· A PARTNERSHIP BETWEEN ·

W. H. FREEMAN & SCIENTIFIC AMERICAN

biology

FOR A CHANGING WORLD

with Physiology

SECOND EDITION

MICHÈLE SHUSTER | JANET VIGNA

MATTHEW TONTONOZ | GUNJAN SINHA

Publisher   Susan Winslow
Acquisitions Editor   Beth Cole
Development Editor   Andrea Gawrylewski
Marketing Director   John Britch
Market Development Manager   Shannon Howard
Marketing Assistant   Tess Sanders
Manager of Digital Development   Amanda Dunning
Associate Media Editor   Betsye Mullaney
Assistant Editor   Jane Taylor
Project Editor   Liz Geller
Copyeditor   Nancy Brooks
Art Director   Diana Blume
Designer   Tom Carling
Art Manager   Matthew McAdams
Artwork   Precision Graphics
Photo Editor   Christine Buese
Photo Researchers   Jacqui Wong, Donna Ranieri, and Teri Stratford
Production Manager   Paul Rohloff
Composition   MPS Limited
Printing and Binding   King Printing Co., Inc.

Library of Congress Control Number: 2013956228

ISBN-13: 978-1-4641-5113-2
ISBN-10: 1-4641-5113-X

Printed in the United States of America

Fourth printing

W. H. Freeman and Company
41 Madison Avenue
New York, NY 10010
Houndmills, Basingstoke RG21 6XS, England
www.whfreeman.com

TO OUR TEACHERS AND STUDENTS

You are our inspiration

About the Authors

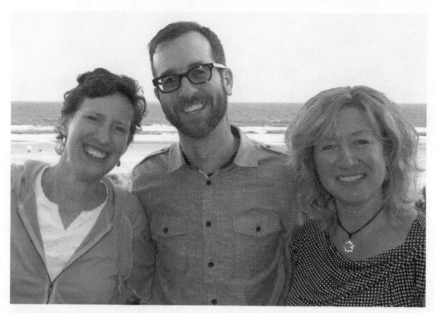

From left to right: Michèle Shuster, Matthew Tontonoz, Janet Vigna

Michèle Shuster, Ph.D., is an associate professor in the biology department at New Mexico State University in Las Cruces, New Mexico. She focuses on the scholarship of teaching and learning, studying introductory biology, microbiology, and cancer biology classes at the undergraduate level, as well as working on several K-12 science education programs. Michèle is an active participant in programs that provide mentoring in scientific teaching to postdoctoral fellows, preparing the next generation of undergraduate educators. She is the recipient of numerous teaching awards, including the Westhafer award for Teaching Excellence at NMSU. Michèle received her Ph.D. from the Sackler School of Graduate Biomedical Sciences at Tufts University School of Medicine, where she studied meiotic chromosome segregation in yeast.

Janet Vigna, Ph.D., is an associate professor in the biology department at Grand Valley State University in Allendale, Michigan. She is a science education specialist in the Integrated Science Program, training and mentoring K-12 science teachers. Janet has 18 years of undergraduate teaching experience, with a special interest in effectively teaching biology to nonmajors. She has recently been recognized with the GVSU Outstanding Teacher Award. Her scholarly interests include biology curriculum development, the effective use of digital media in science education, and research on the effects of biological pesticides on amphibian communities. She received her Ph.D. in microbiology from the University of Iowa.

Matthew Tontonoz has been a development editor for textbooks in introductory biology, cell biology, biochemistry, evolution, and environmental science. After a brief stint in medical school in California, he realized he was better suited to saving sentences than saving lives. Matt received his B.A. in biology from Wesleyan University and his M.A. in the history and sociology of science from the University of Pennsylvania. He is currently staff science writer at the Cancer Research Institute, where he covers advances in cancer immunology and blogs about the history of medicine. He lives in Brooklyn, New York.

Gunjan Sinha has been writing about science for over a decade. Her articles have been published in *Science, Nature Medicine, Nature Biotechnology, Scientific American*, and several other magazines and journals. She holds a graduate degree in molecular genetics from the University of Glasgow, Scotland, and a graduate degree in journalism from New York University. She currently works as a freelance science journalist and lives in Berlin, Germany.

DEAR READER,

Thank you for opening this book! We hope that your journey through it will be as rewarding for you as our journey in writing it has been. When we first came together to collaborate on the development of this text, our biggest goal was to get students interested in biology by showing its relevance to daily life. We wanted to create a textbook that students would actually want to read. Our model and partner in this process has been Scientific American, *a visually stunning magazine that's been successfully bringing science to the public for more than 150 years. The result is a unique textbook that takes a novel approach to teaching biology, one that we think has the potential to greatly improve learning. We hope that this brief introduction will serve as a road map of the book, so that you can get the most out of your experience with it and be as captivated by the wonders of life as we are.*

The main approach of each chapter is the presentation of key science concepts within the context of a relevant and engaging story–a story of discovery, of determination, of human interest, of adventure. From the search for life on Mars to the problem of antibiotic-resistant bacteria, we use stories to bring science to life and to show scientists in action. After all, science is not just a collection of facts, so why would we present it that way? We ask you, our students, to study biology so you can use knowledge to make choices in the real world. We value those stories that will lead you to ask questions about life and how it works and to see the relevance of biology to daily activity. We have seen how stories engage students in our classrooms, and we hope you will be similarly intrigued.

While gripped by a story, you may not even realize how much you are learning. To reinforce the basic learning process, we rely on several strategies:

▸ *Each story is prefaced by a set of Driving Questions. By keeping these in mind as you navigate the story, you will have a good framework for learning the key science concepts.*

▸ *Eye-catching Infographics highlight and drill down into the science of each story. The set of Infographics in a chapter provides a science storyboard for that chapter, illustrating the key scientific concepts and linking them to the story.*

▸ *Key terms are defined in the margins, making it easy to check a definition without having to leave the story.*

▸ *Chapter summaries provide a concise set of bullet points that distill the key scientific concepts.*

▸ *Test Your Knowledge questions at the end of each chapter reinforce basic facts and also allow the facts to be applied through data interpretation and mini cases.*

By taking full advantage of these resources, you will be better able to appreciate how biology affects each and every one of us as well as our close and distant relatives on this planet. We hope that you will talk about biology with your friends and family, and that what you learn here will be applicable to your life. We hope that you will think as critically about every decision you make in the future as we will ask you to do here in these pages.

Welcome to Biology for a Changing World. *We hope that you enjoy your journey, and complete it more prepared for your life in a changing world.*

MICHÈLE SHUSTER   *JANET VIGNA*   *MATTHEW TONTONOZ*   *GUNJAN SINHA*

Science taught through stories

Current, engaging stories capture the imagination and inspire students to read

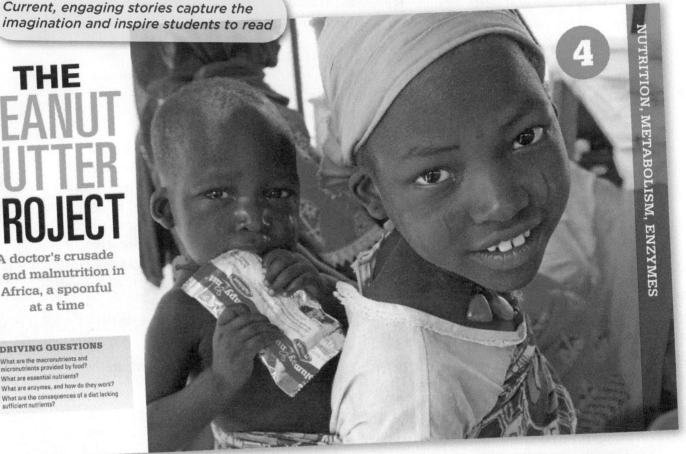

NUTRITION, METABOLISM, ENZYMES

4

THE PEANUT BUTTER PROJECT

A doctor's crusade to end malnutrition in Africa, a spoonful at a time

▸▸DRIVING QUESTIONS

1. What are the macronutrients and micronutrients provided by food?
2. What are essential nutrients?
3. What are enzymes, and how do they work?
4. What are the consequences of a diet lacking sufficient nutrients?

74

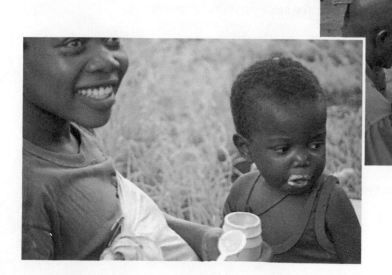

Real people, real voices, and real science bring the story to life

Modern magazine look and feel

Key Terms are easy to find

The science is taught in context of the story

Shorter chapters hold student attention and make the science more manageable

►COFACTOR
An inorganic substance, such as a metal ion, required to activate an enzyme.

►COENZYME
A small organic molecule, such as a vitamin, required to activate an enzyme.

Vitamins and minerals play numerous roles in the body. Some play structural roles–the mineral calcium, for example, is a primary component of bones and teeth. Others play a functional role, helping other molecules to act. Perhaps their most critical role is serving as **cofactors** for enzymes.

Cofactors are accessory or "helper" chemicals that enable enzymes to function. Cofactors include inorganic metals such as zinc, copper, and iron. Cofactors can also be organic molecules, in which case they are called **coenzymes**. Most vitamins are important coenzymes. Without cofactors and coenzymes that bind to enzymes and enable them bind to substrates, cell metabolism would grind to a halt **(INFOGRAPHIC 4.6)**.

By adding extra vitamins and minerals to the milk that children were receiving in the hospital, Manary found that he could get the death rate to go way down, but the recovery rate still wouldn't budge. That's when he knew it was time for a new approach.

INFOGRAPHIC 4.6 VITAMINS AND MINERALS HAVE ESSENTIAL FUNCTIONS

Minerals are micronutrients, nutrients that are essential for health but required in far smaller amounts than macronutrients.

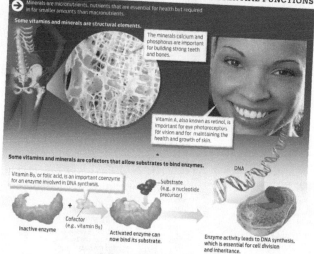

Some vitamins and minerals are structural elements.

The minerals calcium and phosphorus are important for building strong teeth and bones.

Vitamin A, also known as retinol, is important for eye photoreceptors for vision and for maintaining the health and growth of skin.

Some vitamins and minerals are cofactors that allow substrates to bind enzymes.

Vitamin B₉, or folic acid, is an important coenzyme for an enzyme involved in DNA synthesis.

Inactive enzyme

Cofactor (e.g. vitamin B₉)

Activated enzyme can now bind its substrate.

Substrate (e.g. a nucleotide precursor)

DNA

Enzyme activity leads to DNA synthesis, which is essential for cell division and inheritance.

Before and After: For many children in Malawi, peanut butter RUTF has been a life-saver.

EMPTYING THE WARDS

By the time Manary began seriously thinking about home-based therapy, he had been working in Malawi for about 5 years, witnessing the failure of the standard WHO treatment, and he had spent time in villages. In his head, he'd been tossing around the idea of something new and different. Out of the blue, he got an e-mail from a doctor in France, André Briend, who was also thinking about home-based therapy as a treatment for malnutrition.

The two scientists corresponded for about a year by e-mail, weighing the pros and cons of various foods that could be eaten at home yet still pack a nutrient wallop. After considering various options–they considered biscuits, pancakes, even Nutella–the two researchers eventually decided on peanut butter as the best choice. In 2001, they decided to test the idea.

Briend had some of the peanut butter RUTF made up in France and then sealed in foil packets and shipped to Malawi. They then used this product in a carefully designed scientific study. After a brief stabilization phase in the hospital (during which antibiotics were given, if necessary), the children were discharged and sent home on one of three different treatment regimens: (1) ample amounts of traditional food–corn flour and soy; (2) a small amount of peanut butter-based RUTF, to be used as a supplement to the normal diet at home; and (3) the full dose of peanut butter-based RUTF, with sufficient nutrients to meet the total nutritional needs of the children. The goal of the study was to see which treatment regimen was most effective.

Within a few months, the results were astonishing: 95% of the children who received the full peanut butter-based RUTF recovered. Those who received traditional food also did pretty well–about 75% of them recovered, but still the peanut butter was better. And all the home-based treatments were significantly better than standard hospital-based milk therapy, which historically had a 25%-40% recovery rate.

Driving Questions are the roadmap for each chapter

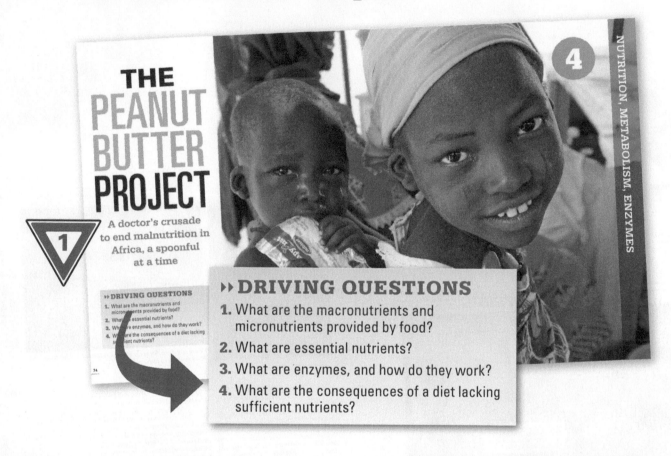

THE PEANUT BUTTER PROJECT

A doctor's crusade to end malnutrition in Africa, a spoonful at a time

NUTRITION, METABOLISM, ENZYMES

4

▶▶ **DRIVING QUESTIONS**

1. What are the macronutrients and micronutrients provided by food?
2. What are essential nutrients?
3. What are enzymes, and how do they work?
4. What are the consequences of a diet lacking sufficient nutrients?

INFOGRAPHIC 4.3 **MACRONUTRIENTS BUILD AN**

▶▶ **DRIVING QUESTIONS**

1. What are the macronutrients and micronutrients provided by food?
2. What are essential nutrients?
3. What are enzymes, and how do they work?
4. What are the consequences of a diet lacking sufficient nutrients?

The four macromolecules that make up cells are car... and nucleic acids. Cells synthesize these macromolec... by the digestion of macronutrients in food.

The subunits released by digestion are taken into the circulation and carried to the body's cells.

Cells use building macromo cellular s

Carbohydrates

Simple sugars

Proteins

Amino acids

Fats

Fatty acids and glycerol

Phospholipids for membranes

Nucleic acids

Nucleotides

DNA tran...

Colorful and engaging infographics teach the most important science

↑ **DRIVING QUESTIONS** ↑

INFO + GRAPHIC = INFOGRAPHIC

Crucial practice built around the Driving Questions

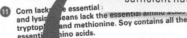

92

UNIT 1 • WHAT IS LIFE MADE OF? CHEMISTRY, CELLS, ENERGY

④

③ A multivitamin supplement is a(n) _____ supplement.
  a. macronutrient
  b. micronutrient
  c. mineral
  d. enzyme
  e. a and b

④ Which of the following foods is a rich source of protein?
  a. lean meat, such as chicken breast
  b. whole grains (e.g., whole wheat bread)
  c. olive oil
  d. leafy greens
  e. berries (e.g., blueberries and raspb...

USE IT

⑤ Explain the difference ... macronutrients and micronutrients.

⑥ A typical multivi... supplement contains vitamin A, vitamin C, vi... D, vitamin E, vitamin K, vitamin B₁, vit... ...c, biotin, calcium, iron, magnes... ...nc, selenium, copper, manganese, and ... Explain your answers to the following qu...
  a. Are all of th... vitamins? If there are ingredients that are not vitamins, what are they?
  ...are all of these m... nutrients?

DRIVING QUESTION 2

What are essential nutrients?

By answering the questions below and studying Infographics 4.3 and 4.5 and Table 4.1, you should be able to generate an answer for the broader Driving...

  c. an essential macronutrient.
  d. a nonessential vitamin.
  e. a nonessential amino a...

⑩ Which component of pean... essential amino acids?
  a. milk powder
  b. peanut butter
  ...vegetable oil
  e. powered vita...ns and...
  f. a and c

⑪ Corn lack... ...e essential ... and lysi... ...eans lack the essential amino acids tryptop... ...and methionine. Soy contains all the essenti... ...mino acids.
  a. Cou... ...eone survive on a diet with a corn-base... ...tein alone? Why or why not?
  b. Wh... ...y traditional diets combine corn (e.g... ...as) with beans?
  c. Why did one of the home-based feeding therapies in Malawi combine soy flour with corn flour?

DRIVING QUESTION 3

What are enzymes, and how do they work?

By answering the questions below and studying Infographics 4.4, 4.5, and 4.6, you should be able to generate an answer for the broader Driving Question above.

KNOW IT

⑫ The substrate of an enzyme is

▸▸ **DRIVING QUESTIONS**

1. What are the macronutrients and micronutrients provided by food?
2. What are essential nutrients?
3. What are enzymes, and how do they work?
4. What are the consequences of a diet lacking sufficient nutrients?

3

INFOGRAPHIC 4.4 ENZYMES FACILITATE CHEMICAL REACTIONS

→ Cells require enzymes to break down and build up macromolecule... ...es are proteins that speed up chemical reactions.

...abolic Reaction: Bonds are broken

Substrates (e.g., carbohydrates)

...nd linking s...rates is brok...

Active site

Enzyme

...drates are ...d by enzymes ...atabolic ...s.

1. Substrates bind to the active site of a specific enzyme.

2. The active site of the enzyme changes shape, stressing the bond, and thereby making it easier to break.

3. The substrate bond breaks and the resulting products are released from the enzyme. The enzyme is ready to be used again.

...abolic Reaction: Bonds are created

...strates (..., amino acids)

Bond linking substrates is formed.

Muscle protein

...tive site

Amino acid subunits are used to build proteins during anabolic reactions.

Enzyme

1. Substrates bind to the active site of a specific enzyme.

2. The active site of the enzyme changes shape, which orients the substrates so they can bind to one another. This facilitates bonding.

3. The enzyme releases the resulting products. The enzyme is ready to be used again.

▸▸ **DRIVING QUESTIONS**

1. What are the macronutrients and micronutrients provided by food?
2. What are essential nutrients?
3. What are enzymes, and how do they work?
4. What are the consequences of a diet lacking sufficient nutrients?

...NTAIN CELLS

...ins, lipids, ...bunits released

...s the ...le the ...y for ...ction.

...y storage and ...surface molecules

...ring and ...instructions

Biology for a Changing World AT A GLANCE

More ways to practice each Driving Question

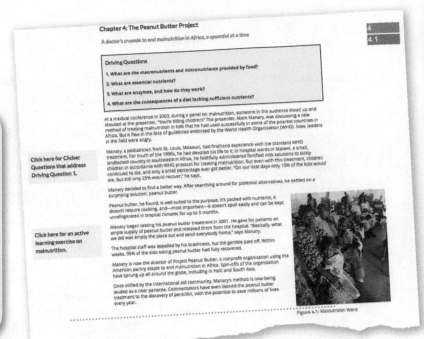

INSTRUCTOR'S GUIDE

The electronic Instructor's Guide is designed to help instructors make the most of our story-based text. Instructional tips from the book's author and other experienced non-majors instructors are integrated into the e-Book, a comprehensive resource for any instructor teaching with our text. Features include active learning exercises, misconception alerts, links to online resources, additional scientific detail, and more.

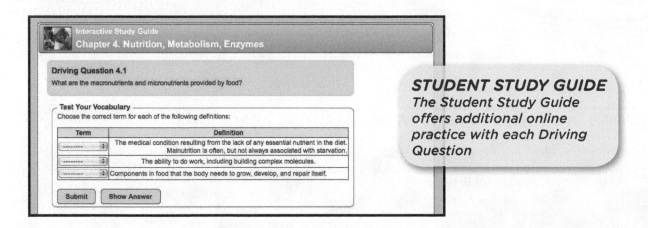

STUDENT STUDY GUIDE
The Student Study Guide offers additional online practice with each Driving Question

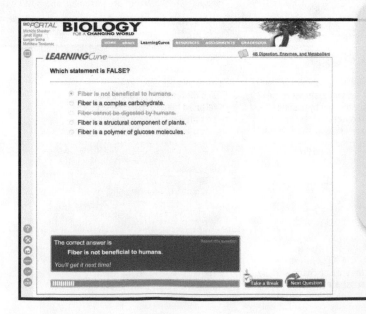

LEARNINGCURVE

LearningCurve is a set of formative assessment activities, built around the Driving Questions, that uses a gamelike interface to guide students through a series of questions tailored to their individual level of understanding. A personalized study plan is generated based upon their quiz results, and students can automatically link to the relevant passages in the e-Book for help.

New data problems, case studies, and Bring it Home activities in each chapter

INTERPRETING DATA

26 The United States currently uses approximately 19 million barrels of oil per day. Of this, about half (9.67 million barrels per day) is imported, and the rest is from U.S. sources (offshore drilling or extraction from shale). The table shows production costs estimated for different oil sources. (The actual cost is driven by a variety of market and geopolitical factors, so we will use production costs as a substitute for actual cost.)

a. Using the data for cost per barrel of various oils, calculate the cost to produce oil to meet current U.S. daily use. Assume that approximately half the imports are from the United Arab Emirates (UAE), and the remainder is split equally between imports from Canada and from Brazil.

b. Let's say that the United States replaces half of its current oil imports from the UAE with domestically produced algal biofuel. What will this do to the cost of produc... our daily needs?

Source	Production Cost per Barrel (Range)	Production Cost per Barrel (Average)
U.S. (offshore)	$70–$80	$75
U.S. (shale oil)	$50–$70	$60
Canada (tar sands)	$70–$90	$80
UAE	$50–$70	$60
Brazil	$60–$80	$70
Algal biofuel	$140–$900	$520

SOURCE: http://insideclimatenews.org/news/20101122/algae-fuel-inches-toward-price-parity-oil; http://www.odac-info.org/newsletter/2011/09/16

c. From what you've read in this chapter, how ...

INTERPRETING DATA

23 The table at right shows the frequencies for STR lengths (repeat number) in different U.S. populations. You can determine the probability of a particular combination of STRs by multiplying the frequencies: for example, the probability of a Hispanic person's having 14 D3S1358 repeats (0.079) and 18 TH01 repeats (0.125) is 0.079 × 0.125 = 0.009

a. What is the probability of a Caucasian American's having a 16, 17 combination for D3S1358?

b. What is the probability of an African American's having a 16, 17 combination for D3S1358?

c. What is the probability of an Hispanic American's having a 16, 17 combination for D3S1358?

d. What is the probability of a Caucasian American's having a 16, 17 combination for D3S1358, and a 5, 9 combination for TH01?

...lity of a Caucasian
...a 16, 17 combination for
...bination for TH01 and an
...for D18S51?

...ers to a, d, and e. Why does
...e many CODIS markers
...two)?

CODIS* STR	NO. OF REPEATS	FREQUENCY (CAUCASIAN)	FREQUENCY (AFRICAN AMERICAN)	FREQUENCY (HISPANIC)
D3S1358	14	0.094	0.089	0.079
	15	0.111	0.186	0.293
	16	0.200	0.248	0.286
	17	0.281	0.242	0.204
	18	0.200	0.155	0.125
TH01	5	0.002	0.004	0
	6	0.232	0.124	0.214
	7	0.190	0.421	0.096
	8	0.084	0.194	0.096
	9	0.114	0.151	0.150
D18S51	10	0.008	0.006	0.004
	11	0.017	0.002	0.011
	12	0.127	0.078	0.118
	13	0.132	0.053	0.111
	14	0.137	0.072	0.139

...or 15 autosomal STR loci on U.S. Caucasian, African American and Hispanic populations. Journal of Forensic Sciences 48(4): 908–911.

...a government database of DNA profiles from offenders, crime scenes, and missing persons.

MINI CASE

21 A college student returns home at the end of the school year. His mother is shocked by the large number of unhealed scrapes and sores on his knees and arms. She also notices that he has put on a few pounds. The student tells his mother that the scrapes are just left over from a skateboarding mishap a few weeks ago and that he guesses he could cut back on some of his snacks. A week after coming home, he goes to the dentist for his yearly checkup. The dentist is alarmed by his bleeding and swollen gums. When asked about his diet, the student notes that he and some of his friends challenged one another to see who could go the longest eating nothing but eggs, mac 'n' cheese, and toast with butter. He proudly announces that he had stayed on this diet for 6 months.

a. Could this student be suffering from malnutrition? Explain your answer.

b. What mineral(s) or vitamin(s) (or both) are you most concerned about, given the symptoms noted by the dentist?

c. What dietary recommendations would you make for this student?

BRING IT HOME

23 A number of concerns have been expressed about GMOs. Search the internet for reliable sources about a particular GMO that you have heard of or in which you are interested (e.g., Golden Rice or genetically modified salmon). List what you consider to be the pros and cons of at least two GMOs. Has what you have read about other genetically modified organisms and the transgenic goats in this chapter changed your opinions about GMOs? What restrictions (if any) would you place on GMOs?

Digital Resources Complete the Package

A robust suite of digital materials provides the tools that instructors and students need to teach and learn with *Biology for a Changing World*. Organized by each chapter's Driving Questions, our instructor and student resources will help ensure that students learn the science behind the stories.

Instructor Resources

LaunchPad

The new standard in online course management, Launch-Pad makes it easier than ever to create interactive assignments, track online homework, and access a wealth of extraordinary teaching and learning tools. Fully loaded with our **customizable e-Book** and all **student** and **instructor resources**, the LaunchPad is organized around a series of prebuilt LaunchPad Units—carefully curated, ready-to-use collections of material for each chapter of *Biology for a Changing World*.

Instructor's Guide

The electronic Instructor's Guide is designed to help instructors make the most of our story-based text. With instructional tips from the book's authors and others who have taught using the text, the Instructor's Guide is a comprehensive resource for any instructor teaching with our text. Features include active learning exercises, misconception alerts, links to online resources, additional scientific detail, and more.

Lecture Tools

Optimized Art

Infographics are optimized for projection in large lecture halls and split apart for effective presentation.

Interactive Animated Infographics

Every piece of art in the text is interactive. All animations are embedded in the text of the LaunchPad e-Book.

PowerPoint Slides

Prebuilt lecture slides provide a great starting point for your lecture and help with the transition to a new textbook.

Clicker Questions

Designed as interactive in-class exercises, these questions reinforce core concepts and uncover misconceptions.

Story Summaries

The story summaries offer a brief synopsis of a chapter's main story, providing interesting details relevant to content found in the book and in online resources.

Activities from the Yale Center for Scientific Teaching

Teaching tidbits created by the Yale Center for Scientific Teaching provide new active learning exercises for the classroom. Each activity is created specifically for *Biology for a Changing World* as a part of Yale's "Theory and Practice of Scientific Teaching" course for graduate science students, and incorporates important tenets of the Scientific Teaching method. Tidbits include PPT slides for class, instructions for teachers, and any additional content needed to implement the activity in your own course. To learn more about Yale's Center for Scientific Teaching, visit http://cst.yale.edu.

Assessment

LEARNINGCurve

LearningCurve is a set of formative assessment activities that uses a gamelike interface to guide students through a series of questions tailored to their individual level of understanding. A personalized study plan is generated based upon their quiz results, and students can automatically link to the relevant passages in the e-Book for help.

Test Bank

A collection of more than a thousand questions organized by chapter and Driving Question and presented in a sortable, searchable environment. The Test Bank includes multiple-choice, short-answer, and matching questions. Questions can also be sorted by type, difficulty level, and important word or concept.

Instructor's Resource Flash Drive

Combines a variety of instructor resources, including art, PowerPoint slides, Clicker Questions, and the Test Bank in one convenient package.

Course Management System

e-Packs are available for Blackboard, Canvas, Desire2Learn, and other course management platforms.

Student Resources

Student Resources reinforce chapter concepts and give students the tools they need to succeed in the course. All student resources are organized by Driving Questions and can be found in the LaunchPad.

LaunchPad

Students have access to a variety of study tools in the LaunchPad along with a complete online version of the textbook. Carefully curated LaunchPad Units provide suggested learning paths for each chapter in the text.

Student Study Guide

Organized by Driving Question, the newly revised electronic Student Study Guide deepens students' understanding of chapter content. The Student Study Guide digs in to each Infographic and develops students' knowledge through a variety of open-ended and multiple-choice self-study questions.

Interactive Animated Infographics

Every piece of art in the text has been animated. All animations are embedded in the text of the LaunchPad e-Book.

*LEARNING*Curve

LearningCurve is a set of formative assessment activities that uses a gamelike interface to guide students through a series of questions tailored to their individual level of understanding. A personalized study plan is generated based upon their quiz results, and students can automatically link to the relevant passages in the e-Book for help.

Pre-Quiz Study Questions with Feedback

Quiz questions with response-specific feedback help explain concepts and correct student misunderstandings.

Art Notebook

The art in each chapter is available in PDF form for students to download and print before lectures or for individual study.

Key Term Flashcards

Students can drill and learn the most important terms in each chapter using interactive flashcards.

Acknowledgments

We are thrilled to introduce the second edition of *Scientific American Biology for a Changing World*. We have replaced six stories and added two new Milestones to keep this book at the forefront of current issues in biology. New Driving Questions target students' reading, and new mini cases and data questions test their knowledge. A fresh, updated design showcases our unparalleled Infographics, and compelling narratives continue to reveal how biology is relevant to daily life.

As with the first edition, we could not have completed this project without the help of a great team at W. H. Freeman. The authors would like to thank Elizabeth Widdicombe, Susan Winslow, and the folks at W. H. Freeman and Company and *Scientific American* for continuing to support this vision for biology education. They recognized our diverse strengths and brought us together to make this vision a reality. We continue to learn so much from one another on this challenging and rewarding professional journey, and none of us has likely worked so hard and so passionately on a project as we all have on this one.

We would like to thank all those who were interviewed and who generously contributed information for the chapters in this edition. Their stories are central to the impact that this book will have on the students we teach. They are authentic examples of biology in a changing world, and they bring this book to life.

A special thank you is required for our Acquisitions Editor, Beth Cole, for her unwavering encouragement and ability to bring stable direction and support to the project. Development Editor Andrea Gawrylewski has spent many hours in the pages of this book, editing the details, managing our chaos, and smoothing our rough edges. Sara Blake and Jane Taylor were invaluable in the scheduling, reviewing, and managing of this edition. We thank them for their dedication, patience, experience, and expertise. Thanks go to Betsye Mullaney and Amanda Dunning for their tireless work on our media and assessment program.

Many thanks to Liz Geller, Nancy Brooks, Diana Blume, Matthew McAdams, Christine Buese, Paul Rohloff, and all the people behind the scenes at W. H. Freeman for translating our ideas into a beautiful, cohesive product. We would like to thank Rachel Rogge and Jan Troutt at Precision Graphics for their outstanding work on the Infographics. We appreciate their patience with the many edits and quick timelines throughout the project. They do amazing work.

We'd like to thank John Britch and Shannon Howard for their enthusiasm and hard work in promoting this book in the biology education community. We thank the enthusiastic group of salespeople who connect with biology educators across the country and do a wonderful job representing this book.

We would like to thank our families and friends who have been close to us during this process. They have been our consultants, served as sounding boards about challenges, celebrated our successes, shared our passions, and supported the extended time and energy we often diverted away from them to this project. We are grateful for their patience and unending support.

And finally, a sincere thank you to our many teachers, mentors, and students over the years who have shaped our views of biology and the world, and how best to teach about one in the context of the other. You are our inspiration.

Media and Assessment Authors

John Harley, Eastern Kentucky University, *Learning Curve Activities*

Beth Collins, Iowa Central Community College, *Learning Curve Activities, Clicker Questions*

Carolyn Wetzel, Holyoke Community College, *Test Bank*

Ryan Shanks, North Georgia College & State University, *Test Bank*

Cindy Malone, California State University, Northridge, *Lecture PowerPoint slides*

Jennifer Cymbola, Grand Valley State University, *Lecture PowerPoint slides*

Ann Aguanno, Marymount Manhattan College, *Pre-Quiz Study Questions and Lecture PowerPoint slides*

Jodi DenUyl, Grand Valley State University, *Pre-Quiz Study Questions*

Brett Macmillan, McDaniel College, *Student Study Guide*

Jenny Knight, University of Colorado-Boulder, *Teaching Guide*

Anne-Marie Hoskinson, University of Colorado-Boulder, *Teaching Guide*

Ben Brammel, Eastern Kentucky University, *Teaching Guide*

Zuzana Swigonova, University of Pittsburgh, *Teaching Guide*

Yvette Gardner, Clayton State University, *Teaching Guide*

Jessamina Blum, Yale University

Annika Moe, University of Minnesota

Elizabeth Hobson, New Mexico State University

Julie Glenn, Gainesville State College

Lisa Strong, Northwest Mississippi Community College

Stephen C. Burnett, Clayton State University

Reviewers

We would like to extend our deepest thanks to the following instructors who reviewed, tested, and advised on the book manuscript at its various stages of development.

Stephanie Aamodt, *Louisiana State University-Shreveport*
Marilyn Abbott, *Lindenwood University*
Ann Aguanno, *Marymount Manhattan College*
Holly Ahern, *SUNY Adirondack*
Kellie Aitchison, *Finger Lakes Community College*
Jon Aoki, *University of Houston-Downtown*
Lisa Appeddu, *Southwestern Oklahoma State University*
Amanda Ashley, *New Mexico State University*
Bert Atsma, *Union County College*
Hernan Aubert, *Pima Community College*
Rao Ayyagari, *Lindenwood University*
Anita Baines, *University of Wisconsin-La Crosse*
Ellen Baker, *Santa Monica College*
Laura Bannan, *USC Upstate*
Verona Barr, *Heartland Community College*
Morgan Benowitz-Fredericks, *Bucknell University*
Christine Bezotte, *Elmira College*
Charles Biles, *East Central University*
Curtis Blankespoor, *Calvin College*
Lanh Bloodworth, *Florida State College-Jacksonville*
Lisa Boggs, *Southwestern Oklahoma State University*
Cheryl Boice, *Florida Gateway College*
Brenda Bourns, *Seattle University*
Bradley Bowden, *Alfred University*
Ben Brammell, *Eastern Kentucky University*
Susan Brantley, *Gainesville State College*
Clay Britton, *Methodist University*
Carole Browne, *Wake Forest University*
Jamie Burchill, *Troy University*
Martha Smith Caldas, *Kansas State University*

Judith Carey, *Springfield Technical Community College*
Jennifer Carney, *Finger Lakes Community College*
Michelle Cawthorn, *Georgia Southern University*
Maitreyee Chandra, *Diablo Valley College*
Genevieve Chung, *Broward College*
Kimberly Cline-Brown, *University of Northern Iowa*
Scott Cooper, *University of Wisconsin-La Crosse*
Erica Corbett, *Southeastern Oklahoma State University*
Richard Cowart, *Coastal Bend College*
Jan Crook-Hill, *University of North Georgia*
Kathleen Curran, *Wesley College*
Karen Curto, *University of Pittsburgh*
Jennifer Cymbola, *Grand Valley State University*
Don Dailey, *Austin Peay State University*
Michael Dann, *Penn State University*
Craig Denesha, *Spartanburg Community College*
Jodi Denuyl, *Grand Valley State University*
Jody Lee Duek, *Pima Community College-Desert Vista Campus*
Jacquelyn Duke, *Baylor University*
Joseph Eagan, *SUNY Adirondack*
Ralph Eckerlin, *Northern Virginia Community College*
Jennifer Ellington, *Belmont Abbey College*
Lianna Etchberger, *Utah State University-Uintah Basin Regional Campus*
Sarah Finch, *Florida College-Jacksonville*
Ryan Fisher, *Salem State University*
Teri Foster, *Graceland University*
Patrick Galliart, *North Iowa Area Community College*
Cynthia Galloway, *Texas A&M-Kingsville*
Richard Gardner, *Southern Virginia University*

ACKNOWLEDGMENTS

Dana Garrigan, *Carthage College*
Thomas Gorczyca, *Northern Essex Community College*
Jen Grant, *University of Wisconsin-Stout*
Sherri Graves, *Sacramento City College*
Andrea Green, *Northeastern State University*
Melissa Greene, *Northwest Mississippi Community College*
Sara Gremillion, *Armstrong Atlantic State University*
Charles Grossman, *Xavier University*
Cheryl Hackworth, *West Valley College*
Kristy Halverson, *The University of Southern Mississippi*
Joyce Hardy, *Chadron State College*
Janelle Hare, *Morehead State University*
Chris Haynes, *Shelton State Community College*
Nathaniel Hemstad, *Inver Hills Community College*
Timothy Henkel, *Valdosta State University*
Jerome Hojnacki, *University of Massachusetts-Lowell*
Thomas Horvath, *SUNY Oneonta*
Anne-Marie Hoskinson, *University of Colorado-Boulder*
Tonya Huff, *Riverside City College*
Kimberly Hurd, *Bakersfield College*
Catherine Hurlbut, *Florida State College-Jacksonville*
Joseph Husband, *Florida State College-Jacksonville*
Virginia Irintcheva, *Black Hawk College*
Evelyn Jackson, *University of Mississippi*
Wendy Jamison, *Chadron State College*
Jamie Jensen, *Brigham Young University*
Wanda Jester, *Salisbury University*
Tanganika Johnson, *Southern University and A&M College*
Laurie Johnson, *Bay de Noc Community College*
John Jones, *Calhoun Community College*
Leslie S. Jones, *Valdosta State University*
Jacqueline Jordan, *Clayton State University*
Arnold Karpoff, *University of Louisville*
Dennis Kitz, *S. Illinois Univ Edwardsville*
Cindy Klevickis, *James Madison University*
Daniel Klionsky, *University of Michigan*
Jennifer Kneafsey, *Tulsa Community College*
Olga Kopp, *Utah Valley University*
Dennis Kraichely, *Cabrini College*
David Krauss, *City University of New York*
Ruhul Kuddus, *Utah Valley University*
Wendy Kuntz, *Kapiolani Community College*
George Labanick, *University of South Carolina-Upstate*
Dale Lambert, *Tarrant County College-Northeast*
Elaine Larsen, *Skidmore College*
Brenda Leady, *University of Toledo*
David Lemke, *Texas State University*
Beth Leuck, *Centenary College of Louisiana*
Harvey Liftin, *Broward College*
Lee Likins, *University of Missouri-Kansas City*
Tammy Liles, *Bluegrass Community and Technical College*
Cayle Lisenbee, *Arizona State University*
Paul Lonquich, *California State University-Northridge*
David Luther, *George Mason University*
Christi Magrath, *Troy University*
Charles Mallery, *University of Miami*
Cindy Malone, *California State University-Northridge*
Donna Maus, *Pittsburg State University*
Kathy McCormick, *Western Piedmont Community College*
Mary McDonald, *University of Central Arkansas*
Terrence McGlynn, *California State University-Dominguez Hills*
Suzanne Meers, *Gainesville State College*
Diane Melroy, *University of North Carolina-Wilmington*
Michael Millward, *Northern Kentucky University*
Brenda Moore, *Truman State University*
Jeanelle Morgan, *University of North Georgia*
Ann Murkowski, *North Seattle Community College*
Necia Nicholas, *Calhoun Community College*
Zia Nisani, *Antelope Valley College*
Cynthia Kay Nishiyama, *California State University-Northridge*

Fidelma O' Leary, *St. Edward's University*
Jorge Obeso, *Miami Dade College*
Peter Oelkers, *University of Michigan, Dearborn*
Joshua Parker, *Clayton State University*
Monica Parker, *Florida State College-Jacksonville South Campus*
Virginia Pascoe, *Mt. San Antonio College*
David Peyton, *Morehead State University*
Joel Piperberg, *Millersville University of Pennsylvania*
Mary Poffenroth, *San Jose State University*
Michelle Priest, *El Camino Compton Educational Center*
Adrianne Prokupek-Pickett, *Nebraska Wesleyan University*
Dianne Purves, *Crafton Hills College*
Logan Randolph, *Polk State College*
Claudia Rauter, *University of Nebraska-Omaha*
Neil Reese, *South Dakota State University*
Nick Reeves, *Mt. San Jacinto College*
Tim Revell, *Mount San Antonio College*
Stanley Rice, *Southeastern Oklahoma State University*
Nelda Rogers, *Brownsville Independent School District*
Troy Rohn, *Boise State University*
Peggy Rolfsen, *Cincinnati State Technical and Community College*
Thomas Rooney, *Wright State University*
Mark Sandheinrich, *University of Wisconsin-La Crosse*
Celine Santiago Bass, *Kaplan University*
Justin Shaffer, *University of North Carolina*
Greg Sievert, *Emporia State University*
Michael Small, *Texas State University*
Sharon Smith, *Florida State College-Jacksonville*
Adrienne Smyth, *Worcester State University*
Stephanie Songer, *University of North Georgia*
Bryan Spohn, *Florida State College-Jacksonville*
Carol St. Angelo, *Hofstra University*
Kurt Stanberry, *University of Houston-Downtown*
Sharon Standridge, *Middle Georgia State College*
John Starnes, *Somerset Community College*
Alicia Steinhardt, *Hartnell Community College*
Mikel Stevens, *Brigham Young University*
Terri Stilson, *North Seattle Community College*
Gail Stratton, *University of Mississippi*
Lisa Strong, *Northwest Mississippi Community College*
Robyn Stroup, *Tulsa Community College*
Zuzana Swigonova, *Univeristy of Pittsburgh*
Don Terpening, *Ulster County Community College*
Janice Thomas, *Montclair State University and Kean University*
Ken Thomas, *Northern Essex Community College*
Rita Thrasher, *Pensacola State College*
Candace Timpte, *Georgia Gwinnett College*
Robert Tompkins, *Belmont Abbey College*
Rick Topinka, *American River College*
Chris Trzepacz, *Murray State University*
Christine Tuaillon, *Nassau Community College*
Linda Twining, *Truman State University*
Eileen Underwood, *Bowling Green State University*
Steve Uyeda, *Pima Community College*
Rani Vajravelu, *University of Central Florida*
Jagan Valluri, *Marshall University*
Dirk Vanderklein, *Montclair State University*
Timothy Wakefield, *John Brown University*
Amanda Waterstrat, *Somerset Community College*
Teresa Weglarz, *University of Wisconsin-Fox Valley*
Michael Wenzel, *California State University Sacramento*
Alexander Werth, *Hampden-Sydney College*
Wayne Whaley, *Utah Valley University*
Alicia Whatley, *Troy University*
Robert Whyte, *California University of Pennsylvania*
Frank Williams, *Langara College*
Christina Wills, *Rockhurst University*
Michael Wolyniak, *Hampden-Sydney College*
Holly Woodruff, *Central Piedmont Community College*
Carol Wymer, *Morehead State University*

Contents

> *Americans burn through 378 million gallons of gasoline a day, enough to fill about 540 Olympic-size swimming pools.*
>
> — CHAPTER 5

UNIT 2
How Does Life Perpetuate? Cell Division and Inheritance

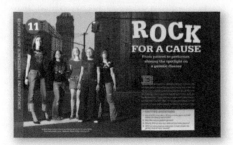

> Mendel provided a new explanation for heredity, decades before the word "genetics" was coined.
>
> — MILESTONE 4

CHAPTER 13

Stem Cells and Cell Differentiation 282

Grow Your Own

Stem cells could be the key to engineering organs

UNIT 3

How Does Life Change over Time? Evolution and Diversity

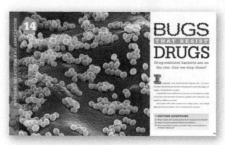

CHAPTER 14

Natural Selection and Adaptation 302

Bugs That Resist Drugs

Drug-resistant bacteria are on the rise. Can we stop them?

M5 MILESTONES IN BIOLOGY

Adventures in Evolution 320

Charles Darwin and Alfred Russel Wallace on the trail of natural selection

Scientists are now seeing bacterial infections that don't respond to any known antibiotics, leading many to fear the day when we run out of treatment options altogether.

— CHAPTER 14

" *[Lost City] is a good example of what we really don't know and what there is to still discover on the seafloor.* **"**

— GRETCHEN FRÜH-GREEN, CHAPTER 18

> The human heart beats more than 2.5 billion times in an average lifetime. — CHAPTER 27

CHAPTER 32

Plant Physiology 716

Q&A: Plants

Exploding seeds, carnivorous flowers, and other colorful adaptations of the plant world

> Plants are among the most "flirtatious" and sexually active organisms on Earth.
>
> — CHAPTER 32

JAVA REPORT

Making sense of the latest buzz in health-related news

IN 1981, A STUDY IN THE PROMINENT *New England Journal of Medicine* made headlines when it reported that drinking two cups of coffee a day doubled a person's risk of getting pancreatic cancer. Drinking five cups a day supposedly tripled the risk. "Study Links Coffee Use to Pancreas Cancer," trumpeted the *New York Times*. "Is there cancer in the cup?" asked *Time* magazine. The lead author of the study, Brian MacMahon of the Harvard School of Public Health, appeared on the *Today* show to warn of the dangers of coffee.

"I will tell you that I myself have stopped drinking coffee," said MacMahon, who had previously drunk three cups a day.

Just 5 years later, MacMahon's research group was back in the news, reporting in the same journal that a second study had found *no* link between coffee and pancreatic cancer. Subsequent studies by other researchers also failed to reproduce the original findings.

A global affair: Customers in the Yunnan province in China line up for their morning coffee fix at a Starbucks.

Today, more than 30 years later, coffee is once again making headlines. Recent studies have suggested that, far from causing disease, coffee may actually help *prevent* a number of conditions—everything from Parkinson disease and diabetes to cancer and tooth decay. "Java Junkies Less Likely to Get Tumors," announced a 2010 CBS News headline. "Morning Joe Fights Prostate Cancer," proclaimed a blog. The September 2010 issue of *Prevention* magazine ran an article titled "Four Ways Coffee Cures," itemizing supposed health benefits for each additional cup consumed.

Not everyone is buying the coffee cure, however. Public health advocates are increasingly alarmed by our love affair with—some might say, addiction to—caffeine. Emergency rooms are reporting more caffeine-related admissions, and poison control centers are receiving more calls related to caffeine "overdoses." In response, politicians are pressuring the Food and Drug Administration (FDA) to force manufacturers to place warning labels on energy drinks. Nevertheless, caffeine's "energizing" effect is advertised on nearly every street corner, where, increasingly, you're also likely to find a coffee shop. As of 2012, according to Foodio54.com, there were 231 Starbucks within a 5-mile radius of a Manhattan zip code; nationally, the average within the same radius is 10.

Why the mixed messages about caffeine? Are researchers making mistakes? Are journalists getting their facts wrong? While both of these possibilities may be true at times, the bigger problem is widespread confusion over the nature of science and the meaning of scientific evidence.

"Consumers are flooded with a firehose of health information every day from various media sources," says Gary Schwitzer, publisher of the consumer watchdog site HealthNewsReview.org and former director of health journalism at the University of Minnesota. "It can be—and often is—an ugly picture: a bazaar of disinformation." Too often, he says, the results of studies are reported in incomplete or misleading ways.

Why might consuming coffee or caffeine be associated with such dramatically different results? The risks or benefits of a caffeinated beverage may depend on the amount a person drinks—one cup versus a whole pot. Or maybe it matters *who* is drinking the beverage. The *New England Journal of Medicine* study, for example, looked at hospitalized patients only. Would the same results have been seen in people who weren't already sick? Sometimes, to properly evaluate a scientific claim, we need to look more closely at how the science was done (INFOGRAPHIC 1.1).

SCIENCE IS A PROCESS

When many people think about science, they think of a body of facts to be memorized: water boils at 100°C; the nucleus is a part of a cell. But that's not the whole story. **Science** is a *way* of knowing, a *method* of seeking answers to questions on the basis of observation and experiment. Scientists draw conclusions from the best evidence they have at any one

▶SCIENCE
The process of using observations and experiments to draw conclusions based on evidence.

time, and the process is not always easy or straightforward. Conclusions based on today's evidence may be modified in the future as other scientists ask different—and sometimes better—questions. Moreover, with improved technology, researchers may uncover better data; new information can cast a new light on old conclusions. Science is a never-ending process.

Perhaps the best way to understand science is to do it. Let's say you want to determine scientifically if coffee has energizing effects. How might you go about investigating this question? A logical place to start would be your own experience. You may notice that you feel more awake when you drink coffee or that coffee helps you concentrate. Such informal personal observations are called **anecdotal evidence.** This is a type of evidence that may be interesting but is often unreliable, since it isn't based on systematic study. A poll of your classmates to find out how they experience coffee would also be anecdotal evidence.

Nevertheless, this anecdotal evidence might lead you to formulate a question: Does coffee improve mental performance? Will it help me study or do better on a test? To find out what information currently exists on the subject, you could read relevant studies that have already been conducted, available in online databases of journal articles or in university libraries. Generally, you can trust the information in scientific journals because it has been subject to **peer review.** The aim of peer review—the review of an article by experts before publication—is to weed out sloppy research as well as overstated claims, and thus ensure the integrity of the journal and its scientific findings. To further reduce the chance of bias, authors must declare any possible conflicts of interest and name all funding sources (for example, pharmaceutical or biotechnology companies). With this information, reviewers and readers can view the study with a more critical eye.

From your perusal of the scientific literature, you would soon discover (if you didn't know it already) that coffee contains caffeine, and that caffeine is a chemical known to stimulate the brain. Brain regions controlling sleep, mood, memory, and concentration are especially sensitive to this chemical, and some researchers believe that caffeine accounts for the general feeling of alertness that many coffee drinkers experience.

▶**ANECDOTAL EVIDENCE**
An informal observation that has not been systematically tested.

▶**PEER REVIEW**
A process in which independent scientific experts read scientific studies before they are published to ensure that the authors have appropriately designed and interpreted the study.

INFOGRAPHIC 1.1 CONFLICTING CONCLUSIONS

 A variety of studies published in peer-reviewed scientific journals report different conclusions about the risks and benefits of coffee. In order for the public to understand and use these outcomes to its advantage, a closer look at the scientific process and the factors that surround coffee drinking is necessary.

Scientific Studies Report That Drinking Coffee...

· May cause pancreatic cancer
· Is linked to infertility and low infant birth weight
· Lowers the risk of Parkinson disease
· Does not cause pancreatic cancer
· Reduces risk of ovarian cancer

Factors That May Influence These Results:

· Chemicals naturally present in coffee, including caffeine
· The climate and soil in which different coffee plants are grown (which in turn influences the chemicals in coffee)
· How the beans are roasted and processed
· How much coffee a person drinks
· The gender, age, and general health of a coffee drinker
· Other social factors, such as whether coffee is consumed with a meal or with a cigarette
· Other unknown factors that just happen to correlate with coffee drinking

INFOGRAPHIC 1.2 SCIENCE IS A PROCESS: NARROWING DOWN THE POSSIBILITIES

Multiple scientists doing multiple experiments narrow down the pool of possible hypotheses. Those that are rigorously tested and supported by other experiments emerge with greatest confidence.

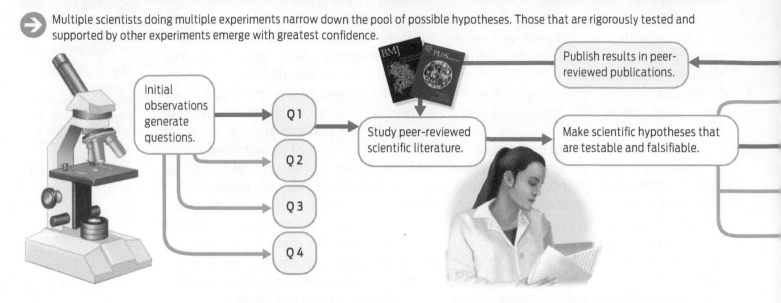

▶**HYPOTHESIS**
A tentative explanation for a scientific observation or question.

▶**TESTABLE**
A hypothesis is testable if it can be supported or rejected by carefully designed experiments or observational studies.

▶**FALSIFIABLE**
Describes a hypothesis that can be ruled out by data that show that the hypothesis does not explain the observation.

▶**EXPERIMENT**
A carefully designed test, the results of which will either support or rule out a hypothesis.

Armed with this information, you could go one step further and formulate a specific **hypothesis** about how coffee affects mental performance. A hypothesis is a possible answer to the question under investigation. For example, one hypothesis about coffee might be that consuming the caffeine in coffee improves memory. Another might be that high levels of caffeine increase concentration. Not all possible explanations will be *scientific* hypotheses, though. A scientific hypothesis must be **testable** and **falsifiable**–that is, it can be established or rejected by experiment; and if it is false, it can be proved wrong. Statements of opinion and hypotheses that depend on supernatural or mystical explanations that cannot be tested or refuted fall outside the realm of scientific explanation. (Some call such explanations "pseudoscience"; astrology is a good example.)

With a clear scientific hypothesis in hand–"caffeinated coffee improves memory"–the next step is to test it, generating evidence for or against the idea. If a hypothesis is shown to be false–if the finding is "caffeinated coffee does not improve memory"–the hypothesis can be rejected and removed from the list of possible answers to the origi-

nal question. You, the scientist, are then forced to consider other hypotheses. On the other hand, if data support the hypothesis, then it will be accepted, at least until further testing and data show otherwise. Because it is impossible to test whether a hypothesis is true in every possible situation, a hypothesis can never be proved true once and for all. The best we can do is support the hypothesis with an exhaustive amount of evidence (**INFOGRAPHIC 1.2**).

There are a number of ways in which a hypothesis can be tested. One is to design a controlled **experiment.** In this case, you might measure the effects of coffee drinking on a group of participants. In 2002, Lee Ryan, a psychologist at the University of Arizona, decided to do just that. Ryan noticed that memory is often optimal early in the morning in adults over age 65 but tends to decline as the day goes on. She also noticed that many adults report feeling more alert after drinking caffeinated coffee. She therefore hypothesized that drinking coffee might prevent this decline in memory, and devised an experiment to test her hypothesis.

First she collected a group of participants–40 men and women over age 65, who were active, healthy, and who reported consuming some form of caffeine daily. She then randomly divided these people into two groups: one that would get caffeinated coffee, and one that would receive decaf. The caffeine group is known as the **experimental group,** since these participants are receiving the factor being tested—in this case, caffeine. The decaf group is the **control group,** which serves as the basis of comparison. Both groups were given memory tests at 8 A.M. and again at 4 P.M. on two nonconsecutive days. The experimental group received a 12-ounce cup of regular coffee containing approximately 220-270 mg of caffeine 30 minutes before each test. The control group received a **placebo:** a 12-ounce cup of decaffeinated coffee containing no more than 5-10 mg of caffeine per serving. No participants knew

The studies reported in scientific journals are reviewed by experts before publication to ensure accuracy.

▶**EXPERIMENTAL GROUP**
The group in an experiment that experiences the experimental intervention or manipulation.

▶**CONTROL GROUP**
The group in an experiment that experiences no experimental intervention or manipulation.

▶**PLACEBO**
A fake treatment given to control groups to mimic the experience of the experimental groups.

INFOGRAPHIC 1.3 ANATOMY OF AN EXPERIMENT

There are many ways to approach a scientific problem. Controlled experiments are one way. As illustrated here, controlled experiments have two groups—the control group and the experimental group—that differ only in the independent variable.

Hypothesis: Drinking caffeinated coffee prevents daily memory decline.

Population of 40 men and women over age 65

	Control Group	Experimental Group
Random placement into equivalent groups (with respect to age, gender, health, activity level, etc.)		
Independent variable (the variable that is changed in a systematic way)	Placebo Treatment 12 oz. decaffeinated coffee (30 minutes prior to test)	Test Treatment 12 oz. caffeinated coffee (30 minutes prior to test)
Dependent variable (the variable that is measured in the experiment)	Memory Test given morning and afternoon on different days	Memory Test given morning and afternoon on different days
Results from data	Memory Test Scores Afternoon scores were <u>worse</u> than morning scores	Memory Test Scores Afternoon scores were <u>the same</u> as morning scores
Evidence-based conclusion	**Caffeinated coffee prevents memory decline in this population.**	

▶**INDEPENDENT VARIABLE**
The variable, or factor, being deliberately changed in the experimental group.

▶**DEPENDENT VARIABLE**
The measured result of an experiment, analyzed in both the experimental and control groups

to which group they were assigned—in other words, the study was "blind." (In a "double-blind" study, neither the investigator nor the participants know who is getting what treatment.)

By administering a placebo, Ryan could ensure that any change observed in the experimental group was a result of consuming caffeine and not just any hot beverage. In addition, all participants were forbidden to eat or drink any caffeine-containing foods or drinks—like chocolate, soda, or coffee—for at least 4 hours before each test. Thus, the con-

trol group was identical to the experimental group in every way except for the consumption of caffeine.

In this experiment, caffeine consumption was the **independent variable**—the factor that is being changed in a deliberate way. The tests of memory are the **dependent variable**—the outcome that may "depend" on caffeine consumption.

Ryan found that participants who drank decaffeinated coffee did worse on tests of memory function in the afternoon compared to the morning. By contrast, the experimental group who drank caffeinated coffee performed equally well on morning and afternoon memory tests. The results, which were reported in the journal *Psychological Science,* support the hypothesis that caffeine, delivered in the form of coffee, prevents the decline of memory–at least in certain people (INFOGRAPHIC 1.3).

Because other factors might possibly explain the link between coffee and mental performance (perhaps coffee drinkers are more active, and their physical activity rather than their coffee consumption explains their mental performance), it's too soon to see these results as proof of coffee's memory-boosting powers. To win our confidence, the experiment must be repeated by other scientists and, if possible, the methodology refined.

SIZE MATTERS

Consider the size of Ryan's experiment–40 people, tested on two different days. That's not a very big study. Could the results have simply been due to chance? What if the 20 people who drank caffeinated coffee just happened to have better memory?

One thing that can strengthen our confidence in the results of a scientific study is **sample size.** Sample size is the number of individuals participating in a study, or the number of times an experiment or set of observations is repeated. The larger the sample size, the more likely the results will have **statistical significance**–that is, they will not be due to chance (INFOGRAPHIC 1.4).

News reports are full of statistics. On any given day, you might hear that 75% of the

▶**SAMPLE SIZE**
The number of experimental subjects or the number of times an experiment is repeated. In human studies, sample size is the number of participants.

▶**STATISTICAL SIGNIFICANCE**
A measure of confidence that the results obtained are "real" and not due to chance.

INFOGRAPHIC 1.4 SAMPLE SIZE MATTERS

 The more data collected in an experiment, the more you can trust the conclusions.

Data from a few participants:

Effect of Caffeine on Memory

Conclusions drawn from these data might suggest that caffeine has only a slight positive influence on memory, a 15% average increase, but are not definitive because of the small sample size.

Data from dozens of participants:

Effect of Caffeine on Memory

These data show a more convincing positive effect of caffeine on memory, a 45% average increase, because the observed effect is supported by more data. A statistical analysis would show that this positive influence is **significant** — in other words, it is not due to chance.

American public opposes a piece of legislation. Or that 15% of a group of people taking a medication experienced a certain unpleasant side effect—like nausea or suicidal thoughts—compared to, say, 8% of people taking a placebo. Are these differences significant, or important? Whenever you hear such numbers being tossed around, it's important to keep in mind the sample size. In the case of the side effects, was this a group of 20 patients (15% of 20 patients is 3 people), or was it 2,000 (15% of 2,000 is 300)? Only with a large enough sample size can we be confident that the results of a given study are statistically significant and represent something other than chance. Moreover, it's important to consider the population being studied. For example, do the people reporting their views on a piece of legislation represent a broad cross section of the public, or are most of them watchers of the same television network, whose views lie at one extreme? Likewise, in Ryan's study, are the 65-year-old self-described "morning people" who regularly consume coffee representative of the wider population?

If you search for "caffeine and memory" on PubMed.gov (a database of medical research papers), you'll see that the memory-enhancing properties of caffeine is a well-researched topic. Many studies have been conducted, at least some of which tend to support Ryan's results. Generally, the more

INFOGRAPHIC 1.5 EVERYDAY THEORY VS. SCIENTIFIC THEORY

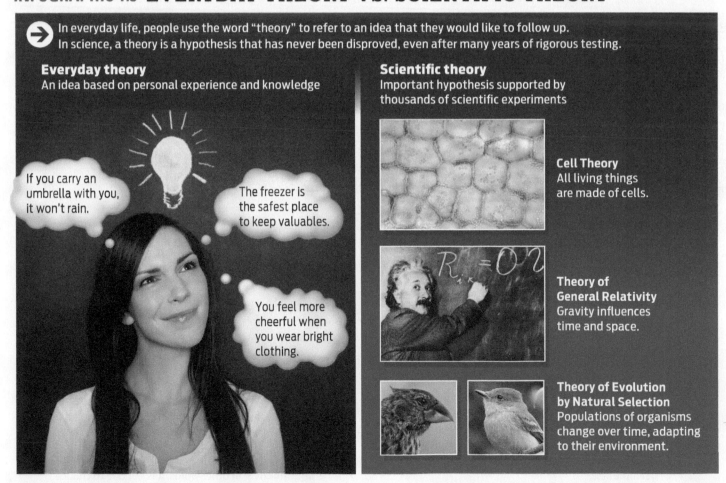

→ In everyday life, people use the word "theory" to refer to an idea that they would like to follow up. In science, a theory is a hypothesis that has never been disproved, even after many years of rigorous testing.

Everyday theory
An idea based on personal experience and knowledge

If you carry an umbrella with you, it won't rain.

The freezer is the safest place to keep valuables.

You feel more cheerful when you wear bright clothing.

Scientific theory
Important hypothesis supported by thousands of scientific experiments

Cell Theory
All living things are made of cells.

Theory of General Relativity
Gravity influences time and space.

Theory of Evolution by Natural Selection
Populations of organisms change over time, adapting to their environment.

> Science is a way of knowing, a method of seeking answers to questions on the basis of observation and experiment.

experiments that support a hypothesis, the more confident we can be that it is true.

Truth in science is never final, however. What is accepted as fact today may tomorrow need to be modified or even rejected, when more evidence comes to light. Nevertheless, scientific knowledge does progress. The highest point of scientific knowledge is what's called a **scientific theory.** The word "theory" in science means something very different from its colloquial meaning. In casual conversation we may say something is "just a theory," meaning it isn't proved. But in science, a theory is an explanation of the natural world that is supported by a large body of evidence compiled over time by numerous researchers. Far from being a fuzzy or unsubstantiated claim, a theory is a scientific explanation that has been extensively tested and has never been disproved (**INFOGRAPHIC 1.5**).

THIS IS YOUR BRAIN ON CAFFEINE

Caffeine is a stimulant. It is in the same class of psychoactive drugs as cocaine, amphetamines, and heroin (although less potent than these, and acting through different mechanisms). Caffeine boosts not just memory and mental performance but physical performance as well. Sports physiologists agree that consuming caffeine before a workout can boost stamina—a fact that is no secret among athletes. A 2004 study found that 33% of 193 track and field athletes and 60% of 287 cyclists said they consumed caffeine to enhance their performance. Recognizing that caffeine is a performance-enhancing drug, the International Olympic Committee prohibited athletes from using it until 2004 (when the committee decided to allow it, presumably because it had become too common a substance to regulate).

While the exact mechanisms are not fully understood, scientists think that caffeine exerts its energizing effect primarily by counteracting the actions of a chemical in the brain called adenosine, which is a type of neurotransmitter. Adenosine is the body's natural sleeping pill—its concentration increases in the brain while we are awake and by the end of the day promotes drowsiness. Caffeine blocks the effect of adenosine in the brain, thereby delaying fatigue and keeping us more alert.

Consumption of caffeinated beverages has skyrocketed in the past 25 years, especially among young people. A 2009 study in the journal *Pediatrics* found that teenagers in a Philadelphia suburb consume anywhere from 23 mg to 1,458 mg of caffeine a day—the equivalent of nearly 10 cups of coffee, and more than three times the recommended safe dose for an adult of 400 mg/day or less. In excess, caffeine can cause anxiety, jitters, heart palpitations, trouble sleeping, dehydration, and more serious symptoms as well—especially in people who are sensitive to it. Of the 4,852 caffeine overdoses reported to poison control centers in the United States in 2008, 49% occurred in those younger than 19, according to a 2011 study in *Pediatrics*. In 2012, a 14-year-old Maryland girl died of cardiac arrhythmia after drinking two 24-ounce Monster Energy drinks (240 mg of caffeine each) in a 24-hour

> **▶SCIENTIFIC THEORY**
> An explanation of the natural world that is supported by a large body of evidence and has never been disproved.

INFOGRAPHIC 1.6 SIDE EFFECTS OF CAFFEINE

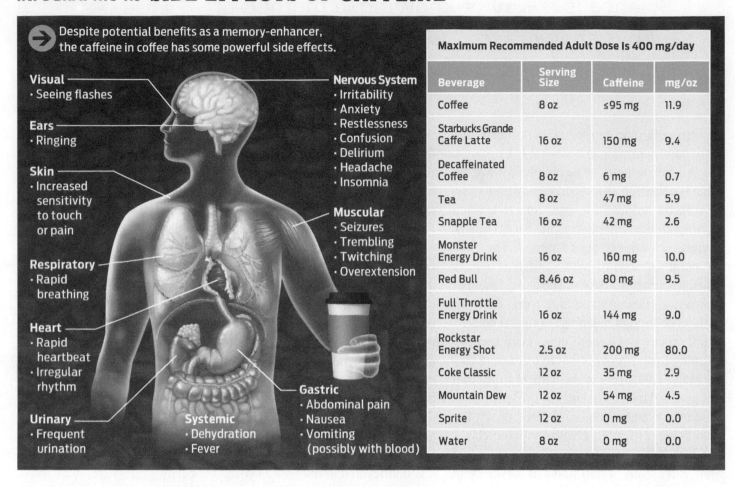

→ Despite potential benefits as a memory-enhancer, the caffeine in coffee has some powerful side effects.

Visual
· Seeing flashes

Ears
· Ringing

Skin
· Increased sensitivity to touch or pain

Respiratory
· Rapid breathing

Heart
· Rapid heartbeat
· Irregular rhythm

Urinary
· Frequent urination

Systemic
· Dehydration
· Fever

Nervous System
· Irritability
· Anxiety
· Restlessness
· Confusion
· Delirium
· Headache
· Insomnia

Muscular
· Seizures
· Trembling
· Twitching
· Overextension

Gastric
· Abdominal pain
· Nausea
· Vomiting (possibly with blood)

Maximum Recommended Adult Dose Is 400 mg/day			
Beverage	Serving Size	Caffeine	mg/oz
Coffee	8 oz	≤95 mg	11.9
Starbucks Grande Caffe Latte	16 oz	150 mg	9.4
Decaffeinated Coffee	8 oz	6 mg	0.7
Tea	8 oz	47 mg	5.9
Snapple Tea	16 oz	42 mg	2.6
Monster Energy Drink	16 oz	160 mg	10.0
Red Bull	8.46 oz	80 mg	9.5
Full Throttle Energy Drink	16 oz	144 mg	9.0
Rockstar Energy Shot	2.5 oz	200 mg	80.0
Coke Classic	12 oz	35 mg	2.9
Mountain Dew	12 oz	54 mg	4.5
Sprite	12 oz	0 mg	0.0
Water	8 oz	0 mg	0.0

period, prompting U.S. Senator Dick Durbin to call for greater federal oversight of energy drinks (**INFOGRAPHIC 1.6**).

For regular coffee drinkers who crave their morning buzz, such side effects are unlikely to persuade them to kick the habit. This may be because, like many other psychoactive substances, caffeine is addictive. Those who drink a significant amount of coffee every day may notice that they don't feel quite right if they skip a day; they may be cranky or get a headache. These are symptoms of withdrawal. In fact, some researchers contend that coffee's mind-boosting effects are an indirect result of the cycle of dependency. Improvement in mood or performance following a cup of coffee, they say, may simply represent relief from withdrawal symptoms rather than any specific beneficial property of coffee.

To test this dependency hypothesis, scientists could conduct an experiment. They could compare the effects of drinking coffee in two groups: one group of regular coffee drinkers who had abstained from coffee for a short period, and another group of non-coffee drinkers. Does coffee give both groups a boost, or only the regular coffee drinkers looking for their fix?

This very experiment was done in 2010 by a group of researchers at the University of Bristol in England. Their study, published in the journal *Neuropsychopharmacology*, looked at caffeine's effect on alertness. Researchers gave caffeine or a placebo to 379 participants and asked them to take a test that rated their level of alertness. The study found that caffeine did not boost alertness in non-coffee drinkers compared to those drinking a placebo (although it did boost their level of anxiety and headache). Heavy coffee drinkers, on the other hand, experienced a steep drop in alertness when given the placebo.

"What this study does is provide very strong evidence for the idea that we don't gain a benefit in alertness from consuming caffeine," the study author, Peter Rogers, said. "Although we feel alert, that's just caffeine bringing us back to our normal state of alertness."

For those of us who rely on coffee to perk us up, the results of this one study are unlikely to change our minds—or our habits. Yet even if confirmed, these results don't really explain why people get hooked on coffee in the first place.

FINDING PATTERNS

Performing controlled laboratory experiments like those discussed above is one way that scientists try to answer questions. Another approach is to make careful observations or comparisons of phenomena that exist in nature, generating data that can then be analyzed systematically. This is the approach taken by scientists who study **epidemiology**—the scientific study of the incidence of disease in populations—or a topic like the movement of stars or the nature of prehistoric life, phenomena that cannot be directly experimented upon.

For example, an epidemiologist who wanted to learn about the relationship between cigarette smoking and lung cancer could compare the rates of lung cancer in smokers and nonsmokers. But an experiment in which participants were asked to smoke cigarettes and were observed to see whether

> **▶EPIDEMIOLOGY**
> The study of patterns of disease in populations, including risk factors.

The famous maker of caffeinated beverages, Coca Cola, has a presence in more than 200 countries around the world.

or not they developed cancer would be highly unethical.

Although epidemiological studies do not provide the immediate gratification of a laboratory experiment, they do have certain advantages. For one thing, they can be relatively inexpensive to conduct, since often the only procedure involved is a participant questionnaire. And you can study factors that are considered harmful, such as excess alcohol or smoking, that you would be unable to test experimentally. Finally, epidemiological studies have the power of numbers and time. The Framingham Heart Study, for example, is a famous epidemiological study that has tracked rates of cardiovascular disease in a group of people and their descendants in Framingham, Massachusetts, in order to identify common risk factors for the disease. Begun in

1948, the study has been going on for decades and has provided mountains of data for researchers in many fields, from cardiology to neuroscience.

Most of the health studies featured in the news are epidemiological studies. Consider a study on coffee and Parkinson disease published in the *Journal of the American Medical Association (JAMA)* in 2000. Researchers examined the relationship between coffee drinking and the incidence of Parkinson disease, a condition that afflicts more than 1 million people in the United States, affecting men and women of all ethnic groups. There is no known cure, only palliative treatments to help lessen symptoms, which include trembling limbs and difficulty coordinating speech and movement.

For more than 30 years, researchers at the Veterans Affairs Medical Center in Honolulu

INFOGRAPHIC 1.7 CORRELATION DOES NOT EQUAL CAUSATION

➔ While the data shown below show a convincing **correlation** between reduced caffeine intake and an increased risk of Parkinson disease, it is impossible to state that less coffee **causes** Parkinson disease. Other factors that were not tested or controlled for could be causing the reduced risk.

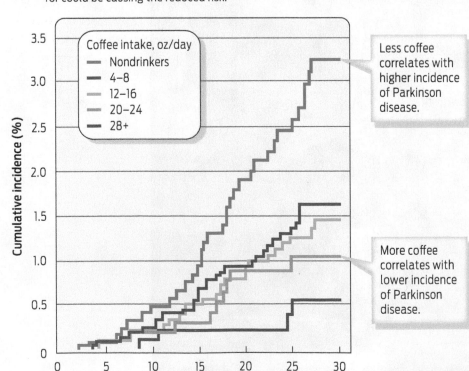

Coffee intake, oz/day
- Nondrinkers
- 4–8
- 12–16
- 20–24
- 28+

Less coffee correlates with higher incidence of Parkinson disease.

More coffee correlates with lower incidence of Parkinson disease.

So, does drinking coffee *cause* Parkinson disease?

Perhaps…but there are many possible explanations of the data:

· Drinking coffee reduces the risk of developing Parkinson disease.
· People who are at risk for developing Parkinson disease are less likely to drink coffee.
· Drinking coffee masks the symptoms of Parkinson disease, thereby reducing the rate of diagnosis of Parkinson disease in coffee consumers.
· Parkinson is a complex disease, so there may be other influential environmental or behavioral factors that are not controlled for in this experiment.
· The results may be specific to Japanese-American men and not translate well to the general population.

SOURCE: Ross, G. W., et al. (2000) Association of coffee and caffeine intake with the risk of Parkinson disease. *Journal of the American Medical Association* 283:2671–2679.

> **❝ Consumers are flooded with a firehose of health information every day from various media sources.❞**
>
> — GARY SCHWITZER

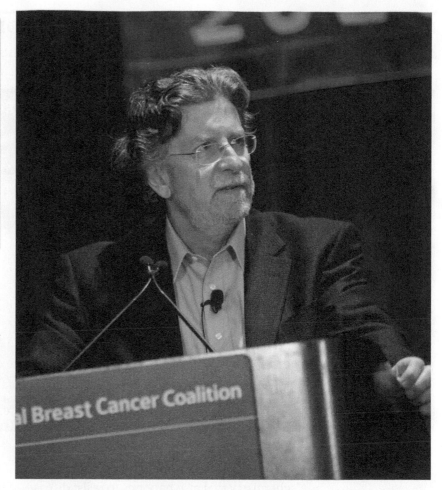

Gary Schwitzer, founder of HealthNewReview.org, lectures to a class at the University of Minnesota School of Journalism and Mass Communication about quality in health care journalism.

followed more than 8,000 Japanese-American men, gathering all sorts of information about them: their age, diet, health, smoking habits, and other characteristics. Of these men, 102 developed Parkinson disease. What did these 102 men have in common? Epidemiologists found that most of them did not drink caffeinated beverages—no coffee, soda, or caffeinated tea.

By contrast, coffee drinkers had a lower incidence of Parkinson disease. In fact, those who drank the most coffee were the least likely to get the disease. Men who drank more than two 12-ounce cups of coffee each day had one-fifth the risk of getting the disease compared to non-coffee drinkers. The same trend was observed for overall caffeine intake that included sources other than coffee.

So does coffee (specifically, caffeine) prevent Parkinson disease? The occurrence and progression of many diseases are affected by a complex range of factors, including age, sex, diet, genetics, and exposure to bacteria and environmental chemicals, as well as lifestyle factors like drinking, smoking, and exercise. Although the study discussed here suggests a link—or **correlation**—between caffeine and lower incidence of Parkinson disease, it does not necessarily show that caffeine prevents the disease. In other words, correlation is not causation. Perhaps

the people who like to drink coffee have different brain chemistry, and it's this different brain chemistry that explains the differing incidence of Parkinson disease among coffee drinkers **(INFOGRAPHIC 1.7)**

Indeed, other studies have found that cigarette smoking also correlates with a lower risk of Parkinson disease. Both coffee drinking and smoking could be considered types of thrill-seeking behavior, observed in people who enjoy the high they get from stimulants such as caffeine or nicotine. The lower risk of Parkinson disease among coffee drinkers might therefore result from thrill-seeking brain

▶ **CORRELATION**
A consistent relationship between two variables.

INFOGRAPHIC 1.8 FROM THE LAB TO THE MEDIA: LOST IN TRANSLATION

The data as reported in peer-reviewed journals are often very complex. Scientists interpret these data in lengthy discussions, but the public receives them as isolated media headlines.

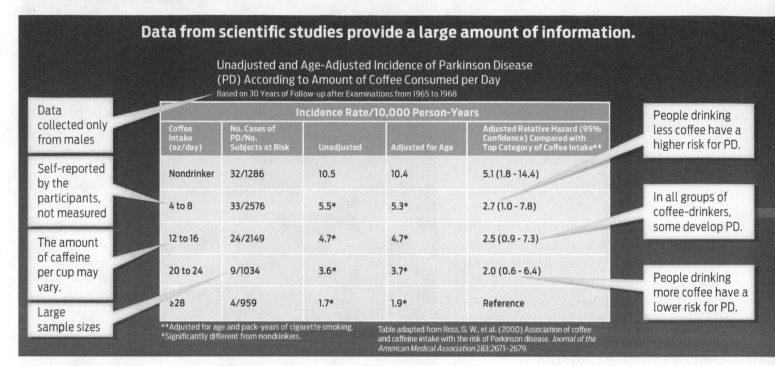

Data from scientific studies provide a large amount of information.

Unadjusted and Age-Adjusted Incidence of Parkinson Disease (PD) According to Amount of Coffee Consumed per Day
Based on 30 Years of Follow-up after Examinations from 1965 to 1968

Coffee intake (oz/day)	No. Cases of PD/No. Subjects at Risk	Incidence Rate/10,000 Person-Years		Adjusted Relative Hazard (95% Confidence) Compared with Top Category of Coffee Intake**
		Unadjusted	Adjusted for Age	
Nondrinker	32/1286	10.5	10.4	5.1 (1.8 - 14.4)
4 to 8	33/2576	5.5*	5.3*	2.7 (1.0 - 7.8)
12 to 16	24/2149	4.7*	4.7*	2.5 (0.9 - 7.3)
20 to 24	9/1034	3.6*	3.7*	2.0 (0.6 - 6.4)
≥28	4/959	1.7*	1.9*	Reference

**Adjusted for age and pack-years of cigarette smoking.
*Significantly different from nondrinkers.

Table adapted from Ross, G. W., et al. (2000) Association of coffee and caffeine intake with the risk of Parkinson disease. *Journal of the American Medical Association* 283:2671–2679.

Data collected only from males

Self-reported by the participants, not measured

The amount of caffeine per cup may vary.

Large sample sizes

People drinking less coffee have a higher risk for PD.

In all groups of coffee-drinkers, some develop PD.

People drinking more coffee have a lower risk for PD.

chemistry that also happens to resist disease—rather than being caused by either smoking or drinking coffee per se.

Moreover, the study followed Japanese-American men. Would the same relationship between caffeine and Parkinson disease be seen in other ethnic groups or in women? Several other epidemiological studies have found a correlation between caffeine consumption and a lower incidence of Parkinson disease in men of other ethnicities. But in women the results have been inconclusive. All in all, there's still no direct evidence that caffeine actually prevents the disease in either men or women.

"While our study found a strong correlation between coffee drinkers and low rates of Parkinson's disease," stated the study's lead author, G. Webster Ross, in a press release, "we have not identified the exact cause of this effect. I'd like to see these findings used as a basis to help other scientists unravel the mechanisms that underlie Parkinson's onset."

Finding a correlation that is not necessarily a cause is a typical result of epidemiological studies, but that does not mean such studies are worthless. In fact, as the author of this study of Japanese-American men indicates, correlations can be a jumping-off point for additional research aimed at identifying root causes. Correlations provide suggestive evidence that merits additional research.

To get a clearer picture of caffeine's role in Parkinson disease, researchers could conduct a type of experiment known as a **randomized clinical trial,** in which the effects of coffee would be measured directly under controlled conditions. One could divide a population into two groups, put one group on coffee and the other on decaf, and then follow both groups for a number of years to see which one had the higher incidence of disease. The problem with such a study is that it is often logistically challenging to conduct, since it can be difficult to get people to stick to

▶RANDOMIZED CLINICAL TRIAL
A controlled medical experiment in which subjects are randomly chosen to receive either an experimental treatment or a standard treatment (or a placebo).

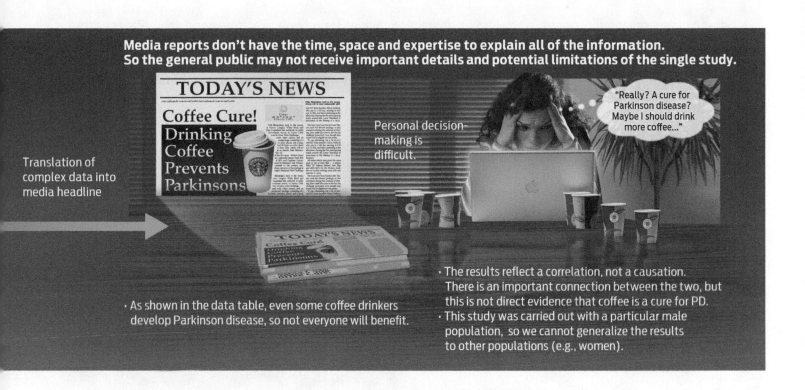

Media reports don't have the time, space and expertise to explain all of the information. So the general public may not receive important details and potential limitations of the single study.

Translation of complex data into media headline

TODAY'S NEWS

Coffee Cure! Drinking Coffee Prevents Parkinsons

Personal decision-making is difficult.

"Really? A cure for Parkinson disease? Maybe I should drink more coffee..."

· As shown in the data table, even some coffee drinkers develop Parkinson disease, so not everyone will benefit.

· The results reflect a correlation, not a causation. There is an important connection between the two, but this is not direct evidence that coffee is a cure for PD.
· This study was carried out with a particular male population, so we cannot generalize the results to other populations (e.g., women).

the regimen for the length of the study. (And such studies are unethical if the experimental treatment is likely to cause harm.)

GETTING BEYOND THE BUZZ

While a lower risk of Parkinson disease represents a potential boon to coffee drinkers, the news for caffeine addicts isn't all good. Over the years, epidemiological studies have also linked caffeine consumption to *higher* rates of various diseases, including osteoporosis, fibrocystic breast disease, and bladder cancer. As with the link to Parkinson disease, however, such correlations do not necessarily prove that caffeine causes any of these diseases. Nevertheless, such studies are often quite influential and newsworthy—like the apparent correlation between coffee and pancreatic cancer that made headlines in 1981. That study was based on a single, small epidemiological study, which was later discounted by further research.

Journalists face unique challenges in covering health news, says Gary Schwitzer of HealthNewsReview.org: "They must cover complex topics, do it quickly, creatively, accurately, completely and with balance—and then be sure they don't 'dumb it down' too much for a general news audience. . . . If they can't do it right, they must realize the *harm* they can do by reporting inaccurately, incompletely, and in an imbalanced way" (INFOGRAPHIC 1.8).

Journalists and scientists aren't the only ones who bear the responsibility of determining what information is trustworthy. As consumers and citizens, we can become more knowledgeable about how science is done and which studies deserve to influence our behavior. Whether it's the latest media report linking cell phones to brain tumors or vaccines to autism, the only way to really judge the value of a study is to sift through the evidence ourselves. Of course, to do that, we might first need a cup of coffee. ■

CHAPTER 1 Summary

▶ Science is an ongoing process in which scientists conduct carefully designed studies to answer questions or test hypotheses.

▶ Scientific hypotheses are tested in controlled experiments or in observational studies, the results of which can support or rule out a hypothesis.

▶ Scientific hypotheses can be supported by data but cannot be proved absolutely, as future studies may provide new findings.

▶ The strength of the conclusions of a scientific study depends on, among other factors, the type of study carried out and the sample size.

▶ Every experiment should have a control—a group that is identical in every way to the experimental group except for one factor: the independent variable.

▶ The independent variable in an experiment is the one being deliberately changed in the experimental group (e.g., coffee intake). The dependent variable is the measured result of the experiment (e.g., effect of coffee on memory).

▶ Often a control group takes a placebo, a fake treatment that mimics the experience of the experimental group.

▶ In epidemiological studies, a relationship between an independent variable (such as caffeine intake) and a dependent variable (such as development of Parkinson disease) does not necessarily mean one caused the other; in other words, correlation does not equal causation.

▶ A randomized clinical trial is one in which test participants are randomly chosen to receive either a standard treatment (or a placebo) or an experimental treatment (e.g., caffeine).

▶ Scientists rely on peer-reviewed scientific reports to learn about new advances in the field. Peer review helps to ensure that the scientific results are valid as well as accurately and fairly presented.

▶ Most of the general public relies on media reports for scientific information. Media reports are not always completely accurate in how they portray the conclusions of scientific studies.

▶ To understand a study properly, it is often necessary to look at how the study was designed and to analyze the data oneself.

▶ Scientific theories are different from everyday theories. A scientific theory has withstood the test of time and extensive testing and is supported by a significant body of evidence.

MORE TO EXPLORE

▶ HealthNewsReview.org

▶ Ryan, L., et al. (2002) Caffeine reduces time-of-day effects on memory performance in older adults. *Psychological Science* 13:68–71.

▶ Seifert, S. M., et al. (2011) Health effects of energy drinks on children, adolescents, and young adults. *Pediatrics* 127(3):511–528.

▶ Rogers, P. J., et al. (2010) Association of the anxiogenic and alerting effects of caffeine with ADORA2A and ADORA1 polymorphisms and habitual level of caffeine consumption. *Neuropsychopharmacology* 35(9):1973–1983.

▶ Ross G. W., et al. (2000) Association of coffee and caffeine intake with the risk of Parkinson disease. *Journal of the American Medical Association* 283(20):2674–2679.

CHAPTER 1 Test Your Knowledge

DRIVING QUESTION 1

How is the scientific method used to test hypotheses?

By answering the questions below and studying Infographics 1.2, 1.3, 1.4, and 1.7, you should be able to generate an answer for the broader Driving Question above.

KNOW IT

1. When scientists carry out an experiment, they are testing a _____.
   - a. theory
   - b. question
   - c. hypothesis
   - d. control
   - e. variable

2. Of the following, which is the earliest step in the scientific process?
   - a. generating a hypothesis
   - b. analyzing data
   - c. conducting an experiment
   - d. drawing a conclusion
   - e. asking a question about an observation

3. In a controlled experiment, which group receives the placebo?
   - a. the experimental group
   - b. the control group
   - c. the scientist group
   - d. the independent group
   - e. all groups

4. In the studies of coffee and memory discussed, the independent variable is _____ and the dependent variable is _____.
   - a. caffeinated coffee; decaffeinated coffee
   - b. memory; caffeinated coffee
   - c. caffeine; memory
   - d. memory; caffeine
   - e. decaffeinated coffee; caffeinated coffee

5. Can an epidemiologist who finds a correlation between the use of tanning beds and melanoma (an aggressive form of skin cancer) in college-age women conclude that tanning beds cause skin cancer?
   - a. yes, as long as the correlation was statistically significant
   - b. yes, but only for college-age women
   - c. yes, but only melanoma skin cancer, not other forms of skin cancer
   - d. no; the study would have to be done with a wider range of participants (males and females of different ages) before it can be concluded that tanning beds cause melanoma
   - e. no; correlation is not proof of causation

USE IT

6. You carry out a clinical trial to test whether a new drug relieves the symptoms of arthritis better than a placebo. You have four groups of participants, all of whom have mildly painful arthritis (rated 7 on a scale of 1 to 10). Each group receives a daily pill as follows: control (group 1) - placebo; group 2- 15 mg; group 3- 25 mg; group 4- 50 mg. At the end of 2 weeks, participants in each group are asked to rate their pain on a scale of 1 to 10. What is the independent variable in this experiment?
   - a. the amount of pain experienced at the start of the experiment
   - b. the amount of pain experienced at the end of the experiment
   - c. the degree to which pain symptoms changed between the start and the end of the experiment
   - d. the drug
   - e. The independent variable could be a, b, or c.

7. You are working on an experiment to test the effect of a specific drug on reducing the risk of breast cancer in postmenopausal women. Describe your control and experimental groups with respect to age, gender, and breast cancer status.

8. Design a randomized clinical trial to test the effects of caffeinated coffee on brain activity. Design your study so that the results will apply to as many people in as many scenarios as possible.

DRIVING QUESTION 2

What factors influence the strength of scientific studies and whether the results of any given study are applicable to a particular population?

By answering the questions below and studying Infographics 1.3 and 1.4, you should be able to generate an answer for the broader Driving Question above.

KNOW IT

9. In which of the following would you have the most confidence?
   - a. a randomized clinical trial with 15,000 subjects
   - b. a randomized clinical trial with 5,000 subjects
   - c. an epidemiological study with 15,000 subjects
   - d. an endorsement of a product by a movie star
   - e. a report on a study presented by a news organization

10. What is the importance of statistical analyses?
    - a. They can reveal whether or not the data have been fabricated.

b. They can be used to support or reject the hypothesis.

c. They can be used to determine whether any observed differences between two groups are real or a result of chance.

d. all of the above

e. b and c

USE IT

11 You carry out a clinical trial to test whether a new drug relieves the symptoms of arthritis better than a placebo. You have four groups of participants, all of whom have mildly painful arthritis (rated 7 on a scale of 1 to 10). Each group receives a daily pill as follows: control (group 1) - placebo; group 2- 15 mg; group 3- 25 mg; group 4- 50 mg. At the end of 2 weeks, participants in each group are asked to rate their pain on a scale of 1 to 10. The mean pain rating of the participants was 6.5 for the placebo, 6.0 for 15 mg of the drug, 4.5 for 25 mg of the drug, and 4.5 for 50 mg of the drug. What is your next step?

a. Invest in the drug company.

b. Conclude that the drug relieves arthritis pain.

c. Run a statistical analysis to determine if the differences are significant.

d. Conclude that the drug doesn't work very well (even the placebo group went down on the pain scale, and there was no difference in results between doses of 25 mg and 50 mg of the drug).

e. a and b

12 Looking at Infographic 1.4 (Sample Size Matters), you see that both graphs show a positive impact of caffeine on memory. However, the data in the graph on the right carry more weight. Why is that? If you read a study that reported only the data in the left graph, would you find the relationship to be compelling? Why or why not?

13 From what you have read in this chapter, would you say a 21-year-old Caucasian female can count on caffeinated coffee to reduce her risk of Parkinson disease?

a. yes, because the results of a peer-reviewed study showed that drinking caffeinated beverages reduced the risk of Parkinson disease

b. no, because participants in that peer-reviewed study were Japanese-American males; it cannot be inferred that the same results would hold for Caucasian females

c. no; she would have to restrict her consumption of coffee to decaffeinated coffee to reduce her risk of Parkinson disease

d. yes; coffee is known to reverse the symptoms of Parkinson disease

e. There are no data on the relationship between drinking caffeinated beverages and Parkinson disease because it would be unethical to conduct such an epidemiological study.

INTERPRETING DATA

14 Most statistical tests report a *p* value that determines whether or not the results are statistically significant (i.e., not produced by chance). Usually the cutoff for *p* values is 0.05: if the *p* value is less than 0.05 the results are considered to be statistically significant. The graph below shows data from a 2012 study published in the *New England Journal of Medicine*. The study examined the impact of the drug Tofacitinib on ulcerative colitis. From the data shown, what dose(s) of Tofacitinib is/are significantly better than the placebo in treating ulcerative colitis?

a. 0.5 mg   d. 15 mg

b. 3 mg   e. both 10 mg and 15 mg

c. 10 mg   f. All doses are more effective than the placebo.

SOURCE: Sandborn, W.J., et al. (2012) Tofacitinib, an oral Janus kinase inhibitor, in active ulcerative colitis. *New England Journal of Medicine* 367:616–624.

DRIVING QUESTION 3

How can you evaluate the evidence in media reports of scientific studies?

By answering the questions below and studying Infographics 1.2, 1.3, 1.4, 1.7, and 1.8, you should be able to generate an answer for the broader Driving Question above.

KNOW IT

15 You hear a news report about a new asthma treatment. What would you want to know before you asked your doctor if this treatment was right for you?

a. Was the drug tested in a randomized clinical trial?

b. How many participants were in the trial?

c. Was there a significant difference between the effect of the new drug and the treatment used in the control group?

d. Did any of the researchers have financial ties to the manufacturer of the new asthma drug?

e. all of the above

16 You are listening to a news report that claims a new study has found convincing evidence that a particular weight-loss product is much more effective than diet and exercise. What can you infer about "convincing" evidence in this case?

a. that it agrees with the hypothesis

b. that statistical tests showed significantly more weight loss in the participants who used the weight-loss product than those who relied on diet and exercise

c. that all the participants lost at least 10 pounds

d. that only the participants who used the weight-loss product lost weight

e. that the participants who used the weight-loss product lost an average of 3 pounds, while the participants that used diet and exercise lost an average of 2 pounds

USE IT

17 How can two different studies investigating the same thing (e.g., the relationship, if any, between caffeinated coffee and memory) come to different conclusions?

a. They may have had different sample sizes.

b. They may have used different types of participants (e.g., participants of different ages or professions).

c. They may have used different amounts of caffeine.

d. They may have evaluated memory differently (e.g., long-term vs. short-term memory).

e. all of the above

18 A scientist who reads an article in a scientific or medical journal can be confident that the report has been peer reviewed.

a. What is a "peer-reviewed" report? Is an article in a daily newspaper a peer-reviewed article?

b. What is the role of a peer reviewer of a scientific article?

c. Why do scientists place so much value on the peer-review process?

19 The mother of a friend is a self-described "coffee addict." She recently received a diagnosis of Parkinson disease. Does her experience negate the results of the *JAMA* study described in this chapter? Why or why not?

20 You may have seen advertisements on television that show beautiful people with clear skin who claim that a specific skin care product is "scientifically proven" to reduce acne. The product reportedly gave these people glowing, clear skin.

a. Is their testimony alone strong enough evidence for you to act on? Why or why not?

b. What kind of scientific evidence would persuade you to spend money on this product? Explain your answer.

DRIVING QUESTION 4

How does the scientific method apply in clinical trials designed to

investigate important issues in human health?

By answering the questions below and studying Infographics 1.2 and 1.3, you should be able to generate an answer for the broader Driving Question above.

USE IT

21 Following the prompts below, design a clinical trial to test the impact of a particular intervention on a specific aspect of human health. You will need to use everything you have learned to do this.

a. From scientific articles or press releases from health organizations you have read, or from your own experiences, what observation(s) can you start with?

b. Do some investigation to generate a testable hypothesis.

c. Design the trial. Consider sample size, whether or not you will use a placebo, what level of blinding you will use, and possible independent and dependent variables.

BRING IT HOME

22 There are many misconceptions about breast cancer and its causes. In the late 1990s, there were rumors that antiperspirants cause breast cancer. There are still catalogues that offer alternative underarm hygiene products that purport to reduce the risk of breast cancer. One viral e-mail claimed that by blocking perspiration, antiperspirants prevent the body from purging toxins, instead forcing the body to store the toxins in lymph nodes in the underarm area near breast tissue. The e-mail stated that men were less likely to develop breast cancer from antiperspirants because their underarm hair trapped most of the product away from direct contact with skin. And as men are less likely to shave their underarms, they are less likely to have shaving nicks through which antiperspirants can enter the body.

a. Read the abstracts of the two articles for which links are provided below.

Darbre, 2005 http://is.gd/pPLxwZ

Harvey and Everett, 2004 http://is.gd/ycqDD8

From the abstracts, and from any other investigation you do, name the components of underarm deodorants and antiperspirants that have been identified as possible culprits in causing breast cancer.

b. Briefly comment on the strengths and weaknesses of each study (consider sample size, control groups, and overall study design).

c. From what you read in the abstracts and from other research you do (cite any additional reliable sources that you consulted), do you think that use of antiperspirants or deodorants or both is a consistent risk factor for breast cancer? Has your opinion about underarm hygiene changed? Explain how and why your opinion has either changed or remained consistent, referring to the abstracts that you have reviewed.

2

MISSION

Prospecting for life on the red planet

▸▸ DRIVING QUESTIONS

1. How is matter organized into molecules of living organisms?
2. What is the definition of life, and how could Martian life be recognized?
3. What is the basic structural unit of life?
4. Why is water so important for life and living organisms?

TO MARS

> **"** *For the landing to succeed, hundreds of events will need to go right, many with split-second timing and all controlled autonomously by the spacecraft.* **"**
> — PETE THEISINGER

SEVEN MINUTES OF TERROR." THAT'S HOW NASA SCIENTISTS DESCRIBED the final moments of its 2012 Mars landing. In that harrowing interval, the speeding spacecraft would need to slow from about 13,200 mph to less than 2 mph as it descended blindly through the Martian atmosphere.

"Those seven minutes are the most challenging part of this entire mission," said Pete Theisinger, project manager at NASA. "For the landing to succeed, hundreds of events will need to go right, many with split-second timing and all controlled autonomously by the spacecraft."

The most frightening part was the last few moments of the descent, when the spacecraft would release its cargo–a 1-ton, SUV-size rover named *Curiosity*–down a floating "sky crane," essentially, a trio of nylon cables suspended from a rocket-powered backpack.

The maneuver had never been attempted before, and even NASA's own scientists had their doubts about it, dubbing it "rover on a rope." But it worked. On August 6, 2012, at 1:32 A.M.

Curiosity is lowered to Mars down the floating sky crane.

Relief and excitement at NASA's Jet Propulsion Laboratory when Curiosity landed safely.

Eastern Standard Time, NASA's Mobile Science Laboratory, aka *Curiosity*, landed successfully on the surface of the red planet.

"Touchdown confirmed," announced NASA engineer Allen Chen. "We're safe on Mars!"

News of the successful landing sent NASA's Jet Propulsion Laboratory into loud cheers as people hugged and high-fived one another. Moments later, *Curiosity* began sending back grainy black and white images of its landing site to a planet full of eager witnesses.

The purpose of NASA's latest trip to Mars–the seventh to date–is to find out whether the planet could have once supported life, and might still support it. NASA calls the mission the "prospecting" stage of its search for life on Mars. What *Curiosity* discovers will allow scientists to begin to answer some fundamental questions, not only about life on Mars, but also about life on Earth. Questions like: How did life begin? Is there more than one type of life? Could life have arrived here on a meteorite from outer space?

These are big questions, but NASA's new rover is nothing if not curious.

THE SEARCH FOR MARTIAN LIFE

This isn't the first time NASA has sent a rover to Mars. Other rovers have moseyed over the red planet, including *Phoenix* in 2008 and *Spirit* and *Opportunity* in 2004. But none has done so with as much flair. With a $2.5 billion price tag and an entire laboratory built into its sleek frame, *Curiosity* is unquestionably the most technologically advanced rover to date.

It's also the most socially connected. Shortly after touching down, *Curiosity* began tweeting news of its progress to its more than 1 million followers, even checking into Mars on Foursquare: "One check-in closer to being Mayor of Mars!" it chirped.

Finding the answer to question of whether there is life on Mars seems straightforward: look and see if anything is growing, or running around, or playing Xbox. By these measures, clearly, there is no life on Mars. The earliest pictures of Mars obtained by NASA's *Mariner 4* spacecraft revealed a dry, rocky landscape, more reminiscent of our lifeless moon than the lush, blue marble we call home. But what

▶HOMEOSTASIS
The maintenance of a relatively constant internal environment.

▶ENERGY
The ability to do work. Living organisms obtain energy either directly from sunlight (through photosynthesis) or from food they consume.

▶METABOLISM
All the chemical reactions taking place in the cells of a living organism that allow it to obtain and use energy.

if Mars harbored microscopic life, invisible to the naked eye? Could life be lurking in the Martian soil?

The first NASA spacecraft to investigate this question was *Viking Lander 1,* which touched down on July 20, 1976. Equipped with mechanical arms that could grab and test Martian soil, *Viking* was designed to look for signs of microscopic life. NASA scientists hypothesized that if life were present in the soil, they should be able to measure its activity. Was anything emitting carbon dioxide, for example, as many organisms on Earth do? They added nutrients to Martian soil and waited to see what would happen.

Initially, the results seemed promising: something in the Martian soil did indeed seem to be breaking down the added nutrients and producing carbon dioxide gas. More intriguing, when the soil was heated to a very high temperature–a temperature that would kill most life–no carbon dioxide was measured. This result seemed like evidence of life.

But scientists now know that such chemical reactions, or reactivity, can occur even in the absence of life–they occur in dry, lifeless deserts on Earth, for example.

"It turns out reactivity is not uniquely biological," says Chris McKay, an astrobiologist with NASA's Ames Research Center in California, who has studied desert analogs of Mars. Still, he says, the experiment was instructive, as it focused attention on a fundamental question in biology: what is life?

Biology is the study of life, so naturally it's important for biologists to know what things

fall under that heading–mosses, for example, are living, but not the rocks they grow on (leave those to the geologists). There are many ways one could define life–on the basis of what it looks like, what it's made of, how it behaves, and so on. And, indeed, scientists and philosophers have offered many definitions of life over the years. But, as the *Viking* experiments show, life can be tricky to search for on the basis of these definitions.

In looking for evidence of chemical reactions, NASA scientists were, in effect, employing a definition of life based on what living things *do*. Biologists generally agree that–on Earth at least–all living things have in common five functional traits. Living things (1) grow and (2) reproduce: they increase in size, and they produce offspring. Living things (3) maintain a relatively stable internal environment in the face of changing external circumstances–producing heat when they're cold, for example–a phenomenon known as **homeostasis** (see Chapter 25). To maintain homeostasis, living things (4) sense and respond to their environment, as when a plant grows toward sunlight. And to carry out these and other life-defining activities, all living organisms (5) obtain and use **energy,** the power to do work. Energy comes from sunlight or food, which living things break down through a series of chemical reactions, the sum total of which is called **metabolism** (INFOGRAPHIC 2.1).

If scientists found an alien being with all these traits, they could make a good case for having found life. But what if the alien had some of, but not all, these traits? Could it be alive?

> **❝** *Our experience with life is limited to familiar Earth life and we know that all known life on Earth descends from a last universal common ancestor.* **❞**
>
> — CAROL CLELAND

"The question 'What is life?' has taken on increasing scientific importance in recent years," says Carol Cleland, a philosopher at the University of Colorado and a member of NASA's Astrobiology Institute. It's become a subject for debate, she says, as molecular biologists attempt to create life from scratch in the lab, and as astrobiologists grapple with how alien life might differ from life on Earth.

"The problem," says Cleland, "is that our experience with life is limited to familiar Earth life and we know that all known life on Earth descends from a last universal common ancestor." Who's to say that all life will look and behave just like it does on this planet?

Even on Earth, our definitions of life don't always hold. For example, mules are clearly alive, but they are sterile and cannot reproduce. Likewise, fire uses energy and grows, but most people would not say fire is alive.

More problematic for NASA, if life once existed on Mars but no longer does, the whole question of what life does is moot, since dead organisms don't *do* anything. That's why, for its latest mission, NASA is relying on different criteria.

CURIOUS ABOUT CHEMISTRY

As one of the scientists behind *Curiosity*, McKay's main job is figuring out the best way to look for life on Mars. He was part of the team responsible for packing *Curiosity's* scientific toolkit.

In many ways, says McKay, *Curiosity* is the sequel to *Viking*, picking up where *Viking* left off. In terms of technological prowess, though, it's light-years ahead. *Curiosity* is equipped with 10 scientific instruments, including a laser that can identify the chemical composition of rocks at a distance. It can also move around, which is a huge advantage over *Viking*.

NASA set *Curiosity* down in a large canyon called Gale Crater. The rover's ultimate destination, which it will travel to over weeks and months, is the slope of a steep peak dubbed Mount Sharp, located at the crater's center. Geologically, the region is similar to the Grand

INFOGRAPHIC 2.1 FIVE FUNCTIONAL TRAITS OF LIFE

Growth
For unicellular (one-celled) organisms, growth is an increase in cell size before reproduction. For multicellular organisms, growth refers to an increase in an organism's size as the number of cells making up the organism increases.

Reproduction
The process of producing new organisms. Offspring are similar, but not necessarily identical, to their parents in general structure, function, and properties.

Homeostasis
Organisms maintain a stable internal environment, even when the external environment changes.

Sense and Respond to Stimuli
Organisms respond to stimuli in many ways. For example, they may move toward a food source or move away from a threatening predator.

Obtain and Use Energy
All living organisms require an input of energy to power their activities. Organisms obtain energy from food (which they either produce themselves by photosynthesis or consume from the environment). Chemical reactions convert that energy into usable forms. The sum total of all these reactions is metabolism.

Canyon, with exposed strata of past eons layered one on top of the other, like a many-layered cake. NASA thinks that this is the best spot to try to reconstruct the past history of Mars, as

INFOGRAPHIC 2.2 TOUCHDOWN, GALE CRATER

→ *Curiosity* set down in Gale Crater, and will navigate up Mount Sharp, where it will look for evidence of life.

Mount Sharp is *Curiosity's* ultimate destination.

The slope of Mount Sharp is similar to that of the Grand Canyon. *Curiosity* will sample the eons of Martian history exposed in the rock layers of this region.

Possible route

Curiosity rover

Curiosity's landing site

Mountains Compared
Mount Sharp is taller than any peak in the continental United States.

MOUNT EVEREST

MOUNT McKINLEY

MOUNT SHARP

MOUNT RAINIER

▶**ELEMENT**
A chemically pure substance that cannot be chemically broken down; each element is made up of and defined by a single type of atom.

▶**MATTER**
Anything that takes up space and has mass.

▶**ATOM**
The smallest unit of an element that cannot be chemically broken down into smaller units.

▶**PROTON**
A positively charged subatomic particle in the nucleus of an atom.

well as to look for signs of fossilized life—much in the way that scientists look for evidence of life in fossils that have been discovered in the clay and sandstone of the Grand Canyon (**INFOGRAPHIC 2.2**).

Unlike Viking, *Curiosity* will not actually test for living organisms (by monitoring carbon dioxide emissions, for example). Instead, it will look for other evidence of life: the chemical building blocks necessary to assemble it.

All life we know of—from amoeba to zebra—uses the same basic chemical recipe: a stew of carbon-based ingredients floating in a broth of water. Carbon (C) is one of approximately 100 different **elements** in the universe. Elements are substances that cannot be broken down by chemical means into smaller substances. They are themselves considered the fundamental components of anything that takes up

space or has mass—the **matter** in the universe. Elements make up both living and nonliving things. The rocky surface of Mars, for example, appears red because of an abundance of the element iron (Fe) that has long since rusted.

The smallest unit of an element that still retains the property of that element is an **atom.** What gives each atom its identity is its specific number of positively charged **protons,** negatively charged **electrons,** and neutral **neutrons.** A carbon atom, for example, has six protons, six electrons, and six neutrons. The relatively heavy protons and neutrons are packed into the atom's dense core, or **nucleus,** while the tiny electrons orbit around it (**INFOGRAPHIC 2.3**).

Carbon is the fourth most common element in the universe and the second most common element in the human body. In fact, just six ele-

> **"** *We don't see any other element that has the sort of flexibility and utility that carbon has.* **"**
>
> — CHRIS MCKAY

ments make up the bulk of you: oxygen (65%), carbon (18%), hydrogen (10%), nitrogen (3%), calcium (1.5%), and phosphorus (1%).

When astrobiologists (and science fiction writers) talk about "carbon-based life forms," they are referring to the fact that carbon forms the backbone of nearly every chemical making up living things. Just as humans have a backbone made of a chain of interconnected vertebrae, the chemicals making up living things have a backbone of interconnected carbon atoms. The backbone can be linear, like our

▶**ELECTRON**
A negatively charged subatomic particle with negligible mass.

▶**NEUTRON**
An electrically uncharged subatomic particle in the nucleus of an atom.

▶**NUCLEUS**
The dense core of an atom.

INFOGRAPHIC 2.3 ALL MATTER IS MADE OF ELEMENTS

The periodic table of elements represents all known elements in the universe. Each element is placed in order on the table by its atomic number, the number of protons found in the nucleus of its corresponding atom.

Periodic Table of Elements

Element
Elements exist as atoms, which have distinct chemical properties determined by the number of subatomic particles the atom has.

Carbon atom

Neutrons are uncharged particles in the nucleus of an atom. A carbon atom has six neutrons.

The atomic mass of an atom is determined by adding the number of protons and neutrons. One type of carbon atom has an atomic mass of 12.

Protons are positively charged particles in the nucleus of an atom. A carbon atom has six protons.

Electrons are negatively charged particles that orbit around the nucleus of an atom in distinct energy shells. A carbon atom has a total of six electrons.

▸**COVALENT BOND**
A strong chemical bond resulting from the sharing of a pair of electrons between two atoms.

▸**MOLECULE**
Atoms linked by covalent bonds.

▸**ORGANIC**
Describes a molecule with a carbon-based backbone and at least one C–H bond.

▸**INORGANIC**
Describes a molecule that lacks a carbon-based backbone and C–H bonds.

▸**CARBOHYDRATE**
An organic molecule made up of one or more sugars. A one-sugar carbohydrate is called a monosaccharide; a carbohydrate with multiple linked sugars is called a polysaccharide.

▸**PROTEIN**
An organic molecule made up of linked amino acid subunits.

▸**LIPIDS**
Organic molecules that generally repel water.

▸**NUCLEIC ACIDS**
Organic molecules made up of linked nucleotide subunits; DNA and RNA are examples of nucleic acids.

▸**MACROMOLECULES**
Large organic molecules that make up living organisms; they include carbohydrates, proteins, and nucleic acids.

▸**MONOMER**
One chemical subunit of a polymer.

spine, or circular–in which case the first carbon in the chain binds to the last carbon in the chain.

"Carbon is very cool, very flexible, very useful," says McKay. "We don't see any other element that has the sort of flexibility and utility that carbon has." That's the main reason why scientists are so interested in it.

SAM I AM

In his lab at NASA's Goddard Space Flight Center in Greenbelt, Maryland, chemist Paul Mahaffy monitors a microwave-size device hairy with cords and wires. When it boots up, the device declares: "Sam I am, I am Sam!" in a playful homage to Dr. Seuss. SAM stands for "Sample Analysis on Mars." It is the tool that *Curiosity* will use to check for life's chemical building blocks. Mahaffy is SAM's principal investigator.

SAM is mounted on *Curiosity* like a backpack. If the powerful cameras on board *Curiosity* are its eyes, then SAM is its extremely sensitive nose. And, according to Mahaffy, the instrument is "getting ready to start sniffing."

The sniffing takes place in a series of interconnected chambers. Once a sample of rock or dirt is scooped up by *Curiosity*, it is popped into SAM, where it is baked to a high temperature. The gases given off from this high-tech crockpot are then analyzed to determine their precise chemical makeup. The whole opera-

tion takes about a day, but interpreting the results can take much longer.

According to Mahaffy, SAM is particularly curious about the carbon on Mars–does it appear linked to other carbon atoms in chains or rings, as it is in living things on Earth? One way that atoms link is by sharing electrons to form a **covalent bond** between them. Atoms linked by covalent bonds form **molecules.** Different atoms can form different numbers of covalent bonds, in different geometries. Carbon atoms have four attachment, or bonding, sites, giving the element enormous versatility in forming molecules.

Living things on Earth are made up of so-called **organic** molecules, which have a backbone of interconnected carbon atoms and at least one carbon attached to a hydrogen atom. Most organic molecules require living things to make them, which is why they are often telltale signs of life. An example of a simple organic molecule is glucose, a type of sugar. Its molecular formula is $C_6H_{12}O_6$. This means that each molecule of glucose has 6 carbon atoms, 12 hydrogen atoms, and 6 oxygen atoms. Glucose is a ring-shaped molecule, with the carbon atoms forming the backbone of the ring.

Nonliving things can also contain carbon, but this carbon is **inorganic:** inorganic molecules do not have a carbon-carbon backbone and a carbon-hydrogen bond. Carbon dioxide

Meet SAM, Curiosity's organic molecule detector.

INFOGRAPHIC 2.4 CARBON IS A VERSATILE COMPONENT OF LIFE'S MOLECULES

Molecules are atoms linked by covalent bonds. The element carbon is a key component of the molecules of living organisms because it can form multiple covalent bonds.

Carbon can form multiple covalent bonds

Electrons

Methane

Covalent bond
Atoms share electrons.

Carbon atoms have four potential binding sites. They can therefore bind up to four different atoms.

Carbon forms both organic and inorganic molecules

Organic molecules have chains of carbon atoms covalently bound to other carbon and hydrogen atoms.

Carbon atoms in inorganic molecules are not covalently bound to other carbon or hydrogen atoms.

Organic molecule: Glucose ($C_6H_{12}O_6$)

Inorganic molecule: Carbon dioxide (CO_2)

(CO_2), for example, is an inorganic molecule, one found in the atmospheres of both Mars and Earth (**INFOGRAPHIC 2.4**).

Just four types of organic molecules make up living things on Earth: **carbohydrates, proteins, lipids,** and **nucleic acids.** Every molecule in the human body can be classified as one or other of these organic molecules. Your skin, for example, is composed of the proteins collagen and elastin, and the protein hemoglobin carries oxygen in your blood. The padding in your soft spots is composed of lipids called triglycerides, also known as fats. And in your liver and muscle cells, a carbohydrate called glycogen helps store energy. All of these

organic molecules have a backbone of interlinked carbon atoms.

Organic molecules can be quite large and are therefore considered **macromolecules.** Macromolecules share a similar organization in that they are composed of subunits called **monomers** linked together in a chain. When two or more monomers join they form a **polymer.** Carbohydrates, for example, are polymers made up of linked monomers called **monosaccharides;** similarly, proteins are made up of subunits called **amino acids** that are bonded together; and nucleic acids are polymers composed of **nucleotides** that form long chains (see **UP CLOSE: MOLECULES OF LIFE**).

▶**POLYMER**
A molecule made up of individual subunits, called monomers, linked together in a chain.

▶**MONOSACCHARIDE**
The building block, or monomer, of a carbohydrate.

▶**AMINO ACID**
The building block, or monomer, of a protein.

▶**NUCLEOTIDE**
The building block, or monomer, of a nucleic acid.

2

UP CLOSE MOLECULES OF LIFE: CARBOHYDRATES, PROTEINS, LIPIDS, NUCLEIC ACIDS

a. Carbohydrates Are Made of Monosaccharides

Carbohydrates are made up of repeating subunits known as monosaccharides, or simple sugars. Carbohydrates act as energy-storing molecules in many organisms. Other carbohydrates provide structural support for cells.

Glucose is an important monosaccharide.

Carbon atoms

Complex carbohydrate

Monosaccharides
The backbone of carbon atoms in monosaccharides is most often arranged in a ring.

Complex Carbohydrates
Monosaccharides like glucose can be bonded together in straight or branching chains called complex carbohydrates.

b. Proteins Are Made of Amino Acids

Proteins are polymers of different small repeating units called amino acids joined together by peptide bonds. Proteins carry out many functions in cells. They help speed up the rate of chemical reactions. They also move things through and around cells and even help entire cells move.

Amino Acid
There are 20 different amino acids found in proteins. Each amino acid shares a common core structure (shown in green).

Amino group

Carboxyl group

Linear Strand of Amino Acids
Different amino acids have different side chains (highlighted in different colors).

Folded, Three-Dimensional Protein
Proteins do not function properly until they fold into their unique three-dimensional shape.

c. Lipids Are Hydrophobic Molecules

There are different types of lipid, each with a distinct structure and function. Lipids are not made up of repeating subunits or building blocks, but they are all hydrophobic molecules, meaning they don't mix with water.

Saturated

Unsaturated

Cholesterol

Fatty Acids
Fatty acids contain long chains of carbon atoms bonded to one another and to hydrogen atoms.

Sterols
Sterols have four connected carbon rings. Cholesterol is a sterol that's an important component of cell membranes. Other sterols may be hormones or color-inducing pigments.

Glycerol   Fatty acids

Choline group   Phosphate group   Glycerol   Fatty acids

Triglycerides
Triglycerides, also known as fat, have three fatty acid chains attached to a glycerol molecule. Fats store large amounts of energy and also provide padding and thermal insulation.

Phospholipids
Phospholipids have two fatty acid chains and a phosphate group attached to a glycerol molecule. Phospholipids are an important component of cell membranes.

d. Nucleic Acids Are Made of Nucleotides

Nucleic acids are polymers of repeating subunits known as nucleotides. There are two types of nucleic acid, DNA and RNA, each of which is made up of slightly different types of nucleotide. DNA and RNA are critical for the storage, transmission, and execution of genetic instructions.

P

Base

Sugar

Nucleotide
Nucleotides share a common core structure, including a phosphate group and a sugar, which varies slightly between DNA and RNA. Each of the five different nucleotides differs by virtue of the individual base.

RNA
RNA molecules consist of only one linear chain of bonded nucleotides.

DNA
A DNA molecule consists of two chains of bonded nucleotides twisted into a helical shape.

▸**CELL**
The basic structural unit of living organisms.

▸**CELL MEMBRANE**
A phospholipid bilayer with embedded proteins that forms the boundary of all cells.

▸**PHOSPHOLIPID**
A type of lipid that forms the cell membrane.

▸**HYDROPHOBIC**
"Water-fearing"; hydrophobic molecules will not dissolve in water.

▸**HYDROPHILIC**
"Water-loving"; hydrophilic molecules dissolve in water.

Using SAM's powerful sensors, *Curiosity* will test Martian soil for these molecules of life or their subunits. While their presence would not absolutely prove the existence of life–a few can be made without life–it would show that Mars has one of the main ingredients necessary for building life. So it is an important first step.

As for what organic molecules, or "organics," he hopes to find, McKay isn't picky. Just to go from no organics to organics, he says, would be an exciting first step. Once scientists locate the organics, they can take the next logical step and ask whether they resemble ones on Earth: "Are any of them sugars? Or amino acids? Or chocolate?" he says. That's when things will really get interesting.

McKay cautions that recognizing evidence of life might be harder than we realize. After all, he says, we have only one example of life–Earth life–so we have to be careful about extrapolating. Nevertheless, the fact that Mars is so close to Earth, and that the early histories of

the two planets were likely very similar, means that any remnants of life on Mars might conceivably resemble those on Earth.

They may even share a common origin.

"MY FAVORITE EARTHLINGS"

When not helping to send rovers into space, McKay spends his time researching more terrestrial habitats–including ones nearly as foreboding as Mars, such as Chile's Atacama Desert and the Dry Valleys of Antarctica. Deserts are intriguing to him, he says, because they are good analogs of Mars: they allow scientists to ask questions about the Martian environment without actually traveling there.

McKay's office is full of rock souvenirs from deserts all over world. Rocks are home to some of his favorite creatures–cyanobacteria–which he calls "my favorite earthlings."

Cyanobacteria form a layer of green beneath desert rocks, where they live on light and moisture that is trapped there as in a greenhouse. Cyanobacteria represent one of the few organisms that can survive in these extreme environments. In fact, cyanobacteria are some of the most ancient organisms on Earth, having first evolved some 2.5 billion years ago. Not only that, through photosynthesis (see Chapter 5) they filled the atmosphere with its first breaths of oxygen, making it habitable for the rest of life.

Not bad for an organism made up of only a single **cell.** Cells are the basic structural unit of life on Earth; they are what enclose life, giving it boundaries. Some organisms, like cyanobacteria, are made up of only a single cell; humans contain trillions of them.

All cells have the same basic structure: they are water-filled sacs bounded by a **cell membrane.** The cell membrane is a two-ply layer of **phospholipids** in which proteins are embedded. Each phospholipid has one **hydrophobic** ("water-fearing") end that repels water and a **hydrophilic** ("water-loving") end that attracts it. What happens when a bunch of

Astrobiologist Chris McKay studies desert analogs of Mars.

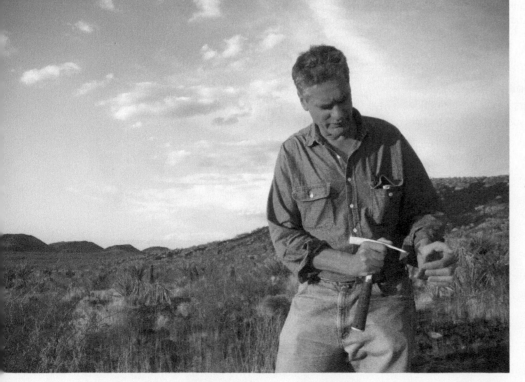

INFOGRAPHIC 2.5 A LAYER RICH IN PHOSPHOLIPIDS DEFINES CELL BOUNDARIES

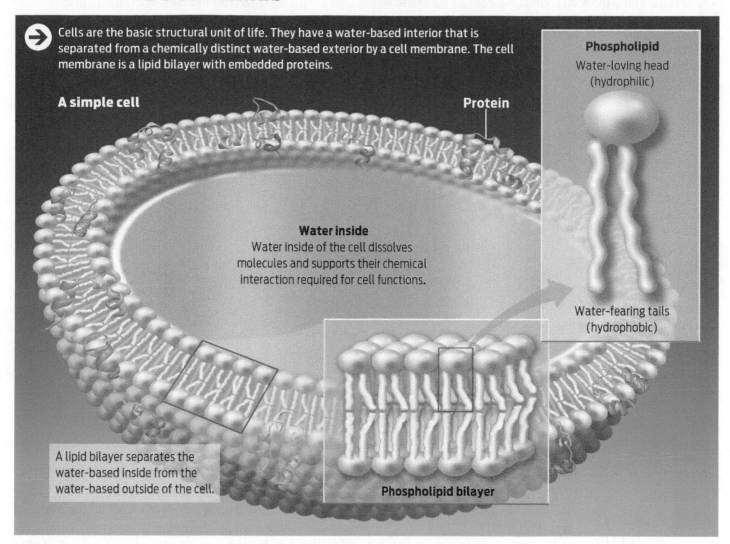

Cells are the basic structural unit of life. They have a water-based interior that is separated from a chemically distinct water-based exterior by a cell membrane. The cell membrane is a lipid bilayer with embedded proteins.

A simple cell

Protein

Phospholipid

Water-loving head (hydrophilic)

Water inside
Water inside of the cell dissolves molecules and supports their chemical interaction required for cell functions.

Water-fearing tails (hydrophobic)

A lipid bilayer separates the water-based inside from the water-based outside of the cell.

Phospholipid bilayer

phospholipids are surrounded by water? They form a lipid sandwich: the hydrophobic tails cluster together, burying themselves in the middle of the membrane, as far away from water as possible, while the hydrophilic heads face out, toward the water (**INFOGRAPHIC 2.5**).

No one knows whether or not cells exist on Mars, but researchers suspect that if life ever did evolve on Mars, it would have been microscopic and unicellular, since that is what the earliest life on Earth was like.

McKay knows of only one desert where it's so dry that even cyanobacteria can't survive,

and that's Atacama Desert in Chile. It is perhaps the most Mars-like of any place on Earth. In fact, as dry as Atacama is, Mars is drier.

It wasn't always that way. Scientists believe Mars was likely a lot moister in its earlier days—3 or 4 billion years ago, when the planet still had an atmosphere to speak of.

FOLLOW THE WATER

In their search for extraterrestrial life, astrobiologists often use this rule of thumb: "Follow the water." Water is viewed as a proxy for life because it is so crucial to life on Earth. Water

INFOGRAPHIC 2.6 WATER IS POLAR AND FORMS HYDROGEN BONDS

Water is a polar molecule because electrons are not shared equally between the oxygen and the hydrogen atoms in each of its polar covalent bonds. Electrons are pulled closer to the oxygen atom than to the hydrogen atoms, creating a slightly negative oxygen atom and a slightly positive hydrogen atom. When many water molecules are near one another, the partially positive hydrogen atoms of some molecules are attracted to the partially negative oxygen atoms of neighboring water molecules. These attractions are hydrogen bonds, weak electrical attractions.

Water Is Polar
One end of the molecule is more positive and the other is more negative.

Partial positive charge

More-positive hydrogen side

Partial negative charge

More-negative oxygen side

The negatively charged electrons in these covalent bonds are more strongly attracted by the oxygen atom.

Water Forms Hydrogen Bonds
Two polar molecules are attracted to each other.

Hydrogen bond between opposite partial charges

Hydrogen bonds

▶POLAR MOLECULE

A molecule in which electrons are not shared equally between atoms, causing a partial negative charge at one end and a partial positive charge at the other; for example, water.

▶HYDROGEN BOND

A weak electrical attraction between a partially positive hydrogen atom and an atom with a partial negative charge.

makes up more than 75% of a cell's weight. All of life's chemical reactions take place in water, and many living things can survive only a few days without it.

A water molecule–H_2O–is shaped like Mickey Mouse's head (the oxygen atom is the face, and the two hydrogens are the ears). Many of water's life-conducive properties are a function of this simple shape. Water is a **polar molecule,** meaning that the electrons in the bonds between oxygen and hydrogen in water are shared unequally; the oxygen atom has a stronger pull on electrons than the hydrogen atoms do, giving water a partial negative charge on one side and a partial positive charge on the other. When water molecules are near one another, the partial negative and partial positive charges of water molecules attract one another, forming weak electri-

cal interactions known as **hydrogen bonds** (INFOGRAPHIC 2.6).

Hydrogen bonds are crucial for many of water's life-sustaining properties. First, they make water "sticky," acting as a kind of glue holding water molecules together. The stickiness of water allows water molecules to cling to one another **(cohesion)** or to a surface **(adhesion).** You can see evidence of water's stickiness wherever you look: a drop of water clinging to a leaf despite the downward pull of gravity, or an insect able to walk on the surface of a pond.

Compared to other molecules its size, water has a large liquid range–freezing at 0°C (32°F) and boiling at 100°C (212°F). That's because water molecules can absorb a lot of energy before they get hot and vaporize (that is, turn into a gas)–again because of their tenacious hydrogen bonds. Water's liquid range can be extended even further: add salt and you can lower the freezing point to -46°C (-50°F); increase the pressure and you can bump up the boiling point to over 343°C (650°F). It's because there is so much salt in seawater that most oceans don't freeze in winter.

Unlike most substances on Earth, water has the unusual property of being less dense as a solid than as a liquid: ice floats. When water freezes, the water molecules form an ordered array, with individual water molecules spaced at equal and fixed "arm's length" from one another. This increased separation between water molecules decreases the density of ice, allowing ice to float on water. And because it does, fish can live beneath frozen lakes in winter and not turn into ice cubes (INFOGRAPHIC 2.7).

Perhaps the most important life-conducive property of water is that it is a universal **solvent,** capable of dissolving just about any substance–even gold. Water transports all of life's dissolved molecules, or **solutes,** from place to place–whether through a cell, a body, or an ecosystem. Life, in essence, is a water-based **solution.** In fact, many biological molecules, like proteins and DNA, have the specific shapes

▶**COHESION**
Water molecules sticking to water molecules through hydrogen bonding.

▶**ADHESION**
Water molecules sticking to other surfaces through hydrogen bonding.

▶**SOLVENT**
A substance in which other substances can dissolve; for example, water.

▶**SOLUTE**
A dissolved substance.

▶**SOLUTION**
The mixture of solute and solvent.

INFOGRAPHIC 2.7 HYDROGEN BONDS GIVE WATER ITS UNIQUE PROPERTIES

Water Is Sticky
Hydrogen bonding allows water molecules to stick to one another (cohesion) and to other surfaces (adhesion), making them wet.

Water Can Absorb a Lot of Energy
It takes a lot of heat energy to disrupt the strong hydrogen bonds between water molecules. That is why water is a liquid at a wide range of temperatures.

Liquid water   Ice

Ice Is Less Dense than Liquid Water
When water freezes, the bonds between the water molecules become rigid and expand, increasing the overall volume of the water. This makes ice less dense than liquid water. That is why ice floats.

INFOGRAPHIC 2.8 WATER IS A GOOD SOLVENT

Because water molecules have partial charges, they can interact with charged ions and other hydrophilic molecules, allowing water to coat and then dissolve these solutes.

Ions Form Ionic Bonds

Sodium and chloride atoms become charged ions when sodium donates an electron to chloride.

Oppositely charged ions are attracted to each other and form ionic bonds.

Sodium atom (Na)

Chlorine atom (Cl)

Sodium ion (Na+)

Chlorine ion (Cl-)

Sodium chloride (NaCl)

Charged Substances Dissolve in Water

Table salt crystals are formed by ionic bonds between sodium and chloride ions.

Polar water molecules disrupt the ionic bonds between Na and Cl, allowing water to dissolve salt into Na⁺ and Cl⁻ ions.

Na⁺   Cl⁻

Water

Salt

Dissolved chloride ion

The more-positive ends of the water molecules interact with the negative chloride ions.

The more-negative ends of the water molecules interact with the positive sodium ions.

Water

Dissolved sodium ion

▶IONIC BOND
A strong electrical attraction between oppositely charged ions formed by the transfer of one or more electrons from one atom to another.

▶ION
An electrically charged atom, the charge resulting from the loss or gain of electrons.

▶pH
A measure of the concentration of H⁺ in a solution.

they do only because of the surrounding water with which they interact.

Water is an excellent solvent for other polar molecules with partial charges and for substances, like salt, that contain **ionic bonds.** Ionic bonds form between atoms that have opposite electrical charges (positive and negative). Such charged atoms, or **ions,** form when one atom loses a negatively charged electron (becoming a positively charged ion) and another atom gains that electron (becoming a negatively charged ion). Ionic bonds are the strong bonds formed between these oppositely charged ions. By surrounding each charged ion, water dissolves the bond between them **(INFOGRAPHIC 2.8).**

When astrobiologists speak about the importance of water for life, they make an important qualification: *liquid* water. Frozen water is found throughout the universe; there are abundant quantities on Mars, for example. But only on Earth does water exist primarily in its liquid form at room temperature. "Liquid water is the key requirement in the search for life," says McKay. "The other worlds of the solar system have enough light, enough carbon, and enough of the other key elements for life. Water in the liquid form is rare."

Though liquid water is not present on the surface of Mars today, scientists suspect that liquid water—lots of it—once covered the planet.

Clues to this ancient water can be seen all over the surface of the planet, which in many places is carved out like sections of the Grand Canyon. *Phoenix Lander* also found telltale signs of liquid water's past on the surface of Mars in the form of salt deposits like those you can see when seawater evaporates. "The case for water, we could say, is tight," says McKay.

Where all this water went, no one knows. But some believe that liquid water may still lurk beneath the surface of the planet, and may even bubble to the surface periodically, as is suggested by photographs of apparent water flows taken in 2004 and 2005 by NASA's *Mars Global Surveyor* satellite.

If there is water within Mars, would it have the properties of Earth water and therefore support life? Depending on what's dissolved in it, water can have a wide range of characteristics—from caustic drain cleaner to calming chamomile tea. The different chemical properties of water-based solutions reflect their **pH,** the concentration of hydrogen ions (H⁺) in a solution, defined as ranging from 0 to 14. In every water-based solution, water molecules (H₂O) split briefly into separate hydrogen (H⁺) and hydroxide (OH⁻) ions. In pure water, the number of separated H⁺ ions is by definition exactly equal to the number of separated OH⁻ ions, and the pH is therefore 7, or neutral—the dead center of the 0 to 14 scale. Acidic solutions, or **acids,** have a higher concentration of hydrogen ions (H⁺) and a pH closer to 0. When acids are added to water, they increase the concentration of hydrogen ions and make the solution more acidic. Basic solutions, or **bases,** on the other hand, have a lower concentration of H⁺ ions and a pH closer to 14. Bases remove H⁺ ions from a solution, thereby increasing the proportion of OH⁻ ions.

Strong acids and bases are highly reactive with other substances, which makes them destructive to the molecules in a cell. Also, many biochemical reactions take place only at a certain pH. Living things are thus extremely sensitive to changes in pH, and most function

best when their pH stays within a specific range. The pH of human blood is about 7.4. If that pH were to fall even slightly, to 7, our biochemistry would malfunction and we would die. *Phoenix Lander* calculated the pH of Mars as 7.7–mild enough to grow asparagus, as the mission's chief chemist put it **(INFOGRAPHIC 2.9).**

▶**ACID**
A substance that increases the hydrogen ion concentration of solutions, making them more acidic.

▶**BASE**
A substance that reduces the hydrogen ion concentration of solutions, making them more basic.

INFOGRAPHIC 2.9 SOLUTIONS HAVE A CHARACTERISTIC pH

The pH of a solution is a measure of the concentration of hydrogen ions (H⁺) in it. Solutions with a low concentration of H⁺ ions have a basic pH (greater than pH 7). Solutions with a high concentration of H⁺ ions have an acidic pH (a pH of less than 7). Both acids and bases can be damaging because they are highly reactive with other substances. A neutral solution has a pH of 7.

▶VIRUS
An infectious agent made up of a protein shell that encloses genetic information.

▶PRION
A protein-only infectious agent.

"WEIRD LIFE"

So far, NASA's search for life on Mars has stuck very closely to our understanding of life on Earth, where living things seem to share certain chemical and structural properties, like carbon-based molecules and cells. Nevertheless, there are a few exceptions, or boundary cases, that seem to bend the rules of life on Earth. **Viruses,** for example, reproduce and pass their genetic information on to new viruses, but they are not made of cells. Instead, they consist of a protein shell that encloses genetic information. Viruses reproduce by infecting a host cell and hijacking its cellular machinery to make copies of itself. Other noncellular, self-reproducing entities include **prions,** infectious proteins that are responsible for mad cow disease and related illnesses. Whether or not viruses and prions are truly alive is hotly debated among scientists (INFOGRAPHIC 2.10).

If viruses and prions bend the rules, then might not Martian life as well? In 2007, the National Academy of Sciences issued a "weird life" report suggesting that NASA should not

INFOGRAPHIC 2.10 ON THE FRINGE

Many organisms defy our criteria for living organisms, and many infectious agents that are not technically alive have powerful impacts on living organisms.

Viruses are not cellular. They infect other cells and use host cell machinery to replicate.

Prions are not cellular. They are infectious proteins that replicate in cells, causing disease.

These bacteria live in a high concentration of arsenic, a chemical that is toxic to most organisms.

These bacteria live in conditions of extreme heat and pressure that would kill most organisms.

> **❝** *We shouldn't lock ourselves into
> a definition that might blind us
> to the presence of unfamiliar forms of life
> should we be so fortunate
> to encounter them.* **❞**
> — CAROL CLELAND

be so narrowly focused on water and organic molecules in its search for life on other planets. True, water may be crucial to life on Earth, but that doesn't mean that other solvents–ammonia or methane, for example–could not support life elsewhere, the report noted. The report also urged the space agency to avoid being "fixated on carbon," even though carbon forms the scaffold of life on Earth. Other elements, for example silicon, could hypothetically provide a functional scaffold for life on other planets.

"Because our experience of life is limited to a single example–familiar Earth life," says philosopher Cleland, "we shouldn't lock ourselves into a definition that might blind us to the presence of unfamiliar forms of life should we be so fortunate to encounter them."

In 2010, researchers with the U. S. Geological Survey reported a very unusual finding that seemed to bear out this warning: they had discovered a bacterium, in a lake full of arsenic, that apparently could substitute arsenic for phosphorus in its DNA. In other words, this bacterium was an apparent exception to the "universal" formula of life. And not only that, it broke the rules by incorporating an element–arsenic–that is wickedly poisonous to most living creatures on Earth. The news made a big splash when it was first reported.

Unfortunately, other researchers were not able to reproduce the controversial findings, and scientists now think that this "exception" does not hold up: the bacteria do need phosphorus, and the universal formula of life stands unbroken.

Nevertheless, the fact that this bug can live happily in a lake full of poison does challenge our notions of what life looks like and where it can survive. Microscopic organisms have been found living just about anywhere on Earth, from radioactive waste and deep-sea vents to frozen Antarctic lakes submerged under miles of ice. Such extreme-loving organisms reveal that life is nothing if not adaptive. Could similarly adaptive organisms have once inhabited Mars? Might they still? NASA researchers are cautiously optimistic.

In December 2012, *Curiosity* ran its first sample of Martian dirt through SAM. No definitive proof of organics yet, but the rover still has many more promising locations to search within Gale Crater. Will *Curiosity* find what she is looking for? McKay, for one, is confident she will.

"I spend my time and energy in the search for evidence of life on Mars," he says. "Obviously, this is because I think there must have been life there and we have a good chance of finding evidence of it." ■

CHAPTER 2 Summary

▶ Life is difficult to define in universal terms because we have only a single example of it to consider: life on Earth.

▶ On Earth, living organisms share a number of functional characteristics: they grow and reproduce, maintain homeostasis, sense and respond to their environment, and rely on energy to carry out their functions.

▶ All matter is composed of elements, of which there are about 100 in the universe. Each element has a unique atomic structure, with a particular number of protons, neutrons, and electrons.

▶ When atoms share pairs of electrons they form covalent bonds, building molecules.

▶ On Earth, living organisms are made up of organic molecules, those containing a backbone of the element carbon.

▶ Four types of carbon-based organic molecule make up living things: proteins, carbohydrates, nucleic acids, and lipids.

▶ Living organisms on Earth are made of cells, which contain water and are surrounded by a cell membrane; cells are the smallest unit of life.

▶ Water is a polar molecule, with a partial positive and a partial negative charge.

▶ Because water molecules have partially positive hydrogen atoms and partially negative oxygen atoms, they can form hydrogen bonds (attractions between these opposite partial charges) with each other and interact with other charged molecules.

▶ Water has many properties that make it a crucial component of life on Earth: it is "sticky," it regulates heat well, it floats when frozen, and it is a good solvent.

▶ When atoms lose or gain electrons, they become ions. Oppositely charged ions can form ionic bonds—strong electrical attractions. Water is a good solvent of substances with ionic bonds.

▶ Substances that, like salt, easily dissolve in water are considered hydrophilic; substances that do not dissolve in water, like lipids, are hydrophobic.

▶ The concentration of H^+ ions in a solution determines its pH. Most chemical reactions in cells take place at a nearly neutral pH.

▶ If life is found on other planets, it may or may not use the chemical framework used by life on Earth.

MORE TO EXPLORE

▶ NASA, Mars Science Laboratory http://www.nasa.gov/mission_pages/msl/index.html

▶ Twitter @MarsCuriosity

▶ McKay, C. P. (2010) An origin of life on Mars. *Cold Spring Harbor Perspectives in Biology*. March 3.

▶ McKay, C. P. (2004) What is life—and how do we search for it in other worlds? *PLoS Biol* 2(9):1260–1263.

▶ Cleland, C. E., and Chyba, C. F. (2007) Does 'life' have a definition? In Sullivan, W. T., III, and Baross, J. A., eds., *Planets and Life: The Emerging Science of Astrobiology*. Cambridge, UK: Cambridge University Press.

▶ *The Limits of Organic Life in Planetary Systems*. (2007) Washington, DC: National Academies Press. http://www.nap.edu/catalog.php?record_id=11919#toc

CHAPTER 2 Test Your Knowledge

DRIVING QUESTION 1

How is matter organized into molecules of living organisms?

By answering the questions below and studying Infographics 2.3, 2.4, and 2.8 and Up Close: Molecules of Life, you should be able to generate an answer for the broader Driving Question above.

KNOW IT

1 What subatomic particles are located in the nucleus of an atom?

- **a.** protons
- **b.** neutrons
- **c.** electrons
- **d.** protons, neutrons, and electrons
- **e.** protons and neutrons

2 When an atom loses an electron, what happens?

- **a.** It becomes positively charged.
- **b.** It becomes negatively charged.
- **c.** It becomes neutral.
- **d.** Nothing happens.
- **e.** Atoms cannot lose an electron because atoms have a defined number of electrons.

3 Glucose (a monosaccharide) has the molecular formula $C_6H_{12}O_6$. How many carbon atoms are in each glucose molecule?

USE IT

4 Consider the types of lipid.

- **a.** How does a sterol, such as cholesterol, differ from a triglyceride?
- **b.** Structurally, what do triglycerides and phospholipids have in common?

5 A cell is unable to take up or make sugars. Which molecule(s) will it be unable to make?

- **a.** carbohydrates
- **b.** proteins
- **c.** lipids
- **d.** nucleic acids
- **e.** all of the above
- **f.** a and d

DRIVING QUESTION 2

What is the definition of life, and how could Martian life be recognized?

By answering the questions below and studying Infographics 2.1 and 2.10, you should be able to generate an answer for the broader Driving Question above.

KNOW IT

6 Which of the following is *not* a generally recognized characteristic of most (if not all) living organisms?

- **a.** the ability to reproduce
- **b.** the ability to maintain homeostasis
- **c.** the ability to obtain energy directly from sunlight
- **d.** the ability to sense and respond to the environment
- **e.** the ability to grow

7 What is homeostasis? Why it is important to living organisms?

8 What does it mean to say a macromolecule is a polymer? Give an example.

9 A collection of amino acids could be used to build a

- **a.** protein.
- **b.** complex carbohydrate.
- **c.** triglyceride.
- **d.** nucleic acid.
- **e.** cell.

USE IT

10 How would you assess whether or not a possibly living organism from another planet were truly alive?

11 Which of the characteristics of living organisms (if any) allow you to distinguish between living and formerly living (that is, dead) organisms? Explain your answer.

12 If, in a mound of dirt, you had evidence that carbon dioxide was being consumed and converted to glucose, what could you conclude about the presence of a living organism? Explain your answer.

2

DRIVING QUESTION 3

What is the basic structural unit of life?

By answering the questions below and studying Infographics 2.5 and 2.10, you should be able to generate an answer for the broader Driving Question above.

KNOW IT

13 The basic building blocks of life are

    a. DNA molecules.

    b. cells.

    c. proteins.

    d. phospholipids.

    e. inorganic molecules.

14 The cell membrane is made of

    a. water.

    b. proteins.

    c. phospholipids.

    d. nucelotides.

    e. b and c.

USE IT

15 What are the arguments for and against considering viruses living organisms?

16 Why do phospholipids form a bilayer in water-based solutions?

DRIVING QUESTION 4

Why is water so important for life and living organisms?

By answering the questions below and studying Infographics 2.5, 2.6, 2.7, 28, and 2.9, you should be able to generate an answer for the broader Driving Question above.

KNOW IT

17 Is olive oil hydrophobic or hydrophilic? What about salt? Explain your answer.

18 The "stickiness" of water results from the _____ bonding of water molecules.

    a. hydrogen

    b. ionic

    c. covalent

    d. acidic

    e. hydrophobic

19 Coffee or tea with sugar dissolved in it is an example of a water-based solution.

    a. What is the solvent in such a beverage?

    b. What is the solute in such a beverage?

    c. Given that the sugar has dissolved in the beverage, are sugar molecules hydrophobic or hydrophilic?

20 As an acidic compound dissolves in water, the pH of the water

    a. increases.

    b. remains neutral.

    c. decreases.

    d. doesn't change.

    e. becomes basic.

21 The bond between the oxygen atom and a hydrogen atom in a water molecule is a(n) _____ bond.

    a. covalent

    b. hydrogen

    c. ionic

    d. hydrophobic

    e. noncovalent

22 How do ionic bonds compare to hydrogen bonds? What are the similarities and differences?

USE IT

23 Why do olive oil and vinegar (a water-based solution) tend to separate in salad dressing? Will added salt dissolve in the oil or in the vinegar? Explain your answer.

24 Which of the following are most likely to dissolve in olive oil?

    a. a polar molecule

    b. a nonpolar molecule

    c. a hydrophilic molecule

    d. a and c

    e. b and c

INTERPRETING DATA

25 Look at Infographic 2.9 and classify each of the following as an acid or a base and state the hydrogen ion concentration relative to a solution with a neutral pH: drain cleaner; coffee; soda.

MINI CASE

26 One approach to finding out if there is life on Mars is to bring Martian dirt samples to Earth for analysis. What are possible considerations for science and society if a Martian life form is released on Earth? Given that Curiosity has landed on Mars, what are the possible consequences if an Earth life form is released on Mars? What steps can mission control take to minimize these risks?

BRING IT HOME

27 Your tax dollars are being invested in projects such as the Curiosity rover project. Investigate the NASA website to learn more about NASA's rationalization for the investment in this mission. Now draft a letter to your congressional representative that expresses your opinion about this expenditure of taxpayer dollars. If you agree, state specific reasons why you think this a good investment of your money. If you disagree, state your reasons, and describe at least two other programs that you would prefer to see funded.

Alexander Fleming
in his lab

WONDER DRUG

How a chance discovery in a London laboratory revolutionized medicine

O N A SEPTEMBER MORNING IN 1928, BIOLOGIST ALEXANDER Fleming returned to his laboratory at St. Mary's Hospital in London after a short summer vacation. As usual, the place was a mess. Flasks were scattered everywhere, and his bench was strewn with the petri plates on which he was growing bacteria. On this day, as Fleming sorted through the plates, he noticed that one was growing a patch of fluffy white mold. It had been contaminated,

▶▶ DRIVING QUESTIONS

1. What structural features are shared by all cells, and what are the key differences between prokaryotic and eukaryotic cells?

2. How do solutes and water cross membranes, and what determines the direction of movement of solutes and water in different situations?

3. How do antibiotics target bacteria, and in what situations is antibiotic therapy indicated?

4. What are some key eukaryotic organelles and their functions?

▶**ANTIBIOTIC**
A chemical that can slow or stop the growth of bacteria; many antibiotics are produced by living organisms.

▶**CELL THEORY**
The concept that all living organisms are made of cells and that cells are formed by the reproduction of existing cells.

▶**PROKARYOTIC CELLS**
Cells that lack internal membrane-bound organelles.

likely by a rogue mold spore that had drifted in from a neighboring laboratory.

Fleming was about to toss the plate in the sink when he noticed something unusual: wherever mold was growing, there was a zone around the mold where the bacteria did not seem to grow. Curious, he looked under a microscope and saw that the bacterial cells near the mold had burst, or lysed. Something in the mold was killing the bacteria.

Fleming scooped a bit of mold from the plate, grew it in culture broth, then tested the liquid in a slew of additional experiments. The results were clear and dramatic: even when diluted 800 times, this "mold juice"–Fleming's term–was a potent killer of many different kinds of bacteria. Moreover, no other fungus that he tested–including one obtained from a pair of moldy old shoes–had this remarkable killing power. Fleming published his results in 1929 in the *British Journal of Experimental Pathology*. He named the antibacterial substance "penicillin," after the fungus that produces it, *Penicillium notatum*. It was the birth of the first **antibiotic** (INFOGRAPHIC 3.1).

Fleming was not the first to notice the bacteria-killing property of *Penicillium*, but he was the first to study it scientifically and publish the results. In fact, Fleming had been looking for bacteria-killing substances for a number of years, ever since he had served as a medical officer in World War I and witnessed soldiers dying from bacteria-caused infections. He had already discovered one such antibacterial agent–the chemical lysozyme–which he detected in his own tears and nasal mucus. Nevertheless, his 1928 discovery was serendipitous: "Penicillin started as a chance observation," Fleming said many years later. "My only merit is that I did not neglect the observation and that I pursued the subject as a bacteriologist."

Most people have seen the *Penicillium* fungus growing on a piece of moldy bread or rotting fruit. It doesn't look very impressive, but the chemical it produces ushered in a whole new age of medicine. For the first time, doctors had a way to treat such deadly bacteria-caused illnesses as pneumonia, syphilis, gonorrhoea, and meningitis. Before penicillin, there was nothing much doctors could do for a patient with a serious bacterial infection. Now, they had a powerful weapon on their side. As physician and author Lewis Thomas recalled in his 1992 memoir, *The Fragile Species,* "We could hardly believe our

INFOGRAPHIC 3.1 HOW PENICILLIN WAS DISCOVERED

A fortuitous observation by Fleming led to the discovery of the first antibiotic. He realized that the fungus on his culture plate was somehow inhibiting the reproduction of bacteria.

A single bacterial cell lands on a culture plate.

Nutrients in the plate support the growth and division of the bacterial cells.

After many rounds of cell division, enough cells accumulate in this spot to be visualized as acolony on the plate.

Bacterial colonies form at locations far away from the mold, but cannot form near the mold.

Staphylococcus bacteria

Penicillium mold under a microscope and on an orange

INFOGRAPHIC 3.2 CELL THEORY: ALL LIVING THINGS ARE MADE OF CELLS

 All living organisms are composed of cells. These cells arise from the reproduction of existing cells. Different cells have different structures and functions.

Diatoms (algae): single-celled eukaryotes

Amoeba (a protozoan):
a single-celled eukaryote

Bacteria: single-celled prokaryotes

Molds (fungi):
single and multicellular eukaryotic cells

Elodea (an aquatic plant):
a multicellular eukaryote

Humans (these are heart cells):
multicellular eukaryotes

eyes on seeing that bacteria could be killed off without at the same time killing the patient. It was not just amazement, it was a revolution."

BUG BULLET

What makes antibiotics special is not just their ability to kill bacteria. After all, bleach kills bacteria just fine. The important thing about antibiotics is that they exert their destructive effects on bacteria without (typically) harming their human or animal host, even if they are taken internally.

Although Fleming didn't know it at the time, penicillin and other antibiotics preferentially kill bacteria because they target what is unique about bacterial cells. According to the **cell theory,** all living things are made of cells, and every new cell comes from the division of a pre-existing one. But not all cells are alike. Cells come in many shapes and sizes and per-

form a variety of different functions. Moreover, they fall into one of two fundamentally different categories: **prokaryotic** or **eukaryotic.** Prokaryotic cells are relatively small and lack internal membrane-bound compartments, called **organelles.** Eukaryotic cells, by contrast, are much larger and contain many such organelles. Bacteria have prokaryotic cells (**INFOGRAPHIC 3.2**).

To understand why antibiotics affect prokaryotic and eukaryotic cells differently, it helps to understand first what the two cell types have in common. All cells, both prokaryotic and eukaryotic, are surrounded by a **cell membrane** composed of phospholipids and proteins (see Chapter 2). This flexible yet sturdy structure forms a boundary between the external environment and the cell's watery **cytoplasm;** it separates inside from outside and literally defines the cell. In addition, all cells have **ribosomes,**

▶**EUKARYOTIC CELLS**
Cells that contain membrane-bound organelles, including a central nucleus.

▶**ORGANELLES**
The membrane-bound compartments of eukaryotic cells that carry out specific functions.

▶**CELL MEMBRANE**
A phospholipid bilayer with embedded proteins that forms the boundary of all cells.

▶**CYTOPLASM**
The gelatinous, aqueous interior of all cells.

▶**RIBOSOME**
A complex of RNA and proteins that carries out protein synthesis in all cells.

▶NUCLEUS
The organelle in eukaryotic cells that contains the genetic material.

▶CELL WALL
A rigid structure enclosing the cell membrane of some cells that helps the cell maintain its shape.

▶OSMOSIS
The diffusion of water across a semipermeable membrane from an area of lower solute concentration to an area of higher solute concentration.

▶HYPOTONIC
Describes a solution surrounding a cell that has a lower concentration of solutes than the cell.

which synthesize the proteins that are crucial to cell function, and all have DNA, the molecule of heredity.

Beyond these common features, however–cell membrane, cytoplasm, ribosomes, and DNA–the two cell types are structurally quite different. In a prokaryotic cell, for instance, the DNA floats freely within the cell's cytoplasm, while in a eukaryotic cell it is housed within a central "command center" called the **nucleus.** The nucleus is one of many organelles found within eukaryotic cells but not in their simpler prokaryotic cousins (**INFOGRAPHIC 3.3**).

Penicillin kills bacteria because of an important difference between prokaryotic and eukaryotic cells. Unlike human and other animal cells, most bacteria are surrounded by a **cell wall.** This rigid structure, which encloses the cell membrane, is what allows bacteria to survive in watery environments– say, the intestines, blood, or a pond.

Water has a tendency to move across cell membranes in order to balance the solute concentrations on each side of the membrane, a process called **osmosis.** Water will predictably move from the solution with the lower solute concentration to the solution with the higher solute concentration. For example, cells placed in a lower-solute, or **hypotonic,** solution will tend to take up water and swell. On the other hand, cells placed in a higher-solute, or **hypertonic,** solution will tend to lose water and shrivel. In an **isotonic** solution, where the solute concentration is the same as the cell's cytoplasm, there is no net movement of water into or out of the cell. In all cases, water moves in a direction that will tend to even out the solute concentrations on each side of the membrane.

The environments that many bacteria find themselves in tend to be hypotonic. Water will enter the bacterial cells by osmosis,

INFOGRAPHIC 3.3 PROKARYOTIC AND EUKARYOTIC CELLS HAVE DIFFERENT STRUCTURES

→ While all cells have a cell membrane, cytoplasm, ribosomes, and DNA, there are specific structural differences between prokaryotic and eukaryotic cells. Eukaryotic cells contain a variety of membrane-enclosed organelles; prokaryotic cells do not.

Basic Prokaryotic Cell

Prokaryotic and eukaryotic cells share these common structures:

Basic Eukaryotic Cell

Cell membrane

Cytoplasm

Ribosome

Genetic material (DNA)

Prokaryotic cells have a cell wall.

Eukaryotic cells have specialized compartments (organelles) for specific cell functions.

A magnified bacterium

A magnified animal cell

INFOGRAPHIC 3.4 WATER FLOWS ACROSS CELL MEMBRANES BY OSMOSIS

 The direction of water movement across the cell membrane is determined by the solute concentration on either side. Water always moves toward the side with the higher solute concentration. In a hypotonic solution, water will flow into the cell, and in a hypertonic solution, water will flow out of a cell. The bacterial cell wall helps protect the cell from lysing in a hypotonic environment. In the presence of some antibiotics the cell wall is disrupted, leaving the cell susceptible to lysis.

Hypotonic Solution
· Higher solute concentration inside cell
· Water flows into cell.

Water flows in.

Cell swells.

Isotonic Solution
· Equal solute concentration in and out of cell
· Water flows equally in both directions.

Water flows equally in both directions.

Hypertonic Solution
· Higher solute concentration outside cell
· Water flows out of cell.

Water flows out.

Cell shrivels.

In the presence of penicillin
1. In a hypotonic solution, water flows into cell, applying pressure on the cell wall.

Water flows in.

2. Antibiotic interferes with cell wall synthesis, weakening the structure.

Cell wall

Cell membrane

3. The bacterial cell wall cannot resist the osmotic pressure of the water. The cell ruptures in this environment.

---

causing them to swell. This swelling would be fatal to bacteria were it not for the cell wall, without which the bacteria would fill up with water and burst. The rigid cell wall counteracts the osmotic pressure, preventing excess water from rushing in. (Eukaryotic cells are protected from osmotic pressure partly by the cholesterol in their cell membrane, which imparts strength.)

What makes the bacterial cell wall rigid is the molecule **peptidoglycan,** a polymer made of sugars and amino acids that link to form a chainlike sheath around the cell. Different bacterial walls can have different structures, but all have peptidoglycan, which

is found only in bacteria. And here's where penicillin comes in: by interfering with the synthesis of peptidoglycan, penicillin weakens the cell wall, which is then no longer able to counteract osmotic water pressure. Eventually, the cell bursts (INFOGRAPHIC 3.4).

Bacteria are not the only organisms with a cell wall (plant cells and fungi have them, too), but they are the only ones that have a cell wall made of peptidoglycan—which is why penicillin is such a selective bacteria killer.

Ironically, despite its remarkable killing powers, penicillin was not immediately recognized as a medical breakthrough when it was first discovered. In fact, Fleming didn't think

▶**HYPERTONIC**
Describes a solution surrounding a cell that has a higher concentration of solutes than the cell.

▶**ISOTONIC**
Describes a solution surrounding a cell that has the same solute concentration as the cell.

▶**PEPTIDOGLYCAN**
The macromolecule found in all bacterial cell walls that confers rigidity.

> **" *I had not the slightest suspicion that I was at the beginning of something extraordinary.* "**
>
> — ALEXANDER FLEMING

his mold had much of a future in medicine. "I had not the slightest suspicion that I was at the beginning of something extraordinary," recalled Fleming years later in a graduation speech to students.

At the time, the idea that an antiseptic agent could kill bacteria without at the same time harming the patient was unheard of, so Fleming never considered that penicillin might be taken internally. Nor was he a chemist, so he lacked the expertise to isolate and purify the active ingredient from the mold. While he found that his mold juice made a "reasonably good" topical antiseptic, he noted in a 1940 paper that "the trouble of making it seemed not worthwhile," and largely gave up working on it.

Ten years would pass before anyone reconsidered Fleming's mold. By then, history had intervened and given new urgency to the search for antibacterial medicines.

FROM FUNGUS TO PHARMACEUTICAL

On September 1, 1939, Germany invaded Poland, plunging the world into war for the second time in a generation. With the horrors of World War I seared into memory, many feared the death toll that would result from the hostilities. Millions of soldiers and civilians had died in World War I, many not as a result of direct combat injuries but from infected wounds. With few other antibacterial medicines available, penicillin suddenly became the focus of research during World War II.

In 1938, Ernst Chain, a German-Jewish biochemist, was working in the pathology department at Oxford University, having fled Germany for England in 1933 when the Nazis came to power. Both Chain and his supervisor, Howard Florey, were interested in the biochemistry of antibacterial substances. Chain stumbled across Fleming's 1929 paper on penicillin and set about trying to isolate and concentrate the active ingredient from the mold, which he succeeded in doing by 1940. Chain's breakthrough allowed Florey's group to begin testing the drug's clinical efficacy. They injected the purified chemical into bacteria-infected mice and found that the mice were quickly rid of their infection. Human trials followed next, in 1941, with the same remarkable result.

As encouraging as these results were, there was one nagging problem: it took up to 2,000 liters (more than 500 gallons) of mold fluid to obtain enough pure penicillin to treat one person. The Oxford doctors used almost their whole supply of the drug treating their first patient, a policeman ravaged by a staphylococcal infection. The team stepped up their

Manufacturing penicillin in 1943: culture flasks are filled with the nutrient solution in which penicillin mold is grown.

Thanks to PENICILLIN
...He Will Come Home!

FROM ORDINARY MOLD—
the Greatest Healing Agent of this War!

Penicillin and the war effort: feelings of wartime patriotism were enlisted to support production of the drug.

purification efforts—even culturing the mold in patients' bedpans and repurifying the drug from patients' urine—but there was no way they could keep up with demand.

The turning point came in 1941, when Oxford scientists approached the U.S. government and asked for help in growing penicillin on a large scale. The method they devised took advantage of something America's heartland had in abundance: corn. Using a by-product of large-scale corn processing as a culture medium in which to grow the fungus, the scientists were able to produce penicillin in much greater quantities.

At first, all the penicillin harvested from U.S. production plants came from Fleming's original strain of *Penicillium notatum*. But researchers continued to look for more potent strains to improve yields. In 1943, they got lucky: researcher Mary Hunt discovered one such strain growing on a ripe cantaloupe in a Peoria, Illinois, supermarket. This new strain, *Penicillium chrysogenum,* produced more than 200 times the amount of penicillin as the original strain. With it, production of the drug soared. By the time the Allies invaded France on June 6, 1944–D-day–they had enough penicillin to treat every soldier who needed it. By the following year, penicillin was widely available to the general public.

The optimism with which patients and doctors greeted the new bacteria-killer cannot be overstated. "Penicillin seemed to justify a carefree attitude to infection," says science historian Robert Bud, principal curator of the Science Museum in London. "In Western countries, for the first time in human history, most people felt that infectious disease was ceasing to be a threat, and sexually infectious disease had already been conquered. For many it seemed cure would be easier than prevention."

Yet, as effective as penicillin was, it was effective only against certain types of bacteria; against others, it was powerless.

STOCKPILING THE ANTIBIOTIC ARSENAL

As Fleming knew, most of the bacterial world falls into one of two categories, **Gram-positive** or **Gram-negative;** these names reflect the way bacterial cell walls trap a dye known as Gram stain (after its discoverer, the Danish scientist Hans Christian Gram). Fleming found that while penicillin easily killed Gram-positive bacteria like *Staphylococcus* and *Streptococcus,* the microbes that cause staph infections and

▶**GRAM-POSITIVE**
Refers to bacteria with a cell wall that includes a thick layer of peptidoglycan that retains the Gram stain.

▶**GRAM-NEGATIVE**
Refers to bacteria with a cell wall that includes a thin layer of peptidoglycan surrounded by an outer lipid membrane that does not retain the Gram stain.

To identify bacteria, researchers treat them with a violet Gram stain. Gram-positive bacteria retain the dye, and so appear purple or blue under a microscope. Gram-negative bacteria appear red or pink.

Figure 2 from Fleming's 1929 paper, showing the response of different bacteria to penicillin.

strep throat, respectively, it had little effect on Gram-negative bacteria like *Escherichia coli* and *Salmonella*, whose cell walls have an extra layer of lipids surrounding them. This extra lipid layer prevents penicillin from reaching the peptidoglycan underneath.

The discovery that penicillin was effective mostly against Gram-positive bacteria led researchers in the 1940s to look for other antibiotics that could kill Gram-negative bacteria. The first such broad-spectrum antibiotic was streptomycin, discovered in 1943 by Albert Schatz and Selman Waksman at Rutgers University. In addition to killing Gram-negative bacteria, streptomycin was the first effective treatment for the deadly bacterial disease tuberculosis. The reason for its effectiveness? Streptomycin

has a chemical structure that allows it to pass more easily through the outer lipid layer of the Gram-negative bacterial cell wall. (Although natural penicillin cannot pass this layer, many modern synthetic varieties of penicillin, known collectively as beta-lactams, can.)

Once inside the cell, streptomycin works by interfering with protein synthesis on bacterial ribosomes. Ribosomes are the molecular "machines" that assemble a cell's proteins. While both eukaryotic and prokaryotic cells have ribosomes, their ribosomes are of different sizes and have different structures. Because streptomycin targets features specific to bacterial ribosomes, it doesn't harm the human who is taking it (**INFOGRAPHIC 3.5**).

Other antibiotics work in different ways—by inhibiting a bacterium's ability to make a critical vitamin or to copy its DNA before dividing, for example. When this happens, the bacterium dies instead of reproducing.

INFOGRAPHIC 3.5 SOME ANTIBIOTICS INHIBIT PROKARYOTIC RIBOSOMES

Ribosomes are responsible for the synthesis of proteins in both prokaryotic and eukaryotic cells, but their structure is slightly different in the two types of cell. Antibiotics that interfere with prokaryotic ribosomes leave eukaryotic ribosomes unaffected.

Prokaryotic Ribosome

Genetic instructions from the nucleus enter the ribosome.

Antibiotic disrupts ribosome function.

Bacterial ribosome

No protein formed

Antibiotics interfere with bacterial ribosomes. Protein synthesis is interrupted.

Eukaryotic Ribosome

Because the human ribosome has a different structure or shape, the antibiotic cannot attach to it and disrupt its function.

Antibiotic

Using these instructions, a new protein chain is formed.

Genetic instructions from the nucleus enter the ribosome.

Human ribosome

Functional protein

Eukaryotic ribosomes are unaffected by antibiotic. Proteins are still produced.

INFOGRAPHIC 3.6 MEMBRANES: ALL CELLS HAVE THEM

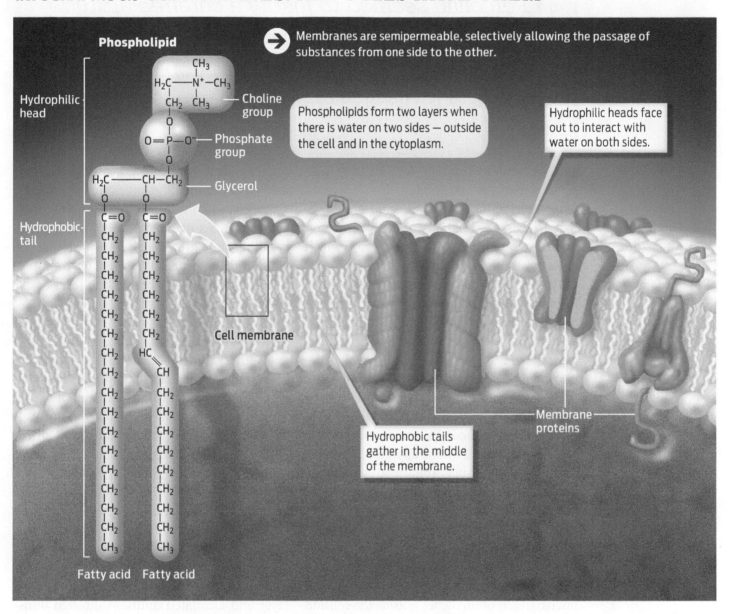

Phospholipid

→ Membranes are semipermeable, selectively allowing the passage of substances from one side to the other.

Hydrophilic head

CH₃
H₂C—N⁺—CH₃
CH₂   CH₃
— Choline group

O=P—O⁻
— Phosphate group

H₂C—CH—CH₂ — Glycerol

Phospholipids form two layers when there is water on two sides — outside the cell and in the cytoplasm.

Hydrophilic heads face out to interact with water on both sides.

Hydrophobic tail

Cell membrane

Hydrophobic tails gather in the middle of the membrane.

Membrane proteins

Fatty acid   Fatty acid

CROSSING ENEMY LINES

For any drug to be effective, it has to reach its designated target. In the case of many antibiotics, that means getting inside the cell to do their work. But the cell membrane, which surrounds all cells, acts as a barrier to the free flow of substances into the cell. How do antibiotics get past this barrier?

Consider the structure of the cell membrane. Phospholipids comprise the bulk of the membrane. They have two main parts: a hydrophilic "head" and a hydrophobic "tail." In the watery context of a cell, the hydrophobic tails cluster together in the middle of the membrane away from water, while the hydrophilic heads face outward, toward water. When arranged in this way, the phospholipids form a two-ply structure called a bilayer. Proteins sit nestled in the lipid bilayer, where they perform a variety of functions such as transporting nutrients in and wastes out. The membrane is semipermeable: only certain substances can cross it freely (INFOGRAPHIC 3.6)

With its densely packed collection of hydrophobic phospholipid tails, the lipid bilayer is largely impermeable to large molecules, like

> **❝** *Penicillin seemed to justify a carefree attitude to infection. In Western countries, for the first time in human history, most people felt that infectious disease was ceasing to be a threat.* **❞**
>
> —ROBERT BUD

▶SIMPLE DIFFUSION
The movement of small, hydrophobic molecules across a membrane from an area of higher concentration to an area of lower concentration; simple diffusion does not require an input of energy.

▶TRANSPORT PROTEINS
Proteins involved in the movement of molecules across the cell membrane.

▶FACILITATED DIFFUSION
The process by which large or hydrophilic solutes move across a membrane from an area of higher concentration to an area of lower concentration with the help of transport proteins; facilitated diffusion does not require an input of energy.

glucose, and hydrophilic (or charged) substances, like sodium ions, and only weakly permeable to water. In fact, the only things that do cross the lipid bilayer easily are small, uncharged molecules like oxygen (O_2), which cross by **simple diffusion.** Diffusion is the natural tendency of dissolved substances to move from an area of higher concentration to one of lower concentration—think of food coloring dispersing in a glass of water. In simple diffusion, the substance moves directly through the phospholipids of the membrane, from the side of the membrane with a higher concentration to the side with a lower concentration, thus balancing the concentrations on both sides. Because substances naturally move from an area of higher concentration to an area of lower concentration, no additional energy is required for this movement beyond what is stored in the concentration difference, or gradient, itself.

Take oxygen, for example. The concentration of oxygen molecules, which are small and uncharged, is often higher outside the cell and lower inside it. This concentration gradient allows oxygen to diffuse easily into the cell—a good thing, because the cell needs oxygen in order to survive. In particular, oxygen is needed to convert chemical energy from the diet into usable energy to carry out cellular work (see Chapter 6).

But the cell also needs some large or hydrophilic molecules in order to survive—one of them is glucose, the cell's energy source. To move such molecules across the membrane the cell makes use of **transport proteins.** Transport proteins sit in the membrane bilayer with one of their ends outside the cell and the

other inside it. By acting as a kind of channel, carrier, or pump, transport proteins provide a passageway for those large or hydrophilic molecules to cross the membrane. They are also very specific: a protein that transports glucose will not transport calcium ions, for example. The cells of the body contain hundreds of different transport proteins.

Transport proteins can move substances either with or against a concentration gradient—either "downhill" or "uphill" across the membrane. When a substance moves downhill by a transport protein from an area of higher concentration to an area of lower concentration, the process is called **facilitated diffusion.** Like simple diffusion, facilitated diffusion requires no additional energy besides that in the concentration gradient. For this reason, facilitated diffusion is also sometimes known as passive transport. Many substances enter the cell by facilitated diffusion—glucose and water, for example. (Osmosis, discussed above, relies on both simple diffusion through the lipid bilayer and facilitated diffusion through transport proteins to passively move water through the membrane.)

Antibiotics move across membranes in a number of ways. Some antibiotics are small hydrophobic molecules that can cross the cell membrane directly by simple diffusion—tetracycline, for example. Others, including penicillin and streptomycin, pass through membranes by facilitated diffusion.

Just because an antibiotic makes it inside a bacterial cell, however, doesn't mean it will stay there. Some bacteria have transport proteins that can actively pump the antibiotic back out of the cell. This bacterial counteroffensive measure is an example of **active transport,** in which proteins pump a substance uphill from an area of lower concentration to an area of higher concentration. Unlike facilitated diffusion, active transport requires an input of chemical energy. In this case, active transport keeps the antibiotic concentration in the bacterial cell low, but the cell must expend energy to keep pumping the antibiotic out against its concentration gradient (INFOGRAPHIC 3.7).

Pumping antibiotics out of the bacterial cell is one way in which bacteria can resist the destructive power of an antibiotic. Others include chemically breaking down the antibiotic with enzymes. Why would bacteria have such built-in mechanisms for counteracting or resisting drugs? Remember that penicillin was

originally isolated from a living organism, a fungus. Streptomycin comes from microorganisms living in soil. Microorganisms like these have evolved chemical defenses as a way to protect themselves from other organisms. In turn, their combatants have evolved countermeasures that give them resistance. Humans thus find themselves embroiled in a battle originally waged solely between microorganisms. We have "amplified a local warfare among microbes in a few grams of soil into a global planetary war between Man and Microbe," wrote Alexander Tomasz, a microbiologist at the Rockefeller University, in the book *Fighting Infection in the 21st Century*. In the early 1980s, Tomasz helped discover how penicillin works, and is now an expert on antibiotic resistance.

▶**ACTIVE TRANSPORT**
The energy-requiring process by which solutes are pumped from an area of lower concentration to an area of higher concentration with the help of transport proteins.

INFOGRAPHIC 3.7 MOLECULES MOVE ACROSS THE CELL MEMBRANE

Simple Diffusion
Small, uncharged molecules cross the cell membrane from areas of higher concentration to areas of lower concentration without the help of transport proteins or the input of energy.

Facilitated Diffusion
Large or hydrophilic molecules cross the cell membrane from areas of higher concentration to areas of lower concentration with the help of specific transport proteins, but without the input of energy.

Active Transport
Large or hydrophilic molecules cross the cell membrane from areas of lower concentration to areas of higher concentration with the help of specific transport proteins and chemical energy to pump molecules against the gradient.

Higher solute concentration

Transport protein

Energy

Lower solute concentration

YOUR INNER BACTERIUM

Antibiotics kill bacteria but leave humans unharmed because their cells have different structures. Of all the ways that prokaryotic and eukaryotic cells differ, the most obvious is the complexity of eukaryotic cells compared to their smaller prokaryotic cousins. In particular, eukaryotic cells—both animal and plant cells—are characterized by the presence of multiple, distinct membrane-bound organelles (INFOGRAPHIC 3.8).

You can think of a eukaryotic cell as a miniature factory with an efficient division of labor. Each organelle is separated from the cell's cytoplasm by a membrane similar to the cell's outer membrane, and each performs a distinct function.

The nucleus is the defining organelle of eukaryotic cells (from the Greek *eu,* meaning "good" or "true," and *karyon,* meaning "nut" or "kernel"). It is surrounded by the **nuclear envelope,** a double membrane made of two lipid bilayers dotted by small openings, called pores. The nucleus encloses the cell's DNA and acts as a kind of control center. Important reactions for interpreting the genetic instructions contained in DNA take place in the nucleus, as well as the manufacture of ribosomes.

Other organelles in a eukaryotic cell perform other specialized tasks. **Mitochondria** are the cell's "power plants"—they use oxygen to extract energy from food and convert that energy into a useful form. Animals and plants both have mitochondria. Humans who inherit or develop defects in their mitochondria usually die—an indication of just how important these organelles are.

The **endoplasmic reticulum (ER)** is a vast network of membranes that serves as a kind of assembly line for the manufacture of proteins and lipids. The "rough" ER is studded with ribosomes making proteins; the "smooth" ER makes lipids. Newly made proteins travel from the ER to the **Golgi apparatus,** which

INFOGRAPHIC 3.8 EUKARYOTIC CELLS HAVE ORGANELLES

→ Humans and other animals, as well as plants, fungi and protozoans, are eukaryotes—they are made up of eukaryotic cells that contain a number of internal organelles.

Animal Cell

Plant Cell

Water vacuole

Nucleus

Endoplasmic reticulum

Ribosome

Mitochondrion

Golgi apparatus

Lysosome

Cellulose cell wall

Chloroplast

Plant cells have a few plant-specific structures, including chloroplasts, a cellulose cell wall, and a central water vacuole.

packs the protein "cargo" into vesicles and then ships them to specific destinations, such as the cell membrane, other organelles, and the bloodstream. The nucleus, ER, and Golgi apparatus thus work together to make and transport proteins to specific locations in and out of the cell.

Other eukaryotic organelles include the **chloroplast,** responsible for photosynthesis in plants, and **lysosomes,** the cell's recycling centers, which digest and recycle molecules. In addition to these membrane-bound structures, a vast network of protein fibers called the **cytoskeleton** allows cells to move and maintain their shape, much the same way the human skeleton does (see **UP CLOSE: EUKARYOTIC ORGANELLES**).

Prokaryotic cells carry out similar functions of energy conversion and protein transport, but they don't contain these processes within separate organelles; everything occurs in the cytoplasm.

How did eukaryotic cells develop their factory-like compartments? That question has long intrigued biologists. One fascinating hypothesis was proposed in the 1960s by biologist Lynn Margulis, who argued that eukaryotic organelles such as mitochondria and chloroplasts were once free-living prokaryotic cells that become incorporated–engulfed–by other free-living prokaryotic cells.

Although many considered this notion of **endosymbiosis** a crazy idea at first, a wealth of evidence now supports it. Mitochondria and chloroplasts are about the same size as bacteria, and to reproduce they divide in a manner similar to prokaryotic cells. Both mitochondria and chloroplasts have circular strands of DNA, just like prokaryotic cells. They also contain ribosomes that are similar in structure to prokaryotic ribosomes–so similar, in fact, that some antibiotics that target prokaryotic ribosomes can affect the ribosomes in eukaryotic mitochondria, which accounts for both the toxicity and the side effects of these antibiotics.

Scanning electron micrograph of Neisseria gonorrhoeae, the bacterium that causes gonorrhea.

▶**CHLOROPLAST**
An organelle in plant and algal cells that is the site of photosynthesis.

▶**LYSOSOME**
An organelle in eukaryotic cells filled with enzymes that can degrade worn-out cellular structures.

▶**CYTOSKELETON**
A network of protein fibers in eukaryotic cells that provides structure and facilitates cell movement.

▶**ENDOSYMBIOSIS**
The scientific theory that free-living prokaryotic cells engulfed other free-living prokaryotic cells billions of years ago, forming eukaryotic organelles such as mitochondria and chloroplasts.

Once perceived as a magic bullet, penicillin is now largely ineffective at treating many infections, including gonorrhea.

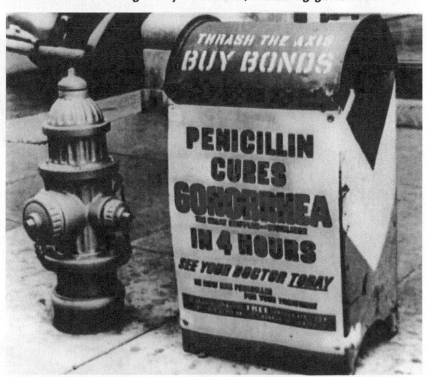

UP CLOSE EUKARYOTIC ORGANELLES

Nucleus

The nucleus is the defining organelle of eukaryotic cells. The nucleus is separated from the cytoplasm by a double membrane (two phospholipid bilayers), known as the nuclear envelope. The nuclear envelope controls the passage of molecules between the nucleus and cytoplasm. The nucleus contains the DNA, the stored genetic instructions of each cell. In addition, important reactions for interpreting the genetic instructions occur in the nucleus.

DNA (genetic material)

Nuclear envelope

Endoplasmic Reticulum

The endoplasmic reticulum (ER) is an extensive, membranous intracellular "plumbing" system that is critical for the production of new proteins. The "rough ER" has a rough appearance because it is studded with ribosomes that are making proteins. The rough ER is contiguous with the "smooth ER," the site of lipid production.

Rough endoplasmic reticulum

Smooth endoplasmic reticulum

Vesicle   Ribosomes

Golgi Apparatus

The Golgi apparatus is a series of flattened membrane compartments, whose purpose is to process and package proteins produced in the rough endoplasmic reticulum. The processed molecules are packaged into membrane vesicles, then targeted and transported to their final destinations.

2. As the proteins make their way through the Golgi apparatus, they are processed to complete their structure and identify them for transport to specific locations in the cell.

3. Proteins are then packaged into transport vesicles which deliver the proteins to their final destination.

1. Transport vesicle delivers proteins from the rough endoplasmic reticulum to the Golgi apparatus.

Transport vesicle

The Nucleus, Endoplasmic Reticulum, and Golgi Apparatus Work Together to Produce and Transport Proteins

Nucleus

2. Proteins are made in the ER and packaged into vesicles for transport to the Golgi apparatus.

3. Proteins receive final modifications in the Golgi apparatus. They are packaged into vesicles for transport to the site of protein function.

1. The nucleus provides instructions for protein production.

Endoplasmic reticulum

Golgi apparatus

Cell membrane

Secreted from cell

Various locations within cell

Mitochondria

Mitochondria are found in almost all eukaryotes, including plants. Mitochondria have two membranes surrounding them. The inner one is highly folded. Mitochondria carry out critical steps in the extraction of energy from food, and the conversion of that "trapped" energy to a useful form. They are the cell's "power plants."

Ribosome

Outer membrane

Inner membrane

Chloroplast

Chloroplasts are organelles found in algae and in the green parts of plants. Chloroplasts have two membranes surrounding them, as well as an internal system of stacked membrane discs. Chloroplasts are the sites of photosynthesis, the reactions by which plants capture the energy of sunlight in a usable form.

Ribosome

Granum

Thylakoid membranes

Inner and outer membranes

Lysosome

Lysosomes are the cell's "recycling centers." Full of digestive enzymes, lysosomes break down worn-out cell parts or molecules so they can be used to build new cellular structures.

Outside of cell

Cell membrane

Cytoplasm

Digestion

Damaged cell parts are digested so their parts can be recycled.

The cell takes in molecules and brings them to a lysosome for digestion. The digested parts are then used in the cell for various functions.

Food and other particles

Lysosome

Golgi apparatus

Lysosome

Digestion

Cytoskeleton

The cytoskeleton is a meshwork of protein fibers that carry out a variety of functions, including cell support, cell movement, and movement of structures within cells. Each type of cytoskeletal fiber has a specific structure and function.

Cell membrane

Microfilament

Intermediate filaments

Microtubule

WINNING THE BATTLE, LOSING THE WAR

To those who first benefited from its healing powers, penicillin seemed like a wonder drug, a magic bullet. A once-lethal bacterial infection could now be cleared in a matter of days with a course of antibiotic. Today, antibiotics are some of the most commonly prescribed drugs.

Antibiotics are so common, in fact, that many people routinely take them when they catch a cold or the flu. But antibiotics are powerless against these illnesses. That's because viruses, not bacteria, cause colds and flu. Since viruses are not made of cells—and according to the cell theory are not even considered to be alive—they can't be killed with an antibiotic.

But that doesn't stop people from trying. The American College of Physicians estimates that of the more than 130 million courses of antibiotics prescribed in the United States each year, as many as 50% are unnecessary, since they are being used to treat colds and other viral infections. What's more, many patients who are prescribed antibiotics for bacterial infections use them improperly. Taking only part of a prescribed dose, for example, can spare some harmful bacteria living in the body, and those bacteria that survive are often heartier and more resistant to the antibiotic than the ones that were killed. Such overuse and misuse of antibiotics have led to an epidemic of such antibiotic-resistance, which the Centers for Disease Control and Prevention calls "one of the world's most pressing public health problems."

Fleming himself foresaw this very danger. In his own research, he found that whenever too little penicillin was used or when it was used for too short a time, populations of bacteria emerged that were resistant to the antibiotic. Fleming warned that improper use of penicillin among patients could lead to the emergence of virulent strains of bacteria that are resistant to the drug. He was right. In 1945, when penicillin was first introduced to the public, virtually all strains of *Staphylococcus aureus* were sensitive to it. Today, more than 90% of *S. aureus* strains are resistant to the antibiotic that once defeated it. (For more on antibiotic-resistant bacteria, see Chapter 14.)

Because of the alarming growth in antibiotic-resistant superbugs, drug companies and researchers are working to develop new antibiotics. One strategy they employ is to tweak the chemical structure of existing antibiotics just enough that a bacterium cannot disable it. Another approach is to look for antibiotics that target other bacterial weaknesses.

But all these efforts would be nothing without the man who gave a moldy petri dish a second glance nearly a century ago. That famous dish now sits in the museum at St. Mary's Hospital in London. For his pioneering research, Alexander Fleming—along with Oxford researchers Howard Florey and Ernst Chain—won a Nobel prize in 1945. ■

As many as 50% of the courses of antibiotics prescribed every year are unnecessary.

CHAPTER 3 Summary

▶ Antibiotics are chemicals, originally produced by living organisms, that selectively target and kill bacteria.

▶ According to the cell theory, all living organisms are made of cells. New cells are formed when an existing cell reproduces.

▶ There are two fundamental types of cells, distinguished by their structure: prokaryotic and eukaryotic. Prokaryotic cells lack membrane-bound organelles; eukaryotic cells have a variety of membrane-bound organelles.

▶ All cells are enclosed by a cell membrane made up of phospholipids and proteins. The cell membrane controls passage of molecules between the exterior of the cell and the cytoplasm.

▶ Small hydrophobic molecules can cross cell membranes by simple diffusion, a process that does not require an input of energy.

▶ Large or hydrophilic molecules are transported across the membrane with the help of membrane transport proteins.

▶ Facilitated diffusion is the transport of molecules down a concentration gradient through a transport protein; it does not require an input of energy. Active transport is the transport of molecules up a concentration gradient through a transport protein; it requires an input of energy.

▶ Water crosses cell membranes by osmosis in order to balance the solutes on each side.

▶ Bacteria are surrounded by a cell wall containing peptidoglycan, a molecule not found in eukaryotes. Some antibiotics, like penicillin, work by preventing peptidoglycan synthesis.

▶ All cells have ribosomes, complexes of RNA and proteins that synthesize new proteins.

▶ Despite their common function, the structures of prokaryotic and eukaryotic ribosomes differ. Some antibiotics, like streptomycin, work by interfering with prokaryotic ribosomes.

▶ Eukaryotic cells contain a number of specialized organelles, including a nucleus, endoplasmic reticulum, Golgi apparatus, mitochondria, chloroplasts, each of which carries out a distinct function.

▶ Eukaryotic cells likely evolved as a result of endosymbiosis, the engulfing of one single-cell prokaryote by another.

▶ Increased and sometimes inappropriate use of antibiotics has led to the emergence of antibiotic-resistant bacteria. Infections caused by these bacteria are very hard to treat.

MORE TO EXPLORE

▶ Fleming, A. (1929) On the antibacterial action of cultures of a penicillium, with special reference to their use in the isolation of *B. influenzae. British Journal of Experimental Pathology* 10:226–236.

▶ Fleming, A. (1945) Penicillin. Nobel Lecture. http://is.gd/ZIPvlu

▶ Bud, R. (2007) *Penicillin: Triumph and Tragedy.* New York: Oxford University Press.

▶ Lax, R. (2005) *The Mold in Dr. Florey's Coat: The Story of the Penicillin Miracle.* New York: Owl Books.

▶ Andrew, P. W., et al., eds. (2000) *Fighting Infection in the 21st Century.* Oxford, U.K.: Blackwell Science.

③

CHAPTER 3 Test Your Knowledge

DRIVING QUESTION 1

What structural features are shared by all cells, and what are the key differences between prokaryotic and eukaryotic cells?

By answering the questions below and studying Infographics 3.2, 3.3, and 3.4, you should be able to generate an answer for the broader Driving Question above.

KNOW IT

1 What does the cell theory state?

2 Which of the following statements best explains why bacteria are considered living organisms?
   a. They can cause disease.
   b. They are made up of biological macromolecules.
   c. They move around.
   d. They are made of cells.
   e. They contain organelles.

3 What are the two main types of cells found in organisms?

4 Which of the following is *not* associated with human cells?
   a. cell membrane
   b. ribosomes
   c. DNA
   d. cell wall
   e. All of the above are associated with human cells.

5 Bacteria have _____ cells, defined by the _____.
   a. prokaryotic; presence of a cell wall
   b. eukaryotic; presence of organelles
   c. eukaryotic; absence of a cell wall
   d. prokaryotic; absence of organelles
   e. eukaryotic; absence of organelles

6 Which of the following is associated with eukaryotic cells but not with prokaryotic cells?
   a. cell membrane
   b. cell wall
   c. DNA
   d. ribosome
   e. nucleus

USE IT

7 According to the cell theory, all living organisms are made of cells. More specifically, what do all living organisms have in common? For example, do all living organisms carry genetic instructions? Do their cells all have a nucleus? What other features do they have in common?

8 You find a single-cell organism with a cell wall in the soil of a forest—can this organism be an animal? Why or why not? Which of the following facts would convince you that the organism is a bacterium and not a plant?
   a. The cell wall is made of cellulose.
   b. The DNA is contained in a nucleus.
   c. The cell wall is made of peptidoglycan.
   d. a and b
   e. b and c

DRIVING QUESTION 2

How do solutes and water cross membranes, and what determines the direction of movement of solutes and water in different situations?

By answering the questions below and studying Infographics 3.5 and 3.7, you should be able to generate an answer for the broader Driving Question above.

KNOW IT

9 The two major components of cell membranes are _____ and _____.
   a. phospholipids; DNA
   b. DNA; proteins
   c. peptidoglycan; phospholipids
   d. peptidoglycan; proteins
   e. phospholipids; proteins

10 If a solute is moving through a phospholipid bilayer from an area of higher concentration to an area of lower concentration without the assistance of a protein, the manner of transport must be
   a. active transport.
   b. facilitated diffusion.
   c. simple diffusion.
   d. any of the above, depending on the solute.
   e. Solutes cannot cross phospholipid bilayers.

11 Consider the movement of molecules across the cell membrane.
   a. What do simple diffusion and facilitated diffusion have in common?
   b. What do active transport and facilitated diffusion have in common?

12 Water is moving across a membrane from solution A into solution B. What can you infer?

  a. Solution A must be pure water.

  b. Solution A must have a lower solute concentration than Solution B.

  c. Solution A must have a higher solute concentration than Solution B.

  d. Solution A and Solution B must have the same concentration of solutes.

  e. Solution B must be pure water.

USE IT

13 Why does facilitated diffusion require membrane transport proteins while simple diffusion does not?

14 Sugars are large, hydrophilic molecules that are important energy sources for cells. How can they enter cells from an environment with a very high concentration of sugar?

  a. by simple diffusion

  b. by osmosis

  c. by facilitated diffusion

  d. by active transport

  e. by using ribosomes

15 Many foods—for example, bacon and salt cod—are preserved with high concentrations of salt. How can high concentrations of salt inhibit the growth of bacteria? (Think about the high solute concentration of the salty food relative to the solute concentration in the bacterial cells. What will happen to the bacterial cells under these conditions?)

MINI CASE

16 Marc, a first-year college student, starts out on a backpacking trip in southern New Mexico. It is September, so the daytime temperatures are quite high, and the desert air is very dry. He has a portable water filter to treat river and stream water that he finds on his planned route through the Gila wilderness. On the second day of his weeklong trip his water filter breaks. He is afraid of contracting giardiasis (a protozoal disease spread through water contaminated by animal feces) so he drinks only the small amount of water that he can boil on his camp stove at night. By the fifth night he is feeling weak and thirsty, and starts to hike out. He makes it to a local highway and collapses. A passing motorist calls 911 for an ambulance.

  a. Given that Marc has sweat a lot, and that sweat causes the loss of more water than solutes, what has happened to the solute concentration of his blood as a result of his dehydration?

  b. From the solute concentration of his blood, what is likely to be happening to his body cells that are in contact with his blood and related fluids (e.g., lymph and cerebral spinal fluid)?

  c. The paramedics have available three saline solutions. One is a "normal" isotonic saline—0.9% NaCl. One is a hypertonic saline (3% NaCl). The last is a "half normal" saline (0.45% NaCl). Which one would you use to treat Marc? Why?

DRIVING QUESTION 3

How do antibiotics target bacteria, and in what situations is antibiotic therapy indicated?

By answering the questions below and studying Infographics 3.1, 3.5 (bottom), and 3.6, you should be able to generate an answer for the broader Driving Question above.

KNOW IT

17 Penicillin interferes with the synthesis of

  _____.

  a. bacterial cell membranes

  b. peptidoglycan

  c. the nuclear envelope

  d. membrane proteins

  e. ribosomes

18 Would phospholipids of the cell membrane be a good target for an antibiotic? Explain your answer.

USE IT

19 If a bacterial infection were treated with two different antibiotics, one that stopped bacterial reproduction and one (penicillin, for example) that inhibited the production of new peptidoglycan, would this use of penicillin or similar drug be effective? Explain your answer.

20 If bacterial cells were placed in a nutrient-containing solution (one that supports their growth) that had the same solute concentration as the cytoplasm, and which also contained penicillin, would the cells burst? Explain your answer. What if the same experiment were repeated with lysozyme? What if the two experiments were repeated in solutions that have lower solute concentrations than the cytoplasm, and did *not* contain growth-supporting nutrients?

21 Fungi are eukaryotic organisms. Scientists have found it more challenging to develop treatments for fungal infections (e.g., yeast infections, athlete's foot, and certain nail infections) than for bacterial infections. Why is this so?

INTERPRETING DATA

22 Bacteria can be characterized as sensitive, intermediately resistant, or fully resistant to different antibiotics. If a strain of bacteria is sensitive to an antibiotic, we can prescribe that antibiotic to treat an infection caused by that strain and have confidence that it will work. If the strain is fully resistant to an antibiotic, that antibiotic cannot treat that infection. In cases of intermediate resistance, it is better to try and find an antibiotic to which the strain is sensitive, as the infection may not respond to antibiotics to which it has intermediate resistance.

Antibiotic	Effective Dose for Sensitive Bacterial Strains (μg/ml)	Effective Dose for Intermediately Resistant Bacterial Strains (μg/ml)	Effective Dose for Fully Resistant Bacterial Strains (μg/ml)
Oxacillin	≤ 2 μg/ml		≥ 4 μg/ml
Vancomycin		8–16 μg/ml	≥ 32 μg/ml
Erythromycin	≤ 0.5 μg/ml	1–4 μg/ml	≥ 8 μg/ml
Tetracycline	≤ 4 μg/ml		≥ 16 μg/ml
Levofloxacin	≤ 2 μg/ml	4 μg/ml	≥ 8 μg/ml

The table shows the concentrations of antibiotics that determine how a bacterial species will respond to those antibiotics. A sensitive strain will be killed by the concentration of antibiotic shown in the "sensitive" column. A strain with intermediate resistance will only be affected by concentrations in the range indicated in the "intermediate" column. And a fully resistant strain requires concentrations shown in the "fully resistant" column.

A hospital patient has a *Staphylococcus aureus* infection. As part of laboratory testing, the *S. aureus* from the patient was grown in different concentrations of various antibiotics. For oxacillin, the lowest concentration that inhibited the growth of the strain was 8 μg/ml; for vancomycin, 4 μg/ml; for erythromycin, 16 μg/ml; for tetracycline, 32 μg/ml; and for levofloxacin, 8 μg/ml. Which antibiotic should be used to treat the infection in this patient?

BRING IT HOME

23 Many patients attempt to pressure their physician to prescribe antibiotics for colds. Is this a good idea? Why or why not?

DRIVING QUESTION 4

What are some key eukaryotic organelles and their functions?

By answering the questions below and studying Infographic 3.8 and Up Close: Eukaryotic Organelles, you should be able to generate an answer for the broader Driving Question above.

KNOW IT

24 Briefly describe the structure and function of each of the following eukaryotic organelles:

a. mitochondrion
b. nucleus
c. endoplasmic reticulum
d. chloroplast

25 Which of the following is *not* a cytoskeletal fiber in eukaryotic cells?

a. macrotubules
b. intermediate filaments
c. microfilaments
d. microtubules
e. All of the above are cytoskeletal fibers.

26 Insulin is a protein hormone secreted by certain pancreatic cells into the bloodstream. Which of the following organelles are involved in the synthesis and secretion of insulin?

a. rough ER
b. Golgi apparatus
c. ribosomes
d. all of the above
e. a and c

USE IT

27 Some inherited syndromes, for example Tay-Sachs disease and MERRF (myoclonic epilepsy with ragged red fibers), interfere with the function of specific organelles. MERRF disrupts mitochondrial function. From what you know about mitochondria, why do you think the muscles and the nervous system are the predominant tissues affected in MERRF? (Think about the activity of these tissues compared to, say, skin.)

28 Which organelle would cause the most damage to cytoskeletal fibers in the cytoplasm if its contents were to leak into the cytoplasm?

a. smooth ER

b. nucleus

c. lysosome

d. Golgi apparatus

e. rough ER

29 Cystic fibrosis is an inherited condition that affects the lungs and digestive tract (see Chapter 11). In many people with cystic fibrosis, a membrane channel protein is found in the rough endoplasmic reticulum instead of the cell membrane. How could a cell membrane protein end up in the rough endoplasmic reticulum? (Hint: Look at the box about the cooperation of the nucleus, endoplasmic reticulum, and Golgi apparatus in Up Close: Eukaryotic Organelles.)

SCIENTIFIC ЯEBEL

Lynn Margulis and the theory of endosymbiosis

An immune cell (green)
engulfs a cell of thrush
fungus (orange).

> **" Life on Earth is such
> a good story you can't afford
> to miss the beginning."**
>
> — LYNN MARGULIS

LYNN MARGULIS NEVER MET A MICROBE SHE DIDN'T LIKE. From the time she first peered through a microscope as a teenager at the University of Chicago and witnessed the vast microcosm of unicellular organisms swimming in a drop of pond water, she was hooked. Other biologists were impressed by the adaptations of plants and animals, but Margulis was smitten with the world of microscopic life. Long before there were animals on the planet, she pointed out, there were microbes, and if it weren't for them, we humans wouldn't be here.

"Life on Earth is such a good story you can't afford to miss the beginning," Margulis said in an interview with the University of Wisconsin alumni magazine. "Do historians begin their study of civilization with the founding of Los Angeles? This is what studying natural history is like if we ignore the microcosm."

In 1966, when she was 28, Margulis used her knowledge of the microcosm to propose a radical hypothesis about how cells had come to be. What distinguishes eukaryotic cells from prokaryotic cells, Margulis knew, was the presence in eukaryotic cells of internal membrane-bound organelles (see

▸▸ DRIVING QUESTIONS

1. What does endosymbiosis say about the origin of organelles such as mitochondria and chloroplasts?
2. What evidence supports the proposed origin of organelles such as mitochondria and chloroplasts?

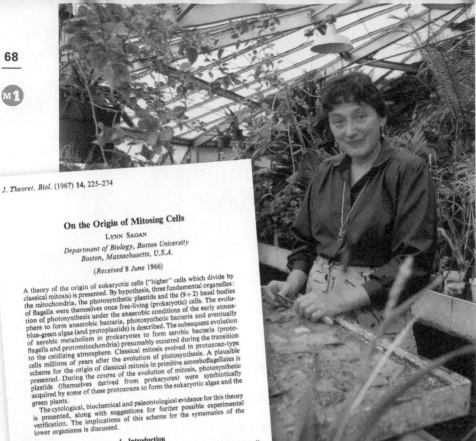

J. Theoret. Biol. (1967) 14, 225–274

On the Origin of Mitosing Cells

LYNN SAGAN

*Department of Biology, Boston University
Boston, Massachusetts, U.S.A.*

(Received 8 June 1966)

A theory of the origin of eukaryotic cells ("higher" cells which divide by classical mitosis) is presented. By hypothesis, three fundamental organelles: the mitochondria, the photosynthetic plastids and the (9+2) basal bodies of flagella were themselves once free-living (prokaryotic) cells. The evolution of photosynthesis under the anaerobic conditions of the early atmosphere to form anaerobic bacteria, photosynthetic bacteria and eventually blue-green algae (and protoplastids) is described. The subsequent evolution of aerobic metabolism in prokaryotes to form aerobic bacteria (proto-flagella and protomitochondria) presumably occurred during the transition to the oxidizing atmosphere. Classical mitosis evolved in protozoan-type cells millions of years after the evolution of photosynthesis. A plausible scheme for the origin of classical mitosis in primitive amoeboflagellates is presented. During the course of the evolution of mitosis, photosynthetic plastids (themselves derived from prokaryotes) were symbiotically acquired by some of these protozoans to form the eukaryotic algae and the green plants.

The cytological, biochemical and paleontological evidence for this theory is presented, along with suggestions for further possible experimental verification. The implications of this scheme for the systematics of the lower organisms is discussed.

1. Introduction

All free-living organisms are cells or are made of cells. There are two basic cell types: *prokaryotic* and *eukaryotic*. Prokaryotic cells include the eubacteria, the blue-green algae, the gliding bacteria, the budding bacteria, the pleuro-pneumonia-like organisms, the spirochaetes and rickettsias, etc. Eukaryotic cells, of course, are the familiar components of plants and animals, molds and protozoans, and all other "higher" organisms. They contain subcellular organelles such as mitochondria and membrane-bounded nuclei and have many other features in common.

"The numerous and fundamental differences between the eukaryotic and prokaryotic cell which have been described in this chapter have been fully recognized only in the past few years. In fact, this basic divergence in cellular structure which separates the bacteria and blue-green algae

16

225

T.B.

Above: Lynn Margulis in her greenhouse. Left: The 1967 paper that made her famous, published under her married name at the time, Lynn Sagan.

Chapter 3). Margulis proposed that eukaryotic cells, with their internal organelles, had formed when one prokaryotic cell engulfed another—essentially eating it for lunch. Instead of being digested, the ingested cell survived and took up residence in its new host. Eventually, the two cells formed a mutually beneficial relationship, and the result was a complex eukaryotic cell. Margulis called her idea endosymbiosis (from the Greek "endo" meaning "within," and "symbiosis" meaning "living together").

Many scientists dismissed the idea outright as a crackpot notion. The paper she wrote proposing the idea, which she titled "On the Origin of Mitosing Cells," was rejected by about 15 scientific journals before finally being accepted by the *Journal of Theoretical Biology,* in 1967. But an interesting thing happened after it was published: people couldn't stop talking about it.

MICROSCOPIC CLUES

In one sense, the idea of endosymbiosis was not entirely new. A few biologists in the late 19th century and early 20th century had suggested it after observing the remarkable resemblance between free-living bacteria and certain organelles of the eukaryotic cell. But in those early days it was not possible to test the idea, and so it was largely ignored. In 1925, Edmund Wilson, a prominent cell biologist, wrote, "To many, no doubt, such speculations may appear too fantastic for present mention in polite biological society." But he went on to suggest that those speculations might "someday call for serious consideration."

In 1960, while working on her master's degree in biology at the University of Wisconsin, Margulis first became intrigued by the notion of endosymbiosis. Her advisors, Walter Plaut and Hans Ris, had recently made the startling discovery that chloroplasts—the tiny green organelles inside plant cells that carry out photosynthesis—had their own DNA, the molecule of heredity. Most DNA in a eukaryotic cell is housed in the nucleus, where it serves as the genetic blueprint for life. What was DNA doing in a chloroplast? To Margulis, the discovery suggested that chloroplasts had once led a separate existence as independent, free-living cells, and thus needed DNA to reproduce, much as bacteria do.

She pursued this idea further as part of her Ph.D. work at the University of California at Berkeley. She studied, in particular, the small unicellular eukaryote called euglena, which lives in water and contains numerous chloroplasts. She used radioactively labeled nucleotides, the building blocks of DNA, to show that the little squiggle inside a euglena chloroplast was indeed DNA, as her advisors had claimed. (Cells incorporate the nucleotides into DNA when they divide, and the radioactivity can be detected with photographic film.) This work lent additional support to the idea of endosymbiosis.

The more Margulis looked, the more evidence she found. Not only did chloroplasts have their own DNA, but they also had ribosomes—the structures that, in the cytoplasm of both prokaryotic and eukaryotic cells, synthesize proteins. Chloroplasts were also about the same size as bacteria, and to reproduce they divided much the same way bacteria divide. Similarly, mitochondria, the cell's "power plants," also had these traits, suggesting that these organelles, too, had once been free-living cells (INFOGRAPHIC M1.1).

In her 1967 paper, Margulis gave credit to the biologists who had previously suggested endosymbiosis, but she did more than simply rehash old ideas. She brought together for the first time all the existing evidence from cell biology and biochemistry and wove it

together into a coherent account. She also put the idea into evolutionary context and offered a rough timeline of when these events happened.

In Margulis's view, the first cells on Earth were prokaryotic bacteria, first evolving some 4 billion years ago. These early cells were anaerobic—they did not use oxygen in their metabolism (which makes sense, since the early Earth had no substantial oxygen). These cells evolved and diversified for billions of years until, about 2.5 billion years ago, some developed the capacity to harvest the energy of sunlight, splitting water and releasing oxygen gas as a by-product. As oxygen built up in

INFOGRAPHIC M1.1 CHLOROPLASTS AND MITOCHONDRIA SHARE TRAITS WITH BACTERIA

Margulis observed that chloroplasts and mitochondria shared several traits with free-living bacteria.

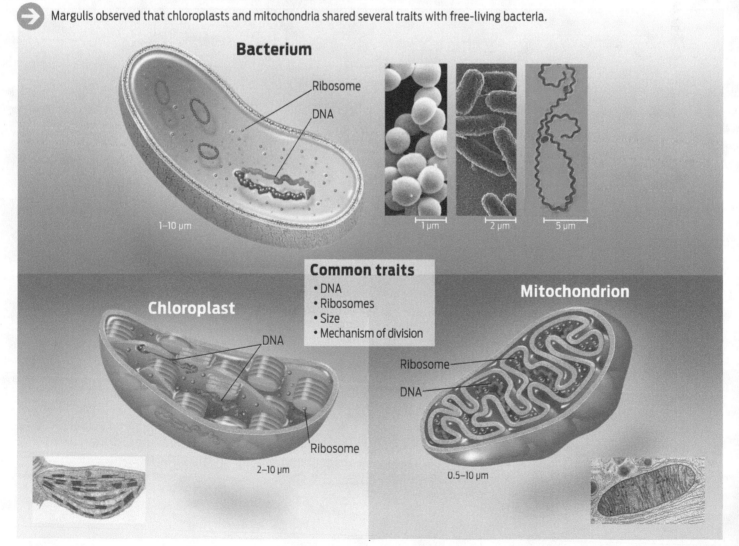

Bacterium

Ribosome

DNA

1–10 μm

1 μm   2 μm   5 μm

Common traits
• DNA
• Ribosomes
• Size
• Mechanism of division

Chloroplast

DNA

Ribosome

2–10 μm

Mitochondrion

Ribosome

DNA

0.5–10 μm

INFOGRAPHIC M1.2 THE FIRST EUKARYOTES WERE PRODUCTS OF ENDOSYMBIOSIS

All eukaryotes are characterized by the presence of a membrane-enclosed nucleus and organelles. While the origin of the nucleus is not yet completely understood, there is good evidence that mitochondria and chloroplasts arose when ancient eukaryotic ancestors engulfed bacteria.

Photosynthetic Eukarya

Photosynthetic protists

Plants

Nonphotosynthetic Eukarya

Animals

Fungi

Nonphotosynthetic protists

Choroplast

Mitochondrion

Mitochondrion

Ancestral non-photosynthetic eukaryote

Bacteria

Endosymbiosis of photosynthetic bacterium

Ancestral eukaryote

Archaea

Endosymbiosis of aerobic bacterium

Anaerobic ancestral archaeon

Photosynthetic bacteria

Aerobic bacteria

Common ancestor (prokaryotic)

the atmosphere as a result of this photosynthesis, it produced an environment that strongly favored bacteria that could use oxygen in their metabolism.

At that point, according to Margulis, one such oxygen-using, or aerobic, bacterium was engulfed by a larger anaerobic cell. The ingested bacterium was not destroyed; instead, the two cells formed a symbiotic, mutually beneficial relationship. The smaller aerobic bacterium enabled the larger anaerobic cell to use oxygen to obtain energy, and the larger anaerobic cell provided the smaller bacterium with a source of sugars to eat. (These ideas bear a strong resemblance to the way that eukaryotic cells still, to this day, metabolize glucose during glycolysis and aerobic respiration; see Chapter 6.)

What began as a mutualistic relationship became, over time, an obligate one: the two cells could no longer survive apart, as each became increasingly codependent on the other. At some point later, this power duo got cozy with a third bacterium—a photosynthetic one—and engulfed it as well. These composite cells, Margulis argued, are the ancestors of eukaryotes (**INFOGRAPHIC M1.2**).

Margulis's paper was highly speculative, but it provided some clear, testable hypotheses. If mitochondria and chloroplasts were descended from free-living bacteria, then it should be possible to determine from their DNA what those free-living bacteria were. At the time Margulis wrote her paper, chemical analysis of DNA was in its infancy, and she couldn't use this technique. But by the mid-70s, analyzing DNA had become routine, and it was DNA evidence that clinched her case: By comparing the sequences of mitochondrial and chloroplast DNA to a wide range of bacterial DNA, researchers discovered that mitochondrial DNA closely resembled DNA from a small bacterium called *Rikettsia*—interestingly, a type of intracellular parasite that burrows inside other cells in order to live. Chloroplast DNA, they discovered, is essentially the same as cyanobacterial DNA.

Electron micrograph of a Rikettsia bacterium burrowing inside a mouse mesothelial cell.

Cyanobacteria, once mistakenly known as blue-green algae, were the first photosynthesizers on Earth, evolving some 2.5 billion years ago. By taking up residence in a larger host cell, these smaller bacteria endowed the host cell with the capacity to photosynthesize and thus paved the way for the evolution of green plants. "Plants are something that hold up cyanobacteria. That's all plants are," Margulis said in a 2006 interview with *Astrobiology Magazine*. "Fundamentally, if you cut them out of the plant cell, and throw away the rest of the plant cell, the little green dot is the only thing that can do that oxygen production. That is the greatest achievement of life on Earth, and it occurred extremely early in the history of life."

FROM HERESY TO ORTHODOXY

As soon as her 1967 paper was published, criticism rolled in. Many of Margulis's colleagues were skeptical, even dismissive, citing a lack of supporting evidence. Most of the evidence that Margulis marshaled in support of her hypothesis was circumstantial rather than direct (the conclusive DNA evidence did not come until 1978). There was philosophical opposition, too. To many people, bacteria were "bad" because they were known to cause disease; the idea that we have bacteria to thank for our very existence was difficult for many to accept. Then, too, the notion of endosymbiosis seemed to go against evolutionary dogma, which held that evolution occurred in small steps as a result of an individualistic "struggle for existence." Endosymbiosis was not a small step, it was a huge one. And it wasn't about competition as much as cooperation. For these reasons, it just didn't sit well with many hard-nosed Darwinists.

For her part, Margulis never shrank from her position or gave up pushing her case. In

Animals live symbiotically with a vast universe of bacteria, collectively known as the microbiome—from bacteria in a cow's digestive tract (top), to bacteria on human skin (bottom).

fact, the resistance she encountered seemed almost to embolden her, and she spent years uncovering many other examples of symbiotic relationships at work in nature.

"Look at a cow," she said in a 2011 interview with *Discover* magazine. "It is a 40-gallon fermentation tank on four legs. It cannot digest grass and needs a whole mess of symbiotic organisms in its overgrown esophagus to digest it."

Or look at your own body. "There are hundreds of ways your body wouldn't work without bacteria. Between your toes is a jungle; under your arms is a jungle. There are bacteria in your mouth, lots of spirochetes, and other bacteria in your intestines. We take for granted their influence. Bacteria are our ancestors."

Much current research is focused on understanding our human microbiome—the population of bacteria that lives on and in our bodies and influences many aspects of our health. In addition to helping us digest food and shaping our immune system, our unique microbiome may even influence our susceptibility to conditions such as diabetes and obesity. The idea that these bacteria are not passive freeloaders but crucial constituents of our bodies is a relatively new way of looking at things—but it would hardly have come as a surprise to Margulis.

When Margulis died in 2011 at the age of 73, she was remembered as the person who most fundamentally changed our view of cells. These days, the idea that mitochondria and chloroplasts started as free-living organisms is accepted nearly as fact by the scientific mainstream, and we now refer to the theory of endosymbiosis, acknowledging the abundant evidence and wide support it has.

"The evolution of the eukaryotic cells was the single most important event in the history of the organic world," said Ernst Mayr, the grandfather of modern evolutionary studies, who died in 2005 at the age of 100. "Margulis's contribution to our understanding the symbiotic factors was of enormous importance." Richard Dawkins, the British don of evolutionary biology, described the theory of endosymbiosis as "one of the great achievements of twentieth-century evolutionary biology." Botanist Peter

Raven said the idea caused "nothing less than a revolution" in our thinking about the cell.

Not all of Margulis's ideas have gained wide support. In particular, her claim that the whip-like tail of a sperm cell derives from a formerly free-living bacterium called a spirochete is not accepted by the scientific establishment. Nor is her idea, put forward in recent years, that AIDS is caused not by a virus but by a cellular parasite, the same one that causes syphilis.

But on one of the most important questions of modern biology, her intellectual daring paid off. Asked by the *Discover* interviewer how she felt about being the source of so many controversial ideas over the years, Margulis responded in typical fashion: "I don't consider my ideas controversial. I consider them right." ■

Margulis was awarded the National Medal of Science by President Bill Clinton in 1999.

MORE TO EXPLORE

▶ Sagan, L. (1967) On the origin of mitosing cells. *Journal of Theoretical Biology* 14:225–274.

▶ Margulis, L., and Sagan, D. (1986) *Microcosmos: Four Billion Years of Evolution from Our Microbial Ancestors*. New York: Summit Books.

▶ Teresi, D. (2011) Interview with Lynn Margulis. *Discover*. http://discovermagazine.com/2011/apr/16

▶ Margulis, L. (2006) *Microbial planet*. Interview with *Astrobiology Magazine*, Part 1. http://www.astrobio.net/interview/2100/microbial-planet

▶ Sapp, J. (2012) "Too Fantastic for Polite Society: A Brief History of Symbiosis Theory." In *Lynn Margulis: The Life and Legacy of a Scientific Rebel*, edited by Dorian Sagan. White River Junction, Vermont: Chelsea Green Publishing.

MILESTONES IN BIOLOGY 1 Test Your Knowledge

1 What is the function of mitochondria? Of chloroplasts?

2 What evidence did Margulis present to support her hypothesis that organelles had once been free-living prokaryotic organisms?

3 On the basis of DNA sequence analysis,

   a. which bacteria are likely the closest relatives of the chloroplast?

   b. which bacteria are likely the closest relatives of the mitochondria?

4 The endosymbiosis in human cells involves

   a. a mutually beneficial relationship.

   b. the growth of bacteria in our gut.

   c. an organelle derived from a photosynthetic bacterium.

   d. a parasitic bacterium infecting our cells during embryonic development.

   e. all of the above

5 From what you have read here about endosymbiosis:

   a. Could you live without your endosymbiotic organelles? Why or why not?

   b. Could you live if plants did not have their endosymbiotic organelles? Explain your answer.

THE PEANUT BUTTER PROJECT

A doctor's crusade to end malnutrition in Africa, a spoonful at a time

▸▸ **DRIVING QUESTIONS**

1. What are the macronutrients and micronutrients provided by food?
2. What are essential nutrients?
3. What are enzymes, and how do they work?
4. What are the consequences of a diet lacking sufficient nutrients?

T A MEDICAL CONFERENCE IN 2003, DURING A PANEL ON MALNUTRI-

TION, someone in the audience stood up and shouted at the presenter, "You're killing children!"

The presenter, Mark Manary, was discussing a new method of treating malnutrition in kids that he

had used successfully in some of the poorest countries in Africa. But it flew in the face of guidelines

endorsed by the World Health Organization (WHO). Now, leaders in the field were angry.

Manary, a pediatrician from St. Louis, Missouri, had firsthand experience with the standard

WHO treatment. For much of the 1990s, he had devoted his life to it: in hospital wards in Malawi,

a small, landlocked country in southeastern Africa, he faithfully administered fortified milk solu-

tions to sickly children in accordance with WHO protocol for treating malnutrition. But even with

this treatment, children continued to die, and only a small percentage ever got better. "On our best

days only 10% of the kids would die. But still only 25% would recover," he says.

Manary decided to find a better way. After searching around for potential alternatives, he settled

on a surprising solution: peanut butter.

Mothers wait in hospitals to get rations of food and milk for their malnourished children.

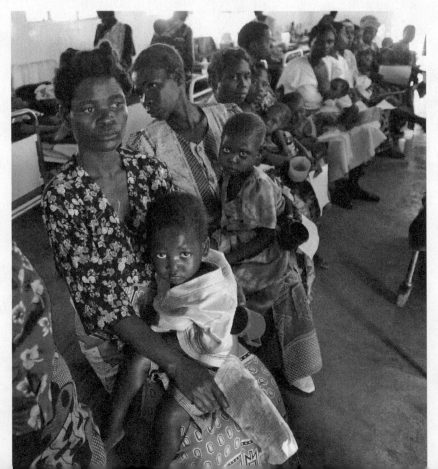

Peanut butter, he found, is well suited to the purpose. It's packed with nutrients, it doesn't re-quire cooking, and—most important—it doesn't spoil easily and can be kept unrefrigerated in tropi-cal climates for up to 3 months.

Manary began testing his peanut butter treatment in 2001. He gave his patients an ample supply of peanut butter and released them from the hospital. "Basically, what we did was empty the place out and send everybody home," says Manary.

The hospital staff was appalled by his brash-ness, but the gamble paid off. Within weeks, 95% of the kids eating peanut butter had fully recovered.

Manary is now the director of Project Pea-nut Butter, a nonprofit organization using the American pantry staple to end malnutrition in Africa. Spin-offs of the organization have sprung up all around the globe, including in Haiti and South Asia.

Once vilified by the international aid commu-nity, Manary's method is now being lauded as a near panacea. Commentators have even likened the peanut butter treatment to the discovery of penicillin, with the potential to save millions of lives every year.

THE ELEPHANT IN THE ROOM

In 1994, Manary became a visiting faculty member at the University of Malawi School of Medicine. When he arrived, he asked the hospital director, "What's your biggest problem?" The director replied, "The malnutrition ward." So that's where Manary chose to work.

It was a dismal and discouraging place: a large room where about 50 starving kids lay close to death on crowded cots, their bodies little more than skin and bone. About a third of the kids would die, despite the doctors' best efforts at rehabilitating them. This poor response rate was, Manary says, "the elephant in the room."

Severe acute **malnutrition**–hunger and starvation medically defined–is the number one killer of children in the world. An estimated 3.5 million children die from malnutrition every year–more than the number who die from AIDS,

tuberculosis, and malaria combined. Most of these deaths occur in sub-Saharan Africa, where grinding poverty is endemic, and food is scarce for large portions of the year (**INFOGRAPHIC 4.1**).

In Malawi, the crisis is particularly acute. The World Food Programme of the United Nations estimates that 1.3 million people, or about 11% of the country's population, will experience food shortages at some point during the year. Historically, more than half of all Malawian children are chronically malnourished, and one in eight dies from lack of food.

Why is food so scarce? Malawi is one of the poorest countries in the world, with a population of mostly subsistence farmers. The primary agricultural crops are corn and soybeans. Farmers also grow commercial cash crops for export, including tobacco, sugar cane, coffee, and tea. But agriculture is not easy in Malawi. Rain comes only once a year, from December to March, when it might rain every day. New crops are

▶**MALNUTRITION**
The medical condition resulting from the lack of any essential nutrient in the diet. Malnutrition is often, but not always, associated with starvation.

INFOGRAPHIC 4.1 HUNGER AROUND THE WORLD

→ While food is the most basic of human requirements for survival, on average 1 in 8 people goes to bed hungry each night.

Proportion of total population undernourished, 2010-12

| Missing or insufficient data | <5% Very low undernourished | 5–14% Moderately low undernourished | 15–24% Moderately high undernourished | 25–34% High undernourished | 35% Very high undernourished |

Data from the World Food Programme (wfp.org) Hunger Map 2012.

INFOGRAPHIC 4.2 FOOD IS A SOURCE OF MACRONUTRIENTS

The most important dietary macronutrients are carbohydrates, proteins, and lipids (commonly referred to as fats). While most foods contain all of these, one or two macronutrients predominate in each food type. A well-balanced diet is one that includes a variety of foods to ensure that the body gets enough of each macronutrient to grow and remain healthy.

Foods rich in carbohydrates
- Fruits and Vegetables
- Grains
- Legumes

Foods rich in proteins
- Meats
- Legumes
- Dairy

Foods rich in fats
- Dairy
- Meats
- Oils

planted during this time, but the harvest won't be available until March. Quite often supplies from the previous year's harvest run out before the next one is in. They call it the hungry season.

Without enough food, people lack adequate **nutrients,** which provide the chemical building blocks our bodies need to live, grow, and repair themselves. All organisms are made up of chemical building blocks such as water, ions, and organic molecules (see Chapter 2). Because humans (and other animals) can't make these components from thin air, we need to obtain them from our diet. Nutrients also provide us with the **energy** needed to power essential life

activities. Both building blocks and energy are crucial components of food, but for simplicity we will discuss them separately. This chapter focuses on food as a source of chemical building blocks; Chapters 5 and 6 consider energy–the fuel component of food–in more detail.

When experts talk about a nutritious diet, they mean one that provides all the nutrients our bodies need in appropriate amounts. Nutrients that the body requires in large amounts are called **macronutrients.** The macronutrients we need in our diet include carbohydrates, proteins, and lipids (commonly known as fats)–three of the four organic macromolecules discussed in Chapter 2. Because most foods contain mixtures of these macronutrients, those of us who eat a varied diet that includes vegetables, oils, grains, meat, and dairy products can easily obtain all the macronutrients our bodies need (INFOGRAPHIC 4.2).

Macronutrients from the diet cannot be used directly by our bodies, in part because they are too large to be absorbed into the bloodstream from our digestive tract. To be useful, macronutrients must first be broken down into smaller subunits by the process of digestion (see Chapter 26). These subunits are small enough to be absorbed from the digestive tract, are taken up by cells, and used to build the macromolecules our cells need. For example, dietary carbohydrates from bread and pasta are broken down into simple sug-

ars, which our bodies use to build an energy-storing carbohydrate called glycogen in liver and muscle tissue. Proteins from a steak are broken down into amino acids, which can be taken up and used to build new proteins, like those making up our muscles. Fats from our diet are broken down into fatty acids and glycerol, which are used to assemble the phospholipids that make up cell membranes.

The food we eat also contains nucleic acids, the fourth macromolecule making up cells. Although not considered macronutrients (because we need them in smaller amounts), nucleic acids are also broken down into smaller subunits, called nucleotides, which are used by cells to build DNA and RNA. Breaking down food to build up our bodies means that, quite literally, we are what we eat (INFOGRAPHIC 4.3).

INFOGRAPHIC 4.3 MACRONUTRIENTS BUILD AND MAINTAIN CELLS

Macronutrients in our diet include the major macromolecules from which all life is made. These macromolecules must be digested to release their smaller subunits.

→ The four macromolecules that make up cells are carbohydrates, proteins, lipids, and nucleic acids. Cells synthesize these macromolecules from the subunits released by the digestion of macronutrients in food.

The subunits released by digestion are taken into the circulation and carried to the body's cells.

Cells use these subunits as the building blocks to assemble the macromolecules necessary for cellular structure and function.

Carbohydrates

Simple sugars

Proteins

Amino acids

Fats

Fatty acids and glycerol

Nucleic acids

Nucleotides

Energy storage and cell-surface molecules

Enzymes and structural proteins

Phospholipids for membranes

DNA and RNA for storing and transmitting genetic instructions

In Malawi, villages are sometimes miles away from the nearest health facility.

▶ESSENTIAL
NUTRIENTS
Nutrients that can't be
made by the body, so
must be obtained from
the diet.

▶STARCH
A complex plant
carbohydrate made of
linked chains of glucose
molecules; a source of
stored energy.

▶ESSENTIAL
AMINO ACIDS
Amino acids the human
body cannot synthesize
and must obtain from
food.

To a certain extent, our bodies can compensate for a deficiency in one or another nutrient by synthesizing it from other chemical components. For example, if a particular amino acid is in short supply, cells may be able to make it from another amino acid that is in excess. But there are some nutrients that our bodies can't manufacture and which must be obtained pre-assembled from our diet. These are called **essential nutrients.**

In Malawi, most families subsist on a single crop: corn. While corn is a good source of complex carbohydrates, like **starch,** it is not a significant source of protein or fat. This leaves out many essential nutrients needed for a healthful diet, including vital amino acids. From starting materials in food, adults can synthesize 11 of the 20 amino acids they need to make proteins. The other nine must be obtained pre-assembled from our diet. Because our body can't manufacture them, these nine amino acids are called **essential amino acids.** (A few more amino acids are considered essential for infants and children.) Animal products such as meat, eggs, fish, and dairy are the richest sources of essential amino acids, but in many places around the globe, these foods are luxuries people can't afford.

In 1999, when Manary first looked around for an alternative treatment for severe hunger,

he considered sending kids home with a bag of ingredients, such as a corn flour and soy, to make traditional meals. He quickly realized, however, that this would not work. First, such foods would need to be cooked, and cooking is extra work for an already overburdened family. Second, severely malnourished children would need to eat many bowls of such foods per day in order to obtain enough nutrients to recover. Third, the food wouldn't keep.

Peanut butter, on the other hand—especially when it has been fortified with additional nutrients—is a highly effective way of delivering the most nutrients per spoonful of food. Technically, it's known as a ready-to-use-therapeutic food (RUTF), which means it is a complete source of nutrition. "If you eat RUTF you don't need to eat anything else," says Manary. "You're getting everything you need—period."

The RUTF that Manary uses is made up of four main ingredients: full-fat milk powder, sugar, vegetable oil, and peanut butter. Peanut butter is a useful treatment for malnutrition partly because it is full of fat, which helps kids put on weight fast. "Peanuts are the only plant I know of that has 50% fat, when you get rid of the water," says Manary. But the fattening quality is only part of peanut butter's appeal. Because it contains very little water and has a pasty consistency that keeps out air, peanut

> **❝ If you eat RUTF you don't need to eat anything else. You're getting everything you need—period.❞**
>
> — MARK MANARY

butter is naturally resistant to spoiling; bacteria can't grow without water. It also doesn't require cooking and is therefore ready to eat. And it's full of protein, a crucial macronutrient for growing children. All these things make peanut butter a near-perfect therapeutic food, supplying children with the necessary fats, proteins, and essential amino acids (from the milk powder) that they would otherwise lack.

Incidentally, many people in industrialized countries have peanut allergies, but in the developing countries such allergies are rare or nonexistent, he says. That has more to do with the way we train our immune systems in our hyperscrubbed world, Manary explains, than with any intrinsic quality of peanuts.

LIFE IN THE VILLAGE

In 1999, having spent 5 years working in a hospital in Malawi, Manary took an unusual step for an American-trained doctor: he left the hospital and went to live in a rural village for 10 weeks. He wanted to get a first-hand view of the way people lived and the obstacles they faced. The thing that struck him the most was the sheer monotony of village life.

"Every day is the same," he says. "It starts with going to the water pump, hauling water. Then it's finding firewood. Then it's cooking food. . . . It's work all day, 365 a year."

Instantly, he understood why childhood malnutrition is such a terrible trap for families. Families live perilously close to the edge of sufficiency, and any slight change in their circumstances can push them over; even one extra mouth to feed can create food shortages. Moreover, for many rural populations, the nearest hospital is miles away. Taking a sick kid to the hospital is a major disruption to family life, so most mothers do it only as a last resort, when the child is already quite sick. Lastly, telling mothers—as they routinely were told by hospital doctors—that they needed to feed their malnourished child seven times a day was not a workable solution. As Manary explains, "It'd be like me telling you 'Hey, you need to walk back and forth [from New York] to Connecticut every day.'" It just wasn't realistic.

Malnutrition problems usually start when children are 6 months old. Kids younger than 6 months old are sustained easily by breastfeeding alone. Breast milk is a perfect food for children at this age, explains Manary. It is nutritionally appropriate for what their bodies can

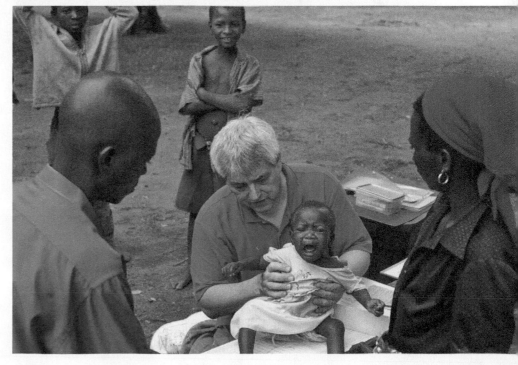

Physician Mark Manary tends to a malnourished child.

INFOGRAPHIC 4.4 ENZYMES FACILITATE CHEMICAL REACTIONS

Cells require enzymes to break down and build up macromolecules. Enzymes are proteins that speed up chemical reactions.

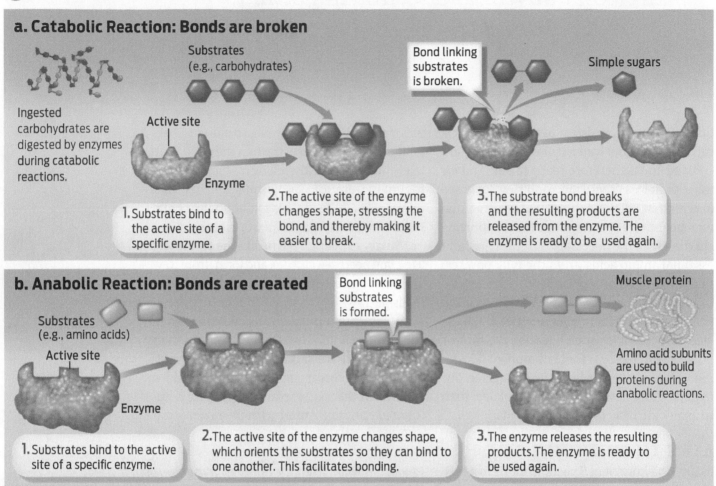

a. Catabolic Reaction: Bonds are broken

Substrates (e.g., carbohydrates)

Bond linking substrates is broken.

Simple sugars

Ingested carbohydrates are digested by enzymes during catabolic reactions.

Active site

Enzyme

1. Substrates bind to the active site of a specific enzyme.

2. The active site of the enzyme changes shape, stressing the bond, and thereby making it easier to break.

3. The substrate bond breaks and the resulting products are released from the enzyme. The enzyme is ready to be used again.

b. Anabolic Reaction: Bonds are created

Bond linking substrates is formed.

Muscle protein

Substrates (e.g., amino acids)

Active site

Enzyme

Amino acid subunits are used to build proteins during anabolic reactions.

1. Substrates bind to the active site of a specific enzyme.

2. The active site of the enzyme changes shape, which orients the substrates so they can bind to one another. This facilitates bonding.

3. The enzyme releases the resulting products. The enzyme is ready to be used again.

▶CATABOLIC REACTION
Any chemical reaction that breaks down complex molecules into simpler molecules.

▶ANABOLIC REACTION
Any chemical reaction that combines simple molecules to build more-complex molecules.

▶METABOLISM
All biochemical reactions occurring in an organism, including reactions that break down food molecules and reactions that build new cell structures.

handle, and it also confers protection against infections and disease. Since 2001, the World Health Organization has recommended that women breast-feed exclusively until a child is 6 months old (and they are encouraged to continue for 2 years and beyond, even after infants begin to eat other food). Virtually all women in Malawi breast-feed.

At 6 months, however, children's growing bodies require more nourishment than breast milk can provide, but no other food is as plentiful or accessible. Malnutrition due to chronic undereating sets in after a period of time, usually between ages 1 and 3.

Scientists sometimes refer to the first 1,000 days of life, from gestation to age 2, as the "golden interval." "Babies go through a huge amount of physical and cognitive development during this period of 1,000 days," says Mary Arimond, a nutrition expert at the University of California, Davis, who studies nutrition in the developing world. "If deficits occur in this 'window,' they are difficult to reverse, especially in continued conditions of poverty."

An underfed child, one that has gone without adequate nutrition for an extended period of time, will lack the raw material for growth and development. That's because growth and development are, essentially, a series of chemical reactions requiring nutrients as starting materials. These reactions begin as soon as the digestion of food begins: reactions that break down larger structures into smaller ones are called **catabolic reactions.** Reactions that build new structures from smaller subunits are called **anabolic reactions.** Together, all the chemical reactions occurring in the body constitute **metabolism.**

To proceed normally, metabolism requires the assistance of helper molecules called **enzymes.** Nearly every chemical reaction in the body requires enzymes. For example, digestive enzymes made by cells in the digestive tract help us digest food molecules into their constituent subunits. When cells divide–for instance in bone marrow to generate new blood cells, or in skin to replace cells lost to death or injury–they rely on a suite of enzymes to carry out a variety of processes, including copying the DNA in cells. Similarly, enzymes produced by cells in bone carry out anabolic reactions that contribute to the formation of new bone.

Enzymes work by speeding up, or catalyzing, chemical reactions (the process is called **catalysis**). In order to accelerate a reaction, the enzyme must bind to molecules involved in the reaction. The molecules that enzymes bind to are called **substrates.** The part of the enzyme that binds to substrates is its **active site.** Each enzyme has an active site that fits only one particular substrate molecule or molecules. An enzyme that breaks down carbohydrates, for example, cannot break down proteins **(INFOGRAPHIC 4.4).**

Enzymes catalyze reactions by lowering the amount of energy required to nudge a chemical reaction into motion. Enzymes substantially reduce this **activation energy,** allowing the reaction to occur more easily. For example, when complex carbohydrates are ingested, enzymes in our digestive tract help break the bonds that hold the molecules together, releasing simple sugars–a reaction that would not otherwise occur **(INFOGRAPHIC 4.5).**

▶**ENZYME**
A protein that speeds up the rate of a chemical reaction.

▶**CATALYSIS**
The process of speeding up the rate of a chemical reaction (e.g., by enzymes).

▶**SUBSTRATE**
A molecule to which an enzyme binds and on which it acts.

▶**ACTIVE SITE**
The part of an enzyme that binds to substrates.

▶**ACTIVATION ENERGY**
The energy required for a chemical reaction to proceed. Enzymes accelerate reactions by reducing their activation energy.

INFOGRAPHIC 4.5 ENZYMES CATALYZE REACTIONS BY LOWERING ACTIVATION ENERGY

 The activation energy is the energy that must be put into a reaction in order to make it "go." Enzymes reduce the activation energy in both anabolic and catabolic reactions, making them occur more rapidly.

Reactants need an energy boost to initiate the reaction.

Activation energy

Once the activation energy has been achieved, the reaction proceeds spontaneously.

Total energy

Reactant (e.g., a complex carbohydrate)

Product (e.g., simple sugars)

Uncatalyzed reaction pathway

Enzyme

Enzyme lowers the activation energy needed to initiate the reaction.

Activation energy

Lowering the activation energy makes it easier for more products to be produced.

Total energy

Reactant (e.g., a complex carbohydrate)

Product (e.g., simple sugars)

Enzyme-catalyzed reaction pathway

> **"** *Babies go through a huge amount of physical and cognitive development during this period of 1,000 days.* **"**
>
> — MARY ARIMOND

If a child doesn't get enough to eat, the enzymes that carry out the anabolic reactions necessary for growth do not have substrates to act on, so these anabolic reactions do not occur. In fact, catabolic reactions often begin to break down existing structures–for example, muscle protein–to obtain amino acids. The results are the telltale signs of malnutrition: thin arms with skin wrinkling over wasted muscle; painful swelling of the legs and feet caused by a buildup of fluid; blond or rust-colored hair resulting from a deficiency of protein; and a distended, bloated stomach, a sign of nutrient deficiency.

Because a healthful diet requires a balance of many different kinds of nutrients, it's not only the sheer quantity of food that is important, but also the specific type of food that matters. Some nutrients can be missing from a diet, even if a child is well fed.

HIDDEN HUNGER

When Manary first arrived at the Malawi hospital in 1994, he made a few changes to the then-standard treatment regimen of fortified milk. One thing he did right away was to add extra potassium to the milk that kids were drinking. Immediately, the fatality rates dropped–from 33% to 10%. Potassium, required for proper muscle contraction and nerve function, is one of many **minerals** that we need to keep healthy.

Because most minerals are needed only in small amounts, they are known as **micronutrients** (as opposed to macronutrients, which are needed in much larger quantities). **Vitamins** are another kind of micronutrient. Just because our bodies require

only small amounts of micronutrients, however, doesn't mean they aren't important. In fact, micronutrient deficiency can have serious health consequences. Iron deficiency causes the blood disease known as anemia, for example, and lack of vitamin C causes a tissue-deteriorating disease called scurvy. (To prevent scurvy while on long sea voyages, British sailors would eat limes and other citrus fruit, which are high in vitamin C; this is why British people are sometimes called "limeys.")

In Malawi, as in many parts of the developing world, people often suffer deficiencies of vitamin A and zinc–which can cause vision and immune problems. Such micronutrient deficiencies are sometimes known as "hidden hunger," because the problem is not lack of food but a lack of necessary micronutrients. Manary says the problem is widespread because most of the world's staple crops–foods like rice, corn, wheat, and cassava–do not contain adequate micronutrients.

Food producers routinely add to foods some micronutrients that are hard to obtain from natural sources. Iodine, for example, is added to table salt (in "iodized" salt) to prevent goiter, an abnormal thickening of the neck caused by an enlarged thyroid gland due to a lack of dietary iodine.

The peanut butter RUTF that Manary uses contains a mineral and vitamin mix that is 1.6% of the paste by weight. Although peanut butter naturally contains many micronutrients, the amounts malnourished children need are greater than what is normally found in most foods. So the peanut paste is deliberately enriched with micronutrients **(TABLE 4.1)**.

▶**MINERAL**
An inorganic chemical element required by organisms for normal growth, reproduction, and tissue maintenance; examples are calcium, iron, potassium, and zinc.

▶**MICRONUTRIENTS**
Nutrients, including vitamins and minerals, that organisms must ingest in small amounts to maintain health.

▶**VITAMIN**
An organic molecule required in small amounts for normal growth, reproduction, and tissue maintenance.

TABLE 4.1 A SAMPLE OF MICRONUTRIENTS IN YOUR DIET

MINERALS: Inorganic elements not synthesized by the body.

MINERAL	FUNCTION	FOOD SOURCES	PROBLEMS OF DEFICIENCY	PROBLEMS OF EXCESS
Calcium	Bone and teeth formation, blood clotting	Dairy products, green vegetables, legumes	Osteoporosis, stunted growth	Kidney stones
Iron	Components of hemoglobin in red blood cells; carries oxygen throughout the body	Green vegetables, beef, liver	Anemia, fatigue, dizziness, headaches, poor concentration	Constipation, risk of type 2 diabetes
Potassium	Electrolyte balance, muscle contraction, nerve function	Fruits, vegetables, meat	Muscle weakness, neurological disturbances	Muscle weakness, heart failure
Sodium	Electrolyte balance, muscle contraction, nerve function	Salt, bread, milk, meat	Muscle cramps, reduced appetite, neurological disturbances	High blood pressure

WATER-SOLUBLE VITAMINS: Organic molecules not synthesized by the body. Excess vitamin is excreted in urine and so does not harm health.

VITAMIN	FUNCTION	FOOD SOURCES	PROBLEMS OF DEFICIENCY	PROBLEMS OF EXCESS
B_1 (thiamine)	Cofactor for enzymes involved in energy metabolism and nerve function	Leafy vegetables, whole grains, meat	Heart failure, depression	None
Folate	Cofactor for enzymes involved in DNA synthesis and cell production	Dark green vegetables, nuts, legumes, whole grains	Neural tube defects, anemia	None
B_{12}	Cofactor for enzymes involved in the breakdown of fatty acids and amino acids and nerve cell maintenance	Meat, milk, eggs	Anemia, neurological disturbances	None
C	Cofactor for enzymes involved in collagen synthesis; improves iron absorption and immunity	Citrus fruits	Scurvy, poor wound healing	None

FAT-SOLUBLE VITAMINS: Organic molecules not synthesized by the body (except vitamin D). Excess vitamin is stored in fat cells and can harm health.

VITAMIN	FUNCTION	FOOD SOURCES	PROBLEMS OF DEFICIENCY	PROBLEMS OF EXCESS
A (retinol)	Component of eye pigment, supports skin, bone, and tooth growth, supports immunity and reproduction	Fruits and vegetables, liver, egg yolk	Skin problems, blindness	Headaches, intestinal pain, bone pain
D	Calcium absorption, bone growth	Fish, dairy products, eggs	Bone deformities	Kidney damage
E	Antioxidant, supports cell membrane integrity	Green leafy vegetables, legumes, nuts, whole grains	Neural tube defects, anemia, digestive-health problems	Fatigue, headaches, blurred vision, diarrhea
K	Supports synthesis of blood clotting factors	Green leafy vegetables, cabbage, liver	Abnormal blood clotting, bruising	Liver damage, anemia

▶**COFACTOR**
An inorganic substance, such as a metal ion, required to activate an enzyme.

▶**COENZYME**
A small organic molecule, such as a vitamin, required to activate an enzyme.

Vitamins and minerals play numerous roles in the body. Some play structural roles–the mineral calcium, for example, is a primary component of bones and teeth. Others play a functional role, helping other molecules to act. Perhaps their most critical role is serving as **cofactors** for enzymes.

Cofactors are accessory or "helper" chemicals that enable enzymes to function. Cofactors include inorganic metals such as zinc, copper, and iron. Cofactors can also be organic mole-

cules, in which case they are called **coenzymes.** Most vitamins are important coenzymes. Without cofactors and coenzymes that bind to enzymes and enable them bind to substrates, cell metabolism would grind to a halt (**INFOGRAPHIC 4.6**).

By adding extra vitamins and minerals to the milk that children were receiving in the hospital, Manary found that he could get the death rate to go way down, but the recovery rate still wouldn't budge. That's when he knew it was time for a new approach.

INFOGRAPHIC 4.6 VITAMINS AND MINERALS HAVE ESSENTIAL FUNCTIONS

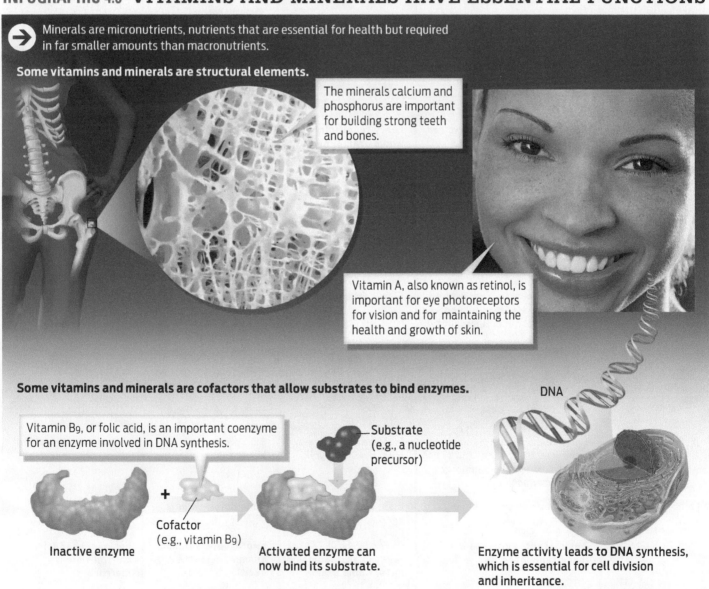

→ Minerals are micronutrients, nutrients that are essential for health but required in far smaller amounts than macronutrients.

Some vitamins and minerals are structural elements.

The minerals calcium and phosphorus are important for building strong teeth and bones.

Vitamin A, also known as retinol, is important for eye photoreceptors for vision and for maintaining the health and growth of skin.

Some vitamins and minerals are cofactors that allow substrates to bind enzymes.

DNA

Vitamin B9, or folic acid, is an important coenzyme for an enzyme involved in DNA synthesis.

Substrate
(e.g., a nucleotide precursor)

Inactive enzyme

Cofactor
(e.g., vitamin B9)

Activated enzyme can now bind its substrate.

Enzyme activity leads to DNA synthesis, which is essential for cell division and inheritance.

Before and After: For many children in Malawi, peanut butter RUTF has been a life-saver.

EMPTYING THE WARDS

By the time Manary began seriously thinking about home-based therapy, he had been working in Malawi for about 5 years, witnessing the failure of the standard WHO treatment, and he had spent time in villages. In his head, he'd been tossing around the idea of something new and different. Out of the blue, he got an e-mail from a doctor in France, André Briend, who was also thinking about home-based therapy as a treatment for malnutrition.

The two scientists corresponded for about a year by e-mail, weighing the pros and cons of various foods that could be eaten at home yet still pack a nutrient wallop. After considering various options—they considered biscuits, pancakes, even Nutella—the two researchers eventually decided on peanut butter as the best choice. In 2001, they decided to test the idea.

Briend had some of the peanut butter RUTF made up in France and then sealed in foil packets and shipped to Malawi. They then used this product in a carefully designed sci-entific study. After a brief stabilization phase in the hospital (during which antibiotics were given, if necessary), the children were discharged and sent home on one of three different treatment regimens: (1) ample amounts of traditional food—corn flour and soy; (2) a small amount of peanut butter-based RUTF, to be used as a supplement to the normal diet at home; and (3) the full dose of peanut butter-based RUTF, with sufficient nutrients to meet the total nutritional needs of the children. The goal of the study was to see which treatment regimen was most effective.

Within a few months, the results were astonishing: 95% of the children who received the full peanut butter-based RUTF recovered. Those who received traditional food also did pretty well—about 75% of them recovered, but still the peanut butter was better. And all the home-based treatments were significantly better than standard hospital-based milk therapy, which historically had a 25%–40% recovery rate.

INFOGRAPHIC 4.7 PEANUT BUTTER–BASED RUTF SAVES MORE CHILDREN

 Studies show that significantly more children recover when treated at home with peanut butter–based RUTF compared to a corn/soy flour diet or their regular diet supplemented with a small amount of the RUTF supplement.

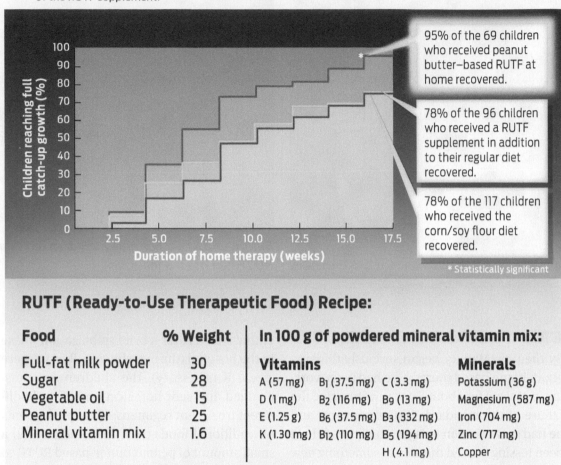

95% of the 69 children who received peanut butter–based RUTF at home recovered.

78% of the 96 children who received a RUTF supplement in addition to their regular diet recovered.

78% of the 117 children who received the corn/soy flour diet recovered.

* Statistically significant

RUTF (Ready-to-Use Therapeutic Food) Recipe:

Food	% Weight
Full-fat milk powder	30
Sugar	28
Vegetable oil	15
Peanut butter	25
Mineral vitamin mix	1.6

In 100 g of powdered mineral vitamin mix:

Vitamins			Minerals
A (57 mg)	B_1 (37.5 mg)	C (3.3 mg)	Potassium (36 g)
D (1 mg)	B_2 (116 mg)	B_9 (13 mg)	Magnesium (587 mg)
E (1.25 g)	B_6 (37.5 mg)	B_3 (332 mg)	Iron (704 mg)
K (1.30 mg)	B_{12} (110 mg)	B_5 (194 mg)	Zinc (717 mg)
		H (4.1 mg)	Copper

SOURCE: Manary, M. J., et al. (2004) Home based therapy for severe malnutrition with ready-to-use food. *Archives of Disease in Childhood* 89:557–561.

Manary couldn't quite believe how effective the treatment was. "I said, 'Damn, this stuff really works!'" (INFOGRAPHIC 4.7).

One of the main reasons that peanut butter RUTF–which locals call chiponde, or "nutpaste"–is more effective than standard therapy is that it can be administered safely at home. When children are malnourished, their immune systems aren't functioning at optimal levels, which means that hospitals are often the worst place for them to be because of the risk of infection from other sick patients. Peanut butter RUTF is also something that children can eat on their own, without help from their parents. And, because they clearly like the taste of it (one doctor described it as tast-ing like the inside of a Reese's Peanut Butter Cup), they gobble it up.

As dramatic as Manary's initial results were, however, not everyone was convinced. In fact, most leaders in nutrition science were vehemently opposed to the approach, which is why Manary was heckled at the conference in 2003.

Manary wasn't the only person in the field to face a backlash. Other doctors working in this area faced similar opposition. "It was pretty nasty," says Steve Collins, a physician and early advocate of home-based treatment, whose work was also lambasted.

At that time, around 2003, the conventional wisdom in humanitarian aid circles was that these children were so sick that they needed to be hospitalized so that doctors and nurses could take care of them. But doctors who, like Manary, were working in the trenches, knew differently. They knew that even with the best available treatments–administered with careful precision by trained professionals–children were recovering only 25%-40% of the time. And, as Manary showed, mothers who were sent home with RUTF could do much better.

To raise awareness of his peanut butter approach and to start making the treatment available on a wider basis, in 2004 Manary started Project Peanut Butter, which would produce the product locally and distribute it to families in Malawi. So far, he says, more than 500,000 kids have been helped by peanut butter RUTF–not only in Malawi, but all across the world, including in Haiti and Sierra Leone.

Thanks in part to Project Peanut Butter, and the extensive body of evidence that Manary and others have collected, the world aid community eventually came around to Manary's view: in 2007, in a dramatic about-face, the joint UN relief agencies–UNICEF, WHO, and the World Food Programme–issued an official statement saying that home-based therapy with peanut butter RUTF is the preferred way to treat acute malnutrition. Manary's approach was vindicated.

PREVENTING HUNGER?

Once made in a broom closet in a Malawi hospital, peanut butter RUTF is now produced in giant industrial factories and used all over the world. The largest producer is the France-based company Nutriset, which sells its product (called Plumpy'Nut) to relief agencies like UNICEF and Doctors without Borders, which then distribute it free to countries in need.

The undeniable success of peanut butter RUTF in the treatment of malnutrition has inspired some aid groups to want to expand its reach into the realm of prevention. Several UN agencies and NGOs (nongovernmental organizations) have experimented with providing related products, known as RUSFs (ready-to-use supplemental foods), to hungry children around the world. These products, which provide less than the full therapeutic dose of nutrients, are

> Within a few months, the results were astonishing: 95% of the children who received the full peanut butter–based RUTF recovered.

meant to supplement local foods and thus help prevent malnutrition before it starts. The jury is still out on whether this approach to prevention is effective. "The state of the evidence is that there is not enough evidence," says nutritionist Arimond, of UC-Davis, who also works with the International Lipid-Based Nutrient Supplements (iLiNS) Project, which is conducting some of this research. But even if malnutrition could be prevented this way, there are those who argue that it should not be.

"Some people feel the private sector can never be trusted and can never be part of the solution," says Arimond. "These concerns have a very legitimate basis, which is the very

Peanut butter RUTF tastes like the inside of a Reese's Peanut Butter Cup.

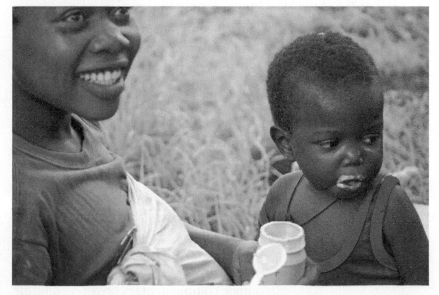

INFOGRAPHIC 4.8 A BALANCED DIET

→ A balanced diet includes all the nutrients needed for full health.

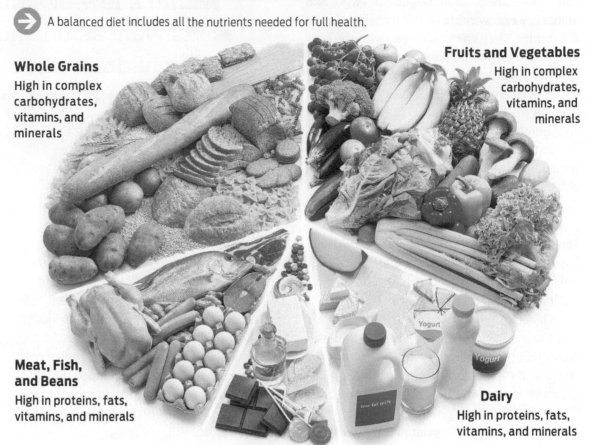

Whole Grains
High in complex carbohydrates, vitamins, and minerals

Fruits and Vegetables
High in complex carbohydrates, vitamins, and minerals

Meat, Fish, and Beans
High in proteins, fats, vitamins, and minerals

Dairy
High in proteins, fats, vitamins, and minerals

Refined Sugars and Fats
High in carbohydrates and fats

negative role infant formula companies have played in developing countries over the years."

For others, the concern is less whether the profit motive is involved, and more who benefits from those profits. "If it's benefiting U.S. agriculture, or European agriculture, or European manufacturing and European transporters, it's not benefiting the right people," says Collins. For that reason, nonprofits like Project Peanut Butter and the Haiti-based Meds & Foods for Kids have made a clear commitment to producing RUTF locally. By using local peanuts, and local farmers, these groups contribute directly to the health and growth of the local economy, and thus help to alleviate the conditions that spawn malnutrition in the first place.

Other critics worry that RUTF will be seen as a substitute for breast-feeding and will thus discourage complementary breast-feeding for children 6 months to 3 years of age, and that it will take the focus off the root problem. The solution to malnutrition, they say, is making sure that people have the resources to eat a healthful, well-balanced diet in the first place (INFOGRAPHIC 4.8).

Manary doesn't disagree. He believes that the ultimate solution to malnutrition is prevention through improved agriculture. To that end, he is also working on a project sponsored by the Bill and Melinda Gates Foundation that aims to use genetic engineering to improve the nutritional quality of subsistence crop plants in Africa. Called BioCassava Plus, the project seeks to make cassava more nutritious by enriching it with additional vitamins and protein. Manary says he has high hopes for the project, and envisions a day when children do not die from hunger. "My professional goal is to fix malnutrition for kids in Africa," he says. ■

CHAPTER 4 Summary

▸ Food is a source of nutrients. Nutrients provide the chemicals required to build and maintain cells and tissues and furnish cells with the energy needed to function.

▸ Nutrients required in large amounts are called macronutrients; nutrients required in smaller amounts are called micronutrients. Both are essential for good health.

▸ Macronutrients include carbohydrates, proteins, and fats; these are among the organic macromolecules that make up our cells.

▸ Digestion breaks down macromolecules into smaller subunits, which are then used by cells to build cell structures and carry out cell functions.

▸ Enzymes are proteins that accelerate the rate of chemical reactions. Nearly all reactions in the body require enzymes, including those required for growth and development.

▸ Enzymes speed up reactions by binding specifically to substrates and reducing the activation energy necessary for a reaction to occur. Enzymes mediate both bond-breaking (catabolic) and bond-building (anabolic) reactions.

▸ Many enzymes require small "helper" chemicals known as cofactors to function. Micronutrients

such as minerals and vitamins, found abundantly in fruits and vegetables, are important cofactors.

▸ Most nutritionists recommend that fruits, vegetables, and whole grains make up the largest portion of our diet.

▸ Malnutrition results when adequate macronutrients and micronutrients are lacking in the diet.

▸ Malnutrition is especially dangerous for children, whose bodies are, or should be, growing rapidly.

MORE TO EXPLORE

▸ **Project Peanut Butter** http://www.project peanutbutter.org

▸ **Meds & Foods for Kids** http://mfkhaiti.org/

▸ **Manary, M. J., et al. (2004) Home based therapy for severe malnutrition with ready-to-use food.** *Archives of Disease in Childhood* 89:557–561.

▸ **Latham, M. C., et al. (2011) RUTF stuff. Can the children be saved with fortified peanut paste?** *World Nutrition* 2(2) 62–85.

▸ **Collins, S. (2001). Changing the way we address severe malnutrition during famine.** *The Lancet* 358(9280):498–501.

▸ **Diamond, J. (1999)** *Guns, Germs, and Steel: The Fates of Human Societies.* New York: W. W. Norton.

CHAPTER 4 Test Your Knowledge

DRIVING QUESTION 1

What are the macronutrients and micronutrients provided by food?

By answering the questions below and studying Infographics 4.2 and 4.3 and Table 4.1, you should be able to generate an answer for the broader Driving Question above.

KNOW IT

1 A macronutrient is

a. a nutrient with a large molecular weight.
b. a nutrient that is abundant in the diet.
c. a nutrient that is required in large amounts.
d. a nutrient that is stored in large amounts in the body.
e. a nutrient that the body makes in large quantities.

2 Which of the following is/are macronutrient(s)?

a. protein
b. iodine
c. vitamin C
d. fats
e. all of the above
f. a and d

3 A multivitamin supplement is a(n) _____ supplement.

a. macronutrient
b. micronutrient
c. mineral
d. enzyme
e. a and b

4 Which of the following foods is a rich source of protein?

a. lean meat, such as chicken breast
b. whole grains (e.g., whole wheat bread)
c. olive oil
d. leafy greens
e. berries (e.g., blueberries and raspberries)

USE IT

5 Explain the difference between macronutrients and micronutrients.

6 A typical multivitamin supplement contains vitamin A, vitamin C, vitamin D, vitamin E, vitamin K, vitamin B_1, vitamin B_2, vitamin B_6, biotin, calcium, iron, magnesium, zinc, selenium, copper, manganese, and chromium. Explain your answers to the following questions.

a. Are all of these vitamins? If there are ingredients that are not vitamins, what are they?
b. Are all of these micronutrients?

DRIVING QUESTION 2

What are essential nutrients?

By answering the questions below and studying Infographics 4.3 and 4.5 and Table 4.1, you should be able to generate an answer for the broader Driving Question above.

KNOW IT

7 What subunits are proteins broken down into during digestion?

a. fatty acids
b. amino acids
c. glycerol
d. nucleotides
e. simple sugars

8 Where (or how) do we obtain essential amino acids?

a. from carbohydrates in our diet
b. by synthesizing them from other amino acids
c. from oils in our diet
d. from bright orange fruits and vegetables
e. from protein in our diet

USE IT

9 Our bodies cannot synthesize vitamin C, but requires it. Therefore, vitamin C is

a. an essential micronutrient.
b. an essential mineral.
c. an essential macronutrient.
d. a nonessential vitamin.
e. a nonessential amino acid.

10 Which component of peanut butter RUTF supplies essential amino acids?

a. milk powder
b. peanut butter
c. sugar
d. vegetable oil
e. powered vitamins and minerals
f. a and c

11 Corn lacks the essential amino acids isoleucine and lysine. Beans lack the essential amino acids tryptophan and methionine. Soy contains all the essential amino acids.

a. Could someone survive on a diet with a corn-based protein alone? Why or why not?
b. Why do many traditional diets combine corn (e.g., tortillas) with beans?
c. Why did one of the home-based feeding therapies in Malawi combine soy flour with corn flour?

DRIVING QUESTION 3

What are enzymes, and how do they work?

By answering the questions below and studying Infographics 4.4 ,4.5, and 4.6, you should be able to generate an answer for the broader Driving Question above.

KNOW IT

12 The substrate of an enzyme is

a. an organic accessory molecule.
b. the molecule(s) released at the end of an enzyme-facilitated reaction.
c. the shape of the enzyme.
d. one of the amino acids that makes up the enzyme.
e. what the enzyme acts on.

13 Compare and contrast enzyme cofactors and coenzymes.

14 Enzymes speed up chemical reactions by

a. increasing the activation energy.
b. decreasing the activation energy.
c. breaking bonds.
d. forming bonds.
e. releasing energy.

15 How is folate (folic acid) best described?

a. as a substrate of an enzyme
b. as a nucleotide
c. as an organic cofactor (coenzyme)
d. as an enzyme
e. a and b

USE IT

16 If the shape of an enzyme's active site were to change, what would happen to the reaction that the enzyme usually speeds up?

17 Considering the function of folate (folic acid) given in Infographic 4.6, why would you say pregnant women (and women who could become pregnant) should ensure that they have adequate levels of folate in their diets?

DRIVING QUESTION 4

What are the consequences of a diet lacking sufficient nutrients?

By answering the questions below and studying Infographics 4.1, 4.6, and 4.7 and Table 4.1, you should be able to generate an answer for the broader Driving Question above.

KNOW IT

18 When vitamins are consumed:

a. Why are there no problems of excess for vitamin C but there are for vitamin E?

b. If you were to take a supplement with a high amount of vitamin C, what would happen to all that vitamin C? Would it all be used? Would some of it be stored in your body?

19 What ingredients in RUTF peanut paste specifically help bone growth? (Hint: Refer to Table 4.1.)

a. calcium   d. all of the above

b. vitamin D   e. a and b

c. potassium

INTERPRETING DATA

20 Infographic 4.7 shows the results of a study examining three different home-based therapies for malnourished children.

a. From the data shown, how many of all the children in the study reached full catch-up growth?

b. What percentage of the children in the study does this number represent? How does this compare to previous recovery rates of 25%–40% for children who had received standard hospital therapy?

The children who received the RUTF were given enough of it to supply 730 kJ of energy per kg of body weight. This is sufficient energy to meet their needs.

c. A malnourished 2-year-old girl weighs a mere 6 kg (~13 pounds). If she had been in the RUTF group in the study, how many daily kJ would she have obtained from the RUTF?

d. If the same malnourished 2-year-old had been in the RUTF supplement group, she would have received 2,100 kJ per day from the supplement. What percentage of her daily energy needs would this represent? (Hint: Use your answer to part c.)

e. Children in the RUTF supplement group ate a traditional diet of corn/soy flour to make up the rest of their diet. Corn/soy flour contains 4 kJ per gram. How many grams of the traditional mix would this 2-year-old need to consume (on top of the RUTF supplement) to meet her daily needs?

MINI CASE

21 A college student returns home at the end of the school year. His mother is shocked by the large number of unhealed scrapes and sores on his knees and arms. She also notices that he has put on a few pounds. The student tells his mother that the scrapes are just left over from a skateboarding mishap a few weeks ago and that he guesses he could cut back on some of his snacks. A week after coming home, he goes to the dentist for his yearly checkup. The dentist is alarmed by his bleeding and swollen gums. When asked about his diet, the student notes that he and some of his friends challenged one another to see who could go the longest eating nothing but eggs, mac 'n' cheese, and toast with butter. He proudly announces that he had stayed on this diet for 6 months.

a. Could this student be suffering from malnutrition? Explain your answer.

b. What mineral(s) or vitamin(s) (or both) are you most concerned about, given the symptoms noted by the dentist?

c. What dietary recommendations would you make for this student?

BRING IT HOME

22 Is malnutrition solely a problem in developing countries? Do some investigative research on at least two food-aid programs, local, federal, or international. What criteria would you consider before deciding to donate money to a food-aid program? Explain your answer.

5

Bassin 2B

THE FUTURE OF
FUEL

Scientists hope to make algae into
the next global energy source

AS AN ENGINEER WORKING FOR the Navy Seals in 1978, Jim Sears took a nighttime scuba dive off the coast of Panama City, Florida—one of many he took to do underwater research. The dive started out routinely, but then, suddenly, glowing phosphorescent algae appeared as if out of nowhere. When Sears put his hands out in front of him, sparkling streamers of microbes came off his fingertips. "It was magical," he recalls.

Bioluminescent algae glow in the tide along a beach.

▶▶ DRIVING QUESTIONS

1. What are the photosynthetic organisms on the planet, and why are they so important?

2. What are the different types of energy and what transformations of energy do organisms carry out?

3. How do plants and algae convert the energy in sunlight into energy-rich organic molecules? (And why can't humans do this?)

4. How do algal biofuels compare to other fuels in terms of costs, benefits, and sustainability?

Sears is an inventor with many and varied devices to his credit. When working for the Navy in the 1970s and 1980s, he built an underwater speech descrambler and a portable mine detector, among other things. Later he moved on to more creative ventures, including a "hump-o-meter" that could tell farmers when their animals were in heat or mating.

But the seeds of his real claim to fame weren't sown until 2004, when Sears was working on agricultural electronics. That's when he began to turn his attention toward what he felt was the world's biggest problem: dwindling fossil fuel reserves. After he did some thinking and a little research, the tiny, glowing organisms that had wowed him during his dive more than two decades earlier came to mind, in part because of a website he stumbled across that discussed the unique properties of algae. He realized suddenly that they might be able to help.

Algae are perhaps best known for the green, red, or brown hues they give to the surfaces of ponds and swimming pools, but they have other unique characteristics, too. The evolutionary ancestors of plants, algae were among the first eukaryotic life forms to appear on our planet. There are large multicellular forms, like seaweed, and microscopic single-cell forms called microalgae. Both types share with green plants the impressive ability to capture the energy of sunlight and convert it into forms of energy usable by other organisms. Even more remarkable, much of that usable energy is in the form of oils ideally suited to making fuel. The oil that microalgae produce is very similar to common vegetable oil. It accumulates inside the microbes' tiny cells, and once extracted, it can be processed to make biodiesel, gasoline, or jet fuel. "The more I looked into them, the more amazing they were," Sears says.

And that's good news, because America is desperate for new fuels. After all, Americans burn through 378 million gallons of gasoline a day, enough to fill about 540 Olympic-size swimming pools. And despite the fact that our demand will likely increase over the course of the next 25 years, the sources of our precious gasoline–oil reserves buried deep underground–are finite, take millions of years to replenish, and largely lie outside U.S. borders (**INFOGRAPHIC 5.1**).

INFOGRAPHIC 5.1 DISTRIBUTION OF RECOVERABLE OIL RESERVES

 The gasoline used to power cars begins as oil formed deep in the ground over millions of years. The United States depends heavily on oil recovered from other countries for its fuel supply.

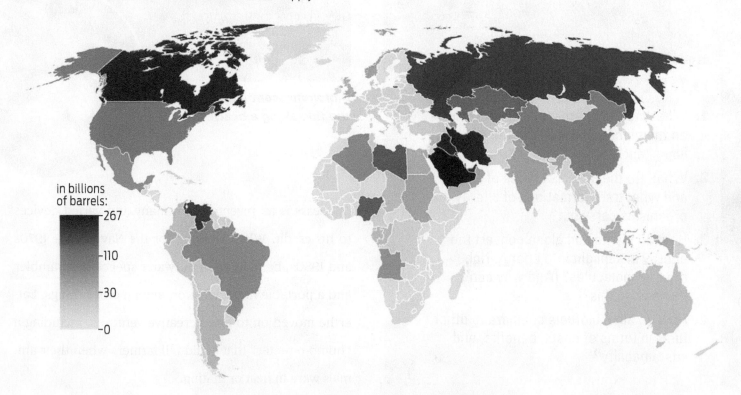

in billions
of barrels:

267

110

30

0

The Toyota Algaeus, which is powered by algae, gasoline, and batteries.

Confronted with this looming crisis, scientists and politicians are now turning toward alternatives such as **biofuels**—renewable fuels made from living organisms. In an effort to end our addiction to oil, in 2007 President George W. Bush signed the Energy Independence and Security Act, which requires the United States to produce 36 billion gallons of renewable fuels by 2022, of which 21 billion gallons must be advanced biofuels (and not corn-based ethanol).

Convinced of the promise of algae biofuels, in 2006, Sears founded Solix, one of the first biotechnology companies working to mass-produce biodiesel from algae. Though he is no longer involved with Solix, the company is still going strong. In 2009, it began commercially producing its algae-based fuel, with the goal of making the equivalent of 3,000 gallons of oil per acre of cultivated algae. Other companies are getting on the algae bandwagon—as of 2012, there were more than 150 companies dedicated to making fuel from algae. In January 2009, Continental Airlines flew its first plane powered in part by jet fuel made from algae. In September of that year, a modified Toyota Prius dubbed Algaeus drove 3,750 miles across the country powered by a fuel mix of algal and conventional gasoline, plus batteries. And in November 2012,

> *Americans burn through 378 million gallons of gasoline a day, enough to fill about 540 Olympic-size swimming pools.*

California gas stations began pumping algae-derived biodiesel. Algae, many say, are the fuel source of the future.

ENERGY BASICS

To power nearly everything in our modern lives we need energy. Americans get their energy from many sources, by far the largest of which are the **fossil fuels**—oil, gas, and coal. The compressed remains of once-living organisms, fossil fuels are considered to be nonrenewable because they take millions of years to form. Currently, only about 9% of our energy comes from renewable sources like wind, solar, and biofuels, but that proportion is growing. For powering our planes and cars, the most useful

▶**BIOFUELS**
Renewable fuels made from living organisms (e.g., plants and algae).

▶**FOSSIL FUELS**
Carbon-rich energy sources, such as coal, petroleum, and natural gas, which are formed from the compressed, fossilized remains of once-living organisms.

INFOGRAPHIC 5.2 U.S. ENERGY CONSUMPTION

 The United States is the largest consumer of fossil fuels. Fossil fuels are considered non-renewable because they take millions of years to create by natural processes. As we continue to deplete fossil fuels, new energy sources are being developed that reduce our demand on petroleum and other fossil fuels.

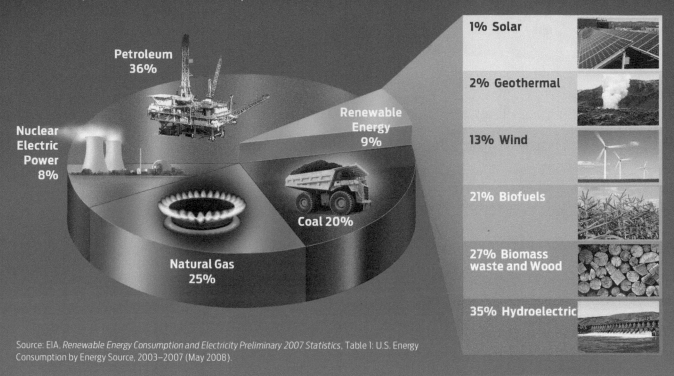

Petroleum 36%

Nuclear Electric Power 8%

Renewable Energy 9%

Coal 20%

Natural Gas 25%

1% Solar

2% Geothermal

13% Wind

21% Biofuels

27% Biomass waste and Wood

35% Hydroelectric

Source: EIA, *Renewable Energy Consumption and Electricity Preliminary 2007 Statistics*, Table 1: U.S. Energy Consumption by Energy Source, 2003–2007 (May 2008).

▶ENERGY
The capacity to do work. Cellular work includes processes such as building complex molecules and moving substances into and out of the cell.

▶CHEMICAL ENERGY
Potential energy stored in the bonds of biological molecules.

fuel sources are those that come in liquid form, such as gasoline and diesel, both derived from oil (petroleum) (**INFOGRAPHIC 5.2**).

Of course, energy isn't just needed to fly planes and drive cars. **Energy**–defined as the capacity to do work–is critical to all life on earth. Energy powers every activity we perform, from the more obvious ones like breathing, thinking, and running to less obvious activities like building the molecules that make up our bodies. Without a source of energy, all life on Earth would grind to a halt, like a cell phone with a dying battery.

Organisms can't simply create energy when they need it, however–energy cannot be created–they must obtain it from an outside source. Humans and other animals obtain the energy they need by eating food. We've already seen that our digestive systems break down food to obtain nutrients (see Chapter 4). As these molecules are further broken down, the energy stored in the molecules is made available to do

work. The bonds that hold molecular subunits together represent a form of stored **chemical energy**; breaking these bonds releases that stored energy, making it available to power cell functions.

Algae and plants, on the other hand, get their energy from the sun. They trap the energy of sunlight and store it in the form of molecules inside their cells. Algae could be the world's next energy source because these tiny organisms are very good at trapping and storing energy–the oil they produce is rich in chemical energy. Even more impressive, all they need to make this oil is sunlight, carbon dioxide, and water (plus a few nutrients). Give them these tidbits, and algae grow rapidly–some strains double their volume in 12 hours–all the while accumulating gobs of oil inside their cells. In addition to this oil, which can be used to make biodiesel, algae also make proteins and carbohydrates (sugars), which can be converted into other biofuels, like ethanol and butanol. These biofuels can be mixed with gasoline to power cars (**INFOGRAPHIC 5.3**).

Sears wasn't the first to consider algae's fuel potential. In 1978—the same year Sears took his fateful night dive—the U.S. Department of Energy started its Aquatic Species Program, with the goal of exploring algae's fuel possibilities. This venture was a direct response to the oil crisis of the 1970s, during which the cost of oil skyrocketed, supplies were rationed, and long lines formed at the pump. But when oil fell to $20 a barrel in 1996, the government abandoned the program, assuming that oil made from algae would always be too expensive. Now, with oil prices much more volatile, and the cost of al-

ternatives coming down, biofuel from algae has become an attractive option again.

With so much talk about dwindling energy reserves, it's tempting to think that energy is something that we simply use up over time. But energy cannot be created or destroyed. When energy is used to power our cars—or our brain cells—that energy is not destroyed, it merely changes form, a principle known as the **conservation of energy**. This principle, one of the laws of thermodynamics, is key to understanding energy use—as applicable to cars as to people.

Consider a cyclist who eats a cereal bar before a ride. The bar contains chemical en-

▶**CONSERVATION OF ENERGY**
The principle that energy cannot be created or destroyed, but can be transformed from one form to another.

INFOGRAPHIC 5.3 ALGAE CAPTURE ENERGY IN THEIR MOLECULES

➡ Algae can use sunlight, carbon dioxide, and nutrients to produce a high volume of oil readily used to produce biofuel, in addition to carbohydrates and proteins that can be useful as additional energy sources.

CO_2 Sunlight   Nutrients

Grow algae in open ponds or closed bioreactors.

Harvest algae cells and separate the components.

Cell Wall Biomass   **Proteins**   **Carbohydrates**   **Oil Lipids**

- Burned for electricity and heat
- Organic fertilizer

- Animal feed

- Fermented for ethanol
- Burned for electricity

- Biodiesel
- Gas

▸**POTENTIAL ENERGY**
Stored energy.

▸**KINETIC ENERGY**
The energy of motion or movement.

▸**HEAT**
The kinetic energy generated by random movements of molecules or atoms.

ergy in the form of chemical bonds that hold the carbohydrate and protein molecules of that bar together. Chemical energy is **potential energy**, meaning that it is stored and waiting for use. When the cyclist eats and digests the bar, digestion breaks those chemical bonds, and the stored energy is released. As the cyclist begins to pedal, his body converts this potential energy into the **kinetic energy** of muscle contraction and **heat**. The kinetic energy of muscle movement is then converted into the kinetic energy of moving wheels. From start to finish, from cereal bar to spinning wheels, energy is converted from one form into another, but is never destroyed (**INFOGRAPHIC 5.4**).

Energy from biofuel has a similar life story. Oil from algae is rich in chemical energy: the lipids in the oil store energy in their bonds. When these bonds are broken—for example, when they are burned—they release large amounts of energy that can be used to power machines. In a car's combustion engine, the chemical energy in biofuel is rapidly and explosively converted to heat energy that warms

the gas molecules inside a chamber, causing them to expand. The expansion of the heated gas molecules pushes against the pistons, causing the wheels to move. The chemical potential energy of biofuel is thus converted into the kinetic energy of car movement.

If energy is never destroyed, only converted, why do we need to keep filling our tanks? It turns out that the conversion of energy from one form to another isn't 100% efficient. With each energy conversion, a bit of energy is "lost" to the environment as heat. This is why our bodies heat up when we exercise and why car engines are warm after being driven. In the case of a car engine, the generation of heat serves a purpose—to push the pistons. But heat lost to the *outside* of the car is wasted—it is no longer available to do useful work. This inefficiency is the reason we need to keep supplying energy to any system. We eat three meals a day to replenish the energy our bodies have lost as heat and converted into the chemical energy of cells and the kinetic energy of movement. Similarly, cars need a new tank of fuel after having

INFOGRAPHIC 5.4 ENERGY IS CONSERVED

➡ Energy in the universe is neither created nor destroyed, but is converted from one form to another. Stored potential energy, for example, can be converted to kinetic energy, as the cyclist below illustrates.

Potential chemical energy in food molecules

Chemical energy is converted into kinetic heat energy that is lost from the body.

Chemical energy is converted into the kinetic energy of muscle and leg movement.

Kinetic energy in wheel movement

1. A cyclist takes in stored (potential) chemical energy in the form of an energy bar.

2. Digestion breaks the chemical bonds and releases energy, which is used to contract muscles and generate heat.

3. Kinetic energy in the muscles is converted into kinetic energy of wheel movement. The bike moves forward.

As energy is converted from one form to another, some of the available energy is not fully converted to the next form. Instead, some energy is converted into heat that escapes into the environment, and some energy is not converted at all.

Unconverted fuel energy removed in exhaust

Heat energy (kinetic) escapes

Heat energy (kinetic) escapes

Heat energy (kinetic) escapes

Fuel Is Chemical Energy

Fuel Combustion
Chemical energy is converted to heat (kinetic) energy.

Pistons Fire
Kinetic energy of heated air molecules is converted to kinetic energy of piston movement.

Tires Roll
Kinetic energy of pistons is converted to kinetic energy of tire movement.

Energy In Car Is Depleted
More chemical energy is required to keep tires rolling.

burned through the previous one to convert the chemical energy of fuel into the heat and kinetic energy of motor movement. Just how well your car converts the chemical energy of gas into the kinetic energy of car speed determines your mpg–your miles per gallon. If an engine doesn't combust efficiently, some of the fuel molecules will undergo chemical reactions and be converted to other molecules–like pollutants–rather than power the pistons as heat. If the pistons can't use the heat efficiently, the heat will leave the car without powering the wheels. At each step of energy transformation, energy is lost from the car system and into the environment, and we're back to the fuel pump once more (**INFOGRAPHIC 5.5**).

SOLAR-POWERED CELLS

Tom Allnutt calls himself an "algae guy." While a student at Virginia Tech in the 1970s, he de-

cided on a whim to take a class in phycology–the study of algae–and immediately he was hooked. Algae fascinate him in part because they can survive pretty much anywhere, from the scalding thermal vents in Yellowstone National Park to the dry bitter cold of Antarctica, where Allnutt later spent 3 years as a scientist diving in freezing lakes and probing rock fissures for signs of the tiny organisms.

Now Allnutt has turned his attention to algae biofuels. He is the Senior Vice President of Research & Development at Phycal, a biotechnology company based in Ohio. In 2010, Phycal received $50 million in federal funding to build a 40-acre pilot facility to grow algae in shallow ponds, extract their oils, and test the oil for its viability as a commercial fuel. The plant is being built in Hawaii, where there is almost constant daytime sunlight–and therefore plenty of energy for the algae to convert into

▶**PHOTOSYNTHESIS**
The process by which plants and other autotrophs use the energy of sunlight to make energy-rich molecules using carbon dioxide and water.

▶**AUTOTROPHS**
Organisms such as plants, algae, and certain bacteria that capture the energy of sunlight by photosynthesis.

fuel. "We're taking the easy stuff first," he says, referring to the company's decision to locate its pilot plant in a sunny spot. It's a common choice: many algae-growing companies have positioned themselves in sunny locations like California.

Why is steady sun so important? For almost all living things on Earth, the ultimate source of energy is the sun. The sun functions like a giant thermonuclear reactor, converting matter into the radiant energy of sunlight. Sunlight is the original source of energy carried in both our fuels and our foods. When you fill your tank, or take a bite of a burger, you are using energy that came originally from the sun.

Humans and other animals can't use the power of sunlight directly, since they have no means to capture it and convert it into a usable form. However, they use it indirectly by eating the products of other organisms that *are* able to capture and convert the energy of sunlight. Through the process of **photosynthesis**, organisms like plants and algae capture the energy of sunlight and convert it into the chemical energy of energy-rich molecules—otherwise known as food.

Photosynthesis is critical to life on Earth because it is the primary mechanism that makes energy available, in the form of food, to almost all living organisms. Photosynthesis is the specialty of **autotrophs**—organisms such as plants, algae, and certain bacteria that can use the energy of sunlight to build organic molecules. Their name means, literally, "self-feeders"—and as you might expect, autotrophs make their own food (**INFOGRAPHIC 5.6**).

INFOGRAPHIC 5.6 AUTOTROPHS CONVERT LIGHT ENERGY INTO CHEMICAL ENERGY

➡ Most autotrophs are organisms that carry out photosynthesis, a process that converts light energy into chemical energy.

There are three basic types of autotrophs:

Plants

Algae

Some bacteria
(e.g., Cyanobacteria)

Light energy
(sunlight)

Photosynthesis

Chemical energy
(glucose sugar)

Immediate energy

Usable Energy
Some of the chemical energy is converted into a form that is available to power cellular functions.

Stored Energy
Some of the chemical energy is stored as potential energy in molecules like oil.

During photosynthesis autotrophs convert light energy into chemical energy

Cell Structures
Some of the chemical energy is used as building blocks for cell structures.

INFOGRAPHIC 5.7 PHOTOSYNTHESIS CAPTURES SUNLIGHT TO MAKE FOOD

 Photosynthesis is the process by which plants and other autotrophs use the energy of sunlight to make food. In plants, photosynthesis occurs in an organelle called the chloroplast, found in cells that make up the green parts of the plant. Photosynthesis has two main steps.

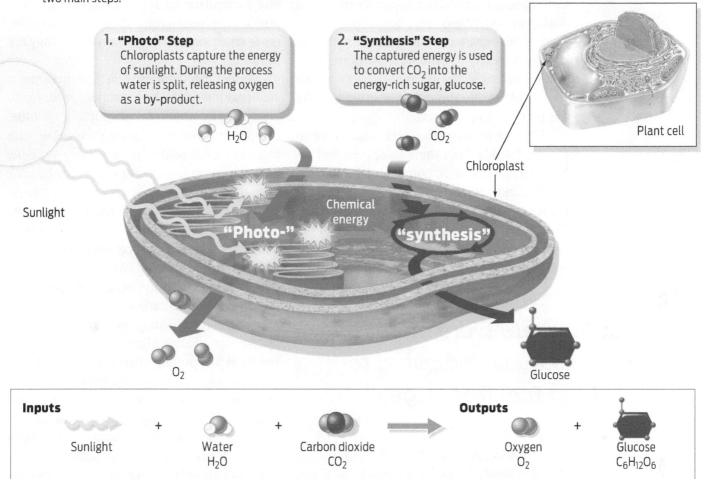

1. "Photo" Step
Chloroplasts capture the energy of sunlight. During the process water is split, releasing oxygen as a by-product.

2. "Synthesis" Step
The captured energy is used to convert CO_2 into the energy-rich sugar, glucose.

Plant cell

Chloroplast

Sunlight

H_2O

CO_2

Chemical energy

"Photo-"

"synthesis"

O_2

Glucose

Inputs

Sunlight + Water H_2O + Carbon dioxide CO_2 →

Outputs

Oxygen O_2 + Glucose $C_6H_{12}O_6$

Autotrophs include not just crop plants like wheat, corn, and soybeans, but also all flowering plants, trees, and bushes. Oceans and lakes have an abundance of autotrophs, too, including much of the planet's algae and photosynthetic bacteria.

CAPTURING ENERGY: PHOTOSYNTHESIS

How exactly do autotrophs use sunlight to create energy-rich molecules? The process of photosynthesis can be summarized in the following equation:

Sunlight + Water + Carbon dioxide → Oxygen + Glucose

That is, plants and other photosynthesizers use the energy of sunlight to make the mole-

cule glucose using carbon dioxide as a source of carbon. In the process, water molecules are split and oxygen is given off as a by-product.

Photosynthesis occurs in two steps: a "photo" step and a "synthesis" step. During the "photo" step, light energy is captured in chemical form. During the "synthesis" step, this chemical energy is used to generate glucose molecules using the carbon atoms of carbon dioxide. This synthesis phase does not directly require sunlight, but it does require the products of the "photo" reactions. The entire process occurs in an organelle called the **chloroplast**, which is present in the leaf cells of plants and the cells of photosynthetic algae (**INFOGRAPHIC 5.7**).

The glucose made from photosynthesis is used by plants (or algae) in a variety of ways:

▶CHLOROPLAST
The organelle in plant and algae cells where photosynthesis occurs.

it can be used for growth—to build new plant parts, like stems and fruit—or as an energy source to power cellular reactions.

While glucose is the major product of photosynthesis, other smaller sugars are produced during the "synthesis" reactions. Glucose and these other sugars provide the building materials for a variety of metabolic reactions in the cell—for example, the assembly of amino acids for protein synthesis and the synthesis of the oils that make up biofuels.

Like all photosynthesizers, algae take in carbon dioxide from the atmosphere and release oxygen. This is convenient for life on Earth, since many creatures—including humans—use this oxygen to breathe (see Chapter 6). Moreover, too much carbon dioxide in the atmosphere can be a bad thing. Carbon dioxide (CO_2) is not just the gas that plants and algae

> As of 2012, there were more than 150 companies dedicated to making fuel from algae.

take in during photosynthesis, it is also the gas that is released by burning fossil fuels. If you think about it, this makes perfect sense: fossil fuels such as coal, petroleum, and natural gas are the compressed remains of once-living photosynthetic organisms that have formed over millions of years; burning these fuels releases this stored carbon dioxide, sending it back into the atmosphere. Carbon dioxide is a greenhouse gas that is accumulating in the atmosphere and is in part responsible for increasing temperatures around the globe (see Chapter 23). By pulling carbon dioxide out of the atmosphere during photosynthesis, plants, algae, and other photosynthetic organisms help mitigate the effects of climate change.

Another boon of using algae as a source of biofuel is that, because algae remove CO_2 from the atmosphere as they grow, they release little net CO_2 when they are burned—they are

basically just returning to the atmosphere the same carbon they had removed. In addition, compared to traditional diesel, studies show that biodiesel releases less carbon monoxide and particulate matter. Algae-based biofuels aren't only renewable, then, they are also more environmentally friendly when they are burned.

Algae's need for carbon also presents a challenge for biofuel production: in order to grow large numbers of these microbes, biofuel companies need to provide their algae with more carbon dioxide than is readily available in the surrounding air. In other words, they have to have another source, which can be costly. Sears's company, Solix, has set up its first biofuel production plant next to a beer manufacturer that produces carbon dioxide as a by-product of brewing (see the discussion of fermentation in Chapter 6). The company simply siphons off this carbon dioxide and feeds it to its algae, thus helping them grow.

Mass-producing algae, however, is not as simple as just putting microbes in a pond with some carbon dioxide, sitting back, and watching them grow. One problem is that algae need to be stirred frequently to incorporate carbon dioxide, a process that requires a lot of extra energy. Aurora Biofuels, an algae company based in Alameda, California, gets around this difficulty by using the carbon dioxide that is pumped into the ponds to drive circulation, rather than using a paddle wheel. This saves about four-fifths of the energy normally required. Another problem for algae growers is that in the summer, sunlight is so intense that it actually oversaturates the microbes, limiting their growth. Algae evolved to live below the surface of the ocean, where light is limited, so they don't handle lots of sun very well.

To overcome this hurdle, in 2007 researchers at the University of California at Berkeley engineered algae with smaller antennae, the body parts algae use to absorb sunlight, similar to the way radio antennas receive electromagnetic signals. These smaller antennae help them avoid oversaturation and increase growth.

Solazyme founders Jonathan Wolfson (left) and Harrison Dillion.

A CHEAPER STRATEGY?

In 2005, Harrison Dillon and Jonathan Wolfson, the founders of Solazyme, a San Francisco-based algae biofuels company, came to a shocking realization: the biofuel they were producing from algae was going to cost approximately $1,000 a gallon. While photosynthesis is free, the cost of maintaining algae is not.

Photosynthetic algae can be expensive to maintain because they are either grown in large transparent tanks called photobioreactors, which are costly to operate, or they are grown in open ponds, which require lots of freshwater and constant monitoring for contamination. Wolfson and Harrison looked for a solution. They knew that, of the more than 100,000 species of algae, there are rare strains that do not rely on sunlight. Instead, they survive by "eating" sugar. If they used these algae to make biofuel, could they save money? Algae that could be grown in closed, nonphotosynthetic vats and fed cane sugar could conceivably be much cheaper.

With support from their investors, Wolfson and Harrison changed their business model and gave the new idea a whirl. It was a good decision: they now believe that they can make oil using these nonphotosynthetic algae that cost only $60 to $80 a barrel (roughly the cost of a barrel of regular oil), in part because they can save money on overhead costs. The sugar is also relatively cheap to obtain since it comes from plant scraps, switchgrass, and other plentiful sources that aren't of direct use to humans as food. Many other algae companies still use photosynthetic algae, and they have developed cost-effective ways to make them work, but for Solazyme, the rare sugar-eating algae are a better choice.

Organisms like ones that Solazyme uses—organisms that can't photosynthesize and must eat molecules produced by other organisms to obtain energy—are called **heterotrophs** ("other-feeders"). It's a group that includes not only sugar-feeding algae but also all animals, fungi, and most bacteria. When humans and other heterotrophs eat plants—or eat animals

▶HETEROTROPHS
Organisms, such as humans and other animals, that obtain energy by eating organic molecules that were produced by other organisms.

that have eaten plants–specialized processes release chemical energy stored in the molecules making up the plant or animal (see the discussion of aerobic respiration in Chapter 6). Heterotrophs can then either use this energy to grow, in which case the energy will become stored in the chemical bonds in their bodies, or to move and power other chemical reactions in cells. The result: energy is converted from one form into another, flows from the sun to autotrophs to heterotrophs, and finally is released back into the environment as heat.

FROM SUN TO FUEL

For heterotrophic humans, the idea of being able to subsist solely on sunlight, carbon dioxide, and water, seems remarkable. How can something as intangible as sunlight carry energy? If you've ever walked barefoot across a sandy beach on a hot summer day, you know that sunlight is a potent source of heat energy. You may also have a sense that certain colors absorb or reflect sunlight better than others–on a sunny day, wearing a reflective white shirt keeps you cooler than a black one that absorbs more of the sun's rays.

These properties reflect the nature of **light energy,** which is part of the electromagnetic spectrum of radiation. Light energy from the sun travels to Earth in waves. These waves of light are made up of discrete packets of energy called **photons.** Photons of different wavelengths contain different amounts of energy, and different objects on earth absorb and reflect different wavelengths of light. Some of these wavelengths, when viewed by the human eye and interpreted by the human brain, appear to us as different colors **(INFOGRAPHIC 5.8).**

When sunlight hits a green plant, for example, its leaves absorb red and blue wavelengths and reflect green wavelengths–which is why plants appear green to our eyes. The molecule that absorbs and reflects these wavelengths of light is the pigment **chlorophyll**–a crucial player in photosynthesis. It is chlorophyll that actually captures the energy of sunlight. During the "photo" reaction, chlorophyll molecules absorb energy from the red and blue wavelengths of sunlight. In addition to chlorophyll, plants and algae contain other pigment molecules that absorb and reflect other wavelengths of light, giving them their distinctive colors. But chlorophyll is the main pigment involved in photosynthesis. When red and blue photons of sunlight hit chlorophyll, the electrons in its atoms become excited. These excited electrons are used to generate an energy-carrying molecule known as **adenosine triphosphate (ATP),** which is used in the "synthesis" part of photosynthesis to make sugar **(INFOGRAPHIC 5.9).** (We'll talk more about ATP, the cell's "energy currency," in Chapter 6.)

Plants use the sugar they make as food–as a source not only of chemical energy to power cellular reactions, but also as the building materials to make all the organic molecules critical

INFOGRAPHIC 5.8 THE ENERGY IN SUNLIGHT TRAVELS IN WAVES

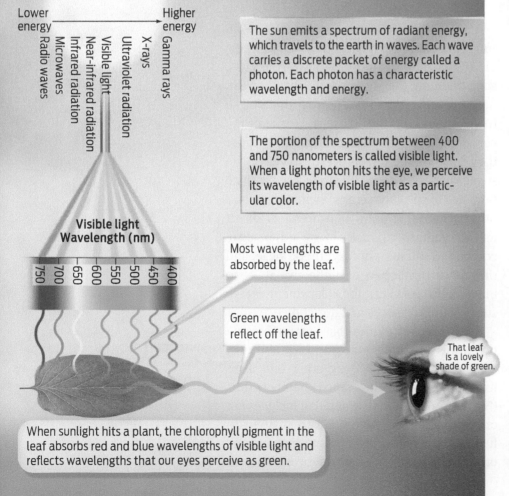

Lower energy → Higher energy

Radio waves
Microwaves
Infrared radiation
Near-infrared radiation
Visible light
Ultraviolet radiation
X-rays
Gamma rays

The sun emits a spectrum of radiant energy, which travels to the earth in waves. Each wave carries a discrete packet of energy called a photon. Each photon has a characteristic wavelength and energy.

The portion of the spectrum between 400 and 750 nanometers is called visible light. When a light photon hits the eye, we perceive its wavelength of visible light as a particular color.

Visible light Wavelength (nm)

750 700 650 600 550 500 450 400

Most wavelengths are absorbed by the leaf.

Green wavelengths reflect off the leaf.

That leaf is a lovely shade of green.

When sunlight hits a plant, the chlorophyll pigment in the leaf absorbs red and blue wavelengths of visible light and reflects wavelengths that our eyes perceive as green.

INFOGRAPHIC 5.9 PHOTOSYNTHESIS: A CLOSER LOOK

1. Light "Photo" Reactions
Chlorophyll pigments within internal chloroplast membranes absorb photons. Chlorophyll electrons (e-) become excited and enter a series of reactions that generate an energy-carrying molecule called ATP.

2. Carbon "Synthesis" Reactions
Energy from the breakdown of ATP is used in the carbon reactions to fix carbon dioxide into organic sugar molecules, a form of stored chemical energy.

Water (H_2O)
Water is split during the light reactions. Split water molecules release electrons that replace electrons lost by excited chlorophyll molecules.

Carbon Dioxide (CO_2)
CO_2 gas enters plant cells from the atmosphere. The carbon atoms are incorporated into organic sugar molecules.

H_2O
(from the roots)

CO_2
(from the air)

Chloroplast

ATP

Sunlight

e^-

Light reactions

Chemical energy from light reactions

Carbon reactions

ATP

Internal chloroplast membranes containing chlorophyll

e^-

O_2
(to the air)

Glucose

Oxygen (O_2)
This gas is a by-product of water splitting during the light reactions.

Glucose ($C_6H_{12}O_6$)
The carbohydrate product of photosynthesis, glucose, contains the chemical energy converted from sunlight and the fixed carbon from atmospheric CO_2.

to building a plant, including proteins, nucleic acids, and lipids (see Chapter 4). Many plants use much of their glucose to produce more complex sugar molecules such as starches, and to form plant products such as wheat grains and fruit. Microscopic, single-cell algae, on the other hand, waste little of their sugar making these elaborate plant parts. Instead, they pour most of it into making oil. And that's what makes them so useful as fuel producers—they pack more bang for the buck. In some algae, oil constitutes more than half of their dry weight. Some algae species secrete the oil in order to adhere to other cells, whereas others accumulate the oil inside their bodies to control buoyancy. Still others use the oil as an efficient way to store energy.

Algae are especially popular right now, but they aren't the only living organisms scientists use to make biofuels. Crop plants such as corn, soybeans, and rapeseed are also used. How do algae stack up against these other biofuels? Compared to corn-based ethanol and fuels made from plants like soybeans and rapeseed, algae take the prize in terms of how much fuel they can produce for the amount of space they take up. According to the U.S. Department of Energy, if American soybean farmers had converted all of their crops in 2007—about 64 million acres' worth of soybeans—into biofuel, they would have provided the country with only enough diesel to meet 5% of the nation's needs. On the other hand, if farmers had grown algae

▶ADENOSINE TRIPHOSPHATE (ATP)
The molecule that cells use to power energy-requiring functions.

> **"** *Algae truly are the foundations of our entire planet.* **"**
>
> — JIM SEARS

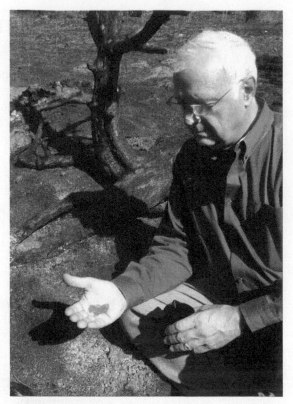

Jim Sears examines a restored soil sample.

on this same amount of land in open ponds or containers, they would have produced enough fuel to supply all our country's diesel needs. And because algae can be cultivated on land that is unsuitable for food crops, production of algal biofuel doesn't compete with or take resources away from food production–a common criticism of corn-derived ethanol.

Extracting fuel from algae is also less energy intensive than extraction from other biofuel sources. To make ethanol, for instance, farmers typically start with corn, which requires more energy to grow than algae do. Then they have to ferment the corn into ethanol, and finally they must harvest the ethanol. It's a multistep process that requires inputs of precious farmland, clean water, and energy. In addition, energy has to be converted first from sunlight into corn, and then from corn into ethanol, which means that some energy is wasted. Algae are far simpler life forms, and therefore more efficient: using sunlight and carbon dioxide, they produce fuel directly.

And they can be grown on poor-quality lands that cannot be cultivated for other crops, using both freshwater and saltwater that is not suitable for agriculture (**TABLE 5.1**).

Which is not to say that all the bugs have been worked out yet, or that algae are a miracle solution to our energy woes. A 2012 report on algae biofuels by the National Academy of Sciences produced at the request of the U.S.

TABLE 5.1 HOW GREEN ARE BIOFUELS?

Biofuels are getting a lot of media attention as potential alternatives to fossil fuels, but not all biofuels are alike in terms of their environmental, energy, and land-use impacts. Compared to conventional crop plants, algae have some of the lowest environmental costs per similar quantity of oil produced.

Crop	Greenhouse Gas Emissions*	Water Use	Fertilizer	Pesticides	Energy Use**	Oil Yield per Acre (gals)
Corn	81–85	High	High	High	High	18
Soybeans	49	High	Low	Medium	Low	48
Rapeseed, canola	37	High	Medium	Medium	Low	127
Algae	183	Medium	Low	Low	High	>5,000

* kg of CO_2 produced per megajoule of energy generated. ** Energy used during growing, harvesting, and refining of fuel.

SOURCE: Groom, M. J., et al. (2008) Biofuels and biodiversity: principles for creating better policies for biofuel production. *Conservation Biology* 22:602–609.

Department of Energy found that scaling up algae biofuel production to meet a significant portion of our energy needs may prove extremely challenging, given the amount of water and nutrients required. Nevertheless, the report stresses that the promise of such fuels makes them worth continued research and development.

BIOFUEL AND BEYOND

In the end, photosynthesis accomplishes two main things. First, it converts light energy from the sun into chemical energy that can be used as food and fuel by plants and animals (including humans). Second, it captures carbon dioxide gas from the air and incorporates those carbon atoms into sugar in a process called **carbon fixation.** By converting inorganic gaseous carbon into an organic form that can be eaten by animals or used by plants to grow and increase their biomass, carbon fixation is ultimately the way carbon enters the global food and energy chain (see Chapter 22).

Carbon fixation is Jim Sears's favorite topic these days. Having left Solix in 2007, Sears is now the president of A2BE Carbon Capture, a company based in Boulder, Colorado, that is looking for ways to reduce carbon dioxide levels in the atmosphere. Since plants, algae, and other photosynthetic organisms all help to temper the effects of global warming by pulling carbon dioxide out of the atmosphere and fixing it into organic sugars, scientists are looking for ways to enhance this natural process.

Believe it or not, Sears says, the healthy soil in your backyard is actually photosynthetic. In soil, tiny bacteria called cyanobacteria thrive, and they perform photosynthesis. These cyanobacteria are good not only for the soil but for the entire planet. Like algae, they perform photosynthesis, which means that in addition to absorbing sunlight, they capture carbon from the atmosphere. They then convert the carbon into forms that provide energy and nutrition to colonies of other microorganisms buried deep within the soil. One square

▶**CARBON FIXATION**
The conversion of inorganic carbon (e.g., CO_2) into organic forms (e.g., sugars).

Algae farms like the one below could be the wave of the future

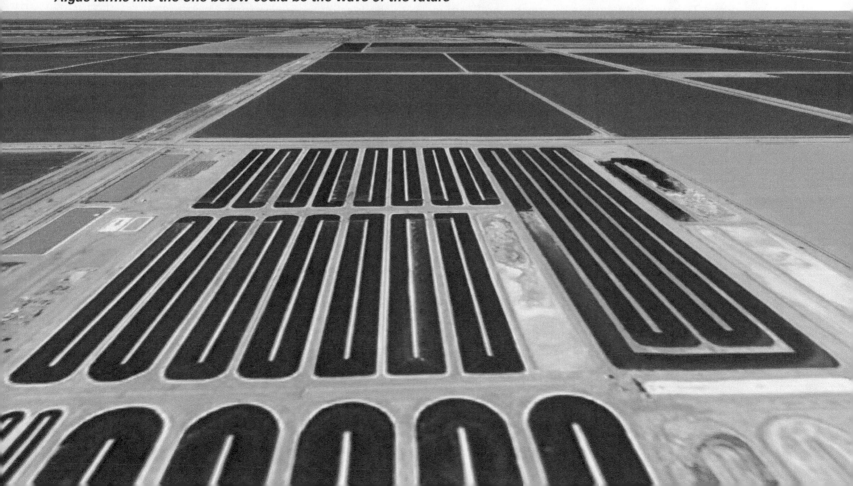

meter of healthy, undisturbed soil can remove 30 g of atmospheric carbon per year.

The problem is that approximately 2 billion out of Earth's 13 billion total hectares of landmass have been damaged by human activity–construction and fires are among the biggest culprits. According to Sears, it can take anywhere from 30 to 3,000 years for soil microorganisms to regenerate after being destroyed–and in the meantime, the damaged soil is unable to remove carbon dioxide from the atmosphere.

Sears, however, has a solution. His new company takes small samples of microorganisms from healthy soil, grows them in a contained facility, and then transplants them to damaged soil, where they spread out and thrive. He estimates that if 1 billion hectares of

land were restored in this way, one-seventh of the world's greenhouse gas problem would be solved because of the vast amounts of carbon dioxide that would be pulled out of the atmosphere by the photosynthetic cyanobacteria in the regenerated soil.

That seemingly simple organisms like algae and cyanobacteria could be so vital to life on Earth is surprising enough–indeed, they provide the planet with much of its breathable oxygen. But soon algae could become the world's most important fuel source, as well as an ally in the fight against climate change. All this from single-cell organisms that have just one major claim to fame: they can convert the energy of sunlight into energy-rich organic molecules. "Algae truly are the foundations of our entire planet," says Sears. ■

CHAPTER 5 Summary

- ▶ All living organisms require energy to live and grow. The ultimate source of energy on Earth is the sun.

- ▶ Photosynthesis is a series of chemical reactions that captures the energy of sunlight and converts it into chemical energy in the form of sugar and other energy-rich molecules. This energy is used by all living organisms to fuel cellular processes.

- ▶ Photosynthesis can be divided into two main parts: a "photo" part, during which the pigment chlorophyll captures light energy, and a "synthesis" part, during which captured energy is used to fix carbon dioxide into glucose.

- ▶ Photosynthetic organisms are known as autotrophs; they include plants, algae, and some bacteria. Animals do not photosynthesize; they are known as heterotrophs.

- ▶ Energy is neither created nor destroyed, but is converted from one form into another, a principle known as the conservation of energy.

- ▶ Kinetic energy is the energy of motion and includes heat energy and light energy. Potential energy is stored energy and includes chemical energy.

- ▶ Energy flows from the sun, is captured and transferred through living organisms, and then flows back into the environment as heat.

- ▶ Energy conversions are inefficient. Some energy is lost as heat with every conversion of energy.

- ▶ The energy-rich molecules produced by some photosynthetic algae include oils that can be used as fuel to power automobiles and aircraft. These biofuels show great promise as alternatives to fossil fuels.

MORE TO EXPLORE

- ▶ U.S. Department of Energy, Office of Energy Efficiency and Renewable Energy, Biomass Program http://www1.eere.energy.gov /biomass/

- ▶ Sheehan, J., et al. (1998) A look back at the U.S. Department of Energy's Aquatic Species Program: biodiesel from algae. National Renewable Energy Laboratory NREL/TP-580-24190.

- ▶ Hunter-Cevera, J., et al. (2012) Sustainable development of algal biofuels. National Academy of Sciences Report.

- ▶ Pienkos, P. T., and Darzins, A. (2009) The promise and challenges of microalgal-derived biofuels. *Biofuels, Bioproducts* and *Biorefining* 3:431–440.

- ▶ Groom, M. J., et al. (2008) Biofuels and biodiversity: principles for creating better policies for biofuel production. *Conservation Biology* 22:602–609.

- ▶ Biello, D. (2011) The false promise of biofuels. *Scientific American* 302:58–65.

CHAPTER 5 Test Your Knowledge

DRIVING QUESTION 1

What are the photosynthetic organisms on the planet, and why are they so important?

By answering the questions below and studying Infographics 5.3 and 5.6, you should be able to generate an answer for the broader Driving Question above.

KNOW IT

1. What do algae, cyanobacteria, and plants have in common?

2. Can animals directly use the energy of sunlight to make their own food (in their own bodies)?

3. What organelle(s) would a nonphotosynthetic alga need to be able to carry out photosynthesis?
   a. mitochondria
   b. nucleus
   c. chloroplast
   d. solar transformer
   e. cell membrane

4. Why do many species of algae appear green?

5. Compare and contrast the ways photosynthetic algae and animals obtain and use energy.

USE IT

6. What would happen to humans and other animals if algae, cyanobacteria, and plants were wiped out? Would we only lose a food source (e.g., plants), or would there be other repercussions?

7. Why would a dark dust cloud that prevented sunlight from reaching Earth's surface be potentially devastating to animal life?

DRIVING QUESTION 2

What are the different types of energy, and what transformations of energy do organisms carry out?

By answering the questions below and studying Infographics 5.4 and 5.5, you should be able to generate an answer for the broader Driving Question above.

KNOW IT

8. The energy of sunlight exists in the form of
   a. glucose.
   b. photons.
   c. gamma rays.
   d. ions.
   e. particles.

9 The energy in a cereal bar is _____ energy. The energy of a cyclist pedaling is _____ energy.

a. light; chemical
b. potential; chemical
c. chemical; kinetic
d. potential; potential
e. kinetic; potential

10 Kinetic energy is best described as

a. stored energy.
b. light energy.
c. the energy of movement.
d. heat energy.
e. any of the above, depending on the situation.

USE IT

11 If you wanted to get the most possible energy from photosynthetic algae, should you eat algae directly or feed algae to a cow and then eat a burger made from that cow? Explain your answer.

DRIVING QUESTION 3

How do plants and algae convert the energy in sunlight into energy-rich organic molecules? (And why can't humans do this?)

By answering the questions below and studying Infographics 5.7, 5.8, and 5.9, you should be able to generate an answer for the broader Driving Question above.

KNOW IT

12 Which of the following photon wavelengths contains the greatest amount of energy?

a. violet
b. red
c. green
d. yellow
e. blue

13 Glucose is a product of photosynthesis. Where do the carbon atoms in glucose come from?

a. starch
b. cow manure
c. molecules in air
d. water
e. soil

14 Mark each of the following as an input (I) or an output (O) of photosynthesis.

Oxygen _____
Carbon dioxide_____
Photons _____
Glucose _____
Water _____

15 Photosynthetic algae are

a. eukaryotic autotrophs.
b. prokaryotic autotrophs.
c. eukaryotic heterotrophs.
d. prokaryotic heterotrophs

USE IT

16 Global warming is linked to elevated atmospheric carbon dioxide levels. How might this affect photosynthesis? If global warming should cause ocean levels to rise, in turn causing forests to be immersed in water, how would photosynthesis be affected?

17 Why are energy-rich lipids from algae more useful as a fuel than energy-rich sugars and other carbohydrates produced by photosynthetic organisms like corn and wheat?

18 Draw a concept map for photosynthesis that includes the following forms of energy and molecules: sunlight; carbon dioxide; glucose (stored chemical energy); water; ATP; heat.

DRIVING QUESTION 4

How do algal biofuels compare to other fuels in terms of cost, benefits, and sustainability?

By answering the questions below and studying Infographic 5.2 and Table 5.1, you should be able to generate an answer for the broader Driving Question above.

KNOW IT

19 Which of the following is/are necessary for biofuel production by algae?

a. sunlight
b. sugar
c. CO_2
d. soil
e. all of the above
f. a and b
g. a and c

20 Why are algae considered more valuable for biofuel than plants (such as corn)?

a. because their photosynthetic products are an oil
b. because they are cheaper to grow
c. because they do not require as much CO_2
d. because they do not require as much fertilizer
e. all of the above

USE IT

21 Many types of algae can divert the sugars they make by photosynthesis into lipids that can be used to make biodiesel. Biodiesel is a promising replacement for fossil fuels. Describe the energy conversions required to make algal lipids for biodiesel and explain why biodiesel might be a more promising fuel than lipids extracted from animals.

22 What do you think are some of the advantages and disadvantages of growing algae in enclosed tubes or bags compared to growing them in open vats? Make a table listing the advantages and disadvantages of each approach and explain your reasoning.

23 Many biofuels require arable land for their production. Discuss competing needs for arable lands in the context of human needs for food and fuel, and how algae may alleviate this tension.

MINI CASE

24 A CEO of a new algal biofuel company is trying to select the site for a production facility. There are three possible options:

- The desert of southern New Mexico (sunny, hot, mild winters, nonarable land, remote)
- Denver, Colorado (sunny, cold winters, urban area with CO_2 emissions from factories and cars)
- Central Washington State (sunny, hot, a rich agricultural zone)

Discuss the pros and cons of each site and make a recommendation to the CEO.

BRING IT HOME

25 Richard Branson has committed his airline, Virgin Atlantic, to using a "green" fuel produced by microbes that use carbon monoxide (CO) from industrial emissions (such as from steel factories) as its carbon and energy source. Through a fermentation process that occurs in a reactor chamber, the microbes convert CO into usable ethanol, a viable green fuel. Consider how this fuel compares and contrasts with algal biofuel and corn ethanol. If this green fuel venture is successful, would Branson's use of green fuel influence your decision to choose Virgin Atlantic over another air carrier? Why?

INTERPRETING DATA

26 The United States currently uses approximately 19 million barrels of oil per day. Of this, about half (9.67 million barrels per day) is imported, and the rest is from U.S. sources (offshore drilling or extraction from shale). The table shows production costs estimated for different oil sources. (The actual cost is driven by a variety of market and geopolitical factors, so we will use production costs as a substitute for actual cost.)

a. Using the data for cost per barrel of various oils, calculate the cost to produce oil to meet current U.S. daily use. Assume that approximately half the imports are from the United Arab Emirates (UAE), and the remainder is split equally between imports from Canada and from Brazil.

b. Let's say that the United States replaces half of its current oil imports from the UAE with domestically produced algal biofuel. What will this do to the cost of production to meet our daily needs?

Source	Production Cost per Barrel (Range)	Production Cost per Barrel (Average)
U.S. (offshore)	$70–$80	$75
U.S. (shale oil)	$50–$70	$60
Canada (tar sands)	$70–$90	$80
UAE	$50–$70	$60
Brazil	$60–$80	$70
Algal biofuel	$140–$900	$520

SOURCE: http://insideclimatenews.org/news/20101122/algae-fuel-inches-toward-price-parity-oil; http://www.odac-info.org/newsletter/2011/09/16

c. From what you've read in this chapter, how are algal biofuel companies working to reduce the costs of algal biofuel?

The biology and culture of our expanding waistlines

SUPERSI

For years, Paul Rozin, a professor of psychology at the University of Pennsylvania, was baffled by this question: How are the French able to eat rich cheeses, butter-laden sauces, flaky croissants and pastries, and still stay slimmer than Americans? To put it more crudely, why doesn't all this fat make them fat?

Rozin is not alone. Researchers have been scratching their heads for years over the weight differences between the populations of the United States and France. As of 2011, 33% of American adults were overweight and 36% were obese, while only 27% of French adults were overweight and 11% obese. Cardiologists have even observed that, despite their rich diet, the French have lower rates of heart disease than Americans and have dubbed the phenomenon "the French paradox." And as the average weight of humans in most

▸▸ **DRIVING QUESTIONS**

1. Why do humans weigh more now than in the past?

2. How does the body use the energy in food?

3. How does aerobic respiration extract useful energy from food?

4. When does fermentation occur, and why can't it sustain human life?

ZE ME?

Paul Rozin, University of Pennsylvania psychology professor, is examining the differences between American and French eating culture.

> **"The idea of what a proper meal is and your own habits are largely instituted by the culture in which you live."**
>
> — PAUL ROZIN

developed nations has increased in the last 30 years, many wonder why waistlines in some countries are expanding faster than in others.

Americans may take the cake for widest waistlines, but they aren't the only ones getting rounder. Globally, obesity rates have more than doubled since 1980. As of 2008, approximately 1.4 billion adults over age 20 were overweight and at least 500 million were obese, according to the World Health Organization (WHO). That means more than 1 in 10 people in the world are obese.

INFOGRAPHIC 6.1 OBESITY IS INFLUENCED BY BIOLOGY AND CULTURE

Biological History
Famine was common. Human bodies evolved to hoard energy in the form of fat to get through times when food was scarce.

Cultural Influence
An abundance of high-fat, processed food is increasingly common in developed countries.

Modern Waistlines Increasing
Today people consume many more Calories than during any other time in history because food is abundant. Our bodies store extra Calories as fat, as they have been evolutionarily programmed to do.

+

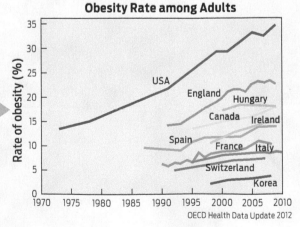

Obesity Rate among Adults

USA
England
Hungary
Canada
Ireland
Spain
France
Italy
Switzerland
Korea

Rate of obesity (%): 0, 5, 10, 15, 20, 25, 30, 35

1970 1975 1980 1985 1990 1995 2000 2005 2010

OECD Health Data Update 2012

Our increasing girth alone wouldn't be a problem were it not for what can come with it: obesity has been linked to increased rates of heart disease, diabetes, and even certain types of cancer. While research into the exact causes is ongoing, many experts agree that obesity poses serious health risks.

So why are people getting heavier? Biologists argue that humans are predisposed to gain weight. Throughout human evolution, famine was the norm, and people had to work hard to grow or hunt and gather their food. Our bodies have adapted by storing extra food as fat for times when food is scarce.

But social scientists peg another weight-gain culprit: culture. Both *how* we eat and *how much* we eat, they argue, are as much to blame for increasing weight as biology. Some societies have remained relatively thin, they point out, despite similar biology. But when these societies adopt American eating habits, which include fast-food, processed snacks, soda, and large portion sizes, they tend to put on pounds (**INFOGRAPHIC 6.1**).

"Culture is underrated as a contributing factor to unhealthy eating," says Rozin. "The idea of what a proper meal is and your own habits are largely instituted by the culture in which you live."

In the United States, there is a growing movement to rein in what has been the cultural norm of unhealthful eating. For example, public health experts have been lobbying the government to pass legislation that would improve people's access to more-healthful foods, such as fresh fruits and vegetables. Many local governments have already banned restaurants from using trans fats to fry foods. In 2010, Congress passed legislation championed by First Lady Michele Obama that sets higher standards for school lunches and limits the amount of junk food that can be sold in schools.

"Unhealthy food choices have become the default food choice," says Kelly Brownell, director of the Rudd Center for Food Policy and Obesity at Yale University. "The question is what can we do about it?"

THE SIZE OF THE PROBLEM

Though the terms "overweight" and "obese" are used casually in popular culture, they actually have specific meanings in scientific circles. They are based on a tool called the **body mass index** (**BMI**). The BMI estimates body fat from the indirect measures of height and weight, and translates it into an easy-to-digest number. People with a BMI between 25 and 29.9 are considered **overweight**; people with a BMI of 30 and above are considered **obese.**

BMI provides a rough estimate of healthy and unhealthy amounts of body fat, and is useful as a general guide to weight. However, it can sometimes be misleading. Athletes and other people with more muscle mass, for example, will sometimes register as overweight when in fact they are a perfectly healthy weight (**INFOGRAPHIC 6.2**).

▶**BODY MASS INDEX (BMI)**
An estimate of body fat based on height and weight.

▶**OVERWEIGHT**
Having a BMI between 25 and 29.9.

▶**OBESE**
Having 20% more body fat than is recommended for one's height, as measured by a body mass index equal to or greater than 30.

INFOGRAPHIC 6.2 BODY MASS INDEX (BMI)

A BMI chart provides an indirect measure of body fat based on the ratio of body height to weight. Because it does not take frame size into account, BMI is only an estimate of body fat. Some people may register as overweight even though they are a healthy weight.

Weight in pounds

Height in feet and inches	120	130	140	150	160	170	180	190	200	210	220	230	240	250
4′6″	29	31	34	36	39	41	43	46	48	51	53	56	58	60
4′8″	27	29	31	34	36	38	40	43	45	47	49	52	54	56
4′10″	25	27	29	31	34	36	38	40	42	44	46	48	50	52
5′0″	23	25	27	29	31	33	35	37	39	41	43	45	47	49
5′2″	22	24	26	27	29	31	33	35	37	38	40	42	44	45
5′4″	21	22	24	26	28	29								43
5′6″	19	21	23	24										40
5′8″	18	20	21	23	24	26								38
5′10″	17	19	20	22	23	24	26	27	29	30	31	33	35	36
6′0″	16	18	19	20	22	23	24	26	27	28	30	31	33	34
6′2″	15	17	18	19	21	22	23	24	26	27	28	30	31	32
6′4″	15	16	17	18	20	21	22	23	24	26	27	28	29	30
6′6″	14	15	16	17	19	20	21	22	23	24	25	27	28	29
6′8″	13	14	15	17	18	19	20	21	22	23	24	25	26	28

 A person 5'6" tall weighing 150 lbs is within the healthy weight range.

☐ Underweight ☐ Healthy weight ☐ Overweight ■ Obese

Even with a few extra pounds here and there, most people still fall within a healthy weight range. Only when our total body fat passes a certain point do the scales tip toward unhealthy. That exact tipping point is the subject of much current research, and there are still many questions and controversies about the exact relationship between health and body fat. For example, is it possible to be fat and relatively healthy? Probably. Does being skinny necessarily mean you are in good cardiovascular health? No. Nevertheless, the preponderance of evidence suggests that there are real health risks to obesity, which is why some researchers have even referred to a global "obesity epidemic."

To get at why Americans are rounding out while their food-loving French counterparts aren't, Rozin and colleagues at the Centre National de la Recherche Scientifique in France decided to look at portion size. In 2003, the team set out to see if Americans do indeed eat more than the French. They compared portion sizes at 11 restaurants in Philadelphia, Pennsylvania, and in Paris, France. Their results, published in the journal *Psychological Science*, were revealing. The average portion size in the Paris restaurants weighed 277 g. By contrast, the average portion size in the Philadelphia restaurants weighed 346 g–25% more. Even restaurant chains in France like McDonald's served smaller portions of certain foods: in Paris, medium fries weighed 90 g, large fries 135 g; in Philadelphia, medium fries weighed 155 g and large fries 200 g.

Once they had surveyed the restaurant scene, Rozin's team went further. They compared the sizes of packaged food in American and in French supermarkets, and they found the same trend: the portions of the majority of food items they tested–from ice cream to chewing gum to yogurt–were smaller in France. Even portion sizes for ingredients in the most commonly used French cookbook were smaller than those in the American favorite, *Joy of Cooking* (**INFOGRAPHIC 6.3**).

INFOGRAPHIC 6.3 AMERICANS EAT LARGE PORTIONS

 Researchers compared portion sizes in restaurants in Philadelphia, U.S., to those in Paris, France. In all but one restaurant, U.S. portions were larger at least half the time. The average portion size in Paris was 277 g; the average size in Philadelphia was 346 g. Philadelphians eat an average of 25% more food than Parisians at every meal.

For sampled menu items, U.S. restaurants consistently serve larger portion sizes.

Restaurant in Paris	Restaurant in Philadelphia	No. of items sampled/ No. larger in U.S.	Mean size ratio (U.S./France)
McDonald's	McDonald's	6/4	1.28
Hard Rock Cafe	Hard Rock Cafe	2/0	0.92
Pizza Hut	Pizza Hut	2/2	1.32
Häagen-Dazs	Häagen-Dazs	2/2	1.42
French: local bistro	French: local bistro	1/1	1.17
Quick	Burger King	5/4	1.36
Local Chinese	Local Chinese	6/4	1.72
Italian: Bistro Romain	Olive Garden	3/2	1.02
Crepes: local	Crepes: local	4/2	1.04
Local ice cream	Local ice cream	2/2	1.24
Pizza: local	Pizza: local	2/2	1.32

Portions at McDonald's in the U.S. are on average 1.28 times larger than the same item in France.

SOURCE: Data from Rozin, P., et al. (2003) The ecology of eating: smaller portion sizes in France than in the United States help explain the French paradox. *Psychological Science* 14:450-454.

Taking it slow: the French have a distinct culture of eating.

That portion size plays a role in weight gain seems pretty obvious: the more we eat, the heavier we get. But why does eating food lead to weight gain? After all, it's not as if we simply increase in weight by the same number of pounds we put on our plate, so something else must be at work. We've already seen, in Chapter 4, that all food consists of mixtures of carbohydrates, fats, proteins, and nucleic acids, and that the relative proportion of each macronutrient varies in different types of food. Meat, for example, contains more protein per unit weight than do potatoes; potatoes have more carbohydrates than meat does. To nourish our bodies, we must eat a balanced diet that includes appropriate amounts of all macronutrients.

But food is not only a source of nutrition; it is also a source of energy—the power to do work. Much in the same way that gasoline powers our cars, food supplies us with chemical energy that our bodies "burn" to power our activities. Food is fuel.

The energy in food comes originally from the sun. Photosynthesizers such as plants capture the energy of sunlight and convert it into chemical energy stored in sugar (see Chapter 5). When we eat this sugar (or eat animals that have eaten this sugar), that stored energy becomes available to us. Not only that, but the energy in food, like all energy, obeys the principle of conservation of energy—energy is never destroyed, it only changes form—a fact with lasting consequences for our waistlines.

Scientists measure food energy in units called calories. A **calorie** (in lower case) is the amount of energy required to raise the temperature of 1 g of water by 1°C (1 degree Celsius). In essence, a calorie is a measurement of energy—the capacity to perform a certain amount of work. On most food labels, the amount of energy stored is listed in kilocalories, which are also referred to as kcals or Calories (the capital "C" indicates kilocalories, not calories). One **Calorie** is equal to 1,000 calories, or 1 kcal.

Different foods contain different amounts of energy, and therefore allow us to perform different amounts of work. Of all the organic molecules, fats are the most energy dense:

▶**calorie**
The amount of energy required to raise the temperature of 1 g of water by 1°C.

▶**Calorie**
1,000 calories or 1 kilocalorie (kcal); the capital "C" in Calorie indicates "kilocalorie." The Calorie is the common unit of energy used in food nutrition labels.

each gram of fat stores approximately 9 Calories in its chemical bonds. Proteins and carbohydrates are about half as energy dense: about 4 Calories per gram. Clearly, a 200-g serving of fatty bacon contains many more Calories than does a 200-g serving of asparagus (**INFOGRAPHIC 6.4**).

All our activities—everything from thinking and digesting to sleeping and running—require energy. So all bodies expend some Calories each day just to stay alive. A person's daily energy needs largely depend on gender, age, body type, and activity levels. A sedentary college-age average-size male, for example, would need to ingest anywhere between 2,200 and 2,400 Calories per day to power his activities and maintain his weight, whereas a football player would need more than 3,200 Calories a day to power and main-

tain his. Exercise or other physical activities require additional energy beyond the basic life-sustaining energy needs of the body. Consequently, athletes, or those who exercise a great deal, generally need to eat more to fuel their activities than do their less-active peers.

Exactly how much more should an athlete eat? Consider the college football player, who must consume 800 to 1,000 Calories more than his sedentary roommate. An average cheeseburger contains anywhere between 400 and 600 Calories depending on its size, so the football player would need about two extra cheeseburgers per day. That's probably less than you thought. Now suppose that same athlete ate a cheeseburger off-season. It would take 1.5 hours of slow swimming, or 2.5 hours of walking,

INFOGRAPHIC 6.4 FOOD POWERS CELLULAR WORK

Food contains the macromolecules that provide the building blocks to make molecules critical to cell function. The subunits released from digested food can also be used as sources of energy for cells.

The body breaks down ingested food into subunits, which pass into the bloodstream. Blood delivers them to the body's cells.

Cells use these subunits either as building blocks to make new macromolecules (Chapter 4) or as energy to fuel cellular work. Fats are the most energy dense molecules.

Carbohydrate

Simple sugars

Energy
4 Calories/gram

Protein

Amino acids

Energy
4 Calories/gram

Fat

Nucleic acid

Fatty acids
and glycerol

Energy
9 Calories/gram

Nucleotides

Not a significant source
of energy for cells

or 3 hours of cycling at 5.5 miles per hour, to burn those extra Calories (**TABLE 6.1**).

Perhaps not surprisingly, not everyone burns Calories at the same rate. There are people who seem to be able to eat to their heart's content and remain slim. And there are those who seem to gain weight just by looking at food. Genetics plays a large role in how much food each one of us actually needs—and, perhaps, wants. There are other factors, too. Men, because their bodies naturally produce more muscle-building hormones than do women's, generally have more muscle mass and therefore need to eat more than women do. Since muscle cells require more Calories than do fat cells, the ratio of muscle mass compared to fat content in our bodies is another factor.

COUNTING CALORIES

Hunger—the feeling that one needs to eat—is a biological urge, and is tightly regulated. When our body senses it needs energy, it sends a signal to the brain, which tells us: *eat!* After we have eaten, our body senses that state, too, and sends an opposing signal: *I'm full!* But we humans eat for many reasons other than hunger—which is when problems can arise. Many of us eat more food than our bodies need, eating either for emotional comfort or for cultural reasons. We also have a natural preference for fatty and sugary foods because such foods are energy dense. For our ancestors, it was likely important to load up on those foods to store energy for times when food was scarce. Today, this ancient taste preference has become a vice that snack food companies have become very good at exploiting. When foods are loaded with sugar and fat, as well as salt, the result is a trifecta of flavors and textures that many people find irresistible.

For the large majority of us, when we eat Calories beyond what our bodies require, the extra energy is stored in one of two places: as glycogen in muscle and liver cells, or as triglycerides in fat cells. **Glycogen** is the energy-storing carbohydrate found in animal cells. You can think of glycogen as a short-term

TABLE 6.1 CALORIES IN, CALORIES OUT

Calories in Select Foods		Calories Burned during Select Activities*	
Food	Calories	Activity	Calories
8 oz. unsweetened green tea	2	Sleeping	55
1 large slice whole wheat bread	79	Sitting	85
½ cup cooked white rice	102	Standing	100
12 oz. nonfat milk	120	Office work	140
12 oz. cola	140	Biking (moderate)	450+
1 glazed doughnut	200	Jogging (5 mph)	500+
1 slice thick-crust cheese pizza	256	Swimming (active)	500+
1 Starbucks grande mocha frappucino with whipped cream	380	Hiking	500+
1 McDonald's Big Mac	540	Power walking	600+
1 Burger King Whopper	670	Cycling (stationary)	650
1 medium apple	80	Squash	650+
10 baby carrots	35	Running	700+

*Approximate number of Calories burned per hour by a 150-pound woman.

storage system. When we require short bursts of energy—in a sprint, for example—the body breaks down glycogen to obtain energy. However, because a gram of glycogen stores only half as many Calories as a gram of fat (about 4 Calories per gram versus 9), our bodies would have to carry around twice as much glycogen to store the same amount of Calories. So our bodies store most excess Calories as **triglycerides** in fat cells, which actually allows us to carry around less weight overall. The body burns this fat only after it has already used up food molecules in the bloodstream and in stored glycogen.

For our ancestors, who lived during times of frequent famine, this system of storing Calories as fat would have come in handy. Their bodies could burn fat for energy to carry them through times of food scarcity. Today, most people in the developed world have plenty of food. But they are more sedentary and eat more Calories than they need—which is why

▶**GLYCOGEN**
A complex animal carbohydrate, made up of linked chains of glucose molecules, that stores energy for short-term use.

▶**TRIGLYCERIDES**
A type of lipid found in fat cells that stores excess energy for long-term use.

INFOGRAPHIC 6.5 GLYCOGEN AND FAT STORE EXCESS CALORIES

When we ingest more Calories than our bodies need, they are stored as glycogen molecules in muscle and liver cells. Once the body's glycogen stores have been replenished, any excess Calories are stored as triglyceride molecules in fat cells.

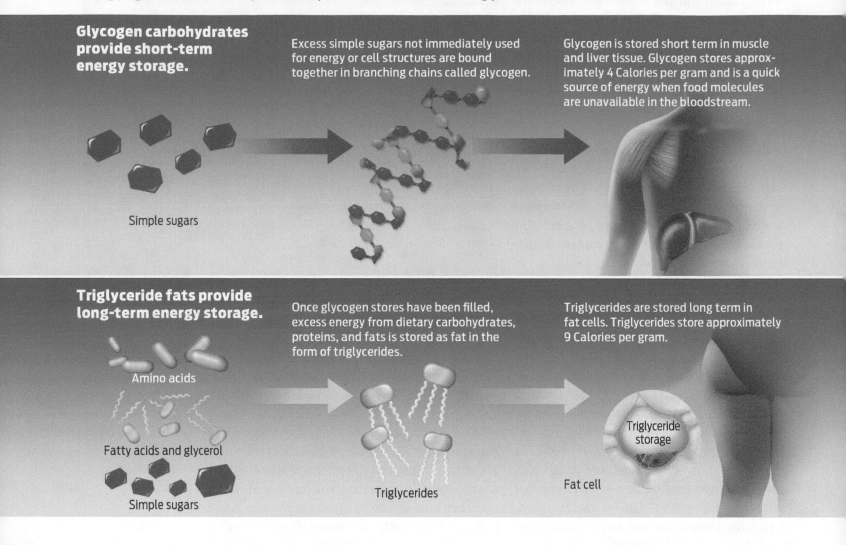

Glycogen carbohydrates provide short-term energy storage.

Excess simple sugars not immediately used for energy or cell structures are bound together in branching chains called glycogen.

Glycogen is stored short term in muscle and liver tissue. Glycogen stores approximately 4 Calories per gram and is a quick source of energy when food molecules are unavailable in the bloodstream.

Simple sugars

Triglyceride fats provide long-term energy storage.

Once glycogen stores have been filled, excess energy from dietary carbohydrates, proteins, and fats is stored as fat in the form of triglycerides.

Triglycerides are stored long term in fat cells. Triglycerides store approximately 9 Calories per gram.

Amino acids

Fatty acids and glycerol

Simple sugars

Triglycerides

Fat cell

Triglyceride storage

they've started to pack on, and keep on, the pounds (**INFOGRAPHIC 6.5**).

THE FRENCH PARADOX

Because each type of energy-rich organic molecule that we ingest—whether protein, carbohydrate, or fat—stores a different amount of energy, it's not only how much we eat but also what we eat that contributes to weight gain. We are more likely to gain weight from a pound of ice cream than a pound of veggies, for example, because ice cream contains more fat—and therefore more Calories.

Some scientists have used this fact to argue that there must be some factor other than small portion size that explains why the French have remained relatively thin. The French may eat less, but they are also world renowned for their love of buttery, creamy sauces, dense desserts, and fatty meats. If the French load up on fat, which has more Calories, how do they manage to keep the weight off? Research such as Rozin's suggests that it may still have to do with portion size: the French may eat fatty foods, but they do so in moderation—savoring smaller amounts of these rich dishes than their American counterparts.

How do the French manage to eat smaller portions? In France and many other European countries, small is the cultural norm. The French don't supersize, and they are less likely to buy in bulk. So the opportunity to eat more may not exist in the first place. Research by other investigators has shown that a person presented with a bigger package of, say, M&M candies will take more from it than when presented with a smaller package. A 2007 study led by Jennifer Fischer at Baylor College of Medicine, for example, found that preschool-age children consumed 33% more Calories when the portion size of the meal was doubled. This behavior combined with meals made with a high ratio of energy-dense ingredients—oil, butter, and sugar—is a significant contributor to childhood obesity, these researchers concluded.

Or it may have to do with the *way* the French eat; not only do they eat smaller portions at each meal, they don't snack, they don't opt for second helpings, and they don't skip meals. Mireille Guiliano in her book *French Women Don't Get Fat* described how she gained 20 pounds during her 5-month stay in America. She snacked, she drank a lot of soda, and she ate standing up without really focusing on her food, she wrote. Guiliano found that she had forgotten how to slow down and enjoy the taste of food the way she used to in France, and so she compensated by eating larger portions. Part of her diet plan when she returned to France, she wrote, was quitting in-between-meals snacking and reacquainting herself with the French culture of eating, which involves slowing down and savoring each bite.

Indeed, evidence suggests that the French do eat more slowly. In his study, Rozin compared the average time people spent eating at McDonald's in Paris and in Philadelphia. In Paris, the average time of the meal was 22 minutes; in Philadelphia, only 14. While scientists don't know for sure how longer meals help people eat less, they speculate that taking it slow may help people enjoy their food more and recognize when they are full.

In addition to having different eating habits, the French may also be more active than Americans. Walking, biking, and taking public transportation to work are more common in Paris than they are in many parts of the United States, which is geographically spread out and more heavily dependent on cars.

However it is that the French remain slim, one thing is clear: they weigh less because they either eat less or burn more Calories. The only way to gain weight is by taking in more Calories than we expend through energy-using activities. In other words, our waistlines obey the principle of conservation of energy (see Chapter 5): energy is neither created nor destroyed but merely converted from one form into another. If more food energy is taken into our bodies than is used to power cellular reactions and physical movement, the excess (minus what is released as heat with every energy conversion) is stored as fat.

PUTTING FOOD TO WORK

Getting energy from food seems simple enough: we eat food and we have energy. But the process of releasing the energy stored in food and putting it to use in cells is a bit more complex. First, food must be broken down into its component subunits—among them, fats and sugars. Then, these breakdown products must go through a series of biochemical reactions that convert the chemical energy stored in these food molecules into a form of fuel we can use. Energy from food is ultimately captured in a molecule called **adenosine triphosphate (ATP)** that our cells use to carry out energy-requiring functions.

You can think of food as a bar of gold: it has a great deal of value, but if you carried that gold bar to your local convenience store, you wouldn't be able to buy even a cup of coffee with it. You would first have to convert your gold bar into bills and coins. ATP is the energetic equivalent of bills and coins; it's

▶ADENOSINE TRIPHOSPHATE (ATP)
The molecule that cells use to power energy-requiring functions; the cell's energy "currency."

UNIT 1 • WHAT IS LIFE MADE OF? CHEMISTRY, CELLS, ENERGY

6

INFOGRAPHIC 6.6 ATP: THE ENERGY CURRENCY OF CELLS

Just as a gold bar must be converted to currency in order to buy merchandise, the energy in food must be converted to ATP before it can be used by the cell.

Money Conversion

Dietary Energy Conversion

Adenosine triphosphate is a nucleotide that stores chemical energy in the bonds between its phosphate groups. Breaking these bonds releases energy.

Adenine

Ribose

Phosphate groups

ATP

Money in a form that can be used to make purchases.

Energy in a form that can be used to power cell functions.

Power Cell Functions

---

currency that your body can actually spend (**INFOGRAPHIC 6.6**).

To make ATP, our bodies first break down food molecules into their smaller subunits through the process of digestion (see Chapter 26). Once released from food, such subunits as fatty acids, glycerol, amino acids, and sugars leave the small intestine and enter the bloodstream, which transports them to the body's cells (see Infographic 6.4). Inside the cells, enzymes break apart the bonds holding these subunits together. The energy stored in those bonds is then captured and converted into the chemical bonds that make up ATP. When cells need energy, they tap these energy-rich bonds in ATP. When ATP bonds are broken, energy is released, allowing cells to "spend" their ATP

currency and carry out normal cellular functions.

The primary process that all eukaryotic organisms, including plants, use to convert food energy into ATP is a form of cellular respiration called **aerobic respiration**. "Aerobic" means "in the presence of oxygen," and this form of cellular respiration predictably requires a continual source of oxygen. Of the subunits released from food, glucose sugar is the most common source of energy for all organisms, from bacteria to humans. The aerobic respiration of glucose can be summarized by this equation:

Glucose + Oxygen → Energy + Carbon dioxide + Water

▶AEROBIC RESPIRATION

A series of reactions that occurs in the presence of oxygen and converts energy stored in food into ATP.

That is, the bonds holding the glucose molecule together are broken. Oxygen from the air we breathe is consumed in the process. When the bonds of glucose are broken, the energy is converted to ATP and some heat is given off as a result of the inefficiency of this conversion (see Chapter 5). Carbon dioxide and water are also given off as waste products of the process (**INFOGRAPHIC 6.7**).

Aerobic respiration is a multistage process that takes place in different parts of the cell. The first stage, known as **glycolysis**, takes place in the cell's cytoplasm. Glycolysis is a series of chemical reactions that splits glucose in half, into two smaller molecules of

pyruvate. Each of these pyruvate molecules has three carbon atoms in its backbone. The pyruvate molecules then enter the cell's mitochondria, where the next two stages of aerobic respiration occur.

During the second stage, the **citric acid cycle**, a series of reactions strips electrons from the bonds between the carbon and hydrogen atoms that were originally in glucose (and are now in pyruvate). The process releases CO_2, which is ultimately exhaled from an organism's lungs.

As the energy-rich bonds in glucose and pyruvate are broken, the electrons from those bonds are picked up by a molecule known as

> **GLYCOLYSIS**
> A series of reactions that breaks down sugar into smaller units; glycolysis takes place in the cytoplasm and is the first stage of both aerobic respiration and fermentation.

> **CITRIC ACID CYCLE**
> A set of reactions that takes place in mitochondria and helps extract energy (in the form of high-energy electrons) from food; the second stage of aerobic respiration.

INFOGRAPHIC 6.7 AEROBIC RESPIRATION TRANSFERS FOOD ENERGY TO ATP

→ During aerobic respiration, our cells use the oxygen we inhale to help extract energy from food. Cells convert the energy stored in food molecules into the bonds of ATP, the cell's energy currency.

1. Blood that circulates throughout the body delivers oxygen from the lungs and food-derived subunits from the small intestines to the body's cells.

Blood vessel

Oxygen (from the lungs)

CO_2

Mitochondrion

Carbon dioxide (to the lungs)

Glucose (from the intestine)

ATP

Water

Animal cell

2. Cells break the chemical bonds of food molecules to release energy, which is used to make ATP. Water and carbon dioxide are produced as waste products.

3. Carbon dioxide exits, travels via the blood to the lungs, and is exhaled as waste.

Inputs			Outputs			
Glucose $C_6H_{12}O_6$	+	Oxygen O_2	→ Energy ATP	+	Carbon dioxide CO_2	+ Water H_2O

INFOGRAPHIC 6.8 AEROBIC RESPIRATION: A CLOSER LOOK

Nearly all eukaryotic organisms carry out aerobic respiration. The three main stages of aerobic respiration occur in specific locations within the cell and yield distinct products.

1. Glycolysis

Glycolysis breaks down food molecules (e.g. glucose) into smaller molecules in the cell's cytoplasm. These molecules then enter the cell's mitochondria. Glycolysis converts some energy into a small number of ATP molecules.

Glucose

Pyruvate molecules

2 ATP

Cytoplasm

Carbon dioxide (CO_2) Carbon atoms released from the chemical bonds of food join with oxygen to form carbon dioxide, which is exhaled from the lungs.

2. Citric Acid Cycle

In a series of reactions, high-energy electrons (e^-) are stripped from the bonds between carbon and hydrogen atoms and carried to the inner membrane of the mitochondria by NADH molecules. During this process, a small amount of ATP is made.

2 ATP

Citric acid cycle

NADH

Electron transport

36 ATP

Oxygen (O_2) Oxygen molecules accept the transported electrons and join with hydrogen to form water.

3. Electron Transport

As the high-energy electrons (e^-) are passed from NADH down a chain of molecules in the mitochondrial membrane, they power a series of reactions that channel energy into the formation of many ATP molecules.

Water (H_2O)

Mitochondrion

▶**NAD⁺**
An electron carrier. NAD^+ can accept electrons, becoming NADH in the process.

▶**ELECTRON TRANSPORT CHAIN**
A process that takes place in mitochondria and produces the bulk of ATP during aerobic respiration; the third stage of aerobic respiration.

NAD^+. When NAD^+ picks up electrons, it becomes NADH (the electron-carrying form of the molecule). NADH then carries the electrons to the inner membrane of the mitochondria, where NADH gives them up (reverting to NAD^+). The electrons then go through the third and last stage of aerobic respiration: the **electron transport chain**.

Electrons stripped from the bonds in glucose contain a lot of potential energy that is temporarily held by NADH. During electron transport, these energetic electrons are passed like hot potatoes from NADH down a chain of molecules in the inner mitochon-

drial membrane. Eventually the electrons are passed to oxygen molecules, which accept the electrons and combine with hydrogen atoms to produce water. As electrons pass down the chain, they supply the energy needed to form ATP. This electron transport chain fuels the bulk of ATP production (INFOGRAPHIC 6.8).

We've focused on glucose, but cells can also burn fats and amino acids for fuel during aerobic respiration. Because fats generally have more carbon-hydrogen bonds than do sugars and amino acids, they have more electrons to be stripped in the citric acid cycle.

More electrons stripped means that more ATP molecules are produced during electron transport (which also explains why a gram of fat contains more Calories than a gram of sugar or protein).

WHEN OXYGEN IS SCARCE

In order to keep performing aerobic respiration, cells need adequate oxygen. If the rate at which cells consume oxygen exceeds the rate at which they take it in when we breathe, aerobic respiration comes to a halt, because the electron transport chain has no oxygen to which it can deliver electrons. Without oxygen, glycolysis still occurs, but its products are shunted into a different process, **fermentation**, which takes place in the cell's cytoplasm (as opposed to the mitochondria).

Fermentation does not actually produce any more ATP beyond what is produced by glycolysis. So why do cells do it? In essence, it's a way to keep glycolysis running. As the cell carries out glycolysis in the absence of O_2, glucose will continue to give up electrons to NAD^+ as it is converted to pyruvate. However, in the absence of O_2, NADH cannot unload its electrons to the electron transport chain, and therefore there is no NAD^+ available to pick up electrons from glucose (since all the NAD will be in the NADH form). Soon glycolysis will grind to a halt unless there is some way for the cell to regenerate NAD^+. This is the purpose of fermentation: during fermentation, NADH unloads its electrons onto pyruvate, thus regenerating NAD^+, which allows glycolysis to keep running and producing small amounts of ATP.

Because fermentation bypasses both the citric acid cycle and the electron transport chain, much less ATP is produced—only about 2 molecules of ATP from each molecule of glucose compared to about 36 ATP produced by aerobic respiration. Instead of carbon dioxide, fermentation produces lactic acid **(INFOGRAPHIC 6.9)**.

▶**FERMENTATION**
A series of chemical reactions that takes place in the absence of oxygen and converts some of the energy stored in food into ATP. Fermentation produces far less ATP than does aerobic respiration.

INFOGRAPHIC 6.9 FERMENTATION OCCURS WHEN OXYGEN IS SCARCE

Glycolysis occurs whether or not oxygen is present. In the absence of oxygen, fermentation reactions follow glycolysis. Fermentation occurs in the cytoplasm and converts the products of glycolysis into lactic acid (or alcohol in some organisms). The only ATP produced is the small amount produced during glycolysis.

1. Glycolysis
Glycolysis occurs identically in the presence and absence of oxygen.

2. Fermentation Reactions
Fermentation converts the products of glycolysis into lactic acid (or alcohol), which leaves the cell.

Blood vessel

NADH unloads electrons from glycolysis.

e^-
NADH

e^-
Pyruvate

Glucose

Glycolysis

Fermentation reactions

Fermentation products

2 ATP

This is the only ATP produced, far less than is produced during aerobic respiration.

Mitochondria are not used in fermentation.

In humans, fermentation takes place primarily during bursts of energy-intensive tasks, such as sprinting or power weight-lifting. It is, in essence, a backup plan for times when oxygen isn't available. The panting you experience on a treadmill is your body's way of trying to obtain more oxygen. (The idea that lactic acid buildup is the cause of muscle cramping is a myth.)

But for many organisms, like certain fungi and bacteria, fermentation is the main way of obtaining energy. In some of these organisms, fermentation produces alcohol rather than lactic acid as a by-product. Humans take advantage of these fermentation reactions when they make alcoholic beverages. Brewer's yeast, for example, is a fungus that ferments sugar in the absence of oxygen, producing alcohol as a result. Humans use brewer's yeast to make beer and wine.

Since fermentation does not break glucose down as completely as does aerobic respiration, there is still quite a bit of carbohydrate energy left in such beverages as beer and wine, about 7 Calories per gram—which explains why most weight-loss diets eliminate alcohol.

Even during aerobic respiration, however, our bodies don't convert every Calorie in food into ATP. The chemical processes aren't 100% efficient, so some energy is always released as heat, which keeps your body warm.

It's important to remember that aerobic respiration does not create energy—it only extracts it from food and converts it into ATP. All the food we eat—whether burger, chicken leg, or Caesar salad—originally gets its energy from the sun, by way of photosynthesis (see Chapter 5). Photosynthesizers such as plants and algae capture the energy of sunlight and convert it into chemical energy stored in sugar. We then eat this sugar (or eat animals that have eaten this sugar), and that stored energy becomes available to us. Plants benefit from the relationship, too: plants use our carbon dioxide waste as raw material for making sugar during photosynthesis. In this way, photosynthesis and respiration form a continual cycle, with the outputs of one process serving as the inputs of the other (**INFOGRAPHIC 6.10**).

In humans, fermentation takes place primarily during energy-intensive tasks.

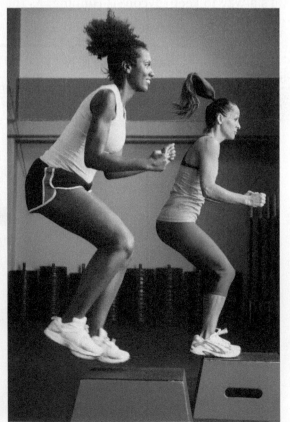

INFOGRAPHIC 6.10: PHOTOSYNTHESIS AND AEROBIC RESPIRATION FORM A CYCLE

→ Photosynthesis and respiration form a continuous cycle, with the outputs of one process serving as the inputs of the other.

Carbon dioxide produced during respiration is used by plants and algae to produce glucose and oxygen during photosynthesis.

Light energy

Carbon dioxide

Photosynthesis

Respiration

ATP

Glucose

Oxygen

Glucose and oxygen produced during photosynthesis are used by plants, animals (that eat the plants), and nearly all other living things to make ATP during aerobic respiration.

CURBING CULTURE

Given the relationship between Calorie intake and our expanding waistlines, many people are interested in altering the way we eat to reduce the amount of Calories that make it into our bloodstream. Food industry scientists, for example, have developed diet drugs that keep the body from absorbing some of the fat molecules in food. (You may have heard of potato chips made with olestra, a fat substitute that is not absorbed by the intestines and so passes through the body as waste, with famously unpleasant effects.) And, of course, many food manufacturers specialize in low-fat and fat-free foods, which are less Calorie dense.

But these efforts to curb obesity offer temporary solutions at best. Studies have shown that 90% of dieters regain most of their lost weight within 2 to 3 years. That's because most people find it terribly difficult to stick to the food restrictions prescribed by diets and tend to revert to their former eating habits. And in some cases, losing weight actually makes people hungrier, and hunger is one of the most powerful biological signals.

A change in eating culture, say some experts, offers a more sustainable fix because it emphasizes a change in eating habits, not the type of food itself. In fact, Americans didn't always consume large portion sizes. Lisa Young and Marion Nestle, both of the Department of Nutrition, Food Studies, and Public Health at New York University, documented in a 2002 study that the current sizes of fries, hamburgers, and soda at restaurants were two to five

The average U.S. portion size is more than twice what it was 20 years ago, and more than 4 times what it was 50 years ago.

20 YEARS AGO	TODAY

CHEESEBURGER

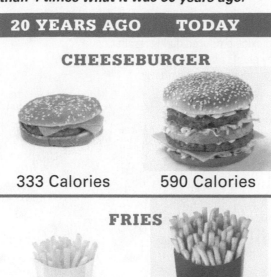

333 Calories	590 Calories

FRIES

210 Calories	610 Calories

TURKEY SANDWICH

320 Calories	820 Calories

BAGEL

140 Calories	350 Calories

SODA

85 Calories	250 Calories

▶**TRANS FAT**
A type of vegetable fat that has been hydrogenated, that is, hydrogen atoms have been added, making it solid at room temperature.

▶**SATURATED FAT**
An animal fat, such as butter; saturated fats are solid at room temperature.

times larger than they were before the 1970s, when portion size began to creep up. Single servings of pasta, muffins, steaks, and bagels today exceed the government-recommended serving sizes by 480%, 333%, 224%, and 195%, respectively. And cookies exceed the standard by a factor of 8, according to Nestle's research. In America, bigger is better.

The French are eating more, too, and the results are evident: the number of obese French people grew from 8.6% in 1997 to 13.1% in 2006, according to a 2008 study by Marie-Aline Charles published in the journal *Obesity*. As American music, movies, and clothing have become pervasive in other countries, so, too, have our eating habits. More and more French people are eating large amounts of nutritionally poor, energy-dense foods. And as jobs increasingly place people in front of computers, they have become less physically active. More French people are eating on the go, eating fast food, and spending less time enjoying formal meals. Much to the dismay of public health experts, French eating culture is tipping toward unhealthful.

Meanwhile, studies suggest that the rate of obesity in the United States may be leveling off. A 2010 study by Katherine Flegal at the Centers for Disease Control and Prevention found no significant increase in the rate of obesity in the United States from 2003 through 2008. Nevertheless, at 30% the prevalence of obesity in the United States is still high and remains higher than in most European countries (and more than double the prevalence in France).

Because of this statistic, there is a growing trend in the United States to legislate changes in the foods people eat. Some cities, for example, have banned the use of **trans fat**, a type of hydrogenated vegetable fat that studies have shown contributes to heart disease. Commercially prepared cookies, French fries, doughnuts, and margarine often contain trans fat, which food manufacturers add to their products to give them a longer shelf life or a pleasing texture. Hydrogenated fat behaves in the body much like **saturated fat**, the type of fat found in butter and other animal products. Studies

> Studies have shown that 90% of dieters regain most of their lost weight within 2 to 3 years.

have shown that eating large amounts of saturated fat can clog arteries (see Chapter 27). By contrast, **unsaturated fats**, which come from plants and are liquid at room temperature, are considered more healthful (although they are still high in Calories). In New York City, restaurants are now required to post their foods' Calorie content, and in 2012, Mayor Michael Bloomberg proposed a ban on the sale of sugary beverages larger than 16 ounces in restaurants, from street carts, and in movie theaters in the city.

"Actions by governments are the only way conditions will change enough to have a major public health impact," says Kelly Brownell. For America's obesity woes, Brownell blames the food industry for its relentless marketing of unhealthful foods, agricultural and trade policies that promote unhealthful diets, and economic policies that make unhealthful foods cheaper than healthful ones.

Not everyone agrees, however, that it is the government's job to restrict our food choices. Americans equate choice in food with democracy, argues Paul Rozin. "We could also over-respond to what many perceive as an obesity epidemic and that could be dangerous. It would restrict individual freedom."

Besides, while such government legislation would restrict *what* we eat, it wouldn't really affect *how* we eat. Rozin hopes people change their behavior voluntarily. For example, people could fit more movement into their daily routines, climbing stairs instead of using

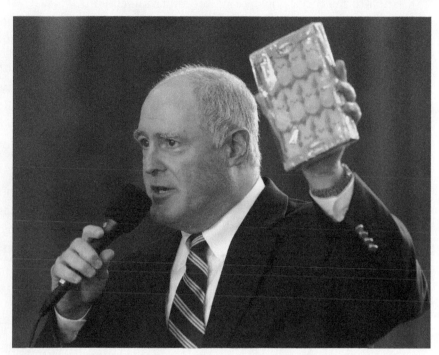

Senator Kel Seliger displays a box of candy in the Texas Senate during a debate on a bill that would ban trans fats in restaurants—he opposed the bill.

an elevator or parking their cars farther away from the entrance to the mall. To combat large portion sizes in restaurants, Rozin advocates ordering less, sharing dishes, or, as Mireille Guiliano recommends, relearning how to savor our food.

The U.S. eating culture can change, Rozin says, but not overnight. Look at cigarette smoking: "It took 50 years to get cigarette smoking to decline and [cigarettes] are much more harmful to health." ■

▶**UNSATURATED FAT**
A plant fat, such as olive oil; unsaturated fats are liquid at room temperature.

CHAPTER 6 Summary

▶ The macronutrients in our food (proteins, carbohydrates, and fats) are sources of dietary energy.

▶ Fats are the most energy-rich organic molecules in our diet. Fats contain twice as many Calories per gram as carbohydrates and proteins.

▶ When we consume more Calories than we use, our bodies store the excess energy in the bonds of glycogen and body fat. Fats store more energy than does glycogen.

▶ Cells carry out chemical reactions that break down food to obtain usable energy in the form of ATP.

▶ In the presence of oxygen, aerobic respiration produces large amounts of ATP from the energy stored in food.

▶ Aerobic respiration occurs in three stages: (1) glycolysis, (2) the citric acid cycle, and (3) electron transport. The first stage occurs in the cytoplasm, the latter two in the mitochondria. Electron transport produces the bulk of ATP.

▶ In the absence of oxygen, fermentation follows glycolysis and produces lactic acid in animals (or, in some organisms, alcohol). Fermentation produces far less ATP than does aerobic respiration.

▶ Exercise helps burn stored Calories. A combination of eating fewer Calories and exercising more will result in weight loss, although hereditary factors play a large role in determining a person's weight.

▶ During exercise, glycogen is used first. Stored fats are tapped only when glycogen stores have been depleted, as might occur during long periods of exercise.

▶ The ultimate source of energy in food is the sun. Photosynthesizers such as plants trap the energy of sunlight and convert it into the chemical energy of sugar. Animals then eat this sugar either directly or indirectly.

▶ Photosynthesis and respiration form a cycle: the carbon dioxide given off by animals, plants, and all organisms that perform aerobic respiration is used by photosynthesizers to make glucose and oxygen during photosynthesis.

MORE TO EXPLORE

▶ Rozin, P., et al. (2003) The ecology of eating: smaller portion sizes in France than in the United States help explain the French paradox. *Psychological Science* 14:450–54.

▶ Guiliano, M. (2007) *French Women Don't Get Fat: The Secret of Eating for Pleasure.* New York: Vintage Books.

▶ Kessler, D. (2010) *The End of Overeating: Taking Control of the Insatiable American Appetite.* Emmaus, PA: Rodale Books.

▶ Yale Rudd Center for Food Policy and Obesity http://www.yaleruddcenter.org/

▶ Let's Move http://www.letsmove.gov/

CHAPTER 6 Test Your Knowledge

DRIVING QUESTION 1

Why do humans weigh more now than in the past?

By answering the questions below and studying Infographics 6.1, 6.2, and 6.3, you should be able to generate an answer for the broader Driving Question above.

KNOW IT

1 A 5'6" female weighs 167 pounds. Use Infographic 6.2 to determine her BMI. Would she be considered underweight, normal weight, overweight, or obese?

2 If a person wants to lose weight, which of the following are viable strategies?

   a. substituting plain water for regular soda

   b. eating the same number of Calories, but eating them all as fruit and veggies instead of burgers and fries

   c. adding more fruits and veggies on top of the current diet

   d. exercising more

   e. a and d

USE IT

3 If you frequently crave French fries, how could you modify your lifestyle to eat fries without gaining weight? Explain your answer.

4 Why do you think that longer meal times translate into fewer Calories consumed?

5 If the government were to issue tax incentives to reduce obesity in the United States, which of the following do you think would be most effective? Explain your choice.

   a. taxing foods high in fat

   b. giving tax breaks to people who join gyms or health clubs

   c. giving rebates for purchasing fresh fruits and vegetables

   d. paying enhanced salaries for teachers in elementary and middle schools to provide education about diet and nutrition

6 If the French eat meals with a higher fat content, why don't the French weigh more on average than Americans?

INTERPRETING DATA

7 Look at the illustration on page 130 showing the increase in size and calorie content for different foods in the past several decades.

Calculate the % increase in Calorie content of a typical serving of the foods shown in the past 20 years. Which food has had the greatest increase in Calories in the past 20 years? Which has had the smallest increase in the past 20 years?

DRIVING QUESTION 2

How does the body use the energy in food?

By answering the questions below and studying Infographics 6.4 and 6.5 and Table 6.1, you should be able to generate an answer for the broader Driving Question above.

KNOW IT

8 Which type of organic molecule stores the most energy per gram?

   a. proteins

   b. starch

   c. nucleic acid

   d. fats (triglycerides)

   e. glycogen

9 A moderately active 21-year-old female has a choice of eating a 2,500-Calorie meal that is primarily protein or a 2,500-Calorie meal that is primarily sugar. What would be the result, in terms of energy, of choosing one over the other?

   a. Nothing; she would burn all these Calories, given her age, gender, and activity level.

   b. She would store the excess Calories as protein, regardless which meal she ate.

   c. She would store the excess Calories as protein if she ate the protein meal, and as glycogen if she ate the sugar meal.

   d. In either case, once her glycogen stores are replenished, she will store the excess Calories as fat.

   e. Regardless of the number of Calories, she will get more energy from the sugar meal.

10 If you exercise for an extended period of time, you will use energy first from _____, then from _____.

a. fats; glycogen
b. proteins; fats
c. glycogen; proteins
d. fats; proteins
e. glycogen; fats

MINI CASE

11 Consider a well-trained 130-pound female marathon runner. She has just loaded up on a carbohydrate meal and has the maximum amount of stored glycogen (6.8 g of glycogen per pound of body weight).

a. How many grams of glycogen is she storing?
b. How many Calories does she have stored as glycogen?
c. If this same number of Calories were stored as fat, how much would it weigh?
d. Suppose she decides to go for a run at a pace of 9 miles per hour (she will be running 6.5-minute miles). Given her weight, she will burn 885 Calories per hour at this pace. How long will it take her to deplete her glycogen stores? How many miles can she run before her glycogen supplies run out? Will she be able to complete a 26.2-mile marathon?
e. Once her glycogen supplies run out, what has to happen if she wants to keep running?

DRIVING QUESTION 3

How does aerobic respiration extract useful energy from food?

By answering the questions below and studying Infographics 6.6, 6.7, 6.8, and 6.10, you should be able to generate an answer for the broader Driving Question above.

KNOW IT

12 Which process is not correctly matched with its cellular location?

a. glycolysis—cytoplasm
b. citric acid cycle—mitochondria
c. glycolysis—mitochondria
d. electron transport—mitochondria
e. none of the above; they are all correctly matched

13 In the presence of oxygen we use _____ to fuel ATP production. What process do plants use to fuel ATP production from food?

a. aerobic respiration; photosynthesis

b. aerobic respiration; aerobic respiration
c. fermentation; aerobic respiration
d. fermentation; photosynthesis
e. glycolysis; photosynthesis

14 Given 1 g of each of the following, which would yield the greatest amount of ATP by aerobic respiration?

a. fat
b. protein
c. carbohydrate
d. nucleic acid
e. alcohol

15 During aerobic respiration, what molecule has (and carries) electrons stripped from food?

a. NAD^+
b. NADH
c. O_2
d. H_2O
e. pyruvate

16 During aerobic respiration, how does NADH give up electrons to regenerate NAD^+?

a. by giving electrons to O_2
b. by giving electrons to pyruvate
c. by giving electrons to glucose
d. by giving electrons to the electron transport chain
e. by giving electrons to another NAD^+

USE IT

17 Draw a carbon atom that is part of a CO_2 molecule such as you just exhaled. Using a written description or a diagram, trace what happens to that carbon atom as it is absorbed by the leaf of a spinach plant and then what happens to the carbon atom when you eat that leaf in a salad.

18 If you ingest carbon in the form of sugar, how is that carbon released from your body?

a. as sugar
b. as fat
c. as CO_2
d. as protein
e. in urine

DRIVING QUESTION 4

When does fermentation occur, and why can't it sustain human life?

By answering the questions below and studying Infographic 6.9, you should be able to generate an answer for the broader Driving Question above.

KNOW IT

19 Compared to aerobic respiration, fermentation produces _____ ATP.

   a. much more
   b. the same amount of
   c. a little less
   d. much less
   e. no

20 What process is most directly prevented in the absence of adequate oxygen?

   a. citric acid cycle
   b. glycolysis
   c. electron transport chain
   d. a, b, and c
   e. glycolysis and the citric acid cycle

21 During fermentation, how does NADH give up electrons to regenerate NAD⁺?

   a. by giving electrons to O_2
   b. by giving electrons to pyruvate
   c. by giving electrons to glucose
   d. by giving electrons to the electron transport chain
   e. by giving electrons to another NAD⁺

22 Where in the cell does fermentation take place?

   a. cytoplasm
   b. mitochondria
   c. nucleus
   d. cytoplasm and mitochondria
   e. Fermentation doesn't occur in cells, it occurs in circulating blood.

USE IT

23 Explain how the presence or absence of oxygen affects ATP production. (The terms *aerobic respiration* and *fermentation* should be in your answer.)

24 Consider fermentation.

   a. How much ATP is generated during fermentation?
   b. How does this compare to aerobic respiration?
   c. In humans, why can't fermentation sustain life? Hint: Think of two reasons—one is related to the product of fermentation and what happens if it accumulates.

BRING IT HOME

25 Losing 1 pound of weight requires a deficit of 3,500 Calories. Design three strategies for someone weighing 150 pounds to lose 1 pound in two weeks. The first should rely on (a healthy) diet alone. The second should rely on exercise alone. The third should be a combination of diet and exercise. Which of the three do you think would be easiest to follow?

BIOLOGICALLY Unique

How DNA helped free an innocent man

ROY BROWN THOUGHT THE POLICE WERE JUST CHECKING up on him when an officer knocked on his door one day in May 1991. Brown, a self-professed hard drinker who earned a living selling magazine subscriptions, had only a week before been released after serving an 8-month prison term. His crime: threatening to kill the director of the Cayuga County Department of Social Services in upstate New York. A caseworker had deemed Brown unfit to care for his 7-year-old daughter. Furious, Brown made a series of threatening phone calls to the director. But he had served his time. What could the officer want from him now?

Three days earlier, police had found the battered body of a woman lying in the grass about 300 feet from the farmhouse where she had lived. Someone had burned the place to the ground. The body was identified as that of Sabina Kulakowski, a social worker at the Cayuga County Department of Social Services. The crime was horrific. The murderer had beaten the 49-year-old Kulakowski, bitten her several times, dragged her outside, and then stabbed and strangled her. It was obvious that Kulakowski had struggled; her body was covered with defensive wounds.

▶▶ **DRIVING QUESTIONS**

1. What is the structure of DNA, and how is DNA organized in cells?

2. How is DNA copied in living cells, and how can DNA be amplified for forensics?

3. How does DNA profiling make use of genetic variation in DNA sequences?

4. How does DNA evidence fit into forensic investigations?

In 1991, Roy Brown was found guilty of homicide and sentenced to 25 years to life in prison.

▶DEOXYRIBONUCLEIC ACID (DNA)

The molecule of heredity, common to all life forms, that is passed from parents to offspring.

▶CHROMOSOME

A single, large DNA molecule wrapped around proteins. Chromosomes are located in the nuclei of most eukaryotic cells.

Although Kulakowski had not been involved in Brown's case, officers arrested Brown that day on suspicion of murder. Eight months later, a jury found Brown guilty of homicide and sentenced him to prison for 25 years to life. The prosecution argued that Brown's motive was revenge against the Department of Social Services. But what really nailed the case was testimony from an expert who stated that bite marks on the victim's body matched Brown's teeth.

Brown, however, maintained his innocence. "I never knew Ms. Kulakowski, and I had nothing to do with that woman's death . . . I am truly innocent," he told the court and onlookers after the verdict had been announced.

Even from prison Brown never stopped trying to prove his innocence. He repeatedly petitioned, in vain, for a retrial.

Then something unexpected happened: Brown uncovered additional evidence that strongly suggested he was not the perpetrator. The evidence was so compelling, in fact, that in late 2004, after Brown had spent 12 years in prison, his lawyers decided to contact the Innocence Project, a nonprofit organization founded in 1992. Its mission: to use DNA evidence to free people wrongly convicted of crimes.

DNA AS EVIDENCE

When the jury convicted Brown in 1992, DNA testing was still in its infancy, so DNA was rarely used as evidence in criminal cases. But over the next decade, DNA testing became a standard part of court cases, as science increasingly showed that it was an extremely accurate way to match crime scene evidence to perpetrators.

How can scientists use DNA to identify a person? The answer lies in the chemical makeup of this molecule, often referred to as the blueprint of life. **Deoxyribonucleic acid, or DNA,** is the hereditary molecule that is common to all life forms—from bacteria to plants to humans—and that is passed from parents to offspring. DNA serves as the instruction manual from which we are built; it's the reason children resemble their relatives—parents, an aunt, perhaps even a grandparent.

Where can you find DNA? The molecule exists inside the nucleus of almost every cell in our body in the form of **chromosomes,** strands of DNA wound around proteins. Humans have 23 pairs of chromosomes in the vast majority of their cells; we inherit one chromosome of each pair from our mother and the other from our father, for a total of 46. The 23rd chromosome pair consists of the sex chromosomes, X and Y, which determine a person's sex. Fathers have an X and a Y and can pass on either. Mothers have two Xs and therefore can pass on only an X. Children who inherit a Y from dad and an X from mom are males. Since nearly all cells contain DNA, scientists can collect evidence such as blood, semen, saliva, or hair from a crime scene and extract DNA from it to identify a perpetrator (INFOGRAPHIC 7.1).

DNA testing has helped the Innocence Project free more than 300 people from prison since 1992, including 18 who served time on death row. The technology has not only given these people their lives back but has also thrown a spotlight on our flawed criminal justice system. Why were people

INFOGRAPHIC 7.1 WHAT IS DNA, AND WHERE IS IT FOUND?

→ Deoxyribonucleic acid, or DNA, is the hereditary molecule common to all living organisms. It is the instruction manual from which an organism is built.

Each human cell has 23 pairs of chromosomes; one chromosome of each pair is inherited from the mother, the other from the father. The 23rd chromosome pair determines a person's sex.

Cell Nucleus

Chromosome

DNA molecules are organized into discrete structures called chromosomes.

If a single DNA molecule were stretched out it would be 1 to 3 m long.

DNA exists in the nuclei of most cells.

DNA

Proteins

Each chromosome consists of a single, long DNA molecule wrapped around proteins.

wrongly convicted and placed on death row? Innocence Project lawyers have found the usual suspects: dishonest witnesses, unscrupulous police officers, apathetic or overburdened lawyers, mistakes in eyewitness identification. But perhaps even more important, DNA technology has helped find ways to improve the system.

"DNA is only one example of how advances in science have made the criminal justice system more reliable," says Peter Neufeld, who, along with Barry Scheck, both from the Benjamin N. Cardozo School of Law in New York City, founded the Innocence Project. "But what we really hope to do now is use DNA as the gold standard of reliability to weed out junk science."

Peter Neufeld and Barry Scheck, founders of the Innocence Project.

UNRELIABLE EVIDENCE

Indeed, it was "junk science" that convicted Roy Brown. The only physical evidence linking Brown to the case was his teeth. A dentist hired by the prosecution testified that the bite marks on Kulakowski's body matched Brown's teeth. But as the defense pointed out, the bite marks came from someone with six upper teeth—Brown had only four. The prosecution's witness argued that Brown could have twisted Kulakowski's skin while biting her and filled in the gaps—an argument that ultimately proved convincing to the jury.

Bite-mark analysis is a particularly troubling form of evidence. No widely accepted rules or standards govern its use, and no government or outside scientific commission has ever validated its claims. In fact, studies show error rates—the rate at which experts have falsely identified bite marks as belonging to a particular person—as high as 91%.

Hair analysis, another common type of evidence, can be equally troublesome. In dozens of cases, Innocence Project lawyers found that forensic scientists had testified that hairs from crime scenes matched the accused, explains Neufeld. But when scientists subsequently tested the DNA inside the follicle cells from those hairs, the DNA didn't match.

The problem is that hair analyses, performed under a microscope, can reveal only certain characteristics: it can distinguish whether

INFOGRAPHIC 7.2 DNA IS MADE OF TWO STRANDS OF NUCLEOTIDES

➔ DNA is a double-stranded molecule. Each of the two strands consists of a chain of subunits called nucleotides that are bonded together. There are four types of nucleotide: adenine (A), thymine (T), guanine (G), and cytosine (C). Nucleotides are the building blocks of DNA.

Nucleotide Structure

Each nucleotide consists of three parts: a sugar, a phosphate, and one of four bases (A, T, C, G)

Phosphate

Sugar

Base

Hydrogen bonding, or base pairing, between the bases of each strand holds the strands together. Bases pair in a specific pattern: A with T, C with G. As and Ts have two hydrogen bonds between them, while Gs and Cs have three.

DNA Sequences Are Unique

Along each strand, the four nucleotides occur billions of times. It is the particular sequence of nucleotides that makes each person unique.

> To identify perpetrators, forensic scientists examine the specific sequence of nucleotide bases along one strand of a person's DNA.

hair is human or not, show a person's ancestry (because of ethnic differences in hair texture), whether the hair has been dyed, cut in a certain way, or pulled out, and where on the body it came from. Hair samples can exclude a suspect, but not positively identify one.

By contrast, each person's DNA is unique. To understand how DNA varies from person to person, consider its structure. DNA is made up of two strands of subunits linked together in long chains. Each subunit—called a **nucleotide**—has three parts: a sugar, a phosphate, and a base. The phosphate group of one nucleotide binds to the sugar group of the next nucleotide to form a chain of interlinked nucleotides. The two strands of linked nucleotides pair up and twist around each other to form a spiral-shaped **double helix.** The sugars and phosphates form the outside "backbone" of the helix and the bases form the internal "rungs," like steps on a twisting ladder. The bases are held together by hydrogen bonds.

The nucleotide rungs, made up of the bases, are most useful in DNA profiling. There are four different possible nucleotide bases: adenine (A), thymine (T), guanine (G), and cytosine (C). In DNA, these four nucleotide bases are repeated over and over, billions of times, in different orders along a DNA strand. The order of nucleotide bases is a key form of genetic information in cells—it provides the instructions for making proteins (see Chapter 9). To identify perpetrators, forensic scientists examine the specific sequence of nucleotide bases along one strand of a person's DNA—the precise order of As, Ts, Gs, and Cs. With the exception of identical twins, no two people share exactly the same order of DNA nucleotides (INFOGRAPHIC 7.2).

BROWN GETS A BREAK

While Brown sat in prison, he never stopped trying to prove his innocence. For 11 years, he petitioned for an appeal and requested specifically that DNA tests be performed on evidence collected at the crime scene. On each occasion, the judge denied his request. Finally, in 2003, Brown filed a Freedom of Information Act request from prison to obtain copies of all documents relating to his case. That's when he made a surprising discovery.

The additional evidence included four affidavits collected by the Cayuga County Sheriff's Department the day after the murder—documents that neither Brown nor his lawyers had ever seen. In the affidavits, four people described the suspicious behavior of another man: Barry Bench. Bench was the brother of Kulakowski's former boyfriend. The Bench family owned the farmhouse in which Kulakowski had been living.

The affidavits, which included sworn testimony from neighbors as well as from Bench's then girlfriend, Tamara Heisner, stated that on the day of Kulakowski's murder, Bench argued with Heisner, went to a local bar, and returned home between 1:30 and 1:45 A.M.—the same time the victim's neighbors alerted the fire department that the farmhouse was ablaze.

The statements further noted that Bench, who came home highly intoxicated, had left the bar at approximately 12:30 A.M. That left 60 to 75 minutes unaccounted for until he arrived home—although he lived only a mile from the bar. When Bench came home, he immediately went inside to "wash up," according to Heisner.

Brown realized that Bench would have had to drive by the farmhouse to get home from the bar and thought it strange that Bench

▶NUCLEOTIDES
The building blocks of DNA. Each nucleotide consists of a sugar, a phosphate, and a base. The sequence of nucleotides (As, Cs, Gs, Ts) along a DNA strand is unique to each person.

▶DOUBLE HELIX
The spiral structure formed by two strands of DNA nucleotides bound together.

▶DNA PROFILE
A visual representation of a person's unique DNA sequence.

▶DNA REPLICATION
The natural process by which cells make an identical copy of a DNA molecule.

▶COMPLEMENTARY
Fitting together; two strands of DNA are said to be complementary in that A always pairs with T, and G always pairs with C.

would not have noticed the raging fire on his own property. While not conclusive, this new evidence was enough to prompt Brown's lawyers to contact the Innocence Project for help.

Meanwhile, Brown decided to write Bench a letter detailing what he had found and urging him to confess. He warned him of his intent to obtain a DNA test on evidence from the murder. "Judges can be fooled and juries make mistakes," he wrote, "[but] when it comes to DNA testing there's no mistakes. DNA is God's creation and God makes no mistakes."

Five days after Brown mailed his letter, Bench threw himself in front of an Amtrak train and died instantly.

Soon after, the Innocence Project team took on Brown's case and filed a motion to have Kulakowski's nightshirt tested for DNA at a New York State crime lab. The nightshirt

had been found in some tall grass near the body. It was not only bloodstained, it was also stained with saliva. Since both saliva and blood contain cells that carry DNA, scientists could chemically extract the DNA from the cells to create a **DNA profile** of the perpetrator.

MAKING MORE DNA

In principle, creating a DNA profile is simple enough, but there is a huge practical hurdle: having enough crime scene DNA to analyze. Although all body fluids and materials contain cells that house our DNA, the amount left at crime scenes is often very small. Without some way to increase the amount of DNA in a saliva stain, for example, DNA would be useless as evidence.

Forensic scientists solve this problem by using a method to increase, or amplify, the

INFOGRAPHIC 7.3 DNA STRUCTURE PROVIDES A MECHANISM FOR DNA REPLICATION

When cells reproduce, they must first replicate their DNA so that each new cell contains a copy of the DNA molecule. Complementary base pairing between DNA strands guides replication of two new strands.

Hydrogen bonds holding nucleotide base pairs together are broken and the helix is unwound.

An enzyme called DNA polymerase adds complementary nucleotides (green) to the new strand using the rules of base pairing, A with T and G with C.

DNA polymerase

New nucleotides (green) are complementary to old nucleotides (blue).

Semiconservative replication:

After replication, two copies of the original DNA exist. Each molecule consists of an original and a new strand.

Kary Mullis receiving a Nobel prize (in 1993) for developing the polymerase chain reaction.

amount of DNA in a sample. The method takes advantage of the way cells normally make more DNA in the process of **DNA replication.** DNA replication occurs throughout our lives whenever cells reproduce. Because each cell comes from the division of a pre-existing cell (see Chapter 3), the DNA of the parent cell must be replicated so that there is one copy for each daughter cell. The process of DNA replication is essentially the same in all organisms.

To understand how DNA replication works, note that the two strands of nucleotides in a DNA helix do not pair up randomly, but in a consistent pattern: A pairs with T, and G pairs with C. These nucleotides pair preferentially because they are the right shape to form stable hydrogen bonds with each other. Because of this patterned pairing, the two strands are said to be **complementary,** meaning that they fit together like pieces of a puzzle (see **Milestones in Biology: The Model Makers**).

During DNA replication, each strand of DNA serves as a template for the creation of a new complementary strand. The new complementary strand will have bases complementary to the original strand, based on the above "rules" of base pairing. For example, if there is an A on the template strand, a T will be added to the complementary strand being formed.

The steps of replication happen in a precise order: First, the helix is unwound and the two strands "unzip" from each other. Then, an enzyme called **DNA polymerase** builds a new strand of DNA along each unzipped strand: free nucleotides floating inside the cell's nucleus are added to each new strand in a sequence that is complementary to the nucleotide sequence on the original template strand, A pairing with T and C with G. The end result is two complete double-stranded molecules of DNA. Because each replicated DNA molecule is made up of one original and one new strand, DNA replication is said to be **semiconservative** (INFOGRAPHIC 7.3).

DNA replication is a remarkably accurate process that happens at mind-boggling speeds,

the polymerase enzyme adding about 1,000 nucleotides per second and rarely making a mistake. On a human scale, that's like a car speeding down the highway at 300 miles per hour, weaving in and out of traffic without hitting any other cars.

Forensic scientists make use of this remarkable cellular process when they need to amplify the DNA in a crime scene sample. The method they use, called the **polymerase chain reaction (PCR),** is similar to DNA replication that occurs naturally–but it takes place in a test tube.

PCR was the brainchild of Kary Mullis, a 39-year-old chemist and amateur surfer who was working for a biotech company in the San Francisco Bay Area in 1983 when he developed the method. Mullis has said that the idea for PCR came to him as "a single lightning bolt"

▶**DNA POLYMERASE**
An enzyme that "reads" the sequence of a DNA strand and helps to add complementary nucleotides to form a new strand during DNA replication.

▶**SEMI-CONSERVATIVE**
DNA replication is said to be semiconservative because each newly made DNA molecule has one original and one new strand of DNA.

▶**POLYMERASE CHAIN REACTION (PCR)**
A laboratory technique used to replicate, and thus amplify, a specific DNA segment.

From a starting sample of just a few molecules of DNA, PCR can make billions of copies.

while driving to his cabin in Mendocino late one Friday night, his girlfriend asleep in the passenger seat next to him. The technology has so transformed the field of biology—not to mention forensics—that Mullis was awarded a Nobel prize in 1993.

Here's how it works: To a small sample of DNA, scientists add nucleotides, the DNA polymerase enzyme, and primers—short segments of DNA that act as guideposts and flag the section to which DNA polymerase should bind to begin replication. The DNA is first heated to separate the strands, and then cooled to allow new nucleotides to be added. From a starting sample of just a few DNA molecules, PCR can make billions of copies of a specific region of the DNA in less than a few hours (INFOGRAPHIC 7.4).

INFOGRAPHIC 7.4 THE POLYMERASE CHAIN REACTION AMPLIFIES SMALL AMOUNTS OF DNA

The polymerase chain reaction (PCR) is similar to naturally occurring DNA replication, except that it occurs in a test tube and only replicates specific regions of a DNA molecule. From a starting sample of just a few molecules of DNA, PCR can make billions of copies.

DNA from a sample
Nucleotides A, T, G, C
DNA polymerase
Primers

DNA region to be replicated

Round 1

Round 2

Round 3

> 1 billion copies (Round 30)

During each round of PCR:
The two nucleotide strands separate, and each strand serves as a template for the addition of nucleotides according to base-pairing rules: A with T, G with C.

1. Heating separates DNA strands.

2. Cooling allows DNA polymerase to pair new nucleotides with the original template strands.

Each person's DNA sequence is unique and can be used as a kind of molecular "fingerprint."

DNA FROM THE CRIME SCENE

In 2006, a New York State crime lab used PCR to amplify DNA from various items of evidence collected by Cayuga County law enforcement officials during their original investigation of Kulakowski's homicide. The evidence included remnants of cotton swabs used to sample bite marks on the victim; the nightshirt stained with saliva and blood; fingernail clippings; and vaginal swabs from the victim.

To extract DNA from a forensic sample, scientists typically use chemicals to separate cells from other material, like fabric. The specific type of chemical used depends on the starting material. Then, a device known as a centrifuge, which spins samples at high speeds to separate materials, is used in combination with other chemicals help to further extract DNA from cells. DNA extraction is usually the most painstaking step of the process because it can be difficult to obtain sufficient cells in a forensic sample to yield enough DNA for PCR. Also, improperly stored samples can degrade

too much to be useful. Samples can also become contaminated with foreign DNA from improper handling, which would render results useless.

In Brown's case, the laboratory's first report on the stained nightshirt was disappointing. Technicians hadn't been able to obtain any DNA from the bite-mark swab. The lab's second report was more conclusive: seven different pieces of the victim's nightshirt contained DNA.

Moreover, the report went on to state that six of the pieces contained DNA from two different people, the victim and another person who was male.

DNA PROFILING: HOW IT WORKS

Once DNA from crime scene evidence is obtained, the next step is to analyze it. Human cells contain vast amounts of DNA–there are on the order of 3 billion nucleotide base pairs contained within the full stretch of the human **genome.** Figuring out the sequence of every nucleotide in the genome would be extremely

▶GENOME
One complete set of genetic instructions encoded in the DNA of an organism.

▶**SHORT TANDEM REPEATS (STRs)**
Sections of a chromosome in which DNA sequences are repeated.

▶**GEL ELECTROPHORESIS**
A laboratory technique that separates fragments of DNA by size.

Human cells contain vast amounts of DNA—there are on the order of 3 billion nucleotide base pairs contained within the full stretch of the human genome.

INFOGRAPHIC 7.5 DNA PROFILING USES SHORT TANDEM REPEATS

 No two people have the same exact nucleotide sequence. The specific regions of DNA that forensic scientists analyze are those that contain short tandem repeats (STRs). STRs are short stretches of repeated DNA sequences. People differ in the number of copies of an STR sequence found along their chromosomes.

This person has the STR sequence repeated six times on each chromosome.

This person has the STR sequence repeated two times on one chromosome and eight times on the other.

time consuming and expensive. So instead, forensic scientists use a shortcut—they employ PCR to amplify specific segments of DNA and analyze just these segments. These segments are known as short tandem repeats.

Short tandem repeats (STRs) are blocks of repeated DNA sequences found at points along our chromosomes. These sequences are noncoding: they do not contain instructions for making proteins. They are a bit like nonsense words in our DNA: the sequence AGCT repeated over and over again, for example. While all of us have STRs in the same places along our chromosomes, the exact length of each STR varies from person to person. At a single STR site, one person may carry the AGCT sequence repeated six times while another person might carry the AGCT sequence repeated four times. Also, since we inherit two copies of every chromosome, every person has two copies of each STR, and they can be of two different lengths—four repeats of AGCT on one chromosome and six repeats of AGCT on the other chromosome, for example. Forensic scientists use these differences in STR lengths to distinguish between individuals (INFOGRAPHIC 7.5).

To create a DNA profile, scientists first employ PCR to increase the amount of DNA at multiple STR regions. Then they use a method called **gel electrophoresis** to separate the replicated STRs according to length. Shorter STRs—those with fewer numbers of repeats—are smaller and travel farther in the gel; longer ones do not travel as far. When visualized with a detection technique known as fluorescence (the same kind of light released from glowing jellyfish), the separated segments of DNA create a specific pattern of bands that is unique to each person. This pattern is a person's DNA profile. Scientists can then compare band patterns of DNA from a crime scene to DNA from a suspect. DNA profiles have other applications, too—paternity or ancestry testing, for example (INFOGRAPHIC 7.6).

INFOGRAPHIC 7.6 CREATING A DNA PROFILE

1. Collect cells from crime scene evidence and extract DNA.

Cheek cell in saliva

STR region 1   STR region 2   STR region 3

2. Amplify multiple STR regions by PCR.

STRs from saliva sample   STRs from suspect   A   B   C

3. Separate STRs by gel electrophoresis.

PCR products are inserted into a gel. An electric current applied to the gel causes polar DNA to migrate through it. Shorter fragments travel farther; longer fragments remain near the top.

4. Compare STR banding patterns.

Different STR lengths result in two bands.   Identical STR lengths result in a single band.

The gel shows the results of three different STR regions (green, red, and blue bands) analyzed from the DNA in a crime scene sample and in three suspects.

Saliva sample   Suspect A   Suspect B   Suspect C (a perfect match)

INFOGRAPHIC 7.7 DNA PROFILING USES MANY DIFFERENT STRs

➡ To create a DNA profile, scientists analyze 15 different STRs (yellow boxes) scattered among our chromosomes. Sharing the same number of repeats at any particular STR is relatively common — typically 5% to 20% of people share the same pattern at any one STR site. But it is the combined pattern of STR repeats at multiple sites that is unique to a person; the more STRs tested, the more discriminating the test becomes.

Location of each STR region on a chromosome

Each STR region has a unique name and chromosome location.

TPOX
D3S1358
D5S818
FGA
CSF1PO
D7S820
D8S1179
TH01
VWA

Chromosome no. 1 2 3 4 5 6 7 8 9 10 11 12

D13S317
D16S539
D18S51
D21S11
AMELX
AMELY

13 14 15 16 17 18 19 20 21 22 X Y

Assuming people share the same pattern at each STR:

Analysis	Calculation	People sharing this profile
1 STR region	0.2	1 in 5
2 STR regions	0.2 x 0.2 = 0.04	1 in 25
5 STR regions	0.2 x 0.2 x 0.2 x 0.2 x 0.2 = 0.00032	1 in 3,125
15 STR regions	0.2 x 0.2 x 0.2 x 0.2 x 0.2 x 0.2 x 0.2 x 0.2 x 0.2 x 0.2 x 0.2 x 0.2 x 0.2 x 0.2 x 0.2 = 3.3×10^{-11}	1 in several quintillion

> ❝ *Judges can be fooled and juries make mistakes, [but] when it comes to DNA testing there's no mistakes.* ❞
>
> — ROY BROWN

DNA PROFILING AND THE LAW

Since 1994, the federal government has been collecting DNA profiles of offenders in the Combined DNA Index System (CODIS), a computer database that contains more than 10 million profiles from criminals convicted of specific crimes in all 50 states. Each profile consists of a banding pattern that represents 15 specific STR regions scattered throughout our genomes. Forensic scientists typically describe the likelihood that any two unrelated people will have the same number of repeated sequences at all 15 regions as 1 in some number of quintillions (billions of billions) (INFOGRAPHIC 7.7).

So far, the database of DNA profiles has proved helpful in more than 190,000 cases. More significantly, DNA evidence is helping to change the criminal justice system for the better. That more than 300 prisoners have already been exonerated by the Innocence Project suggests that many more may have been wrongly convicted but lack the evidence to support their cases. In the majority of criminal cases, there is no DNA evidence.

"How many more wrongful convictions will it take for New York to begin addressing

Roy Brown with his family upon his release from prison.

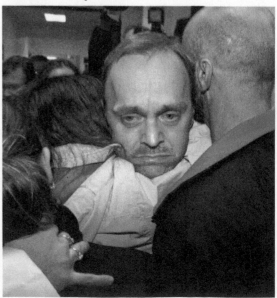

the systemic problems that lead to such miscarriages of justice?" asked Neufeld in 2007, when Brown's case was being reviewed.

Recognizing the flaws in our criminal justice system, Innocence Project lawyers are working with several states to change the way law enforcement operates. For example, studies have shown that witnesses more accurately identify perpetrators if they are shown suspects one at a time instead of in a group line-up. Project lawyers are also helping to force changes in the way interrogations are conducted, calling for them to be videotaped to reduce the possibility of forced confessions. In addition, they are lobbying for legislation to ensure that evidence from crime scenes is properly collected and maintained, since DNA evidence can be ruined or contaminated during collection. They also advocate that anyone convicted of a crime be able to gain access to DNA testing.

"The key is that DNA really gives us an opportunity to start making the other institutions in the system more scientific and reliable as well," says Neufeld.

VINDICATION

The DNA that the New York State crime lab extracted from the victim's nightshirt contained a mixture of DNA from the victim and another person who was male. Analysis showed that this male DNA, however, did not match Roy Brown's. DNA evidence excluded him as Kulakowski's murderer.

Additional testing eventually linked that DNA evidence to Barry Bench. After Bench's suicide, of course, he couldn't provide DNA directly. So lawyers pursued the next best option: a DNA sample voluntarily donated by Bench's biological daughter, Katherine Eckstadt. Because we all receive one set of

Roy Brown with Innocence Project lawyers Nina Morrison and Peter Neufeld after his case was revisited in light of DNA evidence.

chromosomes from our mother and one set from our father, half of Katherine Eckstadt's DNA would have come from her father, and therefore would show great similarity to his. The test yielded dramatic results—a 99.99% probability that the man who deposited his saliva on Sabina Kulakowski's nightshirt was Eckstadt's father, Barry Bench.

To clinch the case, Cayuga County prosecutors eventually agreed to have Bench's body exhumed for DNA tests—which matched the DNA from the saliva stains.

"We've had a lot of crazy cases," says Nina Morrison, the Innocence Project attorney who handled Brown's case, "but this is really up there with the best of them . . . the client solving his own case . . . it's insane." Brown was cleared of all charges and is now putting his life back together. ■

CHAPTER 7 Summary

▸ DNA is the hereditary molecule of all living organisms. DNA contains instructions for building an organism.

▸ DNA sequences determine the genetic uniqueness and relatedness of individuals.

▸ The DNA in a eukaryotic cell is packaged into chromosomes located in the nucleus.

▸ Humans have 23 pairs of chromosomes in their cells—one chromosome of each pair inherited from the mother, the other from the father.

▸ DNA is a double-stranded molecule that forms a spiral structure known as a double helix.

▸ Each strand of DNA is made of nucleotides bonded together in a linear sequence.

▸ There are four distinct nucleotides: adenine (A), thymine (T), guanine (G), and cytosine (C).

▸ The two linear strands of a DNA molecule are bound together by complementary pairing of A with T and G with C.

▸ Complementary pairing of DNA strands guides DNA replication, a fundamental part of cell reproduction.

▸ PCR enables scientists to vastly increase the number of copies of specific DNA sequences.

▸ Forensic scientists use noncoding DNA sequences known as STRs to create a DNA profile.

▸ STRs are blocks of repeated sequences of DNA. People differ in the number of times the sequences are repeated along their chromosomes.

▸ A DNA profile is more accurate and reliable than many other forms of evidence.

MORE TO EXPLORE

▸ **The Innocence Project** http://www.innocenceproject.org/

▸ **National Human Genome Research Institute** http://www.genome.gov

▸ **Scheck B., et al. (2003)** *Actual Innocence: When Justice Goes Wrong and How to Make It Right.* New York: NAL Trade Paperbacks.

▸ **NOVA, Forensics on Trial, October 17, 2012** http://www.pbs.org/wgbh/nova/tech/forensics-on-trial.html

▸ **Mullis, K. (1983) The Polymerase Chain Reaction. Nobel Lecture.** www.nobelprize.org/nobel_prizes/chemistry/laureates/1993/mullis-lecture.html

CHAPTER 7 Test Your Knowledge

DRIVING QUESTION 1

What is the structure of DNA, and how is DNA organized in cells?

By answering the questions below and studying Infographics 7.1 and 7.2, you should be able to generate an answer for the broader Driving Question above.

KNOW IT

1 Which of the following is *not* a nucleotide found in DNA?

a. adenine (A)   c. cytosine (C)   e. uracil (U)
b. thymine (T)   d. guanine (G)

2 If the sequence of one strand of DNA is AGTCTAGC, what is the sequence of the complementary strand?

a. AGTCTAGC   d. GTCGACGC
b. CGATCTGA   e. GCTAGACT
c. TCAGATCG

3 In addition to the base, what are the other components of a nucleotide?

a. sugar and polymerase
b. phosphate and sugar
c. phosphate and polymerase
d. phosphate and helix
e. helix and sugar

④ The_____chromosomes in a human cell from inside the cheek are found in the_____.

a. 46; cytoplasm

b. 23; nucleus

c. 24; cytoplasm

d. 46; nucleus

e. 22; nucleus

⑤ Each chromosome contains

a. DNA only.

b. proteins only.

c. DNA and proteins.

d. the same number of genes and STRs.

e. the entire genome of a cell.

USE IT

⑥ You can detect DNA that is specifically from the X chromosome in a DNA sample from a person. Can you definitively determine the sex of that person (male or female) from the presence of the X chromosome? Explain your answer.

⑦ Human red blood cells are enucleated (that is, they do not have nuclei). Is it possible to isolate DNA from red blood cells? Why or why not?

DRIVING QUESTION 2

How is DNA copied in living cells, and how can DNA be amplified for forensics?

By answering the questions below and studying Infographics 7.3 and 7.4, you should be able to generate an answer for the broader Driving Question above.

KNOW IT

⑧ DNA replication is said to be semiconservative because a newly replicated, double-stranded DNA molecules consists of

a. two old strands.

b. two new strands.

c. one old strand and one new strand.

d. two strands, each with a mixture of old and new DNA.

e. any of the above, depending on the cell type

⑨ Which of the following statements about PCR is true?

a. DNA polymerase is the enzyme that copies DNA in PCR.

b. Primers are not necessary for PCR.

c. PCR does not require nucleotides.

d. PCR does not generate a complementary DNA strand.

e. PCR can make only a few copies of a DNA molecule.

USE IT

⑩ Complete the statements below, and then number them to indicate the order of these two major steps necessary to copy a DNA sequence during PCR.

_____The enzyme _____ "reads" each template strand and adds complementary nucleotides to make a new strand.

_____The two original strands of the DNA molecule can be separated by _____.

⑪ Given this segment of a double-stranded DNA molecule, draw the two major steps involved in DNA replication:

ATCGGCTAGCTACGGCTATTTACGGCATAT
TAGCCGATCGATGCCGATAAATGCCGTATA

DRIVING QUESTION 3

How does DNA profiling make use of genetic variation in DNA sequences?

By answering the questions below and studying Infographics 7.5, 7.6, and 7.7, you should be able to generate an answer for the broader Driving Question above.

KNOW IT

⑫ Which STR will have migrated farthest through an electrophoresis gel?

a. GAAG repeated twice

b. GAAG repeated three times

c. AGCT repeated five times

d. GAAG repeated seven times

e. AGCT repeated seven times

⑬ An individual's STR may vary from the same STR of another individual by

a. the order of nucleotides.

b. the specific bases present.

c. the specific chromosomal location of the STR.

d. the number of times a particular sequence is repeated.

e. the number of coding regions.

⑭ Which of the following represents genetic variation between individuals?

a. whether or not G pairs with C or T

b. the presence of STRs in their genomes

c. the number of chromosomes in the nucleus

d. the sequence of nucleotides along the length of each chromosome

e. the number of chromosomes received from each parent

⑮ A person has an STR with the sequence GACCT repeated six times on one chromosome and eight times on the other chromosome. If this STR were

7

amplified by PCR, and the PCR products run on a gel, which lane would show the corresponding banding pattern (see gel at top right)? The marker lane (M) has fragments starting at 10 nucleotides (at the bottom) and increasing in 10-nucleotide increments.

USE IT

16 A series of statements is presented below. Mark each statement as true (T) or false (F).

a. _____ G pairs with T.

b. _____ Genetic information is passed on to the next generation in the form of DNA molecules.

c. _____ All DNA sequences encode information to produce proteins.

d. _____ Each person carries the same number of STR repeats on both maternal and paternal chromosomes.

e. _____ DNA evidence can be obtained from saliva left in a bite mark.

17 Explain why the statements that you marked as true in Question 16 are in fact true.

18 Rewrite the statements that you marked as false in Question 16 to make them true.

DRIVING QUESTION 4

How does DNA evidence fit into forensic investigations?

By answering the questions below and studying Infographics 7.1, 7.5, 7.6, and 7.7, you should be able to generate an answer for the broader Driving Question above.

INTERPRETING DATA

23 The table shows the frequencies for STR lengths (repeat number) in different U.S. populations. Since we inherit two copies of every chromosome, every person has two copies of each STR. You can determine the probability of a particular combination of STRs by multiplying the frequencies.

If the STR lengths on each chromosome are identical, the probability of this pattern can be determined by multiplying the frequency by itself. For example, the probability of a Hispanic person having 14 D3S1358 repeats (0.079) on both chromosomes is 0.079 x 0.079 = 0.0062

If the STR lengths on each chromosome are different, the probability of this pattern can be determined by multiplying the two frequencies together. Then, because there are two ways of inheriting this pattern (getting a 14 from dad and an 18 from mom or getting the 18 from dad and the 14 from mom), this number must be multiplied by 2.

For example, the probability of a Hispanic person having 14 D3S1358 repeats (0.079) AND 18 D3S1358 repeats (0.125) is 0.079 x 0.125 x 2 = 0.0198.

a. What is the probability of a Caucasian American's having a 16, 17 combination for D3S1358?

b. What is the probability of an African American's having a 16, 17 combination for D3S1358?

c. What is the probability of an Hispanic American's having a 16, 17 combination for D3S1358?

d. What is the probability of a Caucasian American's having a 16, 17 combination for D3S1358, and a 5, 9 combination for TH01?

e. What is the probability of a Caucasian American's having a 16, 17 combination for D3S1358, a 5, 9 combination for TH01 and an 11, 14 combination for D18S51?

f. Consider your answers to a, d, and e. Why does forensic analysis use many CODIS markers (and not just one or two)?

CODIS* STR	NO. OF REPEATS	FREQUENCY (CAUCASIAN)	FREQUENCY (AFRICAN AMERICAN)	FREQUENCY (HISPANIC)
D3S1358	14	0.094	0.089	0.079
	15	0.111	0.186	0.293
	16	0.200	0.248	0.286
	17	0.281	0.242	0.204
	18	0.200	0.155	0.125
TH01	5	0.002	0.004	0
	6	0.232	0.124	0.214
	7	0.190	0.421	0.096
	8	0.084	0.194	0.096
	9	0.114	0.151	0.150
D18S51	10	0.008	0.006	0.004
	11	0.017	0.002	0.011
	12	0.127	0.078	0.118
	13	0.132	0.053	0.111
	14	0.137	0.072	0.139

Data from Butler et al. (2003) Allele frequencies for 15 autosomal STR loci on U.S. Caucasian, African American and Hispanic populations. Journal of Forensic Sciences 48(4): 908–911.

*CODIS stands for "Combined DNA Index System," a government database of DNA profiles from offenders, crime scenes, and missing persons.

KNOW IT

19 The gel at right shows the DNA profile of STRs from four sources: blood from crime scene evidence (E), suspect A, suspect B, and the victim (V). An eyewitness identified suspect A as fleeing the apartment building where the crime occurred. Suspect B was picked up at a local convenience store after using bloodstained money.

E   A   B   V

a. From the DNA profiles shown, can you draw any conclusions about where the crime scene DNA came from?

b. Can you draw any conclusions about relationships among the people profiled? Explain your reasoning.

USE IT

20 Look at Infographic 7.7. From the STRs used in forensic investigations, which STRs on which chromosomes would be particularly useful in determining whether crime scene evidence was left by a female or a male?

21 Explain your response to Question 20, stating the number of STR copies you would expect to see if the perpetrator was female and if the perpetrator was male.

22 The gel below shows a DNA profile using five STRs. The lane labeled W is a mother and the lane labeled C is her child. The lanes labeled M1 and M2 are two men, either of whom, according to the mother, could be the father of the child.

W   C   M1   M2

a. Circle the STR bands that the child (C) inherited from its mother (W).

b. Use the DNA profiles to determine which man is the father of the child.

MINI CASE

24 A female eyewitness has identified a Hispanic American male as the man who stole her car. The eyewitness stated that the man was bleeding profusely from a head wound. Her car was recovered, and male blood with a 16, 17 combination for D3S1358, a 5, 9 combination for TH01, and an 11, 14 combination for D18S51 was found on the driver's seat and steering wheel. Does this finding call the eyewitness evidence into question? Explain your answer.

BRING IT HOME

25 Scientists used DNA from Barry Bench's daughter to pinpoint Bench as a possible suspect, as his DNA was not on file anywhere. Similarly, in cases of disasters, DNA evidence is sometimes required to identify victims. If a victim doesn't have a DNA profile on file, identity must be reconstructed by comparing the victim's DNA profile to that of relatives. These situations illustrate that a DNA profile database has the potential to be useful in cases in which DNA-based identification is required. However, maintenance of such a database is controversial. What arguments can you make for and against banking DNA profiles in a database? If such a database existed, what restrictions would you place on it? Would you choose to register your DNA or your child's DNA in such a database?

THE
MODEL MAKERS

Watson, Crick, and the structure of DNA

IN 1953, WITH INFLATED EGOS, JAMES WATSON AND FRANCIS Crick announced to a crowd in their favorite pub in Cambridge, England, that they had found the secret of life. Given the nature of their discovery, they had every right to boast: in revealing the structure of DNA, they had solved one of the greatest mysteries of science—the chemical basis of inheritance. Their discovery would revolutionize biology and push forward the study of anthropology, evolution, and medicine.

But Watson and Crick's success wasn't merely the result of a marriage between two great minds. Scientific breakthroughs rarely result from single scientists working in a vacuum. Rather, breakthroughs happen when scientists are well positioned to build upon foundations laid down by other scientists. And so it was with DNA. In addition to their own insight, Watson and Crick built on the discoveries of others. They also had luck on their side—they were in the right place at the right time.

Watson and Crick had met in 1951 at Cambridge University, in England. Watson was an American scientist who had just accepted a research position at the university. Crick was then a Ph.D. student, studying protein structure with a technique called x-ray crystallography.

Given their varied backgrounds, the men didn't appear obvious collaborators. Watson was a prodigy. Twelve years younger than Crick, he had earned his Ph.D. at 22. Crick, by contrast, was a late bloomer. He was 38 years old by the time he had his Ph.D. But what they did share was intellectual curiosity. Both had changed their research focus several times. By their

▸▸ DRIVING QUESTIONS

1. Who were the major players in the discovery of the structure of DNA?
2. What pieces of scientific knowledge were assembled in order to elucidate the structure of DNA?

INFOGRAPHIC M2.1 THE DNA PUZZLE

It was known that DNA was made of nucleotides that included a deoxyribose sugar, a phosphate group, and one of four nitrogenous bases. But no one had yet figured out how the nucleotides fit together to produce a DNA molecule.

nucleotides containing a sugar, a phosphate group, and one of four bases: adenine, thymine, cytosine, or guanine. But how did the elements fit together? To answer this question, they took inspiration from the chemist Linus Pauling. Pauling had been studying the structure of proteins and had built a molecular model showing that some proteins exist as a single-stranded, twisting helix. He backed up his model with lab experiments to prove his structure was correct. If an eminent scientist like Pauling could model a structure without first conducting laboratory experiments, Watson and Crick thought they might be able to do the same with DNA (**INFOGRAPHIC M2.1**).

Using wire and metal, Watson and Crick began building scale models of DNA on the basis of existing evidence about the chemical structure of nucleotides. They initially built a three-helix model with the phosphate groups on the inside and the bases radiating outward. But experts who analyzed the structure deemed it chemically unstable.

Then came a crucial finding. In 1951, Watson attended a lecture by a young scientist named Rosalind Franklin. In her laboratory at King's College London, she had been making x-ray diffraction pictures of DNA. X-ray diffraction analyzes the way x-rays bounce off a sample of material in order to determine the chemical structure of that material. Franklin had observed that increasing the humidity of a DNA sample caused it to elongate. To Franklin, this suggested that water molecules were being attracted to the helix, coating it and causing it to stretch out. And, since water is a polar molecule (see Chapter 2), attracted to charged (hydrophilic) molecules, this further suggested that the charged, water-loving phosphate groups of DNA must therefore be on the *outside* of the helix—not, as others had suggested, on the inside.

Franklin's contribution didn't end there. She also discovered other important facts

own admission, both were more interested in solving current hot topics in science—like the structure of DNA—than pursuing the more obscure science that each had trained to do.

Although DNA was first observed in cell nuclei in the late 1860s, it took almost a century before scientists realized its importance. For a long time, the prevailing belief was that proteins carried the genetic information. But by the 1950s it was known that DNA, not protein, was the genetic material. Scientists still, however, didn't understand the structure of DNA, or how it carried genetic information. Solving this problem became a quest for many scientists of the day.

When Crick and Watson came to the problem, they already knew, from the work of other scientists, that DNA was made up of

INFOGRAPHIC M2.2 ROSALIND FRANKLIN AND THE SHAPE OF DNA

Franklin's 1951 x-ray diffraction studies of DNA showed that the structure was likely helical, involving two strands that run in opposite directions, and that the phosphate groups were on the outside of the molecule.

X-ray diffraction images of DNA

X-ray diffraction involves shooting x-rays at a crystallized version of the molecule and recording on film how the x-rays scatter when they bounce off its surface. The imagse shown here are views from the end of a DNA molecule looking down its center.

Rosalind Franklin
July 25, 1920–April 16, 1958

Dry DNA form

Wet DNA form

Humidity

The signature "X" in the middle of this picture suggests a double-stranded helical structure.

The symmetry of the image suggests that the molecule is uniform in width.

DNA elongates in presence of water, allowing the molecule to take on a regular helical shape, shown by the clarity of the image on the right. This evidence suggests that the water-loving phosphate groups are on the outside of the DNA molecule.

about the structure of DNA. Working with a graduate student, Raymond Gosling, she found that her x-ray diffractions confirmed that the elongated form of DNA had all the characteristics of a twisting helix (**INFOGRAPHIC M2.2**).

Maurice Wilkins, who was Franklin's peer and worked in the same laboratory, was also studying DNA structure at the time. In 1953, Wilkins saw Franklin's best unpublished x-ray picture of DNA and showed it to Watson without Franklin's knowledge. "The instant I saw the picture my mouth fell open," Watson recalled in his memoir of the discovery, *The Double Helix*, published in 1968. The sneak preview "gave several of the vital helical parameters."

With that clue in hand, Watson and Crick then took a crucial conceptual step and suggested that the molecule was made of two chains of nucleotides. Each formed a helix, as Franklin had found, but because the DNA molecule was symmetrical—it looked the same when flipped upside down and backward—they realized that the two chains of nucleotides must be oriented in opposite directions.

To construct the model, Watson and Crick also built on a discovery made a few months earlier. In 1952, Erwin Chargaff had found that DNA contained equal amounts of adenine and thymine and equal amounts of guanine and cytosine. This information was critical. As the helix had a smooth shape and a uniform thickness, and as the bases had to point toward the inside of the helix, the different-size bases somehow had to fit together in way that allowed for a consistent length of base pairs. Following a tip about the structure of bases, Watson and Crick were able to use a model to show that A-T pairs and G-C pairs were exactly the same length, explaining the consistent helix shape, and also accounting for Chargaff's finding that the amounts of A and T were equal, as were the amounts of G and C (INFOGRAPHIC M2.3).

The final double-helix model so perfectly fit the experimental data that the scientific community almost immediately accepted it. Watson and Crick published their paper on the structure of DNA in April 1953 in the prominent

INFOGRAPHIC M2.3 ERWIN CHARGAFF'S WORK PROVIDED A CLUE TO BASE PAIRING

Erwin Chargaff studied the nitrogenous bases of DNA. He found that no matter which DNA molecule he analyzed, it always contained equal amounts of adenine and thymine bases and equal amounts of cytosine and guanine bases. Additionally, the width of an A–T pair is the same as the width of a C–G pair. These data suggest that in a double-stranded DNA molecule, adenine must pair with thymine and cytosine must pair with guanine. This pairing is consistent with a uniform base-pair width down the length of the DNA molecule, as Franklin's data suggested.

Erwin Chargaff
August 11, 1905–June 20, 2002

Chargaff's Rule of Base Pairing
Within any DNA molecule:
% adenine = % thymine
% cytosine = % guanine

Adenine and thymine always base-pair.

Cytosine and guanine always base-pair.

Uniform base-pair width

Thymine   Adenine

Cytosine   Guanine

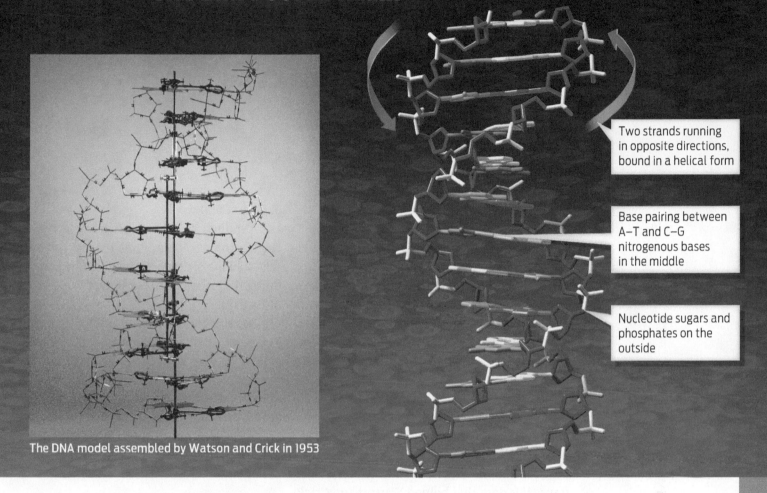

The DNA model assembled by Watson and Crick in 1953

Two strands running in opposite directions, bound in a helical form

Base pairing between A–T and C–G nitrogenous bases in the middle

Nucleotide sugars and phosphates on the outside

journal *Nature*. In it, with considerable understatement, they wrote: "It has not escaped our notice that the specific pairing we have postulated immediately suggests a possible copying mechanism for the genetic material." Indeed, the model that Watson and Crick proposed solved at once both the structure of DNA and how this structure provides a mechanism for DNA replication (see Chapter 7). The mystery of inheritance had finally been unlocked.

Alongside their paper, in the same issue, individual papers by Wilkins and Franklin discussed their respective x-ray diffraction results. In 1962, Watson, Crick, and Wilkins shared the Nobel Prize in Physiology or Medicine (**INFOGRAPHIC M2.4**).

> The final double-helix model so perfectly fit the experimental data that the scientific community almost immediately accepted it.

> **❝ It would have been impossible to give the prize to Maurice [Wilkins] and not to [Franklin because] she did the key experimental work. ❞**
>
> — FRANCIS CRICK

But what about Franklin? Franklin had died of ovarian cancer in 1958, at the age of 37. By Nobel Foundation rules, she was ineligible for nomination since prizes are not awarded posthumously.

Controversy over whether Franklin has been adequately recognized continues. Although Watson and Crick acknowledged her contribution to their research in their article in *Nature,* the extent to which her input helped them build their DNA model was revealed only much later in Watson's 1968 book, published 10 years after Franklin's death. For example, at the time *Nature* published the papers on DNA structure, Franklin's paper was perceived as merely supporting evidence. But it was her data that helped Watson and Crick clinch the structure. Some historians argue that sexist attitudes prevented her from receiving the acclaim she deserved before she died. At the time, female scientists in the biomedical sciences were few and were frequently confronted by negative attitudes from their male peers. "I'm afraid we always used to adopt–let's say, a patronizing attitude towards her," Crick publicly commented after Watson's book was published. He added that if Franklin had lived, "It would have been impossible to give the prize to Maurice [Wilkins] and not to her" because "she did the key experimental work."

Although it was quite normal for colleagues to share data, some have even argued that Wilkins showed Franklin's critical x-ray diffraction photos to Watson out of jealousy or disdain.

Despite controversy, Franklin's contribution to the discovery has never been completely ignored, and she is now recognized as having been a top-notch scientist: her notebooks show that without her thorough scientific research and original ideas, we would have had to wait much longer for what is still considered to be one of the most important discoveries in biology. ■

MORE TO EXPLORE

▸ Watson, J. D., and Crick, F. H. C. (1953) A structure for deoxyribose nucleic acid. *Nature* 171:737–738.

▸ Watson, J. D. (2001 [1968]) *The Double Helix: A Personal Account of the Discovery of the Structure of DNA.* New York: Touchstone Books.

▸ Maddox, B. (2003) *Rosalind Franklin: The Dark Lady of DNA.* New York: Harper Perennial.

1 Which technique did Rosalind Franklin use to examine the structure of DNA?

a. mass spectrometry
b. gel electrophoresis
c. x-ray diffraction
d. model building
e. all of the above

2 Which of the following statements about DNA structure is *not* true?

a. DNA is a double helix.
b. The phosphate groups are on the outside of the helix.
c. The two strands run in opposite orientations.
d. A pairs with A, T with T, C with C, and G with G.
e. The helix has a constant diameter along its length.

3 Describe the scientific contributions of Watson, Crick, Franklin, and Wilkins to the discovery of the structure of DNA.

4 How did differing amounts of water in the DNA crystals help explain the x-ray diffraction patterns?

5 Summarize the structure of a DNA double helix.

8

MEDICINE FROM

Scientists genetically modify animals
to make medicine

MILK

IN A MASSACHUSETTS BARN NESTLED AMONG willow and oak trees, rows of juglike machines drone in a constant hum. Goats, dozens of them, are being milked. But this is no ordinary dairy operation. This farm is one of several worldwide practicing the art of "pharming"– using genetically modified animals to churn out pharmaceutical drugs.

The first drug produced from such genetically modified animals is already available, manufactured by rEVO Biologics, a biotech company based in Framingham, Massachusetts. The drug, called ATryn, consists of a human protein called antithrombin (AT) that was extracted from the goats' milk. ATryn is used to treat patients with inherited antithrombin deficiency, which puts them at risk of developing dangerous blood clots, especially during surgery or childbirth.

For decades, scientists had extracted antithrombin from human blood donations. But blood contains only small amounts of antithrombin, and the supply depends on the number of blood donors. Genetically modified goats,

▶▶ **DRIVING QUESTIONS**

1. What determines the shape of a protein molecule?

2. What are the steps of gene expression, and where do they occur in a cell?

3. How can animals be genetically modified to produce human proteins (with therapeutic uses)?

4. What are some practical applications of genetically modified organisms in treating human disease?

Not your ordinary dairy operation. These goats are being milked to obtain pharmaceutical drugs.

other drugs, for a wide range of diseases, from cancer to hemophilia. The ultimate goal is to make goats the drug factories of the future.

PROFILE OF A PROTEIN

Antithrombin is a **protein.** Recall from Chapter 2 that proteins are one of the four main macromolecules that make up cells. Proteins are the cell's do-it-all molecules. They help us perform countless tasks, everything from contracting our muscles to fighting infection. Some proteins are hormones, important chemical messengers in the body (see Chapter 30); others are enzymes, speeding up the rate of chemical reactions (see Chapter 4); still others act as transporters—for example, the protein hemoglobin transports oxygen throughout the body (see Chapter 28); and the antibodies of our immune system help keep us healthy (see Chapter 31). The list is virtually endless: think of almost any feature of a living organism, and one or more proteins plays a role. Because proteins are so critical to cell function, most drugs are either chemicals that interact with specific proteins or, like antithrombin, are proteins themselves.

All proteins are made of the same building blocks called **amino acids.** There are 20 different amino acids found in proteins. All amino acids have the same basic core structure, but

however, can produce massive amounts of the drug in a relatively short period of time. There is also less risk of transmitting infections such as HIV and hepatitis to healthy people with this method.

Given these and other advantages, some scientists are predicting that genetically modified animals may one day replace human blood donors as a source of therapeutic drugs. "I'm a fan of this technology," says Michael Paidas, co-director of the Yale Women and Children's Center for Blood Disorders, who has conducted some early clinical trials on ATryn. "I think it's going to be a wave of the future for a lot of these difficult-to-get blood products." The Food and Drug Administration (FDA) approved ATryn for use in humans in 2009.

According to Harry Meade, the biologist who conceived the technology, rEVO has about 2,000 goats in its stable, located on a bucolic 180-acre farm in Charlton, Massachusetts. Five hundred of these goats can make enough antithrombin for the whole world, he says.

And antithrombin is just the beginning. The company has plans to produce numerous

▶PROTEIN
A macromolecule made up of repeating subunits known as amino acids, which determine the shape and function of a protein. Proteins play many critical roles in living organisms.

▶AMINO ACIDS
The building blocks of proteins. There are 20 different amino acids.

Biologist Harry Meade first had the idea to make medicine from milk.

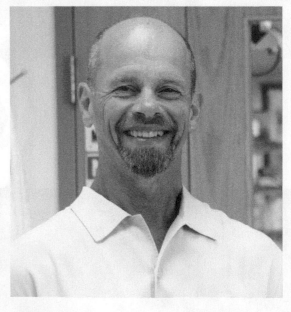

INFOGRAPHIC 8.1 AMINO ACID SEQUENCE DETERMINES PROTEIN SHAPE AND FUNCTION

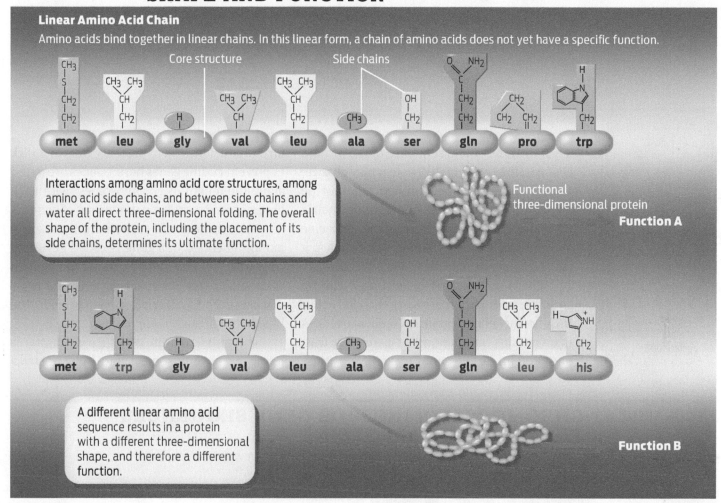

Linear Amino Acid Chain
Amino acids bind together in linear chains. In this linear form, a chain of amino acids does not yet have a specific function.

Core structure

Side chains

met   leu   gly   val   leu   ala   ser   gln   pro   trp

Interactions among amino acid core structures, among amino acid side chains, and between side chains and water all direct three-dimensional folding. The overall shape of the protein, including the placement of its side chains, determines its ultimate function.

Functional three-dimensional protein
Function A

met   trp   gly   val   leu   ala   ser   gln   leu   his

A different linear amino acid sequence results in a protein with a different three-dimensional shape, and therefore a different function.

Function B

each also has a unique chemical side chain that distinguishes one amino acid from another. To form proteins, amino acids bind together in linear chains, much like beads on a string. The human antithrombin protein is a chain of 432 amino acids. Many human proteins are about this size, but chain lengths vary from just a few amino acids to many thousands. The longest human protein, titin (involved in muscle contraction) is a single chain of 34,350 amino acids. The shortest known protein is insulin, which helps regulate blood sugar: it has only 51 amino acids.

In cells, amino acid chains fold into a distinct three-dimensional shape, or conformation. Some proteins, like antithrombin, are made up of just one folded chain. Others, such as the antibodies of our immune system, or the hemoglobin that carries oxygen in our red

blood cells, are made up of more than one folded amino acid chain bound together.

The particular sequence of amino acids in a chain determines how the chain will fold. Interactions between amino acid side chains, and between these side chains and the surrounding water, influence the precise folding pattern. Much like an oil droplet in water, hydrophobic amino acid side chains tend to clump together on the inside of a protein, while hydrophilic amino acids face out toward water, forming a distinct three-dimensional shape. This folded shape is important because shape is what ultimately determines how a protein behaves. In other words, the order of amino acids determines a protein's shape, and the protein's shape determines its specific function (**INFOGRAPHIC 8.1**).

> **❝I think it's going to be a wave of the future for a lot of these difficult-to-get blood products.❞**
>
> —MICHAEL PAIDAS

▶GENE
A sequence of DNA that contains the information to make at least one protein.

▶GENE EXPRESSION
The process of using DNA instructions to make proteins.

▶ALLELES
Alternative versions of the same gene that have different nucleotide sequences.

If the amino acid sequence of the protein changes, the shape of the protein may change. This differently shaped protein may not be able to do its job. For example, if a protein's normal function is to bind to another molecule and the shape of the binding site changes, it may no longer be able to bind to that molecule.

Where do proteins come from? The instructions to make proteins are encoded in our DNA—in our genes. A **gene** is a sequence of DNA that contains instructions for making at least one protein. (A few genes make RNA as their final product instead of proteins, but we'll stick to protein-coding genes here.) Genes are found along the length of chromosomes, with each chromosome carrying a unique set of genes.

The synthesis of a protein from the information encoded in a gene is called **gene expression**. When a cell is making the protein encoded by that gene, the gene is said to be "expressed." In essence, genes provide the blueprint for building proteins.

The antithrombin gene, for example, is found on chromosome 1 and holds instruc-

INFOGRAPHIC 8.2 CHROMOSOMES INCLUDE GENE SEQUENCES THAT CODE FOR PROTEINS

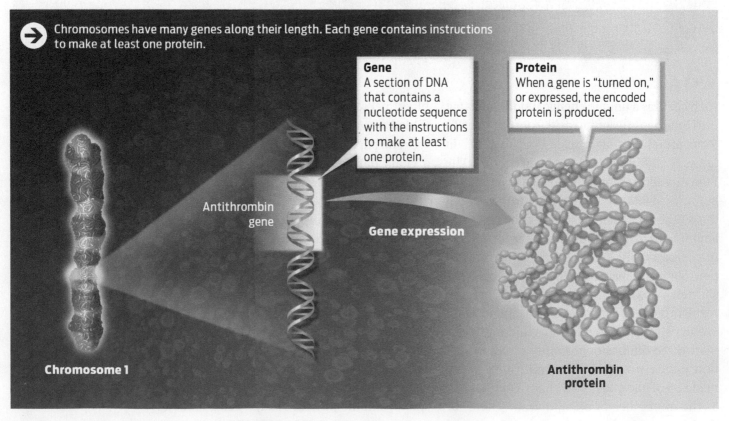

→ Chromosomes have many genes along their length. Each gene contains instructions to make at least one protein.

Gene
A section of DNA that contains a nucleotide sequence with the instructions to make at least one protein.

Protein
When a gene is "turned on," or expressed, the encoded protein is produced.

Antithrombin gene

Gene expression

Chromosome 1

Antithrombin protein

tions to make a chain of 432 amino acids that folds into a three-dimensional protein called antithrombin. When cells express the antithrombin gene, it means they produce the antithrombin protein (**INFOGRAPHIC 8.2**).

In the body, antithrombin protein helps prevent blood from clotting improperly. Antithrombin is produced by cells in the liver and released into the blood. When antithrombin is present, it inactivates enzymes that promote blood clotting. By preventing blood from clotting in the wrong place at the wrong time, antithrombin helps prevent a stroke or a heart attack.

Some people, however, inherit a genetic defect that causes antithrombin deficiency. Inherited antithrombin deficiency isn't rare—about 1 in every 5,000 people in the United States is born with the inability to produce adequate amounts of this protein. People with too little or no functional antithrombin carry a high risk of developing blood clots inside blood vessels, a condition called thrombosis, and consequently they sometimes require antithrombin transfusions to prevent these vessel-blocking clots from forming (**INFOGRAPHIC 8.3**).

When people inherit antithrombin deficiency, it doesn't mean they lack the antithrombin gene. Rather, it means that one or both copies of their antithrombin gene are defective. Remember that most human cells have two copies of every gene (see Chapter 7). These copies can either be identical in nucleotide sequence, or slightly different. Different versions of the same gene, with different nucleotide sequences, are called **alleles**. All genes, including antithrombin, have multiple such alleles.

You can think of alleles as being like words with different spellings. Some of these variant spellings are harmless (for example, color, colour; theater, theatre). Others change the meaning of the word entirely (here, hear; read, red). A change in one letter—one nucleotide—is sometimes enough to alter the function of a protein, or make it nonfunctional. Having a nonfunctional protein is as harmful as not having one at all. In most cases of antithrombin deficiency, only one of the two alleles makes a functional protein. The

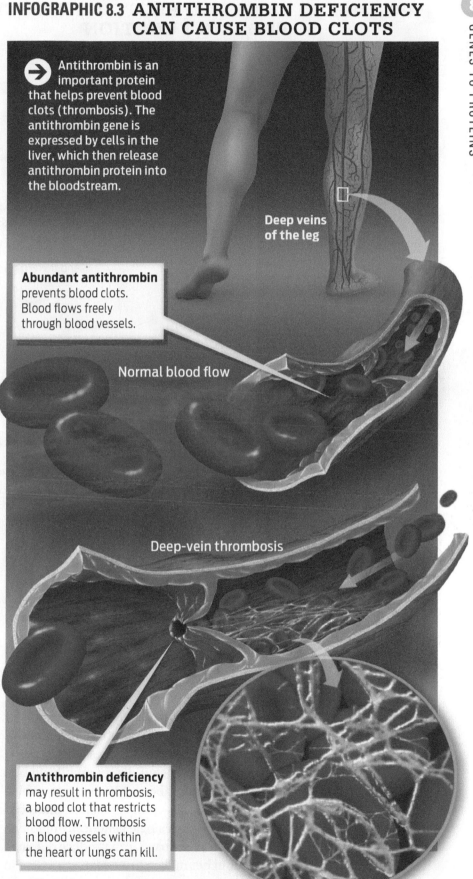

INFOGRAPHIC 8.3 ANTITHROMBIN DEFICIENCY CAN CAUSE BLOOD CLOTS

→ Antithrombin is an important protein that helps prevent blood clots (thrombosis). The antithrombin gene is expressed by cells in the liver, which then release antithrombin protein into the bloodstream.

Deep veins of the leg

Abundant antithrombin prevents blood clots. Blood flows freely through blood vessels.

Normal blood flow

Deep-vein thrombosis

Antithrombin deficiency may result in thrombosis, a blood clot that restricts blood flow. Thrombosis in blood vessels within the heart or lungs can kill.

INFOGRAPHIC 8.4 GENES TO PROTEINS: DIFFERENT ALLELES INFLUENCE PROTEIN FUNCTION

Versions of a gene with different nucleotide sequences are called alleles. Alternative nucleotide sequences may code for a protein with a different shape.

Expression of normal antithrombin allele

Antithrombin gene allele 1

Gene expression

GGCGACCTGAGC
Normal nucleotide sequence

Normal Antithrombin Protein
Protein has normal shape and therefore normal function.

Blood flow is normal when blood contains enough functional antithrombin.

Expression of abnormal antithrombin allele

Antithrombin gene allele 2

Gene expression

GGCGACGTGAGC
Alternative nucleotide sequence

Abnormal Antithrombin Protein
Protein has abnormal shape and therefore doesn't perform its normal function.

Blood flow is restricted by clot when blood lacks enough functional protein.

▶**TRANSGENIC**
Refers to an organism that carries one or more genes from a different species.

▶**REGULATORY SEQUENCE**
The part of a gene that determines the timing, amount, and location of protein production.

amount of protein produced from this single functional allele is simply not enough to prevent improper clotting. Hence, antithrombin deficiency (**INFOGRAPHIC 8.4**).

People with inherited antithrombin deficiency usually take medication to thin their blood and prevent clots. At times when the risk of clots is high–during surgery or pregnancy, for example–they receive antithrombin treatment.

The problem is that it takes 90,000 blood donors to produce 1 kg of antithrombin, enough to treat about 300 patients, which means the drug is often scarce; hospitals can and do run out. A single genetically modified goat, on the other hand, can produce the same amount in her milk in just 1 year, according to rEVO Biologics, ensuring a more stable supply.

GOT MILK?

More than 20 years ago, when Harry Meade was working as a research scientist at a company called Biogen, it occurred to him that producing drugs in a mammal's milk might be more efficient than obtaining proteins from blood. Meade, who grew up on a dairy farm outside Pittsburgh, Pennsylvania, reasoned that the mammary gland is a natural protein factory. To nourish their young, all mammals produce proteins and secrete them into their milk.

At the time of his eureka moment, Meade had been experimenting with putting human genes into mice, which would then produce the human proteins in their blood. The method worked, but in order to obtain the proteins, the mice had to be killed to harvest their blood. Meade realized that if he could get mice or other animals to produce the proteins in their milk, then they wouldn't have to be bled—they could be milked instead.

"I was reminded of the fable about the goose that laid a golden egg each day," wrote Meade in an article for *The Sciences* in 1997. "Killing the goose left the owner with nothing, while caring for it patiently would have yielded a lifetime of steady benefits." Using a larger mammal, like a goat or cow, would be even better than using a mouse, because the yield would be much greater.

To develop this idea, in 1994, Meade co-founded GTC Biotherapeutics (renamed rEVO Biologics in 2012). The company decided to use goats because they produce ample milk but have faster gestation times and are cheaper to maintain than cows.

Animals that have been genetically modified to contain genes from other species are called **transgenic** organisms ("trans" means "across"—in this case, across species, from one to another). The basic idea for making a transgenic goat is simple: isolate the gene of interest from a human chromosome and then insert it into the genome of a goat embryo; the goat grows up containing the human gene in its cells. But in order to make sure the human gene is expressed when the goat produces milk, Meade and his colleagues had to create a hybrid gene that was part human and part goat.

Meade's technique made use of the fact that every gene has two parts: a **regulatory sequence** and a **coding sequence**. Regulatory sequences are like on/off switches for genes; by providing a site where regulatory molecules can bind, they determine when, where, and how much protein a gene makes. Coding sequences determine the identity of a protein: they specify the order, or sequence, of amino acids. Together, the regulatory sequence and the coding sequence ensure that every gene is expressed only at the right time and in the right cells (**INFOGRAPHIC 8.5**).

Meade realized that if he could attach the coding sequence of the human antithrombin gene to the regulatory sequence of a goat gene

INFOGRAPHIC 8.5 THE TWO PARTS OF A GENE

 Genes are organized into two parts. Regulatory sequences determine when and how much protein a gene makes. Coding sequences determine the amino acid sequence of the encoded protein, which determines its shape and function.

Gene

Regulatory sequence
Controls the timing, location, and amount of gene expression

Coding sequence
Determines the sequence of amino acids in the protein

that is expressed in the animal's mammary cells, he could get the goat's mammary cells to make the human protein. In other words, with this gene construct he could dupe the goat's mammary glands into making the human antithrombin protein and secreting it as part of the goat's milk when the animal lactated.

First, the team had to find the right goat gene to work with. That part was easy. About 80% of milk proteins are caseins (from the Latin *caseus,* meaning "cheese"). This family of proteins is produced in the mammary gland and nowhere else in the animal, making these proteins an obvious choice. The team chose one particularly well-characterized casein gene, beta-casein, which had already been studied in mice.

Next, using **genetic engineering** techniques they spliced together the regulatory sequence of this milk gene and the coding sequence of the human antithrombin gene. A genetically engineered hybrid gene like this one is often referred to as a **recombinant gene.** By using the regulatory sequence of the milk gene in their recombinant gene, researchers could ensure that antithrombin is expressed only in the mammary cells, and not in any other tissues. That's important because proteins expressed where they shouldn't be can cause problems.

▶**CODING SEQUENCE**
The part of a gene that specifies the amino acid sequence of a protein. Coding sequences determine the identity, shape, and function of proteins.

▶**GENETIC ENGINEERING**
The process of assembling new genes with novel combinations of regulatory and coding sequences.

▶**RECOMBINANT GENE**
A genetically engineered gene.

If all cells have the same DNA (and they do), why is beta-casein protein–and therefore antithrombin–produced only in mammary cells? It turns out that different cell types contain different suites of proteins; in fact, it is this unique set of proteins that gives each cell type its identity (see Chapter 13). Only mammary cells respond to the signal to produce milk, and only mammary cells have the necessary proteins to bind to the regulatory sequence of the beta-casein gene, turning it on and causing it to be expressed. In other words, the beta-casein regulatory sequence acts as a switch. "So you put the gene in and you put a switch in, and the switch only gets turned on in the mammary gland," explains Simon Lowry, Vice President and Head of Medical Affairs at rEVO.

With the recombinant gene in hand, researchers then began the delicate process of putting this hybrid gene into a goat. Here's how it works: While looking through a microscope, scientists use a needle to inject the recombinant gene into a fertilized single-cell goat embryo. This transgenic embryo is then implanted into a surrogate mother. As the embryo grows and the cells divide, the inserted gene is replicated and passed on to every cell in the developing goat. If this goat is female, it can be used to obtain medicine-rich milk. The goat can also be bred with other goats to create a herd of "pharm" animals.

But the process is harder than it sounds, and took a lot of trial and error. Early attempts had less than a 5% success rate. For that reason, explains Meade, producing a single transgenic founder animal requires implanting 100 to 200 microinjected embryos. The good news is that once a founder goat is produced, it can be bred to create an endless supply of new transgenic goats. From start to finish, it took about 15 years to create a herd of transgenic goats. By 2006, the researchers were in business (**INFOGRAPHIC 8.6**).

Baby goats, or kids, in the nursery at the rEVO Biologics farm.

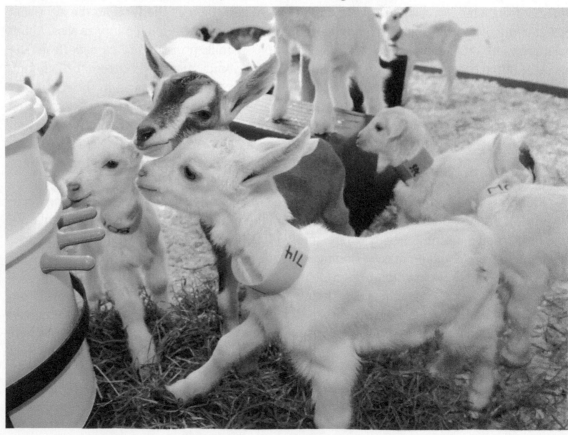

INFOGRAPHIC 8.6 MAKING A TRANSGENIC GOAT

Goat
beta-casein
gene

Enzymes

1. Create recombinant gene
Goat regulatory sequence
and human antithrombin
coding sequence are
cut out of donor cell
chromosomes and joined
together using
specialized enzymes.

Human
antithrombin
gene

Enzymes

Beta-casein
regulatory sequence

Antithrombin
coding sequence

Recombinant gene

Gene expression

Purify protein

2. Microinjection and embryo transfer
The recombinant gene is injected into
fertilized goat embryos. These embryos are
implanted in surrogate goat mothers,
who give birth to transgenic offspring.

3. Harvest milk rich in antithrombin
The recombinant antithrombin gene is
expressed only in mammary tissues of the
transgenic offspring, and the recombinant
antithrombin is secreted into milk.

4. Purify antithrombin from transgenic milk
Antithrombin protein is expressed in the milk
of transgenic females. The protein is isolated
from the milk and used to treat antithrombin-
deficient people.

Animals aren't the only organisms that have been genetically modified by humans. Much of the corn we eat today is transgenic–it contains genes from a soil bacterium. There are strains of transgenic soybeans, transgenic tomatoes, and transgenic insects. Transgenic organisms are also called **genetically modified organisms (GMOs).** Transgenic crops such as corn and soybeans usually contain genes for natural pesticides, which help the plants fight pests and reduce the amount of pesticide a farmer must use. Other varieties contain herbicide-resistance genes, allowing farmers to spray herbicides

on fields to kill weeds without at the same time killing crops.

Such gene-swapping technology also has an important application in medicine: in **gene therapy** scientists attempt to replace a defective human gene with a healthy one, an approach that is already treating, and in some cases curing, debilitating diseases such as severe combined immunodeficiency syndrome (SCID)–a disorder in which babies are born with deficient immune systems. Researchers hope that gene therapy may one day help treat several disorders caused by defective genes, including cystic fibrosis, Huntington disease, and hemophilia.

▶**GENETICALLY MODIFIED ORGANISM (GMO)**
An organism that has been genetically altered by humans.

▶**GENE THERAPY**
A treatment that aims to cure human disease by replacing defective genes with functional ones.

▶TRANSCRIPTION
The first stage of gene expression, during which cells produce molecules of messenger RNA (mRNA) from the instructions encoded within genes in DNA.

Despite the many actual and potential benefits of genetic engineering, mixing and matching genes inspires debate among scientists, environmentalists, and the general public. Many people object to human meddling with the biology of organisms that have evolved naturally. Others worry what might happen to a natural population of organisms—such as corn plants—if their genetically modified cousins (which contain, for example, pesticide-resistance genes) were to escape into the environment and cross with the unmodified population; the consequences are unpredictable.

The fear that meat, cheese, or milk from the transgenic goats might make its way into the human food supply was an explicit concern voiced by environmentalists when ATryn was being evaluated by the FDA. The FDA representative who reviewed the case testified that rEVO has mechanisms in place to make sure that doesn't happen: a double-fenced facility, video cameras on the grounds, and homing devices implanted under the goats' skin to help locate a goat in the unlikely event that one were to escape. But not everyone is convinced. "Humans are fallible and accidents do happen," said Gregory Jaffe, director of the Biotechnology Project of the Center for Science in the Public Interest in an interview with *Scientific American* in 2009. "Even if they're not intended to end up in the food supply, many things do end up in the food supply."

Although the idea of genetically engineering organisms is disquieting to some, others point out that humans have been tampering with the natural evolution of plants and animals for centuries by selectively breeding them for desirable traits. Moreover, from an animal-rights point of view, transgenic goats are no worse off than goats farmed for their milk and meat. (In fact, Meade's goats are treated better—they have generous room to run around, are given toys to play with, stairs to climb, even backscratchers to keep them comfortable.)

Being able to genetically modify organisms for human purposes raises legitimate questions about how to conduct genetic engineering safely and humanely. For example, many people who find nothing ethically troubling about using gene therapy to treat human diseases nonetheless find the prospect of using it to change personality or sexual orientation more problematic. Scientists can now use genetic engineering technology to clone entire animals—growing them from a single cell (Dolly the sheep was the first of these, in 1996). Will humans be next? Many people find that possibility disturbing. With the goats, however, genetic engineering is being used to save human lives—a much less controversial use of the technology.

Karen James, a New Mexico lawyer whose 17-year-old daughter died of a massive blood clot in her brain due to inherited antithrombin deficiency, spoke at the FDA hearing in 2009: "As a wife, as a mother, as someone who would like to be a grandmother someday, I'm here to tell you that antithrombin deficiency is the stuff of nightmares," she said. As long as the animals are not hurt by the practice, James said she has no problems with this method being used to obtain medicine. "If it helps other families avoid tragedies like ours, I think it's a wonderful idea."

> **❝As a wife, as a mother, as someone who would like to be a grandmother someday, I'm here to tell you that antithrombin deficiency is the stuff of nightmares.❞**
> —KAREN JAMES

Gene expression is the process of converting the genetic information of DNA into the sequence of a protein. Gene expression has two main steps: transcription and translation.

mRNA

1. Transcription: DNA to mRNA
Transcription copies the coding sequence of DNA into the complementary messenger RNA (mRNA) sequence.

Nucleus

Cytoplasm

In eukaryotic cells, the mRNA copy leaves the nucleus and enters the cytoplasm where it binds to a ribosome.

mRNA

Ribosome

Protein

The protein is released by the ribosome and folds into its final three-dimensional shape.

2. Translation: mRNA to protein
Translation occurs on a ribosome and uses the mRNA sequence to assemble the appropriate amino acid sequence of the protein.

C C A C A G G A G C G T
DNA sequence

Transcription

G G U G U C C U C G C A
mRNA sequence

Translation

gly | val | leu | ala
Amino acid sequence of protein

MAKING PROTEINS, OR HOW GENES ARE EXPRESSED

Meade's method of making antithrombin takes advantage of the fact that genes provide instructions for making proteins. But what are those instructions? How is the antithrombin protein actually made by goat cells?

In order to get from a gene to a protein, cells carry out two major steps: transcription and translation. **Transcription** is the process of using DNA to make a **messenger RNA (mRNA)** copy of the gene. **Translation** is the process of using this mRNA copy as a set of instructions to assemble amino acids into a protein (**INFOGRAPHIC 8.7**).

Why two separate steps? As the names "transcription" and "translation" imply, the process of gene expression is like copying a text and then converting it into another language. In this case, the text to be translated is a valuable, one-of-a-kind document: DNA. Just as you would be forbidden to borrow a rare manuscript from the library and would instead have to copy the text by hand into your notebook, the cell cannot take DNA out of its "library"–the nucleus. It must first make a copy–the mRNA. The cell can then take this

▶MESSENGER RNA (mRNA)
The RNA copy of an original DNA sequence made during transcription.

▶TRANSLATION
The second stage of gene expression, during which mRNA sequences are used to assemble the corresponding amino acids to make a protein.

INFOGRAPHIC 8.8 TRANSCRIPTION: A CLOSER LOOK

 In eukaryotic cells, transcription occurs in the nucleus and copies a DNA sequence into a corresponding mRNA sequence. RNA polymerase is the key enzyme involved. In prokaryotic cells, transcription occurs in the cytoplasm, where DNA is located.

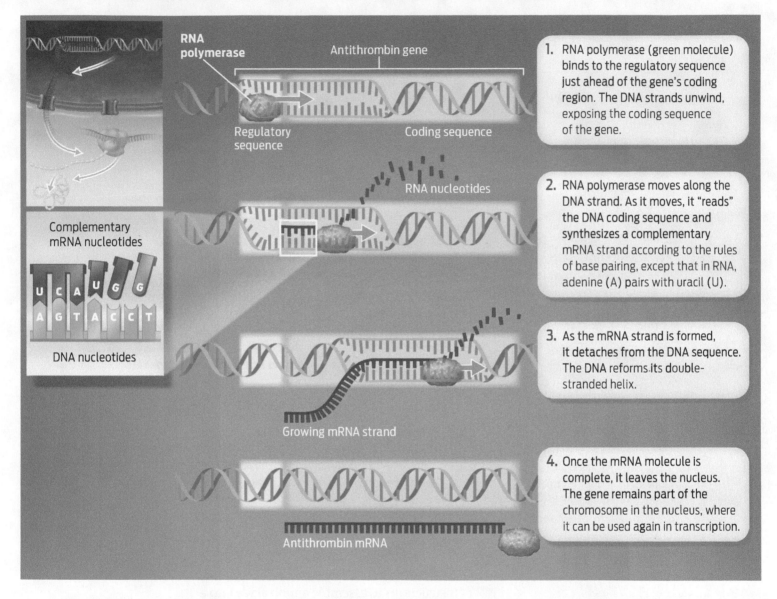

RNA polymerase

Antithrombin gene

Regulatory sequence

Coding sequence

RNA nucleotides

Complementary mRNA nucleotides

UCAUGG

AGTACCT

DNA nucleotides

Growing mRNA strand

Antithrombin mRNA

1. RNA polymerase (green molecule) binds to the regulatory sequence just ahead of the gene's coding region. The DNA strands unwind, exposing the coding sequence of the gene.

2. RNA polymerase moves along the DNA strand. As it moves, it "reads" the DNA coding sequence and synthesizes a complementary mRNA strand according to the rules of base pairing, except that in RNA, adenine (A) pairs with uracil (U).

3. As the mRNA strand is formed, it detaches from the DNA sequence. The DNA reforms its double-stranded helix.

4. Once the mRNA molecule is complete, it leaves the nucleus. The gene remains part of the chromosome in the nucleus, where it can be used again in transcription.

▶**RNA POLYMERASE**
The enzyme that carries out transcription. RNA polymerase copies a strand of DNA into a complementary strand of mRNA.

▶**RIBOSOME**
The cellular machinery that assembles proteins during translation.

mRNA copy into the cytoplasm, where it is translated into a new language: protein.

Transcription begins in the nucleus of a cell when an enzyme called **RNA polymerase** binds to DNA at a gene's regulatory sequence, located just ahead of the coding sequence. At that site, cellular machinery unwinds the DNA double helix and RNA polymerase begins moving along one DNA strand. As it moves, the RNA polymerase "reads" the DNA sequence and synthesizes a complementary mRNA strand according to the rules of base pairing. The same rules of base pairing we discussed in the context of DNA structure apply here, with one difference: RNA nucleotides are made with the base uracil (U) instead of thymine (T). So the complementary base pairs are C with G and A with U **(INFOGRAPHIC 8.8)**.

As its name implies, messenger RNA serves to relay information. Once the mRNA copy is made, it leaves the nucleus and attaches to a complex piece of cellular machinery in the cytoplasm called the **ribosome**. This is the start of translation.

During translation, the ribosome reads the mRNA transcript and translates it into a chain of amino acids. The mRNA transcript specifies which amino acids should be joined together to form chains. Amino acids are specified by groups of three nucleotides; each group is called a **codon.** Each codon is like a word: its letters name a particular amino acid (for example, the codon GGU specifies the amino acid glycine).

The actual building blocks of proteins—amino acids—are delivered to the ribosome by another type of RNA, called **transfer RNA (tRNA),** which physically transports amino acids to the ribosome. Each tRNA molecule is structured like an adaptor: one end binds to an amino acid, the other end binds to mRNA. The part that binds mRNA is called the **anticodon** because it base-pairs in a complementary fashion with an mRNA codon. When the amino acid-toting tRNA finds its codon match, it releases the amino acid to the ribosome, which adds it to the growing amino acid chain (**INFOGRAPHIC 8.9**).

▶**CODON**
A sequence of three mRNA nucleotides that specifies a particular amino acid.

▶**TRANSFER RNA (tRNA)**
A type of RNA that transports amino acids to the ribosome during translation.

▶**ANTICODON**
The part of a tRNA molecule that binds to a complementary mRNA codon.

INFOGRAPHIC 8.9 TRANSLATION: A CLOSER LOOK

In the cytoplasm, the ribosome reads the mRNA sequence and "translates" it into a chain of amino acids to make a protein.

Ribosome
Antithrombin mRNA
C G U U C A U G G A C U C G U U C A U G G

1. The newly transcribed mRNA associates with a ribosome.

ser trp thr arg
U A C A G U A C C U G A G C A tRNA
A U G U C A U G G A C U C G U U C A U G G

2. As the ribosome moves along the mRNA, it "reads" the mRNA sequence in groups of three nucleotides called codons. Each codon specifies a particular amino acid, which is brought to the ribosome by tRNA, which uses its anticodon sequence to find a matching mRNA codon.

Amino acid
Corresponds to the mRNA codon

ser trp
A G U A C C
U C A U G G

Codon
Triplet code on mRNA

Anticodon
Triplet code on tRNA that recognizes a specific codon

Growing chain
trp thr arg ser
A C G U G A G C A A G U
A U G U C A U G G A C U C G U U C A U G G

3. When the correct tRNA is in place, the specified amino acid is added to the growing chain. The ribosome moves on to the next codon.

Antithrombin protein

4. The finished amino acid chain detaches from the ribosome and folds into its three-dimensional shape. mRNAs may be translated several times and tRNAs may pick up another amino acid for reuse in translation.

A U G U C A U G G A C U C G U U C A U G G A C U C G U C C A U G A
G G U A C U

> The genetic code is universal, which means that it is virtually the same in all living organisms.

Although the human genome codes for many thousands of different proteins, each one is pieced together from a starting set of just 20 amino acids. In the same way that the 26 letters in our alphabet can spell hundreds of thousands of words, the basic set of amino acids can make hundreds of thousands of proteins. The set of rules dictating which mRNA codons specify which amino acid is called the **genetic code.** You can think of the genetic code as the key that cracks the mRNA code. Much in the same way the Rosetta Stone provided scholars with the key to decipher what Egyptian hieroglyphics meant, the genetic code tells us what amino acid each codon stands for.

There are two important features of the genetic code. One, the code is redundant: multiple codons specify the same amino acid.

The vast majority of mRNA codons specify a specific amino acid, but there are a few with other functions. Start codons are the first codon of a coding sequence; they tell the ribosome to start translating and begin adding amino acids. Stop codons tell the ribosome to stop translating and not add any more amino acids to the growing chain.

▶GENETIC CODE
The set of rules relating particular mRNA codons to particular amino acids.

INFOGRAPHIC 8.10 THE GENETIC CODE IS UNIVERSAL

Codons are groups of three-nucleotide sequences within chains of mRNA. Most codons specify a particular amino acid. Some codons specify where to start translation (start codons) and where to end (stop codons). There is redundancy in the genetic code, as 64 possible codons code for only 20 different amino acids. Since the genetic code is universal, the same gene will be transcribed and translated into the same protein in virtually all cells and organisms.

Second letter

First letter	U	C	A	G	Third letter
U	UUU UUC Phenylalanine (Phe) / UUA UUG Leucine (Leu)	UCU UCC UCA UCG Serine (Ser)	UAU UAC Tyrosine (Tyr) / UAA UAG Stop codon Stop codon	UGU UGC Cysteine (Cys) / UGA Stop codon / UGG Tryptophan (Trp)	U C A G
C	CUU CUC CUA CUG Leucine (Leu)	CCU CCC CCA CCG Proline (Pro)	CAU CAC Histidine (His) / CAA CAG Glutamine (Gln)	CGU CGC CGA CGG Arginine (Arg)	U C A G
A	AUU AUC AUA Isoleucine (Ile) / AUG Methionine (Met); start codon	ACU ACC ACA ACG Threonine (Thr)	AAU AAC Asparagine (Asn) / AAA AAG Lysine (Lys)	AGU AGC Serine (Ser) / AGA AGG Arginine (Arg)	U C A G
G	GUU GUC GUA GUG Valine (Val)	GCU GCC GCA GCG Alanine (Ala)	GAU GAC Aspartic acid (Asp) / GAA GAG Glutamic acid (Glu)	GGU GGC GGA GGG Glycine (Gly)	U C A G

In many cases, a codon will differ at the third nucleotide position without changing the amino acid that is specified. Two, the genetic code is universal, which means that it is virtually the same in all living organisms. It is because the code is universal that the mammary cells of a goat carrying the human gene for antithrombin can express that gene and produce antithrombin protein in her milk (**INFOGRAPHIC 8.10**).

THE ADVANTAGES OF "PHARMING"

While Meade's goats are currently the talk of the barnyard, this is actually not the first time that living organisms have been used to make drugs to treat humans. Before the era of genetic engineering, medicinally useful proteins were routinely obtained and purified from animal organs or blood. Insulin, for example, which is used to treat diabetes, was originally extracted from pig pancreas. Since the early 1980s, with the rise of genetic engineering, scientists have used bacteria to make recombinant human proteins, including insulin. The problem with this method is that bacteria can make only very simple human proteins. More-complex human proteins, such as antithrombin, require mammalian cells to produce them. For that reason, beginning in the 1990s scientists began using hamster cells grown in culture to make protein drugs. Many currently available biotech drugs, including the cancer drug Avastin and the rheumatoid arthritis drug Humira, are produced by this method.

But even hamster cells grown in culture have limitations. To produce protein drugs from hamster cells, scientists must grow the cells in giant stainless steel vats, which cost millions of dollars to purchase and maintain. The cells secrete proteins into the culture medium, and the proteins are then extracted and purified from this liquid. But typically only very small amounts of protein are obtained this way, so the cost per unit of drug can be quite pricey–thousands of dollars per gram, $50,000-$100,000 per treatment.

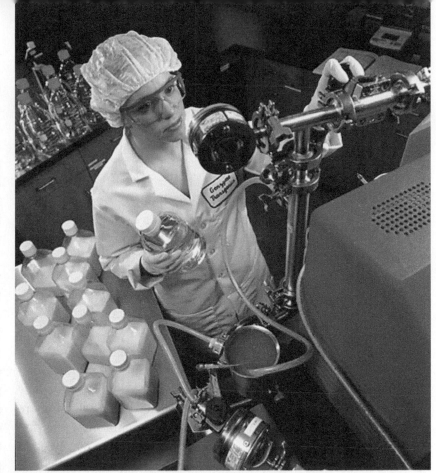

Transgenic goats can produce large quantities of drug in their milk for one-tenth the cost of conventional methods.

As an example, Meade points to the human protein drug factor VII. This protein helps blood clot. Males with the genetic disease hemophilia, a sex-linked trait that is passed from mothers to their sons (see Chapter 12), are born with clotting factor alleles that either encode nonfunctional clotting factor protein or no clotting factor protein at all. Consequently, if a hemophiliac cuts himself, he must be quickly transfused with clotting factor or he may bleed to death. Doctors usually give patients factor VII protein to restore normal coagulation and prevent excessive bleeding.

Companies that sell factor VII produce it by cell culture. But the drug is extremely expensive: 1 mg of factor VII can cost up to $1,000. At the recommended dose of 90 µg/kg (micrograms per kilogram of body weight) given every 2-3 hours to a patient with hemophilia, that's easily $50,000 per day. Transgenic goats, by contrast, can produce human clotting factors in their milk in large volumes for about one tenth of the cost.

The mammary gland is a "natural bioreactor," rEVO scientists like to say. Moreover, it's a bioreactor that lives on inexpensive hay and can be milked twice a day for 7 years. And scaling up is easy. "If I own a giant factory and I want to double capacity, I have to build another factory with all of the industrial issues that that brings," explains Lowry, of rEVO. With goats, however, "If you want to double your capacity, all you have to do is allow boy and girl goats to do what they do."

How do recombinant proteins stack up against ones produced naturally by the body? ATryn has the same amino acid sequence as natural antithrombin, and clinical trials have shown it to have similar efficacy as antithrombin isolated from human donors, with only minor dosing differences. "Clinically, they're equally efficacious in accomplishing the prevention of blood clotting," says physician Paidas, of Yale. It's also cheaper: the wholesale price of ATryn is approximately $4,000 per

vial; the equivalent amount of human-derived antithrombin costs about $5,600.

ATryn is currently only FDA-approved to treat patients with inherited antithrombin deficiency, but as of 2013 researchers were beginning clinical trials of the drug as a treatment for preeclampsia—a life-threatening condition that affects a substantial number of pregnant women and their babies every year.

In addition to ATryn, rEVO is currently working on establishing transgenic animals to produce factor VII and other clotting proteins, as well as a number of other important drugs, including antibodies to treat cancer.

"We have the potential to build an abundant and controlled supply for any plasma protein," says Thomas Newberry, a company spokesperson, who predicts that protein drugs extracted from human blood may become a thing of the past. "In the future, I won't be surprised if people start to think that reinjecting blood products into other people is barbaric." ∎

CHAPTER 8 Summary

- ▸ Genes provide instructions to make proteins. The process of using the information in genes to make proteins is called gene expression.

- ▸ Proteins are folded chains of amino acids that make up cell structures and help cells to function properly.

- ▸ Proteins play an important role in nearly all cellular functions, from muscle movement to metabolism.

- ▸ Many drugs act on proteins in the body or are themselves proteins.

- ▸ Amino acid sequences determine the shape and function of a protein.

- ▸ A change in the DNA sequence of a gene can change the corresponding amino acid sequence, and therefore the function, of a protein.

- ▸ Different versions of the same gene, those with different nucleotide sequences, are called alleles.

- ▸ Every gene has two parts: a coding sequence and a regulatory sequence. The coding sequence determines the identity of a protein; the regulatory sequence determines where, when, and how much of the protein is produced.

- ▸ Gene expression occurs in two stages, transcription and translation, which take place in separate compartments in eukaryotic cells.

- ▸ Transcription is the first step of gene expression, copying the information stored in DNA into mRNA. Transcription occurs in the nucleus.

- ▸ Translation, the second step of gene expression, uses the information stored in mRNA to assemble a protein. Translation occurs in the cytoplasm.

- Proteins are assembled by ribosomes with the help of tRNA, which delivers amino acids to the ribosome.

- The genetic code is the set of rules by which mRNA sequences are translated into protein sequences; the code is redundant and shared by all living organisms.

- Through genetic engineering, genes from one species of organism can be inserted into the genome of another species of organism to make a transgenic organism.

- Transgenic organisms have numerous uses in biotechnology and health.

MORE TO EXPLORE

- Meade, H. (2012) TED Talk: Medicine from milk http://tedxboston.org/speaker/meade

- Meade, H. M. (1997) Dairy Gene. *The Sciences* 37 (5):20–25.

- Echelard, Y., et al. (2006) Production of recombinant therapeutic proteins in the milk of transgenic animals. *BioPharm International* 19:36–46 http://www.biopharminternational.com/biopharm/article/articleDetail.jsp?id=362005

- Stix, G. (2005) "The land of milk and money: the first drug from a transgenic animal may be nearing approval." *Scientific American* 293:102–105.

- Center for Science in the Public Interest http://www.cspinet.org/

CHAPTER 8 Test Your Knowledge

DRIVING QUESTION 1

What determines the shape of a protein?

By answering the questions below and studying Infographics 8.1 and 8.4, you should be able to generate an answer for the broader Driving Question above.

KNOW IT

1. A protein is made up of a chain of _____.

   a. nucleotides
   b. amino acids
   c. lipids
   d. fatty acids
   e. simple sugars

2. What determines a protein's function?

   a. the sequence of amino acids
   b. the three-dimensional shape of the folded protein
   c. the location of its gene on the chromosome
   d. all of the above
   e. a and b

USE IT

3. Heating a protein can cause it to denature, or unfold. What do you think would happen to the function of a protein in a denatured state? Explain your answer.

4. State two ways by which a person can be deficient in antithrombin activity. (Hint: Think of level of expression, which is driven by the regulatory sequence, and the amino acid sequence of the protein, which is determined by the coding sequence.)

DRIVING QUESTION 2

What are the steps of gene expression, and where do they occur in a cell?

By answering the questions below and studying Infographics 8.2, 8.5, 8.7, 8.8, 8.9, and 8.10, you should be able to generate an answer for the broader Driving Question above.

KNOW IT

5. "A gene contains many chromosomes. Each chromosome encodes a protein." Is this statement accurate? If not, explain why not, and rewrite it to make it correct.

6. What is the final product of gene expression?

   a. a DNA molecule
   b. an RNA molecule
   c. a protein
   d. a ribosome
   e. an amino acid

7 For each structure or enzyme listed, indicate by N (nucleus) or C (cytoplasm) where it acts in the process of gene expression in a eukaryotic cell.

RNA polymerase _____

Ribosome _____

tRNA _____

mRNA _____

8 What is the product of transcription?

a. a gene

b. a protein

c. RNA

d. a chromosome

e. RNA polymerase

9 A gene has the sequence ATCGATTG. What is the sequence of the complementary RNA?

a. ATCGATTG

b. AGCTAAC

c. GTTAGCTA

d. UAGCUAAC

e. CAAUCGAU

USE IT

10 If someone has reduced levels of normal functioning antithrombin, would you suspect a problem in the regulatory or in the coding sequence of the antithrombin gene? Explain your answer.

11 If you wanted to use genetic engineering to increase the amount of antithrombin that someone produces, would you modify the regulatory sequence or the coding sequence? Explain your answer.

12 A change in DNA sequence can affect gene expression and protein function. What would be the impact of each of the following changes? How, specifically, would each change affect protein or mRNA structure, function, and levels?

a. a change that prevents RNA polymerase from binding to a gene's regulatory sequence

b. a change in the coding sequence that changes the amino acid sequence of the protein

c. a change in the regulatory sequence that allows transcription to occur at much higher levels

d. a combination of the changes in b and c

DRIVING QUESTION 3

How can animals be genetically modified to produce human proteins (with therapeutic uses)?

By answering the questions below and studying Infographics 8.5, 8.6, and 8.7, you should be able to generate an answer for the broader Driving Question above.

KNOW IT

13 Why is recombinant protein production in milk of transgenic animals an efficient strategy?

a. because milk is secreted, so the protein can be obtained noninvasively

b. because milk is produced in relatively large quantities

c. because mammary glands naturally secrete large quantities of proteins into milk

d. because milk is easier to obtain than other secretions (e.g., urine, sweat, and saliva)

e. all of the above

14 In an antithrombin-producing transgenic goat,

a. is the antithrombin gene construct present in every cell, or only in mammary cells? Explain your answer.

b. is the antithrombin gene construct expressed in every cell, or only in mammary cells? Explain your answer.

15 Explain why scientists used the beta-casein regulatory sequence to express human antithrombin in goats' milk.

USE IT

16 Melanin is a pigment expressed in skin cells; melanin gives skin its color. If you wanted to express a different gene in skin cells, which part of the melanin gene would you use? Why? If you wanted to produce melanin in yeast cells, what part of the melanin gene would you use? Why?

17 Lysozyme is a protein secreted in tears and saliva in all mammals. Amylase is a protein secreted in mammalian saliva.

a. Describe the recombinant gene that you would assemble to express recombinant human insulin in the tears of goats.

b. Describe the recombinant gene that you would assemble to express recombinant human antithrombin in goat saliva.

c. Is either of these approaches as practical as protein production in milk? Why or why not?

DRIVING QUESTION 4

What are some practical applications of genetically modified organisms in treating human disease?

By answering the questions below and studying Infographics 8.3, 8.5, and 8.6, you should be able to generate an answer for the broader Driving Question above.

KNOW IT

18 Goat's milk is used to produce cheese that humans can eat. How is this application of goat's milk for human use different from the one described in this chapter?

19 Why are biotechnology companies eager to design genetically modified organisms to express therapeutic proteins, particularly ones that would otherwise have to be isolated from blood products (e.g., human antithrombin) or animal organs (e.g., insulin, originally isolated from pig pancreas)? (Hint: Consider cost, safety, and practicality.)

USE IT

20 Type 1 diabetes results from a loss of insulin production from the pancreas. People with diabetes take recombinant human insulin expressed in bacteria.

a. Describe the gene construct necessary for expression of human insulin in bacteria.

b. Describe the gene construct necessary to produce human insulin in goat's milk.

c. If you were to attempt gene therapy (genetically modifying the human's genome so that they can produce their own insulin), would you need a recombinant form of the insulin gene? Explain your answer.

INTERPRETING DATA

21 Hereditary antithrombin deficiency occurs in approximately 1 in 5,000 people in the United States. It takes approximately 2.8 g of antithrombin protein to treat one patient with antithrombin deficiency in need of a surgical procedure.

a. How many people in the United States are affected by this disease?

b. A particular transgenic goat produces 2 g of antithrombin per liter of milk. The same goat produces approximately 800 liters of milk over a 10-month lactation period. How much total protein does this goat produce in 10 months? How many surgical patients can be treated with the recombinant protein produced from one goat?

c. Before recombinant antithrombin was available, patients with antithrombin deficiency were treated with antithrombin purified from the blood of blood donors. Each donation of a pint of blood has approximately 0.07 g of antithrombin. How many donations are needed in order to treat a single surgical patient with antithrombin deficiency?

MINI CASE

22 A 75-kg pregnant woman with hereditary antithrombin deficiency requires an emergency cesarean section, a surgical procedure. Her current baseline antithrombin levels are 25% of normal (expressed as International Units (IU) per ml of blood).

a. What are the risks to this mother if she undergoes the procedure?

b. Emergency room doctors must immediately treat her with a "loading dose" of recombinant human antithrombin (ATryn). The loading dose is calculated as [(100% IU – baseline antithrombin (% IU))/2.3] body weight in kilograms. What loading dose, expressed in IU, does this patient require?

c. A vial of ATryn contains 1,750 IU. How many vials are needed for this loading dose?

d. The operation is successful, and both the mother and her new baby boy are fine. Why does mom always tell people that her son's guardian angel is a goat?

BRING IT HOME

23 A number of concerns have been expressed about GMOs. Search the internet for reliable sources about a particular GMO that you have heard of or in which you are interested (e.g., Golden Rice or genetically modified salmon). List what you consider to be the pros and cons of at least two GMOs. Has what you have read about other genetically modified organisms and the transgenic goats in this chapter changed your opinions about GMOs? What restrictions (if any) would you place on GMOs?

IT STARTED WITH A VISIONARY IDEA: THE ENTIRE SEQUENCE of human DNA spelled out, nucleotide by nucleotide, to be read like a book. Such a reference would be an indispensable medical tool. Scientists could, for example, scan the genome for genes that confer susceptibility to disease, which might lead to better treatments. It would enable diagnostic tests that could help predict the risk of developing genetically based diseases. Fields other than medicine would benefit, too. Comparing the human genome to the genomes of other organisms, for example, would shed light on our own evolution. The possible benefits to science were endless.

But when an international group of scientists met in the early 1980s and first floated the idea of sequencing the human genome, they were met with skepticism. Some scientists found the idea absurd, especially given its then-estimated $3 billion price tag. Others thought the potential benefits were illusory or overly hyped. Some simply deemed the task impossible, given the state of the technology at the time.

Over the course of the decade, however, as the fields of molecular biology and genetics progressed, the idea gained both scientific and political support. In 1988, Congress funded both the National Institutes of Health (NIH) and the U.S. Department of Energy to explore the novel concept. By 1990, the collaborative effort to sequence the entire string of more than 3 billion As, Gs, Cs, and Ts that make up a human genome–the Human Genome Project (HGP)–was officially under way.

▶▶ DRIVING QUESTIONS

1. Why is knowing the sequence of the human genome important?

2. What were the similarities and differences between the approaches used by the two research teams that sequenced the genome?

AGACCCACAATA
GTGGGGATGAAG
CT **SEQUENCE** CT
GTGCTGGAAGAT
GCA **SPRINT** CTA
GAAATTTAAAAA
GTTGCTCTCCCA

**Venter and Collins
race to decode
the human genome**

INFOGRAPHIC M3.1 DNA SEQUENCING: HIERARCHICAL SHOTGUN METHOD

The hierarchical shotgun method cuts the genome into large fragments that are mapped to individual chromosomes. These large fragments are then sequenced by the shotgun method. Each fragment is broken into small pieces that are sequenced. Computers are used to assemble the shotgun fragments on the basis of overlaps between different fragments. Because each large fragment is first mapped to a chromosome location, once each large fragment sequence is finished, the entire genome is complete.

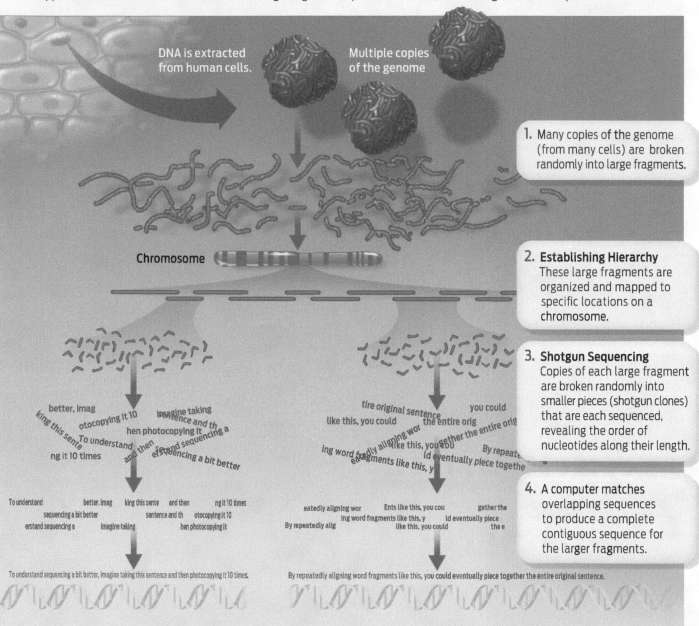

DNA is extracted from human cells.

Multiple copies of the genome

Chromosome

1. Many copies of the genome (from many cells) are broken randomly into large fragments.

2. **Establishing Hierarchy**
   These large fragments are organized and mapped to specific locations on a chromosome.

3. **Shotgun Sequencing**
   Copies of each large fragment are broken randomly into smaller pieces (shotgun clones) that are each sequenced, revealing the order of nucleotides along their length.

4. A computer matches overlapping sequences to produce a complete contiguous sequence for the larger fragments.

To understand sequencing a bit better, imagine taking this sentence and then photocopying it 10 times.

By repeatedly aligning word fragments like this, you could eventually piece together the entire original sentence.

When the project began, the NIH appointed James Watson, the co-discoverer of the structure of DNA, to head and coordinate it. The ambitious and mammoth undertaking involved more than 20 institutions spread around the globe, in China, France, Germany, Japan, the United Kingdom, the United States, and other countries.

Initially, the researchers set about sequencing every nucleotide on every one of the chromosomes that make up the entire genome. But the DNA sequence machines then available could sequence only about 500 nucleotides at a time; the cost and time to sequence the entire human genome 500 consecutive nucleotides at a time would have been prohibitive. So researchers had to develop a shortcut. The approach they came up with is now known as "hierarchical shotgun sequencing"–"hierarchical" because the researchers first chemically broke human chromosomes into large segments, organized them, and mapped each piece to a known physical location on a chromosome; "shotgun" because they then randomly cut those large segments into a rapid-fire slew of even smaller pieces and sequenced the nucleotides of each small piece. They then assembled the final sequence by matching up overlapping pieces.

Here's a useful analogy–imagine photocopying this sentence 10 times. Now imagine taking all the paper copies and tearing them at different points. Depending on where you tear, each copy will contain pieces with different parts of the same word. One copy might contain only the "ima" part of the word "imagine," while another might contain the whole word. You could use the piece containing the entire word to piece together two pieces with "ima" and "gine." By repeatedly aligning word fragments like this, you could eventually piece together the entire original sentence, even if you didn't know at the outset what it said.

In addition to allowing many parts of the genome to be analyzed simultaneously, this approach had the advantage of being able to focus on specific regions of chromosomes that researchers were particularly interested in. But this "shortcut" still took time: by this method, scientists estimated it would take 15 years to finish the entire sequence of 3 billion bases (INFOGRAPHIC M3.1).

The map of sequences grew, and as sequences were assigned to specific chromosomal locations, it became much easier for scientists to find a home for gene sequences they were working on. Since the information was uploaded to a public online database, any scientist could view the map to see where exactly his or her gene of interest was located and apply that information to further research. But after a year, the project wasn't progressing as quickly as planned, and some estimated that if things kept plodding along at the same pace, the total cost–which reflected the cost of labor, laboratory space and equipment, chemical reagents, and computer technology, as well as other expenses–could reach more than $100 billion.

Criticism began to mount over the costs and delays of the HGP. In 1991, Craig Venter, who had helped pioneer automated gene sequencing techniques in his lab at NIH, became especially frustrated with the slow pace of sequencing and publicly proposed

> Comparing the human genome to the genomes of other organisms, for example, would shed light on our own evolution. The possible benefits to science were endless.

an alternative approach. To save time, he proposed focusing on just the protein-coding DNA, skipping the regulatory regions. Using this method, he had already identified thousands more genes than the official project had. But Watson, head of the genome project, dismissed the method as sloppy and inadequate. At one point, he even called Venter's work "brainless," and said that his sequence machines "could be run by monkeys." Angered by his rejection and lured by a hefty salary, Venter left NIH in 1992 to found his own investment-backed institute, called The Institute for Genome Research (TIGR). From his new perch, he continued to explore other sequencing methods.

Meanwhile, in 1992, Watson resigned as head of the HGP after a disagreement with the NIH director over the issue of patenting genes (Watson was opposed). In early 1993, NIH appointed a new head, the geneticist Francis Collins. The idea of speeding up the sequencing using Venter's shortcut didn't sit well with Collins, either, and the decision was made to stay the course.

Then, in 1995, Venter surprised the scientific community again by successfully sequencing the entire genome of a bacterium in a short

INFOGRAPHIC M3.2 DNA SEQUENCING: WHOLE-GENOME SHOTGUN METHOD

The whole-genome shotgun method cuts the genome into different-size fragments that are then sequenced and reassembled without prior information about their position on chromosomes. This approach relies on advanced computer algorithms to assemble all the shotgun fragments on the basis of overlaps between them.

DNA is extracted from human cells.

Multiple copies of the genome

1. Many copies of the genome (from many cells) are broken randomly into different-size fragments (shotgun clones).

2. **Shotgun Sequencing**
These fragments are each sequenced, determining the order of nucleotides along their length.

3. A computer matches overlapping sequences to produce a complete contiguous sequence for the whole genome.

time, using a new sequencing method developed at his institute. This approach involved breaking the entire genome into many small fragments and sequencing them simultaneously–without first mapping them to known chromosome locations, as Collins's group was doing. Venter argued that the same technique could be used on the human genome. But the human genome was much larger and contained many repetitive sequences–how would the multitude of randomly generated sequences be aligned properly? Venter's approach–which he called "whole-genome shotgun sequencing"– would be akin to trying to put a jigsaw puzzle together without a photo of the finished puzzle as a guide. Critics claimed that the sequence data would be riddled with mistakes.

Venter was undeterred. In 1998, he dropped another bomb: he had made a deal with the Perkin-Elmer Corporation, which was about to unveil a new automated sequencing machine. Together they would create a new company, to be called Celera Genomics (from the Latin word *celeritas*, meaning "speed"), that intended single-handedly to sequence the entire human genome in just 3 years for a mere $300 million–a fraction of the cost of the publicly funded consortium. They would use Venter's whole-genome shotgun method (INFOGRAPHIC M3.2).

Collins and other leaders of the public project were troubled. The U.S. Congress might favor Celera's approach and stop funding the public project altogether. Collins was also concerned that Celera was going to try to patent its sequence data, which would have restricted public access to the data.

The race was on. Collins and his colleagues stepped up the pace. Venter wasn't the only one who had access to new automated sequencing machines and powerful computers that could process large amounts of data. Publicly financed scientists, too, had access to these and other new tools. Such technological advances dramatically cut the amount of time it took to sequence each nucleotide and simultaneously cut costs (INFOGRAPHIC M3.3).

About 6 months after Venter's announcement, Collins announced that the public

INFOGRAPHIC M3.3 NEW TECHNOLOGY CUT DNA SEQUENCING TIME AND COST

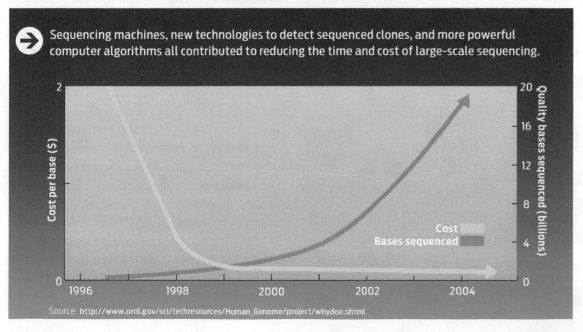

→ Sequencing machines, new technologies to detect sequenced clones, and more powerful computer algorithms all contributed to reducing the time and cost of large-scale sequencing.

Cost

Bases sequenced

Cost per base ($)

Quality bases sequenced (billions)

Source: http://www.ornl.gov/sci/techresources/Human_Genome/project/whydoe.shtml

President Bill Clinton, with J. Craig Venter (left) and Francis Collins (right), announced the completion of the human genome at the White House in June 2000.

consortium would complete sequencing the genome by 2003—2 years ahead of schedule. The consortium also planned to produce a rough draft of the genome by 2001, which was about the same time that Venter planned to finish his draft. Collins justified his decision

Drafts of the human genome sequence were published in 2001.

by stating that scientists were clamoring for the data even in rough form.

For a few years the contest between the privately funded Celera and the publicly funded HGP was bitter, each side criticizing the other's methods. The two sides eventually agreed to share the glory and appeared at a White House press briefing on June 26, 2000, together with U.S. President Bill Clinton and British Prime Minister Tony Blair, to announce publicly that they had completed a rough draft of the human genome sequence. In February 2001, both groups published their drafts of the human genome simultaneously in the journals *Nature* and *Science*.

Just whose genome was in fact sequenced? Geneticists working on the publicly funded project had collected blood samples from anonymous donors. The ultimate sequence is thus a composite pieced together from the gene sequences of several individuals. Celera's sequence data come mostly from the DNA of Venter himself.

Did one group "win" the race? Some say Venter won, because his sequence was further along when the two groups presented their findings. But others were quick to point out that Venter's group had used much of the public HGP data in order to assemble his own sequence—which he wouldn't have been able to do if the data had been kept secret. As the authors of a 2002 scientific report analyzing the two approaches noted, "When speed truly matters, openness is the answer."

Ultimately, when the HGP was officially completed in 2003, 99% of the gene-containing regions of DNA had been sequenced, with an accuracy of 1 in 100,000 nucleotides, and only 341 gaps in 2.85 billion nucleotides. The achievement was hailed as one of the greatest scientific accomplishments of the 20th century. Some even consider it the greatest achievement ever in biology. Not only did it reveal new characteristics of the human genome, it also shed light on how we differ from other organisms.

While the human genome was being sequenced, scientists had also finished sequencing the genomes of some other organisms. Scientists compared our genome to the genomes of other organisms, and what they found astounded them.

Before the sequencing was complete, scientists thought that what made humans such complex creatures was gene number—that the more complex the creature, the more genes it should have. But the HGP showed that humans carry a mere 20,000 to 25,000 genes—about the same number as a lowly roundworm. This evidence suggested that gene number wasn't as important as how those genes were expressed into proteins. Before the project, scientists didn't think noncoding regions of the genome were important. Now they know that noncoding regions actually regulate genes and consequently contribute to the complexity of higher organisms. Moreover, the number of genes an organism has says nothing about the number of proteins that are produced from those genes. Through alternative splicing of mRNA and other mechanisms not yet fully understood, a relatively small number of genes can produce an enormous number of diverse proteins. For example, the human genome may encode

> ## The achievement was hailed as one of the greatest scientific accomplishments of the 20th century.

more than 1 million proteins from fewer than 25,000 genes.

While the full potential of the HGP has yet to be tapped, it has already inspired many useful applications across many fields, from medicine to evolution. For example, the National

Caenorhabditis elegans, a roundworm, was the first animal to have its genome sequenced. It has approximately 24,000 genes.

INFOGRAPHIC M3.4 LINKING GENES TO DISEASE

Having the sequence of the human genome has allowed researchers to identify genes associated with particular diseases. Some of the disease-associated genes on chromosome 1 are shown here.

246 million base pairs

Cataracts
Malignant transformation suppression
Ehlers-Danlos syndrome, type VI
Glaucoma, primary infantile
Hirschsprung disease, cardiac defects
Schwartz-Jampel syndrome
Hypophosphatasia, infantile, childhood
Breast cancer, ductal
Cutaneous malignant melanoma/dysplastic nevus
p53-related protein
Serotonin receptors
Schnyder crystalline corneal dystrophy
Kostmann neutropenia
Oncogene MYC, lung carcinoma-derived
Deafness, autosomal dominant
Porphyria
Epiphyseal dysplasia, multiple, type 2
Intervertebral disc disease
Lymphoma, non-Hodgkin
Breast cancer, invasive intraductal
Colon adenocarcinoma
Maple syrup urine disease, type II
Atrioventricular canal defect
Fluorouracil toxicity, sensitivity to
Zellweger syndrome
Stickler syndrome, type III
Marshall syndrome
Stargardt disease
Retinitis pigmentosa
Cone-rod dystrophy
Macular dystrophy, age-related
Fundus flavimaculatus
Hypothyroidism, nongoitrous
Exostoses, multiple
Pheochromocytoma
Psoriasis susceptibility
Limb-girdle muscular dystrophy, autosomal dominant
Pycnodysostosis
Vohwinkel syndrome with ichthyosis
Erythrokeratoderma, progressive symmetric
Anemia, hemolytic
Elliptocytosis
Pyropoikilocytosis
Spherocytosis, recessive
Schizophrenia
Lupus nephritis, susceptibility to
Migraine, familial hemiplegic
Emery-Dreifuss muscular dystrophy
Cardiomyopathy, dilated
Lipodystrophy, familial partial
Dejerine-Sottas disease, myelin P-related
Hypomyelination, congenital
Nemaline myopathy, autosomal dominant
Lupus erythematosus, systemic, susceptibility
Neutropenia, alloimmune neonatal
Viral infections, recurrent
Antithrombin III deficiency
Atherosclerosis, susceptibility to
Glaucoma
Tumor potentiating region
Nephrotic syndrome
Sjogren syndrome
Coagulation factor deficiency
Alzheimer disease
Cardiomyopathy
Factor H deficiency
Membroproliferative glomerulonephritis
Hemolytic-uremic syndrome
Nephropathy, chronic hypocomplementemic
Epidermolysis bullosa
Popliteala pterygium syndrome
Ectodermal dysplasia/skin fragility syndrome
Usher syndrome, type 2A
Kenny-Caffey syndrome
Diphenylhydantoin toxicity

Homocystinuria
Neuroblastoma (neuroblastoma suppressor)
Rhabdomyosarcoma, alveolar
Neuroblastoma, aberrant in some
Exostoses, multiple-like
Opioid receptor
Hyperprolinemia, type II
Bartter syndrome, type 3
Prostate cancer
Brain cancer
Charcot-Marie-Tooth neuropathy
Muscular dystrophy, congenital
Erythrokeratodermia variabilis
Deafness, autosomal dominant and recessive
Glucose transport defect, blood-brain barrier
Hypercholesterolemia, familial
Neuropathy, paraneoplastic sensory
Muscle-eye-brain disease
Medulloblastoma
Basal cell carcinoma
Corneal dystrophy, gelatinous drop-like
Leber congenital amaurosis
Retinal dystrophy
B-cell leukemia/lymphoma
Lymphoma, MALT and follicular
Mesothelioma
Germ cell tumor
Sezary syndrome
Colon cancer
Neuroblastoma
Glycogen storage disease
Osteopetrosis, autosomal dominant, type II
Waardenburg syndrome, type 2B
Vesicoureteral reflux
Choreoathetosis/spasticity, episodic (paroxysmal)
Hemochromatosis, type 2
Leukemia, acute
Gaucher disease
Medullary cystic kidney disease, autosomal dominant
Renal cell carcinoma, papillary
Insensitivity to pain, congenital, with anhidrosis
Medullary thyroid carcinoma
Hyperlipidemia, familial combined
Hyperparathyroidism
Lymphoma, progression of
Porphyria variegata
Hemorrhagic diathesis
Thromboembolism susceptibility
Systemic lupus erythematosus, susceptibility
Fish-odor syndrome
Prostate cancer, hereditary
Chronic granulomatous disease
Macular degeneration, age-related
Epidermolysis bullosa
Chitotriosidase deficiency
Pseudohypoaldosteronism, type II
Hypokalemic periodic paralysis
Malignant hyperthermia susceptibility
Glomerulopathy with fibronectin deposits
Metastasis suppressor
Measles, susceptibility to
van der Woude syndrome (lip pit syndrome)
Rippling muscle disease
Hypoparathyroidism-retardation-dysmorphism syndrome
Ventricular tachycardia, stress-induced polymorphic
Fumarase deficiency
Chediak-Higashi syndrome
Muckle-Wells syndrome
Zellweger syndrome
Adrenoleukodystrophy, neonatal
Endometrial bleeding-associated factor
Left-right axis malformation
Prostate cancer, hereditary
Chondrodysplasia punctata, rhizomelic, type 2

SOURCE: Genome Management Information System, Oak Ridge National Laboratory, U.S. Department of Energy.

Center for Genome Resources in Santa Fe, New Mexico, uses human genome sequences in an effort to improve human health by improving diagnostics, control, and treatment of disease. They have used genome sequence information to design a screening test that can allow potential parents to know if they are carriers for any of 448 debilitating or fatal diseases. This information can help parents understand their risk of having a child with one of these diseases. They also have developed a test to identify emergency room patients who are at higher risk of developing severe reactions to sepsis (bloodstream infections) and are working on identifying genetic changes that are important in the development of certain types of cancer (INFOGRAPHIC M3.4).

Perhaps the most exciting application of the information is the prospect of "personalized medicine"–treatments based on your genome, tailored specifically to you. It is now possible to sequence a person's genome in a matter of days, and it may soon be possible to use this information to make ever more precise diagnoses and to link specific treatments to your likelihood of responding. Even smart phone app creators are getting into the action: genomic data may one day soon be employed to help you find a compatible mate.

The human genome sequence will remain a crucial scientific tool for years to come. ■

MORE TO EXPLORE

▶ **The Human Genome Project Archive** http://www.ornl.gov/sci/techresources/Human_Genome/home.shtml

▶ **Resnick, R. (2011) TED Talk: Welcome to the genomic revolution** http://www.ted.com/talks/richard_resnick_welcome_to_the_genomic_revolution.html

▶ **Davies, K. (2001) *Cracking the Genome: Inside the Race to Unlock Human DNA*. New York: Free Press.**

MILESTONES IN BIOLOGY 3 Test Your Knowledge

1 **What was the typical sequence length that could be obtained in a single read at the time the HGP began?**

a. ~25 nucleotides

b. ~50 nucleotides

c. ~100 nucleotides

d. ~500 nucleotides

e. ~1,500 nucleotides

2 **Why were computational advances particularly important for the success of the Venter–Celera approach to sequencing the human genome?**

3 **Both Collins and the NIH team and Venter and the Celera team relied on shotgun sequencing at some point. What differed between the approaches of the two teams?**

4 **What were some surprises in the human genome?**

a. The number of nucleotides was much smaller than expected.

b. The number of genes was much smaller than expected.

c. A small new chromosome was discovered.

d. The sequence is extremely similar between different people.

e. all of the above

5 **The National Center for Genome Resources has developed a test that allows couples to determine whether they are carriers for mutations in genes that can cause devastating and lethal genetic diseases in their children. Would you want to take this test before having a child? What do you think you would do if you and your partner were found to be at risk for having a child with a lethal genetic disease?**

9

NATURE'S
PHARMACY

From the bark of an ancient evergreen tree, a cancer treatment blockbuster

▶▶ **DRIVING QUESTIONS**

1. When and how does normal cell division occur in the body?

2. How do normal cells and cancer cells differ with respect to cell division?

3. How are cancer treatment decisions made for a given patient?

4. How are new cancer drugs developed?

I N 1988, DOCTORS AND PATIENTS HAILED THE DISCOVERY OF A powerful new cancer-fighting drug then making its way through clinical trials. Known as taxol, the drug had remarkable effectiveness in treating advanced ovarian cancer. Some 30% of women with ovarian cancer who took taxol saw their tumors shrink, even after other drugs had failed to have any effect. Two years later, taxol was found to be even more effective at treating breast cancer: 50% of patients saw their tumors regress.

For all these women, taxol was a lifesaver, and demand for the drug sky-rocketed. There was just one problem: taxol was isolated from the bark of the Pacific yew tree, found only in old-growth forests of the Pacific Northwest. If every woman with ovarian cancer who needed the drug got it, it would mean harvesting 1.3 million pounds of bark and killing 360,000 trees. Add in the trees required to treat patients with breast cancer, and the numbers were staggering.

"You needed the bark from a fairly good-sized tree to treat one woman with breast cancer," says Susan Band Horwitz, a molecular pharmacologist

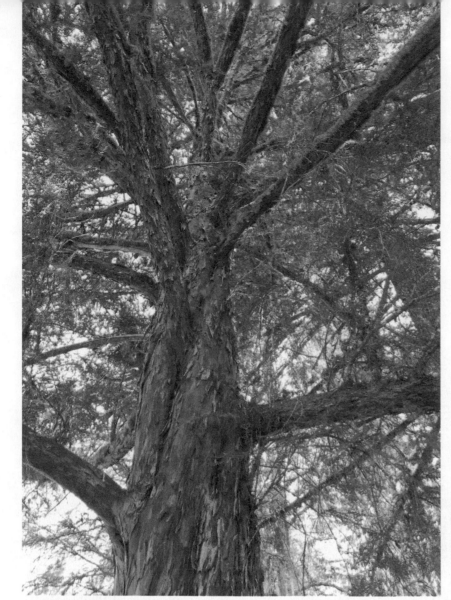

The Pacific yew tree (Taxus brevifolia).

▶CANCER
A disease of unregulated cell division: cells divide inappropriately and accumulate, in some instances forming a tumor.

▶CHEMOTHERAPY
The treatment of disease, specifically cancer, by the use of chemicals.

▶CELL DIVISION
The process by which a cell reproduces itself; cell division is important for normal growth, development, and repair of an organism.

talists countered it was shortsighted to bulldoze a national treasure to harvest a chemical whose source would be depleted in a matter of years. For researchers involved with the project, the message was clear: find a solution to the supply problem or watch a promising new drug go up in smoke.

POISON PILL

Although taxol made a splash when it first entered the clinic, it was no overnight success story. In fact, the roots of the drug extend back more than a half-century. Before 1945, there were basically two ways to treat **cancer:** cut it out or zap it with radiation. These methods worked well for some cancers, but against others they were relatively powerless. Researchers were desperate for a better way.

Then, during World War II, medical officers stumbled upon an important find: soldiers who had been exposed to the poison known as mustard gas had severely lowered white blood cell counts. Autopsies revealed that cells in the normally rapidly dividing bone marrow had stopped dividing. Since cancer of the bone marrow resulted from uncontrolled cell division there, scientists wondered if the deadly poison might be used as a cancer treatment. To their surprise and delight, tests on patients conducted in 1942 proved that it could. **Chemotherapy**–the treatment of disease by chemicals–was born.

Most commonly, chemotherapeutic drugs are used to treat cancer. There are different classes of these drugs, but as researchers soon learned, most work by interfering with a fundamental part of a cell's life: **cell division.**

Although we may think of our bodies as relatively fixed structures, most of our tissues are in constant flux as cells divide periodically to replace cells that have reached the end of their life span. In fact, cell division in our bodies begins long before we are even born. During embryonic development, a single fertilized egg cell divides, and its daughter cells divide again and again, eventually forming trillions of cells by the time a baby is born. As we age, our tissues continually discard old cells and generate

at Albert Einstein College of Medicine in New York, whose work helped bring taxol to the clinic. With 200,000 new cases of breast cancer every year, "it was just untenable to be able to get the amount of drug that was needed."

Coinciding with rising environmental concerns over deforestation, the taxol boom set the stage for a showdown: on the one side were patients and their families clamoring for a lifesaving drug; on the other side, environmentalists seeking to protect the forest. The controversy reached its climax in 1991, when the national news media caught wind of the story: "Save a Life, Kill a Tree?" asked an editorial in the *New York Times.*

Patient advocates argued it was unethical to value trees more than people. Environmennew ones in their place. And when we cut or injure ourselves, cells in the area divide to heal the wound (**INFOGRAPHIC 9.1**).

Certain cells divide more frequently than others—bone marrow cells divide continuously, for example, whereas brain cells divide at a much slower rate or not at all. And cancer cells divide uncontrollably.

To produce new cells, cells pass through a series of stages collectively known as the **cell cycle.** During the cell cycle, one cell becomes two. A cell doesn't simply split in half, however. If it did, each resulting cell would be smaller than the original, and with each division, each cell would lose half its contents. So before a cell divides, it first makes a copy of its contents so that each new cell has the same amount of organelles, DNA, and cytoplasm as the original cell. This preparatory stage of the cell cycle, known

▸**CELL CYCLE**
The ordered sequence of stages that a cell progresses through in order to divide during its life; stages include preparatory phases (G_1, S, G_2) and division phases (mitosis and cytokinesis).

INFOGRAPHIC 9.1 WHY DO CELLS DIVIDE?

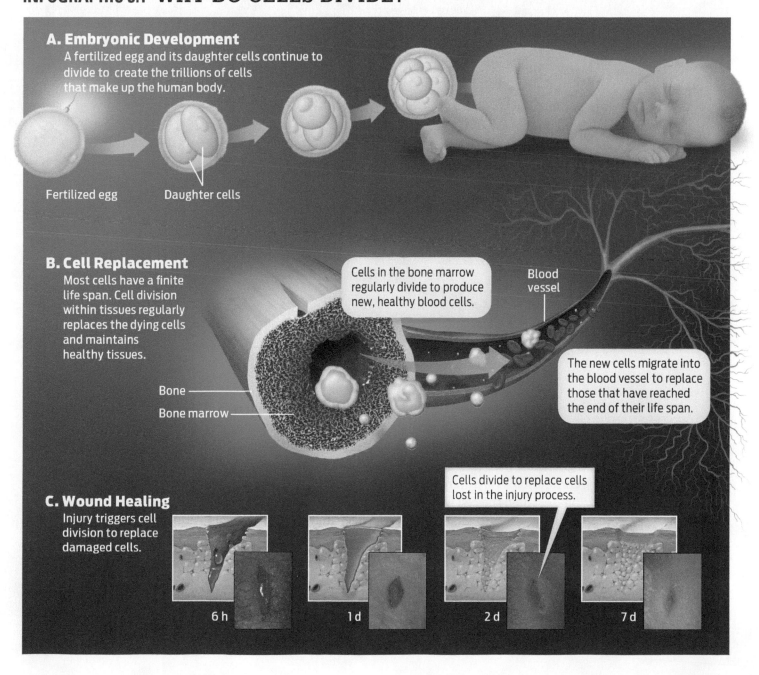

A. Embryonic Development
A fertilized egg and its daughter cells continue to divide to create the trillions of cells that make up the human body.

Fertilized egg

Daughter cells

B. Cell Replacement
Most cells have a finite life span. Cell division within tissues regularly replaces the dying cells and maintains healthy tissues.

Bone

Bone marrow

Cells in the bone marrow regularly divide to produce new, healthy blood cells.

Blood vessel

The new cells migrate into the blood vessel to replace those that have reached the end of their life span.

Cells divide to replace cells lost in the injury process.

C. Wound Healing
Injury triggers cell division to replace damaged cells.

6 h   1 d   2 d   7 d

INFOGRAPHIC 9.2 THE CELL CYCLE: HOW CELLS REPRODUCE

The purpose of the cell cycle is to replicate cells, creating two new daughter cells that are genetically identical to the original parent cell. The cell cycle consists of preparatory phases collectively known as interphase, as well as the division phases, mitosis and cytokinesis.

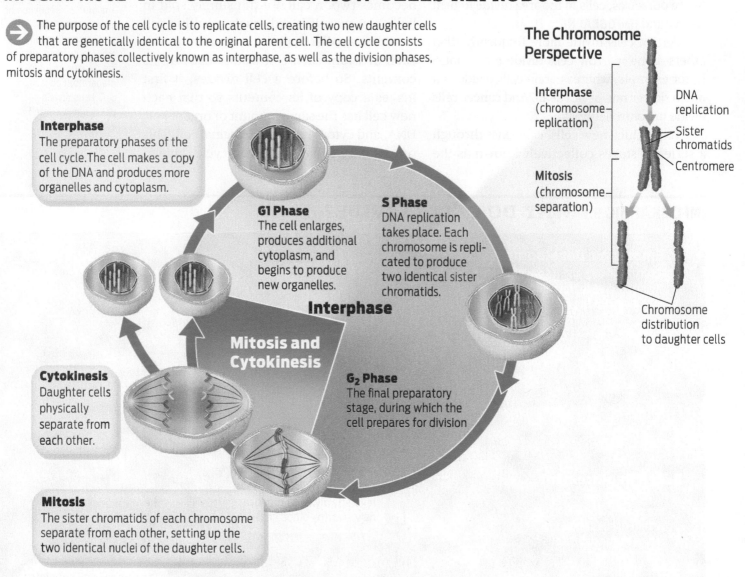

Interphase
The preparatory phases of the cell cycle. The cell makes a copy of the DNA and produces more organelles and cytoplasm.

G1 Phase
The cell enlarges, produces additional cytoplasm, and begins to produce new organelles.

S Phase
DNA replication takes place. Each chromosome is replicated to produce two identical sister chromatids.

G₂ Phase
The final preparatory stage, during which the cell prepares for division

Interphase

Mitosis and Cytokinesis

Cytokinesis
Daughter cells physically separate from each other.

Mitosis
The sister chromatids of each chromosome separate from each other, setting up the two identical nuclei of the daughter cells.

The Chromosome Perspective

Interphase (chromosome-replication)

Mitosis (chromosome-separation)

DNA replication

Sister chromatids

Centromere

Chromosome distribution to daughter cells

▶**INTERPHASE**
The stage of the cell cycle in which cells spend most of their time, preparing for cell division. There are three distinct sub-phases: G₁, S, and G₂.

▶**MITOSIS**
The segregation and separation of duplicated chromosomes during cell division.

▶**CYTOKINESIS**
The physical division of a cell into two daughter cells.

as **interphase,** has separate subphases: G₁ phase, when the cell grows larger and begins to produce more cytoplasm and organelles; synthesis phase (S), when DNA replicates and chromosomes are therefore duplicated; and G₂ phase, when the cell is ready to enter the division phases. In a cell that takes approximately 24 hours to divide, interphase takes about 22 hours to complete.

Once the cell duplicates its contents, it enters the division phases of the cell cycle: **mitosis,** when the chromosomes are evenly divided between the two daughter cells; and **cytokinesis,** when the two daughter

cells physically separate. During mitosis, the duplicated chromosomes line up along the midline of the cell. Each duplicated chromosome is made up of two identical **sister chromatids** connected at a region called the **centromere.** The two sister chromatids of each replicated chromosome are then pulled apart from each other. Each chromatid will form one of two genetically identical chromosomes in a daughter cell. During cytokinesis, the enlarged cell splits into two separate cells, each containing a full complement of organelles and DNA. In this way, one parent divides into two new daughter cells, each of which is identical to the original parent cell (**INFOGRAPHIC 9.2**). Mitosis and cytokinesis take about 2 hours to complete in cells with a 24-hour cell cycle

Most of today's chemotherapy drugs target one or more elements of the cell cycle. Some drugs interfere with cells' ability to copy their DNA and thus to replicate. Other drugs interfere with other steps of the cell cycle, such as the ability of the cells to separate their chromosomes. Unable to complete these important steps of cell division, the cells die.

WITCHES' BREW

Arthur Barclay stood at the foot of a large evergreen tree in a Washington State forest near the base of Mount St. Helens. The year was 1962. A botanist with the U.S. Department of Agriculture (USDA), Barclay was collecting plant samples as part of a new scientific research program sponsored by the National Cancer Institute (NCI). A few years earlier, researchers in Canada had made the accidental discovery that a chemical isolated from the Madagascar periwinkle plant could kill leukemia cells. Inspired by this surprising find, the NCI shifted its focus from synthetic products—the manmade chemicals they had pored over for nearly 20 years—to natural ones, specifically ones from plants.

As part of this plant program, Barclay sliced off a few stems, some needles, and a slab of bark from the tree, then dropped them in his burlap bag. He shipped the samples to a lab in Maryland, where they were screened for cancer-fighting potential.

Plants have long been known to be rich sources of medicinal products. Traditional healers and practitioners of folk medicine have relied on plants to treat illnesses for millennia. And many of the medicines and drugs we use today, including morphine, cocaine, nicotine, and aspirin, come from plants (**TABLE 9.1**). But at the beginning of the 1960s, nothing at all was known about plants that might be useful in fighting cancer. To find a therapeutic needle in the plant haystack, in 1960 the NCI created a screening program to test plant compounds for their potential as chemotherapeutic drugs. Between 1960 and 1981, NCI-affiliated scientists screened more than 100,000 plant extracts.

The overwhelming majority of these chemicals proved to be worthless as drugs. In fact, only a minuscule fraction ever received even so much as a second look. But in a few cases, the effort paid off.

In August 1962, just 2 years into the plant-screening program, the NCI received Barclay's shipment of stems, needles, and bark. As they did with all such specimens, NCI scientists ground up the materials into an extract, which they squirted on cancer cells growing in a dish. Extract from the stems and needles had no effect. But the bark, on the other hand, had a powerful one: it killed the cancer cells.

Taxus brevifolia, the Pacific yew, is one species of a family of related evergreen trees. It grows very slowly and is found only in the forests of Washington, Oregon, and western

The reddish bark of the yew tree is famous for being poisonous.

TABLE 9.1 DRUGS FROM PLANTS

The well-established drugs listed below are among dozens that were developed after scientists began to analyze the chemical constituents of plants used by traditional peoples for medicinal or other purposes.

DRUG	MEDICAL USE	PLANT SOURCE	COMMON PLANT NAME
Aspirin	Reduces pain and inflammation	Filipendula ulmana	Meadowsweet
Codeine	Eases pain, suppresses coughing	Papaver somnifenum	Opium poppy
Ipecac	Induces vomiting	Psychotria ipecacuanha	Ipecacuanha
Pilocarpine	Reduces pressure in the eye	Pilocarpus jaborandi	Jaborandi
Pseudoephedrine	Reduces nasal congestion	Ephedra sinica	Ephedra, ma huang
Quinine	Combats malaria	Cinchona pubescens	Quinine tree
Reserpine	Lowers blood pressure	Rauvolia serpentina	Serpentine wood, snakeroot
Scopalomine	Eases motion sickness	Datura stramonium	Jimson weed
Vinblastine	Chemotherapeutic drug	Catharanthus roseus	Madagascar periwinkle
Paclitaxel	Chemotherapeutic drug	Taxus brevifolia	Pacific yew tree

SOURCE: Cox, P. A., and Balick, M. J. (1994) The ethnobotanical approach to drug discovery. *Scientific American* 270 (6):82–87

Canada. The tree was considered a trash tree by the timber industry, which was more interested in Douglas fir. But people have long found uses for the yew, which has strong but pliant wood. And the reddish-barked tree has a rich history in folklore and literature, most likely because it is poisonous. (In Shakespeare's *Macbeth*, the witches add to their noxious brew "slips of yew, silvered in the moon's eclipse.") When extracts from the bark of *Taxus brevifolia* also turned out to kill cancer cells, it was not surprising that the tree caught the attention of scientists.

Two people who took particular notice were Monroe Wall and Mansukh Wani,

For researchers involved with the project, the message was clear: find a solution to the supply problem or watch a promising new drug go up in smoke.

chemists at the Research Triangle Institute in North Carolina, who spent nearly 10 years, and a mountain of bark, working to identify the specific chemical responsible for *Taxus brevifolia's* exciting effects. Finally, in 1971, they succeeded, and gave the substance a name: taxol.

"A STRUCTURE ONLY A TREE WOULD MAKE"

In 1977, Susan Band Horwitz was a young assistant professor working at Albert Einstein College of Medicine when she received a letter from the NCI asking if she would investigate a potential new drug. At the time, Horwitz had never heard of taxol—only one paper had ever been published on it—but as a molecular pharmacologist interested in the chemistry of natural products, she was immediately smitten with taxol's structure. A strange jumble of rings and tails, it was, she says, "a structure only a tree would make."

Horwitz and a graduate student, Peter Schiff, gave themselves a month to study the chemical to see if it was worth pursuing further. Within only a few weeks, they knew they had something special. They confirmed that very small quantities of taxol did indeed kill cancer cells growing in a dish. But what really caught their attention was what they saw when they looked at the cancer cells under the microscope. The cells were "loaded," Horwitz says, with long fibers known as microtubules.

Microtubules are part of the cell's cytoskeleton (see Chapter 3). Normally, during interphase, microtubules are present as a vast network of fibers running throughout the cytoplasm of the cell. However, as cells enter the early stages of mitosis, the microtubules rearrange to form a structure called the **mitotic spindle.** The microtubules (now known as spindle fibers) begin to attach to sister chromatids of duplicated chromosomes as a cell prepares to enter a phase of mitosis known as metaphase (**SEE UP CLOSE: CELL CYCLE AND MITOSIS**). The microtubules then tug at the sister chromatids from opposite sides of the cell, pulling the chromatids apart during anaphase. Mitosis is such an important part of

the cell cycle that if a cell cannot complete it properly, it self-destructs.

During normal mitosis, microtubules attach to the sister chromatids at the **kinetochore,** a collection of proteins on the centromere. The microtubules pull the sister chromatids apart by shortening at the end attached to the kinetochore. This shortening involves the disassembly of microtubules at the kinetochore end; subunits are removed from this end of the microtubule, shortening it and dragging the chromatid along with it. The shortening pulls the sister chromatids to opposite ends of the cell. Horwitz and Schiff found that taxol both interferes with the normal organization of microtubules and prevents microtubules from shortening. As a result, the cells are unable to

▶**MICROTUBULES**
Hollow protein fibers that are key components of the cytoskeleton and make up the fibers of the mitotic spindle.

▶**MITOTIC SPINDLE**
The structure that separates sister chromatids during mitosis.

▶**KINETOCHORE**
Proteins located at the centromere that provide an attachment point for microtubules of the mitotic spindle.

Susan Band Horwitz (right) and a graduate student in her lab at Albert Einstein College of Medicine.

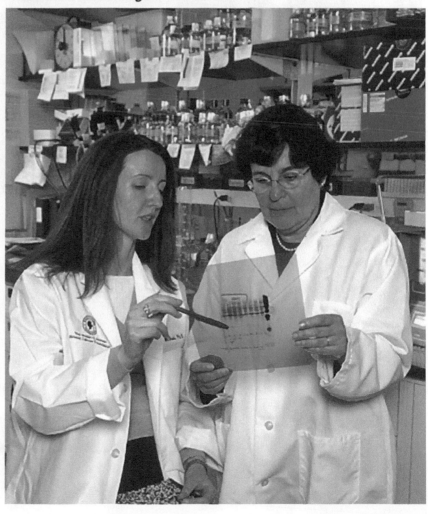

UP CLOSE CELL CYCLE AND MITOSIS

Interphase

- Each chromosome replicates in S phase, resulting in two sister chromatids connected at the centromere.
- Chromosomes are loosely gathered in the nucleus.

Animal Cells

Plant Cells

Mitosis and Cytokinesis

Prophase

- Replicated chromosomes begin to coil up.
- The nuclear membrane begins to disassemble.
- Microtubule fibers begin to form the mitotic spindle.

Metaphase

Microtubules form mitotic spindle.

- Microtubule spindle fibers from opposite ends of the cell attach to the sister chromatids of each chromosome.
- Replicated chromosomes become aligned along the middle of the cell.

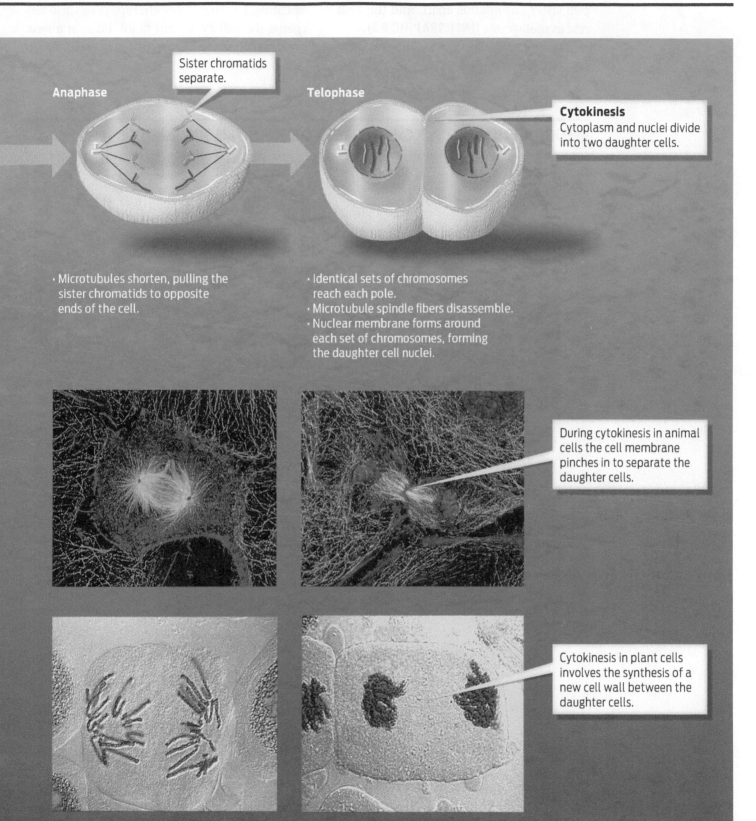

Anaphase

Sister chromatids separate.

Telophase

Cytokinesis
Cytoplasm and nuclei divide into two daughter cells.

· Microtubules shorten, pulling the sister chromatids to opposite ends of the cell.

· Identical sets of chromosomes reach each pole.
· Microtubule spindle fibers disassemble.
· Nuclear membrane forms around each set of chromosomes, forming the daughter cell nuclei.

During cytokinesis in animal cells the cell membrane pinches in to separate the daughter cells.

Cytokinesis in plant cells involves the synthesis of a new cell wall between the daughter cells.

pull sister chromatids apart, and the cells arrest in metaphase (**INFOGRAPHIC 9.3**).

No other chemical worked this way, Horwitz says. And that's what made the discovery so exciting. Existing chemotherapy drugs also targeted the cell cycle, but taxol did it in a new way, which made it a promising new candidate for drug development.

INFOGRAPHIC 9.3 TAXOL INTERFERES WITH MITOSIS

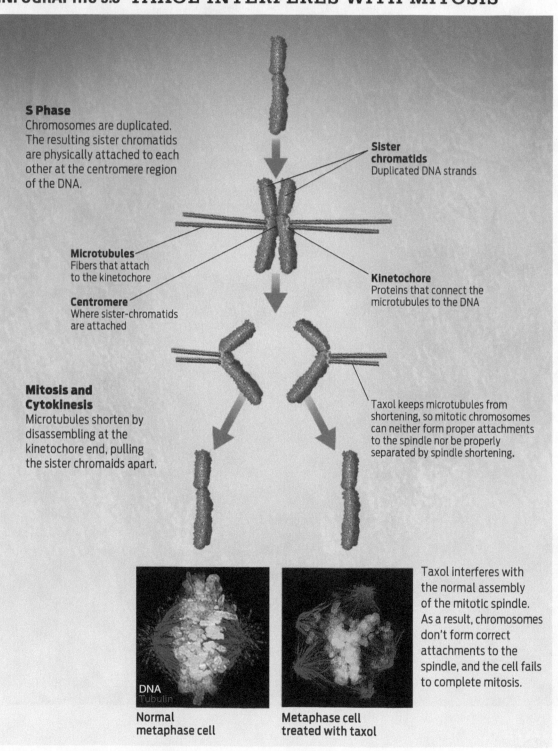

S Phase
Chromosomes are duplicated. The resulting sister chromatids are physically attached to each other at the centromere region of the DNA.

Sister chromatids
Duplicated DNA strands

Microtubules
Fibers that attach to the kinetochore

Kinetochore
Proteins that connect the microtubules to the DNA

Centromere
Where sister-chromatids are attached

Mitosis and Cytokinesis
Microtubules shorten by disassembling at the kinetochore end, pulling the sister chromaids apart.

Taxol keeps microtubules from shortening, so mitotic chromosomes can neither form proper attachments to the spindle nor be properly separated by spindle shortening.

Taxol interferes with the normal assembly of the mitotic spindle. As a result, chromosomes don't form correct attachments to the spindle, and the cell fails to complete mitosis.

DNA
Tubulin

Normal metaphase cell

Metaphase cell treated with taxol

INFOGRAPHIC 9.4 CELL DIVISION IS TIGHTLY REGULATED

 Normal cells have mechanisms to ensure that cell division is carried out precisely and only when necessary. Regulated cell division ensures that adequate cell number and healthy tissue structure are maintained in the body.

Cell Cycle Checkpoints
During the cell cycle, a system of checkpoints regulates a cell's progress. Checkpoints prevent a cell from progressing to the next stage until it accurately finishes the current stage.

Apoptosis
When a normal cell sustains irreparable damage, it undergoes programmed cell death. This cellular suicide prevents cells from producing more damaged daughter cells.

Is there a signal to divide?

G₁

S

Interphase

Mitosis and Cytokinesis

G₂

Irreparable damage?

Apoptotic cell death

Chromosomes aligned for proper chromatid separation?

Have chromosomes replicated properly? Is there DNA damage?

WHEN DIVISION RUNS AMOK: CANCER

Normally, cells divide only on demand, in response to growth signals. When cells no longer need to divide–for example, when a wound has healed or worn-out tissues have been replaced–these growth signals are turned off and cells take a break. The cells pause in their life cycle and stop dividing (although they still carry out other normal cellular functions). By contrast, cancer cells divide haphazardly. Cancer is cell division run amok.

What causes certain cells to "go rogue"? Cancer results when cells accumulate DNA sequence changes. Every time a cell replicates its DNA, for example, there is a small chance that it will make a mistake–insert the wrong nucleotide, for instance. Normally, such

mutations (see Chapter 10) are caught by the cell and fixed at what's known as a **cell cycle checkpoint.** Cells have a series of such checkpoints, which monitor each stage of the cell cycle and check for mistakes. Checkpoints also prevent progression of the cell cycle until previous stages have been successfully completed. At one checkpoint, for example, proteins scan DNA for damage or incorrect base pairing. If problems are detected, one of two things happens: either the cell ramps up DNA repair mechanisms, giving itself time to fix DNA mistakes, or, in cases of severe and irreparable damage, the checkpoints direct a cell to commit suicide, a process called **apoptosis.** Checkpoint mechanisms like these ensure that cells divide accurately and only when necessary **(INFOGRAPHIC 9.4).**

▶**MUTATION**
A change in the nucleotide sequence of DNA.

▶**CELL CYCLE CHECKPOINT**
A cellular mechanism that ensures that each stage of the cell cycle is completed accurately.

▶**APOPTOSIS**
Programmed cell death; often referred to as cellular suicide.

INFOGRAPHIC 9.5 CANCER: WHEN CHECKPOINTS FAIL

➔ Cancer cells have damaged checkpoint mechanisms, which enable them to divide when they should not. This means that DNA damage or errors in chromosome separation are passed on to daughter cells. These cells can bypass apoptosis. With each cell division, the damage is perpetuated and additional errors in DNA accumulate.

Normal Cell Division

Healthy cell

Apoptosis

DNA damage

Healthy cells divide periodically and correct DNA damage and mitotic mistakes.

Unregulated Cell Division

Healthy cell

Unrepaired DNA damage

Mistakes during DNA replication

Mistakes during mitosis

Cancer cells divide in an unregulated manner and accumulate DNA damage and make mitotic mistakes.

Proliferating cancer cells form a tumor.

Even with these repair mechanisms, however, cells with mutations do occasionally manage to complete the cell cycle and divide. This is especially likely if the mutations affect the genes that code for those proteins that function as checkpoints. Defective checkpoint proteins don't do their jobs, so additional mistakes continue to accumulate in these cells. When cells accumulate enough DNA damage to interfere with multiple checkpoints, the result is cancer. Cancer cells plow through the cell cycle uninhibited, divide uncontrollably, and in many cases eventually form a mass of cells called a **tumor (INFOGRAPHIC 9.5)**.

▶TUMOR
A mass of cells resulting from uncontrolled cell division.

Cancer kills by crowding out normal cells and invading other organs, causing them to fail. Cancer cells also secrete a variety of chemicals that wreak havoc on the body's biochemistry, with potentially fatal complications.

Ovarian cancer, which affects more than 20,000 women a year in the United States, is especially insidious. There are no clear symptoms and no effective screening tests. When the condition is eventually discovered, often the cancer has already spread, and the prognosis is poor. Although many patients respond well initially to treatment, the cancer usually recurs. Cure is rare. This is why taxol made such a name for itself in early clinical trials with ovarian cancer–it was effective when everything else had failed.

THE POLITICS OF SUPPLY AND DEMAND

Once taxol's mechanism was understood, the race was on to evaluate its clinical efficacy. Initially, researchers hoped the transition from lab bench to clinic would be straightforward. They soon made a depressing discovery: taxol is unusually hydrophobic and virtually insoluble in water, making it difficult to administer intravenously. Without some way to administer taxol to patients, it would be useless as a chemotherapy drug. At this point, many researchers were ready to abandon taxol altogether, but eventually scientists found a substance in which taxol could dissolve, and by 1984 the drug was ready for clinical testing.

Taxol progressed rapidly through clinical trials, quickly earning a name for itself as a promising new drug for the treatment of advanced ovarian and breast cancer. Taxol is "the most important new drug we have had in cancer for 15 years," Samuel Broder, director of the NCI, told the *New York Times*. "I'm not saying it's a cure, but I will tell you there are women who failed every other treatment who responded."

Yet as effective as taxol was, researchers still had a huge problem on their hands: making enough of the drug for all the patients who needed it. In 1990, just to cover the clinical trials then under way, the NCI estimated that it would need 130 kg of taxol, requiring 1,500,000 pounds of bark.

Making matters worse, researchers soon found themselves facing another hurdle: bad publicity. When, in 1987, the NCI placed an order with the USDA for 60,000 pounds of bark, it caught the attention of environmentalists–as did an even larger order, in 1991, for 750,000 pounds. Concerned that the forest was being quietly shredded, activists protested the government's plans. Working with the Environmental Defense Fund, they launched a series of petitions, with the result that, over the next few years, the NCI found itself increasingly caught up in a fierce battle with an unlikely opponent: the northern spotted owl.

This rare bird made its home in tree nooks in forests where the Pacific yew grew, and environmentalists had for years been arguing that clear-cutting was threatening the birds' habitat. Harvesting of yews for taxol, they claimed, would only exacerbate this problem. Activists lobbied successfully to get the owl classified as a threatened species under the 1963 Endangered Species Act. The owl was just one forest inhabitant, but it became the face of the environmental movement's opposition to deforestation.

Although the number of trees required to obtain taxol paled in comparison to the number of *Taxus* trees that were routinely slashed and burned during normal logging operations, the media presented the controversy as a stark battle between environmentalists and cancer patients–aided in part by the participants themselves: "I love the spotted owl, but I love people more," wrote Bruce Chabner, Chief of the Division of Cancer Treatment at the NCI, in a 1991 *Wall Street Journal* article.

The northern spotted owl.

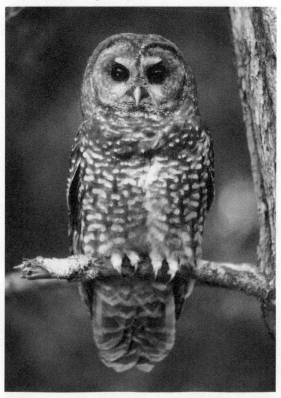

With a political firestorm complicating the already difficult supply problem, scientists at the NCI realized they would need to find an alternative source of the chemical if it was ever to see the light of day as a drug. The director of the NCI's plant program promptly got on the phone with chemists around the country, urging them to find a solution. Robert Holton, a chemist at the University of Florida, remembers the conversation vividly: "He basically told me it was time I got off my butt and did something," Holton later told the journal *Research in Review*. Eighteen months later, Holton delivered the goods. Using a method known as semisynthesis, he isolated the complicated portion of the taxol molecule from needles from the English yew, which is much more common than the Pacific yew. Then he attached the more easily synthesized unique "tail" that gives taxol its distinctive identity.

It was a huge breakthrough. If Holton's method could be developed for use on an industrial scale, then *Taxus* trees would no longer have to be killed, and the spotted owl would be spared as well.

Chemist Robert Holton—whose work led to the development of synthetic taxol—in his lab at the University of Florida.

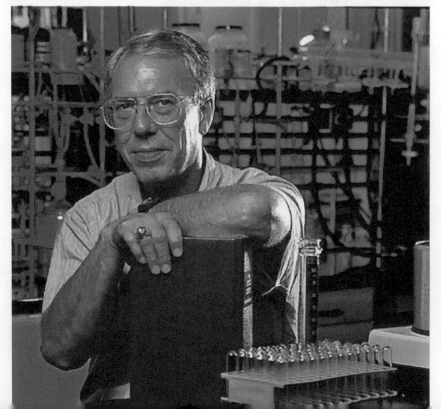

With much of the intellectual work done, and straining under the cost of its by then substantial investment of $32 million, the NCI decided that it was time to turn the project over to a pharmaceutical company equipped to make and sell the drug. (The NCI doesn't market drugs.) The lucky winner of the contract was New Jersey-based Bristol Myers, which acquired rights to taxol in 1991.

In just 2 years, Bristol was able to use Holton's method to bring a plentiful supply of the drug to market. Because much of the clinical research and development had already been completed, the process of getting Food and Drug Administration approval for the drug was speedy. Taxol was approved for the treatment of ovarian cancer in 1992 and for breast cancer in 1994.

What should have been heralded as a therapeutic triumph, however, quickly turned sour when it was discovered that Bristol was charging more than 20 times what it cost the company to produce the drug. Critics of the pharmaceutical company argued that, as a result of its expensive price tag, many of the taxpayers whose money had funded the research and development of the drug would be unable to afford it, while everyone else would effectively be paying for it twice. Bristol countered that its fee for the drug was justified given the size of its investment–$1 billion. Nonetheless, in 1993, the drug's first year on the market, congressional hearings were held to determine whether Bristol's actions were legal. No wrongdoing was found and the matter was ultimately dropped.

Bristol's decision to trademark the chemical name "taxol" also proved controversial. Normally, when a drug goes to market, it gets a new trade name (acetaminophen, for example, became Tylenol). Bristol chose not to create a new trade name, which meant that scientists had to come up with a new generic name for a chemical that had been known as taxol for nearly 30 years. Thus taxol became Taxol®, and the generic name became paclitaxel.

Nevertheless, the company had done what it was supposed to do–solve the supply problems and bring the drug to market. "Bristol did a terrific job in getting the drug to patients," says Horwitz.

INFOGRAPHIC 9.6 TAXOL PROLONGS SURVIVAL IN CANCER PATIENTS

→ Patients treated with Taxol in combination with standard chemotherapies had significantly higher response rate, longer time to progression, and longer survival time compared with standard therapy alone.

SOURCE: Sagent Pharmaceuticals

FIGHTING CANCER

For all its promise as a new chemotherapy drug, Taxol is not a cure for cancer. Battling the disease often means giving a patient more time, rather than completely eradicating the rogue cells. Cancer is a wily disease, and it has ways of outsmarting even the cleverest drug. Because the underlying mutations that led to cancer are not corrected by Taxol, new mutations may arise that allow the cell to evade the drug, leading to drug-resistant forms of the disease. But still, Taxol improves patients' chances, especially when combined with other chemotherapy drugs and other forms of treatment (**INFOGRAPHIC 9.6**).

For many types of cancer, the first line of treatment is often surgery to remove the tumor. Surgery is effective for certain solid tumors, but not for blood cancers (like leukemia), or cancers that have undergone **metastasis,** that is, that have spread to other parts of the body. In these cases, the best option is usually chemotherapy to treat the cancer cells circulating throughout the body.

If the cancer has not yet spread, doctors may also treat a tumor with radiation. In **radia-tion therapy,** beams of high-energy electrons are focused on a tumor. The radiation severely damages cell molecules and causes rampant DNA damage. This DNA damage causes the cancer cells to die, either directly, as a result of the damage itself, or by triggering apoptosis. (While typically multiple checkpoints have failed in cancer cells, others may still be working well enough to trigger apoptosis.)

The downside of both chemotherapy and radiation is that they can cause severe side effects. That's because neither therapy is very specific—both treatments damage all rapidly dividing cells in their path, including healthy ones. Healthy cells lining the intestinal tract, cells in hair follicles, and cells in the bone marrow—all of them cells that normally divide rapidly—are routinely killed by chemotherapy, which leads to nasty side effects such as vomiting, bruising, hair loss, and susceptibility to infection. Radiation therapy, if it is targeted to a part of the body with rapidly dividing cells, can also cause severe side effects, although usually they are more localized than with chemotherapy. While scientists are slowly making progress on cancer therapies that target only

▶**METASTASIS**
The spread of cancer cells from one location in the body to another.

▶**RADIATION THERAPY**
The use of ionizing (high-energy) radiation to treat cancer.

Conventional cancer treatment involves chemotherapy and radiation.

cancerous cells, chemotherapy and radiation remain the mainstays of cancer treatment.

TAXOL AND BEYOND

After elucidating the mechanism of how Taxol killed cancer cells, Horwitz continued to study the drug's effects on cells. Indeed, more than 30 years later, Horwitz's team is still hard at work on Taxol. One puzzle they are working on now is why Taxol works to treat some cancers but not others. It works well for ovarian, breast, and lung cancers, for example, but has no effect on colon or renal cancer. No one knows why.

Another major focus is the basis for acquired resistance to the drug–why cancers tend to become immune to the killing effects of Taxol over time. One possibility is that mutations in the cancer cells change the structure of microtubules so Taxol can no longer bind to them. Resistance is a big problem in cancer treatment, and it's one of the main reasons why it's helpful to have a large and varied arsenal of chemotherapeutic drugs.

In addition to Taxol, Horwitz and her team are also researching other chemicals that may one day find a use in cancer treatment. These chemicals come not from a tree, but from the sea–from invertebrate animals known as sea sponges. Although they live in very different habitats, trees and sea sponges have something very important in common, says Horwitz. They are sessile creatures, meaning they do not move and are stuck where nature put them. Horwitz thinks it is likely that the chemicals these organisms produce are an evolutionary defense mechanism against predation. Unable to run away, they make poisons instead. And it's these very poisons that humans turn to as sources of some of our most powerful medicines. If ever there was an argument for protecting biodiversity, says Horwitz, this is it.

Cervical cancer cells dividing unchecked.

Sitting in her office at Albert Einstein, surrounded by a battery of plaques and awards, Horwitz takes a moment to reflect on the impact of her scientific career. Though smitten with science for its own sake, she says she is happy her work has had such practical importance, improving the lives of so many people with cancer. To date, more than a million patients have received Taxol, which is now the best-selling cancer drug of all time. People are alive today because of the science done in her laboratory, she says. Many have even written to thank her for her work. She pauses a moment, a smile growing across her face: "And that's a wonderful feeling, after all." ▪

CHAPTER 9 Summary

- Cell division is a fundamental feature of life, necessary for normal growth, development, and repair of the body.

- The cell cycle is the sequence of steps that a cell undergoes in order to divide. Stages of the cell cycle include interphase, mitosis, and cytokinesis.

- During mitosis, replicated chromosomes segregate to opposite poles of the dividing cell; during cytokinesis, the cell physically divides into two daughter cells.

- Mitosis takes place in several phases, each of which is important to properly segregate chromosomes into daughter cells.

- Cell cycle checkpoints ensure accurate progression through the cell cycle; repair mechanisms at each checkpoint can fix mistakes that occur, such as DNA damage.

- In the absence of proper checkpoint function, cells can acquire DNA damage during cell division and pass these DNA defects on to daughter cells.

- Mistakes in the course of cell division can lead to cancer, which is unregulated cell division.

- Cancer cells have lost the ability to regulate cell division and reproduce uncontrollably, often eventually forming a tumor.

- Chemotherapy drugs work by interfering with some part of the cell cycle. Taxol works by interfering with separation of sister chromatids by microtubules during mitosis.

- Cancer is often treated with a combination of surgery, chemotherapy, and radiation.

- Chemotherapy and radiation kill all rapidly dividing cells in their path, including both cancer cells and healthy cells, which accounts for the toxicity and side effects of these treatments.

- Many drugs, including some of those used to treat cancer, are extracted from plants.

- Basic scientific research, often funded by the government, is critical to the development of new drugs to treat cancer.

MORE TO EXPLORE

- National Cancer Institute http://www.cancer.gov/

- Horwitz, S. B. (2004) Personal recollections on the early development of Taxol. *Journal of Natural Products* 67:136–138.

- Wall, M. E., and Wani, M. C. (1995) Camptothecin and Taxol: discovery to clinic. *Cancer Research* 55:753–760.

- Goodman, J., and Walsh, V. (2001) *The Story of Taxol: Nature and Politics in the Pursuit of an Anti-Cancer Drug.* Cambridge, UK: Cambridge University Press.

- Stephenson, F. (2002) A Tale of Taxol. *Florida State University Research in Review* 12(3). http://www.rinr.fsu.edu/fall2002/taxol.html

- Hirsch, J. (2006) An anniversary for cancer chemotherapy. *JAMA* 296(12):1518–1520.

CHAPTER 9 Test Your Knowledge

DRIVING QUESTION 1

When and how does normal cell division occur in the body?

By answering the questions below and studying Infographics 9.1 and 9.2 and Up Close: Cell Cycle and Mitosis, you should be able to generate an answer for the broader Driving Question above.

KNOW IT

1 Following mitosis and cytokinesis, daughter cells are

a. genetically unique.
b. genetically identical to each other.
c. genetically identical to the parent cell.
d. contain half of the parent cell's chromosomes.
e. b and c

2 In the cell cycle, DNA is replicated during

a. mitosis.
b. G_1.
c. S.
d. G_2.
e. cytokinesis.

3 What process is critical for embryonic development, wound healing, and replacement of blood cells? (Hint: All these processes require new cells.)

4 During which stage of the cell cycle do sister chromatids separate from each other?

5 During which stage of the cell cycle are sister chromatids initially produced?

USE IT

6 If a cell fails to replicate its DNA completely, what will happen?

a. It will progress through G_2 and mitosis.
b. It will die by apoptosis.
c. It will pause to allow DNA replication to complete.
d. It will stop in S phase and never progress further through the cell cycle.
e. It will stay in interphase indefinitely.

7 Many drugs interfere with cell division. Why shouldn't pregnant women take these drugs?

8 What would be the result if a cell completed interphase and mitosis but failed to complete cytokinesis—how many cells would there be at that point, and how many chromosomes relative to the parent cell would those cells have?

DRIVING QUESTION 2

How do normal cells and cancer cells differ with respect to cell division?

By answering the questions below and studying Infographics 9.2, 9.3, 9.4, and 9.5 and Up Close: Cell Cycle and Mitosis, you should be able to generate an answer for the broader Driving Question above.

KNOW IT

9 A normal cell that sustains irreparable DNA damage will most likely

a. divide out of control.
b. die by apoptosis.
c. arrest in G_2.
d. immediately go back to S phase.
e. stop in S phase and never progress through the cell cycle.

10 Which checkpoint prevents a normal cell from completing its cell cycle if it has not accurately replicated its DNA?

a. interphase (G_1–S)
b. interphase (G_2–M)
c. mitotic (M)
c. a and b
e. all of the above

USE IT

11 After a bad sunburn, skin usually peels. What process best describes what has happened to the burned skin cells?

a. skin cancer
b. metastasis
c. apoptosis
d. checkpoint failure
e. cytokinesis

12 Cancerous cells may not peel after a bad sunburn. What process has failed in this case?

a. DNA replication
b. signaling to divide
c. cytokinesis
d. checkpoints
e. mitosis

13 Complete the table below by placing a checkmark to indicate which cells will divide in which conditions. Growth factors (GFs) are molecules that signal normal cells to divide.

CELL TYPE	NO GFs	GFs PRESENT
NORMAL SKIN CELL		
MELANOMA SKIN CANCER CELL		

DRIVING QUESTION 3

How are decisions about treatment made for a given cancer patient?

By answering the questions below and studying Infographics 9.6 and 9.7, you should be able to generate an answer for the broader Driving Question above.

KNOW IT

14 A patient has metastatic melanoma—skin cancer that has spread throughout the body. Is surgery a viable option for this patient? Why or why not?

15 Which of the following properties should a promising new ovarian cancer drug have?
   a. blocks mitosis in noncancerous cells
   b. blocks mitosis in cancerous cells
   c. prevents entry into S phase in normal cells
   d. enhances the activity of cell cycle checkpoints in noncancerous cells
   e. a and b

16 Explain why chemotherapy can cause nausea, diarrhea, and hair loss.

USE IT

17 Liver cells and neurons rarely, if ever, divide in normal circumstances. The cells lining the digestive tract are replaced by cell division on a regular basis. Explain why chemotherapy frequently causes digestive symptoms but less frequently causes cognitive symptoms.

INTERPRETING DATA

18 Refer to Infographic 9.6 to answer the following questions.
   a. If patients with stage 3 ovarian cancer have Taxol added to their chemotherapy regimen, what proportion of them will be alive after 4 years, relative to patients who do not have Taxol added to their chemotherapy regimen?
   b. For patients with breast cancer, what proportion will be disease free in 1 year if Taxol is added to their chemotherapy? What proportion will be disease free at 1 year in the absence of Taxol? How do these proportions compare to the same patients at 4 years?

MINI CASE

19 Rosa has recently been diagnosed with ovarian cancer. Tests show that her cancer has metastasized to her liver, lungs, and brain. For each of the following, discuss how it works to treat cancer and whether or not it would be feasible in Rosa's case.
   a. surgery
   b. radiation
   c. Taxol
   d. chemotherapy without Taxol

DRIVING QUESTION 4

How are new cancer drugs developed?

By studying Infographics 9.4, 9.5, and 9.6 and answering the questions below, you should be able to generate an answer for the broader Driving Question above.

KNOW IT

20 In the production of taxol:
   a. What plant was at risk from early demand?
   b. What animal was consequently put at risk?

21 If a medically valuable drug is isolated from a rare plant, which of the following steps can be taken to ensure that the drug is made available to patients in need and that the plant is protected from overuse?
   a. Try to cultivate the plant on a farm to ensure that there is sufficient supply independent of the plant in its native setting.
   b. Try to find a related (and less rare) plant that produces a related compound with similar pharmaceutical activity.
   c. Use the structure of the compound to try to make a completely synthetic version of the chemical compound in a lab.
   d. Learn more about what the drug is doing in the body and try to design synthetic molecules that might have the same activity.
   e. all of the above

USE IT

22 You are a senator sitting on a committee to review cancer drug development. You have heard testimony from a patient with drug-resistant ovarian cancer, an environmentalist, a college student, and a cancer researcher. Summarize their testimony and explain how each witness contributed to your position on the conflict between saving trees and saving lives.

BRING IT HOME

23 Your mother's best friend has been diagnosed with ovarian cancer. She was treated with a generic version of Taxol, but her cancer was resistant, and continued to grow. Her doctors have let her know that there aren't a lot of options for her beyond standard chemotherapy. She wonders why there aren't more anticancer drugs for her doctors to choose from. What can you tell her about the process of drug discovery and development that would help to explain why there are few novel anticancer drugs?

FIGHTING FATE

When cancer runs in the family, ordinary measures are not enough

▸▸ **DRIVING QUESTIONS**

1. What are mutations, and how can they occur?

2. How does cancer develop, and how can people reduce their risk?

3. Why do people with "inherited" cancer often develop cancer at a relatively young age?

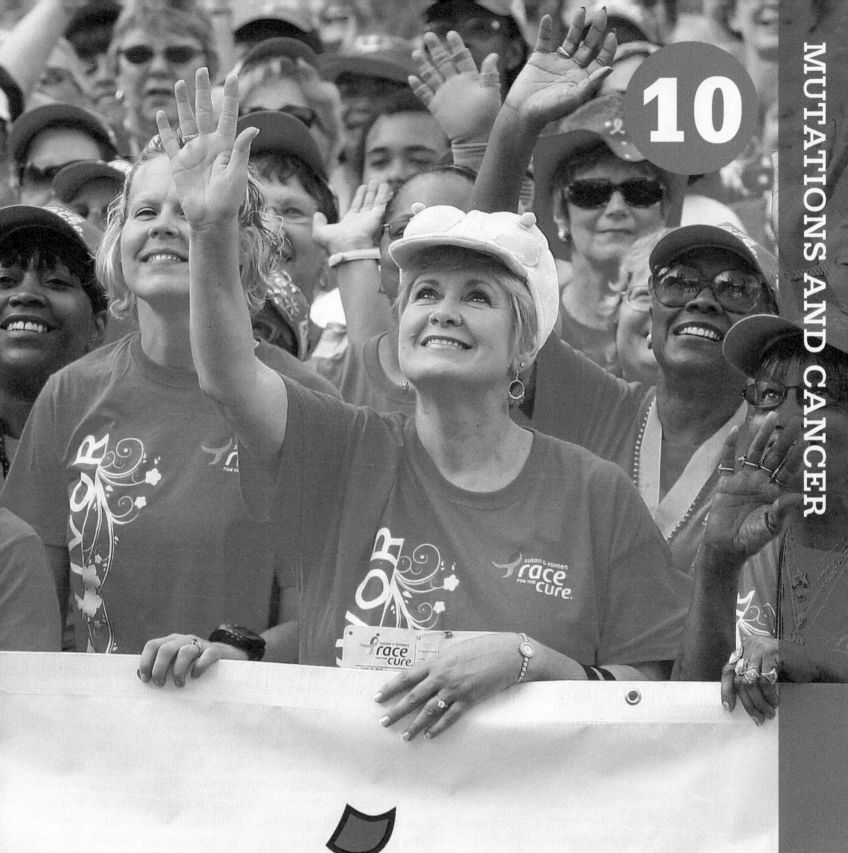

Lorene Ahern wasn't totally surprised when she tested positive for breast cancer. "Half of me was expecting it all my life and part of me was saying, 'No, this won't happen to me,'" says the 47-year-old mother of two in Twinsburg, Ohio. She knew that her risk of cancer might be higher than average—her mother had died of cancer at 49. But until the day she learned the test result, Ahern, who took good care of herself and had a healthy lifestyle, had never fully believed she would develop cancer.

There was more bad news in store for Ahern. About a year after she received the diagnosis of breast cancer, Ahern had DNA extracted from her blood and tested for mutations in two genes—*BRCA1,* located on chromosome 17, and *BRCA2,* located on chromosome 13 ("BRCA" stands for "breast cancer"). Women who are born with mutations in either of these two genes have an exceptionally high risk of developing breast and ovarian cancers; men with these mutations are at higher risk for breast and prostate cancers. This test, too, was positive: Ahern had a mutation in one of her copies of the *BRCA1* gene, which meant that she was at high risk for other cancers as well. Moreover, she could have passed on this mutation to her two children.

> **❝ Half of me was expecting it all my life and part of me was saying, 'No, this won't happen to me.' ❞**
>
> — LORENE AHERN

Aside from nonmelanoma skin cancer, breast cancer is the most common cancer to affect women. Breast cancer affects more than 200,000 women in the United States a year, according to the National Cancer Institute. For most women, the risk of developing breast cancer is about 12%, or 1 in every 8 women. For women with mutations in *BRCA1* or *BRCA2,* however, the risk is much higher: a 40% to 80% lifetime risk of developing breast cancer and a 20% to 50% risk of developing ovarian cancer, depending on the particular *BRCA* alleles they carry.

The good news is that studies have shown that diet and lifestyle changes can dramatically cut a woman's risk of getting cancer—just quitting smoking cuts the risk by 30%. The bad news is that prevention is not that simple for women with inherited predispositions to breast cancer—for this group, diet and lifestyle changes don't necessarily make a difference. "Their cancers just behave differently," says Thomas Sellers, Executive Vice President of the H. Lee Moffitt Cancer Center and Research Institute in Tampa, Florida. Even with traditional treatments like chemotherapy and radiation, hereditary breast cancers are more likely to recur in the same tissue or other tissue in the body. But these women do have treatment options that can drastically reduce their risk of getting cancer or of having it recur.

"HEREDITARY CANCER"

Cancer is a disease of uncontrolled cell division (see Chapter 9). It results from **mutations** in DNA. The mutations affect the structure and function of proteins that are

▶MUTATION
A change in the nucleotide sequence of DNA.

important for the normal functioning of the cell cycle. Often, these mutations occur in the cells of the tissue where the cancer develops—for example, the skin. Because these skin cell mutations are not passed from parent to child, neither is the skin cancer; just because your parent got skin cancer from spending too much time in the sun doesn't mean you will also pay the price.

Women who, like Ahern, have a genetic predisposition to getting cancer are often said to have "hereditary" or "inherited" cancer. This does not mean that their cancer was passed from parent to child. It means that they have inherited a mutation, from one or both parents, that makes the development of cancer much more likely. In other words, the cancer itself is not inherited, but the risk of getting it is.

In Ahern's case, an inherited mutation in one of the copies of her *BRCA1* gene predisposed her to cancer. The mutation causes the *BRCA1* gene to make a dysfunctional BRCA1 protein, which in its normal form helps to regulate the cell cycle so that a cell can repair DNA damage. One mutated gene, inherited from one parent, is enough to significantly increase the risk of getting cancer.

Recall from Chapter 8 that alternative versions of the same gene are called alleles. How does a mutated gene differ from an allele? In fact, they are essentially the same thing: an allele is a mutated version of a gene, and scientists often use the terms interchangeably. Ahern's mutation is just one of more than 600 mutations found in the *BRCA1* gene. Another way of saying this is that there are more than 600 alleles of the *BRCA1* gene in the population. Each allele has a different nucleotide sequence. These different alleles arose because mutations resulted in the substitution, deletion, or insertion of one or more nucleotides within the gene. Some of these alleles put women at high risk for developing cancer (**INFOGRAPHIC 10.1**).

To understand how there can be so many versions of a single gene in a population, remember that every time a cell replicates its DNA, mistakes can occur. About once every 10,000 to 100,000 times, DNA polymerase, the enzyme responsible for adding nucleotides to a new strand of DNA, will add the wrong nucleotide, adding for instance a guanine (G) instead of a thymine (T) to pair with an adenine (A), or adding too many or too few bases in a specific location. Similarly, a variety of chemicals—both in the environment and naturally produced in the body—can alter the base composition.

INFOGRAPHIC 10.1 MUTATIONS IN THE *BRCA1* GENE INCREASE THE RISK OF CANCER

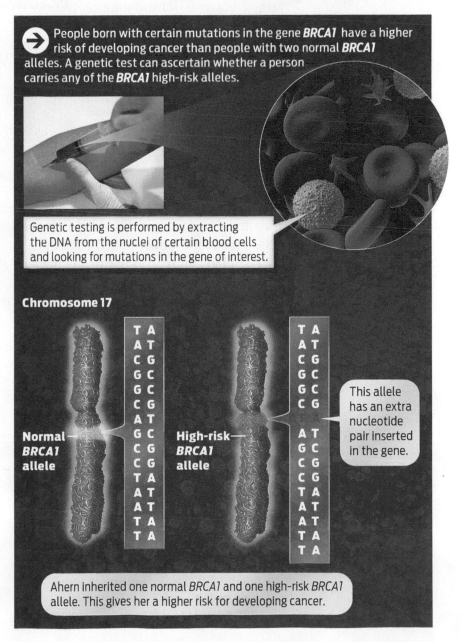

→ People born with certain mutations in the gene *BRCA1* have a higher risk of developing cancer than people with two normal *BRCA1* alleles. A genetic test can ascertain whether a person carries any of the *BRCA1* high-risk alleles.

Genetic testing is performed by extracting the DNA from the nuclei of certain blood cells and looking for mutations in the gene of interest.

Chromosome 17

Normal *BRCA1* allele

High-risk *BRCA1* allele

This allele has an extra nucleotide pair inserted in the gene.

Ahern inherited one normal *BRCA1* and one high-risk *BRCA1* allele. This gives her a higher risk for developing cancer.

INFOGRAPHIC 10.2 REPLICATION ERRORS AND OTHER DNA DAMAGE CAN PRODUCE NEW ALLELES

DNA replication errors and other alterations can lead to permanent mutations if they are not repaired or are repaired incorrectly.

Each old strand is replicated by the addition of new nucleotides.

The wrong nucleotide is added during replication, creating a mutation.

Original sequence

Mutated sequence with a nucleotide mismatch

DNA is not repaired

New allele
Replication establishes a new nucleotide sequence in one of the replicated molecules.

DNA is repaired correctly

Original allele
Sequence is restored to the original allele sequence.

DNA is repaired incorrectly

New allele
Sequence is changed (incorrectly) to create a new nucleotide pair in both replicated molecules.

▶**GERM-LINE MUTATION**
A mutation occurring in gametes; passed on to offspring.

▶**SOMATIC MUTATION**
A mutation that occurs in a body (nongamete) cell; not passed on to offspring.

Most of the time, these mistakes in the nucleotide order are repaired by a cell's error correction machinery. Groups of enzymes "proofread" the nucleotides added to the new DNA strand. If a newly added nucleotide is not complementary to its partner on the template strand–if a G is incorrectly paired with an A, for example–these enzymes replace that nucleotide with the correct one. DNA damage is usually repaired at cell cycle checkpoints (see Chapter 9), points in the cycle at which the cell is monitored for such mistakes, and where, if necessary, the cell's progress through the cycle is delayed until DNA repair has taken place. If the damage is irreparable, the cell self-destructs through apoptosis. On average, less than one mistake in 10 billion nucleotides makes it through this system of checkpoints. But think how often a given gene is replicated in a population and you can see how over time a rate of 1 in 10 billion can produce quite a number of mutations. This is how new alleles are introduced into the population (**INFOGRAPHIC 10.2**).

Not all mutations that arise in cells are passed on to offspring. Mutations are inherited by offspring only if they occur in sperm or egg cells during meiosis. That's because sperm and egg cells are the only parent cells that go into making an embryo.

Some of these mutations are so detrimental that they aren't compatible with life, and the embryo spontaneously aborts. Others aren't severe enough to harm a fetus or prevent birth, but they impair health after birth–as with diseases like cystic fibrosis and Huntington disease. Some mutations are neutral in their effects: they may change a nucleotide here and there, but don't seriously affect health. And some mutations are actually beneficial: a mutation that enables the blood to carry more oxygen, for example, might be an advantage to someone who lives in high altitudes (see Chapter 28).

Mutations we inherit from our parents are present in all the cells of our body. Moreover, these mutations are faithfully copied every time our cells divide. Such inherited mutations are also called **germ-line mutations** because the gene changes are found in the sperm and egg cells—the germ cells—and can be passed from parent to child each generation.

By contrast, **somatic mutations**—those that occur in cells in the rest of the body—are not passed on to future generations, although they can cause disease in the affected individual. A person who acquires a mutation in a skin cell from too much sun exposure, for example, will not pass this mutation on to his or her children. That's because the mutation did not occur in sperm or egg cells, nor will it affect those cells. This mutation can, however, be passed by mitotic cell division to daughter cells of the mutated cell and cause disease in the affected person. In the case of mutations caused by UV exposure, for example, some of these may occur in skin cells and lead to the development of skin cancer. This is one way nonhereditary (or "sporadic") cancers develop (**INFOGRAPHIC 10.3**).

Now imagine germ-line mutations accumulating over thousands of years in a population. As long as a mutation does not affect the

An MRI showing several malignant tumors in the breast of a 32-year-old woman.

ability to reproduce, it will be passed on to future generations through sexual reproduction. The result is that a single gene such as *BRCA1* can have hundreds of different nucleotide sequences, or alleles, in a population.

INFOGRAPHIC 10.3 MUTATIONS CAN BE HEREDITARY OR NONHEREDITARY

Mutations that occur in sperm or egg cells are germ-line mutations that can be passed on to offspring. These inherited mutations are then found in every cell of the offspring, including egg or sperm cells, and can be passed on to subsequent generations. Somatic mutations are those that occur in any cell in the body other than the sperm or egg. These mutations are not inherited but may cause disease in the individual that acquires them.

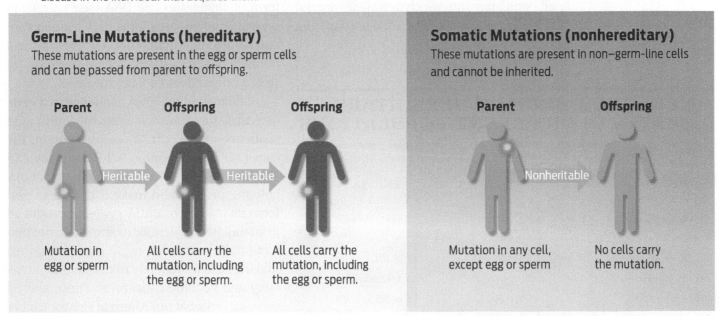

Germ-Line Mutations (hereditary)
These mutations are present in the egg or sperm cells and can be passed from parent to offspring.

Parent	Offspring	Offspring
Heritable →	Heritable →	
Mutation in egg or sperm	All cells carry the mutation, including the egg or sperm.	All cells carry the mutation, including the egg or sperm.

Somatic Mutations (nonhereditary)
These mutations are present in non–germ-line cells and cannot be inherited.

Parent	Offspring
Nonheritable	
Mutation in any cell, except egg or sperm	No cells carry the mutation.

ETHNIC GROUPS AND GENETIC DISEASE

Ahern was not being irrationally fearful about getting cancer. She knew that both her family history and ethnic background put her at significantly increased risk. Ahern descends from a subgroup of Jews known as the Ashkenazi; the term generally refers to Jews of German and Eastern European descent. Ahern's father was born in Germany, immigrating to the United States in 1939; her maternal grandfather was born in Russia.

The history of this Jewish subgroup extends much further back than modern Europe. The Ashkenazi Jews are a subgroup that left the Middle East and began populating parts of Europe more than 2,000 years ago. The majority of Ashkenazis migrated into Europe in the 10th century from the region of present-day Israel, settling in the Rhineland, the valley of the Rhine River, in Germany; from there, many migrated east into Poland and Russia.

A number of historical factors have made the Ashkenazi Jewish population more susceptible than others to genetic diseases. First, they descend from a small group of people. Second, that population has expanded and contracted over time. Third, and most important, members of the population usually marry within the community. In other words, Ashkenazi Jews have many of the characteristics of an isolated population into which new alleles are not frequently introduced by people immigrating into the population.

Consequently, Ashkenazi Jews are an example of an ethnic group that has a more homogeneous genetic background than the general population, and is more likely to pass on certain genetic diseases to future generations. Scientists have discovered more than 1,000 recessive diseases in the general population, but most of them are rare. In Ashkenazi Jews, however, the prevalence of some recessive diseases is increased 100-fold or more. Tay-Sachs disease, Gaucher disease, and Bloom syndrome are genetic diseases that all occur more frequently in this ethnic group than in the general population; approximately 1 in 25 Ashkenazi Jews carries disease alleles for at least one of these disorders (**TABLE 10.1**).

Ashkenazi Jews are not the only ethnic group to have a higher incidence of certain genetic diseases than occurs in the general population. For example, people from Mediterranean, African, and Asian countries have higher rates of thalassemias–blood disorders that cause anemia. Sickle-cell anemia, another type of hereditary anemia, is more common among people of African descent.

Ashkenazi Jews are also more likely than the general population to carry mutations in *BRCA1* and *BRCA2*. Some studies have found that more than 8% of Ashkenazi women carry a mutated *BRCA1* gene, compared to only 2.2% of other women. These alleles can take the form of changes in one DNA base pair, or in several. In some cases, large DNA segments are rearranged. In mutated *BRCA2* genes, a small number of additional DNA base pairs are inserted into or deleted from the gene.

A change in the DNA sequence of a gene can cause problems if it alters the amino acid sequence of the corresponding protein. Because an altered amino acid sequence can also alter the shape of the protein, it can disable the protein and make it unable to perform its usual job. BRCA proteins produced from mutated alleles, for example, do not perform their job as cell cycle regulators, thus making cells more likely to divide uncontrollably and become cancerous. These altered proteins are what put Ahern at risk for cancer (**INFOGRAPHIC 10.4**).

TABLE 10.1 INCIDENCE OF HEREDITARY DISEASES IN DIFFERENT POPULATIONS

HEREDITARY DISEASE	CARRIER RATE IN ASHKENAZI JEWISH POPULATION	CARRIER RATE IN GENERAL POPULATION
Tay-Sachs disease	1 in 25	1 in 250
Canavan disease	1 in 40	Rare/unknown
Niemann-Pick disease, type A	1 in 90	1 in 40,000
Gaucher disease, type 1	1 in 14	1 in 100
Bloom syndrome	1 in 100	Rare/unknown
BCRA mutation	1 in 40	1 in 350–1,000
Familial dysautonomia	1 in 30	Rare/unknown

INFOGRAPHIC 10.4 MUTATIONS IN DNA CAN ALTER PROTEIN FUNCTION AND CAUSE CANCER

Mutations alter the nucleotide sequence of DNA. If a mutation changes the coding region of any gene, the corresponding protein may be dysfunctional. When the protein in question helps regulate the cell cycle, cancer may result.

Normal allele sequence

DNA
T A C G G C A G C C T A C C T
A T G C C G T C G G A T G G A

Mutated allele sequence

T A C G G C A T G C C T A C C T
A T G C C G T A C G G A T G G A

This sequence has an insertion of one base pair (shown in red).

Transcription

Transcription

mRNA
AUG CCG UCG GAU GGA ...

AUG CCG UAC GGA UGG A ...

This mRNA has a nucleotide insertion that shifts the nucleotides after it to the right.

Translation

Translation

Amino acid sequence

Met Pro Ser Asp Gly

Met Pro Tyr Gly Trp

Amino acids coded from sequences after the mutation are different from the original.

Protein

Functional protein
In its normal form, this protein regulates the cell cycle.

Nonfunctional protein
This protein has an altered shape, rendering it dysfunctional and unable to regulate the cell cycle. Cancer may result.

THE MULTI-HIT MODEL

Inheriting certain *BRCA* alleles dramatically increases the risk of cancer, but it doesn't mean that the disease will necessarily develop. In most cases, hereditary cancer occurs only when additional, nonhereditary mutations in a cell accumulate. Similarly, the accumulation of harmful somatic mutations in a cell can lead to cancer, even in someone with no genetic predisposition. This is known as the multi-hit model of cancer, where each "hit" is a mutation, and multiple hits are needed to cause the disease.

What causes mutations, other than DNA replication mistakes? Environmental insults such as chemicals, ultraviolet light, radiation, and other factors can damage DNA and cause it to mutate. Exposure to ultraviolet light, for example, damages the DNA in our skin cells and can lead to skin cancer.

> Environmental insults such as chemicals, ultraviolet light, radiation, and other factors can damage DNA and cause it to mutate.

INFOGRAPHIC 10.5 WHAT CAUSES MUTATIONS?

→ When genes become mutated, their DNA sequence is altered. There are several ways that DNA mutations can arise: they may be inherited; they may occur during DNA replication if errors are not corrected; they can also arise from environmental insults.

Mutagens and carcinogens

Heredity

A mutation in a cancer-associated gene can be inherited from either parent.

DNA replication mistakes

Some errors are not corrected, causing new mutations.

Smoking

Pollutants and pesticides in food

Alcohol

Pollution

Char (blackened bits) on meats grilled or fried at high temperatures

UV light

Many compounds in the environment and in our diet are mutagens, producing mutations in cancer-associated genes.

▶**MUTAGEN**
Any chemical or physical agent that can damage DNA by changing its nucleotide sequence.

▶**CARCINOGEN**
Any chemical agent that causes cancer by damaging DNA. Carcinogens are a type of mutagen.

▶**PROTO-ONCOGENE**
A gene that codes for a protein that helps cells divide normally.

▶**TUMOR SUPPRESSOR GENE**
A gene that codes for proteins that monitor and check cell cycle progression. When these genes mutate, tumor suppressor proteins lose normal function.

Physical or chemical agents that cause mutations are called **mutagens.** Chemicals and other factors such as pesticides and pollutants that can cause cancer are a class of mutagen known as **carcinogens** because they damage DNA in a way that can lead to cancer. Not all mutagens originate outside the body. Some of the reactions that occur in the mitochondria during cellular respiration (see Chapter 6), for example, produce DNA-damaging molecules known as free radicals (**INFOGRAPHIC 10.5**).

Normally cells are able to repair such DNA damage. But very rarely a mistake remains uncorrected, and over time and with age, if enough mutations accumulate in the same cell, that cell may begin to divide abnormally and become cancerous. Such acquired somatic mutations can develop throughout a person's life as he or she is exposed to carcinogenic environmental insults and as cells divide.

Mutations that cause cancer typically occur in two categories of genes that influence the cell cycle: **proto-oncogenes** and **tumor suppressor genes.** Normal proto-oncogenes promote cell division and cell differentiation, but only in response to appropriate signals. When proto-oncogenes are mutated, they can become permanently "turned on," or activated, stimulating cells to divide all the time. In this state they are called **oncogenes**–genes that cause cancer. In other words, oncogenes are proto-oncogenes that have mutated to become overexpressed or permanently activated. *Her2*, a gene overexpressed in certain types of breast cancer, is an example of a proto-oncogene.

Tumor suppressor genes, also known as tumor suppressors, normally pause cell division, repair damaged DNA, and initiate apoptosis, or cell death. Tumor suppressor genes cause cancer when they are inactivated by mutation. "You can think of the suppressors as brakes and the oncogenes as the accelerators," Sellers of the H. Lee Moffitt Cancer Center and Research Institute explains. "They are sort of the yin and yang of each other." If a proto-oncogene is mutated, it's as if the accelerator is stuck down and the car–cell division–keeps going; if a tumor suppressor gene is mutated, it's as if the brakes don't work and the car cannot be stopped. Both *BRCA1* and *BRCA2* are tumor suppressors that, when normally expressed, code for proteins that help the cell progress normally through the cell cycle and respond to DNA damage (**INFOGRAPHIC 10.6**).

It usually takes more than a single mutation in a cell to cause cancer. In most cases,

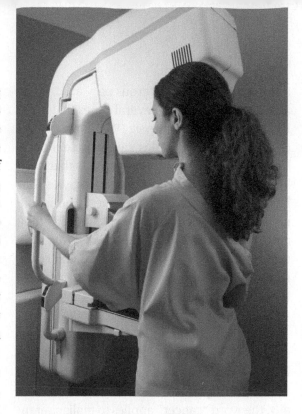

Mammograms (x-rays of the breast) are the best way to detect cancer.

▶**ONCOGENE**
A mutated and overactive form of a proto-oncogene. Oncogenes drive cells to divide continually.

INFOGRAPHIC 10.6 MUTATIONS IN TWO TYPES OF CELL CYCLE GENES CAUSE MOST TYPES OF CANCER

→ During the cell cycle, proteins regulate whether the cell is ready to continue to the next stage or if the cell requires additional time to repair DNA damage before progressing. The proteins that regulate these checkpoints are made by proto-oncogenes and tumor suppressor genes. Accumulated mutations in these types of genes cause cancer.

GO **Proto-oncogenes** signal cells to progress through the cell cycle at the appropriate time. Mutations in these genes cause them to be overactive, causing too much cell division.

STOP **Tumor suppressor genes** signal cells to pause the cell cycle to fix mistakes. Mutations in these genes cause them to be underexpressed or nonfunctional, allowing damaged cells to divide inappropriately.

Some proto-oncogenes, like *Her2*, produce proteins that enable the cell to respond to external signals that tell the cell to divide.

BRCA1 and *BRCA2* are tumor suppressor genes that produce DNA repair proteins.

Tumor suppressor genes like *p53* produce proteins that can induce apoptosis (cell death) instead of allowing the cell to progress through the cell cycle.

GO

G_1

S

Interphase

Mitosis and Cytokinesis

G_2

STOP STOP

STOP

Apoptotic cell death

Some tumor suppressor genes make proteins that suppress the cell cycle when there is DNA damage.

▶BENIGN TUMOR
A noncancerous tumor that will not spread throughout the body.

▶MALIGNANT TUMOR
A cancerous tumor that spreads throughout the body.

a cell will become cancerous only after it has acquired mutations in several genes that regulate the cell cycle or repair DNA damage. The collective mutated genes can include a combination of tumor suppressor genes that have lost their function *and* proto-oncogenes that have been activated to oncogenes. This is one reason why cancer affects people more as they age: as cells accumulate mutations over time through exposure to carcinogens and repeated rounds of cell division, the chances increase that a cell will accumulate enough mutations to become cancerous.

In the multi-hit model, tumors develop in stages following mutations. After one or two hits, a **benign tumor** may form. After several more, a **malignant tumor** may result. Generally, the word "cancer" applies to malignant tumors, which have the capacity to invade other tissues and spread to other parts of the body, or metastasize (**INFOGRAPHIC 10.7**).

People who have inherited high-risk mutations start life with at least one cancer-predisposing mutation, so they require fewer additional mutations to develop cancer. For example, Ahern was born with a predisposing mutation in one of her *BRCA1* alleles. If a second mutation in one of her somatic cells disables her second *BRCA1* allele, that cell and all its descendants will no longer be able to respond effectively to repair DNA damage. Consequently, the cells of women with *BRCA* mutations accumulate DNA damage at a faster rate, which is why hereditary breast cancer often strikes women who are in their 30s and 40s—much younger than women who

INFOGRAPHIC 10.7 TUMORS DEVELOP IN STAGES AS MUTATIONS ACCUMULATE IN A CELL

It takes more than a single mutation to cause cancer. Individuals who have inherited high-risk mutations require fewer additional mutations to get to cancer, and therefore develop cancer at a much earlier age.

A possible scenario for cancer progression (from A to D):

A

Inherit *BRCA1* mutation
A single mutation in remaining *BRCA1* alleles causes loss of *BRCA* function. DNA is not efficiently repaired. Additional mutations are more likely to occur because of these uncorrected mutations.

B

Benign tumor
(noncancerous)

DNA replication mistake not corrected, producing an oncogene
Additional mutations activate proto-oncogenes into oncogenes. The cell divides more often without normal checks, and cells begin to pile up.

C

Smoking mutates *p53* gene
Cells fail to correct mutations or to enter apoptosis. Mutations are propagated through additional cell divisions. Cells do not stop dividing even when crowded.

D Malignant tumor
(cancerous)

Additional mutations permit invasion, new blood vessel growth, and spread
The tumor contains malignant cells that invade surrounding tissues, promote new blood vessel growth, and spread to both nearby and distant locations.

have no inherited predisposition to cancer (**INFOGRAPHIC 10.8**).

Since *BRCA* genes are expressed in many cell types in addition to breast tissue, mutations in either gene raise the risk of cancers in other organs, too. Scientists have linked mutations in both genes to a higher than average risk of prostate, colon, and pancreatic cancers, among others. But the breasts and ovaries are at especially high risk of developing cancers because they respond to the hormone estrogen, which causes cells in these organs to divide more often. In breast tissue, for example, the rise in estrogen during a woman's menstrual cycle signals cells lining the milk glands to divide to prepare to produce milk should a woman become pregnant.

The *BRCA* genes aren't the only ones that predispose women to breast cancer. Scientists now think that genes other than *BRCA* cause up to half of all hereditary breast cancers. Other inherited mutations in tumor suppressor genes and proto-oncogenes have been linked to various other cancers as well.

PREVENTATIVE MEASURES

A decade ago, women diagnosed with hereditary breast cancer had to face the continuing fear of developing new cancers. Their breast cancer could be treated with chemotherapy and radiation (see Chapter 9), but because they were born with genetic predispositions that made their cells less able to repair DNA damage from the start, the likelihood that the cancer would recur or that a new cancer would develop was high. To reduce the risk, they could choose to have their breasts or ovaries surgically removed—a drastic option, and the evidence supporting surgery as means to reduce the risk of repeat cancers was slim.

But over the years, studies have shown that preventative surgery to remove both breasts or ovaries can reduce risk of cancers in those organs by as much as 90%. "The difference is that now we have empirical data," Sellers comments. There is a small risk that

INFOGRAPHIC 10.8 *BRCA* **MUTATIONS INCREASE THE RISK OF BREAST AND OVARIAN CANCER**

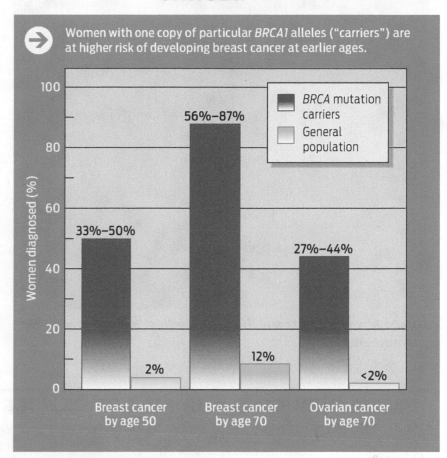

Women with one copy of particular *BRCA1* alleles ("carriers") are at higher risk of developing breast cancer at earlier ages.

Legend:
- BRCA mutation carriers
- General population

Y-axis: Women diagnosed (%)

- Breast cancer by age 50: 33%–50% (carriers), 2% (general population)
- Breast cancer by age 70: 56%–87% (carriers), 12% (general population)
- Ovarian cancer by age 70: 27%–44% (carriers), <2% (general population)

A decade ago, women diagnosed with hereditary breast cancer had to face the continuing fear of developing new cancers.

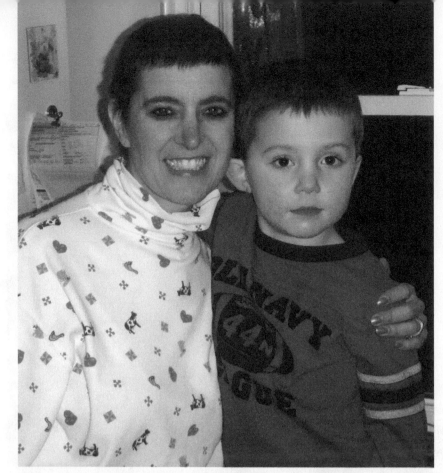

Lorene Ahern with her son. Her hair was just growing back after chemotherapy.

the cancer will recur because breast tissue is distributed across the chest wall and can be found near the armpit, above the collarbone, and as far down as the abdomen; it is impossible for a surgeon to remove all breast tissue. And removal of the breasts or ovaries will not reduce the increased risk of developing cancer in other areas of the body.

Ahern consulted a genetic counselor and decided to have both ovaries removed, a year after doctors diagnosed her breast cancer. Ahern is also considering a mastectomy to remove her breasts.

Though Ahern says she feels "pretty good" right now, there was a time when she was visiting online breast cancer discussion groups every evening after work and all weekend long. They not only helped her cope emotionally but also helped to inform her about her disease and her treatment options. Patient-oriented support groups can be a valuable source of information for women, in addition to what health professionals provide.

TABLE 10.2 REDUCING THE RISK OF CANCER

WHAT?	HOW?
Wear sunscreen.	Sunscreen helps prevent UV-induced DNA damage in skin cells. Women who use tanning beds increase their risk of skin cancer by 55%.
Avoid tobacco (both smoking and chewing).	Agents in tobacco can break DNA, causing mutations and many different cancers. Smoking causes 80%–90% of lung cancer deaths.
Avoid or reduce alcohol consumption.	Excessive consumption of alcohol increases the risk of oral, liver, and breast cancers.
Maintain a healthy weight.	Obesity and lack of physical exercise increases the risk of breast cancer in postmenopausal women. Between 14% and 20% of cancer deaths result from overweight and obesity.
Get screened.	Screening helps detect cancers early and improves the odds of successful treatment.

SPECIAL MEASURES FOR HEREDITARY CANCERS

WHAT?	HOW?
Consider genetic counseling and testing.	Genetic counseling and testing enables better-informed decisions about prevention and treatment.
Screen early.	Cancer screening at an earlier age than recommended for the general population can aid in prevention.
Consider prophylactic surgery.	Removal of tissue, mastectomy, or removal of the ovaries, for example, reduces the risk of developing cancer.
Involve other members of the family.	Genetic testing may help others in the family also make better-informed decisions.

However, this may not be the best route to support for everyone.

Unfortunately, according to Sue Friedman, executive director of Facing Our Risk of Cancer Empowered (FORCE), there is still a lot of misinformation out there about hereditary breast cancer. People still assume that diet and lifestyle changes will cut the risk of cancer in people with hereditary cancer. "Those factors may help, but not enough in our community," says Friedman, who has had cancer herself. Women also have options regarding prophylactic surgery—when to have it and how much is necessary—that aren't always effectively communicated by health professionals (**TABLE 10.2**).

Scientists admit that surgery isn't the most palatable treatment. "Surgery cuts your risk substantially, but it's still pretty traumatic," says Sellars. "It would be nice to say we've got a medication you can take and you'll have the same effect. But we just don't have that kind of treatment right now." ■

CHAPTER 10 Summary

▸ Cancer is uncontrolled cell division caused by mutations in DNA.

▸ Mutations occur spontaneously during DNA replication. They can also be caused by environmental triggers such as tobacco or UV radiation and by chemicals naturally produced by the body.

▸ Mutations in certain genes can lead to cancer if they damage the normal function of the proteins those genes code for.

▸ Mutations in two types of gene, proto-oncogenes and tumor suppressors, cause most cancers.

▸ Multiple mutations must occur in the same cell for it to become cancerous.

▸ Mutations that occur in somatic (body) cells, for example skin or breast cells, are not inherited by offspring. Mutations that occur in germ (sperm or egg) cells are inherited by offspring.

▸ People with "hereditary" cancer inherit predispositions to the disease in the form of specific genetic mutations. These mutations are present in all body cells and can serve as the first mutation that may lead to cancer.

▸ Women with *BRCA* mutations have a much higher risk of developing cancer, and at an earlier age, than women without these mutations.

▸ Mutations introduce new alleles into the population. These alleles may code for proteins that have advantageous, neutral, or harmful effects on an individual.

▸ Certain alleles are more common in ethnic groups that have been reproductively isolated for long periods of time.

MORE TO EXPLORE

▸ **National Cancer Institute FactSheet: BRCA1 and BRCA2** http://www.cancer.gov/cancertopics/factsheet/Risk/BRCA

▸ **American Cancer Society: Heredity and Cancer** http://www.cancer.org/cancer/cancercauses/geneticsandcancer/heredity-and-cancer

▸ **Susan G. Komen for the Cure** http://ww5.komen.org/

▸ **Genetic Disease Foundation** http://www.geneticdiseasefoundation.org/

▸ **Center for Jewish Genetics** http://www.jewishgenetics.org/

CHAPTER 10 Test Your Knowledge

DRIVING QUESTION 1

What are mutations, and how can they occur?

By answering the questions below and studying Infographics 10.1, 10.2, 10.3, and 10.4, you should be able to generate an answer for the broader Driving Question above.

KNOW IT

1 In an otherwise normal cell, what happens if one mistake is made during DNA replication?

a. Nothing; mistakes just happen.
b. A cell cycle checkpoint detects the error and pauses the cell cycle so the error can be corrected.
c. The cell will begin to divide out of control, forming a malignant tumor.
d. A checkpoint will force the cell to carry out apoptosis, a form of cellular suicide.
e. The mutation will be inherited by the individual's offspring.

2 Why does wearing sunscreen reduce cancer risk?

a. Sunscreen can repair damaged DNA.
b. Sunscreen can activate checkpoints in skin cells.
c. Sunscreen can reduce the chance of mutations caused by exposure to UV radiation present in sunlight.
d. It doesn't; sunscreen causes mutation and actually increases cancer risk.
e. Sunscreen can prevent cells with mutations from being destroyed.

3 Are all mutations bad? Explain your answer.

USE IT

4 The mutation illustrated in Infographic 10.4 inserted an A in the third codon of the mRNA shown. Use the genetic code (Infographic 8.10, p. 176) to match each mutation below (all are mutations of the normal mRNA sequence shown in Infographic 10.4) with its effect on the protein.

Mutation

Substitution of an A for the C in the third codon
Substitution of a C for the U in the fourth codon
Substitution of an A for the first C in the second codon

Effect

_____ Protein will have an incorrect amino acid in its sequence.
_____ No impact on the protein
_____ Protein will be shorter than normal.

DRIVING QUESTION 2

How does cancer develop, and how can people reduce their risk?

By answering the questions below and studying Infographics 10.3, 10.4, 10.5, and 10.7, you should be able to generate an answer for the broader Driving Question above.

KNOW IT

5 What are some differences and some similarities between tumor suppressor genes and oncogenes?

6 What is the role of *BRCA1* in normal cells?

7 Which of the following can cause cancer to develop and progress?

a. a proto-oncogene
b. an oncogene
c. a tumor suppressor gene
d. a mutated tumor suppressor gene
e. b and d
f. b and c

USE IT

8 What would you say to a niece if she asked you how she could reduce her risk of breast cancer? (Assume there is no family history of breast cancer.) How might each of your suggestions reduce her risk?

9 Why is age a risk factor for cancer?

DRIVING QUESTION 3

Why do people with "inherited cancer" often develop cancer at a relatively young age?

By answering the questions below and studying Infographics 10.1, 10.6, and 10.7, you should be able to generate an answer for the broader Driving Question above.

KNOW IT

10 A woman with a *BRCA1* mutation

a. will definitely develop breast cancer.
b. is at increased risk of developing breast cancer.
c. must have inherited it from her mother because of the link to breast cancer.
d. will also have a mutation in *BRCA2*.
e. b and c

11 Which of the following family histories most strongly suggests a risk of inherited breast cancer due to *BRCA1* mutations?

a. many female relatives who were diagnosed with breast cancer in their 70s

b. many relatives with skin cancer

c. many relatives diagnosed with skin cancer at an early age

d. many female relatives diagnosed with breast cancer at an early age

e. many female relatives with both early breast cancer and ovarian cancer

USE IT

12 Lorene Ahern was born with an inherited predisposition to cancer. At the cellular and genetic level, what was she born with? At birth, were cells in her breast genetically identical to cells in her liver? Now that she has breast cancer, are her cancer cells genetically identical to her normal breast cells? Explain your answers.

13 Which of the following women would be most likely to benefit from genetic testing for breast cancer?

a. a 25-year-old woman whose mother, aunt, and grandmother had breast cancer

b. a healthy 75-year-old woman with no family history of breast cancer

c. a 40-year-old woman who has a cousin with breast cancer

d. a 55-year-old woman whose older sister was just diagnosed with breast cancer

e. All women can benefit from genetic testing for breast cancer.

14 People like Lorene Ahern have inherited a mutated version of *BRCA1*. Why does this mutation pose a problem? Why are these people at high risk of developing breast cancer when they still have a functional *BRCA1* allele? Describe how the protein encoded by normal *BRCA1* compares to that encoded by mutant alleles of *BRCA1*.

MINI CASE

15 Nellie has a family history similar to Lorene Ahern's. Nellie's mother died at an early age from breast cancer, as did her maternal aunt (her mother's sister). Nellie is not yet 35 but has started having annual mammograms. She has also been tested for *BRCA1* and *BRCA2* mutations. She has a *BRCA2* mutation and is considering prophylactic surgery. Her younger sister, Anne, doesn't want to know the results of Nellie's genetic testing because if Nellie has a *BRCA2* mutation, then there is a chance that Anne could have inherited the same mutation from their mother. Does Nellie or Nellie's doctor have an obligation to tell Anne about the test results? What about Nellie's older brother? Should he be told? There are personal and medical benefits and risks to consider here.

INTERPRETING DATA

16 Refer to Infographic 10.8 to answer the following questions.

a. What is the average percentage of *BRCA1* carriers who develop breast cancer by age 50?

b. Exactly how many times more likely are *BRCA1* carriers to develop breast cancer by age 50 compared to the general population?

c. What is the average percentage of *BRCA1* carriers who develop breast cancer by age 70%

d. Exactly how many times more likely are *BRCA1* carriers to develop breast cancer by age 70 than the general population?

e. Why are both carriers and noncarriers more likely to have developed breast cancer by age 70 than by age 50?

BRING IT HOME

17 If you wanted to change your lifestyle to reduce your risk of developing cancer, what specific steps could you take with respect to each of the following? Be as specific as you can. Take your age and gender into consideration as you consider each factor.

a. alcohol consumption

b. sun exposure

c. tobacco use

d. exposure to pesticides

e. meat preparation (cooking method)

11

In 2003, Emily and her friends launched an all-girl band called Hellen.
From left to right: Katie, Charmain, Becca, Emily, and Amanda.

ROCK
FOR A CAUSE

From patient to performer, shining the spotlight on a genetic disease

EMILY SCHALLER HAD NO IDEA THAT HER DETROIT-BASED rock band would be strumming its way to fame one day. Back in 2003, she and her friends were goofing around singing songs when someone floated the idea of forming a band. None of them could play guitar, bass, or drums, but that didn't stop them. "We just went out and bought a bunch of instruments," says Emily, who chose to play drums. The friends practiced in Emily's parents' basement. Her older brother overheard the original mix of punk and classic

> ### ▸▸ DRIVING QUESTIONS
>
> 1. How does the organization of chromosomes, genes, and their alleles contribute to human traits?
> 2. How does meiosis produce gametes?
> 3. Why do different traits have different inheritance patterns?
> 4. What are some practical applications of understanding the genetic basis of human disease?

rock songs that the five girls put together and was so impressed he asked them to open for his own band on New Year's Eve. The five-girl rock 'n' roll band called Hellen was born.

"It's getting really huge," says Emily excitedly. Hellen performs almost every weekend in

Detroit and is now taking its show on the road, with gigs scheduled in Cleveland, Chicago, and soon, Emily hopes, New York City. The band's unplanned success has given Emily an opportunity to raise awareness about a cause very close to her heart.

Emily may look like a typical rock 'n' roller, with her bleached blond hair and tattooed forearms, but her carefree appearance masks a serious underlying condition. Emily has cystic fibrosis (CF), a genetic disease she inherited from her parents, and each day she takes a cocktail of drugs and vitamins to help cope with the disease. CF has many symptoms, the most dangerous of which is mucus that clogs airways in the lungs and makes it difficult to breathe. People with CF also can't digest food well—mucus blocks the passageways through which the digestive enzymes travel to the intestines. So Emily must swallow enzymes before each meal to ensure that her body gets enough nutrients. She's grown accustomed to the pills—upward of 40 per day—but having to take such meticulous care of her health is hardly routine for most 25-year-olds.

Every year, approximately 2,500 babies are born with CF, making this the most common fatal genetic disease in the United States. In 1989, a team of scientists led by Lap Chee Tsui at Toronto's Hospital for Sick Children and Francis Collins, then at the University of Michigan, discovered that the disease is caused by mutations in a specific gene that sits on chromosome 7. A **mutation** is a change in the nucleotide sequence of DNA, which creates a new allele of a gene. As we saw in Chapter 8, alleles are alternative nucleotide sequences of the same gene. Most genes have not just one allele but several, each created by genetic mutation. Tsui and Collins (who would later head the Human Genome Project and the National Institutes of Health) discovered that CF is caused by mutations in a gene called *Cystic Fibrosis Transmembrane Conductance Regulator*. This gene codes for the protein known as the cystic fibrosis transmembrane regulator (CFTR), which shuttles ions in and out of cells throughout the body.

INFOGRAPHIC 11.1 CF IS CAUSED BY MUTATIONS IN THE *CFTR* GENE

→ Cystic fibrosis (CF) is caused by a variety of mutations in the cystic fibrosis transmembrane regulator (*CFTR*) gene that sits on chromosome 7. One such mutation consists of a deletion of three consecutive nucleotides, which creates a CF-associated mutant allele. Consequently, the mRNA expressed from this allele has a missing codon and the resulting protein lacks an amino acid in a specific location, rendering the protein nonfunctional.

Chromosome 7

CFTR gene

Normal allele

ATCATCTTGGTGTT

Normal CFTR protein

Chromosome 7

CFTR gene

Mutant allele

ATCAT TGGTGTT

Three nucleotides deleted in CF allele

CTT

Nonfunctional CFTR protein that lacks a critical amino acid in its sequence

> Every year, approximately 2,500 babies are born with CF, making this the most common fatal genetic disease in the United States.

The discovery was a milestone. Now that they knew the gene responsible, scientists could study how mutations in the gene make people sick. Because genes provide instructions for making proteins, a change in gene sequence can change the function or shape of a protein. A variety of mutations in the *CFTR* gene can cause CF. In the most common CF allele, three nucleotides within the *CFTR* gene are deleted. People who carry this allele produce a defective CFTR protein. This slight change wreaks havoc on the body: the lungs, sweat glands, and pancreas no longer function normally (**INFOGRAPHIC 11.1**).

Today, more than 20 years later, scientists understand the disease better, and this has led to better drugs and therapies to treat symptoms; people with CF are living longer than ever. But despite scientific advances, there is still much to learn. One aspect of the disease that scientists are studying intensively is that people with identical CF alleles vary in the course of their disease–some have worse symptoms and live shorter lives than others. In recent years, scientists have discovered that there are other genes that contribute to a patient's overall health–so-called modifier genes. That discovery is leading to exciting new therapies that may extend Emily's life and the lives of thousands of other people with CF.

HOW IS CF INHERITED?

When Emily's mother, Debbie, learned that her daughter had CF, she was shocked. She and her husband, Lowell, were both healthy, and they already had two healthy sons. How did their daughter develop a genetic disease that neither Debbie nor her husband had?

The answer has to do with how genes are inherited–how they are passed down from generation to generation. Genes, which provide instructions for making proteins, are the units of inheritance, physically transmitted from parents to children. The particular alleles of genes you received from your parents are the reason you resemble your mother and father, and possibly also an uncle or a grandparent. But not every child of a couple receives exactly the same set of parental alleles, and so children can and do differ from their parents and from one another.

Consider Emily's parents. Like all humans, they are **diploid** organisms, meaning that each of their body cells carries two copies of each chromosome–one inherited from mom, the other from dad. Such paired chromosomes are called **homologous chromosomes**. Because chromosomes come in pairs, we have two copies of nearly every gene in our body cells. (Genes located on the X and Y chromosome in males do not have a second copy.) While the two gene copies have the same general function, the nucleotide sequences of the two copies can differ. In other words, a person can carry two different alleles of the same gene, one of which functions differently from the other. In the case of the gene *CFTR*, a person can have one CF-associated allele and remain healthy if his or her other chromosome has a normal allele to make up for the defective copy. This is why Emily's parents, Debbie

▶**MUTATION**
A change in the nucleotide sequence of DNA.

▶**DIPLOID**
Having two copies of every chromosome.

▶**HOMOLOGOUS CHROMOSOMES**
A pair of chromosomes that both contain the same genes. In a diploid cell, one chromosome in the pair is inherited from the mother, the other from the father.

▸**PHENOTYPE**
The visible or measurable features of an individual.

▸**GENOTYPE**
The particular genetic makeup of an individual.

▸**GAMETES**
Specialized reproductive cells that carry one copy of each chromosome (that is, they are haploid). Sperm are male gametes; eggs are female gametes.

▸**HAPLOID**
Having only one copy of every chromosome.

▸**MEIOSIS**
A specialized type of nuclear division that generates genetically unique haploid gametes.

and Lowell, are healthy: they each have one normal allele that masks the defective copy (**INFOGRAPHIC 11.2**).

But as Debbie and Lowell also illustrate, it's not always possible to know what genes a person has just from outward appearance. In fact, geneticists make a distinction between a person's observable or measurable traits, or **phenotype,** and his or her genes, or **genotype.** Both Debbie and Lowell have normal phenotypes, but they both also carry a disease allele as part of their genotype. They each inherited one CF allele from one of their parents and therefore can pass that defective allele along to their children—as they did to Emily.

But not all the Schaller children have the disease—Debbie and Lowell also have two healthy boys. Why didn't these children inherit CF?

Sexual reproduction is a bit like shuffling the genetic cards. Before parents pass their genes to their offspring, those genes are first mixed up and then the two copies of each gene are separated from each other, so that not every child receives the same combination of alleles. It is the unique combination of maternal and paternal alleles that come together during fertilization that determines a person's genotype and contributes to his or her phenotype.

To reproduce sexually, organisms must first create sex cells called **gametes.** In humans, these are the egg and sperm cells. Unlike the rest of the body's cells, which are diploid, gametes carry only one copy of each chromosome, which makes them **haploid.** To become haploid, the cells that form gametes go through a unique kind of cell division that includes **meiosis,** which halves the number of chromosomes from 46 to 23. When a haploid sperm fertilizes a haploid

INFOGRAPHIC 11.2 HUMANS HAVE TWO COPIES OF NEARLY EVERY GENE

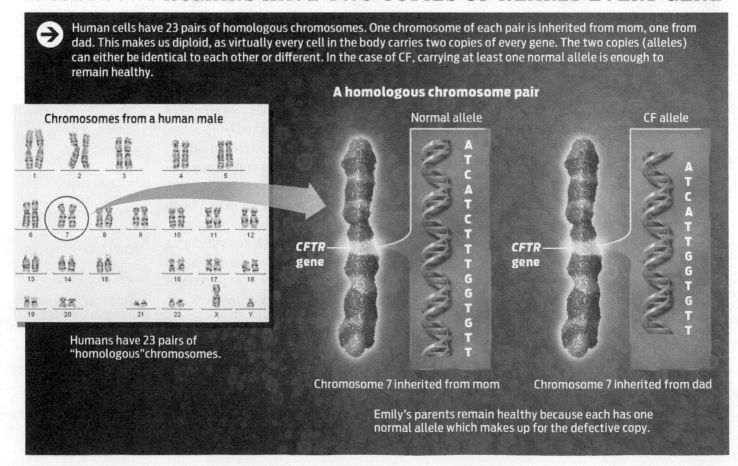

➔ Human cells have 23 pairs of homologous chromosomes. One chromosome of each pair is inherited from mom, one from dad. This makes us diploid, as virtually every cell in the body carries two copies of every gene. The two copies (alleles) can either be identical to each other or different. In the case of CF, carrying at least one normal allele is enough to remain healthy.

Chromosomes from a human male

1 2 3 4 5
6 7 8 9 10 11 12
13 14 15 16 17 18
19 20 21 22 X Y

Humans have 23 pairs of "homologous" chromosomes.

A homologous chromosome pair

Normal allele

CF allele

CFTR gene
ATCATCTTTGGTGTT

CFTR gene
ATCATTGGTGTT

Chromosome 7 inherited from mom

Chromosome 7 inherited from dad

Emily's parents remain healthy because each has one normal allele which makes up for the defective copy.

INFOGRAPHIC 11.3 GAMETES PASS GENETIC INFORMATION TO THE NEXT GENERATION

→ To reproduce sexually, diploid organisms produce specialized sex cells called gametes, which are haploid—they carry only one copy of each chromosome. When a sperm fertilizes an egg the resulting diploid zygote divides by mitotic cell division, eventually generating enough cells to form a baby. The baby is diploid.

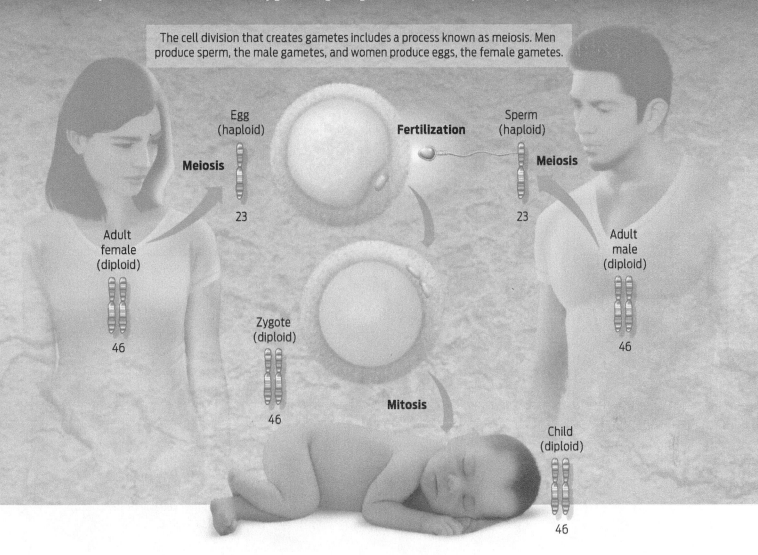

The cell division that creates gametes includes a process known as meiosis. Men produce sperm, the male gametes, and women produce eggs, the female gametes.

Egg (haploid) Fertilization Sperm (haploid)

Meiosis Meiosis

23 23

Adult female (diploid) Adult male (diploid)

46 46

Zygote (diploid)

46

Mitosis

Child (diploid)

46

---

egg, the result is a diploid **zygote** that now carries two copies of every gene on 46 chromosomes. This zygote will divide by mitosis to become an **embryo,** which will eventually grow into a human child (INFOGRAPHIC 11.3).

Meiotic cell division, which produces haploid sperm and egg, is similar to mitotic cell division (see Chapter 9), except that in meiosis there are two separate divisions. The first division separates homologous chromosomes; the second division separates sister chromatids (INFOGRAPHIC 11.4).

Because it unites haploid egg and sperm from two people, sexual reproduction is the primary reason that children don't look and behave exactly like one parent in particular; they inherit alleles from both parents and consequently are genetically a combination of the two.

Besides forming haploid sex cells, meiosis contributes to the genetic diversity of offspring in other ways as well. No two gametes produced by the same parent are identical, and that is because of two major events during

▶ZYGOTE
A cell that is capable of developing into an adult organism. The zygote is formed when an egg is fertilized by a sperm.

▶EMBRYO
An early stage of development reached when a zygote undergoes cell division to form a multicellular structure.

INFOGRAPHIC 11.4 MEIOSIS PRODUCES HAPLOID EGG AND SPERM

Humans produce egg and sperm through meiosis, which takes place in the ovaries and the testes. Meiosis halves the chromosome number from 46 to 23.

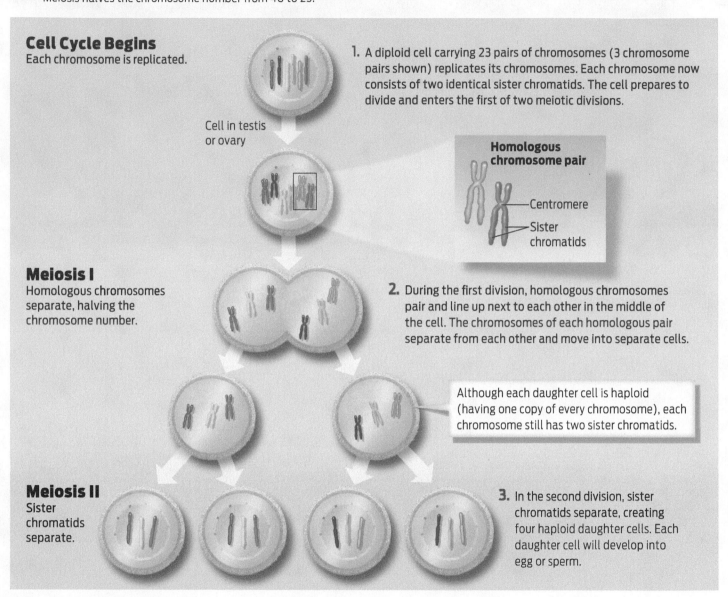

Cell Cycle Begins
Each chromosome is replicated.

Cell in testis or ovary

1. A diploid cell carrying 23 pairs of chromosomes (3 chromosome pairs shown) replicates its chromosomes. Each chromosome now consists of two identical sister chromatids. The cell prepares to divide and enters the first of two meiotic divisions.

Homologous chromosome pair

Centromere

Sister chromatids

Meiosis I
Homologous chromosomes separate, halving the chromosome number.

2. During the first division, homologous chromosomes pair and line up next to each other in the middle of the cell. The chromosomes of each homologous pair separate from each other and move into separate cells.

Although each daughter cell is haploid (having one copy of every chromosome), each chromosome still has two sister chromatids.

Meiosis II
Sister chromatids separate.

3. In the second division, sister chromatids separate, creating four haploid daughter cells. Each daughter cell will develop into egg or sperm.

▶**RECOMBINATION**
An event in meiosis during which maternal and paternal chromosomes pair and physically exchange DNA segments.

▶**INDEPENDENT ASSORTMENT**
The principle that alleles of different genes are distributed independently of one another during meiosis.

meiosis that contribute to the huge variation we see among parents, children, and siblings. The first is **recombination,** in which homologous maternal and paternal chromosomes pair up and physically swap genetic information by exchanging DNA segments. As a result of recombination, also called crossing over, maternal chromosomes actually contain segments (and therefore alleles) from paternal chromosomes and vice versa.

The second vitally important aspect of meiosis is **independent assortment,** the

principle that alleles of different genes are distributed independently of one another, not as a package. (What alleles for hair color you inherit, for example, have no bearing on what alleles for height you inherit.) The physical basis of independent assortment is the random way that maternal and paternal chromosomes are distributed to sex cells in meiosis. During the first division of meiosis (known as meiosis I), maternal and paternal chromosomes line up next to each other along the midline of the cell and segregate into newly forming cells. Because each maternal and paternal chromosome pair can line up in two different ways (sometimes the mother's chromosome is on the left, sometimes the father's is), the exact combination of maternal and paternal chromosomes that each sperm or egg inherits differs every time meiosis occurs. In fact, as a result of recombination and independent assortment, no two sperm or egg cells are exactly alike (**INFOGRAPHIC 11.5**).

INFOGRAPHIC 11.5 MEIOSIS PRODUCES GENETICALLY DIVERSE EGG AND SPERM

Meiosis produces haploid gametes that are genetically unique. Each egg and sperm has its own distinct combination of alleles. The two events that create this diversity are (1) recombination and (2) independent assortment.

1. Recombination

Before separating at meiosis I, the maternal and paternal chromosomes line up next to each other and physically exchange segments of DNA. Consequently, maternal chromosomes contain segments (and thus alleles) from paternal chromosomes, and vice versa.

Cell in meiosis I

Maternal chromosome   Paternal chromosome

Homologous chromosomes

Sister chromatids

Crossing over between nonsister chromatids

Chromatids after exchange

Recombinant daughter chromosomes

2. Independent Assortment

Maternal and paternal chromosome pairs separate according to how they have randomly lined up in the cell. Each time meiosis occurs, the chromosome pairs lines up differently, and thus a different chromosome combination is produced in the resulting gametes. When all 23 chromosome pairs are considered, there are more than 8 million unique chromosome combinations possible.

D = dominant allele
d = recessive allele

Maternal and paternal chromosomes from each homologous pair line up randomly.

Possible allele combinations in the gametes.

When meiosis is complete, each gamete has only 23 chromosomes, which are a mixture of maternal and paternal alleles—and this is the reason that not everyone in the Schaller family has CF. Because alleles randomly distribute into each gamete, some of the Schallers' gametes will carry the CF allele and others will not. If by chance a sperm that carries a CF allele fertilizes an egg that also carries a CF allele, the resulting child will have CF.

After Emily's cystic fibrosis was diagnosed, both Debbie and Lowell learned that their parents had relatives who had died at a very young age. At the time, the cause of death was thought to be a respiratory illness such as pneumonia. But these relatives most likely had CF, Debbie now thinks; doctors at the time simply did not have the tools to diagnose the disease.

The Schallers now knew that the disease ran in both sides of the family. But they could still not help Emily. "They told us she would only live to be about 12 years old," Debbie recalls, adding, "We just put ourselves in the hands of medical professionals."

LIVING WITH THE DISEASE

Growing up, Emily was scarcely aware of her own disability. The visits to doctors and periodic stays in the hospital were just a part of life. All her teachers and friends knew that she had CF. "My family and friends were all so supportive," she says. In high school she played

INFOGRAPHIC 11.6 THE CFTR PROTEIN AND CYSTIC FIBROSIS

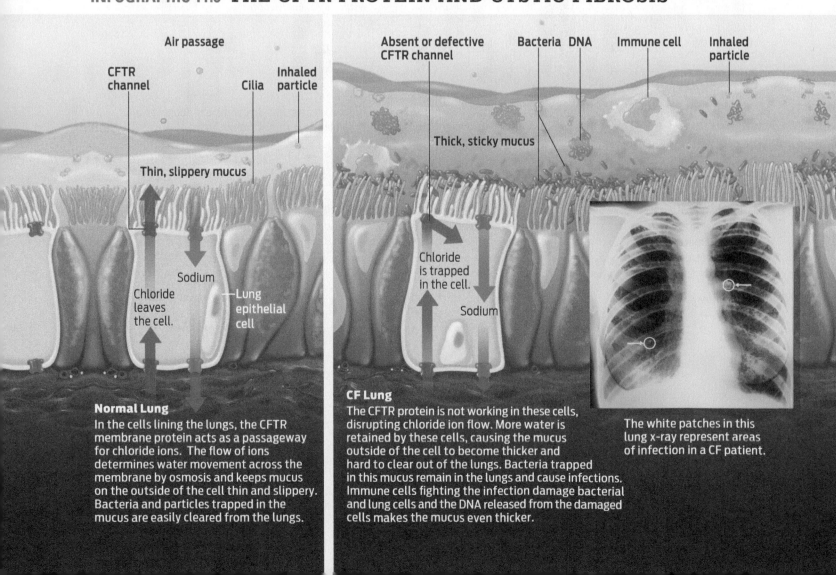

Normal Lung
In the cells lining the lungs, the CFTR membrane protein acts as a passageway for chloride ions. The flow of ions determines water movement across the membrane by osmosis and keeps mucus on the outside of the cell thin and slippery. Bacteria and particles trapped in the mucus are easily cleared from the lungs.

CF Lung
The CFTR protein is not working in these cells, disrupting chloride ion flow. More water is retained by these cells, causing the mucus outside of the cell to become thicker and hard to clear out of the lungs. Bacteria trapped in this mucus remain in the lungs and cause infections. Immune cells fighting the infection damage bacterial and lung cells and the DNA released from the damaged cells makes the mucus even thicker.

The white patches in this lung x-ray represent areas of infection in a CF patient.

volleyball, basketball, and soccer, and participated in many walkathons to raise money for CF research. Thanks to medical progress, Emily has outlived doctors' original expectations by almost two decades.

But she deals daily with the legacy of her genetic inheritance. In healthy people, the CFTR protein acts as a channel through a cell's membrane that allows certain ions to move in and out of the cell, keeping the cell's chemistry in balance. But in people with CF, the channel is distorted or absent altogether, and the mechanism goes awry. The result is that mucus—a slippery substance that lubricates and protects the linings of the airways, digestive system, reproductive system, and other tissues—becomes abnormally thick and sticky.

This abnormal mucus blocks ducts throughout the body. The most problematic symptom, however, is that thick mucus builds up in the lungs. Patients have trouble breathing, and the mucus provides fertile ground for bacteria and other organisms. Over time, repeated infections permanently damage the lungs. As a result, people with CF may slowly lose their ability to breathe, eventually dying of suffocation (**INFOGRAPHIC 11.6**).

To avoid lung damage, every morning Emily straps on an inflatable vest that vibrates to loosen mucus in her lungs. For 30 minutes she inhales a saltwater mist and another medication to thin her mucus, which she then coughs out periodically. To that regime she adds two other medications three times a week to keep her lungs from becoming inflamed and to kill infections. But despite her best efforts, Emily has been hospitalized more frequently in recent years because of serious lung infections that hinder her ability to breathe.

Emily remains undaunted. "I just live each day at a time," she says. She works about 30 hours a week at a retail shop in downtown Detroit, runs to stay fit, and spends her evenings practicing with her band, performing at concerts, playing guitar, or hanging out with friends. She hopes her band's fame and success

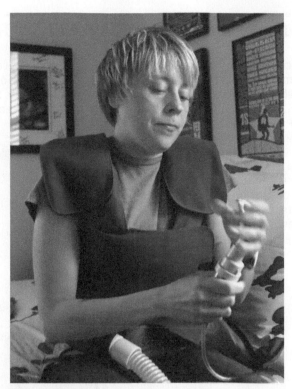

Cystic fibrosis patients like Emily wear vibrating vests to loosen the mucus in their lungs while inhaling a saltwater mist to thin out the mucus.

will grow—if the band's following expands beyond Detroit, she hopes to tour Europe. Emily hasn't ruled out having a family of her own one day. Even though she has CF, her children will not necessarily have the disease.

Emily with her parents, Debbie and Lowell.

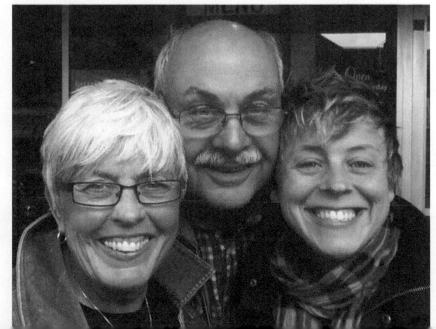

INFOGRAPHIC 11.7 HOW RECESSIVE TRAITS ARE INHERITED

Cystic fibrosis is a recessive trait, which means that the disease phenotype is caused by inheriting two recessive alleles, as in Emily's case. Emily's parents do not have CF because they each possess one dominant allele, but they each also carry one recessive CF allele, making them both heterozygous carriers. To calculate the probability that Debbie and Lowell will have a child with CF, we can determine the possible alleles in their gametes and then join all possible combinations of these sperm and egg in a **Punnett** square.

During meiosis I, maternal and paternal chromosomes separate. Consequently, half of Debbie and Lowell's gametes carry the dominant allele *A* and half carry the recessive allele *a*.

Normal father (heterozygous)

Cell in the testis

A = dominant allele
a = recessive allele

Meiosis

Sperm

Normal mother (heterozygous)

Meiosis

Cell in the ovary

Eggs

Unaffected child
AA

Unaffected child
Aa (carrier)

Unaffected child
Aa (carrier)

Affected child
aa

For heterozygous parents such as Debbie and Lowell, the chance that a child will be born with CF is 1 in 4, or 25%.

▶**RECESSIVE ALLELE**
An allele that reveals itself in the phenotype only if a masking dominant allele is not present.

▶**DOMINANT ALLELE**
An allele that can mask the presence of a recessive allele.

▶**HETEROZYGOUS**
Having two different alleles.

▶**HOMOZYGOUS**
Having two identical alleles.

Why not? Remember that since Emily has CF, her parents, Lowell and Debbie, both must carry disease alleles. But as neither of them has the disease, the CF alleles must be "hidden." When one allele masks the effect of another, the hidden allele is described as **recessive** (designated by a lower-case letter, e.g., *a*). The normal allele, which conceals the effect of the recessive allele, is known as the **dominant allele** (designated by a capital letter, e.g., *A*). Debbie and Lowell are healthy because they each have a dominant normal allele that compensates for their defective recessive CF allele. Geneticists call their genotype **heterozygous.** Their two healthy sons are either heterozygous like their parents, or have two normal alleles–that is, their genotype is **homozygous** for the normal allele. A genotype made up of two dominant alleles is known as homozygous dominant. Emily's genotype, however, is homozygous recessive: she inherited one recessive CF allele from each parent, which is why she has the disease.

What were the chances that Debbie and Lowell would have a child with CF? To figure out the likelihood that parents will have a child with a particular trait, we can plot the possibilities on a **Punnett square** (a tool named for the geneticist Reginald C. Punnett, who devised it). A Punnett square matches up the possible parental gametes and shows the likelihood that particular parental alleles will combine. As heterozygous individuals, Debbie and Lowell each have a 50% chance of passing on their CF allele to a child, which means they have a 25% chance of having a child with CF and a 75% chance of having a healthy child. The chance that a child will be a heterozygous **carrier**–that is, that the child will carry the recessive allele for CF but will not have the disease because the allele's effect is masked by the dominant allele–is 50% (INFOGRAPHIC 11.7).

Just as Emily's genotype is different from her parents' genotype, Emily's children will have different genotypes from her own. Whether or not her children develop CF depends on the father's genotype. Since Emily is homozygous, she can contribute only recessive CF alleles to her children. If Emily were to have children with a man who had two normal alleles, for example, none of her children would have the disease–they would all have a heterozygous genotype but a normal phenotype. But as carriers they could pass on the disease to their children. If Emily had children

with a man who was heterozygous for the CF gene, then her children would have a 1 in 2, or 50%, chance of having CF.

Not all recessive alleles cause disease. Many physical traits are the result of inheriting two recessive alleles of a gene. For example, people with blue eyes or red hair have inherited recessive alleles that prevent the deposition of dark pigment. And not all genetic diseases are caused by recessive alleles: some, such as the neurodegenerative disorder Huntington disease, are determined by dominant alleles. Diseases caused by dominant alleles have a higher probability of showing up in the next generation because it takes only one disease allele to cause the trait (INFOGRAPHIC 11.8).

In all cases, anyone with a genetic disease is at risk for passing it on to his or her children. The risk merely varies, depending

> **▸PUNNETT SQUARE**
> A diagram used to determine probabilities of offspring having particular genotypes, given the genotypes of the parents.

> **▸CARRIER**
> An individual who is heterozygous for a particular gene of interest, and therefore can pass on the recessive allele without showing any of its effects.

INFOGRAPHIC 11.8 HOW DOMINANT TRAITS ARE INHERITED

 Some genetic conditions, such as Huntington disease, a degenerative neurological disease, and polydactyly, having more than five fingers or toes per limb, are caused by dominant alleles. Many common traits such as dark eyes and dimples are also determined by dominant alleles. In these cases, inheriting one copy of the dominant allele is sufficient to display the trait.

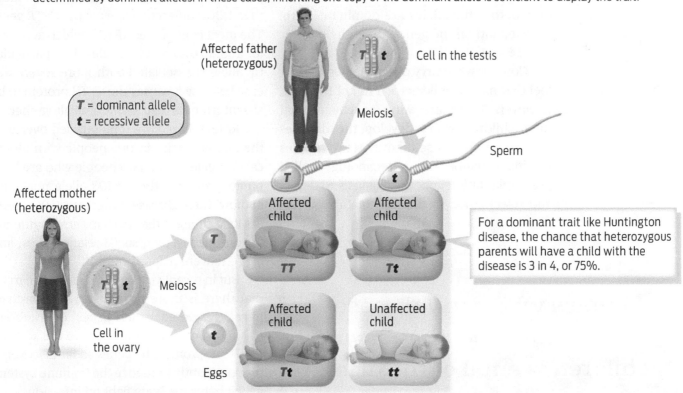

TABLE 11.1 INHERITED GENETIC CONDITIONS IN HUMANS

RECESSIVE TRAIT	PHENOTYPE
Albinism	Lack of pigment in skin, hair, and eyes
Cystic fibrosis	Excess mucus in lungs, digestive tract, and liver; increased susceptibility to infections
Sickle-cell disease	Sickled red blood cells; damage to tissues
Tay-Sachs disease	Lipid accumulation in brain cells; mental deficiency, blindness, and death in childhood

DOMINANT TRAIT	PHENOTYPE
Huntington disease	Mental deterioration and uncontrollable movements; onset at middle age
Freckles	Pigmented spots on skin, particularly on face and arms
Polydactyly	More than five digits on hands or feet
Dimples	Indentations in the skin of the cheeks
Chin cleft	Indentation in chin

on whether the alleles are dominant or recessive, and on the genotype of the partner (TABLE 11.1).

Couples who carry disease genes needn't feel that having children is a roll of the dice, however. There are ways to ensure that their children won't develop the diseases they could otherwise inherit. Many couples in this situation use a technology called pre-implantation genetic diagnosis to detect and select embryos that do not carry defec-

> Couples who carry disease genes needn't feel that having children is a roll of the dice.

tive alleles. Through in vitro fertilization, a man's sperm can fertilize a woman's eggs outside the body (see Chapter 28). The genes of each resulting embryo are then examined for specific alleles, and then only embryos that don't contain defective alleles are implanted into the mother. Hundreds of thousands of babies have been born by this technique. Some couples, however, may choose not to undergo assisted reproduction because of religious or other personal reasons.

NEW RESEARCH IN THE PIPELINE

For children who do inherit a genetic disease, there may be ways to treat the symptoms or reduce their severity. In the case of CF, new treatments are in the pipeline that could help Emily, and her children and grandchildren, too. Furthest along are a class of medications that, when inhaled, can restore the balance of ions inside affected cells in the lungs. Scientists are presently testing more than 20 experimental drugs in humans, and a few are already on the market.

Through basic research, scientists continue to learn more about cystic fibrosis. Over the past 25 years, scientists have discovered more than 1,000 different alleles of the *CFTR* gene. The most common is $\triangle F508$, which accounts for about 70% of all CF alleles. This particular CF allele is associated with more severe disease because it causes the CFTR protein to be absent from the cell membrane altogether.

Researchers have long puzzled over why the disease varies in two people with identical CF alleles—even two people who are both homozygous for the $\triangle F508$ allele may vary in how their disease progresses. Researchers long thought that perhaps environmental factors such as diet, social relationships, and exercise were responsible.

But in recent years, scientists have learned that there is more to the story. Researchers have discovered other genes on different chromosomes that contribute to the severity of CF symptoms. The genes so far discovered predominantly influence the immune system, which helps the body fight off infections.

For example, scientists have found that one allele of a gene called *TGFβ1*, located on chromosome 19, is associated with more-severe lung disease in CF patients. This gene influences the immune response to infection. Scientists suspect that CF patients with certain *TGFβ1* alleles mount a more vigorous response to infections than those with other alleles. Such a heightened immune response can cause lung tissue to scar. So if a CF patient also inherited this specific allele of *TGFβ1* his or her lungs are more likely to scar in response to infections. The impact of such "modifier genes" on the CF phenotype makes it more complicated to assess how disabling any particular person's CF disease will be–but it is not impossible.

Parents who are heterozygous carriers of CF, for example, have a 1 in 4, or 25%, chance of having a child who has CF. If these two parents are also heterozygous for *TGFβ1*, then the probability that their child will be homozygous recessive for *TGFβ1* is also 1 in 4 (25%). The chance of two independent events occurring together is calculated by multiplying the two independent chances together. So the probability of being homozygous recessive for both *CFTR* and *TGFβ1* is 1/4 × 1/4, or 1 in 16. This probability can also be calculated by using a Punnett square (**INFOGRAPHIC 11.9**).

INFOGRAPHIC 11.9 TRACKING THE INHERITANCE OF TWO GENES

People with CF differ in the severity of their disease. Some of this variability is influenced by alleles of other genes that sit on other chromosomes. One such gene, *TGFβ1*, is located on chromosome 19, shown here with symbol *D*. We can also use a Punnett square to follow the inheritance of two genes, as in the example below.

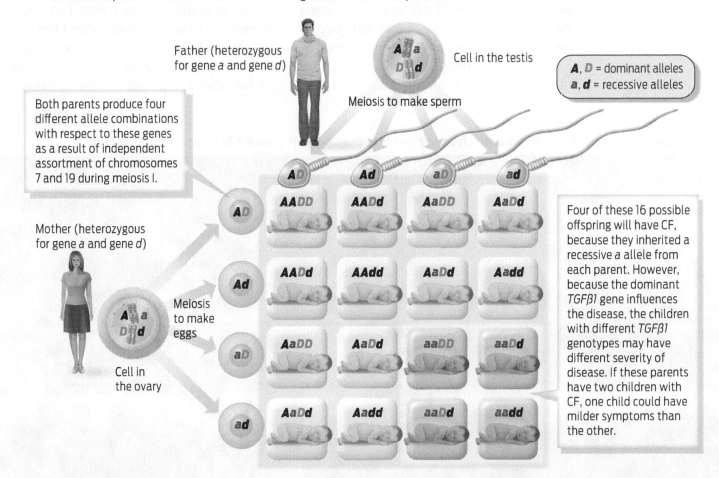

Father (heterozygous for gene *a* and gene *d*)

Cell in the testis

Meiosis to make sperm

A, *D* = dominant alleles
a, *d* = recessive alleles

Both parents produce four different allele combinations with respect to these genes as a result of independent assortment of chromosomes 7 and 19 during meiosis I.

Mother (heterozygous for gene *a* and gene *d*)

Meiosis to make eggs

Cell in the ovary

	AD	Ad	aD	ad
AD	AADD	AADd	AaDD	AaDd
Ad	AADd	AAdd	AaDd	Aadd
aD	AaDD	AaDd	aaDD	aaDd
ad	AaDd	Aadd	aaDd	aadd

Four of these 16 possible offspring will have CF, because they inherited a recessive *a* allele from each parent. However, because the dominant *TGFβ1* gene influences the disease, the children with different *TGFβ1* genotypes may have different severity of disease. If these parents have two children with CF, one child could have milder symptoms than the other.

> Over the past 25 years, scientists have discovered more than 1,000 different alleles of the *CFTR* gene.

Understanding how these modifier genes contribute to the disease may point the way to even more therapies. In some cases, existing drugs may prove useful. Drugs that reduce inflammation by targeting the *TGFβ1* protein, for example, may help reduce scarring in the lungs.

Emily recently had her genotype tested. While she doesn't carry △*F508*, she does carry another allele associated with severe disease, *G551D*. This allele is much rarer than △*F508*—only about 4% of the U.S. population has it. People with this mutation have a CFTR channel that doesn't open properly. The good news for patients with this mutation is that a new FDA-approved drug, called Kalydeco, is remarkably effective at restoring function to their malfunctioning CFTR channel. Unlike other drugs, which merely treat the symp-

toms of the disease, this one addresses the underlying cause: the malfunctioning protein.

Kalydeco is not a cure for CF—the genetic mutation is still present in a person's cells—but the drug allows patients to better control their symptoms. And that is enough for Emily. "Everyone talks about curing a disease—cure CF, cure these other diseases. [But] Kalydeco controls CF at the basic defect, so I'm OK with the other 'c' word, control, because I'm living it and I've never felt better in my life," Emily recently told NPR. She nicknamed the drug "blue lightning," because of how quickly the small blue pill helped her breathing.

Keenly aware of how medical progress has extended her life, Emily conducts her own share of fund raising and education. After that fateful New Year's Eve when she and her friends opened for her brother's

Hellen performing at the 2006 annual "Just Let me Breathe" benefit concert in Royal Oak, Michigan.

band, South Normal, Hellen's following grew. The band wrote more songs and refined their sound. "I favor the old stuff, AC-DC, Led Zeppelin, the Ramones," Emily remarks. And fans just couldn't seem to get enough of the girls' sound. Each year, the number of fans has grown.

Sparked by a South Normal hit single called "Just Breathe," Emily had the idea of organizing a concert to benefit CF research. The song has nothing to do with CF, but Emily thought that it might be a good theme for a concert. Besides, she says, "We were tired of walkathons and black tie events with tickets that cost $300 each."

Emily and her brother organized the first benefit concert in 2004. Called "Just Let Me Breathe," it featured four Detroit bands. The concert sold out and raised about $9,000. Inspired by that success, Emily decided to make the concert an annual event, and even persuaded *Spin* magazine to sponsor it. The benefit concert recently celebrated its seventh anniversary. With all her fund-raising activities, Emily has raised more than $300,000 so far. But she doesn't plan on stopping there. In 2007, she started the Rock CF Foundation, a nonprofit devoted to improving the lives of

Through all her fund-raising activities, Emily has raised more than $300,000 to benefit research on cystic fibrosis.

people with cystic fibrosis and raising awareness of the disease. She continues to work actively with the foundation, even sponsoring and running in an annual half marathon. Her ultimate goal is simple, she says: to keep raising money and, one day, "Rock CF" for good. ■

CHAPTER 11 Summary

▶ Genes, which code for proteins, are the units of inheritance, physically passed down from parents to offspring.

▶ An organism's physical traits constitute its phenotype; its genes constitute its genotype. A person's genotype can't always be determined from his or her phenotype.

▶ Humans are diploid organisms, meaning they have two copies of each chromosome in their cells. Because chromosomes come in pairs, we have two copies of nearly every gene in our body cells. These copies can be the same or different from each other.

▶ Different versions of the same gene are called alleles. Alleles arise from mutations that change the nucleotide sequence of a gene.

▶ Alleles may be dominant or recessive. Dominant alleles can mask the effects of recessive alleles, which can be hidden.

▶ Many traits result from carrying two recessive alleles; others result from carrying one dominant allele.

▶ Meiosis is a type of cell division that produces genetically distinct sperm and egg.

▶ Homologous chromosomes recombine and assort independently during meiosis to generate genetically diverse sperm and eggs. No two sperm or egg cells produced by the same person will be exactly alike.

▶ Haploid gametes fuse randomly during fertilization, generating genetically unique diploid zygotes.

▶ A Punnett square can help predict a child's genotype and phenotype when the pattern of inheritance, dominant or recessive, is known.

▶ Cystic fibrosis (CF) is a recessively inherited genetic disease. Alterations in the gene *CFTR* cause disease by interfering with ion and water balance in cells, especially in the lungs.

MORE TO EXPLORE

▶ Cystic Fibrosis Foundation www.cff.org/

▶ Rock CF Foundation http://letsrockcf.blogspot.com/

▶ FUCF (song) http://www.myspace.com/schallyeah

▶ Pearson, H. (2009) Human genetics: One gene, twenty years. *Nature* 460:164–169. http://www.nature.com/news/2009/080709/full/460164a.html

▶ Palca, J. (January 2, 2013) NPR: "Drug Fulfills Promise Of Research Into Cystic Fibrosis Gene" http://is.gd/wO3v2l

▶ Ramsey, B. W., et al. (2011) A CFTR potentiator in patients with cystic fibrosis and the *G551D* mutation. *New England Journal of Medicine* 365:1663–1672.

CHAPTER 11 Test Your Knowledge

DRIVING QUESTION 1

How does the organization of chromosomes, genes, and their alleles contribute to human traits?

By answering the questions below and studying Infographics 11.1, 11.2, and 11.3, you should be able to generate an answer to the broader Driving Question above.

KNOW IT

1 How do the two alleles of the *CFTR* gene in a lung cell differ?

a. They were inherited from different parents.

b. One is on chromosome 7 and one is on chromosome 3.

c. Only one is expressed.

d. all of the above

e. There is no difference because they are both the same gene.

2 Consider a liver cell.

a. How many chromosomes are present?

b. How many alleles of each gene are present?

3 Consider a gamete.

a. How many chromosomes are present?

b. How many alleles of each gene are present?

USE IT

4 A diploid cell of baker's yeast has 32 chromosomes. How many chromosomes are in each of its haploid spores?

a. 32   c. 8   e. 1

b. 16   d. 64

5 In diploid organisms, having two homologues of each chromosome can be beneficial if one allele of a gene encodes a nonfunctional protein. Can haploid organisms survive the presence of nonfunctional alleles? Explain your answer.

6 From which parent did Emily inherit cystic fibrosis? Explain your answer.

DRIVING QUESTION 2

How does meiosis produce gametes?

By answering the questions below and studying Infographics 11.3, 11.4, and 11.5, you should be able to generate an answer to the broader Driving Question above.

KNOW IT

7 A human female has _____ chromosomes in each skin cell and _____ chromosomes in each egg.

a. 46; 46   c. 46; 23   e. 92; 46

b. 23; 46   d. 23; 23

8 A woman is heterozygous for the CF-associated gene (the alleles are represented here by the letters *A* and *a*). Assuming that meiosis occurs normally, which of the following represent eggs that she can produce?

a. *A*   d. *AA*   g. *A, a,* or *Aa*

b. *a*   e. *aa*

c. *Aa*   f. *A* or *a*

9 Draw a maternal version of chromosome 7 in one color and a paternal version of chromosome 7 in another color. Maintaining this color distinction, now draw a possible version of chromosome 7 that could end up in a gamete following meiotic division.

USE IT

10 An alien has 82 total chromosomes in each of its body cells. The chromosomes are paired, making 41 pairs. If the alien's gametes undergo meiosis, what are the number and arrangement (paired or not) of chromosomes in one of its gametes? Give the reason for your answer.

11 Describe at least two major differences between mitosis (discussed in Chapter 9) and meiosis.

12 If meiosis were to fail and a cell skipped meiosis I, so that meiosis II was the only meiotic division, how would you describe the resulting gametes?

DRIVING QUESTION 3

Why do different traits have different inheritance patterns?

By answering the questions below and studying Infographics 11.7, 11.8, and 11.9, you should be able to generate an answer to the broader Driving Question above.

KNOW IT

13 What is the genotype of a person with CF?

a. homozygous dominant

b. homozygous recessive

c. heterozygous

d. any of the above

e. none of the above

14 Strictly on the basis of the following *CFTR* genotypes, what do you predict the phenotype of each person to be?

   a. heterozygous

   b. homozygous dominant

   c. homozygous recessive

15 How many copies of the CF-associated allele does a person with CF have in one of his or her lung cells? How does this compare to someone who is a carrier for CF? How does it compare to someone who is homozygous dominant for the gene *CFTR*?

16 Women can inherit alleles of a gene called *BRCA1* that makes puts them at higher risk for breast cancer. The alleles associated with elevated cancer risk are dominant. Of the genotypes listed below, which carries the lowest genetic risk of developing breast cancer?

   a. *BB*

   b. *Bb*

   c. *bb*

   d. *BB* and *Bb* carry less risk than *bb*.

   e. All carry equal risk.

USE IT

17 A person has a heterozygous genotype for a disease gene and no disease phenotype. Does this disease have a dominant or a recessive inheritance pattern?

18 Assume that Emily (who has CF) decides to have children with a man who does not have CF and who has no family history of CF.

   a. What combination of gametes can each of them produce?

   b. Place these gametes on a Punnett square and fill in the results of the cross.

   c. On the basis of the Punnett square results, what is the probability that they will have a child with CF?

   d. On the basis of the Punnett square results, what is the probability that they will have a child who is a carrier for CF?

MINI CASE

19 You are a genetic counselor. A 21-year old college student has scheduled an appointment because his 47-year old mother has Huntington disease, and he is worried about developing this disease. You ask about other family members. The student's maternal grandmother (his mother's mother) does not have Huntington disease. The student's father is 62 years old and does not have Huntington disease. Based on this information, you draw a Punnett square to determine the probability that the student will develop Huntington disease.

   a. What could you tell the student about his risk?

   b. The student has a half sister. She is 19 years old and has the same mother but a different father. Her father is 45 and does not have Huntington disease. However, the father's mother died of Huntington disease. How does the half sister's risk compare to her brother's (the student's) risk? Could you give her a definitive answer about her risk? Why or why not?

20 From the discussion in this chapter, why might a person with a homozygous recessive *CFTR* genotype have a somewhat different phenotype from someone who also has a homozygous recessive *CFTR* genotype?

21 Phenylketonuria is considered to be an inborn error of metabolism. It is a recessive genetic condition in which the enzyme that breaks down the amino acid phenylalanine is defective or missing. Testing of all newborns allows this condition to be detected at birth. A special diet that severely minimizes phenylalanine (e.g., by avoiding diet sodas and most usual sources of protein) can treat the condition. If two carriers of both cystic fibrosis and phenylketonuria were to have a child, what is the probability that the child will have

   a. both cystic fibrosis and phenylketonuria?

   b. cystic fibrosis and be a carrier for phenylketonuria?

   c. neither condition?

   d. neither condition and not be a carrier for either?

DRIVING QUESTION 4

What are practical applications of understanding the genetic basis of human disease?

By answering the questions below and studying Infographics 11.6, 11.7, 11.8, and 11.9, you should be able to generate an answer to the broader Driving Question above.

KNOW IT

22 Can cystic fibrosis be diagnosed prenatally by examining the fetal chromosomes (as shown in the inset in Infographic 11.2)? Why or why not?

INTERPRETING DATA

23 Ivacaftor (trade name Kalydeco) is a drug designed to enhance the activity of the CFTR protein encoded by the *G551D* allele. The protein encoded by this allele is present in the cell membrane, but it is not very active. The drug has been shown in laboratory studies to enhance the activity of this protein. To determine if ivacaftor has an impact on disease symptoms in people with cystic fibrosis, the drug was tested in a randomized, double-blind clinical trial.

a. From the information provided here and in the graph below, describe the experimental design of the trial. Include who the participants were, how participants were assigned to the experimental and the control group, and what treatment(s) were given to participants in each group. Why was it important that this was a blind trial?

b. Before treatment started, baseline measurements of lung function were obtained. One of these measurements is the amount of air that a participant can forcibly exhale from the lungs in 1 second (expressed as % of total lung volume). This measurement is known as the predicted FEV. The drug was given to the experimental group every 12 hours, and the predicted FEV was measured. The results are shown in the graph. The average value of *N* (the number of subjects) is

shown for each group, as well as the variability of the measurement (indicated by error bars, vertical lines extending above and below the data point for each average value).

How soon did the experimental group experience an improvement in lung function as measured by predicted FEV?

How long was the improvement sustained?

By the end of the study, what was the absolute improvement of the experimental group relative to baseline?

c. The *ΔF508* allele causes the CFTR protein to be absent from the cell membrane. Is ivacaftor likely to be a viable treatment for patients whose CF is caused by this allele? Why or why not?

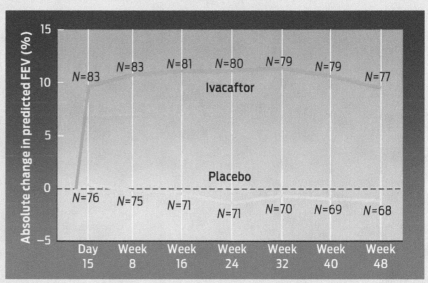

SOURCE: Ramsey, B. W., et al. (2011) A CFTR potentiator in patients with cystic fibrosis and the *G551D* mutation. Figure 1A. *New England Journal of Medicine* 365:1663–1672.

BRING IT HOME

24 Emily took a genetic test to determine which CF alleles she inherited. The results revealed she has a *G551D* allele, making her a candidate for ivacaftor. Since taking this drug, Emily's breathing and lung function has improved. In this case, the genetic test opened up a treatment option for a patient. In other cases (e.g., cystic fibrosis patients with the *ΔF508* allele) treatments are limited, and for some genetic diseases, such as Huntington disease, treatment is limited and there is no cure. If you were faced with the decision to take a genetic test, especially for a disease for which there is no cure, would you take the test? Why or why not?

Mendel's Garden

An Austrian monk lays the foundation for modern genetics

The garden outside the Augustinian Abbey in Brno, where Gregor Mendel performed his experiments.

4

Mendel provided a new explanation for heredity, decades before the word "genetics" was coined.

REGOR JOHANN MENDEL WAS AN UNLIKELY FATHER OF genetics. He was a melancholy Austrian monk who by all accounts suffered from debilitating test-taking anxiety, failing his teaching exam twice. Mendel nevertheless collected the first research suggesting that each parent passes discrete "elements," or hereditary particles, to each child that determine specific traits. These elements remain intact and can be passed on indefinitely to future generations without being diluted. Although he couldn't say at the time what these elements were, Mendel had in fact discovered what came to be called genes. We now know that genes come in pairs, and that they exist in multiple discrete forms we call alleles.

Many scientists of Mendel's day—the mid-19th century—believed that parental traits were blended together in offspring, like mixing paint colors. For example, a tall mother and short father would have a son or daughter of medium height, who would then pass on that trait—medium height—to their children. Other scientists clung to the old idea that a sperm or egg contained

▸▸ DRIVING QUESTIONS

1. How was Mendel able to recognize the transmission of alleles before the discovery of DNA?
2. What do Mendel's two laws state about how offspring inherit alleles from their parents?

249

INFOGRAPHIC M4.1 THEORIES OF INHERITANCE BEFORE MENDEL

Preformation theorie of inheritance, popular in the mid-19th century, posited that the next generation of life already existed fully formed in miniature inside the egg or sperm. It was thought that these tiny individuals needed only to grow before being born. Other ideas of inheritance speculated that substances from the mother and father blend together during conception to produce the traits of the offspring.

Preformationist Ideas

Blending Ideas

Fluid

Sperm

Fluid

Egg

Offspring traits are the result of blending these parent substances.

Spermist theory maintained that sperm held a homunculus inside.

Ovist theory proposed that the preformed human was inside the egg.

Semen and menstrual fluid contain the substance that contributes to offspring traits

Particles from different parts of the body migrate to the egg and sperm.

Gregor Mendel at work in his laboratory.

a miniature adult waiting to be born. But through a series of simple yet elegant experiments conducted in a monastery garden, Mendel provided a new explanation for heredity, decades before the word "genetics" was coined (**INFOGRAPHIC M4.1**).

In 1843, Mendel became a monk at the Augustinian Abbey of St. Thomas in Brünn (now Brno, in the Czech Republic). He studied theology and was ordained a priest in 1847. When he failed his teaching exam (the Augustinians were a teaching order), the abbot at St. Thomas sent Mendel to the University of Vienna to brush up. For 2 years he studied math, physics, zoology, and botany, but once again he flunked the test. The depressing result encouraged him to turn from teaching to research.

Mendel returned to the monastery in 1853, and a year later began researching a topic that had sparked his interest in school: hybrids, the offspring of two different breeds or varieties. Mendel was interested in how hybrids form, and he hoped to explain what he and many others had observed: that physical traits (size, color, etc.) can skip a generation.

He began his research by breeding mice but, as Robin Henig wrote in her 2001 book *The Monk in the Garden*, the local bishop found "toying with the reproduction of animals simply too vulgar an undertaking for a priest." So Mendel decided to work instead with pea plants, which proved a better model organism anyway. The plants grew quickly, and he could better control their environment and breeding.

Mendel began by choosing specific traits that he could see and study, among them seed shape, seed color, pod shape, pod color, flower color, and stem length. Each of the traits he chose to study appeared in two forms. For example, seed shape was either round or wrinkled; seed color was either green or yellow. Because, as he and others had observed, in several types of organism some traits seemed to disappear in one generation only to show up again in the next, he started his breeding experiments with plants that "bred true"–plants with offspring that carried the same traits as the parents, generation after generation. Only then could he study what happened to particular traits when purebred plants of one variety were mated with purebred, or true-breeding, plants of another variety.

Pea plants can self-pollinate, which means that the pea flower contains both male and female sexual organs and a single plant can fertilize itself to produce offspring. To produce true-breeding plants, Mendel covered pea flowers with a small bag so that he could control fertilization, manually fertilizing plants with their own pollen and preventing pollen from another plant from entering. Once he had established true-breeding plants, he could then set up a cross between two different plants. What would happen if he crossed a true-breeding green-seeded plant with a true-bred plant that produced yellow seeds? Or a purple-flowered plant with a white-flowered plant? For each cross, Mendel painstakingly pollinated individual

The view through a window of the Abbey of St. Thomas to the garden where Mendel ran his experiments with pea plants.

flowers from the two plants by hand. He also prevented self-pollination by removing the male reproductive parts from the plants to be fertilized.

Mendel noticed that when he bred a true-breeding white-flowering plant with a true-breeding purple-flowering plant, the first generation of offspring (what we now call the F_1 generation) all had purple flowers. That the flowers were true purple rather than pale purple suggested that parental traits were not blended, as earlier hypotheses of inheritance would have predicted. But the trait for white flowers did not disappear completely, either. When Mendel randomly selected two F_1 purple-flowering plants to breed, he found that on

INFOGRAPHIC M4.2 MENDEL'S EXPERIMENTS

True Bred: self-pollinated; parents and offspring share traits

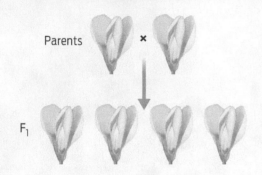

Parents

F₁

> Mendel recognized that there are alternative "elements" for each trait. In this example, there are two elements for flower color—one element for purple, and one for white.

Parents

F₁

Cross-pollination: between two different true-bred parents

Parents

F₁

F₂

> **100% of the F₁ plants have purple flowers.**
> · This suggests that inherited material is not blended to produce traits in offspring because there are no pale purple offspring.

> **75% of the F₂ plants have purple flowers, and 25% have white flowers.**
> · This suggests that the white element was not lost in the F₁ generation, but rather was hidden by the purple element.
> · This result also suggests that each individual has two elements for one trait. The F₁ offspring must have displayed the purple element and carried the hidden white element.

average 1 out of every 4 plants of the second generation of offspring (the F₂ generation) had white flowers. Mendel reasoned that a hidden white element must be present in the purple F₁ plants. So each F₁ plant must have two such elements, one representing purple (the trait that appeared) and the other representing white (the hidden trait) (**INFOGRAPHIC M4.2**).

If these results sound familiar, there's a good reason for that: they reflect dominant and recessive patterns of inheritance, which we discussed in Chapter 11. Purple flower color is dominant over white, which is recessive. Mendel was the first to gather evidence showing that traits could be inherited in a dominant or recessive fashion, and was in fact the one

> ## Mendel studied thousands of pea plant crosses and discovered the basic principles of inheritance.

who coined these terms. While earlier scientists had noticed that traits could disappear in one generation and reappear in later generations, Mendel was the first to offer a coherent explanation of *why* they did.

Over 7 years, Mendel studied thousands of pea plant crosses and discovered the basic principles of inheritance. He published his results in 1866.

Today we know that Mendel's "elements" are alleles of genes, and that genes are located on chromosomes. The principles he discovered have been formalized into two laws. The first of these is Mendel's law of segregation, which states that for any diploid organism, the two alleles of each gene segregate separately into gametes. That is, every gamete receives only one of the two alleles and the specific allele that any one gamete receives is random (INFOGRAPHIC M4.3).

The pea plant, **Pisum sativum.**

INFOGRAPHIC M4.3 MENDEL'S LAW OF SEGREGATION

Mendel's experiments enabled him to formulate the law of segregation. This law has held up over time, although today we call Mendel's "elements" alleles. Mendel's law of segregation states that when an organism produces gametes, the two alleles for any given trait separate so that each gamete receives only one allele. Consequently, each parent donates only one of any two alleles to any offspring. The alleles don't blend, but remain as discrete pieces of information as they pass from one generation to the next.

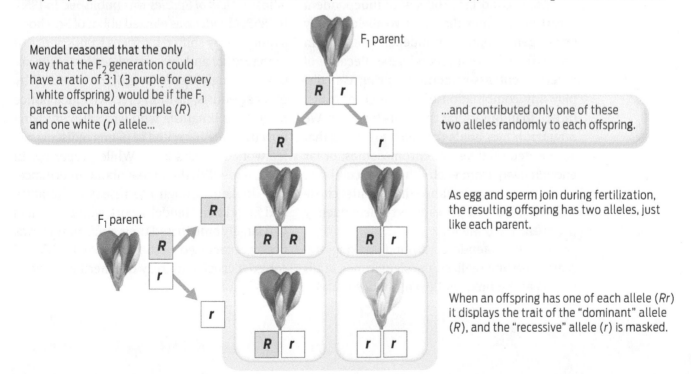

Mendel reasoned that the only way that the F$_2$ generation could have a ratio of 3:1 (3 purple for every 1 white offspring) would be if the F$_1$ parents each had one purple (*R*) and one white (*r*) allele...

F$_1$ parent

...and contributed only one of these two alleles randomly to each offspring.

F$_1$ parent

As egg and sperm join during fertilization, the resulting offspring has two alleles, just like each parent.

When an offspring has one of each allele (*Rr*) it displays the trait of the "dominant" allele (*R*), and the "recessive" allele (*r*) is masked.

INFOGRAPHIC M4.4 MENDEL'S LAW OF INDEPENDENT ASSORTMENT

Mendel went on to study how multiple traits are inherited. For example, he studied plants that had different seed color and seed texture and how those traits passed to the next generation. Tracing two traits at a time helped him formulate the law of independent assortment. This law posits that two alleles for any given trait will segregate independently from any other alleles when passed on to gametes. Consequently, each gamete may acquire any possible allele combination and traits.

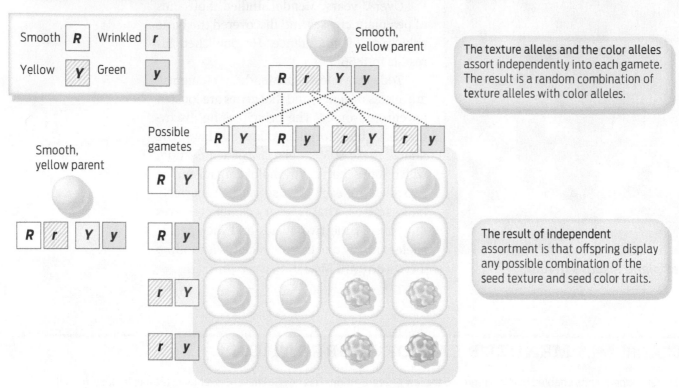

Smooth **R** Wrinkled **r**
Yellow **Y** Green **y**

Smooth, yellow parent

Smooth, yellow parent

Possible gametes

The texture alleles and the color alleles assort independently into each gamete. The result is a random combination of texture alleles with color alleles.

The result of independent assortment is that offspring display any possible combination of the seed texture and seed color traits.

The second law, the law of independent assortment, states that the two alleles of any given gene segregate independently from any two alleles of a second gene. Because of independent assortment, offspring can display any combination of the different traits, rather than inheriting the traits together. We now know this holds true only for genes that are located on different chromosomes, or far enough away from each other to recombine. It was mere happenstance that Mendel chose traits for which the genes assort independently (**INFOGRAPHIC M4.4**).

Despite Mendel's groundbreaking research, no one realized the significance of his results at the time—not even Charles Darwin, whose *Origin of Species* was published in 1859. In 1868, Mendel was elected abbot of St. Thomas and largely shifted his focus from science to monastic life and the administration of the abbey. Although Mendel's research was cited by other scientists, he didn't receive much notice until three botanists who were also studying how traits are inherited in plants rediscovered his work 30 years later. While preparing, in 1900, to publish their ideas about inheritance, they looked through the research literature and found that Mendel's work with pea plants had largely anticipated their own. Mendel was finally recognized as the researcher who had solved a crucial mystery of inheritance many years before. ■

MORE TO EXPLORE

▶ Henig, R. M. (2001) *The Monk in the Garden: The Lost and Found Genius of Gregor Mendel, the Father of Genetics.* New York: Mariner Books.

▶ Miko, I. (2008) Gregor Mendel and the principles of inheritance. *Nature Education* 1(1) http://www.nature.com/scitable/topicpage/gregor-mendel-and-the-principles-of-inheritance-593

▶ Griffiths, A. J. F. et al. (2012) *Introduction to Genetic Analysis.* New York: W. H. Freeman.

MILESTONES IN BIOLOGY 4 Test Your Knowledge

1 **You have a tall pea plant.**

a. What are its possible genotypes?

b. What crosses could you do to determine if it is true breeding or not? For each cross, give the expected phenotypes of the offspring.

2 **In crossing pea plants:**

a. If a true-breeding tall pea plant with purple flowers is crossed with another true-breeding tall pea plant with purple flowers, what will the offspring pea plants look like?

b. Will the offspring be true-breeding? Explain your answer.

3 **Half of the gametes of a heterozygous parent will carry the dominant allele, and half will carry the recessive allele. Which of Mendel's laws explains this?**

4 **Consider Mendel's laws:**

a. Which of Mendel's laws would be violated if the offspring of two heterozygous tall, purple-flowered pea plants (denoted as *TtPp*) were only tall, purple-flowered plants or short, white-flowered plants?

b. What would such a violation suggest about the two genes?

Genetics

Complexities of human genetics, from sex to depression

 What makes a man?

ANSWER: A botched circumcision in 1966 on a little boy named Bruce Reimer in time became a landmark example of how biology shapes sexual identity. Doctors at the hospital where Reimer was circumcised used an experimental procedure that involved burning off the foreskin. The procedure went awry, and Bruce's penis was singed nearly completely off, beyond surgical repair. On the advice of John Money, a well-known psychologist and sex researcher at Johns Hopkins University in Baltimore, Maryland, who had written much about the importance of environment in determining a person's sexual identity, Bruce's parents decided to have their little boy surgically turned into a little girl and rear him as "Brenda."

But Brenda never behaved like a girl. She didn't like playing with girls' toys, didn't enjoy wearing dresses, and often got into fistfights at school. By the time Brenda reached puberty, her behavior became so troublesome that her father broke down and told her what had happened to her. Brenda felt relieved rather than angry. "All of a sudden everything clicked," she later said. "For the first time things made sense and I understood who and what I was."

Brenda eventually had reconstructive surgery to create a penis, took hormones, and changed her name to David. David told his story in a book published in 2000, *As Nature Made Him: The Boy Who Was Raised as a Girl*, by John Colapinto. By that time, it had become

▸▸ **DRIVING QUESTIONS**

1. How do chromosomes determine sex, and how does sex influence the inheritance of certain traits?

2. Some traits are not inherited in simple dominant or recessive inheritance patterns. What are some complex inheritance patterns?

3. How do numerical abnormalities of chromosomes occur, and what are the consequences of these abnormalities?

David Reimer in 2000.

increasingly clear that sexual identity–the experience of oneself as male or female–is significantly influenced by biology. Studies had shown that prenatal exposure to fetal sex hormones such as testosterone not only determine whether a fetus will develop female or male genitalia, but also act on a developing baby's brain, influencing behavior.

Sex hormones are produced by sex organs known as the **gonads**–ovaries in females, testes in males. Sex hormones include androgens (from *andros*, Greek for "man"), such as testosterone, and estrogen (from the Greek *oistros*, meaning "mad passion"). Both males and females produce testosterone and estrogen, but in most cases males produce higher levels of testosterone and females produce higher levels of estrogen. In a developing fetus, these hormones shape the development of both internal and external sexual anatomy.

Whether a fetus develops male or female gonads, and produces male or female hormones, depends on the set of chromosomes it receives from its parents. Humans have 23 pairs of chromosomes; 22 pairs are **autosomes** and

INFOGRAPHIC 12.1 X AND Y CHROMOSOMES DETERMINE HUMAN SEX

Males and females differ by virtue of a pair of sex chromosomes. Females have two X chromosomes and males have a single X and a single Y chromosome. Every person must have at least one X chromosome, but it's the presence of a gene on the Y chromosome that initiates male development.

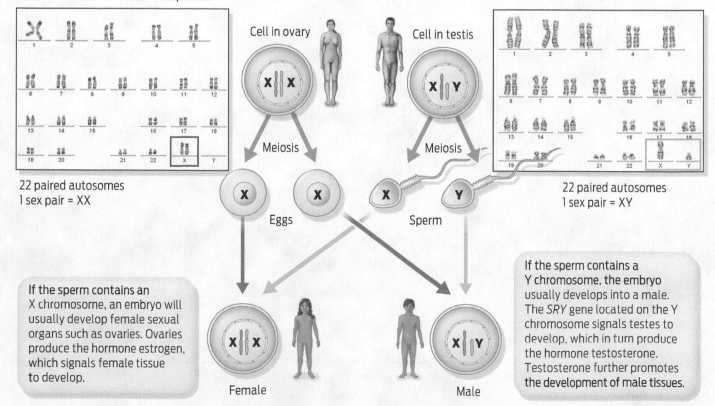

22 paired autosomes
1 sex pair = XX

22 paired autosomes
1 sex pair = XY

Cell in ovary

Cell in testis

Meiosis

Meiosis

Eggs

Sperm

If the sperm contains an X chromosome, an embryo will usually develop female sexual organs such as ovaries. Ovaries produce the hormone estrogen, which signals female tissue to develop.

If the sperm contains a Y chromosome, the embryo usually develops into a male. The *SRY* gene located on the Y chromosome signals testes to develop, which in turn produce the hormone testosterone. Testosterone further promotes the development of male tissues.

Female

Male

1 pair are the **sex chromosomes,** X and Y. Sons inherit one Y chromosome from their father and one X chromosome from their mother. Daughters inherit two X chromosomes, one each from mother and father. So males are XY and females are XX. In the absence of a Y chromosome, a fetus will develop into a female. Thus, fathers determine the sex of a baby, the determination based on whether the sperm fertilizing a mother's egg carries an X or a Y sex chromosome. The Y chromosome contains a gene called *SRY* that signals testes to develop. (*SRY* stands for "*s*ex-determining *r*egion on the *Y* chromosome.") The testes, in turn, produce masculinizing hormones such as testosterone that mold a male body (**INFOGRAPHIC 12.1**).

X-linked traits are passed from mothers to children on their X chromosome.

Although most individuals have internal and external genitalia that are either clearly male or clearly female, there are exceptions. Each year about 1 in every 1,600 babies born in America falls into an intermediate sex category termed "intersex." An intersexual person is someone whose external genitalia do not match his or her internal sex organs or genetic sex—for example, a person with an XX chromosome pair who has internal ovaries but external genitalia that appear male. Often intersex babies are born with ambiguous genitalia.

Debate over the experience of David Reimer, and similar cases, as well as research showing how strongly biology influences sexual identity, has changed the care of intersex babies. Today, such babies are often assigned a sex by parents and doctors only after a period of observation to assess behavior patterns.

Surgeons then perform surgery to create either male or female genitalia. Or, parents may forgo surgery, preferring that their child remain as is.

Some instances of intersex have genetic causes. For example, if the Y chromosome has a mutation in the *SRY* gene, the embryo is likely to have undeveloped gonads with external female genitalia, even though it carries an XY chromosome pair.

There are also cases of people with XY sex chromosomes who have mutations in genes that code for androgen receptors on cells. So even though they carry a functional *SRY* gene and have internal testes, their cells fail to respond to androgens like testosterone. As a result, complete male external genitalia do not develop, and these people appear to be female.

Similarly, there are people with XX sex chromosomes who have male genitalia. In some cases, this is caused by a condition called congenital adrenal hyperplasia. These individuals have one or more mutations in genes on autosomal chromosomes. One result is excessive production of androgens. These people have ovaries but may have genitals that appear more male than female.

Some people have only one sex chromosome, while others have three. Every person must have at least one X chromosome (having none is fatal), but because of errors in chromosome segregation during meiosis, a variety of other X and Y combinations are possible: XXY men, women with only a single X chromosome, XXX females, and XYY males. In many of these cases, a person's physical traits and genitalia reveal that they do not

▸GONADS
Sex organs: ovaries in females, testes in males.

▸AUTOSOMES
Paired chromosomes present in both males and females; all chromosomes except the X and Y chromosomes.

▸SEX CHROMOSOMES
Paired chromosomes that differ between males and females, XX in females, XY in male

TABLE 12.1 BETWEEN MALE AND FEMALE: VARIETIES OF SEX AND INTERSEX

SEX CATEGORY	CHROMOSOMES	GONADS	GENITALIA	OTHER CHARACTERISTICS
Female	XX	Ovaries	Female	
Male	XY	Testes	Male	
Female pseudo-hermaphroditism	XX	Ovaries	Male	Infertile
Male pseudo-hermaphroditism	XY	Testes	Female or ambiguous	Infertile
True gonadal Intersex	XX and/or XY	Ovaries and testes	Male, female, or ambiguous	Infertile; historically called true hermaphrodites
Triple X syndrome	XXX	Ovaries	Female	Fertile, taller than average, learning disabilities
Klinefelter syndrome	XXY	Testes	Male	Infertile, enlarged breast tissue
47, XYY syndrome	XYY	Testes	Male	Fertile, taller than average, elevated risk for learning and emotional disabilities
Turner syndrome	X	Ovaries	Female	Infertile, broad chest, webbed neck

have the usual makeup of sex chromosomes, but not always (**TABLE 12.1**).

As you can see, sex isn't so easy to define; there are genetic, hormonal, anatomical, and behavioral aspects to sex, and these elements don't always align. Defining what counts as "masculine" and "feminine" is even more complicated. For example, some men have characteristics that we typically identify as female, such as a high voice and sparse body hair, yet they are genetically and anatomically male. And many women have what are considered to be more masculine features, such as angular faces and more muscle as compared to body fat. Yet, they are genetically and anatomically female. In fact, there are very few physical or mental characteristics that are entirely male or entirely female. In addition, some people—for example, transgender individuals—may mentally identify with one sex even though their genitalia and chromosomal makeup classify them as the other.

For biologists, the story of David Reimer—known in the medical literature as the "John/Joan case"—put the nail in the coffin of the idea that children are born psychosexually "neutral." Contrary to what sex researcher Money and others had claimed, it was not possible to mold someone into whichever sexual identity you wanted through surgery and child rearing—at least not without damaging psychological fallout. Of this stubborn fact, David Reimer was living proof.

David eventually married a woman and adopted her three children. Though he could not have children of his own (since his testes had been removed), he took the responsibilities of fatherhood seriously. "From what I've been taught by my father," he told Colapinto, "what makes you a man is: You treat your wife well. You put a roof over your family's head. You're a good father. . . That, to me, is a man."

David said he told his story so that others would be spared the nightmarish experience he went through. Tragically, that experience may have led to the depression that cost him his life. David killed himself in 2004. He was 38.

SEX-LINKED INHERITANCE

Q Why do some genetic conditions affect sons more often than daughters?

ANSWER: Some 10 million American men—about 7% of the male population—either cannot distinguish red from green, or they perceive these hues differently from the way other people do. Such red-green color blindness affects only 0.4% of women. Similarly, 1 in 5,000 boys worldwide is born with hemophilia, a blood-clotting disorder, yet hemophilia rarely afflicts girls.

Why this disparity? These conditions are caused by genes found on the X chromosome. When a gene is located on either of the sex chromosomes, daughters and sons don't share the same probability of inheriting it.

Take the neuromuscular condition Duchenne muscular dystrophy (DMD), for example. DMD is a disease in which muscles slowly de-generate, leading to paralysis. About 1 in 2,400 boys worldwide is born with the condition each year. Most affected boys are in wheelchairs by the time they become teenagers, and they rarely live past 30.

Why does DMD primarily affect men? Recall that a female has two X chromosomes. For a recessive trait like DMD, a normal allele on one X chromosome masks the recessive disease allele on the other X chromosome. A male, on the other hand, has a single X chromosome, and will show the effects of any recessive alleles on that X chromosome. Because females can carry the disease allele without showing it, they may not even know they are carriers who can pass it on to their sons. Diseases and other traits such as DMD that are inherited on X chromosomes are called **X-linked traits** (INFOGRAPHIC 12.2).

So, for example, if you are female and DMD runs in your family—say, a male cousin has the disease—you could have inherited one DMD allele from your mother and not even know it. In this case, since your male cousin

> **▶X-LINKED TRAIT**
> A phenotype determined by an allele on an X chromosome.

INFOGRAPHIC 12.2 X-LINKED TRAITS ARE INHERITED ON X CHROMOSOMES

→ Duchenne muscular dystrophy (DMD) is an example of an X-linked trait. Recessive mutations of the *dystrophin* gene on the X chromosome cause the disease. DMD primarily affects males because they inherit only one copy of the X chromosome (from their mothers). Therefore, the single DMD allele they inherit determines their phenotype. Since females have two X chromosomes, they may carry the DMD allele but have a healthy phenotype.

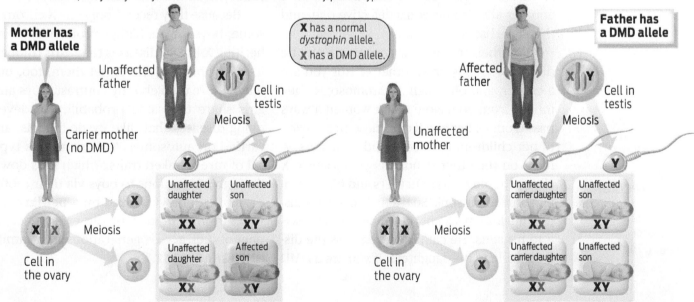

INFOGRAPHIC 12.3 FEMALE CARRIERS CAN PASS DISEASE ALLELES TO THEIR CHILDREN

The following diagram, known as a pedigree, shows how an X-linked trait (in this case Duchenne muscular dystrophy, or DMD) passes through generations.

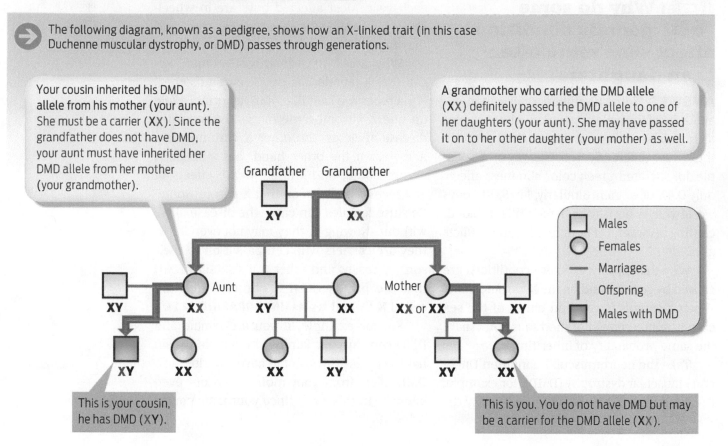

Your cousin inherited his DMD allele from his mother (your aunt). She must be a carrier (**X**X). Since the grandfather does not have DMD, your aunt must have inherited her DMD allele from her mother (your grandmother).

A grandmother who carried the DMD allele (**X**X) definitely passed the DMD allele to one of her daughters (your aunt). She may have passed it on to her other daughter (your mother) as well.

Grandfather Grandmother
XY XX

Aunt
XY XX
XY XX

Mother
XX or XX XY

XY XX

Males
Females
Marriages
Offspring
Males with DMD

This is your cousin, he has DMD (X**Y**).
XY XX

XX XY

XY XX

This is you. You do not have DMD but may be a carrier for the DMD allele (**X**X).

---

is affected, your aunt was a DMD carrier. Her son, your cousin, inherited her DMD-carrying X chromosome. Because males have only one X chromosome, your cousin doesn't have another allele to mask his defective one, and therefore has the disease.

Whether or not your sons will have the disease depends on whether or not you are a carrier, and on which X chromosome they inherit from you. Note that a woman always passes one of her two X chromosomes to each of her children, both sons and daughters. A man, on the other hand, passes his single X chromosome to his daughters and his Y chromosome to his sons. So if a male carries a disease allele on his X chromosome, he can't pass it to his sons. He can, however, pass the disease allele to his daughters. If you are a DMD

carrier, a son who inherits a DMD allele from you will have the disease. Your daughters who inherit your DMD allele will be carriers (assuming their father is healthy).

Because they receive only one X chromosome, boys have a higher probability of inheriting X-linked diseases than girls (though it is possible for girls to get them, too, but much more unlikely). By contrast, boys and girls share the same probability of developing diseases that, like cystic fibrosis, are carried on autosomes. This pattern is typical of rare X-linked traits, which pass down through generations to boys via their mothers and can be visualized on a **pedigree,** a visual representation of the occurrence of phenotypes across generations in a family (**INFOGRAPHIC 12.3**).

▶PEDIGREE
A visual representation of the occurrence of phenotypes across generations.

Y-CHROMOSOME ANALYSIS

Q Did Thomas Jefferson father children with a slave?

ANSWER: Thomas Jefferson was the third president of the United States, the principal architect of the Declaration of Independence, and founder of the University of Virginia. He was also a slave holder. Historians have long debated the meaning of this and other seeming contradictions in the founding father's life and politics. For example, although Jefferson's writings clearly show that he did not believe in the institution of slavery, he owned at least 200 slaves. He made disparaging comments about slaves, yet maintained close relationships with those living in his house.

In some cases, very close. Jefferson was rumored to have fathered at least six children with Sally Hemings, a slave who tended to his family. For decades, historians discredited the rumor as unreliable oral history. In 1998, scientists attempted to find out by testing the DNA of both Hemings's and Jefferson's descendants with a technique called **Y-chromosome analysis.**

Y-chromosome analysis is commonly used to study ancestry and to identify paternity. It is just one of several ways in which science can complement history. Scientists can use Y-chromosome analysis to verify or discredit accepted history, or to fill in missing pieces of historical information.

How does Y-chromosome analysis work? As the name implies, in this technique re-

A slave named Lucy, born at Monticello in 1811. Slaves at Monticello often cared for the Jefferson children.

searchers examine the Y-chromosome, which is very small and contains few genes. Sons inherit their Y chromosome from their fathers. These Y chromosomes are passed through generations, from fathers to sons, largely unchanged. That's because Y chromosomes have no homologous partner chromosome with which to pair and exchange DNA during meiosis (see Chapter 11). In other words, the Y chromosome rarely undergoes genetic recombination. Consequently, the Y chromosome that a son inherits from his father is almost

During Jefferson's own time, people commented on the resemblance of Hemings's children to the president.

identical to the Y chromosome that his father inherited from his father (**INFOGRAPHIC 12.4**).

Comparing DNA sequences on Y chromosomes can reliably establish paternity and help to establish a line of ancestry. For example, scientists have used Y-chromosome analysis to show that 90% of the Cohanim—members of the historic Jewish priesthood—are related, supporting oral and written histories claiming the Cohanim (the Hebrew plural from which we derive the name "Cohen") are all descended from Aaron, brother of Moses. They've also used Y-chromosome analysis to support the

oral history of the Lemba, an African tribe, who claim they are descended from Jews. And Y-chromosome analysis suggests that about 8% of Eastern European and Asian men are descended from Genghis Khan or his family.

During Jefferson's own time, people commented on the resemblance of Hemings's children to the president. But later historians either explained the resemblance away or proposed other explanations—for example, that one of Jefferson's nephews had fathered her children.

To set the record straight, in 1998 a team of geneticists led by Eugene A. Foster compared

INFOGRAPHIC 12.4 Y CHROMOSOMES PASS LARGELY UNCHANGED FROM FATHERS TO SONS

→ Y-chromosome analysis for paternity testing relies upon the fact that the Y chromosome does not undergo recombination during meiosis and so passes unchanged from the father to his sons.

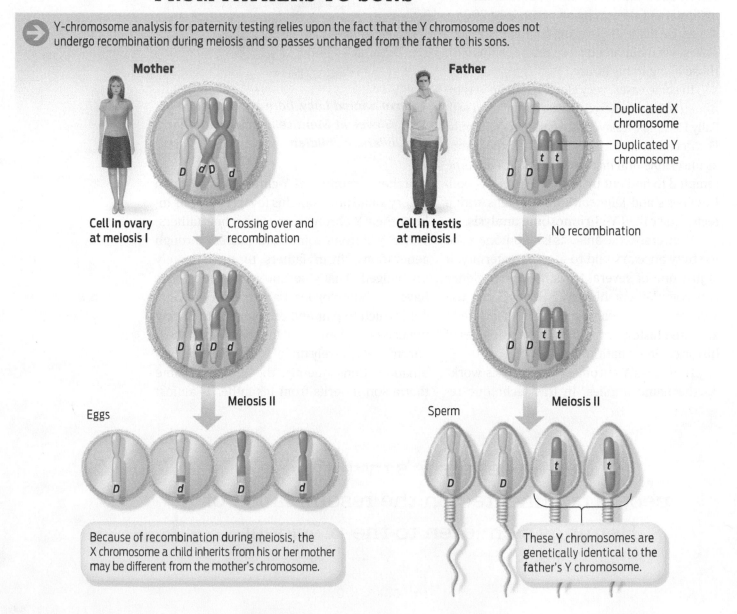

Mother

Cell in ovary at meiosis I

Crossing over and recombination

Meiosis II

Eggs

Because of recombination during meiosis, the X chromosome a child inherits from his or her mother may be different from the mother's chromosome.

Father

Duplicated X chromosome

Duplicated Y chromosome

Cell in testis at meiosis I

No recombination

Meiosis II

Sperm

These Y chromosomes are genetically identical to the father's Y chromosome.

INFOGRAPHIC 12.5 DNA LINKS SALLY HEMINGS'S SON TO JEFFERSON

➡️ Scientists compared DNA sequences on the Y chromosome of Sally Hemings's and Thomas Jefferson's grandfather's descendants. The DNA sequences match at the 11 different STR locations analyzed.

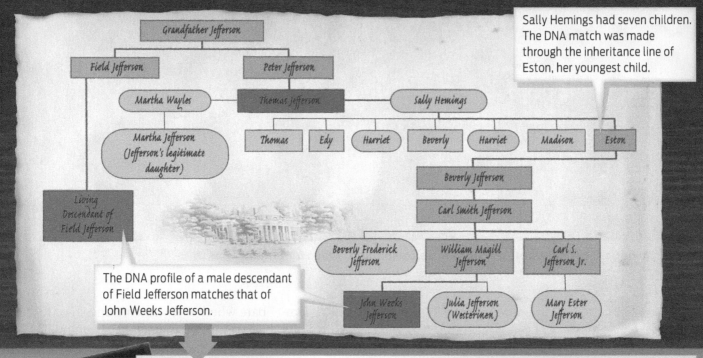

Sally Hemings had seven children. The DNA match was made through the inheritance line of Eston, her youngest child.

The DNA profile of a male descendant of Field Jefferson matches that of John Weeks Jefferson.

STR analysis This table shows the number of repeats at each STR location. Red numbers denote matching profiles between individuals. John Weeks Jefferson's profile matches that of the living descendant of Field Jefferson at every STR. However, John Weeks Jefferson's profile does not perfectly match that of an unrelated individual.

STR location	1	2	3	4	5	6	7	8	9	10	11
Living descendant of Field Jefferson	15	12	4	11	3	9	11	10	15	13	7
John Weeks Jefferson	15	12	4	11	3	9	11	10	15	13	7
Unrelated individual	14	12	5	11	3	10	11	13	13	13	7

the Y chromosomes of three groups of men: descendants of Thomas Jefferson's paternal uncle Field Jefferson; one male descendant of Eston Hemings, Sally Hemings's son; and descendants of Jefferson's sister's sons. Since Jefferson's only surviving child from his wife was a daughter, he did not have any direct male descendants, which is why scientists tested descendants of Jefferson's uncle.

The team analyzed 11 short tandem repeats (STRs) on the Y chromosome. (Recall from Chapter 7 that STRs are short regions of DNA that are repeated a different number of times in different people, and are used in DNA profiling.) They made some startling discoveries: the results clearly showed that descendants of Thomas's nephews have a different Y chromosome from the man descended from Eston Hemings, thus ruling out Thomas's nephews as the father of Sally's children. The results also clearly showed that Eston's descendant—John Weeks Jefferson—has the same Y chromosome as the descendants of Field Jefferson. Consequently, Thomas Jefferson *could* have fathered Eston Hemings. The study does not *prove* that he is the father, but it does show that the father was definitely a male Jefferson (**INFOGRAPHIC 12.5**).

Are these people descendants of Thomas Jefferson and Sally Hemings?

Some historians have argued that Thomas's younger brother Randolph Jefferson could have fathered Eston. (Or, indeed, that any of the other seven Jefferson males who periodically visited Monticello, where Thomas and Sally lived, could be the father.) But other experts have argued that historical evidence—for example, records of the president's travels—place him rather than Randolph under the same roof as Sally at the time of her conceptions. A 2000 report by the Thomas Jefferson Foundation concludes that the preponderance of historical and biological evidence points to a "strong likelihood" of a sexual relationship between Thomas and Sally.

For the descendants of Eston Hemings, the DNA study was powerful vindication, even if debate continues. They had long argued that they were descended from Thomas Jefferson, but without hard evidence, most historians disregarded their claims. "I feel wonderful about it," Julia Jefferson Westerinen, a Staten Island artist and Eston's great-great-granddaughter told the *New York Times* when the study results were published. "I feel honored."

As for the relationship between Jefferson and Sally Hemings, historians continue to de-

bate whether it was consensual or forced. "I was a history major," said Jefferson Westerinen, "And we learned not to say, 'I feel this, I think that,' without knowing the facts. They had a relationship of 38 years. I would like to think they were in love, but how would I know?"

We'll likely never know the truth. Neither Thomas Jefferson nor Sally Hemings left any written evidence of their relationship.

INCOMPLETE DOMINANCE

Q **A woman has straight hair, a man has curly hair. What hair texture will their children have?**

ANSWER: Their children will all have an intermediate hair texture: wavy hair. Like flower color in some plants, where for example red-flowering plants mated with white-flowering plants produce plants with pink flowers, hair texture is an example of **incomplete dominance,** a form of inheritance in

▶**INCOMPLETE DOMINANCE**
A form of inheritance in which heterozygotes have a phenotype that is intermediate between homozygous dominant and homozygous recessive.

which heterozygotes have a phenotype intermediate between homozygous dominant and homozygous recessive.

There are two versions of the gene that specify hair texture: straight (*h*) and curly (*H*). People with straight hair are homozygous recessive (*hh*); they don't produce any of the protein that causes hair to curl. People who are homozygous dominant (*HH*) express a full amount of hair curl protein and consequently have curly hair. Heterozygotes (*Hh*) express half the amount of curl protein relative to

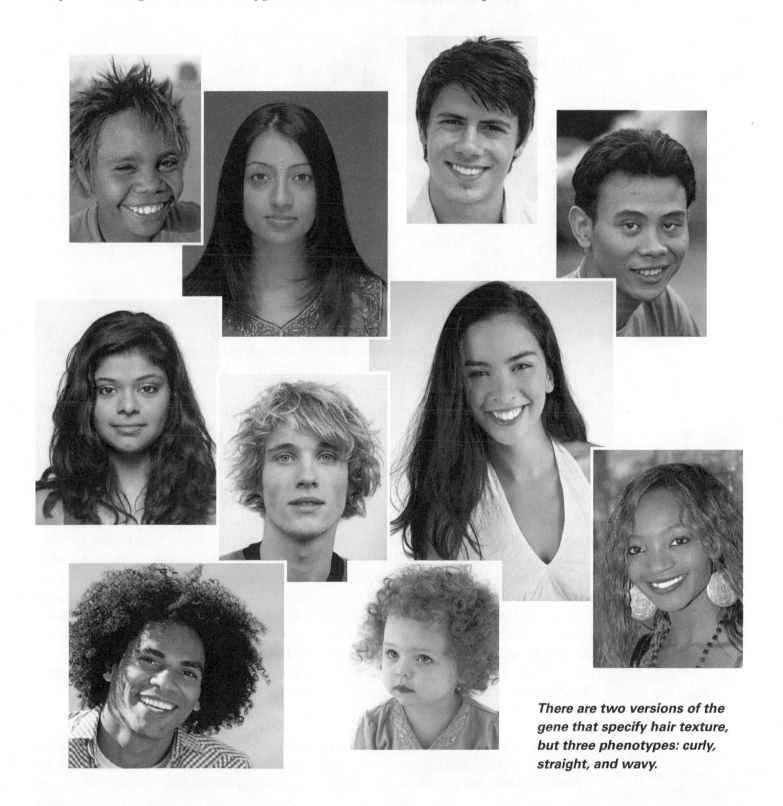

There are two versions of the gene that specify hair texture, but three phenotypes: curly, straight, and wavy.

INFOGRAPHIC 12.6 HAIR TEXTURE EXHIBITS INCOMPLETE DOMINANCE

In incomplete dominance, **heterozygotes** display a phenotype intermediate between homozygous dominant and homozygous recessive. Hair texture is an example. There are two alleles of a gene that determine the texture of hair. One allele *(H)* results in curly hair, and one allele *(h)* results in straight hair.

Curly-haired father

People who are homozygous dominant (*HH*) express only the protein that contributes to curly hair, resulting in curls.

Cell in the testis

Meiosis

Straight-haired mother

People who are homozygous recessive (*hh*) do not produce any of the protein that contributes to curly hair, so their hair is straight.

Meiosis

Cell in the ovary

| Wavy *Hh* | Wavy *Hh* |
| Wavy *Hh* | Wavy *Hh* |

Wavy-haired offspring

Heterozygotes (*Hh*) express half the amount of the "curly" protein relative to people with curly hair, so their hair is less curly (wavy).

someone with curly hair, which makes their hair wavy (**INFOGRAPHIC 12.6**).

CODOMINANCE

Q Who can be a universal blood donor?

ANSWER: When someone needs a blood transfusion, the donated blood cannot come from just anyone. The transfused blood must be compatible in ways that are determined by genetics. The two most important genetic attributes are ABO blood type and Rhesus (Rh) factor, both of which must be compatible between donor and recipient. Mixing incompatible blood causes blood cells to clump, a life-threatening condition.

Your blood type indicates the presence or absence of specific protein markers, or "flags," on your red blood cells. For example, if you have type A blood, your cells display A markers. There are three basic blood type alleles: *A, B,* and *O*. Since we inherit one allele

from each parent, the possible combinations of the three alleles are *OO, AO, BO, AB, AA,* and *BB*.

Blood type is an example of **codominance**—both maternal and paternal alleles contribute equally and separately to the phenotype. Unlike incomplete dominance, in which heterozygotes have an intermediate phenotype, codominant alleles share the limelight: heterozygotes express both traits.

Blood type alleles *A* and *B* are codominant, while *O* is recessive to both *A* and *B*. Consequently, if you have blood type A, your genotype will be either *AA* homozygous or *AO* heterozygous. The same goes for blood type B: you will have a genotype of either *BB* or *BO*. People with type AB blood have an *AB* genotype and express both A and B markers on their cells (**INFOGRAPHIC 12.7**).

Your Rh status, (+) or (−), indicates the presence or absence of Rh proteins on the surface of your red blood cells. Rh factor genes are inherited in a dominant and recessive fashion: the positive Rh factor allele (*Rh*+) is dominant over the negative Rh factor allele

▶**CODOMINANCE**
A form of inheritance in which both alleles contribute equally to the phenotype.

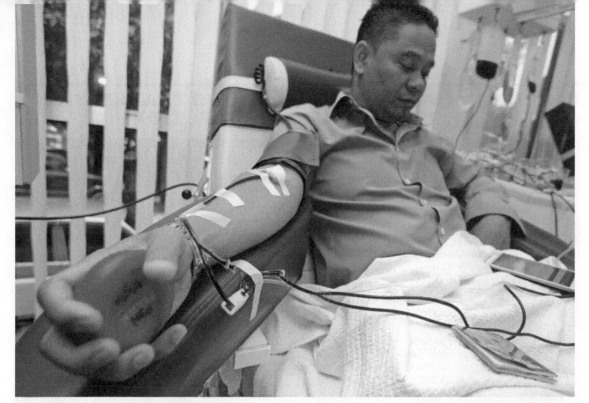

Blood from donors can be used for transfusions, but must be compatible with the recipient.

(*Rh-*). So if a person carries one positive and one negative allele, the positive allele will dominate and that person will have an Rh-positive phenotype.

Type O Rh-negative donors are known as universal donors because their blood can be transfused to patients of any other blood type; red blood cells from these donors lack

INFOGRAPHIC 12.7 HUMAN BLOOD TYPE IS A CODOMINANT TRAIT

In codominant inheritance, heterozygotes display the effects of both alleles in their phenotype. Human blood type is an example. Alleles for blood type code for different surface markers on red blood cells. A person with type AB blood, for example, displays both A and B markers, while type O blood displays no surface markers. A person's blood type must be considered when he or she gives or receives blood.

Blood Transfusions

The ability to donate or receive blood is based on immune rejection. If two people have the same surface markers, then their blood will be compatible. People with type O blood have no surface markers to provoke an immune response in a recipient (so O is the universal donor). People with type AB blood will not recognize either marker as foreign, so can receive blood from any donor.

	Type A markers	Type B markers	Type A and B markers	No blood group markers
Red blood cell type	A	B	AB	O
Genotype	*AA* or *AO*	*BB* or *BO*	*AB*	*OO*
Can donate to	Type A or AB recipient	Type B or AB recipient	Type AB recipient	Type A, B, AB, or O recipient
Can receive from	Type A or O donor	Type B or O donor	Type A, B, AB, or O donor	Type O donor

INFOGRAPHIC 12.8 A MISMATCHED BLOOD TRANSFUSION CAUSES IMMUNE REJECTION

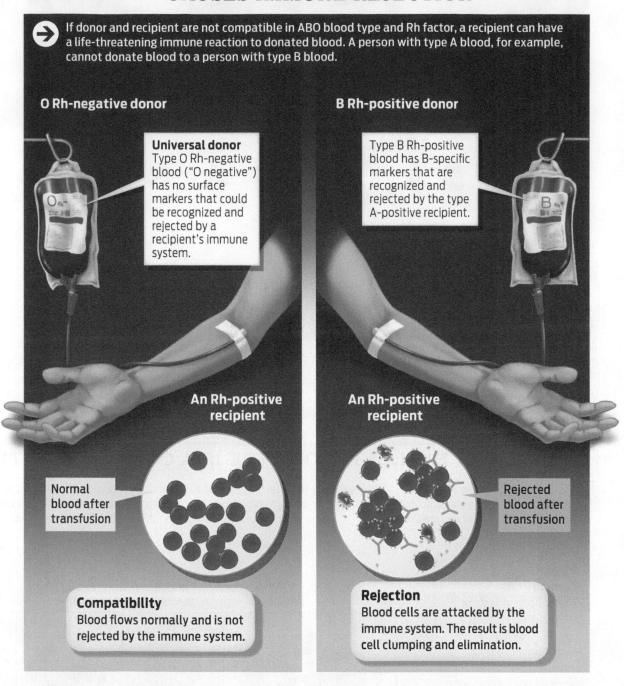

→ If donor and recipient are not compatible in ABO blood type and Rh factor, a recipient can have a life-threatening immune reaction to donated blood. A person with type A blood, for example, cannot donate blood to a person with type B blood.

O Rh-negative donor

Universal donor
Type O Rh-negative blood ("O negative") has no surface markers that could be recognized and rejected by a recipient's immune system.

An Rh-positive recipient

Normal blood after transfusion

Compatibility
Blood flows normally and is not rejected by the immune system.

B Rh-positive donor

Type B Rh-positive blood has B-specific markers that are recognized and rejected by the type A-positive recipient.

An Rh-positive recipient

Rejected blood after transfusion

Rejection
Blood cells are attacked by the immune system. The result is blood cell clumping and elimination.

surface markers and so will not trigger an immune response in a recipient. Because any patient can receive O Rh-negative blood, "O negative" donors are always in demand. Blood banks can fall short of O negative blood during such disasters as earthquakes or hurricanes in which many people are hurt and require blood (**INFOGRAPHIC 12.8**).

CONTINUOUS VARIATION

 How much of human height is genetic?

ANSWER: The short answer is, a lot. Geneticists estimate that height is 60% to 80% genetic. That is, genes account for 60% to 80%

of the difference in height you see from person to person. But there isn't one single gene that determines height–there are many: more than 20 different genes are thought to influence a person's height, which is one reason why we see such a range of heights among us.

In the United States, most people are between 5 feet and 6.2 feet tall, and women tend to be shorter than men. If height is plotted on a graph, the result resembles a bell curve, with a range of heights and the heights of most people falling near the middle of the curve at the top of the bell. In other words, height is a trait that shows **continuous variation** in the population. This is in contrast to the discrete, or discontinuous, traits we've encountered, in which individuals have one of only two or three possible phenotypes for a given trait– Mendel's round or wrinkled peas, or AB blood type, for example. Why does height show an unbroken range of phenotypes rather than discrete categories like tall or short? The main reason is that human height is a **polygenic trait**–one that is influenced by more than one gene. When multiple genes act together, their effects add up to produce a range of phenotypes. Other examples of polygenic traits are skin color and eye color.

The fact that height is largely genetic means that, all other things being equal, two tall parents are very likely to have a child who is also tall; same with two short parents. If you were to plot the height of students in your class against the average of their parents' height, most tall children would come from tall parents, and most short children would come from short parents. But genes aren't the whole story.

Even though 60% to 80% of the variation we see in height is due to genes, another 20% to 40% is due to environmental factors such as nutrition. Why do these estimates of environmental influence vary so much? The answer is that it depends on what environment you're talking about. In developed countries, where most people have access to adequate nutrition, height is more than 80% inherited, or heritable. This means that when scientists compare the height of a person to his or her relatives, they find that height varies less than 20% among close relatives. In developing countries, where many people are still malnourished, environment plays a larger role. Another way of looking at this is that more people in developed countries have reached their genetic potential than people in developing countries because most of

▶**CONTINUOUS VARIATION**
Variation in a population showing an unbroken range of phenotypes rather than discrete categories.

▶**POLYGENIC TRAIT**
A trait whose phenotype is determined by the interaction among alleles of more than one gene.

Height is an example of a trait that shows continuous variation in any given population.

INFOGRAPHIC 12.9 HUMAN HEIGHT IS BOTH POLYGENIC AND MULTIFACTORIAL

Multiple genes as well as environmental factors such as diet, nutrition, and overall health act together to determine how tall we become.

Polygenic: multiple genes working together

Many genes act together to determine one's height. The combination of alleles a person inherits (*aabbcc*, *AabbCc*, etc.) predicts a distinct height phenotype. Within a population, the result of these polygenic interactions is a range of height phenotypes, from very short to very tall.

Multifactorial: genes and environment working together

Human populations show a continuous range of heights (red line), rather than a genetically predicted number of phenotypes (blue bars). Environmental influences blur the genetic boundaries, resulting in a seamless continuity across the phenotype range.

A mating between two people of medium height (where three genes control height: *AaBbCc × AaBbCc*) produces seven distinct phenotypes, determined by the number of dominant genes inherited.

Male gametes

	ABC	ABc	AbC	aBC	Abc	aBc	abC	abc
ABC	AABBCC	AABBCc	AABbCC	AaBBCC	AABbCc	AaBBCc	AaBbCC	AaBbCc
ABc	AABBCc	AABBcc	AABbCc	AaBBCc	AABbcc	AaBBcc	AaBbCc	AaBbcc
AbC	AABbCC	AABbCc	AAbbCC	AaBbCC	AAbbCc	AaBbCc	AabbCC	AabbCc
aBC	AaBBCC	AaBBCc	AaBbCC	aaBBCC	AaBbCc	aaBBCc	aaBbCC	aaBbCc
Abc	AABbCc	AABbcc	AAbbCc	AaBbCc	AAbbcc	AaBbcc	AabbCc	Aabbcc
aBc	AaBBCc	AaBBcc	AaBbCc	aaBBCc	AaBbcc	aaBBcc	aaBbCc	aaBbcc
abC	AaBbCC	AaBbCc	AabbCC	aaBbCC	AabbCc	aaBbCc	aabbCC	aabbCc
abc	AaBbCc	AaBbcc	AabbCc	aaBbCc	Aabbcc	aaBbcc	aabbCc	aabbcc

Female gametes (left axis label)

Genetic prediction
An individual inheriting the genotype *aaBbCc* falls in the 5.0 ft phenotype.

Actual phenotype
That same individual may actually grow to a height of 5.3 ft depending upon environmental influences.

Number of individuals: 800, 600, 400, 200, 0

Human height (feet): 4.0, 4.5, 5.0, 5.5, 6.0, 6.5, 7.0

---

us in the developed world have access to adequate nutrition. In developing countries, access to nutrition varies much more, and this variation is reflected in larger variations in height between a given person and his or her relatives. The average height of the U.S. population has almost leveled off in the past decade, suggesting that the environment has almost maximized the genetic potential of height in this country.

When both genes and environment work together to influence a given trait, the trait is described as **multifactorial**. So height is both polygenic and multifactorial. Many polygenic traits, such as height and skin color, are influenced by both genes and the environment and thus may be considered multifactorial (**INFOGRAPHIC 12.9**).

▶**MULTIFACTORIAL INHERITANCE**
An interaction between genes and the environment that contributes to a phenotype or trait.

MULTIFACTORIAL INHERITANCE

Q Are some people genetically predisposed to depression?

ANSWER: In the early 1990s, Stephen Suomi and Dee Higley, researchers at the National Institute of Child Health and Human Development, were studying how stress affects the mental development of infant monkeys. They were investigating whether certain alleles of a gene that encodes the serotonin transporter protein made infant monkeys more vulnerable to stress early in life.

Other researchers had shown that the serotonin transporter gene influences the levels of serotonin present in the spaces between brain cells, and that low levels of serotonin in these spaces is one biological hallmark of anxiety and depression in people (**INFOGRAPHIC 12.10**).

The serotonin transporter gene is located on chromosome 17. There are two common alleles: a short version and a long version. The long version contains about 44 extra base pairs. Previous research had suggested that people who had at least one copy of the short version of this gene were much more likely to have an anxiety disorder.

Higley and Suomi showed that infant monkeys exposed to stress, such as being deprived of their mothers, and who carried short versions of this allele, behaved differently from their counterparts: they were more anxious,

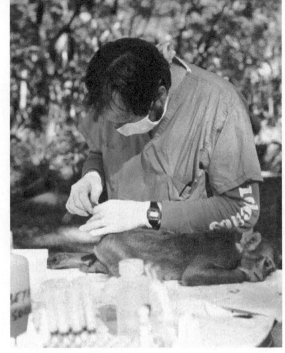

Researcher Dee Higley checks a monkey for wounds and scars—an indication of stressful, violent encounters.

INFOGRAPHIC 12.10 SEROTONIN TRANSPORTER FUNCTION IS LINKED TO DEPRESSION

→ Serotonin is an important signaling molecule in the brain. The serotonin transporter influences the levels of serotonin available for signaling. People with depression have lowered levels of serotonin in the spaces between cells.

Area of serotonin activity (red)

Nerve cell

Serotonin Reuptake
Serotonin that doesn't immediately act on neighboring cells is taken back up into the cell via serotonin transporter proteins (shown in green). A change in the serotonin transporter can cause too much serotonin to be taken back up, leaving little serotonin to act on cells and increasing the likelihood of depression.

Adjacent nerve cell

Serotonin Release
Serotonin is released by a brain cell into the space adjacent to another brain cell.

Serotonin

Serotonin Signal
When serotonin exerts its effects on the adjacent cell, that cell responds to regulate mood. Adequate levels of serotonin are required for a person to maintain a normal mood.

Positive mood signal

aggressive, and some even became alcoholics as adults.

Despite this finding in monkeys, it quickly became clear that having short versus long alleles alone could not explain why some people become severely depressed while others are more resilient. Researchers could not find a clear association between any particular allele and depression in people.

Rather than continue to hunt for depression-related alleles, other researchers took a different tack. In 2003, Terrie Moffitt and Avshalom Caspi, a husband-and-wife team of psychologists at King's College London, decided to test whether genes and environment interact to produce depression in people. Moffitt and Caspi turned to a long-term study of almost 900 New Zealanders, identified these participants' serotonin transporter alleles, and interviewed them about traumatic experiences in early adulthood—experiences such as a major breakup, a death in the family, or serious injury—to see if these difficulties brought out an underlying genetic tendency toward depression.

The results were striking: clinical depression was diagnosed in 43% of participants who had two copies of the short allele and who had experienced four or more tumultuous events. By contrast, only 17% of participants who had two copies of the long allele and who had endured four or more stressful events had become depressed—this was no more than the rate of depression in the general population. Participants with the short allele who experienced no stressful events fared pretty well, too—they also became depressed at no more than the average rate. Clearly, it was the combination of hard knocks and short alleles that more than doubled the risk of depression (**INFOGRAPHIC 12.11**).

Caspi and Moffitt's study was one of the first to examine the combined effects of genes and experience on a specific mental trait—psychiatrists were delighted. While scientists had been trying to tease apart environmental and genetic influences on physical diseases like cancer, this was the first study to investigate this relationship in a mental disorder. Moreover, the findings reinforced the emerging view that the majority of mental illnesses and other complex diseases cannot be explained by genetic or environmental factors alone, but often arise from an interaction between the two. In other words, mental illnesses exhibit multifactorial inheritance.

Multifactorial inheritance is a common pattern of inheritance. Diseases such as asthma, diabetes, and heart disease are all caused by a combination of several genes and their interaction with the environment. For example, some studies have found that

INFOGRAPHIC 12.11 DEPRESSION IS A MULTIFACTORIAL TRAIT

➔ In 2003, Terrie Moffitt and Avshalom Caspi showed that a specific allele of the serotonin transporter gene—a gene that influences levels of the signaling molecule serotonin in the brain—in combination with stressful life events can cause depression. The gene comes in long and short versions.

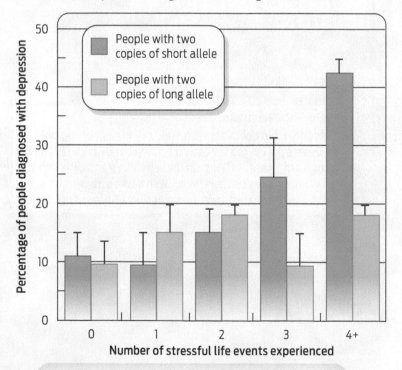

People with two copies of short allele who also experienced four or more stressful events were more than twice as likely to become depressed as those with two copies of the long allele who also experienced stressful events.

SOURCE: Caspi, A., et al. (2003) Influence of life stress on depression: moderation by a polymorphism in the 5-HTT gene. *Science* 301(5631):386-389.

The majority of mental illnesses and other complex diseases cannot be explained by genetic or environmental factors alone, but often arise from an interaction between the two.

cigarette smoke, air pollution, and ozone can exacerbate asthma. Other studies have shown that people who carry the *E4* allele of a gene called *APO*, which codes for a protein found in the cholesterol-carrying lipoproteins in blood (see Chapter 27), have an increased chance of developing heart disease if they smoke and don't exercise compared to people with other *APO E* alleles who smoke and don't exercise. Even for traits that are largely genetically determined, the environment plays a very important role in influencing our phenotype.

cell division. During a menstrual cycle, one egg resumes meiosis and is ovulated. In older women, when these eggs complete meiosis and are ovulated, they are more likely to have an error in chromosome segregation, leading to a chromosomal abnormality.

A chromosomal abnormality means that a developing fetus carries a chromosome number that differs from the usual 46. The most common abnormalities in humans are called **aneuploidies,** deviations from the normal number of chromosomes because single

▶ANEUPLOIDY

An abnormal number of one or more chromosomes (either extra or missing copies).

NONDISJUNCTION

Q Why does the risk of having a baby with Down syndrome go up as a woman ages?

ANSWER: At age 25, a woman's risk of having a baby with Down syndrome is 1 in 1,250 births (that is, 1 out of every 1,250 babies born to women who are 25 is likely to have Down syndrome). At age 40 her risk skyrockets to 1 in 100 births.

As women age, the risk of giving birth to a baby with any chromosomal abnormality increases. That's because as a woman ages, so do her eggs. All the eggs that a woman will ever have were formed before she was born, and they have been aging like the rest of the cells in her body. Until puberty, a woman's eggs are "paused" in the middle of meiosis (at meiosis I); they haven't yet completed their

Most Down syndrome children have learning disabilities that range from mild to moderate.

▶**NONDISJUNCTION**
The failure of chromosomes to separate accurately during cell division; nondisjunction in meiosis leads to aneuploid gametes.

▶**TRISOMY 21**
Carrying an extra copy of chromosome 21; also known as Down syndrome.

chromosomes are either duplicated or deleted. Most aneuploidies arise during meiosis in the parents' sex cells. If chromosomes segregate improperly during meiosis, an occurrence called **nondisjunction,** the resulting gamete will either lack a chromosome or carry an extra copy. When that egg is fertilized by a normal gamete, the resulting zygote can have an abnormal number of chromosomes. In most cases, the abnormality is so severe the zygote spontaneously aborts (**INFOGRAPHIC 12.12**).

There are, however, cases in which the abnormality is not life threatening but does cause severe disability—the most common is **trisomy 21,** also called Down syndrome. Trisomy 21 results when an embryo inherits an extra copy of chromosome 21. Anyone can conceive a child with the abnormality, but older women are at much higher risk.

Most Down syndrome children have learning disabilities that range from mild to moderate, but some have profound mental disability. They are also at higher risk for other diseases and typically don't live beyond 50 years of age.

Down syndrome, as well as other chromosomal abnormalities, can be diagnosed by **amniocentesis.** This procedure is usually

INFOGRAPHIC 12.12 CHROMOSOMAL ABNORMALITIES: ANEUPLOIDY

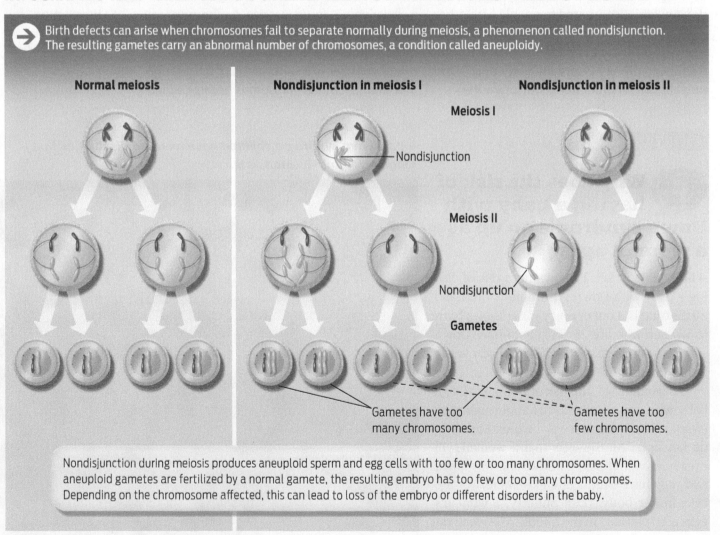

Birth defects can arise when chromosomes fail to separate normally during meiosis, a phenomenon called nondisjunction. The resulting gametes carry an abnormal number of chromosomes, a condition called aneuploidy.

Normal meiosis

Nondisjunction in meiosis I

Nondisjunction in meiosis II

Meiosis I

Nondisjunction

Meiosis II

Nondisjunction

Gametes

Gametes have too many chromosomes.

Gametes have too few chromosomes.

Nondisjunction during meiosis produces aneuploid sperm and egg cells with too few or too many chromosomes. When aneuploid gametes are fertilized by a normal gamete, the resulting embryo has too few or too many chromosomes. Depending on the chromosome affected, this can lead to loss of the embryo or different disorders in the baby.

INFOGRAPHIC 12.13 AMNIOCENTESIS PROVIDES A FETAL KARYOTYPE

Doctors perform a procedure called amniocentesis to obtain fetal cells and diagnose chromosomal abnormalities such as Down syndrome. A karyotype analysis is done on the fetal cells to look for chromosomal abnormalities, in particular missing or extra chromosomes.

Fetal karyotype:

Amniotic fluid with cells from the fetus

Fetus

Uterus (womb)

Placenta

Cervix

Down syndrome results from having three copies of chromosome 21.

performed between 14 and 20 weeks of pregnancy, although some medical centers may perform it as early as 11 weeks. The procedure is quick. A long, thin, hollow needle is inserted through a woman's abdominal wall and into her uterus. Through the needle, the equivalent of 2 to 4 teaspoons of amniotic fluid, which surrounds the growing fetus, is removed. This fluid contains fetal cells that contain the fetus's DNA. From that fluid, technicians analyze the fetal **karyotype**–that is, the chromosomal make-up in its cells (**INFOGRAPHIC 12.13**).

The reasons to undergo amniocentesis vary from couple to couple. But if a test comes back positive, couples have options: they can begin to plan for a disabled child, or make the decision not to carry the child to term.

Although scientists have linked some of the most obvious birth defects to the age of a woman's eggs, recent research also shows that a man's age affects his sperm quality. Men who father children after age 45 are more likely to have children with cognitive disorders such as autism, for example. Male fertility declines over time, too, although much more gradually than does female fertility. Research shows that the older the man, the more likely he is to produce sperm with genetic defects. ∎

▶**AMNIOCENTESIS**
A procedure that removes fluid surrounding the fetus to obtain and analyze fetal cells to diagnose genetic disorders.

▶**KARYOTYPE**
The chromosomal makeup of cells. Karyotype analysis can be used to detect trisomy 21 prenatally.

CHAPTER 12 Summary

▸ Humans have 23 pairs of chromosomes. One of these pairs is the sex chromosomes: XX in females and XY in males. It is the presence of the Y chromosome that determines maleness, and therefore fathers determine the sex of a baby.

▸ Sex determination has genetic, hormonal, anatomical, and behavioral aspects that all interact to produce one's sense of sexual identity; variations can lead to intersex.

▸ Because the Y chromosome in a male does not have a homologous partner, it does not recombine during meiosis. The Y chromosome a son inherits from his father is essentially identical to the Y chromosome his father inherited from his father (the grandfather), a fact that can be used to establish paternity.

▸ Disorders and other traits inherited on X chromosomes are called X-linked traits, and are more common in males than in females.

▸ Hair texture is an example of incomplete dominance, a form of inheritance in which heterozygotes have a phenotype intermediate between homozygous dominant and homozygous recessive.

▸ ABO blood type is an example of a codominant trait—both maternal and paternal alleles contribute equally and separately to the phenotype.

▸ Many traits are polygenic—that is, they are influenced by the additive effects of multiple genes. Polygenic traits often show a continuous, bell-shaped distribution in the population. Human height is an example.

▸ In many cases, a person's phenotype is determined by both genes and environmental influences; this type of inheritance is described as multifactorial. Depression and cardiovascular disease are examples of multifactorial illnesses.

▸ Some genetic disorders result from having a chromosome number that differs from the usual 46. Down syndrome, or trisomy 21, is caused by having an extra copy of chromosome 21.

MORE TO EXPLORE

▸ Colapinto, J. (2000) *As Nature Made Him: The Boy Who Was Raised as a Girl.* New York: HarperCollins.

▸ Foster, E. A., et al. (1998) Jefferson fathered slave's last child. *Nature* 396:27–28.

▸ Griffiths, A. J. F., et al. (2012) *Introduction to Genetic Analysis.* New York: W. H. Freeman.

▸ Caspi, A., et al. (2003) Influence of life stress on depression: moderation by a polymorphism in the 5-HTT gene. *Science* 301(5631):386–9.

▸ Allen, E. G., et al. (2009) Maternal age and risk for trisomy 21 assessed by the origin of chromosome nondisjunction: a report from the Atlanta and National Down Syndrome Projects. *Human Genetics* 125(1):41–52.

CHAPTER 12 Test Your Knowledge

How do chromosomes determine sex, and how does sex influence the inheritance of certain traits?

By answering the questions below and studying Infographics 12.1–12.5, you should be able to generate an answer for the broader Driving Question above.

KNOW IT

1 Which of the following most influences the development of a female fetus?

a. the presence of any two sex chromosomes

b. the presence of two X chromosomes

c. the absence of a Y chromosome

d. the presence of a Y chromosome

e. either b or c

2 Why are more males than females affected by X-linked recessive genetic diseases?

3 If a man has an X-linked recessive disease, can his sons inherit that disease from him? Why or why not?

USE IT

4 Which of the following couples could have a boy with Duchenne muscular dystrophy (DMD)?

a. a male with Duchenne muscular dystrophy and a homozygous dominant female

b. a male without Duchenne muscular dystrophy and a homozygous dominant female

c. a male without Duchenne muscular dystrophy and a carrier female

d. a and c

e. none of the above

5 Predict the sex of a baby with each of the following pairs of sex chromosomes. Use your answer to check your answer to Question 1.

a. XX　　　b. XXY　　　c. XY　　　d. X

6 Consider your brother and your son.

a. If you are female, will your brother and your son have essentially identical Y chromosomes? Explain your answer.

b. If you are male, will your brother and your son have essentially identical Y chromosomes? Explain your answer.

7 A wife is heterozygous for Duchenne muscular dystrophy alleles and her husband does not have DMD. Neither has any other notable medical history. What percentage of their sons, and what percentage of their daughters, will have:

a. Duchenne muscular dystrophy (which is determined by a recessive allele on the X chromosome)

b. an X-linked dominant form of rickets (a bone disease)

Some traits are not inherited as simple dominant or recessive inheritance patterns. What are some of these complex inheritance patterns?

By answering the questions below and studying Infographics 12.6–12.11, you should be able to generate an answer for the broader Driving Question above.

KNOW IT

8 What aspects of height make it a polygenic trait?

9 Which of the following inheritance patterns includes an environmental contribution?

a. polygenic

b. X-linked recessive

c. X-linked dominant

d. multifactorial

e. none of the above

10 What is the difference between polygenic inheritance and multifactorial inheritance?

11 How does incomplete dominance differ from co-dominance?

12 If you are blood type A-positive, to whom can you safely donate blood? Who can safely donate blood to you? List all possible recipients and donors and explain your answer.

USE IT

13 If two women have identical alleles of the suspected 20 height-associated genes, why might one of those women be 5 feet 5 inches tall and the other 5 feet 8 inches tall?

14 Look at Infographic 12.11. How do the data given support the hypothesis that both genes and the environment influence at least some cases of clinical depression?

15 Look at Infographic 12.11. At approximately how many stressful experiences does the homozygous short-allele genotype begin to influence the depression phenotype?

16 From what you have read in this chapter, how can you account for two people with the same genotype for a predisposing disease allele having different phenotypes?

MINI CASE

17 A serious car crash on a freeway has resulted in multiple injuries causing substantial blood loss in three members of a family—a mother, a father and their 2-year-old daughter. The local blood bank will be challenged to supply blood, as their supplies of every blood type were drained after the roof of a shopping plaza collapsed the week before and many transfusions were required.

   a. The EMTs must give blood immediately to all three members of the family. What blood type should they use (consider both ABO blood type and Rh factor)? Explain your answer.

   b. Both parents have a blood donor card in their wallets. The mother is O-negative and the father is A-positive. From this information, what (if any) additional blood types (beyond your answer to part a) can be given to either parent? Explain your answer.

   c. Does knowing the parents' blood types allow you to infer enough information about the daughter's possible blood type to use a different blood type for her transfusion? Why or why not? (Hint: Consider possible blood types for the daughter and the implications of, for example, using A-negative donor blood. Could you guarantee that this would be safe?)

DRIVING QUESTION 3

How do numerical abnormalities of chromosomes occur, and what are the consequences of these abnormalities?

By answering the questions below and studying Infographics 12.12 and 12.13, you should be able to generate an answer for the broader Driving Question above.

KNOW IT

18 What is the normal chromosome number for each of the following?

   a. a human egg

   b. a human sperm

   c. a human zygote

19 When looking at a karyotype, for example to diagnose trisomy 21 in a fetus, is it possible to use that analysis also to tell if the fetus has inherited a cystic fibrosis allele from a carrier mother?

USE IT

20 Which of the following can result in trisomy 21?

   a. an egg with 23 chromosomes fertilized by a sperm with 23 chromosomes

   b. an egg with 22 chromosomes fertilized by a sperm with 23 chromosomes

   c. an egg with 24 chromosomes, two of which are chromosome 21, fertilized by a sperm with 23 chromosomes

   d. an egg with 23 chromosomes fertilized by a sperm with 24 chromosomes, two of which are chromosome 21

21 From what you have read in this chapter, which of the possibilities in Question 20 is most likely? Explain your answer.

BRING IT HOME

22 What factors would lead you to consider prenatal genetic testing? In your opinion, what is the value of having this information?

INTERPRETING DATA

23 The graph at right shows the average ("Mean") age of women who had children with trisomy 21 ("Cases"), of those who did not ("Controls"), and the average age of women giving birth in the population. The data are presented for 15 years.

a. In general, how does the age (at time of birth) of women giving birth to a baby with Down syndrome compare to the age of women giving birth to a baby without Down syndrome?

b. During which year was the average age of the cases closest to the average age of the controls? How close were the average ages in this year?

c. During which year was the average age of the cases the most different from the average age of the controls? How different were the average ages in this year?

d. Using the data points for each year over this 15-year period, calculate the overall average age of women having babies with Down syndrome, and the overall average age of women having babies who do not have Down syndrome.

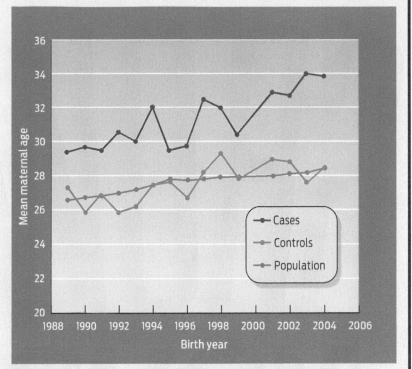

SOURCE: Graves Allen, E., et al. (2009) Maternal age and risk for trisomy 21 assessed by the origin of chromosome nondisjunction: a report from the Atlanta and National Down Syndrome Projects. *Human Genetics* 125(1):41–52.

GROW YOUR OWN

Stem cells could be the key to engineering organs

▸▸ **DRIVING QUESTIONS**

1. What is the structure of tissues and organs, and how can organs be repaired or replaced?

2. What are the properties of specialized cells in tissues, and how do stem cells differentiate into these specialized cells?

3. How do stem cells contribute to regenerative medicine, and how can we obtain or produce stem cells for this purpose?

When Luke Masella was 10 years old, he was in and out of the hospital nearly every week. Doctors didn't know what was wrong with him. Toxins were building up in his blood, and he had lost a quarter of his body weight. After a multitude of tests, the doctors realized that Luke was in kidney failure and traced the problem to a faulty bladder. Urine was backing up inside the kidney and making it malfunction.

In 2001, Luke became one of the first patients in the world to receive an experimental treatment for such bladder problems: an engineered human bladder, one grown from his own cells, was implanted to fix his faulty one. "They take a piece of your bladder out. They grow it in a lab for 2 months into a new bladder that's your own. And they put it back in," Luke explained on ABC television in 2012.

This organ-growing technique is the brainchild of Anthony Atala, director of the Wake Forest University Institute for Regenerative Medicine. In 2006, Atala announced that he and his colleagues had successfully transplanted engineered human bladders into several children and teenagers, surprising the medical community. Although scientists had for years been transplanting organs like hearts and kidneys, the bladders were the first transplanted organs made with a person's own cells.

"It was very significant work," William Wagner, deputy director of the McGowan Institute for Regenerative Medicine at the University of Pittsburgh, says of Atala's accomplishment. "He's overcome a huge number of challenges."

The potential applications of the technique are enormous. Each year, the demand for transplant organs such as hearts, livers, and kidneys vastly exceeds supply. In 2012, for example, surgeons transplanted about 30,000 organs, according to the Organ Procurement and Transplantation Network. Meanwhile, there are about 100,000 people waiting for an organ transplant. And even when an organ does become available, the recipient's body may reject the organ because the donor and recipient immune systems are not compatible–leaving the patient sicker than before the transplant.

Growing organs from a person's own cells not only sidesteps organ rejection, it also eliminates the need for donors. A decade ago, most scientists considered such a feat a pipe dream. After all, many human organs are complex three-dimensional structures made up of millions of cells–how would scientists manage to build such a structure by hand? But

> **❝They take a piece of your bladder out. They grow it in a lab for 2 months into a new bladder that's your own. And they put it back in.❞**
>
> — LUKE MASELLA

since the 2000s, advances in our understanding of **stem cells** have brought this fantasy closer to reality. Stem cells are immature cells that divide repeatedly and give rise to more-specialized cell types. They are important for development, growth, and repair of the body. Stem cells in the bone, heart, and brain, for example, help to regenerate those tissues and organs (**INFOGRAPHIC 13.1**).

With their regenerative properties, stem cells may hold the key to healing damaged or diseased body parts–they can do much of the construction work themselves. Already, scientists are using stem cells to construct new organs for transplant. One particularly eye-catching method is something dubbed "bioprinting," a technique that uses computer graphics and cellular "ink" to manufacture organs from scratch. Need a new bladder? Just hit print.

ENGINEERING ORGANS

The effort to engineer human organs for transplants dates back more than 40 years and is based on the knowledge that organs made from a patient's own cells would have advantages over ones obtained from human or nonhuman donors. Not only would they be genetically compatible, they also wouldn't rely on a supply of donors. Important strides in the field were made in the late 1980s, when Joseph Vacanti of Boston Children's Hospital teamed up with Robert Langer at MIT to engineer tissues. The pair wanted to design synthetic biodegradable scaffolds that could be molded into particular shapes–a human ear, for example–and then coated with cells that would grow into a **tissue.** The scaffold itself would never need to be removed–it would in time dissolve.

Atala, who had collaborated with Vacanti and Langer, applied this research on biodegradable scaffolds to his own work on engineered bladders. Atala, a surgeon, sought to help his patients whose bladders were not functioning normally because of cancer, injury, or a birth defect. For more than a century, doctors have treated such patients by using pieces of their intestine or bowel to reconstruct their bladders. But because intestine and bowel are built to do different things than the bladder, this treat-

INFOGRAPHIC 13.1 STEM CELLS IN TISSUES HAVE REGENERATIVE PROPERTIES

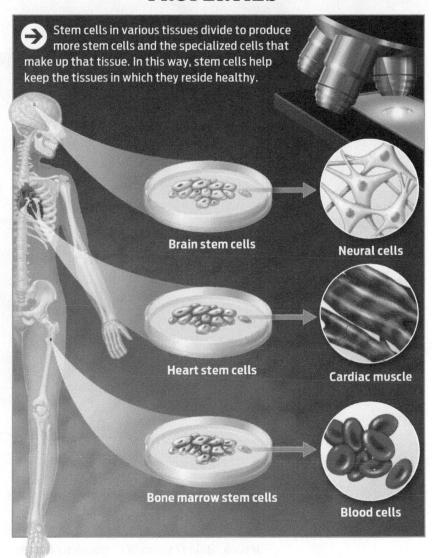

→ Stem cells in various tissues divide to produce more stem cells and the specialized cells that make up that tissue. In this way, stem cells help keep the tissues in which they reside healthy.

Brain stem cells — Neural cells

Heart stem cells — Cardiac muscle

Bone marrow stem cells — Blood cells

ment is not ideal–it can lead to many problems later, including leaks and even cancer. Growing a piece of new bladder to repair the organ, Atala reasoned, would be much better.

Although it has been possible for decades to grow human skin outside the body to treat burn victims, growing more-complex organs like bladders has been challenging. Skin is a thin, flat organ made up of mostly one cell type. A bladder, by contrast, is a round, hollow organ made up of multiple cell types working together. To grow a bladder, scientists must grow several layers of tissue, including muscle and epithelial tissue. (Epithelial cells are flat

▶**STEM CELLS**
Immature cells that can divide and differentiate into specialized cell types.

▶**TISSUE**
An organized group of different cell types that work together to carry out a particular function.

INFOGRAPHIC 13.2 CELLS ARE ORGANIZED INTO TISSUES, ORGANS, AND SYSTEMS

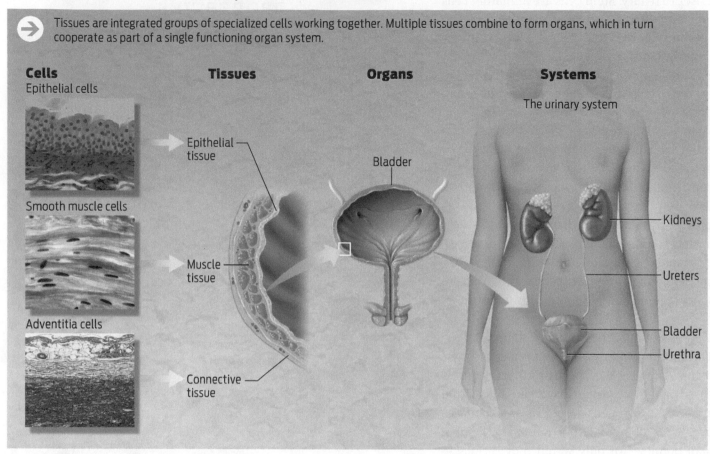

→ Tissues are integrated groups of specialized cells working together. Multiple tissues combine to form organs, which in turn cooperate as part of a single functioning organ system.

Cells
Epithelial cells

Smooth muscle cells

Adventitia cells

Tissues
Epithelial tissue

Muscle tissue

Connective tissue

Organs
Bladder

Systems
The urinary system

Kidneys

Ureters

Bladder

Urethra

cells that line many organs.) Moreover, the thicker the tissue, the more blood vessels required to nourish it (**INFOGRAPHIC 13.2**).

It took Atala 17 years of research to achieve success. He spent the most of those years figuring out how to get bladder cells to divide outside the body, and then how to grow them on a biodegradable scaffold in the shape of a bladder. Next, he had to test the bladders in laboratory animals to make sure they worked. Finally, in 1999, he was ready to test the bladders in patients.

For each patient, Atala cut a piece of tissue smaller than a postage stamp from inside the bladder and extracted two types of cells—muscle stem cells and bladder epithelial stem cells. He then mixed these stem cells with chemicals that promoted cell division, producing a stockpile of millions of cells. Next, he layered these

INFOGRAPHIC 13.3 ENGINEERING

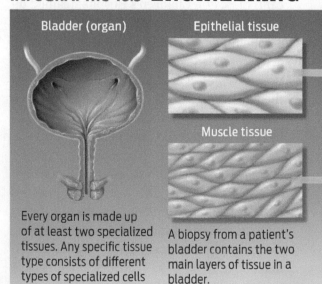

Bladder (organ)

Epithelial tissue

Muscle tissue

Every organ is made up of at least two specialized tissues. Any specific tissue type consists of different types of specialized cells that together have a specific function.

A biopsy from a patient's bladder contains the two main layers of tissue in a bladder.

stem cells onto the biodegradable scaffold, which he had sculpted to resemble a human bladder. He bathed the scaffold in nutrients to stimulate the cells to divide, and then placed the scaffold with nutrients and other growth factors in an incubator to simulate conditions inside the human body. The cells went through several cell divisions, attached to the scaffold, and grew into the tissue layers that make up a bladder. Two months later, surgeons reconstructed the patient's bladders using the new bladder tissue (INFOGRAPHIC 13.3).

The treatment, so far, seems remarkably successful. Not only does the technique improve bladder function, it avoids the complications that result from using tissues from other organs—like the bowel—for bladder repair. "Doing bowel-for-bladder replacements in children really got to me. It's one thing to put them into an adult, but putting them in a child with a 70-plus life expectancy didn't make sense when you knew there would be trouble down the line," Atala told the *New York Times* in 2006, soon after he had published his results.

Atala's technique also avoids the immune rejection that can result from using bladder tissue transplanted from a donor. Just as a food

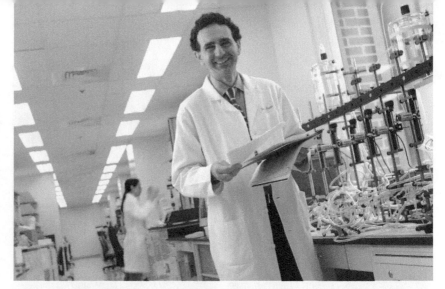

Anthony Atala, in his lab at Wake Forest University Institute for Regenerative Medicine.

allergy can cause shock when the offending food is ingested, a person's immune system can react similarly to a donated organ and reject it. By contrast, tissue grown from a person's own cells poses no such risk of rejection because the tissue is genetically identical to the source.

As of 2012, a total of about 30 patients have received the experimental bladder treatment, which is being evaluated in an ongoing clinical trial. Thanks to the engineered bladder he received in 2001, Luke Masella is now

AN ORGAN USING STEM CELLS

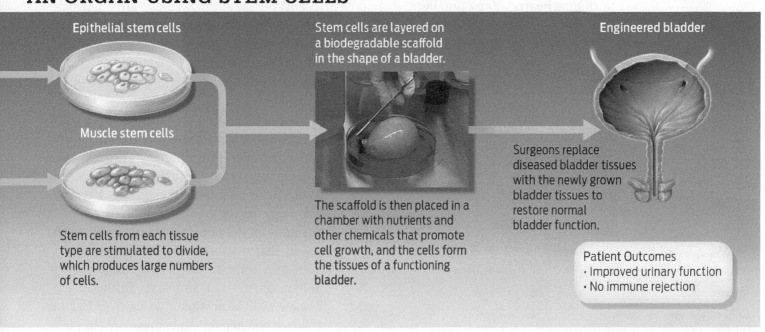

Epithelial stem cells

Muscle stem cells

Stem cells from each tissue type are stimulated to divide, which produces large numbers of cells.

Stem cells are layered on a biodegradable scaffold in the shape of a bladder.

The scaffold is then placed in a chamber with nutrients and other chemicals that promote cell growth, and the cells form the tissues of a functioning bladder.

Engineered bladder

Surgeons replace diseased bladder tissues with the newly grown bladder tissues to restore normal bladder function.

Patient Outcomes
· Improved urinary function
· No immune rejection

An engineered human urethra, grown in Atala's lab at Wake Forest.

a healthy young man with functioning bladder and kidneys. The 21-year-old is currently a senior at the University of Connecticut, studying media, television, and communications.

Atala is using his organ-growing technique to make other organs for transplant as well. He has used it to grow replacement urethras for children whose urethras were damaged by trauma to their pelvis, for example. The urethras are working normally in all five boys who received them, Atala reported in *The Lancet* in 2011. And these organs are just the beginning. Atala's group at Wake Forest is working on developing a whole host of organs for transplant, including heart, kidney, and liver.

▶**ADULT (SOMATIC) STEM CELLS**
Stem cells located in tissues that help maintain and regenerate those tissues.

TABLE 13.1 **HOW OLD ARE YOU?**

Your body is younger than you think. Each kind of tissue has its own turnover time, depending in part on the workload endured by its cells. Only the lens cells of the eye, the neurons of the cerebral cortex, and perhaps the muscle cells of the heart last a lifetime.

TISSUE TYPE	TURNOVER RATE
Epidermis (skin surface)	2 to 3 weeks
Red blood cells	120 days
Liver	300 to 500 days
Bones	More than 10 years
Gut	15.9 years
Rib muscle	15.1 years
Lens of the eye	Never replaced
Neurons of the cerebral cortex	Never replaced

REGENERATIVE MEDICINE

For scientists working to grow new organs, inspiration comes from an unlikely place: salamanders. These four-legged amphibians have a remarkable ability to regenerate body parts that have been injured or even severed completely. In presentations, Atala often shows a time-lapse video of a salamander growing back an entire front limb, complete with webbed digits.

If a salamander can regenerate tissues, Atala asks, why can't a person? In fact, he says, they can. "The human body is constantly regenerating," he explains. The whole superficial layer of our skin turns over every couple of weeks, he points out, while the cells lining our intestines turn over about every week. Even something as seemingly solid as bone is completely replaced about every 10 years. "Basically you don't have any cells left over that were there 10 years prior," Atala says (**TABLE 13.1**).

Scientists have known for decades that some cells, like skin and blood cells, divide continually to repair and replace these tissues. But only within the past 20 years have they discovered that stem cells are found in many organs of the body: they are found in bladders, for example, and in many–perhaps all–tissues. Stem cells found in mature tissues are known as **adult stem cells** or **somatic stem cells** ("somatic" means "of the body"). Scientists are still searching for exactly where adult stem cells are located in various tissues. They suspect these cells reside in particular niches where they wait, not dividing, until disease or injury triggers them to divide.

Both humans and salamanders have adult stem cells. The main difference between humans and salamanders is that humans are genetically programmed to form scar tissue at the site of a wound, whereas salamanders grow new tissues. Scar tissue serves to protect us by sealing off a wound from the outside world, but the downside is that it interferes with regeneration. If there were some way to control scar formation in humans, researchers reason, further regeneration might be possible. "We have not lost the ability to regenerate," Atala stresses. "The question is how can we harness that potential again?"

To heal tissue damaged by injury or disease, stem cells must do more than simply divide repeatedly. The new cells must also go through a process of specialization to develop into the specific cell types appropriate to the tissue in need of healing. Remember that during embryonic development a single cell becomes millions as the embryo grows. These dividing cells eventually become specialized as muscle cells, kidney cells, heart cells, and more than 200 other cell types in the body by the time we are born. This process, in which a cell develops from an immature cell type into a more specialized one, is called **cellular differentiation.** Cells become specialized by turning some genes "on" and others "off," a process called **differential gene expression.** So while every cell in our body carries the exact same DNA, it is a cell's pattern of gene expression–and therefore the proteins produced from those genes (see Chapter 8)–that defines it as one cell type or another.

Take, for example, two cell types with very distinct characteristics: muscle cells and B cells of the immune system. Muscle cells, which are long and slender, allow the body to move by contracting and relaxing. B cells are round, with antibody receptors protruding from their surfaces that detect foreign objects like viruses and bacteria, and thus help the body fight off infection. A cell's physical shape and function are determined by the kinds of protein found within it. Muscle cells contain proteins that allow them to contract and cause body movement; B cells contain proteins that allow them to fight infection. Both muscle cells and B cells contain exactly the same DNA, and therefore contain the same genes. But only a subset of those genes is turned on in each cell type. As a result, each cell type produces a unique set of proteins that distinguish one cell type from another (INFOGRAPHIC 13.4).

▸**CELLULAR DIFFERENTIATION**
The process by which a cell specializes to carry out a specific role.

▸**DIFFERENTIAL GENE EXPRESSION**
The process by which genes are "turned on," or expressed, in different cell types.

INFOGRAPHIC 13.4 SPECIALIZED CELLS EXPRESS DIFFERENT GENES

Every cell in your body has the same genes, or genome. What distinguishes one cell type from another is the pattern of gene expression and, consequently, the proteins each cell makes. A muscle cell makes a different set of proteins than a B cell, a type of immune-system cell. Muscle cells, for example, express large amounts of actin and myosin proteins, which help muscles contract, whereas B cells express high levels of antibody proteins, which help the body fight infections.

Muscle stem cell — Mitosis

Expression of muscle-cell-specific genes

Specialized muscle cell

In any person, each of these cells has the same genome. But each cell type, be it muscle or B-cell, has a unique pattern of genome expression.

The result is that each cell type makes a unique set of proteins and consequently has a unique function in the body. Such cells are said to be "specialized."

Bone marrow stem cell — Mitosis

Expression of B-cell-specific genes

Specialized B cell

Some fully differentiated cells do not divide. As cells become more mature and specialized–becoming, for example, cardiac muscle cells or nerve cells–they lose the ability to undergo cell division, or they divide less frequently. Stem cells, on the other hand, retain the ability to divide almost indefinitely. When a stem cell divides, one of the daughter cells remains an immature stem cell, while the other one "grows up," differentiating into a more specialized cell. In this way, stem cells contribute to the maintenance of tissues, as daughter cells replace cells that have reached the end of their life span, while also retaining the ability to supply additional cells in the future.

When scientists discovered that many tissues had their own pool of stem cells, the search began for ways to stimulate these stem cells to divide and differentiate when they otherwise would not, and thus help to repair tissues and organs. This field of research is called regenerative medicine.

The idea of using stem cells to regenerate damaged or diseased tissues isn't entirely new. Doctors have been treating leukemia–a type of white blood cell cancer–with stem cell transplants for decades. Like other blood cells, white blood cells, called leukocytes, are derived from stem cells in the bone marrow. Leukemia is caused by defective leukocytes that divide uncontrollably. To treat leukemia, doctors administer chemotherapy to kill the patient's bone marrow cells and then replace those cells with marrow cells from an immune-

INFOGRAPHIC 13.5 REGENERATIVE MEDICINE: FOUR APPROACHES

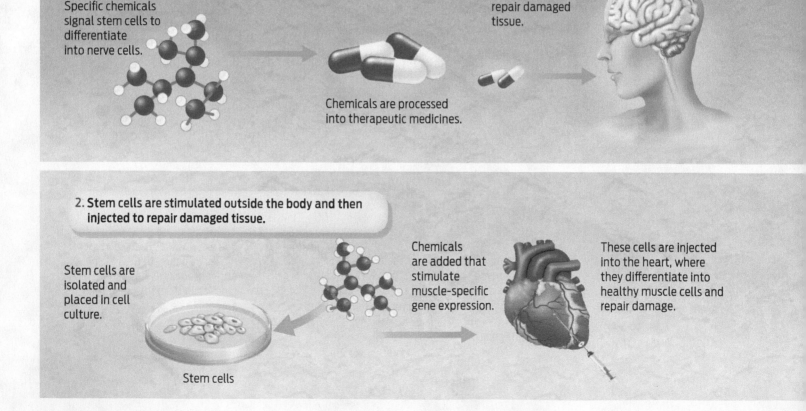

1. **Therapeutic drugs stimulate stem cells within the body to repair damaged tissue.**

Specific chemicals signal stem cells to differentiate into nerve cells.

Chemicals are processed into therapeutic medicines.

Patients can take the chemicals in the form of medicines that stimulate somatic stem cells in the brain to differentiate into nerve cells and repair damaged tissue.

2. **Stem cells are stimulated outside the body and then injected to repair damaged tissue.**

Stem cells are isolated and placed in cell culture.

Stem cells

Chemicals are added that stimulate muscle-specific gene expression.

These cells are injected into the heart, where they differentiate into healthy muscle cells and repair damage.

matched donor. Stem cells in the new marrow repopulate the patient's bloodstream with healthy blood cells. The goal of regenerative medicine is similar: scientists want to use stem cells to heal damaged or diseased tissues. The difference is that regenerative medicine seeks to prod stem cells to divide and differentiate when they normally wouldn't, stimulating them to make cell types that they wouldn't normally make.

One approach in regenerative medicine would be the use of therapeutic drugs to stimulate specific stem cells in the body to grow and differentiate into the specialized cell types of the tissues that need repairing. Another approach would involve removing stem cells from the body, chemically inducing them to reproduce and differentiate, and then re-implanting the cells into a patient with a damaged tissue or organ.

A third approach, already in use, involves implanting the body with biodegradable scaffolds that encourage new tissue growth. Atala's group has shown, for example, that stem cells can repair damaged urethras from inside the body. Give these cells a urethra-shaped scaffold to grow on, and new cells will migrate to the scaffold and regenerate a fully functioning urethra.

The fourth approach is one that Atala has used to create new bladders, ears, and other organs: seeding cells on a biodegradable scaffold outside the body, growing a new organ, then implanting it in a patient (**INFOGRAPHIC 13.5**).

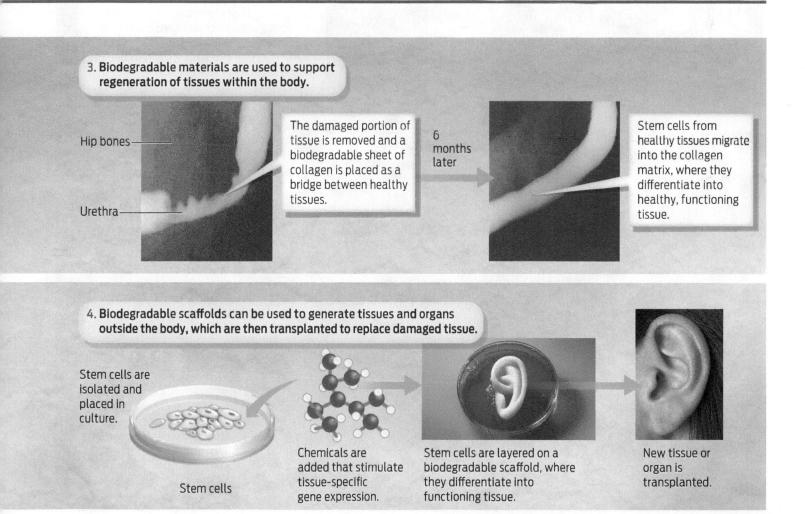

3. Biodegradable materials are used to support regeneration of tissues within the body.

Hip bones

Urethra

The damaged portion of tissue is removed and a biodegradable sheet of collagen is placed as a bridge between healthy tissues.

6 months later

Stem cells from healthy tissues migrate into the collagen matrix, where they differentiate into healthy, functioning tissue.

4. Biodegradable scaffolds can be used to generate tissues and organs outside the body, which are then transplanted to replace damaged tissue.

Stem cells are isolated and placed in culture.

Stem cells

Chemicals are added that stimulate tissue-specific gene expression.

Stem cells are layered on a biodegradable scaffold, where they differentiate into functioning tissue.

New tissue or organ is transplanted.

▶**MULTIPOTENT**
Describes a cell with the ability to differentiate into a limited number of cell types in the body.

▶**EMBRYONIC STEM CELLS**
Stem cells that make up an early embryo and which can differentiate into nearly every cell type in the body.

TURNING BACK THE CLOCK

Because of their remarkable capacity to produce new cells that can differentiate and specialize, stem cells show great promise in regenerative medicine. But it turns out that not all stem cells are created equal. Adult stem cells—those found in mature tissues—typically can differentiate only into one or a few cell types. Most bone marrow stem cells, for example, in normal circumstances differentiate only into blood cells—but not into neurons or bladder cells. Such cells with restricted ability to differentiate are described as **multipotent.**

Other stem cells—those active during embryonic development, for example—can differentiate into many more cell types. Such **embryonic stem cells** are found in early embryos at what's known as the **blastocyst** stage, when the embryo is mostly a hollow ball of cells. Unlike adult stem cells, which differentiate only into certain cell types, em-

INFOGRAPHIC 13.6 EMBRYONIC VS. ADULT STEM CELLS

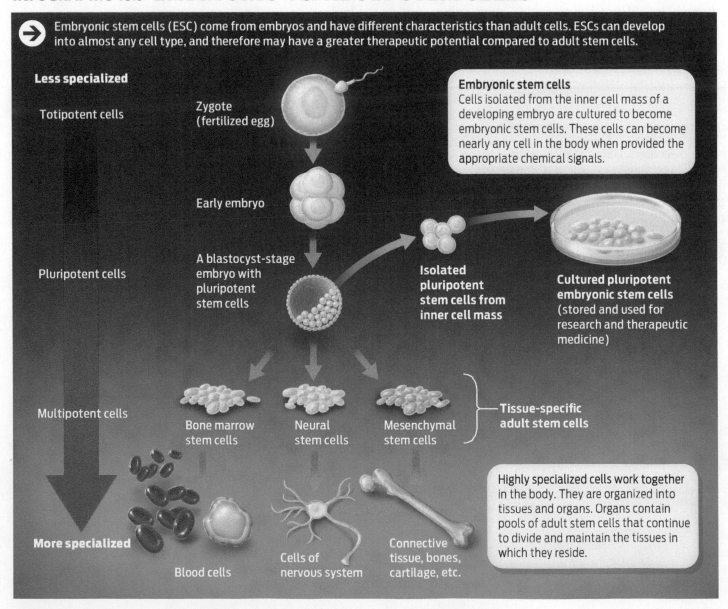

→ Embryonic stem cells (ESC) come from embryos and have different characteristics than adult cells. ESCs can develop into almost any cell type, and therefore may have a greater therapeutic potential compared to adult stem cells.

Less specialized

Totipotent cells

Zygote (fertilized egg)

Embryonic stem cells
Cells isolated from the inner cell mass of a developing embryo are cultured to become embryonic stem cells. These cells can become nearly any cell in the body when provided the appropriate chemical signals.

Early embryo

Pluripotent cells

A blastocyst-stage embryo with pluripotent stem cells

Isolated pluripotent stem cells from inner cell mass

Cultured pluripotent embryonic stem cells (stored and used for research and therapeutic medicine)

Multipotent cells

Bone marrow stem cells

Neural stem cells

Mesenchymal stem cells

Tissue-specific adult stem cells

Highly specialized cells work together in the body. They are organized into tissues and organs. Organs contain pools of adult stem cells that continue to divide and maintain the tissues in which they reside.

More specialized

Blood cells

Cells of nervous system

Connective tissue, bones, cartilage, etc.

bryonic stem cells can give rise to nearly any cell type in the body. For this reason, they are referred to as **pluripotent.** At even earlier stages of embryonic development, embryonic cells can differentiate into every cell type, and are therefore described as **totipotent** (INFOGRAPHIC 13.6).

Some scientists argue that embryonic stem cells hold greater potential in treating disease because they are not as specialized as adult stem cells and can therefore differentiate into many more cell types. With embryonic stem cells, scientists can in principle make any cell type they need.

The challenge has always been how to obtain them. One source of embryonic stem cells is discarded human embryos from fertility clinics. Scientists can extract cells from an embryo and place them in media

Harvesting embryonic stem cells from an 8-cell embryo.

that stimulates them to divide. These cells are then stored, used in research, and could potentially be used in treatment. But that became much harder to do when, in 2001, President George W. Bush put in place severe restrictions on the use of federal funds to support research using stem cells derived from embryos. By this mandate, only a small number of existing cell lines were eligible for federal funding, and funding for making new stem cell lines was prohibited. In 2009, by executive order, President Barack Obama removed some of these restrictions, but funding remains controversial with many members of Congress.

Another way to obtain embryonic stem cells is by a technique called cloning. In this method, scientists replace the nucleus of a haploid unfertilized human egg with the diploid nucleus taken from another cell (a skin cell, for example). The chemical "soup" inside the egg turns on specific genes in the donated nucleus, resetting it to an embryonic state. This technique, known as somatic cell nuclear transfer (SCNT), produces a new embryo with the same genes as the donor cell. This is how Dolly, the first cloned sheep, was created in 1996 (INFOGRAPHIC 13.7).

If scientists were to implant a cloned embryo in a woman's womb, it would develop into a fetus that has the same nuclear DNA as the person who donated the cells from which the diploid nucleus was taken. While such reproductive cloning is prohibited in the United States, scientists funded by nongovernmental sources are allowed to produce cloned embryos for research—so-called "therapeutic cloning." Embryonic stem cells made by SCNT can be extracted from the new embryo and grown in a petri dish to form a population of stem cells that are essentially genetically identical to the donor. Consequently, any differentiated cells derived from these stem cells could be transplanted back into the donor without fear of an immune response against the transplanted cells.

The main ethical difficulty with both sources of embryonic stem cells—frozen embryos from fertility clinics and cloned embryos generated in the lab—is that human embryos are destroyed in the process. To date, most embryonic stem cell lines in the United States have been derived from embryos stored at

▶**BLASTOCYST**
The stage of embryonic development in which the embryo is a hollow ball of cells. Researchers can derive embryonic stem cell lines during the blastocyst stage.

▶**PLURIPOTENT**
Describes a cell with the ability to differentiate into nearly any cell type in the body.

▶**TOTIPOTENT**
Describes a cell with the ability to differentiate into any cell type in the body.

INFOGRAPHIC 13.7 SOMATIC CELL NUCLEAR TRANSFER PRODUCES CLONED EMBRYONIC STEM CELLS

→ Somatic cell nuclear transfer, or cloning, involves replacing the nucleus of an egg with a nucleus from a specialized cell, creating an embryo that is genetically identical to the donor cell.

Egg — Nucleus (genetic material)

Skin cell

Nucleus removed from egg

Enucleated egg

Skin cell nucleus

This cell has the chemical environment of the donated egg and the genome of the donated skin cell.

Very early embryo

If the cloned embryo is implanted in the uterus of a surrogate animal, it will grow into an individual that is the genetic twin of the skin donor.

Blastocyst-stage embryo with the inner cell mass

Inner cell mass removed for stem cell culture

If the inner cell mass of the embryo is cultured, the cloned embryonic skin cells are a genetic match to the skin donor. If used for regenerative therapy in the skin donor, they will not be rejected.

fertility clinics and subsequently donated to research. If not donated to science, these extra embryos would be destroyed after a period of time anyway. Nevertheless, many people find the idea of deliberately interfering with human embryos troubling.

Keenly aware of the ethical challenges, the scientific community has been searching for other methods to generate embryonic stem cells. In 2007, scientists accomplished a feat that could make the ethical controversy obsolete. They discovered a way to produce embryonic-like stem cells without touching an embryo. They did this by reprogramming mature cells, converting them into cells that to all appearances behave exactly like embryonic stem cells, capable of dividing and differentiating into any cell type.

In 2007, James Thompson at the University of Wisconsin and Shinya Yamanaka of Kyoto University in Japan independently showed that they could turn a mature human cell—one that has already differentiated—into an "embryonic" stem cell by inserting four human genes into its genome. They called the new cells **induced pluripotent stem cells** to reflect

▶INDUCED PLURIPOTENT STEM CELL

A pluripotent stem cell that was generated by manipulation of a differentiated somatic cell.

the fact that mature cells had been genetically manipulated to become embryonic-like stem cells. Yamanaka, for example, took an adult skin cell and inserted four human genes into it–genes that are normally switched off in skin cells. The added genes produced proteins that were able to turn off the expression of genes associated with the differentiated state and turn on genes associated with pluripotency. By changing gene expression in this way, the researchers were able to "de-differentiate" the skin cell–to turn back the clock, in a sense–and return the cell to its embryonic state. It was a major technological breakthrough. This promising technique may offer a way to pro-

duce transplantable cells that are genetically matched to a patient without depending on embryos. For his pioneering work, Yamanaka won a Nobel prize in 2012 (**INFOGRAPHIC 13.8**).

Atala's group has also shown that it is possible to isolate embryonic-like stem cells from amniotic fluid, another potentially valuable way of obtaining them.

Research using both embryonic and adult stem cells in regenerative medicine is forging ahead. In 2012, scientists at Advanced Cell Technology, a biotechnology company in Massachusetts, reported that they had partially restored vision to two legally blind patients suffering from an eye disease called macular

INFOGRAPHIC 13.8 INDUCED STEM CELLS

One method of creating embryonic stem cells is to induce adult stem cells to de-differentiate. Scientists have done just that by inserting either specific genes or proteins into adult cells. The expression of these genes causes the differentiated cells to act like pluripotent stem cells. This technology offers the potential to create immune matched stem cells for therapy, just like cloning, but it does not involve destroying an embryo.

"Even the dumbest stem cell is smarter than the smartest neuroscientist."

— EVAN SNYDER

degeneration. The stem cells were injected directly into the patients' eyes, where they differentiated into new retina cells. And researchers at Emory University and the University of Michigan are conducting a clinical trial of embryonic stem cells in patients with ALS (amyotrophic lateral sclerosis, or Lou Gehrig's disease), with some initial positive results: a few patients have regained the ability to move some previously paralyzed muscles.

As for *how* the stem cells are working on a chemical level to heal damaged tissue, researchers are the first to admit they don't know. "Even the dumbest stem cell is smarter than the smartest neuroscientist," says Evan Snyder, program director at the Burnham Institute for Medical Research in La Jolla, California. "The cells are making stuff that we might not be able to identify for centuries." But that hasn't stopped researchers from attempting to use stem cells for therapeutic purposes.

BUILDING ORGANS IN 3-D

Atala's group has been experimenting with engineering other organs besides bladders. At Wake Forest, approximately 300 researchers are working on engineering more than 30 different body parts. They have had the most success so far with simple structures, such as ears, which don't have a lot of blood vessels or nerve connections. These organs are currently being used to treat wounded veterans returning from the Iraq and Afghanistan wars.

In partnership with the U.S. military, Atala's group is also working to make more complex body parts for soldiers whose pelvic regions were harmed by explosive devices. While the treatments are years away from the doctor's office, the group has had some promising success in animal models. For example, scientists have successfully created functional rabbit penises that are recognized and used by the bunny recipients. The approach could one day help the scores of soldiers returning from Iraq and Afghanistan with genital injuries.

A major hurdle in building more-complex organs is the intricacy of the design. A bladder is essentially a balloon. But solid organs, such has hearts and kidneys, are much more complicated. One way that scientists are attempting to solve that problem is by using computers to design the delicate structures on-screen and then print the results.

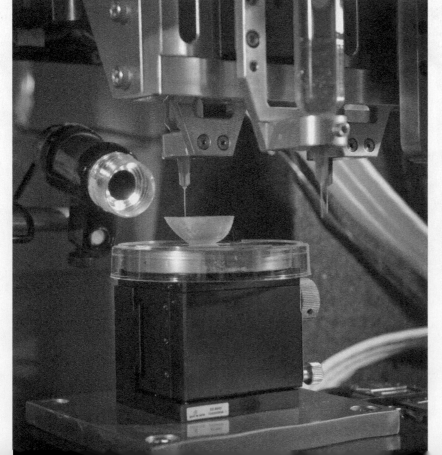

A bioprinter filled with cellular "ink," printing a human kidney.

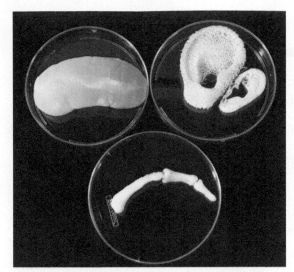

Replacement kidneys, ears, and fingers are among the organs being engineered at Wake Forest University's Institute for Regenerative Medicine.

Until recently, most bioprinters were simply modified ink jet printers with cartridges filled with cells instead of ink. The printers that now do this work are specifically designed and built for the purpose. Bioprinters are being used to build new blood vessels as well as more-complex organs like kidneys. Researchers are even developing printers that can scan a wound and print an appropriately shaped skin "bandage." The first commercial bioprinter was created by the San Diego-based company Organovo in 2009.

The idea behind bioprinting is really no different from other types of three-dimensional printing. Commercially available 3-D printers can make just about anything these days, from plastic toys and jewelry to chocolate candies to dental crowns and prosthetic limbs. All these technologies rely on 3-D graphics software that is much like the programs architects have used for years. The designs are fabricated by a printer that uses gel-like substances to build the structure in three dimensions. Bioprinting does the same thing, but instead of plastic, metal, or chocolate, the construction material is your own differentiated cells, produced from stem cells.

Atala and his group at Wake Forest have used organ printing to build miniature kidneys

that have been shown to be functional—able to filter blood and produce urine. He printed one of the organs before a live audience during a TED Talk in 2011. These experimental prototypes aren't ready to be used in people, but the technology is progressing rapidly.

The main difficulty with engineering more-complex solid organs like livers, hearts, and kidneys is ensuring that all the cells in the organ are linked to an adequate blood supply, which in normal tissues is accomplished by an intricate web of capillaries (see Chapter 27). The organs have to be properly hooked up to nerves, too, so that they can send and receive signals from the brain. "It's a tough challenge, but I think it's doable," Atala told *Scientific American* in 2011.

Researchers in this field look forward to the day, in the not-too-distant future, when patients will be able to order just about any organ they need, designed to work perfectly in their own body. No longer will patients die while waiting for an organ donor.

"With the field of regenerative medicine," says Atala, "the hope is to be able to have tissues and organs available for patients that wouldn't otherwise have them." ∎

A biodegradable scaffold in the shape of a trachea can be seeded with cells and transplanted into a patient.

CHAPTER 13 Summary

- Tissues are integrated groups of specialized cells that perform specific functions.

- Stem cells are relatively unspecialized cells that can divide and differentiate into different cell types.

- Adult stem cells, also known as somatic stem cells, are found in mature tissues; embryonic stem cells make up early embryos.

- Stem cells can be used therapeutically to engineer or regenerate tissues and organs.

- Making new tissues requires both cell division and cell differentiation. Cell differentiation is the process by which an unspecialized cell becomes a specialized cell with a unique function.

- All cells in the body have the same genome but express different genes. Such differential gene expression causes each cell type to produce different proteins and to have different functions.

- Adult stem cells are multipotent, capable of differentiating into a limited number of different cell types.

- Embryonic stem cells are pluripotent, capable of differentiating into nearly any cell type in the body.

- Embryonic stem cells can be obtained from human embryos or from cloned embryos. They may also be created by inducing adult cells to "de-differentiate."

- In somatic cell nuclear transfer (SCNT), also known as cloning, the nucleus of an unfertilized egg is replaced by the nucleus of a donor cell to produce an embryo with the same genes as the donor cell.

- Both adult and embryonic stem cells are being investigated as possible therapies to restore damaged tissue in humans.

MORE TO EXPLORE

- Catalyst, ABC television (Oct, 2012): Organ Bioprinting http://www.abc.net.au/catalyst/stories/3618385.htm

- Atala, A. (2011) TED Talk: Printing a human kidney http://www.ted.com/talks/anthony_atala_printing_a_human_kidney.html

- Atala, A., et al. (2006) Tissue-engineered autologous bladders for patients needing cystoplasty. *The Lancet* 367:1241–46.

- Raya-Rivera, A., et al. (2011) Tissue-engineered autologous urethras for patients who need reconstruction: an observational study. *The Lancet* 377:1175–82.

- Junying, Y., et al. (2007) Induced pluripotent stem cell lines derived from human somatic cells. *Science* 318:1917–1920.

CHAPTER 13 Test Your Knowledge

DRIVING QUESTION 1

What is the structure of tissues and organs, and how can organs be repaired or replaced?

By answering the questions below and studying Infographics 13.1, 13.2, 13.3, and 13.5, you should be able to generate an answer for the broader Driving Question above.

KNOW IT

1 Which of the following statements applies to tissues?

   a. Only one cell type is present.

   b. Multiple cell types are present.

   c. Each tissue has a specific function.

   d. Cells within a tissue cooperate.

   e. all of the above

   f. b, c, and d

2 The brain and spinal cord are made up of nervous tissue. This tissue includes neurons—cells that fire electrical impulses that communicate information— as well as glial cells. Some glial cells enable the electrical impulse to travel faster in a neuron. How are the roles of neurons and glial cells consistent with a functional tissue?

3 You shed skin cells every day. How are those cells replaced?

   a. by mitotic division and specialization of embryonic stem cells

   b. by differentiation of neighboring neurons into skin cells

   c. by differentiation of red blood cells that leave the circulation and migrate into deeper layers of the skin

   d. by mitotic division and differentiation of tissue stem cells

   e. all of the above

USE IT

4 What are the pros and cons of receiving an organ transplant versus growing a replacement organ from one's own cells?

5 Why does a recipient of a liver transplant have a high risk of bacterial infections?

   a. because the liver plays a critical role in the immune response

   b. because donor livers are often contaminated with disease-causing bacteria

   c. because transplant recipients have to take drugs that suppress their immune system

   d. because the surgery poses a high risk for introducing bacteria into the recipient

   e. because the immune system may reject the liver

6 From the information provided in Question 2, would it be sufficient to just replace the neurons in someone who suffered nervous tissue damage? Why or why not?

7 Why is engineering a bladder more challenging than engineering skin?

DRIVING QUESTION 2

What are the properties of specialized cells in tissues, and how do stem cells differentiate into these specialized cells?

By answering the questions below and studying Infographic 13.4, you should be able to generate an answer for the broader Driving Question above.

KNOW IT

8 Relative to one of your liver cells, one of your skin cells

   a. has the same genome (that is, the same genetic material).

   b. has the same function.

   c. has a different pattern of gene expression.

   d. a and c

   e. b and c

9 A muscle cell does not have keratin--the protein that gives skin its elasticity and "waterproofing." Why would muscle cells not have keratin?

10 Is the genome of stem cells larger than that of specialized cells?

   a. yes, because they need the genes found in every cell type, whereas specialized cells need only a subset of all the genes

   b. yes, because they express more genes than do specialized cells

   c. no, because all cells in a person have the identical set of genes in their genome

   d. no; they have a smaller genome, because stem cells are equivalent to gametes (which are haploid) in that they can potentially create an entire individual

   e. no; they have a smaller genome because stem cells express only a subset of genes

13

USE IT

11 Different cells have different functions. Muscle cells contract because of the sliding action of actin and myosin proteins in muscle cells; a protein called retinal is important for the function of the light-detecting photoreceptor cells in the retina of the eye; helper T cells of the immune system have a protein on their surface called CD4 that participates in the immune response. From this information, complete the following table.

	PHOTORE-CEPTORS OF THE RETINA	MUSCLE CELLS OF THE HEART	T CELLS
Myosin gene present?			
Myosin protein present?			
Retinal gene present?			
Retinal mRNA present?			
Retinal protein present?			
CD4 gene present?			
CD4 mRNA present?			
CD4 protein present?			

12 A woman has had a heart attack, and her heart muscle is damaged. What genes would a possible replacement cell have to express in order to begin to take over the function of cardiac muscle? For each gene, explain briefly why its expression would be important in this situation.

DRIVING QUESTION 3

How do stem cells contribute to regenerative medicine, and how can we obtain or produce stem cells for this purpose?

By answering the questions below and studying Infographics 13.3, 13.5, 13.6, 13.7, and 13.8, you should be able to generate an answer for the broader Driving Question above.

KNOW IT

13 Compare and contrast embryonic stem cells and somatic (that is, adult) stem cells in regard to at least two features.

14 An adult stem cell from bone marrow is most useful in treating

a. a heart attack.

b. a large burn on the upper thigh.

c. a disorder affecting the development of white blood cells.

d. a degenerative eye disease affecting the retina.

e. a degenerative nervous system disease such as Alzheimer disease.

USE IT

15 List and then describe some of the successes and challenges associated with using adult stem cells for stem cell therapy in comparison with embryonic stem cells.

MINI CASE

16 History was made in 2008 when a 30-year-old woman whose airway was severely damaged from tuberculosis was treated with an engineered trachea that was a perfect match. A trachea consists of a cartilage tube, chondrocytes (cells that help maintain the cartilage by producing proteins that make up cartilage), and an epithelial lining. Physicians used the following components in order to accomplish the procedure: an intact donor trachea (from a woman who had died from cardiac complications); the patient's own bone marrow stem cells; and the patient's own airway epithelial cells.

The engineered trachea was surgically implanted in the patient. She did not take any drugs to suppress her immune system, and after 3 months, she had normal airway function, no evidence of rejection of the engineered trachea, and a much improved quality of life.

From the information provided, describe the steps that the physicians must have followed in order to accomplish this successful procedure.

SOURCE: Macchiarini, P., et al. (2008) Clinical transplantation of a tissue-engineered airway. *The Lancet* 372:2023–2030.

INTERPRETING DATA

17 Specific proteins expressed by specific cell types can be used as markers to both identify and isolate specific cell types from a population of cells. The table at the right provides a list of markers specifically associated with different cell types.

(i) A scientist has isolated a population of cells that express telomerase.

  a. Can these cells differentiate into neurons? What marker would you look for to determine if they had differentiated into neurons?

  b. Can these cells differentiate into white blood cells? What marker would you look for to determine if they had differentiated into white blood cells?

(ii) Experiments in mice have shown that hematopoietic stem cells can be coaxed out of the bone marrow and into the circulation and then differentiate into cardiac muscle in mice that have experienced a heart attack. What markers must the scientists have followed to confirm this result?

MARKER	CELL TYPE
CD34	Hematopoietic stem cell in bone marrow
Collagen type II	Cartilage cell (chondrocyte)
MAP2	Neuron
CD8	White blood cell
Myosin heavy chain	Cardiac muscle cell
CD133	Neural stem cell
Telomerase	Embryonic stem cell
Nestin	Pancreatic progenitor

(iii) In the Mini Case of the engineered trachea (Question 16), what marker was most likely on the bone marrow stem cells isolated from the patient? What marker would the scientists have looked for to confirm that the cells were differentiating into cartilage cells?

BRING IT HOME

18 Your roommate's best (dog) friend recently died. This beloved pet had been with your roommate since she was in the second grade, and she is very upset about the loss. She has stated that she is going to try and have her pet cloned. You have heard about cloning, and that many pets have been cloned. But you have also heard that cloning is controversial. You decide to do some research. What can you find out about the differences between reproductive and therapeutic cloning? Do you think that reproductive cloning should be allowed? Does your opinion differ for pets vs. humans?

BUGS

THAT RESIST

DRUGS

Drug-resistant bacteria are on the rise. Can we stop them?

IN JANUARY 2008, SIXTH-GRADER CARLOS DON, AN ACTIVE footballer and skateboarder, boarded a bus headed for a class trip, happy and healthy. A month later he was dead.

In April 2006, 17-year-old Rebecca Lohsen was a model student at her high school; she was on the honor roll and was a member of the swim team. Four months later Rebecca was dead.

In December 2003, Ricky Lannetti was a college senior, a star football player and all-around athlete. A few weeks later Ricky was dead.

▶▶ DRIVING QUESTIONS

1. What is staph, and can it be present in the absence of an infection?
2. How do bacteria resist the effects of antibiotics?
3. How do populations evolve, and what is the role of evolution in antibiotic resistance?

> Scientists are now seeing bacterial infections that don't respond to any known antibiotics, leading many to fear the day when we run out of treatment options altogether.

The list of sudden deaths like these goes on and on. But surprisingly, these young people weren't killed in accidents or by violence; they were all killed by an infectious bacterium known as methicillin-resistant *Staphylococcus aureus* (MRSA), which has become widespread in recent years and which is difficult to treat with many existing antibiotics.

MRSA (pronounced "mer-sa") sickens some 94,000 people in the United States each year and kills almost 19,000, according to a 2007 study by Monina Klevens and her colleagues at the Centers for Disease Control and Prevention (CDC). Formerly, outbreaks of MRSA were confined mainly to hospitals. But since the late 1990s, a growing number of healthy people are becoming infected outside hospitals. In addition, new high-risk groups are emerging—children in day care, the prison population, men who have sex with men, and certain ethnic groups—whose MRSA infection rates are higher than those in the general population. Schools nationwide have been reporting outbreaks and young, healthy people are getting sick.

For more than 60 years, bacteria-caused infections have been successfully combated with antibiotics—those "wonder drugs" that Alexander Fleming first discovered in a moldy petri plate nearly a century ago (see Chapter 3). But many of our most trusted antibiotics are no longer effective at killing the bacteria they once defeated. Over time, these bacteria have changed genetically—evolved—to become resistant. As a result, scientists are now seeing bacterial infections that don't respond to any known antibiotics, leading many to fear the day when we run out of treatment options altogether.

"This is a major public health imperative," says Robert Daum, professor of microbiology at the University of Chicago and a member of the Infectious Diseases Society of America. "We need a plan of attack now."

STAPH THE MICROBE

Bacteria are everywhere: in the air, in food, on toothbrushes and computer keyboards–even on and inside the human body. Indeed, the body is host to a wide variety of bacteria, ranging from helpful ones like *Lactobacilli* and *Bifidobacteria* that live in the gut and aid digestion, to harmful ones like *Salmonella* that cause food poisoning. In fact, there are more bacteria living on and in you than there are human cells making up your body. The helpful bacteria on your skin produce acids that make your skin inhospitable to other, less helpful bacteria, generally protecting you from invaders. At any given moment, a fair number of us harbor potentially dangerous bacteria on our skin. Most of the time, this isn't a problem. But in certain circumstances, these bacteria can overgrow your helpful skin bacteria and cause serious infections.

MRSA infection is caused by the *Staphylococcus aureus* bacterium–often simply called "staph." Although several species of staph bacteria can cause human disease, the medical community is especially concerned about staph species, such as *S. aureus,* that have developed resistance to antibiotic drugs that once effectively killed them. "MRSA" is actually a misnomer because the antibiotic methicillin is no longer used to treat staph infections. Drug-resistant strains of staph are usually resistant to several different types of antibiotics, including penicillins and cephalosporins.

INFOGRAPHIC 14.1 THE BACTERIUM *STAPHYLOCOCCUS AUREUS*

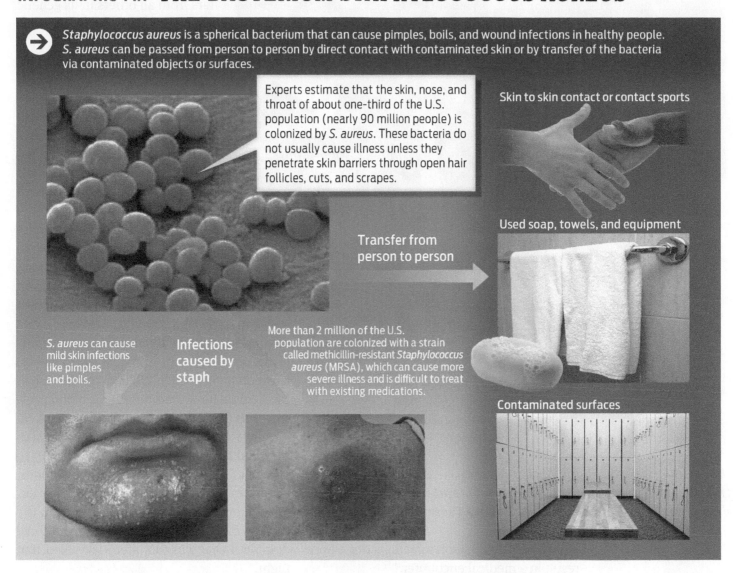

Staphylococcus aureus is a spherical bacterium that can cause pimples, boils, and wound infections in healthy people. *S. aureus* can be passed from person to person by direct contact with contaminated skin or by transfer of the bacteria via contaminated objects or surfaces.

Experts estimate that the skin, nose, and throat of about one-third of the U.S. population (nearly 90 million people) is colonized by *S. aureus*. These bacteria do not usually cause illness unless they penetrate skin barriers through open hair follicles, cuts, and scrapes.

Skin to skin contact or contact sports

Transfer from person to person

Used soap, towels, and equipment

S. aureus can cause mild skin infections like pimples and boils.

Infections caused by staph

More than 2 million of the U.S. population are colonized with a strain called methicillin-resistant *Staphylococcus aureus* (MRSA), which can cause more severe illness and is difficult to treat with existing medications.

Contaminated surfaces

Surprisingly, staph bacteria cause no harm to most of the people who carry them. Between 30% and 40% of the population carries staph on their skin or in their noses, and about 1% of the population carries drug-resistant strains, according to the CDC. If you carry staph of any strain but aren't sick, you are "colonized" but not infected. Staph spreads from person to person through skin-to-skin contact or through shared, contaminated items such as towels and bars of soap.

Most healthy people can be colonized with any staph strain, including MRSA, and not become ill—for the most part the skin and our immune systems protect us. But infections can occur if staph bacteria come into contact with a wound or otherwise enter the body. For example, athletes who have cuts and scrapes may acquire a staph infection in locker rooms or during contact sports. In otherwise healthy people, staph infection usually causes only minor skin eruptions such as boils or pustules that can resemble spider bites [INFOGRAPHIC 14.1].

"Every one of us has probably had a staph infection at some point," explains Daum. "Staph ranges from the commonest cause of infected fingernails all the way to a severe syndrome with rapid death, and everything

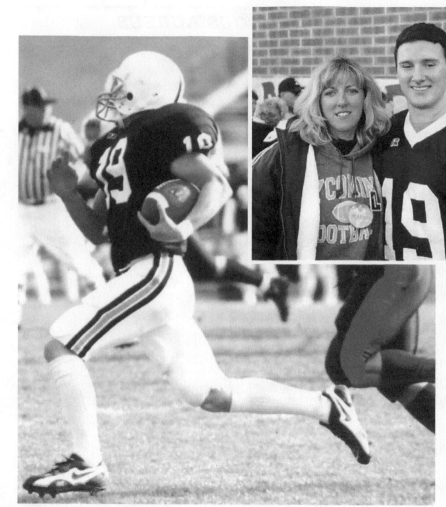

Ricky Lannetti was a healthy 21-year-old football player. Top right: Ricky with his mother, Theresa Drew.

strains of staph. Each strain differs from all others in its genetic makeup. MRSA, for example, is composed of a number of unique strains of staph bacteria, and some cause more serious disease than others. In recent years there have been several cases of healthy people becoming severely ill from MRSA infection, most likely because they were infected by an especially deadly strain of drug-resistant staph.

Ricky Lannetti, for example, was a healthy 21-year-old football player at Lycoming College in Williamsport, Pennsylvania. "He was strong as an ox and he ran like a deer," says his mother, Theresa Drew. In early December 2003, Ricky came down with a bout of flu. He wasn't recovering, and on the morning of December 6, Drew drove her son to Williamsport Hospital. By the time he was admitted, his blood pressure was dangerously low and his body temperature was erratic. As each hour passed, his condition worsened. His lungs began to fail. Doctors tried five different antibiotics, all in vain. When his heart began to weaken, his doctors prepared him to be flown to the cardiac center at a bigger hospital in Philadelphia. But it was too late. Ricky died that night.

It was only after an autopsy was performed that doctors discovered what had killed him: MRSA had infected Ricky's bloodstream and ravaged his insides. Although doctors couldn't be sure how Ricky contracted MRSA, they suspect that it entered his body through a pimple on his buttocks. From there, it spread to his internal organs.

"Doctors tried every antibiotic imaginable, including vancomycin," says his father, Rick Lannetti. But the treatment was too late. Ricky's immune system was already weak because of the flu. When he contracted MRSA, his body was unable to fight back as well as it otherwise would have. "In the end," his father says, "MRSA had broken every one of his organs beyond repair."

in between. Most staph infections don't even result in a medical encounter."

People with a weakened immune system are at risk for more-severe staph infections, especially if staph finds a way to enter the bloodstream, for example through a surgical wound. The elderly, whose immune systems are weakened, and the young, whose immune systems are still developing, are at especially high risk. And people who are already sick and fighting off other infections can be at high risk, too. When bacteria such as staph do cause illness, they do so by multiplying on or in human tissues. They can also secrete toxic substances that harm human cells or interfere with essential cellular processes. MRSA infections in the lungs and bloodstream are especially dangerous.

Staph bacteria can cause such a range of disease because there are many different

ACQUIRING RESISTANCE

Bacterial infections were a common cause of death before the 1940s, when antibiotics first became widely available. **Antibiotics** are chemicals that either kill bacteria or slow their growth by interfering with the function of essential bacterial cell structures. Since Alexander Fleming's discovery of penicillin in 1928 (see Chapter 3), many additional antibiotics have been discovered or synthesized in the lab.

For decades, these antibiotics have been used to treat most common bacterial infections, including staph, and have saved thousands of lives. But almost immediately after antibiotics were introduced, bacteria that could survive antibiotics–drug-resistant bacteria–began to emerge. Within the last decade drug-resistant bacterial strains have become much more common. Most people infected with drug-resistant bacterial strains are still treatable, but they have fewer treatment options. And sometimes–as in Ricky Lannetti's case–existing drugs are completely ineffective.

Drug-resistant strains of staph, for example, are typically resistant to an entire class of antibiotic drugs called the beta-lactams. Beta-lactams include penicillin and the cephalosporin antibiotics, such as methicillin and cephalexin. Beta-lactams are the most commonly prescribed class of antibiotics. They work by interfering with a bacterium's ability to synthesize cell walls (**INFOGRAPHIC 14.2**).

A variety of non-beta-lactam classes of antibiotics can treat MRSA infections, and vancomycin, a non-beta-lactam drug, is the antibiotic of choice when a serious MRSA infection is confirmed. But even vancomycin isn't always effective; there are now staph strains resistant to vancomycin, too.

Ricky Lannetti did not respond to vancomycin. Nor did Rebecca Lohsen, the 17-year-old high school swimmer. She was diagnosed with MRSA 2 days after she was admitted to the hospital for pneumonia. But antibiotics were ineffective in controlling the MRSA that

INFOGRAPHIC 14.2 HOW BETA-LACTAM ANTIBIOTICS WORK

➡ Antibiotic medications are grouped into classes, one of which is called the beta-lactams. Beta-lactam antibiotics interfere with a bacterium's ability to synthesize cell walls.

Antibiotic interferes with proper synthesis of the wall.

Water flows in by osmosis.

The wall breaks apart.

Bacterial cell wall

Bacterial cell membrane

Cytoplasm inside cell

In most bacterial cells, a rigid cell wall surrounds the cell membrane. The cell wall helps keep cells intact, despite the flow of water into the cells.

Beta-lactam antibiotics interfere with the synthesis of new cell wall material, causing a weak wall.

Without a strong wall, the force of excessive water inside the cell is strong enough to rupture the cell.

14

INFOGRAPHIC 14.3 HOW BACTERIA REPRODUCE

→ Bacteria reproduce through a process called binary fission. Binary fission is a form of asexual reproduction in which a single parent cell replicates its contents and then divides into two daughter cells. Note that each daughter cell inherits all its DNA from the single parent cell.

Cell wall — — DNA
— Cell membrane

1. Cell elongates and DNA is replicated.

2. Cell wall and plasma membrane begin to divide.

3. Cross-wall forms completely around divided DNA.

4. Daughter cells separate.

grows in size, and then splits into two daughter cells, each with a copy of the parental DNA. Each time DNA is replicated, however, there is a chance that genetic mutations will occur, and the new alleles will then be carried into each daughter cell. And because bacteria reproduce much more rapidly than other organisms—one generation of bacteria can double in as little as 20 minutes—they accumulate mutations at a relatively high rate. An entire population of bacteria that is genetically different from the original cell can arise very quickly (INFOGRAPHIC 14.3).

Mutation isn't the only way that populations of bacteria can acquire new genetic variation. Bacteria can acquire new alleles, and even new genes, through a mechanism called gene transfer, in which pieces of DNA pass from one type of bacteria to another. (By contrast, populations of sexually reproducing organisms acquire additional genetic diversity by recombining parental chromosomes during meiosis and then mating with genetically distinct individuals; see Chapter 11) (INFOGRAPHIC 14.4).

Staph bacteria, for example, became resistant to drugs either by mutations in their own genes or by picking up "resistance" genes from

Bacteria can pick up genes from other bacteria by the process of gene transfer. The tube-like transfer apparatus is called a pilus.

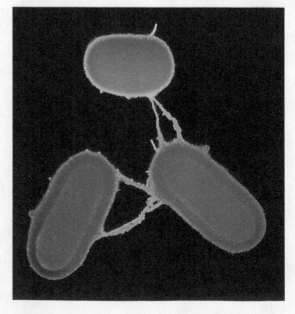

attacked her lungs and then her heart. Twelve-year-old Carlos Don, who skateboarded and played football, also did not respond to vancomycin; he, too, died.

What's behind this rise in drug-resistant bugs? Like all organisms, bacteria can acquire mutations when their DNA replicates during reproduction (see Chapter 10). These random mutations create new alleles in the bacterial population, including ones that may confer antibiotic resistance.

Bacteria reproduce asexually by a process called **binary fission.** Unlike sexual reproduction, in which gametes from two parents fuse, asexual reproduction does not require a partner. In binary fission, a single parental cell simply replicates its single chromosome,

▶BINARY FISSION
A type of asexual reproduction in which one parental cell divides into two.

HOW BACTERIAL POPULATIONS ACQUIRE GENETIC VARIATION

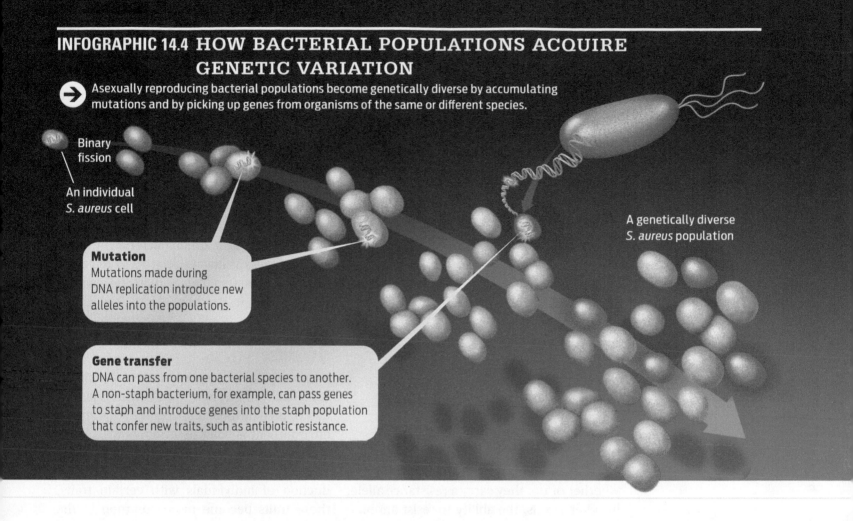

Asexually reproducing bacterial populations become genetically diverse by accumulating mutations and by picking up genes from organisms of the same or different species.

Binary fission

An individual *S. aureus* cell

A genetically diverse *S. aureus* population

Mutation
Mutations made during DNA replication introduce new alleles into the populations.

Gene transfer
DNA can pass from one bacterial species to another. A non-staph bacterium, for example, can pass genes to staph and introduce genes into the staph population that confer new traits, such as antibiotic resistance.

other drug-resistant bacteria. The genetic changes ultimately altered bacterial proteins in ways that helped staph dodge antibiotic drugs. Specifically, the altered or acquired genes either code for proteins that can disable antibiotics, or they code for proteins with altered shapes to which antibiotics can no longer bind. Some bacteria produce enzymes called beta-lactamases that chew up beta-lactam antibiotics. Because different strains of bacteria have developed antibiotic resistance independently, several strains of genetically unique drug-resistant staph circulate through human communities at the same time.

AN EVOLVING ENEMY

While an individual bacterium—or any individual organism—can undergo genetic changes that give it new traits, this doesn't entirely explain how bacterial populations such as staph develop resistance to drugs. An entire popula-

tion of organisms with a new trait can arise only when the environment favors that trait—that is, when carrying the specific trait is advantageous to the organisms carrying it. A **population** is a group of individuals of the same species living together in the same geographic area. Geographic area is relative; it could be an open prairie or a drop of pond water. A population of bacteria can exist anywhere. In the case of staph, populations exist in people's noses and on other parts of their skin.

When a population's environment favors some traits over others, the frequencies of the alleles that code for those traits in the population change over time. Take the trait for drug resistance, for example. A genetically diverse population of bacteria will contain some individuals possessing alleles that confer resistance. In an environment free of antibiotics, individual bacteria will have about an

▶**POPULATION**
A group of organisms of the same species living together in the same geographic area.

INFOGRAPHIC 14.5 AN ORGANISM'S FITNESS DEPENDS ON ITS ENVIRONMENT

The term "fitness" describes the relative ability of an organism to reproduce in a particular environment. Fitness is determined by the interaction between phenotype and environment. Antibiotic-resistant bacteria, for example, have high fitness in the presence of antibiotics.

Reproductive fitness in the absence of antibiotic	Reproductive fitness in the presence of antibiotic	
Similar to the rest of the population	Antibiotic-sensitive variants	Low
Similar to the rest of the population	Moderately antibiotic-resistant variants	Medium
Similar to the rest of the population	Highly antibiotic-resistant variants	High

▶**EVOLUTION**
Change in allele frequencies in a population over time.

▶**FITNESS**
The relative ability of an organism to survive and reproduce in a particular environment.

▶**NATURAL SELECTION**
Differential survival and reproduction of individuals in response to environmental pressure that leads to change in allele frequencies in a population over time.

▶**ADAPTATION**
The process by which populations become better suited to their environment as a result of natural selection.

equal chance of surviving and reproducing, whether or not they carry a resistance allele; in other words, the ability to resist antibiotics will confer neither an advantage nor a disadvantage, since there are no antibiotics around. But in the presence of an antibiotic, bacteria with an allele for resistance may survive, whereas other bacteria die. The surviving bacteria, which are drug resistant, reproduce and pass their alleles for drug resistance on to future generations. Consequently, the frequency of the resistance trait increases. This is how populations evolve. **Evolution** is defined as a change in the frequency of alleles in the population over time.

An organism's ability to survive and reproduce in a particular environment is called its **fitness.** The higher an organism's fitness, the more likely that alleles carried by that organism will be passed on to future generations and increase in frequency. In an environment in which antibiotics are abundant, drug-resistant bacteria are more fit than non-resistant bacteria (**INFOGRAPHIC 14.5**).

And in the United States, antibiotic use is widespread: in 2010, antibiotics were prescribed at the rate of 801 prescriptions per 1,000 people. This makes the United States a top consumer of antibiotics on a global scale. The abundance of antibiotics in the environment has created the perfect breeding ground for antibiotic-resistant staph.

In a different environment, however, one in which antibiotics are less common, these same resistant bacteria will not necessarily have an edge over other bacteria. In other words, fitness is always relative to the environment; organisms can be fit in one environment and not in another.

SELECTING FOR SUPERBUGS

Ultimately, the interplay between an organism's traits, or its phenotype (which is largely determined by its genes, or genotype), and its environment is what determines what traits will predominate in a population. When the environment favors the survival and reproduction of individuals with certain traits, those traits become more common in the population. This process of differential survival and reproduction of individuals within a population in response to environmental pressure is known as **natural selection.** Much in the way plant and animal breeders have for centuries practiced artificial selection to produce individuals with desired traits, the environment also, in a sense, selects individuals with certain traits.

When natural selection acts on a population over time, advantageous traits become more common, and the population becomes better suited to its environment. In other words, evolution by natural selection leads to **adaptation.** This is what we see with antibiotic-resistant bacteria: the population has become better suited, or adapted, to an environment in which antibiotics are abundant because individual bacteria carrying resistance genes are more fit in this environment. Charles Darwin was one of the first people to figure out how natural selection works and to study its results—for example, in birds on the Galápagos Islands, where various finch species had evolved different sizes of beak as adaptations to different food sources (see **Milestones in Biology: Adventures in Evolution**).

Note that evolution by natural selection occurs in populations, not individuals. Individual organisms do not experience a change in allele frequencies over time. Therefore, individual organisms do not evolve (**INFOGRAPHIC 14.6**).

Natural selection doesn't always affect populations in the same way. By studying how natural selection has shaped populations, scientists have defined three major patterns of natural selection. When the predominant phenotype in the population has shifted in one particular direction, we say that **directional selection** has occurred. For example, when bacterial populations evolve from populations sensitive to drugs into ones that resist drugs—that is, toward antibiotic resistance—they are exhibiting directional selection.

When the phenotype of the population settles around the middle of the phenotypic spectrum, we call this **stabilizing selection.** Or, a population can "spread out," so that the

▶**DIRECTIONAL SELECTION**
A type of natural selection in which organisms with phenotypes at one end of a spectrum are favored by the environment.

▶**STABILIZING SELECTION**
A type of natural selection in which organisms near the middle of the phenotypic range of variation are favored by the environment.

INFOGRAPHIC 14.6 EVOLUTION BY NATURAL SELECTION

→ In any genetically diverse population, individual fitness varies. When an organism's environment favors specific genetic variants to survive and reproduce over others, natural selection occurs. Those with high fitness tend to reproduce more successfully. Over generations, the frequency of alleles that confer higher fitness increase while those that confer lower fitness decrease. This non-random change in allele frequencies over generations is called evolution by natural selection.

For example, in the *absence* of antibiotic

Individual bacteria in the population have about the same fitness.

Individuals reproduce at about the same rate and therefore pass on their alleles in similar number to the next generations.

Reproduction

Over time, the frequency of alleles remains the same.

In the *presence* of antibiotic

Resistant bacteria have higher fitness (sensitive individuals cannot reproduce at all).

Resistant bacteria reproduce more often and therefore pass on their resistance alleles in greater number to the next generations.

Sensitive bacterium

Moderately resistant bacterium

Resistant bacterium

Antibiotic

Reproduction

Over time, the frequency of resistant alleles increases dramatically.

▶DIVERSIFYING SELECTION
A type of natural selection in which organisms with phenotypes at both extremes of the phenotypic range are favored by the environment.

population shows extremes of the phenotypic spectrum; this pattern is known as **diversifying selection** (INFOGRAPHIC 14.7).

The particular pattern of natural selection a population follows depends on the interaction of phenotypes with the environment. So, for example, in the absence of antibiotics, populations of staph bacteria might have followed stabilizing or diversifying selection. Instead, directional selection led to the MRSA that killed Carlos Don, Rebecca Lohsen, and Ricky Lannetti.

MRSA IN THE COMMUNITY

Drug-resistant staph strains first emerged in hospitals during the early 1960s, partly as a result of selection pressure from antibiotics. Since then, hospitals have remained hot spots for staph infections. The combination of heavy antibiotic use, lots of sick patients, and close quarters makes hospitals a fertile environment for the emergence of resistant bugs.

In response, many hospitals have implemented measures to reduce infections. For

---

INFOGRAPHIC 14.7 PATTERNS OF NATURAL SELECTION

Directional selection
occurs when a single phenotype predominates in a particular environment.

The environment selects for phenotypes toward one end of the spectrum.

— Before natural selection
— After natural selection

Number of bacteria

Low   High
Antibiotic resistance

Example
Antibiotic-containing environments favor resistant strains of bacteria.

Stabilizing selection
occurs when phenotypes at each end of the spectrum are less suited to the environment than organisms in the middle of the phenotypic range.

The environment selects for a very narrow range of phenotypes.

Number of humans

Low   High
Human birth weight

Example
Human babies with very low birth weights do not survive as well as larger babies, and very large human babies are not easily delivered through the birth canal. Midrange babies are favored.

Diversifying selection
typically occurs in a "patchy" environment, in which extremes of the phenotypic range do better than middle range individuals.

The environment selects for phenotypes at both ends of the spectrum.

Number of finches

Small   Large
Finch beak size

Example
The African finch *Pyrenestes* lives in an environment where only large, hard seeds and small, softer seeds are available. Birds with either large or small beak sizes are selected for, while medium beaks, which are not as successful at cracking either type of seed, are selected against.

example, studies have shown that measures as simple as requiring all health care workers to wash their hands before handling each patient can dramatically reduce the number of infections. However, the rate of hand washing among health care workers remains dismally low. A 2010 study published by researchers at the Atlanta Veterans' Affairs Medical Center found that fewer than 50% of health care workers followed guidelines for hand hygiene in the hospitals observed in the study. Other studies have shown similar numbers. "It is really important that people do low tech things that make a high difference," says Ruth Lynfield, state epidemiologist and medical director of the Minnesota Health Department. "Washing hands well and often is absolutely critical."

Another common strategy for reducing the spread of infections is to include hand-sanitizing stations at the exit of patient rooms and in other high-traffic areas. These devices typically dispense a 62% alcohol gel (for example, Purell), which kills most bacteria. Because the gel is mostly alcohol and evaporates quickly, it can be used in place of hand washing. Since 2002, the CDC has recommended the use of alcohol-based hand sanitizers in hospital set-

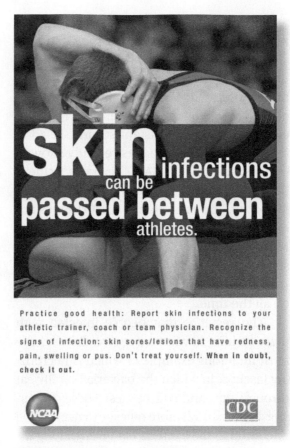

Practice good health: Report skin infections to your athletic trainer, coach or team physician. Recognize the signs of infection: skin sores/lesions that have redness, pain, swelling or pus. Don't treat yourself. When in doubt, check it out.

The CDC teamed up with the National Collegiate Athlete Association to create a series of educational posters to raise awareness about infections.

Hand washing can reduce the spread of infections.

tings as an effective way to reduce the spread of germs when hands are not visibly soiled; when hands are visibly soiled, washing with soap and water is the preferred method of sanitization.

More alarming than MRSA infections in hospitals are MRSA infections in the community at large. Though MRSA has been around for more than 40 years, in the mid-1990s the rate of community-acquired infections in the United States began to soar, explains Daum, of the University of Chicago. Currently, about 85% of MRSA infections occur in people who are hospitalized or who have come in contact with the health care system. The other 15% occur in people in the community with no hospital connection. In some groups, such as day care children and prisoners, 60%-90% of those who show up at a hospital or clinic because of a skin infection are infected by some strain of drug-resistant staph, he says.

What has happened in the last 20 years? A new strain of *Staphylococcus aureus* emerged and flourished in an environment where antibiotic use is rampant. Daum thinks that drug-resistant staph strains circulating in the community evolved separately and more recently than other resistant strains. In addition, he and his colleagues recently showed that a strain called USA300 is more virulent than other MRSA strains. "It appears to be juiced up," he says. Many USA300 genes are expressed at high levels. One gene in particular that controls expression of a number of staph toxins, which can damage cells and tissues as well as cause toxic shock, is turned on all the time.

The result is that people infected with these strains have more-severe disease, the most infamous example of which is necrotizing fasciitis, in which the bacteria literally eat through skin and soft tissues. These "superbugs" can also kill more quickly. In necrotizing pneumonia, the bacteria eat through lung tissue and kill the victim. Symptoms can appear so suddenly that, according to Daum, "You could be healthy at 1:00 in the afternoon and be dead by 1:00 in the morning."

It was likely USA300 or a related strain that killed Carlos Don, Rebecca Lohsen, and Ricky Lannetti. "I've been an infectious disease guy for over 20 years now and we didn't talk about staph necrotizing pneumonia like we do now," says Daum.

Even more troubling, staph is continuing to evolve. There is evidence that when strains that are prevalent in the community mix with strains that are prevalent in hospitals, the risk increases that an even more virulent staph strain will emerge.

STOPPING SUPERBUGS

Staph bacteria aren't the only ones that have grown resistant to antibiotics. It is getting harder to treat patients with severe *Salmonella* food poisoning caused by drug-resistant strains. *Neisseria gonorrhoeae*, the bacterium that causes gonorrhea, has become resistant to another important group of antibiotics, the

fluoroquinolones. And there are now forms of pneumonia caused by strains of *Klebsiella* that are resistant to every available antibiotic.

Because the very use of antibiotics can drive bacterial populations to evolve resistance, antibiotic resistance is inevitable. But humans have hastened the emergence of drug-resistant strains of bacteria by the haphazard use and overuse of antibiotics. From the moment antibiotics were first introduced, physicians began prescribing them for colds, coughs, and earaches, most of which are caused by viruses that aren't killed by antibiotics anyway. Antibiotics are frequently overused or misused for many other ailments as well, contributing to an environment that promotes resistance.

Doctors aren't the only culprits. Farmers give antibiotics in low doses to poultry, swine, and cattle to promote growth. This practice can cause food-borne pathogens such as *Salmonella* or *Campylobacter* to develop antibiotic resistance. Undigested antibiotics in animal manure can contaminate the environment through groundwater or when manure is used as fertilizer. In this environment drug-resistant bacteria are more fit and will therefore be selected for and become more prevalent over time.

"We really have to be careful about how we use antibiotics because antibiotic use is the biggest driver of antibiotic resistance," says Lynfield, of the Minnesota Health Department.

Some public health officials are concerned that increased use of so-called antibacterial soaps that contain the bacteria-killing chemical triclosan might be contributing to the emergence of antibiotic-resistant superbugs. Although triclosan is not used as an antibiotic, some researchers worry that bacteria that become resistant to triclosan might also be able to resist antibiotics by a phenomenon known as cross-resistance. So far, there is no conclusive evidence that triclosan leads to antibiotic resistance in populations. Nevertheless, most experts agree that using antibacterial soaps is overkill: multiple studies have failed to find any benefit to washing with antibacterial soaps over and above washing with plain old soap.

Alcohol-based sanitizers are not thought to contribute to cross-resistance because the alcohol quickly evaporates, leaving no bacteria-killing residue. In contrast, soaps that contain triclosan can linger on surfaces like sinks, providing a possible opportunity for bacteria to adapt to the presence of the chemical.

Clearly, developing stronger antibiotics isn't the only or the best solution to the problem of resistance because bacteria will ultimately adapt to those, too. Perhaps the best way to control resistance, say experts, is to change practices that enable resistant strains to thrive. Careful hygiene and prevention of infection through vaccination are important tools. It is also important that when an antibiotic is prescribed it is taken precisely as prescribed, for the full course of treatment, no matter how much better the patient may be feeling. If bacteria are exposed to antibiotics at low levels or for short durations, the entire population may not be eradicated. The remaining bacteria may be resistant to the antibiotic and proliferate. And anyone taking antibiotics exposes all the bacteria in his or her body to the antibiotics, possibly enabling other drug resistant bacteria to emerge. These drug-resistant bacteria might then be transmitted to other people.

At the community level, the more antibiotics that are used, the more resistance that will emerge. So doctors are heavily discouraged from prescribing antibiotics unnecessarily. And efforts are being made to crack down on the practice of feeding livestock low levels of antibiotics (INFOGRAPHIC 14.8).

INFOGRAPHIC 14.8 PREVENTING AND TREATING INFECTION BY ANTIBIOTIC-RESISTANT BACTERIA

Reduce antibiotics in livestock feed.
Excessive antibiotics in the environment create continuous selective pressure for all bacteria.

Keep locker rooms and sports equipment clean.
Protect young athletes from contact with contaminated surfaces.

Disinfect common surfaces.
Disinfection reduces the transmission of infection by contact with contaminated surfaces, especially in facilities that serve a lot of people.

Wash hands frequently.
Both regular and antibacterial soap, as well as hand sanitizer, are effective at preventing the spread of bacterial infection.

Research new vaccines.
Vaccines prevent resistant bacterial strains from making people ill.

Do not take antibiotics for viral infections.
Viruses are not killed by antibiotics. Overuse of antibiotics causes resistant bacterial strains to become widespread.

UNIT 3 • HOW DOES LIFE CHANGE OVER TIME? EVOLUTION AND DIVERSITY

Of course, these measures won't fight resistant strains that are already circulating. But there are ways to reduce and perhaps prevent infections in this case, too. Because MRSA is more prevalent in certain environments, they present opportunities for health care workers to intervene. Prisons and jails, for example, are hot spots for infection because inmates often have limited access to soap and other forms of hygiene, are housed close together, and may not receive prompt medical care. When infected prisoners are released, they can then spread their germs to relatives and friends. "People go in, they pick up MRSA, they take it home, and then I see the kids come in sick," says Daum. By preventing transmission in correctional settings, health care workers may be able to prevent infections in the larger community.

A vaccine would be another way to prevent staph infections. A vaccine for children against *Streptococcus pneumoniae* introduced in 2000 caused the rate of infection–and especially the rate of drug-resistant infections–to drop dramatically. And not only did the rate of infection drop in vaccinated children, but other age groups benefited as well because the bacteria were not being transmitted as frequently. As another example of the impact that vaccines have on infections, Daum points to the bacterium *Haemophilus influenzae*, which frequently caused pneumonia, meningitis, and other serious diseases in children. Today, children are vaccinated against it. "When I was an intern we used to see 60 to 80 *Haemophilus* infections a month," he says. "Today we see none, it's gone. And MRSA needs to be gone, too." ■

CHAPTER 14 Summary

▶ Populations are groups of individuals of the same species living together in the same geographic area.

▶ Bacteria populations exist nearly everywhere, including on and in our bodies; most are harmless or even beneficial, but some can cause disease.

▶ Within any population, there is genetic variation among individuals.

▶ Bacterial populations, which reproduce asexually, acquire genetic variation by mutation and gene transfer; populations of sexually reproducing organisms acquire genetic variation by mutation and by meiosis and fusion of gametes.

▶ Genetic variation in a population gives rise to corresponding phenotypic variation in the population.

▶ Individuals with different phenotypes will have differing ability to survive and reproduce in a population; that is, they will differ in fitness.

▶ The differential survival and reproduction of individuals in a population over time in response to environmental pressure is called natural selection.

▶ Natural selection is one cause of evolution, which is defined as a change in the allele frequency of a population over time.

▶ Individuals with higher fitness in a given environment reproduce and pass on their alleles more frequently than do individuals with lower fitness, resulting in evolution by natural selection.

▶ Over time, natural selection leads to adaptation: advantageous traits become more common in the population, which as a result becomes more suited to its environment.

▶ Natural selection can shift the allele frequencies in a population in one or other of several patterns: directional selection, diversifying selection, or stabilizing selection.

▶ Antibiotic-resistant populations of bacteria emerge by directional selection in the presence of antibiotics.

MORE TO EXPLORE

▶ Infectious Diseases Society of America: Patient Stories http://www.idsociety.org/Patient_Stories/

▶ Klevens, R. M., et al. (2007) Invasive methicillin-resistant *Staphylococcus aureus* infections in the United States. *Journal of the American Medical Association* 298(15):1763–1771.

▶ Aiello, A. E., et al. (2007) Consumer antibacterial

soaps: effective or just risky? *Clinical Infectious Diseases* 45:S137–S147

▶ Aiello, A. E., et al. (2005) Antibacterial cleaning products and drug resistance. *Emerging Infectious Diseases* 11(10)1565–1570.

▶ Kallen, A. J. (2010) Health care–associated invasive MRSA infections, 2005–2008. *Journal of the American Medical Association* 304(6):641–647.

CHAPTER 14 Test Your Knowledge

DRIVING QUESTION 1

What is staph, and can it be present in the absence of an infection?

By answering the questions below and studying Infographic 14.1, you should be able to generate an answer for the broader Driving Question above.

KNOW IT

1. Can *S. aureus* be present in or on a person who has no evidence of an infection?

   a. no; *S. aureus* is associated only with infections
   b. yes, but only non-MRSA strains are present in the absence of an infection
   c. yes, but only for very short periods of time (between touching a contaminated surface and washing your hands)
   d. yes; *S. aureus* is a common skin bacterium
   e. yes; *S. aureus* is a common bacterium found in the bloodstream

2. The term "MRSA" as it is used today refers to

   a. *S. aureus* bacteria that are resistant to many antibiotics.
   b. a collection of skin and other infections caused by a type of bacteria.
   c. *S. aureus* bacteria that are found only in humans with certain types of skin infections.
   d. *S. aureus* bacteria that are normal residents of human skin in the vast majority of the human population.
   e. all bacteria that are resistant to antibiotics.

3. What is the difference between an *S. aureus* colonization and an *S. aureus* infection?

4. MRSA is most likely to be problematic if found

   a. on the surface of the skin.
   b. in nasal passages.
   c. in the bloodstream.
   d. on the fingernails.
   e. The presence of MRSA in any of those locations indicates a serious infection.

USE IT

5. A young athlete has a nasty skin infection caused by MRSA. How might this infection have been contracted?

6. For the patient in Question 5, which general kinds of antibiotics would you choose (or avoid) in treating the infection? What other measures would you recommend to prevent spread of MRSA to the athlete's teammates and family? Explain your answer.

DRIVING QUESTION 2

How do bacteria resist the effects of antibiotics?

By answering the questions below and studying Infographics 14.2, 14.3, and 14.4, you should be able to generate an answer for the broader Driving Question above.

KNOW IT

7. In the presence of penicillin:

   a. What happens to a sensitive strain of *S. aureus*?
   b. What happens to a resistant strain of *S. aureus*?

8 How do beta-lactam antibiotics kill sensitive bacteria?

a. by attracting water into cells

b. by destabilizing the cell membrane

c. by preventing DNA replication during bacterial reproduction

d. by destabilizing the cell wall

e. all of the above, depending on the specific strain of bacteria

USE IT

9 Why do the beta-lactam antibiotics affect sensitive bacterial cells but not eukaryotic cells? (You may want to review cell structure, discussed in Chapter 3, to answer this question.)

10 A sensitive *S. aureus* bacterium acquires a new gene that allows it to resist the effects of beta-lactam antibiotics (that is, the bacterium is now resistant). What might the protein encoded by that gene do?

a. synthesize beta-lactam antibiotics

b. digest beta-lactam antibiotics

c. produce a toxin

d. enhance colonization of human skin

e. enhance entry into the bloodstream

DRIVING QUESTION 3

How do populations evolve, and what is the role of evolution in antibiotic resistance?

By answering the questions below and studying Infographics 14.5, 14.6, and 14.7, you should be able to generate an answer for the broader Driving Question above.

KNOW IT

11 What are the two major mechanisms by which bacterial populations generate genetic diversity?

a. mutation and meiosis

b. binary fission and evolution by natural selection

c. gene transfer and mutation

d. mutation and binary fission

e. gene transfer and replication

12 What is the environmental pressure in the case of antibiotic resistance?

a. the growth rate of the bacteria

b. how strong or weak the bacterial cell walls are

c. the relative fitness of different bacteria

d. the presence or absence of antibiotics in the environment

e. the temperature of the environment

13 What is the evolutionary meaning of the term "fitness"?

14 The evolution of antibiotic resistance is an example of

a. directional selection.

b. diversifying selection.

c. stabilizing selection.

d. random selection.

e. steady selection.

15 In humans, very-large-birth-weight babies and very tiny babies do not survive as well as midrange babies. What kind of selection is acting on human birth weight?

a. directional selection

b. diversifying selection

c. stabilizing selection

d. random selection

e. steady selection

USE IT

16 Binary fission is asexual. What does this mean? How could two daughter cells end up with different genomes at the end of one round of binary fission?

17 In what sense do bacteria "evolve faster" than other species?

INTERPRETING DATA

18 A single *S. aureus* cell gets into a wound on your foot. *S. aureus* divides by binary fission approximately once every 30 minutes.

a. Thirty minutes after the initial infection, how many *S. aureus* cells will be present?

b. In 1 hour, how many *S. aureus* cells will be present?

c. In 12 hours, how many *S. aureus* cells will be present? (Hint: The general formula is $2^{number\ of\ generations}$; you need to figure out how many generations occurred in 12 hours.)

d. Mutations occur at a rate of 1 per 10^{10} base pairs per generation. *S. aureus* has 2.8×10^6 base pairs in its genome. Therefore, approximately 0.0028 mutations will occur per cell in the population. At the end of 12 hours, how many mutations will be present in the population of *S. aureus* in the wound in your foot? What are the implications of this genetic diversity in the context of treating a possible infection?

19 If we take the most fit bacterium from one environment—one in which the antibiotic amoxicillin is abundant, for example—and place it in an environment in which a different antibiotic is abundant, will it retain its high degree of fitness?

a. yes; fitness is fitness, regardless of the environment

b. yes; once a bacterium is resistant to one antibiotic it is resistant to all antibiotics

c. not necessarily; fitness depends on the ability of an organism to survive and reproduce, and it may not do this as well in a different environment

d. no; what is fit in one environment will never be fit in another environment

20 If a single bacterial cell that is sensitive to an antibiotic—for example, vancomycin—is placed in a growth medium that contains vancomycin, it will die. Now consider another single bacterial cell, also sensitive to vancomycin, that is allowed to divide for many generations to become a larger population. If this population is placed into vancomycin-containing growth medium, some bacteria will grow. Why do you see growth in this case, but not with the transferred single cell?

21 Imagine that a genetically diverse population of garden snails occupies your backyard, in which the vegetation is many shades of green with some brown patches of dry grass.

a. If birds like to eat snails, but they can see only the snails that stand out from their background and don't blend in, what do you think the population of snails in your backyard will look like after a period of time? Explain your answer.

b. Suppose you move the population of snails to a new environment, one with patches of dark brown pebbles and patches of yellow ground cover. Will individual snails mutate to change their color immediately? As the population evolves and adapts to the new environment, what do you predict will happen to the phenotypes in your population of snails after several generations in this new environment? How did this occur? Include the terms *gametes, mutation, fitness, phenotype,* and *environmental selective pressure* in your answer.

MINI CASE

22 Your friend has had a virus-caused cold for 3 days and is still so stuffy and hoarse that he is hard to understand. He seems to be telling you that his doctor called in a prescription for an antibiotic for him to pick up at his pharmacy. You hope that you misunderstood him, but you realize that you heard him perfectly well.

a. Will the antibiotic help your friend's cold?

b. What are the risks to your friend if he takes the antibiotic? (Think about what might happen if he should develop a wound infection in the future.)

BRING IT HOME

23 Your roommate has been prescribed an antibiotic for bacterial pneumonia. She is feeling better and stops taking her antibiotic before finishing the prescribed dose, telling you that she will save the remainder to take the next time she becomes sick. What can you tell your roommate to convince her that this is not a good plan?

It was on a short-cut through the hospital kitchens that Albert was first approached by a member of the Antibiotic Resistance.

ADVENTURES IN
Evolution

**Charles Darwin and
Alfred Russel Wallace
on the the trail of
natural selection**

5

> **❝** *The voyage of the Beagle has been by far the most important event in my life and has determined my whole career.* **❞**
>
> —CHARLES DARWIN

DURING HIS LAST TERM AT CAMBRIDGE, CHARLES DARWIN faced a dilemma: what to do with himself after graduation. He'd considered becoming a physician, like his father. But the sight of blood made him queasy and he hated rote memorization. He changed his focus to theology, intending to become a clergyman, but his real passion was bug collecting. Only that wasn't going to pay the bills.

Then a professor told him about the internship of a lifetime: a 5-year, around-the-world trip as a naturalist aboard a British surveying ship. The ship's captain, Robert FitzRoy, wanted a travel companion who would also collect specimens along the way. Unsure what he wanted to do with his life but eager to see the world, the 22-year-old Darwin jumped at the chance. He later said of the trip, "The voyage of the *Beagle* has been by far the most

▸▸ DRIVING QUESTIONS

1. What observations did Darwin make about nature that helped shape his thinking about evolution?
2. What works by other scientists shaped Darwin's thoughts about evolution?

important event in my life and has determined my whole career."

Yet he almost didn't go. Darwin's father, Robert Darwin, thought his son should buckle down and prepare to enter the clergy. A trip around the world seemed to him a useless distraction—a "wild scheme," he called it—and he refused at first to let his son go. But eventually, at the cajoling of his family, he relented. Charles packed his bags, said goodbye to his girlfriend, Emma, and set sail for South America. It was December 1831.

The passage aboard the 90-foot vessel was frequently harrowing, and Darwin suffered debilitating bouts of seasickness, but his journey aboard the *Beagle* set in motion one of the great revolutions in science. What he saw on that trip planted the seeds of ideas that have completely changed the way we view the world and our place in it. As the evolutionary biologist Stephen Jay Gould put it, "The world has been different ever since Darwin."

JOURNEY TO AN IDEA

Though Darwin is the most famous of the figures associated with the theory of evolution, he did not invent the idea. Nor was he alone among his contemporaries in studying it. In fact, the notion that species change gradually over time had been around for generations. To be sure, most people in the 1830s—Darwin included—still assumed that species were fixed and unchanging, created perfectly by God. But evidence to the contrary had been accumulating for some time. Explorers and naturalists were traveling to faraway lands, and finding unusual plants

and animals they had never seen before. Fossils were being uncovered, providing evidence that some species no longer seen on Earth had lived in the past. And anatomists were noting uncanny physical resemblances between different species, including chimpanzees and humans. Evolution was in the air when Darwin began thinking about it.

However, the ideas that people in Darwin's time had proposed to explain *how* species changed were flawed. One common misconception was Lamarckianism, named after the French naturalist Jean-Baptiste Lamarck, who suggested that species could change through the inheritance of acquired characteristics. In the Lamarckian view, giraffes, for example, developed their long necks by continually stretching them to feed on tall trees. Once it acquired its long neck, a giraffe could then pass that advantageous trait on to its offspring. This idea of the inheritance of acquired characteristics, while incorrect, was a popular one in Darwin's time—one that even Darwin himself found it hard to fully shake off in his writings (INFOGRAPHIC M5.1).

Though it would be years before Darwin proposed his own theory of evolution, his trip around the world provided him with an indispensable foundation. He kept a diary of his adventures, which was later published as *The Voyage of the Beagle* (1839).

While at sea, Darwin had plenty of time to read and think about the ideas then being discussed in scientific circles. He read, for instance, the work of Charles Lyell whose *Principles of Geology* (1830-1833) argued that

> ❝*The world has been different ever since Darwin.*❞
>
> —STEPHEN JAY GOULD

INFOGRAPHIC M5.1 LAMARCKISM: AN EARLY IDEA ABOUT EVOLUTION

→ Lamarck hypothesized that traits acquired in one's lifetime are passed on to offspring.

Jean-Baptiste Lamarck
August 1, 1744–December 28, 1829

Reproduction

If a giraffe strained its neck to eat leaves from higher branches, then over time it would develop a longer neck.

It would then pass this acquired trait on to its offspring, and they would be born with long necks.

Earth was much older than the 6,000 years popularly accepted at the time (a figure based on a literal reading of the Bible), and that its geology had been shaped entirely by incremental forces operating over a vast expanse of time. Valleys, for example, were formed by the slow grinding forces of wind and water, not by catastrophic floods; mountains were pushed up gradually by the action of volcanoes and earthquakes. Lyell's view of incremental change producing dramatic results over great spans of time left an indelible impression on Darwin.

Not long into his trip, he saw something that seemed to confirm Lyell's view. While docked at the Cape Verde islands, 300 miles off the west coast of Africa, he noticed a white layer of compressed seashells and corals embedded in a rock face 30 feet above sea level. This layer had clearly been formed in the sea, but had somehow been lifted up out of the water. Later on his trip, Darwin witnessed how this might happen: after an earthquake devastated the town of Concepción, Chile, he noticed that the coastline had been pushed up several feet, as shown by the rim of mollusks and barnacles that hovered out of the water around the bay. Lyell was right.

With such thoughts of an ancient, slowly changing Earth on his mind, Darwin studied the plants, animals, and geology at each stop on his trip, collecting fossils and specimens

INFOGRAPHIC M5.2 DARWIN'S VOYAGE ON THE BEAGLE

of local flora and fauna wherever he went (**INFOGRAPHIC M5.2**).

While exploring the shore of Argentina in August 1833, Darwin unearthed a particularly prized find: the fossilized remains of several large mammals embedded in a sea cliff, including one that looked like a giant armadillo and another that resembled a giant sloth. These animals had clearly lived long ago and were now extinct, since such oversize mammals no longer roamed the plains of Argentina. And yet the ancient creatures bore a striking resemblance to the much smaller armadillos and sloths that were indigenous to the area. Why should animals separated by such vast epochs of time share such similar anatomical structures? To Darwin, this suggested that these creatures were ancestrally related to one another, and also that the species had changed gradually over time. Darwin was beginning to question the conventional wisdom of his day.

In 1835, the young naturalist stepped ashore on the Galápagos Islands, off the coast of Ecua-

dor. On this archipelago, Darwin observed and collected many creatures, among them a variety of small birds. Months later, while studying the specimens back in England, he learned that they were all species of finch. Each species was distinguishable by a different size and shape of beak, but they all bore a family resemblance. He later wrote in *The Voyage of the Beagle*, "One might really fancy that, from an original paucity of birds in this archipelago, one species had been taken and modified for different ends."

This notion of one species giving rise over time to new species Darwin came to call "descent with modification," which he represented in his notebooks by a diagram that looked like a branching tree. (Darwin himself didn't like using the term "evolution" because he thought it gave a mistaken idea of progress or direction toward a goal, but "evolution" is the term that stuck.)

After 5 years traveling the globe, Darwin returned home to England in October 1836, but his intellectual journey was only just beginning. He began to think more seriously about how species might change over time. A key insight came to him in September 1838 while reading the work of the political economist Thomas Malthus, whose pessimistic book *An Essay on the Principle of Population* (1798) described how hunger, starvation, and disease would ultimately limit human population growth. The same must be true of plant and animal species, Darwin realized. If every individual in a population reproduced, even in a slowly reproducing population such as elephants, the world would be completely overrun with elephants in not that many generations. Since Earth is not overrun with elephants, factors must be limiting their population growth. Such limitations, Darwin reasoned, would lead to competition for resources that would put weaker individuals at a disadvantage. "It at once struck me," Darwin later wrote in his autobiography "that under these circumstances favourable variations would tend to be preserved, and unfavourable ones to be destroyed. The result of this would be the formation of new species."

These favorable variations needn't be very pronounced, he realized. All that was needed was for certain individuals to have a slight edge over others in the competition to survive and reproduce. These helpful variations would then be inherited by offspring and become more common in the population. In effect, the environment was "selecting" for favorable traits, much as plant and animal breeders selected and perpetuated desirable varietals—a plant with especially large fruit, for instance, or the many breeds of dogs we see today.

This idea of "natural selection" was Darwin's original contribution to the theory of evolution. Others had speculated at length about species change—most notably Robert Chambers in *Vestiges of the Natural History of Creation* (1844)—but Darwin was the first to provide a clear *mechanism* of evolution. The philosopher of science Daniel Dennett has called natural selection "the single best idea anyone has ever had."

Among other things, natural selection provided a powerful explanation for the apparent design seen in nature. Before Darwin, the con-

ventional view was that provided by William Paley, whose book *Natural Theology* (1802) Darwin had read in college. In that work, Paley made his famous "argument from design": that is, many creatures are so finely constructed and well adapted to their environment that their existence implies the existence of a designer—much in the way that the existence of a watch implies the existence of a watchmaker. To many people in the 19th century, that designer was God. With the idea of natural selection, Darwin did away with the need for a watchmaker: natural selection could accomplish the same thing without any input from a designer.

> **❝** *It at once struck me that under these circumstances favourable variations would tend to be preserved, and unfavourable ones to be destroyed. The result of this would be the formation of new species.* **❞**
>
> —CHARLES DARWIN

By 1844, Darwin had developed his ideas into a 200-page manuscript that he hoped would be the definitive word on the subject. He did not rush his ideas about natural selection into print, however. He knew that his ideas would be controversial, contradicting as they did strongly held beliefs about God and the special creation of all animals, including humans. Other scientists with evolutionary ideas were causing quite a stir in England and being openly ridiculed (for this reason Robert Chambers had published his book anonymously). Even sharing his theory of evolution by natural selection with trusted colleagues, Darwin said, was "like confessing a murder." To withstand challenges, he knew he would need more detailed evidence.

INFOGRAPHIC M5.3 THE EVOLUTION OF DARWIN'S THOUGHT

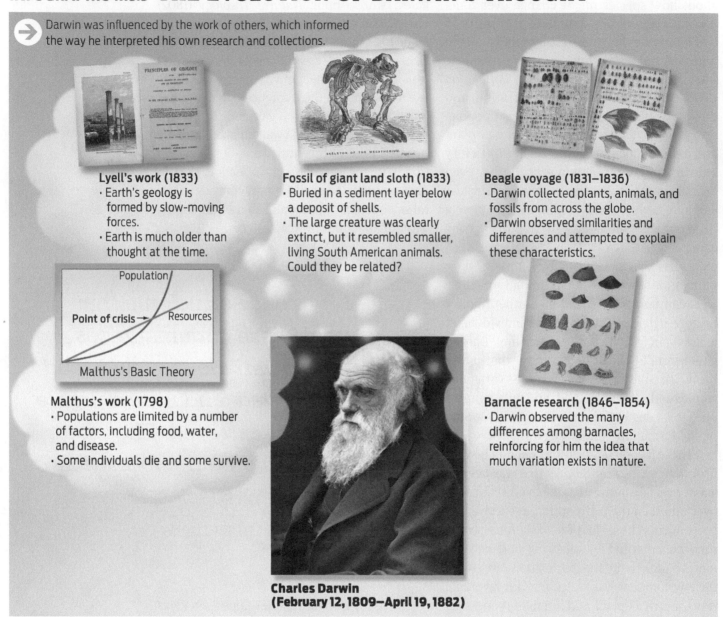

Darwin was influenced by the work of others, which informed the way he interpreted his own research and collections.

Lyell's work (1833)
· Earth's geology is formed by slow-moving forces.
· Earth is much older than thought at the time.

Fossil of giant land sloth (1833)
· Buried in a sediment layer below a deposit of shells.
· The large creature was clearly extinct, but it resembled smaller, living South American animals. Could they be related?

Beagle voyage (1831–1836)
· Darwin collected plants, animals, and fossils from across the globe.
· Darwin observed similarities and differences and attempted to explain these characteristics.

Population

Point of crisis → Resources

Malthus's Basic Theory

Malthus's work (1798)
· Populations are limited by a number of factors, including food, water, and disease.
· Some individuals die and some survive.

Barnacle research (1846–1854)
· Darwin observed the many differences among barnacles, reinforcing for him the idea that much variation exists in nature.

Charles Darwin
(February 12, 1809–April 19, 1882)

And so, at age 37, Darwin began to investigate closely one large group of animals: barnacles, the small invertebrates that cling to ships or marine life. Darwin spent 8 years, from 1846 to 1854, carefully cataloguing the barnacles' tiny features, comparing and contrasting them with those of other known invertebrates. It was tedious work, leading Darwin to write, "I hate a Barnacle as no man ever did before." Yet the work proved valuable: it reinforced his idea that a great deal of variation exists in

nature–barnacles are nothing if not diverse– and it provided ample evidence of descent with modification since the creatures clearly shared adaptations with other invertebrates (INFOGRAPHIC M5.3).

In the summer of 1858, Darwin was hard at work on his "species" book when he received a letter from a young naturalist with whom he had a casual acquaintance, a collector named Alfred Russel Wallace who made a living selling rare butterflies and birds to other collectors and museums. The envelope was postmarked from an island in Indonesia. Inside was a 20-page manuscript describing the author's bold new idea about how species change over time, which he wanted Darwin to read and have published. Darwin, it seemed, had been scooped.

IN DARWIN'S SHADOW

Although we often credit Darwin with the discovery of natural selection, he was not alone in charting this intellectual territory. Another British naturalist was also hot on the trail. Like Darwin, Wallace was fascinated by natural history and had a thirst for adventure. In other ways, though, the two men couldn't have been more different. Darwin came from a wealthy family and had received a prestigious Cambridge education. He was greeted as a minor celebrity when he returned from his trip around the world and was accepted into the scientific establishment. Wallace, on the other hand, was a man of more humble origins, for whom nothing in life had come easily.

The eighth of nine children, Wallace could not afford a university education. He attended night school and supported himself as a builder and railroad surveyor. His budding fascination with natural history, though, led him to read widely. Like Darwin, he read Lyell's work on geology, Malthus's work on human population, and Chambers's *Vestiges*. And, of course, he devoured Darwin's travel account, *The Voyage of the Beagle*.

In 1848, having scrimped and saved, the 25-year-old Wallace set sail for Brazil, to the mouth of the Amazon River. There he hoped to earn his reputation as a respectable scientist by understanding the origin of species. Exploring the rain forest of the Amazon, Wallace was struck by the distribution of distinct yet similar-looking (what he called "closely allied") species, which were often separated by a geographic barrier such as a canyon or river. For example, he noted that different species of sloth monkey were found on different banks of the Amazon River. Over the course of his 4-year trip, Wallace scoured the Amazon and collected thousands of specimens.

Wallace was on his way home to London with his specimens in 1852 when disaster struck: his ship caught fire and sank. Wallace survived, but he lost everything—his notes, sketches, journals, and all his specimens. In spite of this catastrophe, Wallace was undeterred. Less than 2 years later, he was off on another collecting expedition, this time to the Malay archipelago (what is now Singapore, Malaysia, and Indonesia).

Wallace's first paper, "On the Law Which Has Regulated the Introduction of New Species," was published in September 1855. Based on his island work, it focused on the similar geographical distribution of "closely allied" species. For example, "the Galápagos Islands," he wrote "contain little groups of plants and animals peculiar to themselves, but most nearly allied to those of South America." From these observations, Wallace deduced this law, as he called it: "Every species has come into existence coincident both in space and time with a pre-existing closely allied species."

Wallace's article was groundbreaking, foreshadowing Darwin in a number of ways, but it lacked an explanation—a mechanism—of exactly how one species might have evolved from another.

Wallace continued his travels, but in early 1858 disaster struck again: he contracted malaria. Confined to bed, he let his mind wander. He thought about what Malthus had written about disease and how it kept human populations in check. How might these forces of disease and death, multiplied over time, influence animal populations, he wondered? Then came the flash of insight—like "friction upon the specially-prepared match," he recalled: in every generation, weaker individuals will die while those with the fittest variations will survive and reproduce; as a result species will change and adapt to their surroundings, eventually forming new species. Wallace had worked out the mechanism for evolution that was missing from his earlier work. He quickly wrote out his idea and sent it to the one naturalist he thought might be able to appreciate it. This was the 20-page manuscript that arrived on Darwin's doorstep on June 18, 1858 (**INFOGRAPHIC M5.4**).

Darwin was stunned. For 20 years he had been working diligently on the same idea and now it seemed someone else might get credit for it. "All my originality will be smashed," he wailed to his friend Lyell. Recognizing Darwin's predicament, Lyell and other colleagues devised a plan that would clearly establish Darwin's intellectual precedence: they would arrange to have papers by both men presented at a meeting of the Linnaean Society in London. The meeting took place on July 1, 1858. The papers were dutifully read, but there was no discussion or fanfare. In fact, neither of the authors was even present: Wallace was still traveling in Malaysia and Darwin was mourning the recent death of his young son and too distraught to attend.

The scientific meeting secured Darwin's reputation, but still he was unsettled. Wallace's communication had lit a fire under his feet.

INFOGRAPHIC M5.4 THE EVOLUTION OF WALLACE'S THOUGHT

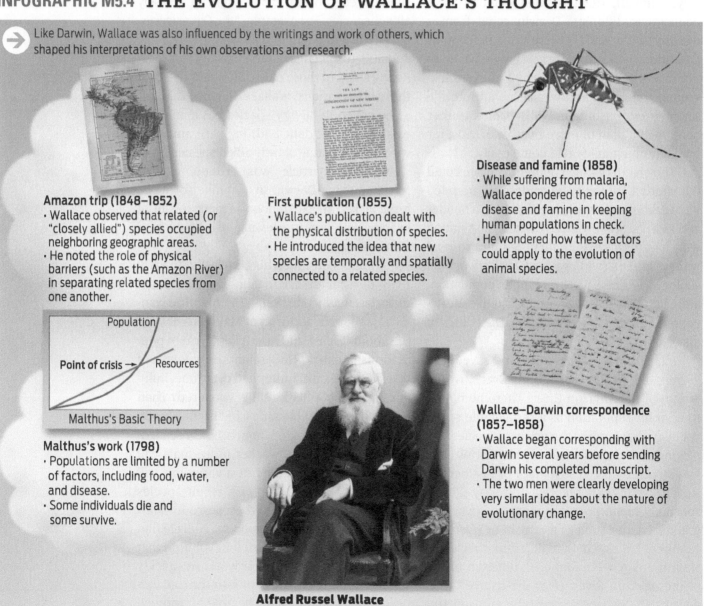

Like Darwin, Wallace was also influenced by the writings and work of others, which shaped his interpretations of his own observations and research.

Amazon trip (1848–1852)
· Wallace observed that related (or "closely allied") species occupied neighboring geographic areas.
· He noted the role of physical barriers (such as the Amazon River) in separating related species from one another.

First publication (1855)
· Wallace's publication dealt with the physical distribution of species.
· He introduced the idea that new species are temporally and spatially connected to a related species.

Disease and famine (1858)
· While suffering from malaria, Wallace pondered the role of disease and famine in keeping human populations in check.
· He wondered how these factors could apply to the evolution of animal species.

Malthus's work (1798)
· Populations are limited by a number of factors, including food, water, and disease.
· Some individuals die and some survive.

Population
Point of crisis → Resources
Malthus's Basic Theory

Wallace–Darwin correspondence (185?–1858)
· Wallace began corresponding with Darwin several years before sending Darwin his completed manuscript.
· The two men were clearly developing very similar ideas about the nature of evolutionary change.

Alfred Russel Wallace
(January 8, 1823–November 7, 1913)

He needed to finish his book. That work, *On the Origin of Species by Means of Natural Selection*, was published in November 1859. It would become one of the most famous books of all time, going through six editions by 1872.

Although it may seem that Wallace was cheated of his rightful recognition as a discoverer of evolution by natural selection, he was never bitter. On the contrary, he was delighted when he heard about his copublication with Darwin. He fully accepted that Darwin had formulated a more complete theory of natural selection before he did, and there is no trace of resentment in his later writings. In fact, Wallace titled his major work *Darwinism*, in recognition of the other man's intellectual influence.

After the presentation of 1858, Wallace stayed in the Malay archipelago for 4 more years, systematically recording its fauna and flora and securing his reputation as both the greatest living authority on the region and an expert on speciation. In fact, Wallace is responsible for our modern-day definition of "species." In work on butterflies, he defined "species" as groups of individuals capable of interbreeding with other members of the group but not with individuals from outside the group. This idea–known today as the biological species concept–remains one of the most important in evolutionary theory. ■

MORE TO EXPLORE

▸ Darwin, C. (1909 [1839]) *The Voyage of the Beagle*. New York: P. F. Collier and Son http://books.google.com/books?id=MDILAAAAIAAJ&q

▸ Beagle Voyage, Natural History Museum http://is.gd/RRNI9T

▸ Browne, J. (2008) *Darwin's Origin of Species*. New York: Grove Press.

▸ Wallace, A. R. (1869) *The Malay Archipelago*. London: Macmillan http://ebooks.adelaide.edu.au/w/wallace/alfred_russel/malay/

▸ Secord, J. (2003) *Victorian Sensation: The Extraordinary Publication, Reception, and Secret Authorship of Vestiges of the Natural History of Creation*. Chicago: University of Chicago Press.

MILESTONES IN BIOLOGY 5 Test Your Knowledge

1 What did the discovery of a fossil sloth in a sea cliff on the coast of Argentina suggest to Darwin?

2 Why was Thomas Malthus's book critical to Darwin's thinking about descent with modification?

3 How did Wallace use Thomas Malthus's book to inform his ideas about species?

4 What did the field experiences Darwin and Wallace had in observing the natural world at first hand add to their understanding of evolution that perhaps reading and thinking alone couldn't provide?

URBAN EVOLUTION

How cities are altering the fate of species

O N A SOGGY DAY IN 2012, BIOLOGISTS JASON MUNSHI-SOUTH and Stephen Harris traipse through the underbrush of Highbridge Park in the Washington Heights section of Manhattan, some 8 miles north of Times Square. They brush back a swatch of green to reveal a small, shoebox-size trap, with one unhappy camper inside: a tiny white-footed mouse. This diminutive rodent, only about 2 inches long, is one of a few urban species that scientists have begun to look at more closely for answers to questions about evolution.

Studying evolution in Manhattan–arguably the most unnatural place on the planet–might seem an odd choice. But Munshi-South and Harris belong to a new breed of biologist, one fascinated by the nature right under our noses. From rodents in parks to ants on median strips to the cockroaches and bedbugs inside buildings–even the bacteria brewing inside our belly buttons–no location is too mundane for this scientific crew.

What's the advantage of staying local? Besides being quite convenient–Munshi-South and Harris work at Baruch College in midtown Manhattan–

▸▸ **DRIVING QUESTIONS**

1. What is a gene pool (and can you swim in it)?

2. How do different evolutionary mechanisms influence the composition of a gene pool?

3. How does the gene pool of an evolving population compare to the gene pool of a nonevolving population?

4. How do new species arise, and how can we recognize them?

▶**POPULATION GENETICS**
The study of the genetic makeup of populations and how the genetic composition of a population changes.

cities like New York offer some unique opportunities for the evolutionary biologist.

"There isn't really anywhere on the planet that isn't impacted by human activity now," says Munshi-South. "If we want to understand how species are actually evolving now, we need to understand the inputs from human activity."

And nowhere is that input more evident than in the Big Apple. With a population of 8 million people packed into just 300 square miles of land, New York City is the largest and most densely populated city in the United States. All those restless urbanites have left a profound mark on the landscape, changing it in both dramatic and subtle ways. Just look at Manhattan, New York City's most densely populated borough: Once an island of thick forest, Manhattan is now a sliver of concrete, interspersed with oases of green. The skyscrapers, the bridges, the subway system–not to mention the bars and coffee shops–have all transformed a once wild place beyond recognition. In the process, says Munshi-South, the city's wildlife has been subject to "a grand evolutionary experiment."

Biologist Jason Munshi-South lays a trap for white-footed mice.

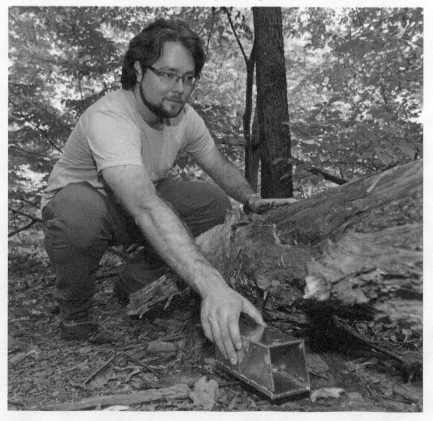

SEX AND THE CITY

To study evolution, Munshi-South has traveled to some pretty farflung locations: Southeast Asia, to study how the mating behavior of small mammals is affected by logging operations, and Africa, to research how the migration patterns and stress levels of forest elephants are affected by the petroleum operations there.

Mice in Manhattan might seem a far cry from that earlier research, but it's really not, he says. "Urbanism is basically one of these large-scale dramatic transformations of landscapes," he says. "And maybe one of the most complete transformations from a seminatural state to a use that's dominated by human activity." The common thread to this work, he says, is understanding how animals cope with a rapidly changing environment.

Urbanization isn't the only human-caused environmental change that animals face, of course–climate change is another big one–but it's one that has taken on pressing urgency in recent years.

"Already 50% of people live in cities," says Munshi-South. "It's going to be 60% in like 15 years, and it's just going to keep going up and up." In fact, demographers predict that the human population will reach 9 billion in 2050, and that by then 70% of us–roughly 6 billion people–will live in cities. All those urban dwellers represent a significant evolutionary force to be reckoned with. So it's important to understand how our fellow animals are adapting–or not adapting–to city life.

If you're an evolutionary biologist, and you want to understand how a group of organisms is coping with environmental changes, you need to know something about their underlying genetics–and not just the genetics of individuals, but of the population as a whole. For that, you need the tools of **population genetics.** Population genetics allows scientists to understand the nature of evolutionary change as it is reflected in the genes of a population. Essentially, it's a way to take stock of who's reproducing, who isn't, and the consequences for the population as a whole.

From a population genetics perspective, each distinct population of organisms–whether

INFOGRAPHIC 15.1 POPULATION GENETICS

 Population geneticists study the gene pools of populations. If a gene pool changes (that is, if the allele frequencies have changed) over the course of generations, then evolution has occurred.

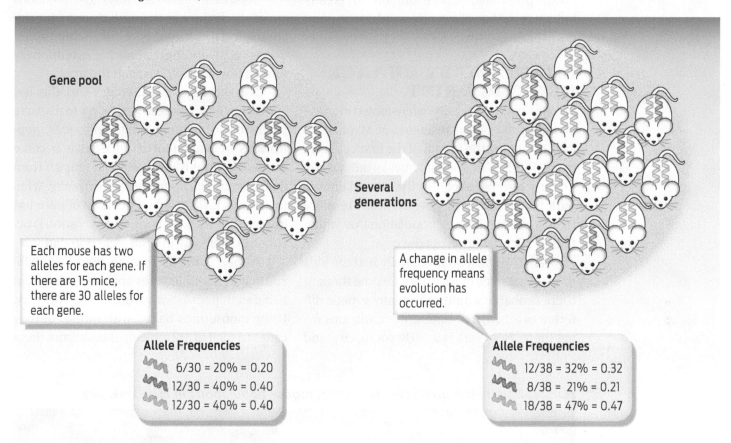

Gene pool

Each mouse has two alleles for each gene. If there are 15 mice, there are 30 alleles for each gene.

Several generations

A change in allele frequency means evolution has occurred.

Allele Frequencies

6/30 = 20% = 0.20
12/30 = 40% = 0.40
12/30 = 40% = 0.40

Allele Frequencies

12/38 = 32% = 0.32
8/38 = 21% = 0.21
18/38 = 47% = 0.47

mice in Manhattan or elephants in Africa—has its own particular collection of alleles, which together constitute its **gene pool.** Within the gene pool, each allele is present in a certain proportion, or **allele frequency,** relative to the total number of alleles for that gene in the population. For example, 50 alleles out of a total of 1,000 alleles would be an allele frequency of 0.05. Over time, several forces can change the frequency of alleles—that is, how common they are in the population. When the frequency of alleles changes over time, a population evolves. Recall from Chapter 14 that this is the definition of evolution (**INFOGRAPHIC 15.1**).

Evolutionary changes in a gene pool can have lasting consequences for a population. They can, for example, result in the population becoming more adapted to its environment—think of the antibiotic-resistant bacteria we met in Chapter 14. The evolution-

ary mechanism that results in adaptation is natural selection.

But natural selection isn't the only mechanism of evolution. Mutation, which introduces new alleles into a population (see Chapter 10), is another important evolutionary mechanism. Because it is rare, mutation by itself does not dramatically change allele frequencies. But that doesn't mean it isn't important—after all, mutation is the source of variation in a population, on which natural selection acts. However, mutation is a fundamentally random process that does not by itself lead to a population becoming more adapted to its environment. In other words, mutation is a type of **nonadaptive evolution.** The other types of nonadaptive evolution are genetic drift and gene flow.

Nonadaptive evolution isn't necessarily "bad," or maladaptive. If mutations didn't introduce variation into a population, there

▶**GENE POOL**
The total collection of alleles in a population.

▶**ALLELE FREQUENCY**
The relative proportion of an allele in a population.

▶**NONADAPTIVE EVOLUTION**
Any change in allele frequency that does not by itself lead a population to become more adapted to its environment; the causes of nonadaptive evolution are mutation, genetic drift, and gene flow.

would be no evolution at all. And many non-adaptive changes in allele frequency can be considered "neutral"–neither "good" nor "bad." But nonadaptive evolution can greatly influence the fate of a species, and so researchers are keen to study it.

CHANGING BY CHANCE: GENETIC DRIFT

Peromyscus leucopus–the white-footed mouse– is one of the oldest residents of Manhattan, long predating the arrival of the first colonists in the 1600s. The rodent squeaks out a living in the green spaces of the city–essentially any park that has canopy cover. There are more than a dozen distinct populations of white-footed mice living across the city.

When Munshi-South initially had the idea of studying mice in New York City, he thought there probably wouldn't be many genetic differences among the different populations living there; New York is a fairly young city, and

evolutionary change generally happens slowly. But the results of his study clearly showed that was not the case.

"It turned out that actually the populations are fairly distinct in the parks," says Munshi-South. And these differences have come about over a relatively short period of evolutionary time–a few hundred years at the most.

Over three centuries, as New York City has gone from a rich tapestry of forest to a sparse patchwork of fragmented green spaces, populations of white-footed mice have become trapped in little islands of green, cut off from their distant cousins in the rest of the city. What this means is that each population of mice has its own distinct gene pool, and evolution is occurring differently in each local population.

To peer into these gene pools, Munshi-South and his colleagues analyzed mice from 15 different populations around New York City. Using mousetraps baited with birdseed, they caught a total of 312 individual mice from these

Habitat fragmentation has led to distinct mouse populations in New York City.

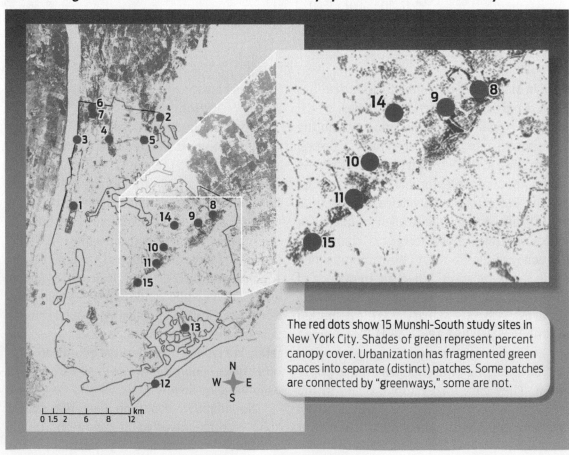

The red dots show 15 Munshi-South study sites in New York City. Shades of green represent percent canopy cover. Urbanization has fragmented green spaces into separate (distinct) patches. Some patches are connected by "greenways," some are not.

INFOGRAPHIC 15.2 GENE POOLS OF NEW YORK CITY MOUSE POPULATIONS

Researchers collected tail DNA from 312 mice at 15 locations in New York City. Once they analyzed each mouse's DNA, the researchers wanted to get a sense of how related the populations were. They assigned mice with similar genotypes particular colors and sorted all the mice by location. They found that mice within a population shared more alleles with one another than they did with mice from other populations.

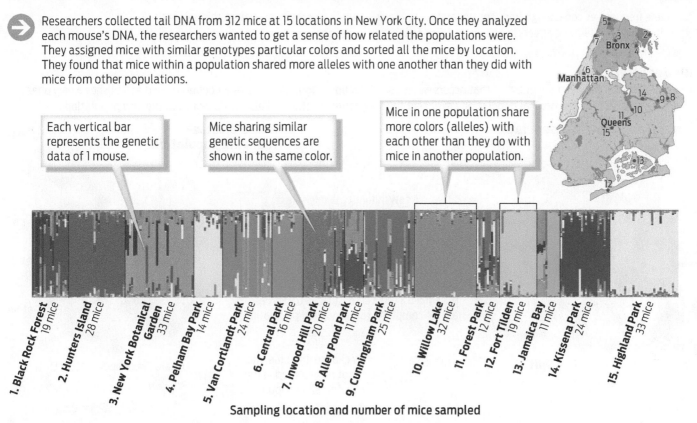

Each vertical bar represents the genetic data of 1 mouse.

Mice sharing similar genetic sequences are shown in the same color.

Mice in one population share more colors (alleles) with each other than they do with mice in another population.

Sampling location and number of mice sampled

1. Black Rock Forest 19 mice
2. Hunters Island 28 mice
3. New York Botanical Garden 33 mice
4. Pelham Bay Park 14 mice
5. Van Cortlandt Park 24 mice
6. Central Park 16 mice
7. Inwood Hill Park 20 mice
8. Alley Pond Park 11 mice
9. Cunningham Park 25 mice
10. Willow Lake 32 mice
11. Forest Park 12 mice
12. Fort Tilden 19 mice
13. Jamaica Bay 11 mice
14. Kissena Park 24 mice
15. Highland Park 33 mice

SOURCE: Munshi-South, J., and Kharchenko, K. (2010) Rapid, pervasive genetic differentiation of urban white-footed mouse (*Peromyscus leucopus*) populations in New York City. *Molecular Ecology* 19:4242 -4254.

populations. After catching each mouse, researchers cut off the tip of its tail and put it in ethanol; DNA from the piece of tail could later be extracted and sequenced at specific chromosome locations (see Chapter 7). The mice, which were not seriously harmed by this procedure, were then released back into the wild.

When DNA sequences at 18 different chromosome locations from all 312 mice were assessed, the results showed distinct clustering of alleles, with mice within one population tending to share more alleles with one another than with mice from other populations. In fact, says Munshi-South, you can accurately predict where a mouse is from just by looking at its DNA (**INFOGRAPHIC 15.2**).

How did these genetic differences among populations come about? One possibility is that each population of mice evolved by natural selection as a result of local differences in the environment—perhaps each different green

space has different predators or food sources, for example, which selected for individuals with different alleles. Given how close these green spaces are to one another, however—in some cases, less than a mile up the road—and given also how similar the environments are, this explanation isn't the most likely one. More likely, says Munshi-South, is **genetic drift.**

Genetic drift is a bit like rolling the evolutionary dice. By simple chance, some individuals survive and reproduce, and others do not. Those that pass on their genes aren't necessarily more fit or better adapted; they're just lucky—perhaps their nest or burrow wasn't swept away in a flash flood, for example.

Over time, genetic drift tends to decrease the genetic diversity of a population, as some alleles are lost completely and others sweep to 100% frequency. Genetic drift will have more dramatic effects in smaller populations than in larger ones, because in a population with few

▶**GENETIC DRIFT**
Random changes in the allele frequencies of a population between generations; genetic drift tends to have more dramatic effects in smaller populations than in larger ones.

INFOGRAPHIC 15.3 GENETIC DRIFT REDUCES GENETIC DIVERSITY

Allele frequencies can change from one generation to the next purely as a result of chance: this is genetic drift.
Drift has more dramatic effects in smaller populations than in larger ones.

Founder Effect

The founder effect is a type of genetic drift that occurs when a small group of "founders" leaves a population and establishes a new one.
If, by chance, alleles from the original population are absent from the founders, they will also be absent from the new population.

A large, diverse original population

A less-diverse founding population

A few individuals migrate

A few founders migrate to previously
uninhabited territory. By chance, these
mice have a reduced frequency for some
alleles compared to the original population.

Population
after several
generations

By chance, the blue allele
is not represented in this
new founding population.

Allele Frequencies before Drift

6/30 = 20% = 0.20
12/30 = 40% = 0.40
12/30 = 40% = 0.40

Allele Frequencies after Drift

10/22 = 45% = 0.45
12/22 = 55% = 0.55

Bottleneck Effect

Genetic bottlenecks occur when a population loses a large proportion of its members. If the original population is large, the reduced
population is likely to retain the same alleles present in the original population. But in a small starting population, bottlenecks are
more consequential: the loss of individuals is more likely to result in the loss of alleles from the population.

A more-diverse original population

A less-diverse bottleneck population

A bottleneck event, like rapid habitat loss,
eliminates a large percentage of the original
population. By chance, the surviving mice
have a different (reduced) allele frequency
than the original population.

Population
after several
generations

By chance, red alleles are
lost from the gene pool.

Allele Frequencies before Drift

6/30 = 20% = 0.20
12/30 = 40% = 0.40
12/30 = 40% = 0.40

Allele Frequencies after Drift

6/18 = 33% = 0.33
12/18 = 67% = 0.67

individuals any single individual that does not reproduce could spell the loss of alleles from the population. But all populations experience some measure of genetic drift, since chance is a fact of life.

Biologists refer to two general types of genetic drift: founder effects and bottlenecks. A **founder effect** occurs when a small group of settlers ("founders") splits off from a main population and establishes a new one. Because a founding population is by definition small, there is a good chance that the particular alleles it carries will not be fully representative of the population it left. Thus founder effects tend to reduce the genetic diversity of the new population.

Because white-footed mice are native to the New York region, it is unlikely that they have experienced founder effects; the populations that are here have all been here for a very long time.

More applicable to these mice is what's known as a **bottleneck.** When a population is cut down sharply—forced through a "bottleneck"—there's a good chance that the remaining population will possess a less-diverse gene pool. Bottlenecks can occur from natural causes—say, a flood that sweeps through the city, killing many individuals—or from human interference, such as the cutting down of a forest. Either way, a population that is forced through a genetic bottleneck usually contains a fraction of the original starting diversity in the population (**INFOGRAPHIC 15.3**).

As an example of a genetic bottleneck, consider the cheetah (*Acinonyx jubatus*), the fastest land animal. Cheetahs almost became extinct 10,000 years ago when harsh conditions of the last ice age claimed the lives of many large vertebrates on several continents. Ultimately, a few cheetahs survived and reproduced, but the more than 12,000 individuals alive today are now so genetically similar that skin grafts between unrelated individuals do not cause immune rejection; nearly all genetic diversity has been eliminated from the population.

Why does genetic diversity matter? You can think of a gene pool as a population's portfolio of financial assets. Having a diverse array of investments is a better strategy for long-term success than having all your money tied up in one kind of stock—especially if that stock loses value in changed economic times.

For example, say a population of mice suddenly finds itself in a more crowded and polluted environment than it did before. If the population carries with it a rich variety of alleles, then some of these alleles (ones associated with stronger immune systems, for example) may help that population survive and reproduce in the altered conditions. Individuals with these alleles will be more fit in this environment, and the population will adapt by natural selection. With less diversity in the population, the opportunity for adaptation will be more limited, and the population may shrink. Preserving genetic diversity is thus a prime concern of conservation biologists interested in protecting natural populations from extinction, especially as humans encroach on more and more wild habitat.

THE DAILY COMMUTE: GENE FLOW

Once a population has lost genetic diversity because of genetic drift, there are only two ways that genetic diversity can be reintroduced: (1) by mutation, which as we saw in Chapter 10 continually introduces new alleles into the population, and (2) by **gene flow,** in which alleles move between populations. Like genetic drift, gene flow is a type of nonadaptive evolution that does not lead to a population becoming more adapted to its environment. Unlike genetic drift, gene flow tends to increase the genetic diversity of a population (**INFOGRAPHIC 15.4**).

Munshi-South and his colleagues found that different New York City mice populations had different levels of gene flow, meaning that some populations exchanged alleles with other populations more than others. They wanted to understand why, so they built a statistical model and tested whether it could explain the data they collected at their study sites. Their idea was this: gene flow should be possible between populations where there is a corridor of tree canopy connecting them. So they

▸FOUNDER EFFECT
A type of genetic drift in which a small number of individuals leaves one population and establishes a new population; by chance, the newly established population may have lower genetic diversity than the original population.

▸BOTTLENECK EFFECT
A type of genetic drift that occurs when a population is suddenly reduced to a small number of individuals, and alleles are lost from the population as a result.

▸GENE FLOW
The movement of alleles from one population to another, which may increase the genetic diversity of a population.

Migration and interbreeding of individuals move alleles between populations. Populations that can interbreed with other populations have higher allele diversity than isolated populations.

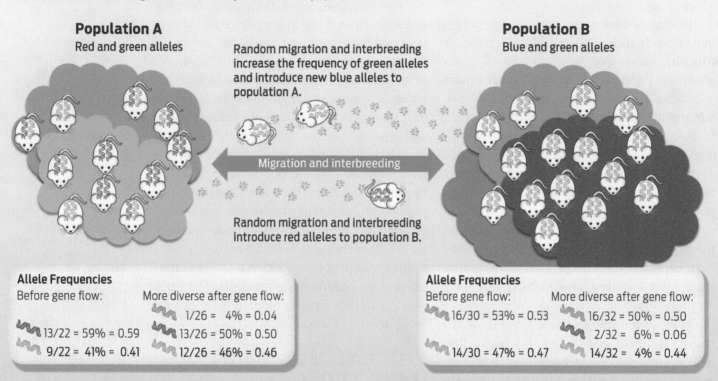

Population A
Red and green alleles

Random migration and interbreeding increase the frequency of green alleles and introduce new blue alleles to population A.

Migration and interbreeding

Random migration and interbreeding introduce red alleles to population B.

Population B
Blue and green alleles

Allele Frequencies

Before gene flow:

13/22 = 59% = 0.59
9/22 = 41% = 0.41

More diverse after gene flow:

1/26 = 4% = 0.04
13/26 = 50% = 0.50
12/26 = 46% = 0.46

Allele Frequencies

Before gene flow:

16/30 = 53% = 0.53
14/30 = 47% = 0.47

More diverse after gene flow:

16/32 = 50% = 0.50
2/32 = 6% = 0.06
14/32 = 4% = 0.44

divided the city into categories based on the percentage of "green" landscape (tree cover) or "gray" landscape (concrete), and then tested how closely their actual data matched the hypothetical results. It was a simple idea, but it worked surprisingly well. "That pretty much explained the variation we saw in gene flow and drift between the populations," he says. In fact, the "green corridor" model explained the amount of gene flow even more than the absolute distance between the populations.

Putting it all together, the picture looks like this: urbanization has led to habitat fragmentation that has isolated and bottlenecked dozens of mice populations, leaving each with a distinct gene pool. Populations that were completely isolated (e.g., sites 10 and 12 in Infographic 15.2) have continued to diverge by genetic drift, while populations connected to other populations by a green corridor (e.g., sites 8, 9, and 11 in Infographic 15.2) have re-

mained more similar to one another because of gene flow between them. Furthermore, mutation randomly adds new alleles to these populations, contributing to diversity between them. "So it's basically this interplay between genetic drift, gene flow, and mutation," says Munshi-South, that is influencing the level of genetic diversity in the populations.

One reason gene flow is important is that small, isolated populations can be damaged by lack of genetic diversity. Take the Florida panther (*Puma concolor*), for example. In the past, Florida panthers mated with puma populations from neighboring states where their ranges overlapped. This interbreeding–breeding among populations–fostered an exchange of alleles that continually enriched the local populations' genetic diversity. By the mid-20th century, however, hunting and development had squeezed the Florida panther population into an isolated region at the state's southernmost tip. By 1967, only 30 panthers remained, and the U.S. Fish and Wildlife Service listed them as endangered. By 1980, the panthers showed unmistakable signs of ill health–birth defects, low sperm count, missing testes, and bent tails–that resulted from **inbreeding,** mating between closely related members of a population.

Inbreeding can have dangerous consequences for a population. Because closely related individuals are more likely to share the same alleles, the chance of two recessive harmful alleles coming together during mating is high. When that happens, homozygous recessive genotypes are created, and previously hidden recessive alleles start to affect phenotypes in negative ways. This effect is called **inbreeding depression.**

To counteract this dangerous trend, the U.S. Fish and Wildlife Service, in 1995, brought in eight female pumas from Texas to mate with Florida's male panthers and thereby introduce genetic diversity. The program was successful: the hybrid kittens–30 in all–showed no signs of inbreeding depression. By 2007, more than 100 healthy panthers were roaming the swamps and grasslands of Florida.

For the time being, says Munshi-South, mice in Manhattan seem to be doing just fine in maintaining adequate genetic diversity. While the populations are clearly distinct from one an-other, each also has within it a fair amount of genetic diversity. This is probably because each population is still quite large and so the drift that is occurring has not dramatically reduced the number of alleles in each population. That fact, combined with mutation and occasional episodes of gene flow between some of the populations, has allowed these populations to maintain significant genetic diversity.

The same cannot be said of other species that Munshi-South and his colleagues are studying in New York City. The northern dusky salamander (*Desmognathus fuscus*), which makes its home in freshwater streams seeping out of the ground, was once extremely common throughout much of the city just 60 years ago. Today, it clings to a single hillside in northern Manhattan and to parts of Staten Island–a severe bottleneck. Not only that, but this single hillside has been bisected not once, but twice, by bridges connecting the boroughs of the Bronx and Manhattan, thus dividing this small, isolated population even further. Munshi-South and his colleagues have found that the populations living on either side of these bridges are genetically distinct–evidence that significant gene flow is not happening between them. Whether or not the dusky salamanders will be able to retain their tenuous hold along this hillside, only time will tell.

▶**INBREEDING**
Mating between closely related individuals. Inbreeding does not change the allele frequency within a population, but it does increase the proportion of homozygous individuals to heterozygotes.

▶**INBREEDING DEPRESSION**
The negative reproductive consequences for a population associated with having a high frequency of homozygous individuals possessing harmful recessive alleles.

Urbanization has altered the distribution of green spaces in New York and is changing the gene pools of nonhuman populations. Below: High Bridge connecting the Bronx and Manhattan. Right: High Line park.

CITY MOUSE, COUNTRY MOUSE

Fifteen miles north of Manhattan, on farms and apple orchards tucked away in the New York countryside, white-footed mice lead slower-paced lives, free from the stresses of urban living.

Have city mice and country mice evolved differently as a result of living in different environments? To answer this question, Munshi-South and a graduate student, Stephen Harris, are collecting DNA samples from country mice in order to compare the gene pools of city and country mice. "The sort of low-hanging fruit that we hope to find are the genes that are consistently different between urban and rural populations," says Munshi-South.

They've only just begun this work, but already they are finding some interesting results. A number of genes do seem to differ consistently between urban and rural populations. Perhaps not surprisingly, they have to do with things like immunity and response to stress. There are basically two types of stress that city mice face, says Munshi-South: stress associated with pollution, such as higher levels of heavy metals in the water and soil, and stress associated with living in crowded quarters and

INFOGRAPHIC 15.5 CITY MOUSE AND COUNTRY MOUSE

Country mice and urban mice differ most often in genes associated with immunity and response to stress. These genes may be responding to selective pressures in the urban environment, such as exposure to pollution and competition for food and mates.

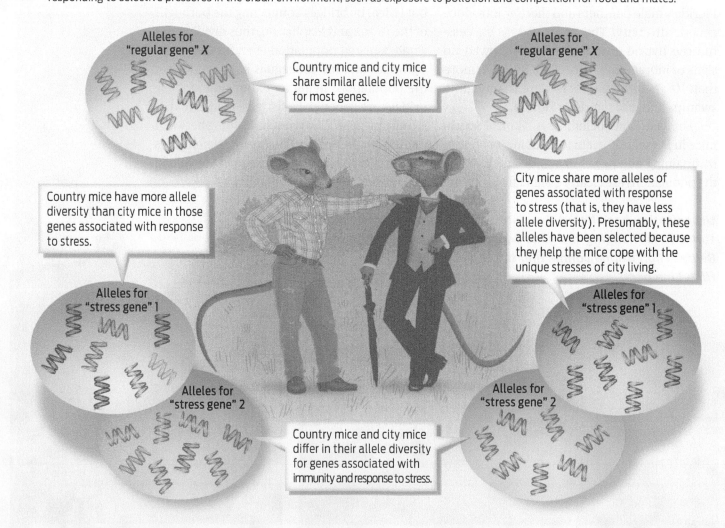

Alleles for "regular gene" X

Alleles for "regular gene" X

Country mice and city mice share similar allele diversity for most genes.

City mice share more alleles of genes associated with response to stress (that is, they have less allele diversity). Presumably, these alleles have been selected because they help the mice cope with the unique stresses of city living.

Country mice have more allele diversity than city mice in those genes associated with response to stress.

Alleles for "stress gene" 1

Alleles for "stress gene" 2

Alleles for "stress gene" 1

Alleles for "stress gene" 2

Country mice and city mice differ in their allele diversity for genes associated with immunity and response to stress.

TABLE 15.1 ADAPTIVE AND NONADAPTIVE MECHANISMS OF EVOLUTION

MECHANISM OF EVOLUTION	HOW ALLELE FREQUENCIES CHANGE	ADAPTIVE OR NONADAPTIVE?	HOW GENETIC DIVERSITY IS AFFECTED
Natural selection	Individuals with favorable alleles reproduce preferentially, increasing the frequency of these alleles.	Adaptive	Usually decreases—unfavorable alleles may be eliminated from the population
Mutation	New alleles are created randomly.	Nonadaptive	Increases—new alleles are introduced into the population
Genetic drift	Allele frequencies change due to chance events.	Nonadaptive	Usually decreases—alleles may be eliminated from the population
Gene flow	Alleles move from one population to another.	Nonadaptive	Increases—new alleles are added to the population

having to compete with lots of other mice for food and sex.

City mice have higher frequencies of certain "stress alleles" than do country mice, which makes perfect sense, he says. The city mice and the country mice have experienced different selective pressures, and these selective pressures have led to differences in the gene pools of these two large populations. Overall, city mice seem a bit better able to cope with these pressures than their country mouse cousins; they are New Yorkers born and bred (INFOGRAPHIC 15.5).

While city mice have experienced selection for urban traits, this does not mean they are all the same. Many genetic differences still exist among the local populations of urbanized mice. That's because mutation, drift, and gene flow continue to occur even as natural selection is happening. In fact, for most natural populations, all four mechanisms of evolution–selection, mutation, gene flow, and genetic drift–are continually operating at the same time, fostering unique evolutionary outcomes (TABLE 15.1).

How do biologists identify genes that have changed because of evolutionary mechanisms? To answer this question, it helps to know how alleles behave in a nonevolving population. Allele frequencies in a nonevolving population behave in a predictable way: by definition, they do not change over time. Furthermore, in a nonevolving population, genotype frequencies remain unchanged from one generation to the next, a condition known as **Hardy-Weinberg equilibrium.** By describing the default pattern of how alleles behave from one generation to the next when evolution is *not* occurring, the Hardy-Weinberg equilibrium provides a baseline from which to judge if a population *is* evolving. (Scientists call this a null hypothesis.)

The **Hardy-Weinberg equation** is a mathematical formula that relates allele frequencies to genotype frequencies in a population at Hardy-Weinberg equilibrium. If, using this equation, we find that the *actual* genotype frequencies are different from the *expected* frequencies, then we know that evolution has occurred (see UP CLOSE: CALCULATING HARDY-WEINBERG EQUILIBRIUM).

Hardy-Weinberg can also help researchers figure out, say, if drift or selection is operating in a given population. Let's say biologists obtain samples of DNA from a random sampling of mice in a population and they look at the

▶HARDY-WEINBERG EQUILIBRIUM
The principle that, in a nonevolving population, both allele and genotype frequencies remain constant from one generation to the next.

▶HARDY-WEINBERG EQUATION
A mathematical formula that calculates the frequency of genotypes and phenotypes one would expect to find in a nonevolving population.

UP CLOSE CALCULATING THE HARDY-WEINBERG EQUILIBRIUM

How do we know if a population is evolving? To find out, we use a mathematical formula called the Hardy-Weinberg equation, which calculates the frequency of genotypes you would expect to find in a nonevolving population from simple rules of probability. For a gene with two alleles, B and b, with allele frequencies p and q, this formula can be written as:

$$p^2 \quad + \quad 2pq \quad + \quad q^2 \quad = 1$$

Frequency of homozygotes BB Frequency of heterozygotes Bb Frequency of homozygotes bb

By definition, a population is not evolving (and is therefore in Hardy-Weinberg equilibrium) when it has stable allele frequencies and stable genotype frequencies from generation to generation. This can be achieved only when all five of the following conditions are met:

1. No mutation introducing new alleles into the population

2. No natural selection favoring some alleles over others

3. An infinitely large population size (and therefore no genetic drift)

4. No gene flow between populations

5. Random mating of individuals

In nature, no population can ever be in strict Hardy-Weinberg equilibrium, since it will never meet all five conditions. For example, because no real population is infinitely large, genetic drift will always occur. In other words, all natural populations are evolving. Nevertheless, by describing the pattern of genotypes in a nonevolving population, Hardy-Weinberg equilibrium provides a baseline from which to measure evolution.

To see how the Hardy-Weinberg equation can be used to detect evolutionary change, consider the following example. Say you have a population of mice with two alleles (B and b) and three possible phenotypes for fur color, gray (BB), brown (Bb), and white (bb). As every individual in the population has two alleles for the fur-color gene (one maternal and one paternal), there

are twice as many alleles as there are members of the population. So a population of 500 mice has 1,000 alleles of the gene for fur color.

Now let's say we sample the DNA of our mice population and determine that there are 800 B alleles in the population and 200 b alleles. We would then say that the frequency of the B allele is 0.8 (800/1,000) and the frequency of the b allele is 0.2 (200/1,000). Since there are only two alleles in the population, their combined frequencies will add up to 1. If we use p to denote the frequency of B and q to denote the frequency of b, then we can say that $p + q = 1$.

Suppose we want to use those allele frequencies to calculate the expected frequency of white-furred (bb) individuals in the population if the population is indeed in Hardy-Weinberg equilibrium. If the frequency of b in the population is q, then we know from the Hardy-Weinberg equation that the expected frequency of bb is $q^2 = (.2)(.2) = .04$, and that this frequency will remain constant over generations. Thus, in our population of mice, 4%, or 20 mice, would be expected to have white fur, if the population is in Hardy-Weinberg equilibrium. If we find out that the actual percentage of white mice in the population is more or less than this number, then we know that our population is evolving, and we can begin to investigate why.

The Hardy-Weinberg equation also has important applications in public health. It can be used, for example, to estimate the frequency of carriers (heterozygotes) of rare recessive diseases in a population (see Question 17 in Test Your Knowledge for an example).

The Hardy-Weinberg Equation:

In a nonevolving population mice will be mating randomly, with no mutation, natural selection, or genetic drift to change allele frequencies.

Male parent sperm

pq

Female parent eggs

pq

p = frequency of
q = frequency of

p q

p p^2 pq

q pq q^2

When we add up the genotype frequencies in this nonevolving population, we get the Hardy-Weinberg equation:
$p^2 + 2pq + q^2 = 1$

Starting population

DNA testing of 500 mice reveals the number of B and b alleles:

800 B alleles
200 b alleles

From this, allele frequencies can be calculated:

Original allele frequencies:
p = frequency of B 800/1000 = 80% = **0.80**
q = frequency of b 200/1000 = 20% = **0.20**
$p + q = 1$
$0.8 + 0.2 = 1$

Frequency of genotypes in a nonevolving population:
p^2 = frequency of B B genotype = 0.8×0.8 = **0.64**
$2pq$ = frequency of B b genotype = $2 \times 0.8 \times 0.2$ = **0.32**
q^2 = frequency of b b genotype = 0.2×0.2 = **0.04**
$p^2 + 2pq + q^2 = 1$
$0.64 + 0.32 + 0.04 = 1$

frequencies of genotypes at 10 different regions of DNA. Nine of those regions have genotype frequencies predicted by the Hardy-Weinberg equation, but one does not—it is far from Hardy-Weinberg equilibrium. Researchers then know that something interesting is happening at that one DNA location—some force of evolution is acting. In fact, this is how Munshi-South and his colleagues identified the candidate genes to compare between city and country mice.

"You can use certain deviations from Hardy-Weinberg equilibrium to find parts of the genome that are under selection," he says. "So, if they strongly deviate from Hardy-Weinberg, whereas the rest of the genome roughly fits it, those outliers are likely to have something interesting going on, like natural selection."

By understanding how city life has changed mice genetically, researchers will have a better understanding of how human activity is influencing mice evolution. That might not sound like a hugely important goal, especially if you're not a fan of mice. But there are larger

> # Manhattan offers a preview of what human activity will do to many other species in the coming years.

lessons to take away. According to Munshi-South, Manhattan offers a preview of what human activity will do to many other species in the coming years.

"Even just global warming alone is going to drive a lot of these processes in the future," he says. Indeed, some urban animals are already feeling the heat.

After several generations...

Nonevolving population
If no evolution is happening, genotype and allele frequencies will stay the same over several generations.

= 10 mice

Evolving population
When the population has a change in its genotype and allele frequencies, it is evolving.

= 10 mice

Observed genotype frequencies:
320 gray mice: BB (p^2) = 320/500 = **0.64**
160 brown mice: Bb ($2pq$) = 160/500 = **0.32**
20 white mice: bb (q^2) = 20/500 = **0.04**

Observed allele frequencies:
Frequency B allele (p) = 800/1000 = **0.8**
Frequency b allele (q) = 200/1000 = **0.2**

Compare with starting population frequencies:
Genotype and allele frequencies in this population have not changed. Therefore, this population is not evolving. When parent and offspring genotype and allele frequencies remain constant, these populations are said to be in Hardy-Weinberg equilibrium.

Observed genotype frequencies:
190 gray mice: BB (p^2) = 190/500 = **0.38**
250 brown mice: Bb ($2pq$) = 250/500 = **0.50**
60 white mice: bb (q^2) = 60/500 = **0.12**

Observed allele frequencies:
Frequency of B allele (p) = 630/1000 = **0.63**
Frequency of b allele (q) = 370/1000 = **0.37**

Compare with starting population frequencies:
Both allele and genotype frequencies have changed. Therefore this population is evolving.

BIODIVERSITY ON BROADWAY

A few years ago, in 2006, just up the road from where Munshi-South works, researchers at Columbia University decided to look at what is perhaps the most urban of all green spaces: median strips. Their idea was to look at these median strips as "islands" of wilderness within the city and to explore the diversity of ant species there. Were any species new to Manhattan, for example, or previously unidentified?

The researchers collected a total of 6,619 individual ants from 44 sites along three different avenues in the New York—Broadway, Park Avenue, and the West Side Highway. Amidst these crawling masses, they identified 13 different species of ant, including both native and introduced species. The most common ant species, found on nearly all medians, was an introduced species known as the pavement ant (*Tetramorium caespitum*), which hails originally from Europe, but ants from as far away as Japan were found. Somehow, despite the close quarters, all these different species are able to coexist alongside one another while maintaining their distinct lifestyles. As the researchers noted in their study, "Manhattan is, if not quite a melting pot of ant species, at least a mixing bowl."

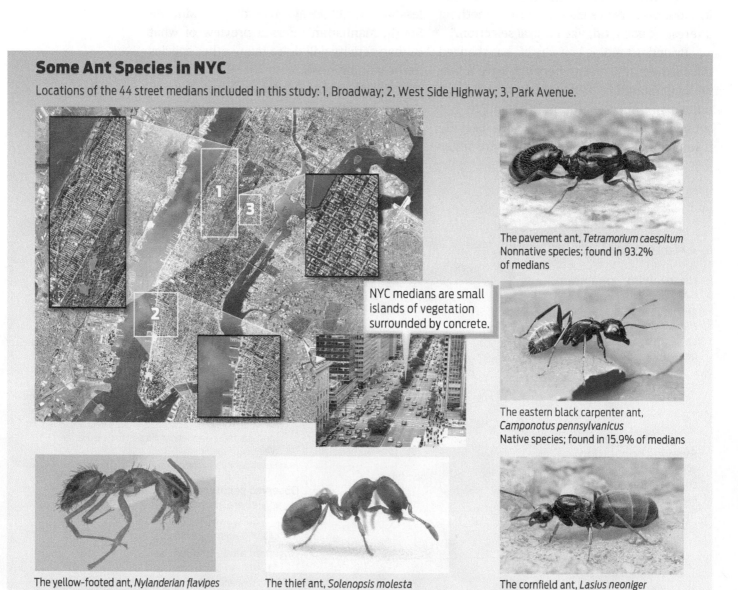

Some Ant Species in NYC

Locations of the 44 street medians included in this study: 1, Broadway; 2, West Side Highway; 3, Park Avenue.

NYC medians are small islands of vegetation surrounded by concrete.

The pavement ant, *Tetramorium caespitum*
Nonnative species; found in 93.2% of medians

The eastern black carpenter ant, *Camponotus pennsylvanicus*
Native species; found in 15.9% of medians

The yellow-footed ant, *Nylanderian flavipes*
Nonnative species; found in 52.3% of medians

The thief ant, *Solenopsis molesta*
Native species; found in 63.6% of medians

The cornfield ant, *Lasius neoniger*
Native species; found in 11.4% of medians

> **"** *Basically, we're mapping what's actually out there. But instead of going off to the rain forest we're looking in people's backyards.* **"**
>
> — ANDREA LUCKY

Rob Dunn, an entomologist at North Carolina State University, was one of the investigators in this ant study. From studying many urban ant populations, he has found evidence that that the ant species that seem to do best in urban settings are ones that hail from warmer climates, like the Southwest. Because of all the heat-absorbing concrete, cities on average tend to be warmer than rural and suburban areas. If cities are any indication, global climate change is going to alter the ranges and diversity of ant species in an area, with unpredictable consequences.

It was while studying ants in New York City that Dunn first had the idea for a larger project: a nationwide study of ants, using ordinary citizens as research assistants to collect specimens. Now in its second year, the project is called, appropriately, School of Ants.

Anyone can enroll. All you need are index cards, ziplock bags, and a pecan sandy–the bait of choice for seasoned ant collectors. Once the ants are collected, they're put in the freezer for a night and then shipped off to a lab at the North Carolina State University, where they will be studied and catalogued.

The aim of School of Ants is to create a map of ant species across the country, especially in urban areas, where ants have been little studied. "Basically, we're mapping what's actually out there," says Andrea Lucky, an entomologist at the University of Florida who, along with Dunn, now runs the project. "But instead of going off to the rain forest we're looking in people's backyards."

The most exciting part of the project, says Lucky, is discovering new ant species–as one

School of Ant researcher did recently in Durham, North Carolina. The new species, which has extremely tiny eyes, behaves as a social parasite, taking over the worker ants of other colonies and making them into "slaves." The unusual species is currently in the process of being studied and named.

How did they know the ant species was new? Basically, says Lucky, you have to know a lot about what is already known. You look in books, in museum collections, and in research papers to see if your ant has been previously described. Then you compare your specimen to the existing ones. "You end up looking at a lot of ants," Lucky says.

Mostly, you're considering physical appearance–shape of the head, number of spines and segments of antennae, even how furry the creature is–but DNA evidence is important for recognizing more subtle differences. The idea is that these physical and genetic differences reflect adaptations to different environments that in turn prevent the species from interbreeding. In essence, that is what defines a species.

The term species comes from the Latin word for "kind" or "appearance." According to the **biological species concept**–which states the more formal definition that biologists use– a species is a population of individuals whose members can interbreed and produce fertile offspring.

Members of different species cannot mate and produce fertile offspring with each other because their populations are reproductively isolated. Such **reproductive isolation** can be caused by a number of factors. For example,

▶BIOLOGICAL SPECIES CONCEPT
The definition of a species as a population whose members can interbreed to produce fertile offspring.

▶REPRODUCTIVE ISOLATION
Mechanisms that prevent mating (and therefore gene flow) between members of different species.

15

UNIT 3 • HOW DOES LIFE CHANGE OVER TIME? EVOLUTION AND DIVERSITY

INFOGRAPHIC 15.6 SPECIES ARE REPRODUCTIVELY ISOLATED

→ Species maintain their reproductive isolation in a variety of ways.

Ecological Isolation
Different environments.
The Arctic Fox and the Desert Fox live in such different places, they never encounter each other.

Temporal Isolation
Mating behavior or fertility at different times.
The Leopard Frog mates in early spring and the Bullfrog mates in early summer.

Behavioral Isolation
Different mating activities.
The Prairie Chicken is not attracted to the mating display of the Ring-Necked Pheasant.

Mechanical Isolation
Mating organs are incompatible.
Plants pollinated by the hummingbird do not receive pollen from plants pollinated by the Black Bee.

Gametic Isolation
Gametes cannot unite.
The gametes from a dog and a cat cannot unite to form a zygote.

Hybrid Inviability
Gametes unite but viable offspring cannot form.
The goat and sheep can mate, but the zygote formed does not survive.

Hybrid Infertility
Viable hybrid offspring cannot reproduce.
Zebras and horses are different species because their hybrid offspring, zebroids, cannot produce offspring of their own.

the two species may have different mating times, locations, or mating rituals—so, like ships passing in the night, they may never have the opportunity to meet. This is true of many ant species, for example, which breed at different times of year. Or, two species may be able to mate—as zebras and horses can—but the hybrid offspring they produce is infertile (**INFOGRAPHIC 15.6**).

Reproductive isolation explains why species remain separate—as do all the species of ants that share a median strip—but how did the species form in the first place? New species form when a strong barrier to gene flow occurs between populations. That barrier could be physical—like a road or river that divides a forest in two—or climatic, like the different temperatures that occur at different elevations on a mountainside. Once this barrier forms, the separated gene pools will evolve independently by the mechanisms we have already encountered: mutation, genetic drift, and selection. Eventually, if enough genetic changes accumulate between populations of the same species to make them reproductively isolated, the two populations may diverge into separate species, a process called **speciation.**

Speciation is happening all the time in nature, but it can be hard to see because it occurs so slowly. It generally takes many thousands of years for species to diverge. We see the *results* of speciation whenever we look at the diversity of nature—there are more than 12,000 known ant species, for example—but observing speciation as it happens is much harder. The classic, and still the best, example of speciation comes from Darwin himself: the finches he observed on the Galápagos Islands, near Ecuador, while traveling aboard the *Beagle* (see **Milestones in Biology: Adventures in Evolution**). The original finch species came from the mainland of Ecuador. Descendants of this mainland population then island-hopped through the archipelago, creating founder populations on each island that then adapted locally, by natural selection, to the unique environments they encountered. From one ancestral species of finch have

INFOGRAPHIC 15.7 SPECIATION: HOW ONE SPECIES CAN BECOME MANY

The Galápagos archipelago is a series of islands off the coast of South America. Finches first came to the Galápagos from a population on the mainland of South America. As they spread from island to island, they encountered different environments, including available food sources, which influenced bill size and shape in each new island population. As separated finch populations evolved in different food environments, they diverged from their ancestral population to have smaller, pointed bills for insects, longer bills for cactus fruit and flowers, or thick, strong bills for hard seeds. In addition, they evolved such that the separated populations could not interbreed. At least 13 finch species have diverged from the original South American species.

Insect eater–thin pointed beak
Mangrove Finch
Camarhynchus heliobates

Seed eater–strong, cone-shaped beak
Large Ground Finch
Geospiza magnirostris

Flower nectar eater–long, tubular beak
Cactus Ground Finch
Geospiza scandens

South American ancestor
Blue-black Grassquit
Volatina jacarina

PACIFIC OCEAN

Pinta

Marchena

Tower (Genovesa)

Santiago (James or San Salvador)

Is. S. Chino

Jervis (Rabida)

North Seymour
Baltra

Fernandina

Pinzon

Santa Cruz (Indefatigable)

Isabela (Albermarle)

Santa Fe (Barrington)

San Cristobal (Chatham)

Tortuga

Floreana (Charles)

Hood (Española)

Ecuador

Galápagos Islands

PACIFIC OCEAN

---

emerged 13 different species of finch, each with a distinctive beak size and shape adapted to a unique environment **(INFOGRAPHIC 15.7)**.

Ants on median strips resemble Darwin's finches in some respects: they are geographically separated and do not typically interbreed. Given enough time, ants from different median strips might very well form different species, much like finches on different islands. The same goes for New York City's mice populations: they, too, might one day form new species, provided the populations remain isolated and continue to diverge genetically. But the timescale involved makes it hard to predict. It took thousands of years for Darwin's finches to

evolve into different species; the median strips and parks themselves may be long gone before that happens for ants and mice.

EMPIRE STATE OF MIND

If cities are the new evolutionary laboratory, then researchers Munshi-South, Dunn, and Lucky are like modern-day versions of Charles Darwin—mapping uncharted biological territory, albeit in places found right under our noses.

"I think a lot people are fairly shocked to find out that everything in North America isn't already named and catalogued and placed a drawer somewhere," says Lucky. "There are multiple levels of discovery just waiting to

▶**SPECIATION**
The genetic divergence of populations, leading over time to reproductive isolation and the formation of new species.

happen." And as she's shown, you don't have to be a professional scientist to contribute to the process of discovery.

Beyond basic curiosity, researchers have an additional reason for studying evolution in urban settings—what Dunn calls the "pigeon paradox." If people are going to care about nature enough to protect it, he says, they're going to have to connect with the nature that's around them. For an increasing number of people in the world, this is urban nature: pigeons, rodents, and ants. While urban species may seem less interesting or exotic than ones in tropical rain forests, they are the ones that more and more people are actually likely to encounter and the ones their actions affect on a daily basis.

As Munshi-South points out, cities are active contributors to the evolutionary process, shaping the wildlife around us in profound ways. Along with climate change, urbanism is one of the main ways that humans are directly altering the face of the planet and thereby shaping the fate of species. Scientists are only just beginning to understand how cities drive evolutionary change, but they will have ample opportunity for continued research. The trend of urbanism shows no signs of stopping.

Those curious about the future can catch a glimpse of what lies in store for us by studying those places that are already intensely urbanized, like New York City. In a way, say biologists, we are all New Yorkers now. ∎

CHAPTER 15 Summary

- From a genetic perspective, a population is identified by the particular collection of alleles in its gene pool.

- Genetic diversity, as reflected by the number of different alleles in a population's gene pool, is important for the continued survival of populations, especially in the face of changing environments.

- Evolution is a change in allele frequencies in a population over time. Evolution can be adaptive or nonadaptive. Mutation, genetic drift, and gene flow are nonadaptive forms of evolution.

- The founder effect is a type of genetic drift in which a small number of individuals establishes a new population in a new location, with reduced genetic diversity as a possible result.

- The bottleneck effect is a type of genetic drift that occurs when the size of a population is reduced, often by a natural disaster, and the genetic diversity of the remaining population is reduced.

- Inbreeding of closely related individuals may occur in small, isolated populations, posing a threat to the health of a species.

- Gene flow is the movement of alleles between different populations of the same species, often resulting in increased genetic diversity of a population.

- Genetic diversity can be assessed by using DNA sequences to determine allele frequency.

- Hardy-Weinberg equilibrium describes the frequency of genotypes in a nonevolving population. The Hardy-Weinberg equation can be used to detect evolutionary change in a population.

- According to the biological species concept, a species is a population of individuals that can interbreed to produce fertile offspring.

- Speciation can occur when gene pools are separated, gene flow is restricted, and populations diverge genetically over time.

MORE TO EXPLORE

- Munshi-South, J. (2012) TED Talk: Evolution in a big city http://ed.ted.com/lessons/evolution-in-a-big-city

- School of Ants http://schoolofants.org/

- Munshi-South, J. (2012) Urban landscape genetics: canopy cover predicts gene flow between white-footed mouse (*Peromyscus leucopus*) populations in New York City. *Molecular Ecology*. 21:1360–1378.

- Munshi-South, J., and Kharchenko, K. (2010) Rapid, pervasive genetic differentiation of urban

white-footed mouse (*Peromyscus leucopus*) populations in New York City. *Molecular Ecology* 19: 4242–4254.

- Menke, S. B., et al. (2010) Urban areas may serve as habitat and corridors for dry-adapted, heat tolerant species; an example from ants. *Urban Ecosystems* 13(2):135–163.

- Dunn, R. R., et al. (2006) The pigeon paradox: dependence of global conservation on urban nature. *Conservation Biology* 20(6):1814–1816.

CHAPTER 15 Test Your Knowledge

DRIVING QUESTION 1

What is a gene pool (and can you swim in it)?

By answering the questions below and studying Infographics 15.1 and 15.2, you should be able to generate an answer for the broader Driving Question above.

KNOW IT

1 Genetic diversity is measured in terms of allele frequencies (the relative proportions of specific alleles in a gene pool). A population of 3,200 mice has 4,200 dominant *G* alleles and 2,200 recessive *g* alleles. What is the frequency of *g* alleles in the population?

2 Of the three populations described below, each of which has 1,000 members, which population has the highest genetic diversity? Note that only one gene is presented, and that this gene has three possible alleles: *A1, A2,* and *a*.

Population A: 70% have an *A1/A1* genotype, 25% have an *A1/A2* genotype, and 5% have an *A1/a* genotype.

Population B: 50% have an *A1/A1* genotype, 20% have an *A2/A2* genotype, 10% have an *A1/A2* genotype, 10% have an *A2/a* genotype and 10% have an *a/a* genotype.

Population C: 80% have an *A1/A1* genotype, and 20% have an *A1/a* genotype.

USE IT

3 A small population of 26 individuals has five alleles, *A* through *E*, for a particular gene. The *E* allele is represented only in one, homozygous individual:

Five individuals are *D/A* heterozygotes.
Five individuals are *A/A* homozygotes.
Five individuals are *A/B* heterozygotes.
Five individuals are *C/D* heterozygotes.
Five individuals are *C/C* homozygotes.
One individual is an *E/E* homozygote.

If five *A/E* heterozygotes migrate into the population, what will be the impact on the allele frequencies of each of the five alleles?

4 Some populations, for example cheetahs, have gene pools with very few different alleles. What approach(es) could be taken to try and introduce new alleles into these kinds of population?

5 From their gene pool and population size, which of the four populations in the accompanying table would you be most concerned about from a conservation perspective? Why would you be concerned?

POPULATION	NUMBER OF INDIVIDUALS	NUMBER OF ALLELES, GENE 1	NUMBER OF ALLELES, GENE 2	NUMBER OF ALLELES, GENE 3
1	50	1	7	5
2	1,000	1	5	7
3	50	3	2	2
4	1,000	1	1	2

6 The global human population continues to grow, and more people than ever are living in crowded cities. Given this situation, what selective pressures might the human population be currently facing or be expected to face in the near future?

DRIVING QUESTION 2

How do different evolutionary mechanisms influence the composition of a gene pool?

By answering the questions below and studying Infographics 15.3, 15.4, and 15.5 and Table 5.1, you should be able to generate an answer for the broader Driving Question above.

KNOW IT

7 Which of the following are examples of genetic drift?

a. founder effect

b. bottleneck effect

c. inbreeding

d. a and b

e. a, b, and c

8 A bottleneck is best described as

a. an expansion of a population from a small group of founders.

b. a small number of individuals leaving a population.

c. a reduction in the size of an original population followed by an expansion in size as the surviving members reproduce.

d. the mixing and mingling of alleles by mating between members of different populations.

e. an example of natural selection.

9 A population of ants on a median strip has 12 different alleles, *A* through *L*, of a particular gene. A drunk driver plows across the median strip, destroying most of the median strip and 90% of the ants. The surviving ants are all homozygous for allele *H*.

a. What is the impact of this event on the frequency of alleles *A* through *L*?

b. What type of event is this?

USE IT

10 Question 2 looked at the allele frequencies of three populations, A, B and C. From your answer to that question, which population would you predict to have the greatest chance of surviving an environmental change? Explain your answer.

11 In humans, founder effects may occur when a small group of founders immigrates to a new country, for example to establish a religious community. In this situation, why might the allele frequencies in suc-

ceeding generations remain similar to those of the founding population rather than gradually becoming more similar to the allele frequencies of the population of the country to which they immigrated?

12 Why is genetic drift considered to be a form of evolution? How does it differ from evolution by natural selection?

INTERPRETING DATA

13 The figure below shows a structure bar plot of moles from different parks in New York City. As in Infographic 15.2, each vertical bar represents genotypes from 18 genomic locations in one animal. The bars are color coded, with similar genotypes represented by the same color. From the data presented in the figure:

a. Are these three populations genetically isolated from one another? Explain your answer. What factors could explain their isolation, or lack thereof?

b. Is one of the populations potentially experiencing gene flow with another population? If so, which one, and how do you know?

Population A   Population B   Population C

DRIVING QUESTION 3

How does the gene pool of an evolving population compare to the gene pool of a nonevolving population?

By answering the questions below and studying Infographic 15.5 and Up Close: Calculating Hardy-Weinberg Equilibrium, you should be able to generate an answer for the broader Driving Question above.

KNOW IT

14 Which of the following statements is/are true about a nonevolving population?

a. Allele frequencies do not change over generations.

b. Genotype frequencies do not change over time.

c. Individuals choose mates with whom they share many alleles.

d. all of the above

e. a and b

15 A starting population of bacteria has two alleles of the *TUB* gene, *T* and *t*. The frequency of *T* is 0.8 and the frequency of *t* is 0.2. The local environment undergoes an elevated temperature for many generations of bacterial reproduction. After 50 generations of reproduction at the elevated temperature, the frequency of *T* is 0.4 and the frequency of *t* is 0.6. Has evolution occurred? Explain your answer.

16 Why is inbreeding detrimental to a population?

USE IT

17 Phenylketonuria (PKU) is a rare, recessive genetic condition that affects approximately 1 in 15,000 babies born in the United States. (You may have noticed on products that contain aspartame the statement "Phenylketonurics: contains phenylalanine," a warning for people with PKU that they should avoid consuming that product.) Calculate the expected frequency of carriers (that is, of heterozygotes) in the U.S. population, based on the information provided about rates of PKU among U.S. births, assuming that the population is in Hardy-Weinberg equilibrium for this gene.

18 Assume a population of 100 individuals. Five are homozygous dominant (*AA*), 80 are heterozygous (*Aa*), and 15 are homozygous recessive (*aa*) for the *A* gene. Determine *p* and *q* for this gene for this population. Now use those values for *p* and *q* and plug them into the Hardy-Weinberg equation. Is this population in Hardy-Weinberg equilibrium? Why or why not?

DRIVING QUESTION 4

How do new species arise, and how can we recognize them?

By answering the questions below and studying Infographics 15.7 and 15.8, you should be able to generate an answer for the broader Driving Question above.

KNOW IT

19 The biological species concept defines a species

a. on the basis of similar physical appearance.

b. on the basis of close genetic relationships.

c. on the basis of similar levels of genetic diversity.

d. on the basis of the ability to mate and produce fertile offspring.

e. on the basis of recognizing one another's mating behaviors.

20 How does geographic isolation contribute to speciation?

USE IT

21 Two populations of rodents have been physically separated by a large lake for many generations. The shore on one side of the lake is drier and has very different vegetation from that on the other side. The lake is drained by humans to irrigate crops, and now the rodent populations are reunited. How could you assess if they are still members of the same species?

22 If geographically dispersed groups of a given species all converge at a common location during breeding season, then return to their home sites to bear and rear their young, what might happen to the gene pools of the different groups over time?

MINI CASE

23 More than 50% of the global human population now lives in urban areas, and it is predicted that 70% will live in urban areas by 2050. Researchers have hypothesized that the emotional health of urbanites is influenced positively by interaction with nature. Given this information, and what you have read in this chapter, write a compelling paragraph on the need to conserve urban species and approaches to such conservation based on population genetics.

BRING IT HOME

24 The School of Ants is a citizen-scientist project to document the distribution and diversity of ants across the United States. Similarly, the Audubon Christmas Bird Count is a conservation-related project carried out by volunteer citizen scientists. The Audubon Society uses the data collected by these volunteers to evaluate the health of bird populations and make informed decisions about conservation. There are numerous other citizen-scientist projects (e.g., Project Squirrel and the Gravestone Project). Carry out an Internet search to find a citizen-scientist project that you find interesting. Now take the next step: enroll, collect some data, and contribute to science!

A Fish With FINGERS?

A transitional fossil fills a gap in our knowledge of evolution

FOR 5 YEARS, BIOLOGISTS NEIL SHUBIN AND TED DAESCHLER spent their summers trekking through one of the most desolate regions on Earth. They were fossil hunting on remote Ellesmere Island, in the Canadian Arctic, about 600 miles from the north pole. Even in summer, Ellesmere is a forbidding place: a windswept, frozen desert where sparse vegetation grows no more than a few inches tall, where sleet and snow fall in the middle of July, and where the sun never sets. Only a handful of wild animals survive here, but

▸▸ DRIVING QUESTIONS

1. How does the fossil record reveal information about evolutionary changes?
2. What features make *Tiktaalik* a transitional fossil, and what role do these types of fossil play in the fossil record?
3. What can anatomy and DNA reveal about evolution?

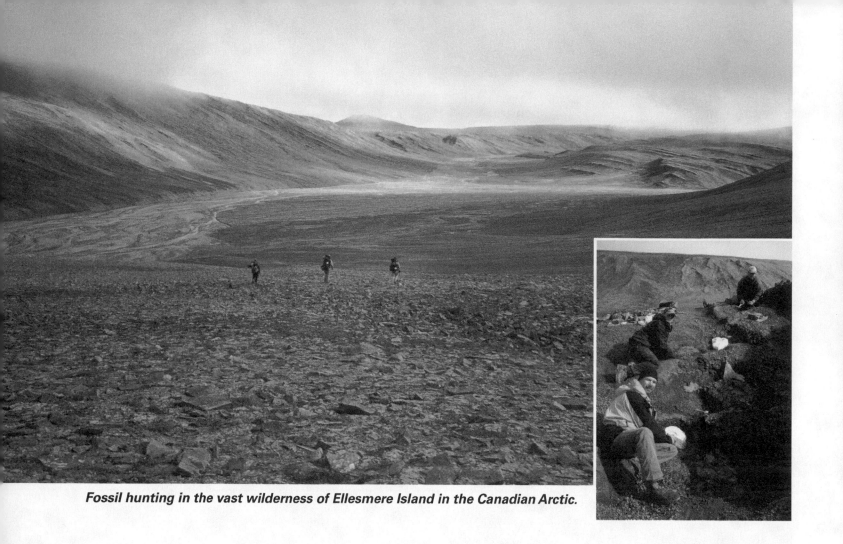

Fossil hunting in the vast wilderness of Ellesmere Island in the Canadian Arctic.

▶**VERTEBRATE**
An animal with a bony or cartilaginous backbone.

▶**DESCENT WITH MODIFICATION**
Darwin's term for evolution, combining the ideas that all living things are related and that organisms have changed over time.

▶**FOSSILS**
The preserved remains or impressions of once-living organisms.

those that do make for dangerous working conditions: hungry polar bears and charging herds of muskoxen are known hazards of working in the Arctic, says Daeschler, who carried a shotgun for protection.

When not looking over their shoulders, the researchers drilled, chiseled, and hammered their way through rocks looking for fossils. Not just any rocks and fossils, but ones dating from 375 million years ago, when animals were taking their first tentative steps on land. For three summers, they scoured the site of what was once an active streambed but found little of interest. Then, in 2004, the team made a tantalizing discovery: the snout of a curious-looking creature protruding from a slab of pink rock. Further excavation revealed the well-preserved remains of several flat-headed animals between 4 and 9 feet long. In some ways, the animals resembled giant fish—they had fins and scales. But they also had traits that resembled those of land-dwelling amphibians—notably, a neck, wrists, and finger-like bones. The researchers named the new species *Tiktaalik roseae*; *tiktaalik* (pronounced tic-TAH-lick) is a native word meaning "large freshwater fish." This ancient hybrid animal no longer exists, but it represents a critical phase in the evolution of four-legged, land-dwelling **vertebrates**—including humans.

Tiktaalik "splits the difference between something we think of as a fish and something we think of as a limbed animal," says Daeschler, a curator of vertebrate zoology at the Academy of Natural Sciences in Philadelphia. "In that sense, it is a wonderful transitional fossil between two major groups of vertebrates."

Today, of course, four-legged animals roam far and wide over land. But 400 million years ago it was a different story. Life was mostly aquatic then, restricted to oceans and freshwater streams. How life made the jump

"" *Tiktaalik "splits the difference between something we think of as a fish and something we think of as a limbed animal."* **""**

— TED DAESCHLER

from water to land is a question that has long intrigued evolutionary biologists. In fact, scientists have been searching for evidence of this milestone ever since Charles Darwin first proposed that all life on the planet is related by a tree of common descent. According to Shubin, a professor of biology at the University of Chicago and the Field Museum of Natural History, *Tiktaalik* is the most compelling example yet of an animal that lived at the cusp of this important transition. Not only does it fill a gap in our knowledge, the discovery also provides persuasive evidence in support of Darwin's theory.

READING THE FOSSIL RECORD

The theory of evolution—what Darwin called **descent with modification**—draws two main conclusions about life on Earth: that all living things are related, and that the different species we see today have emerged over time as a result of natural selection operating over millions of years. Many lines of evidence support this theory (remember that in science a "theory" is an idea supported by a tremendous amount of evidence and which has never been disproved). One of the most compelling lines of evidence for evolution comes from **fossils,** the preserved remains or impressions of once-living organisms. Fossils

are like snapshots of past life, capturing what life was like at particular moments in time.

They are formed in a number of ways: an animal or plant may be frozen in ice, trapped in amber, or buried in a thick layer of mud. The entombed organism is thereby protected from being eaten by scavengers or rapidly decomposed by bacteria. Over time, if conditions are right, the organism's shape is preserved. Not all organisms are equally likely to form fossils, however: animals with bones or shells are more likely to be preserved than animals without such hard parts (think earthworms or jellyfish) that decay quickly. And conditions permitting fossilization

A **Tiktaalik roaseae** *fossil.*

INFOGRAPHIC 16.1 FOSSILS FORM ONLY IN CERTAIN CIRCUMSTANCES

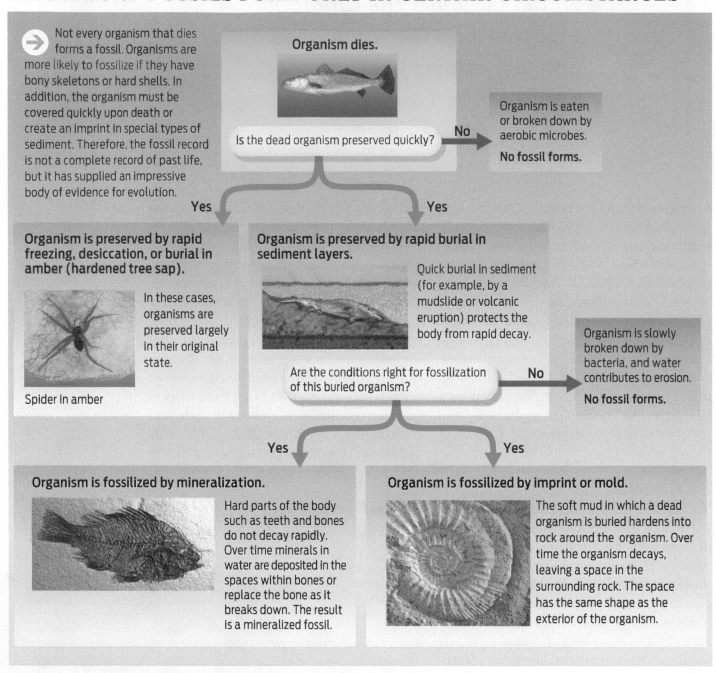

→ Not every organism that dies forms a fossil. Organisms are more likely to fossilize if they have bony skeletons or hard shells. In addition, the organism must be covered quickly upon death or create an imprint in special types of sediment. Therefore, the fossil record is not a complete record of past life, but it has supplied an impressive body of evidence for evolution.

Organism dies.

Is the dead organism preserved quickly? — **No** → Organism is eaten or broken down by aerobic microbes. **No fossil forms.**

Yes

Yes

Organism is preserved by rapid freezing, desiccation, or burial in amber (hardened tree sap).

In these cases, organisms are preserved largely in their original state.

Spider in amber

Organism is preserved by rapid burial in sediment layers.

Quick burial in sediment (for example, by a mudslide or volcanic eruption) protects the body from rapid decay.

Are the conditions right for fossilization of this buried organism? — **No** → Organism is slowly broken down by bacteria, and water contributes to erosion. **No fossil forms.**

Yes

Yes

Organism is fossilized by mineralization.

Hard parts of the body such as teeth and bones do not decay rapidly. Over time minerals in water are deposited in the spaces within bones or replace the bone as it breaks down. The result is a mineralized fossil.

Organism is fossilized by imprint or mold.

The soft mud in which a dead organism is buried hardens into rock around the organism. Over time the organism decays, leaving a space in the surrounding rock. The space has the same shape as the exterior of the organism.

▶**FOSSIL RECORD**
An assemblage of fossils arranged in order of age, providing evidence of changes in species over time.

▶**PALEONTOLOGIST**
A scientist who studies ancient life by examining the fossil record.

are rare: the organism has to be in just the right place at just the right time (**INFOGRAPHIC 16.1**).

Because not all organisms are preserved, the **fossil record** is not a complete record of past life. Nevertheless, the existing fossil record is remarkably rich and offers an exciting window into the past. **Paleontologists,** scientists who study ancient life, have uncovered hundreds of thousands of fossils throughout the world, from many evolutionary time peri-

ods. When fossils are arranged in order of age, they provide a tangible history of life on Earth. The fossil record is extensive enough to show the overall arc of life and provides compelling evidence in support of Darwin's theory.

For example, if all organisms have descended from a single common ancestor billions of years ago, as the theory of evolution concludes they did, then we would expect the fossil record to show an ordered succession of evolutionary stages as organisms evolved and diversified. And, indeed, that is exactly what you see: prokaryotes appear before eukaryotes, single-cell organisms before multicellular ones, water-dwelling organisms before land-dwelling ones, fish before amphibians, reptiles before birds.

Moreover, we would expect to see changes over time in a family of organisms, and we do. One exceptionally well studied example is hors-es. Comparisons of modern-day horse bones with fossils of horse ancestors reveal how, in the course of evolution, horses have lost most of their toes. The fossils show a continual series of changes over time, with the most recent fossils being the most similar to modern organisms, and the more ancient fossils being the most different. But they all clearly share a family resemblance (INFOGRAPHIC 16.2).

INFOGRAPHIC 16.2 FOSSILS REVEAL CHANGES IN SPECIES OVER TIME

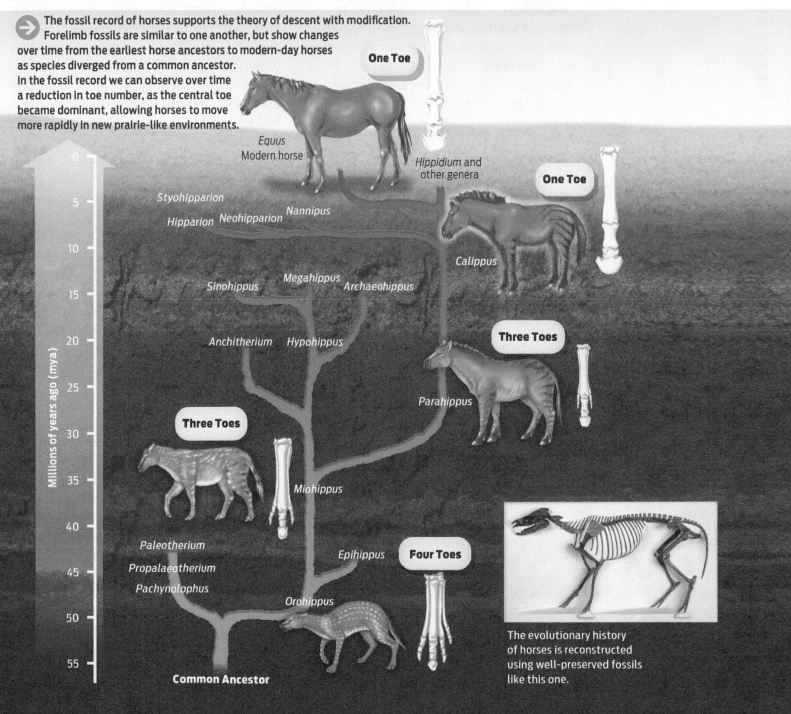

The fossil record of horses supports the theory of descent with modification. Forelimb fossils are similar to one another, but show changes over time from the earliest horse ancestors to modern-day horses as species diverged from a common ancestor. In the fossil record we can observe over time a reduction in toe number, as the central toe became dominant, allowing horses to move more rapidly in new prairie-like environments.

The evolutionary history of horses is reconstructed using well-preserved fossils like this one.

Descent with modification also predicts that the fossil record should contain evidence of intermediate organisms—those with a mixture of "old" and "new" traits. Darwin acknowledged in *The Origin of Species* that the fossil record of his day did not provide many examples of such intermediate organisms—a state of affairs he described as "probably the gravest and most obvious of all the many objections which may be urged against my views." Yet Darwin knew that if his hypothesis were correct, then such intermediate fossils would eventually be found. And indeed they have been. Scientists have discovered animals with mixtures of reptile and bird characteristics, and animals with mixtures of reptile and mammal characteristics. But the transition between fish and amphibians has remained more obscure.

THE FOSSIL HUNT

Shubin and Daeschler began their hunt for fossils in the Canadian Arctic in 1999, after stumbling upon a map in an old geology textbook.

Ellesmere Island, nearly as large as Great Britain, contains Canada's most northern point.

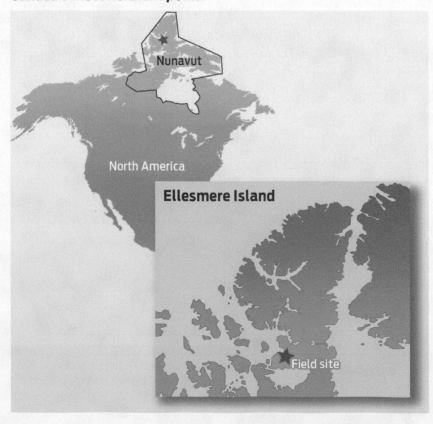

The map showed that the region contained large swaths of exposed rock dating back 375-380 million years—just the period of time the researchers were interested in.

Why was this period so important to the scientists? They knew that there are no land-dwelling vertebrates in the fossil record before 385 million years ago. By 365 million years ago, organisms easily recognizable as amphibians are well documented in the fossil record. The scientists hypothesized that if they looked at rocks sandwiched in between these two time periods—around 375 million years old—they might find one of Darwin's elusive "intermediates." Moreover, Ellesmere is one of only three places on Earth where rocks of this time period are exposed, yet to Shubin and Daeschler's knowledge, no other paleontologists had explored the area, which meant it was a potential fossil gold mine.

Knowing exactly where to look for fossils was tricky, since Ellesmere Island covers 75,000 square miles. To locate the most promising dig site, the scientists first studied aerial photographs. Once on the ground, the scientists and their team split up and spent the first two seasons just walking the rocky exposures, prospecting for bits and pieces of fossils that had eroded out from the rock. When they found something interesting on the surface, they would start to dig.

It was while walking these rocky exposures in 2002 that Daeschler and his team found the first piece of what would turn out to be a *Tiktaalik* fossil—"basically part of the snout," he says. At first, they didn't think much of the find, but collected it anyway along with other fossil pieces. Back in Philadelphia, researchers cleaned the fossil, removing the remaining rock. Even then, says Daeschler, it wasn't clear what the snout belonged to. Not until a visiting graduate student remarked on the resemblance of the skull to one from the earliest known amphibians did the researchers realize what they had found. If ever there was a "lightbulb" moment, he says, this was it. But, alas, they had only one small piece of the creature.

The team returned to Ellesmere in 2004 for another round of hunting and digging. It

INFOGRAPHIC 16.3 HOW FOSSILS ARE DATED

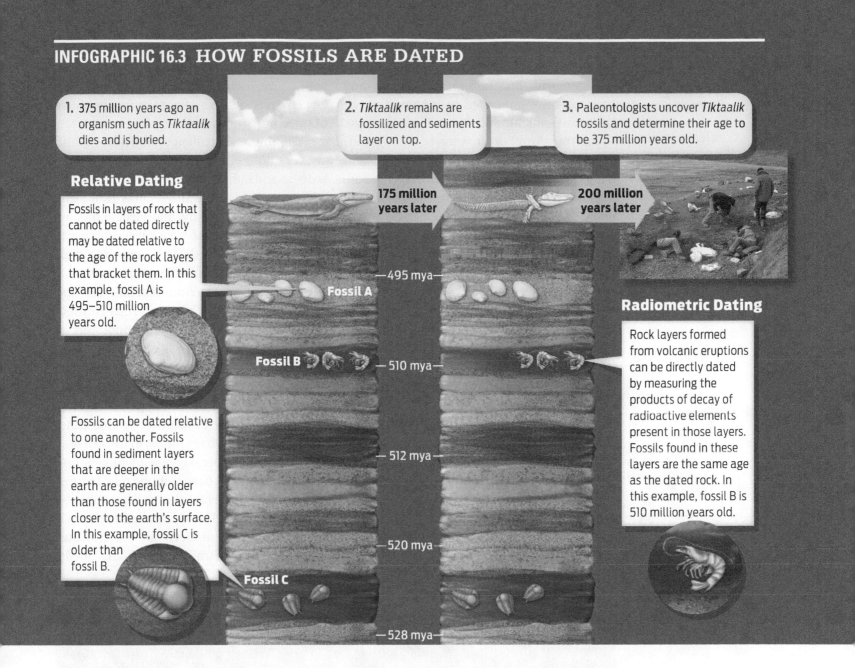

1. 375 million years ago an organism such as *Tiktaalik* dies and is buried.

2. *Tiktaalik* remains are fossilized and sediments layer on top.

3. Paleontologists uncover *Tiktaalik* fossils and determine their age to be 375 million years old.

Relative Dating

Fossils in layers of rock that cannot be dated directly may be dated relative to the age of the rock layers that bracket them. In this example, fossil A is 495–510 million years old.

Fossils can be dated relative to one another. Fossils found in sediment layers that are deeper in the earth are generally older than those found in layers closer to the earth's surface. In this example, fossil C is older than fossil B.

175 million years later

200 million years later

Fossil A
—495 mya—

Fossil B
—510 mya—

—512 mya—

—520 mya—

Fossil C

—528 mya—

Radiometric Dating

Rock layers formed from volcanic eruptions can be directly dated by measuring the products of decay of radioactive elements present in those layers. Fossils found in these layers are the same age as the dated rock. In this example, fossil B is 510 million years old.

---

didn't take long for their patience to be rewarded: "Literally inches," Daeschler says, from where they'd been excavating before, they hit pay dirt.

The fossils they found looked like the elusive intermediate creature the team had been hunting for. But how could they be sure it was the right age? Logically, fossils are at least as old as the rocks that encase them, so if you know the age of the rocks, then you know the age of the fossils, too. Some types of rocks can be dated directly by a method known as **radiometric dating,** which uses the proportion of certain radioactive isotopes in rock crystals as a geologic clock (see Chapter 17). Fossils found in or near these layers can be dated quite precisely. If fossils are found in rock layers that cannot be directly dated, they can be dated indirectly by their position with respect to rocks or fossils of known age, either deeper or shallower, a technique called **relative dating.** Generally speaking, the deeper the fossils, the older they are. Using a combination of both methods, scientists have determined that the rocks where *Tiktaalik* was found are 375 million years old, which means *Tiktaalik* is that old as well (**INFOGRAPHIC 16.3**).

▶**RADIOMETRIC DATING**
The use of radioactive isotopes as a measure for determining the age of a rock or fossil.

▶**RELATIVE DATING**
Determining the age of a fossil from its position relative to layers of rock or fossils of known age.

16

▶INVERTEBRATE
An animal without a backbone.

▶TETRAPOD
A vertebrate animal with four true limbs, that is, jointed, bony appendages with digits. Mammals, amphibians, birds, and reptiles are tetrapods.

SETTING THE STAGE FOR LIFE ON LAND

The geologic time period that Shubin and Daeschler are interested in is known as the Devonian—roughly 400-350 million years ago. Great transformations were occurring during the Devonian: jawed fishes, sharks, land plants, and insects all diversified in this period. Because sea levels were high worldwide, and much of the land lay submerged under water, the Devonian Period has been called the age of fishes.

Back then, what is now the Canadian Arctic had a warm, wet climate and a landscape veined by shallow, meandering streams. Early in the Devonian Period there was little plant growth, and the world would have looked fairly brown and empty. By the middle of the Devonian, if you were standing on the bank of a stream you would have seen some of the first land plants, the first forests, as well as the first **invertebrates**—spiderlike creatures and millipedes, for example—crawling on land. Still, there would have been no land-dwelling vertebrates at this time: nothing with bony limbs, nothing with a backbone or skull.

By the late Devonian, things were changing quickly. By then, says Daeschler, "you had

a green floodplain, a green world." It was this green world—a rich and productive ecosystem, with energy-rich leaf litter flowing into shallow streams—that set the stage for the move of vertebrates onto land.

The physical challenges of living on land are very different from those in water. Water is dense and difficult to move through, but fish glide smoothly through it thanks to a sleek shape, a muscular body, and flexible fins. By contrast, animals that walk on land have to cope with gravity. Air doesn't support animals as they move, so the bodies of land animals need a sturdier structure. Animals on land can also dry out, which is dangerous for them because cells need water to function. And, of course, taking in oxygen is different on land and in water.

Of the many features that distinguish land animals from fish, biologists have singled out one as a key evolutionary milestone: limbs. Fish do not have limbs, in the sense of jointed, bony appendages with fingers and toes. Instead, they have webbed fins. In most fishes, the fin bones are thin and fan out away from each other. These so-called ray-finned fishes include the modern-day perch, trout, and bass. By contrast, amphibians, birds,

An artist's representation of what Tiktaalik *may have looked like.*

A physical model of Tiktaalik's *forelimb fin, made from replicas of fossil bones fused with flexible wire.*

most reptiles, and mammals all have two pairs of limbs, defining them as **tetrapods** (from the Greek for "four-footed").

While having limbs is a key feature distinguishing tetrapods from fish, one small group of fish—the lobe-finned fish—seems to blur this distinction. First appearing in the fossil record about 400 million years ago, lobe-finned fish have fleshy fins supported by a stalk of bones that resemble primitive limb bones.

Lobe-finned fish are thought to have evolved in shallow streams, where rich plant material lured small fish and other creatures close to the water's edge. The lobe-finned fish likely used their sturdy fins to touch the bottom of the streambed while maneuvering to catch prey. As Daeschler explains, it was the unique ecological opportunity afforded by shallow streams that enabled the lobe-finned fish to start developing features that were adaptive in shallow water. But lobe-finned fish were still very far from being true tetrapods. *Tiktaalik* is a step closer: "It looks like a fish in that it has scales and fins," said Shubin in a 2006 interview at the University of Chicago, "but when you look inside the skeleton you see how special it really is."

THE FISH THAT DID PUSHUPS

Shubin and Daeschler were lucky: the fossils they found were so well preserved that they were able to study *Tiktaalik*'s skeletal anatomy in detail, even seeing how the bones interacted and where muscles attached. From these fossil bones, they determined that *Tiktaalik* was a predatory fish with sharp teeth, scales, and fins. In addition to these fishy attributes, it had a flat skull reminiscent of a crocodile head and a flexible neck. To Shubin and Daeschler, the neck was one of the most surprising finds. Having a flexible neck meant that, unlike a fish, *Tiktaalik* could swivel its head independently of its body, perhaps enabling it to catch a glimpse of predators sneaking

> **❝** *It looks like a fish in that it has scales and fins, but when you look inside the skeleton you see how special it really is.* **❞**
>
> — NEIL SHUBIN

Tiktaalik *likely used its sturdy forelimb fins to pull itself out of the water for short excursions on land.*

up on it from behind or to snap its jaws sideways like a crocodile. It also had the full-fledged ribs of a modern land animal, sturdy enough to support the animal's trunk out of water even against the force of gravity.

But it is *Tiktaalik*'s fins that have justly made it famous. While possessing many features of a lobe-finned fish, *Tiktaalik* appears also to have had a jointed elbow, wrist, and fingerlike bones. From the fossil pieces, Shubin and Daeschler were able to create a model of how the bones would have moved relative to one another, and they have modeled these movements digitally. The models show that the bones and joints were strong enough to support the body and worked like those of the earliest known tetrapods–the early amphibians. "This animal was able to hold its fin below its body, bend the fin out toward what we think of as a wrist, and bend the elbow," explains Daeschler. In other words, it was a fish that could do a push-up.

With this hybrid anatomy, *Tiktaalik* was not galloping on land, of course. It probably lived most of the time in water, but Shubin and Daeschler suspect that *Tiktaalik* may have used its supportive fins to pull itself out of the water for brief periods. "This is a fish that can live in the shallows and even make short excursions

onto land," said Shubin, in the same 2006 interview. The ability to crawl onto land would certainly have been a useful trait in the Devonian, when open water was a brutal fish-eat-fish world, whereas land was a predator-free paradise, full of nourishing bugs.

Like other fish living at the time, *Tiktaalik* is thought to have had both lungs and gills, which explains how it could breathe out of water for these short excursions. Most modern fish have retained their gills but lost their lungs over time (the lungs have evolved into the balloon-like structure known as a swim bladder, which helps fish float). Some modern fish known as lungfish, however, have retained this ancestral trait. Lungfish are lobe-finned fish closely related to the lobe-finned fish from which *Tiktaalik* is believed to have descended.

There was, of course, no forethought involved in this process of limb evolution. Fish did not develop limbs for the purpose of walking on land. Rather, limbs first evolved in shallow water, where they proved adaptive and were thus retained in the descendants of the organisms who first developed them. Then, when there was an opportunity to take advantage of a tantalizing new habitat–land–the amphibious creatures already had the skeletal "toolkit."

For all its amphibian-like adaptations, *Tiktaalik* is still considered a fish because its limbs lack the true jointed fingers and toes that characterize tetrapod limbs (in other words, they're still fins). But it's by far the most tetrapod-like of all the fishes so far discovered. Scientists have jokingly referred to it as a "fishapod" (**INFOGRAPHIC 16.4**).

And that's what makes *Tiktaalik* such an important find: it occupies a midpoint between fish and tetrapods. "It very much fits in that gray area between things we typically call fish and things we typically call limbed animals," says Daeschler. Such intermediate, or transitional, fossils document important steps in the evolution of life on Earth. They help biologists understand how groups of organisms evolved, through natural selection, from one form into another. And they confirm that Darwin's theory of descent with modification—which predicts such intermediate forms—is correct.

INFOGRAPHIC 16.4 *TIKTAALIK*, AN INTERMEDIATE FOSSILIZED ORGANISM

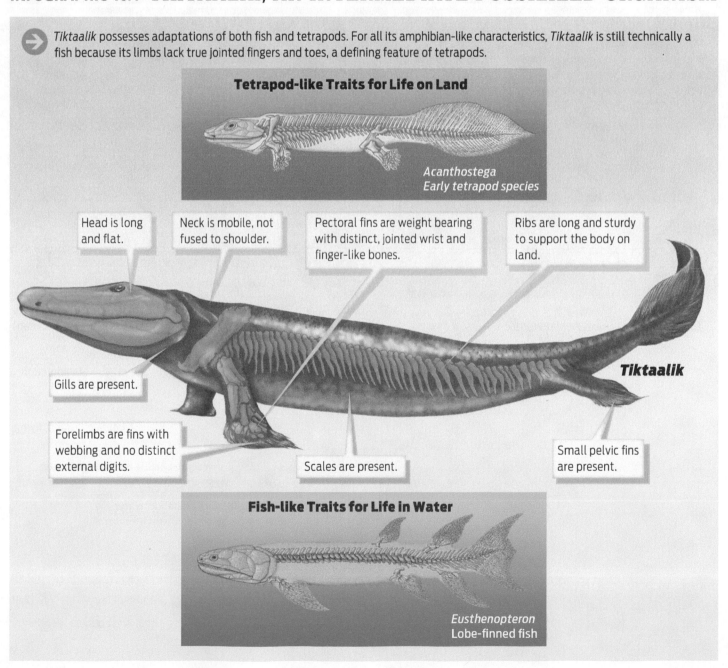

→ *Tiktaalik* possesses adaptations of both fish and tetrapods. For all its amphibian-like characteristics, *Tiktaalik* is still technically a fish because its limbs lack true jointed fingers and toes, a defining feature of tetrapods.

Tetrapod-like Traits for Life on Land

Acanthostega
Early tetrapod species

Head is long and flat.

Neck is mobile, not fused to shoulder.

Pectoral fins are weight bearing with distinct, jointed wrist and finger-like bones.

Ribs are long and sturdy to support the body on land.

Gills are present.

Tiktaalik

Forelimbs are fins with webbing and no distinct external digits.

Scales are present.

Small pelvic fins are present.

Fish-like Traits for Life in Water

Eusthenopteron
Lobe-finned fish

▶**HOMOLOGY**
Anatomical, genetic, or developmental similarity among organisms due to common ancestry.

A FIN IS A PAW IS AN ARM IS A WING

In *The Origin of Species*, Darwin asked, "What can be more curious than that the hand of a man, formed for grasping, that of a mole for digging, the leg of the horse, the paddle of the porpoise, and the wing of the bat, should all be constructed on the same pattern, and should include similar bones, in the same relative positions?" To Darwin, this uncanny similarity was evidence that all these organisms were related–that they share a common ancestor in the ancient past.

The fact that all tetrapods share the same forelimb bones, arranged in the same order, is an example of **homology**–a similarity due to common ancestry. Before Darwin, comparative anatomists had identified many such similarities in anatomy; what they lacked was a satisfactory explanation for why such similarity should exist.

INFOGRAPHIC 16.5 FORELIMB HOMOLOGY IN FISH AND TETRAPODS

The number, order, and underlying structure of the forelimb bones are similar in all the groups illustrated below. The differences in the relative width, length, and strength of each bone contribute to the specialized function of each forelimb. This anatomical homology is strong evidence that these organisms all have a common ancestor at some time in the distant past. The variations in bone shape and function reflect evolutionary adaptations to different environments.

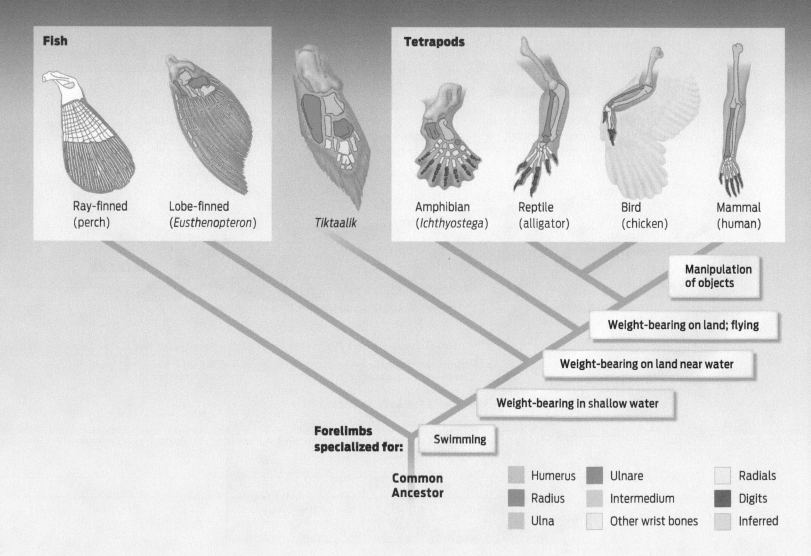

Fish

Ray-finned (perch)

Lobe-finned (*Eusthenopteron*)

Tiktaalik

Tetrapods

Amphibian (*Ichthyostega*)

Reptile (alligator)

Bird (chicken)

Mammal (human)

Manipulation of objects

Weight-bearing on land; flying

Weight-bearing on land near water

Weight-bearing in shallow water

Forelimbs specialized for:

Swimming

Common Ancestor

Humerus · Ulnare · Radials
Radius · Intermedium · Digits
Ulna · Other wrist bones · Inferred

INFOGRAPHIC 16.6 VERTEBRATE ANIMALS SHARE A SIMILAR PATTERN OF EARLY DEVELOPMENT

We can identify homologous structures by tracing their embryological development. Some of our middle ear bones, for example, are homologous with the jaw bones of reptiles and bones supporting gills in fish. We know this because all of these structures develop from the pharyngeal pouches that appear in all vertebrate embryos early in development. This developmental homology is strong evidence that all vertebrate animals are related by common ancestry. Genetic changes over time have introduced modifications in later stages that give rise to distinct species with vast physical differences.

Early Embryos

Human	Cat	Fish	Snake	Chicken

Pharyngeal pouches

Post-anal tail

Early-stage embryos of related organisms share common structures.

Adult Organisms

Later in development, these structures take on species-specific shape and function.

Darwin provided that explanation: homologous structures are ones that are similar because they are inherited from the same ancestor–in this case, an amphibious creature like *Tiktaalik*. Why is this significant? Think of it this way: every time you bend your wrist back and forth–to swipe a paint brush or hold a cell phone to your ear, for example–you are using structures that first evolved 375 million years ago in fish. As Shubin points out, "This is not just some archaic, weird branch of evolution; this is *our* branch of evolution" (INFOGRAPHIC 16.5).

If they have the same bones, why then do a human arm and a bird wing look so different? Remember that during the process of inheritance mutations are continually introduced into the DNA of genes. Such mutations can produce subtle changes in the proteins encoded by those genes–proteins involved in constructing the bones that make up an arm or a wing, for example. Changes in bone proteins can result in slightly altered bones, for instance making

them longer or thinner. When these modified bones are helpful to an organism's survival and reproduction, the advantageous traits are passed on to the next generation, and populations emerge that have these adaptations. This "descent with modification" (Darwin's phrase again) results in diverse organisms sharing common–homologous–structures and putting them to different uses.

We can see homology not only in adult anatomy, but in early development as well. Take a look at early embryos of vertebrate animals as diverse as humans, fish, and chickens and you'll see that they all look remarkably similar. Why should the embryonic stage of a human resemble the embryonic stage of a fish when the adults of each species look so different? Similar embryological structures are further evidence that all vertebrates have a common ancestor (INFOGRAPHIC 16.6).

Development helps us solve other evolutionary conundrums as well, such as why reptiles like snakes don't have limbs like other tetrapods. In fact, snake embryos *do* possess the beginnings of limbs, but these limb buds remain rudimentary and do not develop into full-fledged limbs (although you can still see stubby hindlimbs in some species of snake today). Such **vestigial structures,** which serve no apparent function in an organism, are strong evidence for evolution: these now apparently useless features are inherited from an ancestor in whom they *did* serve a function.

Zooming in even further, to the molecular level, we find still more examples of homology–and thus more evidence of common an-

> In essence, DNA serves as a kind of molecular clock: each additional sequence difference is like a tick of the clock, showing the amount of time that has elapsed since the two species shared a common ancestor.

cestry. Scientists have known since the 1950s that DNA is the molecule of heredity, and that it is shared by all living organisms on Earth. Every molecule of DNA–whether from fish, maple tree, bacterium, or human–is made of the same four nucleotides (A, C, T, and G), and all organisms use the information encoded by those nucleotides to make proteins in the same basic way, using the universal genetic code (see Chapter 8). Why should all living things use the same system of decoding genetic information? The best explanation is that this system was the one used by the ancient ancestor of all living organisms, passed on to all of its descendants, and preserved throughout billions of years of evolution.

▶**VESTIGIAL STRUCTURE**
A structure inherited from an ancestor that no longer serves a clear function in the organism that possesses it.

DNA AND DESCENT

While all living organisms share DNA and the genetic code, no two species share the exact same sequence of DNA nucleotides. That's because (as described in Chapter 10) errors in DNA replication and other mutations are continually introducing variation into DNA sequences (and the proteins they encode). Over time, neutral and advantageous mutations will tend to be preserved, while harmful mutations will tend to be selected against and eliminated. In addition, much of our DNA consists of long stretches of noncoding sequences with no known function. Because mutations in these regions have no effect on an organism, they accumulate over time. As mutations are passed on to descendants, the number of sequence differences between the ancestor and its descendants grows–slowly in the case of sequences coding for critical proteins whose structures are well adapted to their functions, and more rapidly in the case of noncoding DNA not involved with making proteins. Closely related species will therefore have more similar DNA sequences than species that are more distantly related.

For example, when scientists looked at one specific region of DNA–the cystic fibrosis transmembrane regulator (or *CFTR*) region–they discovered that human DNA in this region is 99% identical to chimpanzee DNA. The fact that the DNA of the two species is nearly identical reflects the fact that humans and chimps share a common ancestor that lived relatively recently–just 5-7 million years ago. By contrast, human DNA is 85% identical to the DNA of a mouse at this same region, which makes sense given that humans and mice share a common ancestor that lived between 60 and 100 million years ago. Less sequence identity would be seen between a human and a toad, whose common ancestor– a lobed-finned fish–lived roughly 375 million years ago. The more distantly related two species are, the more sequence differences in DNA sequences you will see. In essence, DNA serves as a kind of molecular clock: each additional sequence difference is like a tick of the clock, showing the amount of time that

INFOGRAPHIC 16.7 RELATED ORGANISMS SHARE DNA SEQUENCES

→ Related organisms share DNA sequences inherited from a common ancestor. Over time, the sequence in each species acquires independent mutations. The more time that has passed, the greater the number of sequence differences that will be present. Thus, the percentage of nucleotides that differ between two species gives an indication of the evolutionary distance between them.

Sequence homology between species

Species A	GGTATCGAGGTTCTACATTGCAACTTCTAC
Close relative	GGAAACGAGGTTCTACATTGCCACTTCTAC
Distant relative	GGAAACGAGGTTCGACATAGCCACTTCTAC

3 differences in 30 nucleotides
3/30 = 10%; or 90% similarity

5 differences in 30 nucleotides
5/30 = 17%; or 83% similarity

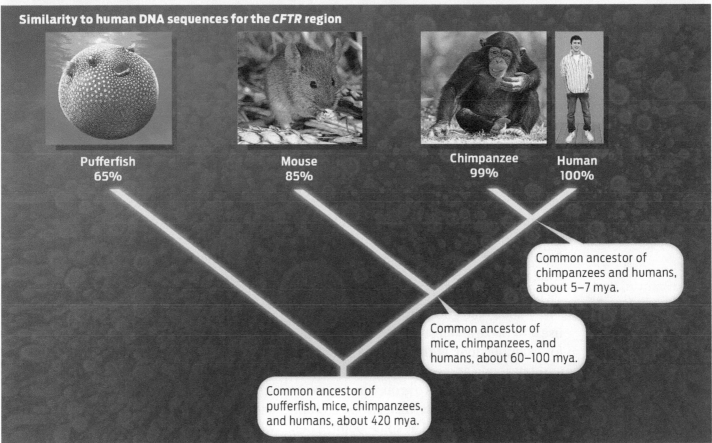

Similarity to human DNA sequences for the *CFTR* region

Pufferfish
65%

Mouse
85%

Chimpanzee
99%

Human
100%

Common ancestor of chimpanzees and humans, about 5–7 mya.

Common ancestor of mice, chimpanzees, and humans, about 60–100 mya.

Common ancestor of pufferfish, mice, chimpanzees, and humans, about 420 mya.

SOURCE: Data for the *CFTR* coding region from Thomas, J. W., et al. (2003) Comparative analyses of multi-species sequences from targeted genomic regions. *Nature* 424:788–893.

has elapsed since the two species shared a common ancestor (**INFOGRAPHIC 16.7**).

When combined with evidence from the fossil record, anatomy, and development, molecular data become a powerful tool for understanding evolution. As we'll see in Chapter 17, DNA evidence is often a more reliable clue to common ancestry than physical appearance, and can serve as a check on conclusions derived from the fossil record or anatomy. As well, DNA is deepening our knowledge of

how limbs are constructed at the molecular level. Scientists working in Shubin's lab have shown that the same genes orchestrate limb development in animals as diverse as sharks, chickens, and humans. Learning how these genes work and how changes in their DNA sequences can produce large-scale changes in body plan or limb structure is a hot area of biology right now, informally known as "evo-devo" (short for "evolutionary developmental biology").

Fossils like this well-preserved sample give scientists a glimpse into evolutionary history.

FILLING IN THE GAPS

Asked what he thinks is most interesting about the discovery of *Tiktaalik,* Daeschler homes in on what he says is a popular misconception about the fossil record—that it's "spotty" and "chaotic." But that's simply not true, he says. Despite the fact that it does not record *all* past life, the fossil record is still "very good"—so good, in fact, that you can use it to make and test predictions. You can, for example, look at the fossil record of fish and tetrapods and—suspecting on the basis of anatomy that the two groups are related—hypothesize that an intermediate-looking animal must have existed at some point. Then you can go look for it. Daeschler refers to this process as "filling in the gaps," and it's exactly what he and Shubin did with *Tiktaalik*. They knew, based on the existing fossil record, *when* such a creature was likely to have existed, so then it was just a question of *where* to look for it.

For Shubin and Daeschler, *Tiktaalik* is exciting mostly because it shows that our understanding of evolution is correct: "It confirms that we have a very good understanding of the framework of the history of life," says Daeschler. "We predicted something like *Tiktaalik*, and sure enough, with a little time and effort, we found it." ■

CHAPTER 16 Summary

▶ The theory of evolution—what Darwin called "descent with modification"—draws two main conclusions about life: that all living things are related, sharing a common ancestor in the distant past; and that the species we see today are the result of natural selection operating over millions of years.

▶ The theory of evolution is supported by a wealth of evidence, including fossil, anatomical, and DNA evidence.

▶ Fossils are preserved remains or impressions of once-living organisms that provide a record of past life on Earth. Not all organisms are equally likely to form fossils.

▶ Fossils can be dated directly or indirectly on the basis of the age of the rocks they are found in, or their position relative to rocks or fossils of known ages.

▶ When fossils are dated and placed in sequence, they show how life on Earth has changed over time.

▶ As predicted by descent with modification, the fossil record shows the same overall pattern for all lines of descent: younger fossils are more similar to modern organisms than are older fossils.

▶ Descent with modification also predicts the existence of "intermediate" organisms, such as *Tiktaalik*, that possess mixtures of "old" and "new" traits. *Tiktaalik* has features of both fish and tetrapods (four-limbed vertebrates).

▶ An organism's anatomy reflects adaptation to its ecological environment. Changed ecological circumstances provide opportunities for new adaptations to evolve by natural selection.

▶ Homology—the anatomical, developmental, or genetic similarities shared among groups of related organisms—is strong evidence that those groups descend from a common ancestor.

▶ Homology can be seen in the common bone structure of the forelimbs of tetrapods, the similar embryonic development of all vertebrate animals, and the universal genetic code.

▶ Many genes, including those controlling limb development, are shared among distantly related species, an example of molecular homology owing to common ancestry.

▶ DNA can be used as a molecular clock: more-closely related species show greater DNA sequence homology than do more-distantly related species.

MORE TO EXPLORE

▶ *Tiktaalik roseae* http://tiktaalik.uchicago.edu

▶ Shubin, N. (2008) *Your Inner Fish: A Journey into the 3.5-Billion-Year History of the Human Body.* New York: Random House.

▶ Shubin, N. H., et al. (2006) The pectoral fin of *Tiktaalik roseae* and the origin of the tetrapod limb. *Nature* 440:764–771.

▶ Daeschler, E. B., et al. (2006) A Devonian tetrapod-like fish and the evolution of the tetrapod body plan. *Nature* 440:757–763.

▶ Shubin, N., et al. (2009) Deep homology and the origins of evolutionary novelty. *Nature* 457:818–823.

▶ Thomas, J. W., et al. (2003) Comparative analyses of multi-species sequences from targeted genomic regions. *Nature* 424:788–793.

CHAPTER 16 Test Your Knowledge

How does the fossil record reveal information about evolutionary changes?

By answering the questions below and studying Infographics 16.1, 16.2, and 16.3, you should be able to generate an answer for the broader Driving Question above.

KNOW IT

1 Which of the following is most likely to leave a fossil?

a. a jellyfish

b. a worm

c. a wolf

d. a sea sponge (an organism that lacks a skeleton)

e. All of the above are equally likely to leave a fossil.

2 Generally speaking, if you are looking at layers of rock, at what level would you expect to find the newest—that is, the youngest—fossils?

3 You are examining a column of soil that contains vertebrate fossils from deeper to shallower layers. Would you expect a fossil with four limbs with digits to occur higher or lower in the soil column relative to a "standard" fish? Explain your answer.

4 What can the fossil shown below tell us about the structure and lifestyle of the organism that left it? Describe your observations.

USE IT

5 You have molecular evidence that leads you to hypothesize that a particular group of soft-bodied sea cucumbers evolved at a certain time. You have found a fossil bed with many hard-shelled mollusks dating to the critical time, but no fossil evidence to support your hypothesis about the sea cucumbers. Does this find cause you to reject your hypothesis? Why or why not?

6 A specific type of oyster is found in North American fossil beds dated to 100 million years ago. If similar oyster fossils are found in European rock, in layers along with a novel type of barnacle fossil, what can be concluded about the age of the barnacles? Explain your answer.

BRING IT HOME

7 Do an Internet search to find out about fossils discovered in your home state. Determine what kinds of organisms they represent, how old they are, and where in your state you would need to go in order to have a chance of finding fossils in the field.

What features make *Tiktaalik* a transitional fossil, and what role do these types of fossil play in the fossil record?

By answering the questions below and studying Infographics 16.4 and 16.5, you should be able to generate an answer for the broader Driving Question above.

KNOW IT

8 Which of the following features of *Tiktaalik* is not shared with other bony fishes?

a. scales

b. teeth

c. a mobile neck

d. fins

e. none of the above

9 *Tiktaalik* fossils have both fishlike and tetrapod-like characteristics. Which characteristics are related to supporting the body out of the water?

USE IT

10 *Tiktaalik* fossils are described as "intermediate" or "transitional" fossils. What does this mean? Why are transitional organisms so significant in the history of life?

11 *Tiktaalik* has been called a "fishapod"—part fish, part tetrapod. Speculate on the fossil appearance of its first true tetrapod descendant—what features would distinguish it from *Titkaalik*? How old would you expect those fossils to be relative to *Titkaalik*?

12 If some fish acquired modifications that allowed them to be successful on land, why didn't fish just disappear? In other words, why are there still plenty of fish in the sea if the land presented so many favorable opportunities?

DRIVING QUESTION 3

What can anatomy and DNA reveal about evolution?

By answering the questions below and studying Infographics 16.5, 16.6, and 16.7, you should be able to generate an answer for the broader Driving Question above.

KNOW IT

13 Compare and contrast the structure and function of an eagle wing with the structure and function of a human arm.

14 Vertebrate embryos have structures called pharyngeal pouches. What do these structures develop into in an adult human? In an adult bony fish?

15 You have three sequences of a given gene from three different organisms. How could you determine how closely the three organisms are related?

USE IT

16 What is the evolutionary explanation for the fact that both human hands and otter paws have five digits?

17 Could you use the presence of a tail to distinguish a human embryo from a chicken embryo? Why or why not?

18 If, in humans, the DNA sequence TTTCTAGGAATA encodes the amino acid sequence phenylalanine–leucine–glycine–isoleucine, what amino acid sequence will that same DNA sequence specify in bacteria?

19 Gene *X* is present in yeast and in sea urchins. Both produce protein *X*, but the yeast protein is slightly different from the sea urchin protein. What explains this difference? How might you use this information to judge whether humans are closer evolutionarily to yeast or to sea urchins?

INTERPRETING DATA

20 The gene responsible for hairlessness in Mexican hairless dogs is called corneodesmosin (*CDSN*). This gene is present in other organisms. Look at the sequence of a portion of the *CDSN* gene from pairs of different species, given below. For each pair, determine the number of differences.

From the variations in this sequence, which organism appears to be most closely related to humans? Which organism appears to be least closely related to humans?

SPECIES	SEQUENCE
Homo sapiens (human)	ACTCCGGCCCCTACATCCCCAGCTCCCA
Canis lupus familiaris (dog)	ATTCTGGCTCCTACATTTCCAGCTCCCA
Homo sapiens (human)	ACTCCGGCCCCTACATCCCCAGCTCCCA
Pan troglodytes (chimpanzee)	ACTCCGGCCCCTACATCCCCAGCTCCCA
Homo sapiens (human)	ACTCCGGCCCCTACATCCCCAGCTCCCA
Sus scrofa (pig)	AGTCTGGCTCCTACATCTCCAGCTCCCA
Homo sapiens (human)	ACTCCGGCCCCTACATCCCCAGCTCCCA
Macaca mulatta (rhesus monkey)	ACTCTGGCCCCTACATCCCCAGCTCCCA

MINI CASE

21 Fossils allow us to understand the evolution of many lineages of plants and animals. They therefore represent a valuable scientific resource. What if *Tiktaalik* (or an equally important transitional fossil) had been found by amateur fossil hunters and sold to a private collector? Do you think there should be any regulation of fossil hunting to prevent the loss of valuable scientific information from the public domain?

Evolution

From moon rocks to DNA, clues to the history of life on Earth

 How old is Earth, and how do we know?

ANSWER: When the *Apollo 11* astronauts Neil Armstrong and Buzz Aldrin returned to Earth from their historic 1969 moon walk, they carried with them a cargo of lunar rock chipped from the moon's surface. Embedded within these hunks of shimmering anorthosite lay clues to the earliest history of our solar system, including the planet we call home.

According to the nebular hypothesis, the most widely accepted explanation for the formation of our solar system, the planetary objects in our solar system are the result of a single event: the collapse of a swirling solar nebula, which formed both the sun and the planets out of cosmic dust. Since all the planets were formed at roughly the same time, we can date the age of the solar system by dating any planetary object within it. Of the many moon rocks obtained over the course of the six *Apollo* missions, the oldest have been

▶▶ DRIVING QUESTIONS

1. What do we know about the history of life on Earth, and how do we know it?

2. What factors help to explain the distribution of species on Earth?

3. What are the major groups of organisms, and how are organisms placed in groups?

Visitors admire a display in the Hall of Biodiversity at the American Museum of Natural History in New York City.

MAMMALS

MAMMALS   CHELICERATES   CRUSTACEANS   INSECTS AND MYRIAPODS   MENTED W

The Genesis Rock, a sample of lunar crust from about the time the moon was formed, was retrieved by Apollo 15 astronauts James Irwin and David Scott in 1971.

calculated to be some 4.4 to 4.5 billion years old, which means that Earth is at least that old as well.

Why go to the moon to date Earth? With the exception of a few meteorite battle scars, the moon's surface has remained largely intact over the course of its existence. In contrast, Earth is a swirling ball of molten lava that continuously churns and digests its rocky outer crust. Because of this perpetual changing, it is difficult to find original, undisturbed rocks from Earth's earliest period. The oldest known rocks on Earth's surface, the Acasta Gneiss in a remote region of northern Canada, are 4.0 billion years old. Other ancient Earth rocks, from Western Australia, contain single mineral crystals that are 4.4 billion years old.

While these values do not establish an absolute age of Earth, they do provide a lower limit: Earth is at least as old as the materials that make it up. From these Earth minerals and moon rocks, as well as material from meteorites that have fallen to Earth, scientists estimate that the age of Earth—and of the solar system more generally—is 4.54 billion years, give or take a few million years.

How are such rocks, extraterrestrial or earthly, dated? The most important method is **radiometric dating,** in which the amount of radioactivity present in a rock is used as a geologic clock. When rocks form, the minerals in them contain a certain amount of **radioac-** tive isotopes—atoms of elements such as uranium-238, potassium-40, and rubidium-87—that are unstable and decay into other atoms.

Radioactive isotopes decay by releasing high-energy particles from the nucleus, a change that causes one element literally to transform into another. For example, an atom of the radioactive isotope uranium-238 eventually decays into a stable atom of lead-206. The time it takes for half the isotope in a sample to break down is called its **half-life.**

Different radioactive elements decay at different rates. Uranium-238 has a half-life of 4.5 billion years, whereas potassium-40 has a half-life of 1.3 billion years. The half-life of carbon-14 (useful for dating once-living, organic remains) is relatively short: it decays to nitrogen-14 in just 5,730 years. Because the isotopes decay at a known and constant rate, they can be used to determine the age of the materials in which they're found (**INFOGRAPHIC 17.1**).

As wind and water washed over rocks throughout Earth's history, they stripped off, or eroded, particles and carried them to other places. Sometimes the deposited particles were compressed over many years into new rock layers by water or by additional particles. Such rock, called sedimentary rock, can be seen in the distinctive striations, or stripes, marking successive layers of sandstone and limestone found in former riverbanks like those surrounding the Grand Canyon in Arizona. Most fossils are found in sedimentary rocks.

Rocks can also form suddenly as erupting volcanoes spew lava and ash over an area. When this molten debris cools and hardens, it forms what is called igneous rock ("igneous" is from the Latin word for "fire").

Radiometric dating is typically performed on igneous rocks, with the uranium-lead method used to date the oldest igneous rocks. Here's how it works: When igneous rocks form, crystals of the mineral zircon are produced within the rock. Zircon crystals have a highly ordered structure that incorporates uranium but excludes lead; when these crystals first form, there is no lead present and the radioactive clock is set to zero. Over time, the uranium decays at a constant rate into lead-206.

▶RADIOMETRIC DATING
The use of radioactive isotopes as a measure for determining the age of a rock or fossil.

▶RADIOACTIVE ISOTOPES
An unstable form of an element that decays into another element by radiation, that is, by emitting energetic particles.

▶HALF-LIFE
The time it takes for one-half of a sample of a radioactive isotope to decay.

By measuring the ratio of uranium to lead in these crystals, scientists can calculate the age of the rock (INFOGRAPHIC 17.2).

Sedimentary rocks cannot be dated by radiometric methods because they are made up of particles from rocks of various ages. But when fossils are found in sedimentary rock between layers of igneous rock, the dating of the igneous layers gives an accurate estimate of the age range of the fossils sandwiched in between.

Dating rocks by radioactive isotopes is quite precise and can be confirmed by cross-checking with different methods. For example, scientists used three different methods to date minerals taken from layers of rock in Saskatchewan, Canada: the potassium-argon method yielded an age of 72.5 million years; the uranium-lead method, an age of 72.4 million years; and the rubidium–strontium method, an age of 72.54 million years.

By radioactive dating, scientists have established that Earth is indeed quite old–old enough for evolution to have been acting for billions of years. They have also been able to establish precisely the dates of key events in the timeline of life on Earth as represented in the fossil record.

INFOGRAPHIC 17.1 UNSTABLE ELEMENTS UNDERGO RADIOACTIVE DECAY

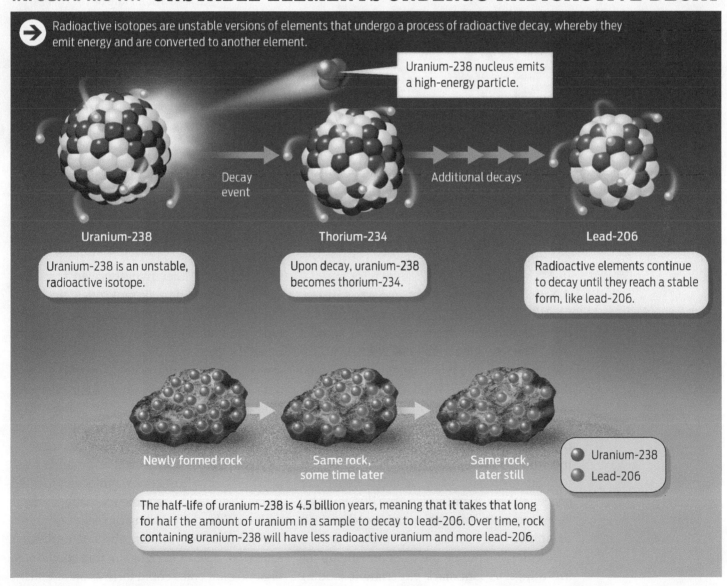

Radioactive isotopes are unstable versions of elements that undergo a process of radioactive decay, whereby they emit energy and are converted to another element.

Uranium-238 nucleus emits a high-energy particle.

Decay event

Additional decays

Uranium-238

Thorium-234

Lead-206

Uranium-238 is an unstable, radioactive isotope.

Upon decay, uranium-238 becomes thorium-234.

Radioactive elements continue to decay until they reach a stable form, like lead-206.

Newly formed rock

Same rock, some time later

Same rock, later still

● Uranium-238
● Lead-206

The half-life of uranium-238 is 4.5 billion years, meaning that it takes that long for half the amount of uranium in a sample to decay to lead-206. Over time, rock containing uranium-238 will have less radioactive uranium and more lead-206.

Volcanic eruptions deposit rock and ash that contain zircon crystals that are 100% uranium-238 0% lead.

→ Some rock types, like those produced during volcanic eruptions, contain radioactive minerals like zircon that can be used to determine the age of the rock. Because the isotope uranium-238 decays to lead at a constant rate, the age of rock layers containing these minerals can be calculated by measuring the ratio of uranium-238 to lead-206 in the mineral sample.

1.3 billion years ago, a volcanic eruption deposited this layer of rock and ash.

2.4 billion years ago, a volcanic eruption deposited this layer of ash.

Percent uranium-238

1.3 billion-year-old zircon has decayed to 85% uranium-238:15% lead.

2.4 billion-year-old zircon has decayed to 65% uranium-238:35% lead.

Constant rate of uranium-238 decay

y-axis: 100%, 85%, 75%, 65%, 50%, 25%

x-axis: Today, 2.25, 4.5, 6.75, 9.0, 11.25, 13.5

Billions of years ago (bya)

BIOCHEMISTRY

Q When and how did life begin?

ANSWER: At some point in Earth's distant past, life did not exist. Then, at a later point, it did. Where did this life come from? How did it start? The transition from nonliving to living occurred more than 3 billion years ago and left no discernible evidence. We can now only hypothesize how that transition might have occurred.

Scientists have offered a number of hypotheses to explain how life began on Earth, including the idea that it arrived here fully formed on an asteroid or meteorite from outer space. Others hypothesize that life emerged in stages over time, as inorganic chemicals combined into successively more complex molecules, including the ones making up living things. A landmark experiment lending

support to this hypothesis was performed by University of Chicago chemist Harold Urey and his 23-year-old graduate student Stanley Miller in 1953.

Urey and Miller hypothesized that they could synthesize organic molecules–the building blocks of life–by replicating the chemical environment of the early Earth. They combined the gases hydrogen (H_2), methane (CH_4), ammonia (NH_3), and water vapor (H_2O) in a flask filled with warm water–their best estimate of what the "primordial soup" was like. They then replicated lightning by discharging sparks into the chamber with an electrode.

As the gases condensed and rained back into the flask, a host of new molecules formed from the basic ingredients–among them amino acids, the building blocks of proteins. This landmark experiment showed for the first time that, given the right starting conditions, organic molecules could form spontaneously, on their own, from the inorganic materials believed to be present in the primordial soup.

Since Urey and Miller's experiment, other researchers have confirmed and extended their results, showing that it is possible, by varying the composition of the starting materials, to produce from inorganic precursors essentially all the organic molecules used by living organisms, including all 20 amino acids, as well as sugars, lipids, nucleic acids, and even ATP–the molecule that powers almost all life on Earth. These results are significant because organic molecules are the building blocks of life, and without them life could not exist. Their synthesis in the primordial past would therefore have been a necessary first step for life to emerge. The importance of organic molecules for life is the main reason that NASA's *Curiosity* rover is currently looking for evidence of them on Mars–to see if Mars could have once supported or harbored life (see Chapter 2).

Although organic molecules are a prerequisite for life, they are not themselves alive. To be alive, something must, among other things, be able to grow and reproduce. Today, of course, cells carry out these life-sustaining functions. How then did living cells come about? Recall from Chapter 2 that cell boundaries are defined by a lipid membrane. Researchers hypothesize that at a certain point the lipid molecules in the primordial soup formed bubbles, a reasonable conclusion since lipids are hydrophobic and naturally form bubbles in water. The other organic molecules were incorporated into these bubbles and, over the course of millions of years, these membrane-bound bubbles filled with organic molecules eventually became cells, capable of reproducing. While these ideas are highly speculative, research on microorganisms living today in such unlikely places as hydrothermal vents at the bottom of the ocean are giving us concrete insights into how life might have begun (see Chapter 18).

PALEONTOLOGY

Q What was life like millions of years ago?

ANSWER: Humans weren't around millions of years ago, so we have no cave paintings or other records to help us picture what life on Earth was like. Most of what we know about past life on Earth comes from fossils–the preserved remains of once-living organisms, such as *Tiktaalik*, discussed in Chapter 16.

Stanley Miller recreates the experiment he first performed in 1953 with Harold Urey.

While each fossil find is a treasure, any single specimen reveals only a tiny slice of evolutionary history. What paleontologists really want to understand is how each fossil fits into the larger story told by the fossil record. By dating the rock layers, or strata, near where fossils are buried, scientists can determine when different organisms lived on Earth. Combined with geological evidence, the fossil record has enabled scientists to construct a timeline of life on Earth (INFOGRAPHIC 17.3).

INFOGRAPHIC 17.3 GEOLOGIC TIMELINE OF LIFE ON EARTH

LIFE IN WATER | **LIFE ON LAND**

Precambrian

- 3,500 mya — First evidence of prokaryotic cells.
  Prokaryotes dominate the oceans. — **Archean**
- 2,500 mya
  First eukaryotes appear. Soft-bodied invertebrates develop. — **Proterozoic**
- 545 mya
  Expansion of ocean animal diversity. — **Cambrian**
- 495 mya
  First coral reefs. Primitive fish. Ocean plants diversify. First fungi appear. — **Ordovician**
- 439 mya — Mass extinction

Paleozoic Era

  Seedless plants, primitive insects and soft-bodied animals appear on land. — **Silurian**
- 408 mya
  Fish species diversify. First insects and seed-bearing plants appear on land. — **Devonian**
- 354 mya — Mass extinction
  Amphibians appear and begin to diversify. — **Carboniferous**
- 290 mya
  Oceans abundant in coral species. Reptiles appear on land. — **Permian**
- 251 mya — Mass extinction
  Ocean life diversifies in recovery from Permian extinction. Dinosaurs and mammals appear on land. — **Triassic**
- 206 mya — Mass extinction

Mesozoic Era

  First flowering plants and bird species appear. Large dinosaurs are plant-eaters. — **Jurassic**
- 144 mya
  Dinosaurs diversify. Cone-bearing and flowering plants dominate many habitats. — **Cretaceous**
- 65 mya — Mass extinction

Cenozoic Era

  Mammals, birds and flowering plants diversify. Grasses appear. First primates and early humans appear. — **Tertiary**
- 1.8 mya
  Many large mammals become extinct. Appearance of modern organisms present today. — **Quaternary**
- Today

The timeline shows that during the 4.6 billion years that Earth has been around, its geography and climate have gone through dramatic changes. For the first few hundred million years or so Earth was a fiery inferno, coursing with seas of molten lava and bombarded by meteorites. Not until things simmered down and the surface cooled a bit, around 3.8 billion years ago, could it support life.

The oldest known fossils date from some 3.5 billion years ago, when Earth's climate was very different from what it is today, notably because of the absence of substantial amounts of oxygen (O_2). In this oxygenless world, the only organisms that could thrive were unicellular prokaryotes that did not rely on oxygen for their metabolic reactions. With the emergence and proliferation of unicellular photosynthetic organisms, between 3.0 and 2.5 billion years ago, oxygen began to accumulate in the atmosphere, opening the door for more-complex eukaryotic organisms to evolve.

The first multicellular, eukaryotic organisms to make use of this oxygen were green algae, which appeared 1.2 billion years ago. Soft-bodied aquatic animals followed, about 600 million years ago, but it is only since about 545 million years ago, during the Cambrian Period, that we see fossil evidence of a truly diverse animal world. During the Cambrian explosion, as this event is known, ocean life swelled with a mind-boggling array of strange-looking creatures, including *Opabinia*, an organism with five eyes and a snout resembling a vacuum-cleaner hose, discovered in fossils from this period.

The first organisms to colonize land were primitive plants, appearing roughly 450 million years ago. By 350 million years ago, forests of seedless plants covered the globe.

Then, 250 million years ago, life was drastically cut down: roughly 95% of living species were extinguished in a mass die-off known as the Permian **extinction.** Scientists do not know what caused the Permian extinction, but some hypothesize that massive volcanic activity filled the atmosphere with heat-trapping gases that led to a rapidly changing climate. The Permian extinction wasn't bad for all organisms, though; some flourished as space and resources opened

The Burgess Shale, in the Canadian Rockies, is famous for the fine state of preservation of the soft parts of its fossils.

up for the survivors, who spread and diversified as a result in a phenomenon known as **adaptive radiation.** Among these were reptiles, who thrived in the hot, dry climate of the following Triassic Period. The most famous group of reptiles, the dinosaurs, dominated the land for nearly 200 million years, thanks to a combination of drought-resistant skin and fast-moving legs, until they died out in another **mass extinction** at the end of the Cretaceous Period, 65 million years ago.

The reason for the extinction of the dinosaurs was a mystery for many years. Evidence now suggests that what killed off the dinosaurs (and 60% of the other species living at the time) was a massive 6-mile-wide asteroid that plowed into Earth with almost unimaginable force, sending a thick layer of soot and ash into the atmosphere and blocking out the sun for months. A crater 110 miles wide in Mexico's Yucatán peninsula, near the town of Chicxulub, is the likely impact site.

With the extinction of the dinosaurs, it was mammals' chance to spread and diversify on land and thus give rise to many of the species of organisms we see on the planet today. This pattern of sudden change—extinctions followed by adaptive radiations—is seen in the fossil record and is an example of **punctuated equilibrium,** in which most evolutionary change occurs in sudden bursts related to environmental change rather than taking place gradually.

▶**EXTINCTION**
The elimination of all individuals in a species; extinction may occur over time or in a sudden mass die-off.

▶**ADAPTIVE RADIATION**
The spreading and diversification of organisms that occur when they colonize a new habitat.

▶**MASS EXTINCTION**
An extinction of between 50% and 90% of all species that occurs relatively rapidly.

▶**PUNCTUATED EQUILIBRIUM**
Periodic bursts of species change as a result of sudden environmental change.

▶**BIOGEOGRAPHY**
The study of how organisms are distributed in geographical space.

BIOGEOGRAPHY

Q Why are there no penguins at the north pole, and no polar bears at the south pole?

ANSWER: In terms of habitat, the north pole and the south pole are pretty similar: cold, snow-capped, and surrounded by ocean. Yet polar bears are found only in the north and penguins are found only in the south. Why? If environmental conditions were the only fac-

tors, then you might expect that polar bears and penguins would do equally well in either habitat and should therefore coexist in the same location (as they sometimes do in TV commercials). But the distribution of organisms around the world reflects the details of their evolutionary history—their connection in time and space to the species they are related to and descended from. In fact, this was one of the main pieces of evidence that Darwin used to support his theory of descent with modification. His voyage on the *Beagle* was in a sense the first work in **biogeography**, the study of the natural geographic distribution

INFOGRAPHIC 17.4 MOVEMENT OF EARTH'S PLATES INFLUENCES CLIMATE AND BIOGEOGRAPHY

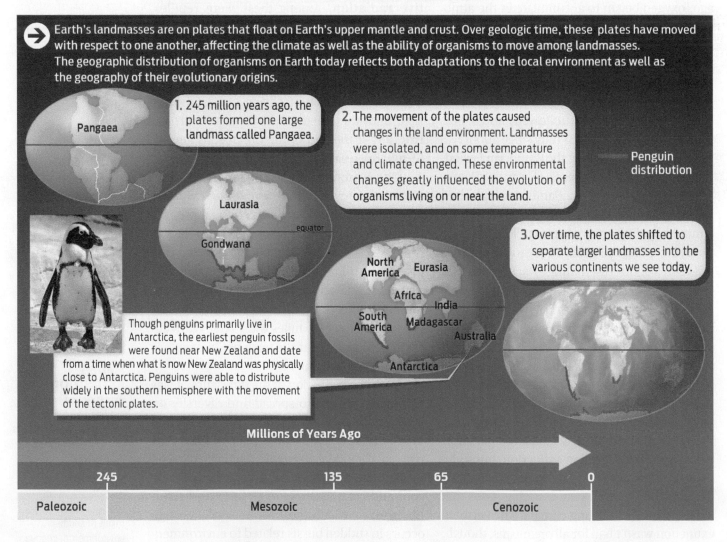

Earth's landmasses are on plates that float on Earth's upper mantle and crust. Over geologic time, these plates have moved with respect to one another, affecting the climate as well as the ability of organisms to move among landmasses. The geographic distribution of organisms on Earth today reflects both adaptations to the local environment as well as the geography of their evolutionary origins.

1. 245 million years ago, the plates formed one large landmass called Pangaea.

2. The movement of the plates caused changes in the land environment. Landmasses were isolated, and on some temperature and climate changed. These environmental changes greatly influenced the evolution of organisms living on or near the land.

3. Over time, the plates shifted to separate larger landmasses into the various continents we see today.

Though penguins primarily live in Antarctica, the earliest penguin fossils were found near New Zealand and date from a time when what is now New Zealand was physically close to Antarctica. Penguins were able to distribute widely in the southern hemisphere with the movement of the tectonic plates.

Penguin distribution

Millions of Years Ago

245   135   65   0

Paleozoic | Mesozoic | Cenozoic

of species. Biogeography seeks to explain why particular organisms are found in some areas and not others.

Penguins, for example, make their home primarily in the southern hemisphere, especially in the coastal regions of Antarctica. According to fossil evidence, penguins first appeared about 65 million years ago near what is now southern New Zealand. Polar bears, on the other hand, live only in the Arctic. From fossil and DNA evidence, scientists conclude that polar bears evolved from brown bears roughly 150,000 years ago in Siberia, when that region became isolated by glaciers. Both penguins and polar bears have thus lived in their respective habitats for a long time, and it's easy to understand why they haven't migrated from one pole to the other—obstacles include great distance and the warmer oceans separating their icy habitats. But how did they get to their homes in the first place?

Though today they are at opposite ends of Earth, the Arctic and Antarctic landmasses weren't always so far apart. In fact, 250 million years ago, the continents were bound together in one large landmass that geologists named Pangaea. At that time it was hypothetically possible for populations of land-dwelling animals to roam far and wide over the entire land surface. But because of a geologic process called **plate tectonics,** over time this giant landmass split and split again, forming the continents of the northern and southern hemispheres. In the process, the ancestors of penguins and polar bears were isolated from each other, as if on different lifeboats cast out to sea. Because the animals we know today as penguins and polar bears evolved from their ancestors after the split of the northern and southern landmasses, they are found today at different ends of Earth. Plate tectonics also helps explain how a flightless bird such as a penguin could spread to four distant continents in the south hemisphere: the landmasses were much closer then (INFOGRAPHIC 17.4).

Tectonic plates continue to move, occasionally causing dramatic events such as earthquakes and volcanic eruptions. GPS measurements, used to track the direction and velocity of plate movement, show that the rate of movement for different plates ranges from 2.5 to 15 cm per year. Earth is a work in progress.

CONVERGENT EVOLUTION

Q Are creatures that look alike always closely related?

ANSWER: Polar bears share many traits with their brown-bear cousins—both species are recognizable as bears despite obvious differences in color. The fact that polar bears resemble brown bears is persuasive evidence that the two species share a recent common ancestor. But common ancestry is not the only reason that two species might appear similar. Even species that are not closely related may share similar adaptations as a result of independent episodes of natural selection, a phenomenon called **convergent evolution.**

Cold-dwelling fish provide a good example. In the frigid waters of the Antarctic Ocean, fish have a unique adaptation that keeps them from becoming ice cubes: their blood is pumped full of "antifreeze." Fish blood contains molecules called glycoproteins that lower the temperature at which body fluids freeze; the glycoproteins surround any tiny ice crystals that may form in the blood and stop them from growing. Arctic fish, at Earth's other pole, also have "antifreeze proteins," but the genes that code for them are different.

Arctic and Antarctic fish diverged from their common ancestor long before each species developed antifreeze proteins, which means these adaptations must have evolved more than once. In other words, at least two separate, independent episodes of evolution occurred with the same functional results. Sometimes similar environmental challenges will favor the same adaptations time and time again.

▶**PLATE TECTONICS**
The movement of Earth's upper mantle and crust, which influences the geographical distribution of landmasses and organisms.

▶**CONVERGENT EVOLUTION**
The process by which organisms that are not closely related evolve similar adaptations as a result of independent episodes of natural selection.

17

▶**TAXONOMY**
The process of identifying, naming, and classifying organisms on the basis of shared traits.

DIVERSITY AND TAXONOMY

Q How many species are there on Earth, and how do scientists keep track of them?

ANSWER: Current estimates of the total number of species on Earth range anywhere from 5 to 30 million, of which 1.5 million or so have been formally described. Many of these species are found in diversity hot spots like rain forests, and new species are constantly being discovered—on the order of 17,000 new species a year (**INFOGRAPHIC 17.5**).

With so many species out there, how do scientists keep track of them all? The process by which scientists systematically identify, name, and classify organisms is called **taxonomy.** (Taxonomy is part of the broader study of systematics, the study of biological diversity of life on Earth.)

Taxonomy is an attempt to impose a human sense of order on this vast array of species, categorizing them on the basis of features they have in common, such as whether their cells are eukaryotic or prokaryotic, whether or not they photosynthesize, whether or not they have four legs and fur.

To sort organisms, taxonomists use a system of seven progressively narrower categories: kingdom, phylum, class, order, family, genus, species. As you move down the list, from kingdom to species, the categories are increasingly exclusive, until finally only one member is included. The genus and species names provide a useful scientific identifier for every living organism. Because that scientific name is in Latin, it can be easily recognized in many languages.

Take humans, for example. Humans are animals, members of the kingdom Animalia. Within the animal kingdom, they belong to the phylum Chordata, a group that includes the

INFOGRAPHIC 17.5 HOW MANY SPECIES ARE THERE?

→ The numbers of species in each group below represents only those that have been formally characterized and classified. The true number of species is likely to be much higher (estimated numbers are shown in parentheses). Prokaryotic diversity may be immeasurable because of their size and ability to live in just about every environment on the planet. Therefore, it is not currently possible to make reliable estimates for the true number of prokaryotic species.

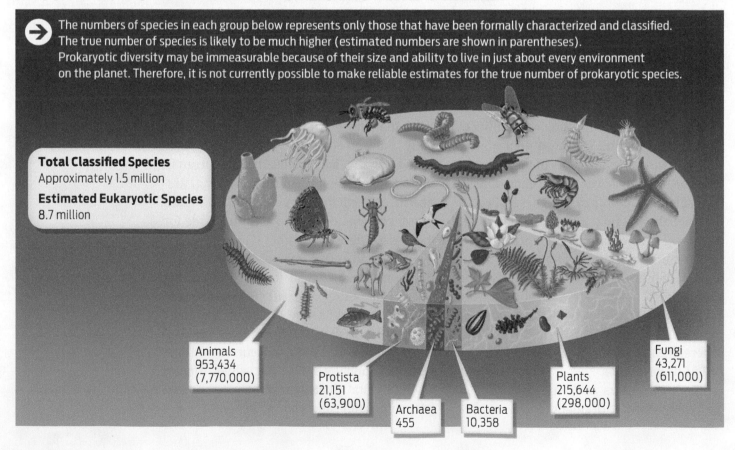

Total Classified Species
Approximately 1.5 million

Estimated Eukaryotic Species
8.7 million

Animals
953,434
(7,770,000)

Protista
21,151
(63,900)

Archaea
455

Bacteria
10,358

Plants
215,644
(298,000)

Fungi
43,271
(611,000)

INFOGRAPHIC 17.6 CLASSIFICATION OF SPECIES

 Organisms are classified into groups that are increasingly exclusive. In the broadest category (animal kingdom), all animals are included. Closely related organisms are grouped based on morphological, nutritional, and genetic characteristics. There are far fewer organisms in an order than in a phylum.

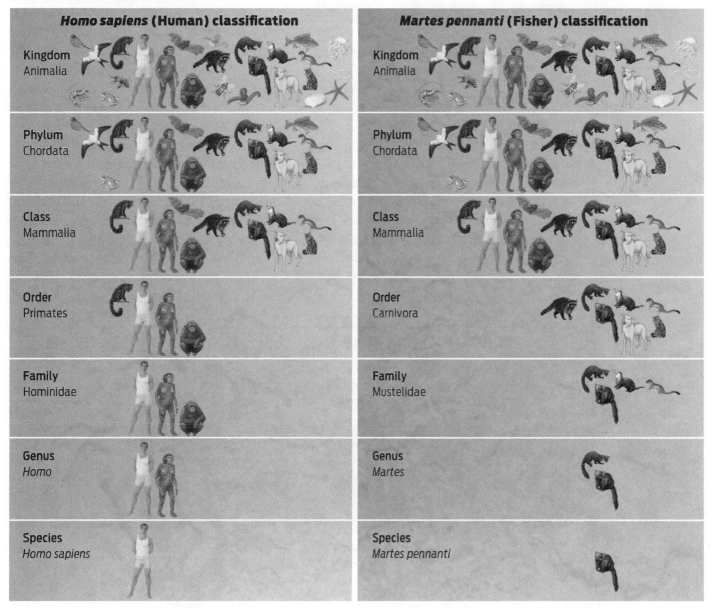

Homo sapiens (Human) classification	*Martes pennanti* (Fisher) classification
Kingdom Animalia	**Kingdom** Animalia
Phylum Chordata	**Phylum** Chordata
Class Mammalia	**Class** Mammalia
Order Primates	**Order** Carnivora
Family Hominidae	**Family** Mustelidae
Genus *Homo*	**Genus** *Martes*
Species *Homo sapiens*	**Species** *Martes pennanti*

vertebrates, animals with a rigid backbone. Further, humans are **mammals,** members of the class Mammalia; they share with all members of this class mammary glands and a body that is covered with hair. Humans belong to the Primate order, which also includes monkeys, apes, and lemurs. And humans are members of the Hominidae family, and so are closely related to their fellow hominids: chimpanzees, gorillas, and orangutans. Our scientific name—made up of our genus and species names—is *Homo sapiens* ("wise human") **(INFOGRAPHIC 17.6).**

Classification would seem to be a simple matter—just observe, measure, and sort. But deciding which category an organism belongs

▶**VERTEBRATES**
Animals with a rigid backbone.

▶**MAMMALS**
Members of the class Mammalia; all members of this class have mammary glands and a body covered with hair.

▶**PHYLOGENY**
The evolutionary history of a group of organisms.

▶**PHYLOGENETIC TREE**
A branching diagram of relationships showing common ancestry.

in can sometimes be tricky, as the example of convergent evolution has shown. Sometimes, to properly classify organisms, scientists have to look a little deeper.

READING PHYLOGENETIC TREES

 Is a crocodile more closely related to a bird or to a lizard?

ANSWER: The fact that all land vertebrates have four limbs and the same forelimb bones indicates that they all share a common ancestor. But how precisely are they related? In other words, who's more closely related to whom? Biologists want not only to categorize organisms, but also to have those categories reflect **phylogeny,** the actual evolutionary history of the organisms. This history is represented

visually by a diagram called a **phylogenetic tree,** which is similar in some respects to a family tree.

Phylogenetic trees can be drawn in a number of ways, but most have certain features in common. At the base, or root, is the common ancestor shared by all organisms on the tree. Over time, and with different selective pressures, different groups of organisms diverged from that common ancestor and from one another, leading to separate branches on the tree. The points on the tree at which these branch points occur are called nodes. A node represents the common ancestor shared by all organisms on the branches above that node. At the very tips of the branches we find the most recent organisms in that lineage, including living organisms and organisms that became extinct. We can thus establish relationships between living organisms (at the tips of the

INFOGRAPHIC 17.7 HOW TO READ A PHYLOGENETIC TREE

→ Evolutionary history, or phylogeny, is represented visually by a phylogenetic tree. Trees have a common structure, with a root, nodes, and branch points. To determine evolutionary relationships among living or extinct organisms, consider the most recent common ancestors.

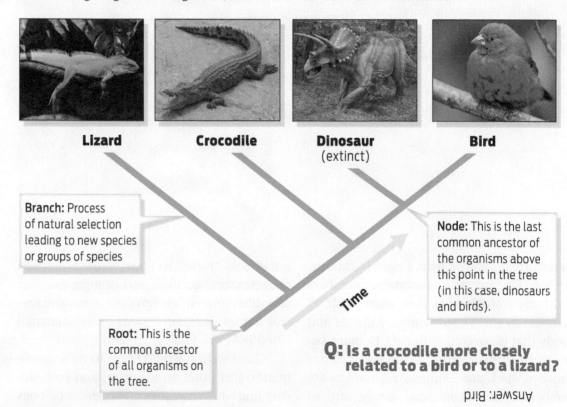

Lizard   Crocodile   Dinosaur (extinct)   Bird

Branch: Process of natural selection leading to new species or groups of species

Node: This is the last common ancestor of the organisms above this point in the tree (in this case, dinosaurs and birds).

Root: This is the common ancestor of all organisms on the tree.

Time

Q: Is a crocodile more closely related to a bird or to a lizard?

Answer: Bird

branches) on the basis of the ancestors they share. The more recently two groups share a common ancestor, the more closely they are related (**INFOGRAPHIC 17.7**).

A phylogenetic tree is a visual representation of the best hypothesis we currently have for how species are related, based on a shared evolutionary history. The evidence for a phylogenetic tree comes from many sources, including the fossil record, physical traits, and shared DNA sequences. For many years, biologists relied solely on observable physical or behavioral features to construct phylogenetic, or evolutionary, trees. But with the genetic revolution, it's become common to include DNA evidence. Typically, researchers compare sequence differences in a gene that is found in all living organisms, such as a ribosomal RNA (rRNA) gene.

Sometimes the new genetic information yields surprises. Modern genetic evidence shows, for example, that crocodiles are more closely related to birds than they are to lizards, appearances notwithstanding. Genetics, you might say, is shaking the evolutionary tree.

CLASSIFICATION AND PHYLOGENY

How many branches does the tree of life have?

ANSWER: Since each living species sits on its own branch of a phylogenetic tree, the complete tree of life has as many branches as there are species in the world. Today's species are like thin twigs in the upper branches of an enormous oak tree. Closer to the bottom of the tree, nearer to the ancient trunk, however, we find significant forks. Just how many forks there are at the bottom of the tree is a question that has been debated for decades.

Before the 18th century, biologists divided living things into just two main categories: animals and plants. This classification was based on whether an organism moved around and ate or did not move around and eat. By the mid-19th century, the microscope had revealed

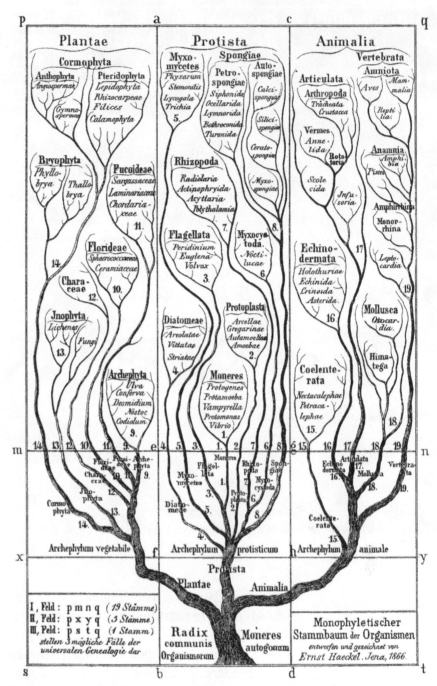

An early version of the tree of life, drawn by Ernst Haeckel in 1866. Haeckel's tree was the first to include microscopic organisms known as protists.

a whole new world of organisms, and so a third branch was added to life's tree. The German biologist and artist Ernst Haeckel, in 1866, was the first to draw a phylogenetic tree that included these microscopic organisms, which he termed Protista, or protists.

By the 1960s, taxonomists realized that even three branches did not fully capture

INFOGRAPHIC 17.8 GENETICS DEFINES THREE DOMAINS OF LIFE: BACTERIA, ARCHAEA, EUKARYA

→ All living organisms have evolved from a common ancestor. On the basis of genetic evidence, we can group living things into one of three domains of life, each with a distinct evolutionary history. While the Bacteria and Archaea both have prokaryotic cells, they have distinct evolutionary histories, with Archaea being genetically more closely related to Eukarya than to Bacteria. The domain Eukarya encompasses protists, plants, fungi, and animals, including humans (see Chapter 19).

Bacteria
- Spirochetes
- Green filamentous bacteria
- Gram-positive bacteria
- Proteobacteria
- Cyanobacteria
- Planctomyces
- Bacteroides Cytophaga
- Thermotoga
- Aquifex

Archaea
- Methanobacterium
- Methanosarcina
- Methanococcus
- Halophiles
- T. celer
- Thermoproteus
- Pyrodicticum

Eukarya
- Slime molds
- Animals
- Fungi
- Plants
- Ciliates
- Flagellates
- Trichomonads
- Microsporidia
- Diplomonads

Common Ancestor

▶ **DOMAIN**
The highest category in the modern system of classification; there are three domains— Bacteria, Archaea, and Eukarya.

the diversity of life; many organisms—such as fungi—didn't fit neatly into any of these groups, and so another classification scheme was proposed. This one grouped all living organisms into five large kingdoms on the basis of how they obtained their food (by eating, photosynthesizing, or decomposing) and whether they had eukaryotic or prokaryotic cells. The five kingdoms were Animalia, Plantae, Fungi, Protista, and Monera. Protista included mostly single-cell eukaryotic organisms (such as the amoeba), and Monera included all prokaryotic organisms (such as bacteria).

Yet even this revised classification scheme eventually had to be overhauled as more information became available. In the 1970s, genetic studies by Carl Woese revealed that, on the basis of genetic relatedness, not all prokaryotes could be lumped together; likewise, protists were too genetically diverse to be put in one category. Consequently, scientists now group organisms into one of three large **domains**— Bacteria, Archaea, and Eukarya—which represent three fundamental branch points in the trunk of the evolutionary tree. The original kingdom Monera is now divided into two domains, Archaea and Bacteria. Within the domain Eukarya, Animalia, Plantae, and Fungi remain recognized kingdoms, but the protists (members of the former kingdom Protista) are dispersed across the eukaryotic domain of life on the basis of genetic evidence (**INFOGRAPHIC 17.8**).

CHAPTER 17 Summary

▶ The age of Earth and its rock layers can be determined by measuring the amount of radioactive isotopes present in rocks, a method known as radiometric dating.

▶ Life on Earth may have emerged in stages, as inorganic molecules combined to form organic ones in the primordial soup, and as these were incorporated into lipid bubbles to form cells.

▶ From geological evidence and the fossil record, paleontologists have been able to construct a geologic timeline of life on Earth.

▶ Earth's history can be divided into important eras and periods. Dinosaurs, for example, lived primarily from 250 to 65 million years ago, during the Mesozoic Era, from the Triassic through the Cretaceous periods.

▶ The history of life on Earth is marked by repeated extinctions and adaptive radiations, a phenomenon of intermittent rather than steady change known as punctuated equilibrium.

▶ Ancient movement of Earth's major landmasses affected the eventual distribution of species around the globe, the study of which is known as biogeography.

▶ Convergent evolution is the evolution of similar adaptations in response to similar environmental challenges in groups of organisms that are not closely related.

▶ Life is astoundingly diverse. Current estimates of the total number of species on Earth range anywhere from 5 to 30 million, of which 1.5 million have been formally described.

▶ Biologists sort organisms into a series of nested categories based on shared anatomical and genetic features: domain, kingdom, phylum, class, order, family, genus, species.

▶ The scientific name of an organism is given by its genus and species names (the scientific name of humans is *Homo sapiens*).

▶ Both physical evidence and genetic evidence are used to understand evolutionary history, or phylogeny. Branching trees of common ancestry are used to represent that history visually.

▶ On the basis of genetic evidence, all living organisms can be classified into one of three domains: Bacteria, Archaea, or Eukarya.

MORE TO EXPLORE

▶ Newman, W. L. (1997) "Age of the Earth." In *Geologic Time*. U.S. Geological Service. http://pubs.usgs.gov/gip/geotime/age.html

▶ Gould, S. J. (1989) *Wonderful Life: Burgess Shale and the Nature of History*. New York: W. W. Norton.

▶ Mora, C., et al. (2011) How many species are there on Earth and in the ocean? *PLOS Biology* 9(8):e1001127. doi:10.1371/journal.pbio.1001127. http://www.plosbiology.org/article/info:doi/10.1371/journal.pbio.1001127

▶ TimeTree, The Timescale of Life http://www.timetree.org/index.php

▶ Woese, C.R., et al. (1990) Towards a natural system of organisms: proposal for the domains Archaea, Bacteria, and Eucarya. *Proceedings of the National Academy of Sciences* 87:4576–4579

CHAPTER 17 Test Your Knowledge

DRIVING QUESTION 1

What do we know about the history of life on Earth, and how do we know it?

By answering the questions below and studying Infographics 17.1, 17.2, and 17.3, you should be able to generate an answer for the broader Driving Question above.

KNOW IT

1 What do uranium-238, carbon-14, and potassium-40 have in common?

2 To date what you suspect to be the very earliest life on Earth, which isotope would you use: uranium-238, carbon-14, or potassium-40? Explain your answer.

3 Place the following evolutionary milestones in order from earliest (1) to most recent (7), providing approximate dates to support your answer.

_____ The first multicellular eukaryotes

_____ The first prokaryotes

_____ The Permian extinction

_____ The Cambrian explosion

_____ The first animals

_____ The extinction of dinosaurs

_____ An increase in oxygen in the atmosphere

USE IT

4 Consider a rock formed at about the same time as Earth was formed.

a. How old is this rock?

b. How much of the original uranium-238 is likely to be left today in that rock?

5 Diverse animal fossils are found dating from the Cambrian Period, not earlier. Why might these organisms have made their first appearance in the fossil record only then, even though their ancestors may have been living, and evolving, for a long time before the Cambrian? (Think about what kinds of new structures might have evolved during the Cambrian Period that would have allowed these organisms to leave fossils.)

MINI CASE

6 Along the banks of a river, some sedimentary rock strata have been revealed by erosion. By radiometric dating, the layer above these strata is determined to be ~290 million years old, and the layer beneath has been dated to ~354 million years ago. A paleontologist starts to uncover fossils in the sedimentary rock strata. The fossils are clearly land-dwelling vertebrates. Are they more likely to be reptiles or amphibians? Explain your answer.

INTERPRETING DATA

7 You have carried out radiometric analysis on four igneous rocks uncovered at several sites you are exploring. From the % lead you determine in each case, what is the approximate age of the rock?

Rock A: 75% lead _____

Rock B: 50% lead _____

Rock C: 30% lead _____

Rock D: 10% lead _____

DRIVING QUESTION 2

What factors help to explain the distribution of species on Earth?

By answering the questions below and studying Infographic 17.4, you should be able to generate an answer for the broader Driving Question above.

KNOW IT

8 If two organisms strongly resemble each other in their physical traits, can you necessarily conclude that they are closely related? Explain your answer.

9 What did the arrangement of landmasses on Earth look like between 135 and 65 million years ago? What happened to these landmasses, and how does this change help explain the distribution of organisms found on the planet?

USE IT

10 A cactus called ocotillo (*Fouquieria splendens*), which grows in New Mexico, looks very much like *Alluaudia procera,* a species of plant that grows in the deserts of Madagascar. These two plant species are not closely related—they are in different orders in the kingdom Plantae. Why then do they look so alike?

11 If penguins and polar bears had evolved before Pangaea split into northern and southern continents, what might you predict about their geographic distribution today?

12 Both bats and insects fly, but bat wings have bones and insect wings do not. Would you consider bat and insect wings to be a result of convergent evolution, or of homology—evolution based on inheritance of similar structures from a common ancestor? Explain your answer.

DRIVING QUESTION 3

What are the major groups of organisms, and how are organisms placed in groups?

By answering the questions below and studying Infographics 17.5, 17.6, 17.7, and 17.8, you should be able to generate an answer for the broader Driving Question above.

KNOW IT

13 Which of the following is *not* a domain of life?

a. Animalia
b. Eukarya
c. Bacteria
d. Archaea
e. Plantae
f. Neither a nor e is a domain of life.

14 Put the following terms in order from most inclusive (1) to least inclusive (5).

_____ Domain

_____ Species

_____ Kingdom

_____ Genus

_____ Phylum

15 A phylogenetic tree represents

a. a grouping of organisms on the basis of their shared structural features.
b. a grouping of organisms on the basis of their cell type.
c. a grouping of organisms on the basis of their complexity.
d. a grouping of organisms on the basis of their evolutionary history.
e. a grouping of organisms on the basis of where they are found.

USE IT

16 Why was the classification of the kingdom Monera split into two domains? What are these two domains?

17 On the tree below, which number represents the most recent common ancestor of humans and corn?

a. 1   b. 2   c. 3   d. 4

e. Humans and corn do not share any ancestors.

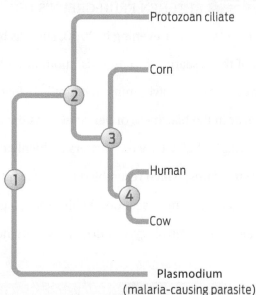

Protozoan ciliate

Corn

Human

Cow

Plasmodium
(malaria-causing parasite)

BRING IT HOME

18 Carry out some online research on the fossils found in your home state. What groups of organisms are represented in the statewide fossil record? What is the oldest fossil found in your state? What do the fossils in your state suggest about the pattern(s) of evolution in your state?

GRETCHEN FRÜH-GREEN'S HEART HAD NEVER BEAT SO fast. Late on a December evening in 2000, she was hunkered down in the control room of the research ship *Atlantis,* monitoring live video streaming up to the ship from a camera swimming half a mile below. Strange white shapes began to appear in the blackness of her screen. As the camera panned, an underwater landscape of ghostly white towers suddenly came into focus. The huge limestone structures, which resembled the stalagmites found in caves, loomed above an otherwise empty seafloor. Früh-Green, a geologist with the Institute for Mineralogy and Petrology in Zurich, Switzerland, knew immediately that

Lost City sits 2,600 feet down on an undersea mountain in the middle of the Atlantic Ocean.

▸▸ DRIVING QUESTIONS

1. What are the prokaryotic domains of life?
2. What are the features of bacteria and of archaea?
3. What are the challenges faced by organisms living at Lost City, and how do they face them?

ST CITY

Probing life's origins at
the bottom of the sea

she was looking at something special and raced to tell her colleagues. "It was really quite exciting," she said. "It was late at night and kind of woke us all up."

The next day, the mission's chief scientist, Deborah Kelley, and two members of the team climbed into a submersible craft that dived 2,600 feet down to the site, which sits on an undersea mountain in the middle of the Atlantic Ocean. There they found a dense "cityscape" of rocky spires, stretching across roughly six football fields of dark seafloor. It was an uncharted world no one had known existed. Kelley named the undersea world "Lost City." The tallest tower, measuring 200 feet, she dubbed "Poseidon," after the Greek god of the sea.

Subsequent research has shown that each tower is a type of underwater chimney, or spring, known as a hydrothermal vent. As rocks beneath Earth's crust come in contact with seawater, they react chemically, giving off a steady stream of heat and combustible gas that seeps out of each vent. The resulting fluid is highly caustic, with a pH of 9 to 11–similar to that of drain cleaner–and temperatures as hot as 100°C (212°F).

INFOGRAPHIC 18.1 HYDROTHERMAL VENTS OF LOST CITY PROVIDE AN EXTREME ENVIRONMENT FOR LIFE

The hydrothermal vents of Lost City are huge rock chimneys. Scalding fluid with an extremely basic pH flows out of the tops of these chimneys. The towering size and range of environments within and around a single Lost City spire mean it can host a variety of unique microbial communities.

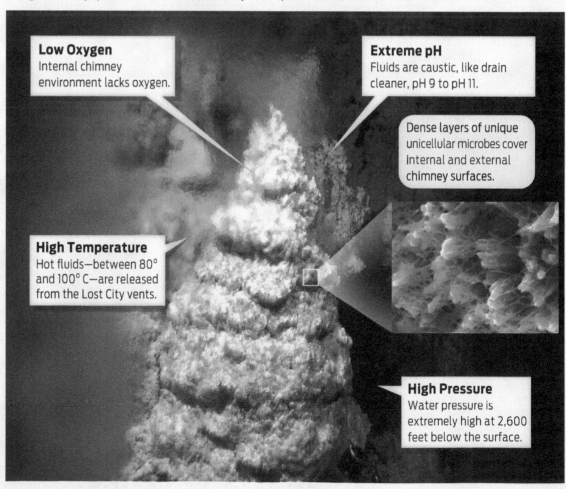

Low Oxygen
Internal chimney environment lacks oxygen.

Extreme pH
Fluids are caustic, like drain cleaner, pH 9 to pH 11.

Dense layers of unique unicellular microbes cover internal and external chimney surfaces.

High Temperature
Hot fluids—between 80° and 100° C—are released from the Lost City vents.

High Pressure
Water pressure is extremely high at 2,600 feet below the surface.

The submersible craft Alvin is deployed from the research ship Atlantis.

Though deep-sea hydrothermal vents had been discovered before, the ones at Lost City were unique in their chemistry and in the type of life they support. Nothing like them had ever been seen. "Rarely does something like this come along that drives home how much we still have to learn about our own planet," said Kelley, an oceanographer at the University of Washington in Seattle, who now leads the international team of scientists who are exploring Lost City's mysteries.

An extreme and seemingly inhospitable environment, the towers of Lost City are nonetheless home to a surprising number of life forms. The most prevalent living creatures are dense layers of unicellular microbes that coat the towers, inside and out (INFOGRAPHIC 18.1). Microbial life exists pretty much everywhere on Earth—in frozen glaciers, in radioactive dirt, in the intestines of animals—yet the microbes at Lost City have earned the fascination of scientists.

"They're living in boiling toilet bowl cleaner full of flammable gas," says Bill Brazelton, a postgraduate researcher with NASA and the University of Washington, who studies the microbes. The extreme conditions of this environment resemble what researchers think the ancient Earth was like, some 3.8 billion years ago, when life was getting started. Scientists have long wondered how life managed to emerge and flourish in such inhospitable conditions. At Lost City, they are finding some tantalizing clues.

EXPLORING THE DEEP

Since Lost City was discovered in 2000, researchers have organized three exploratory trips to the site. During each of these month-long expeditions, a team of more than 50 people—scientists, pilots, engineers, and ship crew—works around the clock to orchestrate dives and obtain specimens for research. A squad of sophisticated robotic assistants aids the effort.

> It was an uncharted world no one had known existed. Kelley named the undersea world "Lost City."

> ❝ *All the time you're looking at something that nobody's ever looked at before, and that's really cool.* ❞
>
> — BILL BRAZELTON

Researchers dive to the site in *Alvin*, a tiny three-person submersible craft. Each trip to the murky depths takes about 30 minutes and is a risky descent into dark, uncharted waters. This isn't flat seafloor, after all; the Lost City towers are like tall buildings, some as high as 18 stories. Members of the team compare the journey to flying through New York City in a helicopter at night with no lights. It's well worth the effort, though: "All the time you're looking at something that nobody's ever looked at before, and that's really cool," says Brazelton.

To collect rock samples, researchers use a pair of remotely operated mechanical vehi-

INFOGRAPHIC 18.2 INVESTIGATING LIFE IN LOST CITY

Collecting Samples

Various types of life are collected from the surface of the Lost City spires and surrounding fluid. Collection is painstakingly completed with the use of robotic claws and suction devices that move samples into collection boxes for transport to the surface.

Robotic arms grasp rock samples for further analysis on *Atlantis*.

A "slurp gun" sucks up samples containing smaller organisms.

Processing Microbes

Once Lost City microbe samples arrive in the laboratory on board *Atlantis* they are processed so that each organism can be grown in a laboratory culture and then identified.

A scientist processes microbial samples in an anaerobic glove bag that eliminates oxygen that may be harmful to Lost City microbes.

Lost City microbes are placed in tubes with specialized energy sources, like hydrogen or methane, and placed in the incubator on board *Atlantis*.

Left: Gretchen Früh-Green examines rock samples collected from Lost City.
Right: Deborah Kelley monitors a video feed from a robotic camera.

cles, *Jason* and *Hercules,* each equipped with robotic arms. The arms enable researchers to reach out and grab rocks. But that's easier said than done: it's a bit like attempting to grab a toy with a shaky mechanical claw at the arcade, only under water with strong ocean currents whipping the claw around. The chalky limestone prizes can also be quite brittle, crumbling if squeezed too tightly. Often the researchers grab something only to have it fall down between the spires.

To collect living specimens, the researchers use what they call a "slurp gun." A robotic arm aims the gun, which then gulps a sample from the spires. Everything is caught on camera by *Hercules.* Eight hours later, the vehicles return to *Atlantis* with their cargo of samples.

Because many of the Lost City microbes cannot tolerate oxygen, the biologists have to be careful not to expose them to air. Using a special airtight bag with built-in gloves, the researchers transfer samples of microbes into test tubes without introducing oxygen into their environment. The microbes are then put in warm incubators and coaxed to grow (INFOGRAPHIC 18.2).

WHAT ARE THEY?

The microbial life in Lost City poses many riddles for the scientists trying to understand this mysterious deep-sea world: What are these creatures? How are they related to known organisms? What adaptations allow them to survive?

Lost City houses a community of life forms ranging from mats of microbes to translucent 1-cm-long animals and the larger fish that eat them. But most of the living things at Lost City are **prokaryotes**—unicellular organisms whose single cell lacks internal membrane-bound organelles and whose ribbon of DNA floats freely in the cytoplasm (rather than being housed in a nucleus, as in eukaryotes; see Chapter 3). Prokaryotic organisms are microscopic, on the order of 1-10 microns, which is about 1/10 the thickness of a human hair (INFOGRAPHIC 18.3).

What prokaryotes lack in size they make up for in numbers. Prokaryotes occupy virtually every niche on the planet, and most scientists agree that we have barely scratched the surface in cataloguing their numbers and diversity. There are more prokaryotes in a handful of dirt than there are plants and animals in a rain forest. More prokaryotic organisms live on and in you right now than there are human cells in your body. At Lost City, up to 1 billion such prokaryotic organisms inhabit each gram of chimney rock, forming a mucuslike biofilm several centimeters thick. It looks like the chimneys got sneezed on, says Brazelton.

▶**PROKARYOTE**
A usually unicellular organism whose cell lacks internal membrane-bound organelles and whose DNA is not contained within a nucleus.

→ Prokaryotic cells are much smaller than eukaryotic cells and do not have the same internal organization. Prokaryotic cells lack organelles, instead carrying out all cellular functions in one central space. The DNA molecule floats freely in the cytoplasm.

Prokaryotic Organisms
· Typically single cells
· No organelles
· No nucleus; DNA floats freely

Prokaryotes are about the size of a eukaryotic mitochondrion...

...and 1/10 the diameter of a human hair.

Ribosomes

Chromosome

As their numbers testify, prokaryotes are an extraordinarily successful product of evolution. From fossil evidence, we know that prokaryotes were the first colonizers of our planet, and for nearly 2 billion years its only life form. Having first evolved nearly 4 billion years ago, prokaryotic organisms have had plenty of time to adapt to a wide range of environments, including many that would kill most eukaryotes. In fact, prokaryotes can thrive just about anywhere. At another type of deep-sea hydrothermal vent, off the Oregon coast–one with very acidic, extremely hot fluids–scientists have discovered more than 40,000 different kinds of prokaryote (INFOGRAPHIC 18.4).

How do the prokaryotic organisms found at Lost City compare to those living elsewhere? When biologists want to identify a prokaryote, they can't always rely on physical appearance,

since many prokaryotic organisms look similar under a microscope. Nor can prokaryotes necessarily be grown in the laboratory. Many of the unusual prokaryotes at Lost City were almost impossible to culture in the lab, making it hard to study their features and behavior. Instead, biologists generally rely on DNA to identify prokaryotic organisms. The number of DNA sequence similarities between the new species and known ones establishes their degree of relatedness. Finding a unique DNA sequence in a sample means the researchers have discovered a new organism.

By looking at DNA sequences, researchers have discovered several new species of prokaryotic organisms living at Lost City. Two of these species fall into a group of prokaryotes known as archaea. Archaea aren't the only prokaryotes present at Lost City–the site is rich in bacterial populations, too–but it's the archaea that are most interesting to researchers. That's because the archaea are doing things that even bacteria can't do.

At Lost City, bacterial populations congregate on the outsides of the active vents, where temperatures are relatively mild and where oxygen is present in the seawater. Archaea, by contrast, are found inside the vents, where temperatures are hottest and where there is no oxygen. So far, just two species of archaea have been detected in this environment. "The conditions are so extreme that they're the only thing that has been able to survive," says Brazelton. Because of their preference for such extreme environments, archaea have been nicknamed "extremophiles."

As intriguing as these extreme-loving organisms are, however, they weren't even recognized as a distinct evolutionary group until quite recently. For many years all prokaryotes were lumped together into one large group—the kingdom Monera—a classification based largely on their cell structure. Then, in the late 1970s, Carl Woese and his colleagues at the University of Illinois made the surprising discovery that not all prokaryotic organisms are genetically similar enough to be classified as a single group. His work established an entirely new branch of prokaryotic organisms, now known as the archaea. While most archaea don't look that different from bacteria under the microscope—

INFOGRAPHIC 18.4 PROKARYOTES ARE ABUNDANT AND DIVERSE

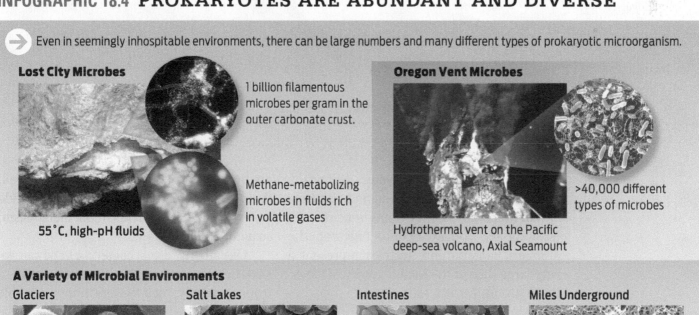

→ Even in seemingly inhospitable environments, there can be large numbers and many different types of prokaryotic microorganism.

Lost City Microbes

1 billion filamentous microbes per gram in the outer carbonate crust.

Methane-metabolizing microbes in fluids rich in volatile gases

55°C, high-pH fluids

Oregon Vent Microbes

>40,000 different types of microbes

Hydrothermal vent on the Pacific deep-sea volcano, Axial Seamount

A Variety of Microbial Environments

Glaciers	Salt Lakes	Intestines	Miles Underground
Environment: freezing, high-pressure, low-nutrient, low-oxygen Organism trait: produces antifreeze chemicals	Environment: high-salt, low-nutrient Organism trait: Photosynthesizes	Environment: no oxygen, moist, high growth rate at 37°C Organism trait: metabolizes sugars	Environment: dry, high-pressure, low-nutrient Organism trait: metabolizes coal

INFOGRAPHIC 18.5 BACTERIA AND ARCHAEA, LIFE'S PROKARYOTIC DOMAINS

Two of the domains of life, Bacteria and Archaea, have prokaryotic cells, but each has a distinct evolutionary history, with Archaea being genetically closer to Eukarya than to Bacteria. The genetic differences between Bacteria and Archaea translate into a variety of structural and functional adaptations.

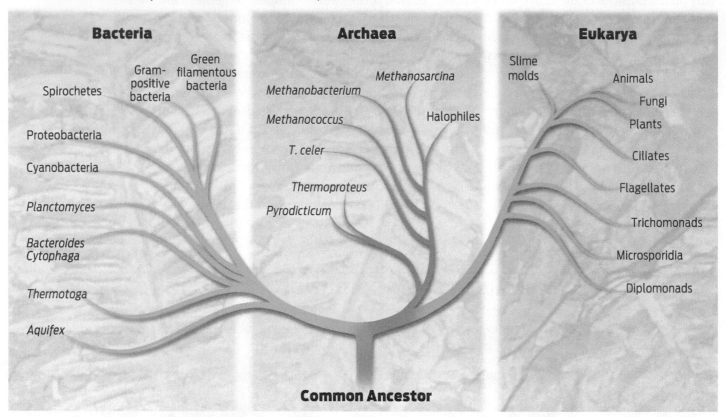

both are unicellular prokaryotes—genetically they are as different from bacteria as humans are. In other words, they represent a distinct evolutionary domain of life. Together, the domains Bacteria and Archaea represent a very large slice of the total diversity of life on Earth (**INFOGRAPHIC 18.5**).

BOUNTEOUS AND (MOSTLY) BENEFICIAL: BACTERIA

Bacteria are the invisible workhorses of the planet. They include commonly encountered microbes such as the beneficial strain of *Escherichia coli* present in the human gut and *Staphylococcus aureus,* which causes skin infections, as well as not so commonly encountered ones like the fluorescence-emitting bacteria living in the head of a sea squid. While all bacteria

▶BACTERIA
One of the two domains of prokaryotic life; the other is Archaea.

are prokaryotic, and most possess a cell wall, their genetic diversity translates into a wide variety of differences in nutrition, metabolism, structure, and lifestyle.

Like all organisms, bacteria can be categorized by what they eat. Some bacteria are autotrophs (literally, "self-feeders"): they are able to make their own food directly, using material from the nonliving environment, from carbon dioxide to rocks. Others are heterotrophs (literally, "other feeders"): they must rely on other living organisms to provide them with food.

One of the largest and most important groups of autotrophic bacteria is the cyanobacteria, which are found in oceans and freshwater, as well as on exposed rocks and soil—virtually everywhere sunlight can reach them. Cyanobacteria use the energy of sunlight to carry out photosynthesis in a manner similar to plants, taking in CO_2 and generating much of the oxygen that other organisms, including humans, rely on. Cyanobacteria are thought to be the oldest photosynthetic organisms on Earth, dating back roughly 2.5 billion years and playing a pivotal role in making the atmosphere breathable for the rest of us. Many cyanobacteria also perform the ecologically useful task of converting nitrogen from the atmosphere into a form that plants can use to grow. This process, called **nitrogen fixation,** is indispensable for the survival of life on Earth (see Chapter 23).

Not all autotrophic bacteria rely on sunlight for energy. Some, including those living at Lost City, can obtain energy directly from geological sources such as inorganic gases pouring out of hydrothermal vents—making them among the few organisms on Earth that do not rely on the sun's energy to survive.

Heterotrophic bacteria include those that obtain food by consuming material from living or dead organisms. Such heterotrophic bacteria play an important role in decomposition, allowing carbon and other elements—which would otherwise be trapped in dead organisms, sewage, or landfills—to be recycled. They are also useful in bioremediation projects. For example, some types of bacteria metabolize droplets of oil, much the way humans digest butter, so they naturally help clean up oil spills.

Bacteria break down their food molecules through a variety of metabolic pathways, some of which require oxygen, some of which do not. For example, many bacteria employ the anaerobic process known as fermentation (see Chapter 6) to get energy from food. The products of fermentation can be valuable (and tasty) to humans. You may have seen "L. bulgaricus" listed as an ingredient of yogurt; live *Lactobacillus bulgaricus* bacteria are present and at work in the yogurt, fermenting sugars into lactic acid, which helps the milk solidify into yogurt and which gives yogurt its tangy taste. Other bacteria use oxygen to break down organic molecules, like the aerobic bacteria that feast on an oil spill.

Highly resourceful, bacteria have a diverse range of living arrangements with other creatures. Many live in close association, or **symbiosis,** with other organisms—often to the benefit of one or both partners. Naturally occurring lactobacilli in the female vaginal tract, for example, obtain nourishment by fermenting naturally occurring sugars to lactic acid. The resulting acidity of the vaginal tract suppresses the growth of yeast, preventing yeast infections. Antibiotics taken for a bacterial infection are likely to kill the resident lactobacilli

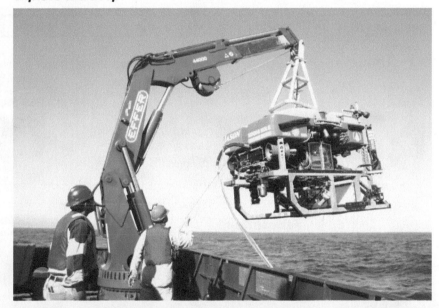

Jason, a remotely operated mechanical assistant, is used to explore the deep.

▶**NITROGEN FIXATION**
The conversion of atmospheric nitrogen into a form that plants can use for growth.

▶**SYMBIOSIS**
A relationship in which two different organisms live together, often interdependently.

▶PATHOGEN
A disease-causing agent, usually an organism.

as well as the invaders, and a yeast infection is often an unhappy side effect.

Another example of beneficial bacterial symbiosis is *Vibrio fischeri,* a bioluminescent bacterium that lives and feeds inside the light organs of certain species of squid. The glow-in-the-dark *Vibrio* produces light beneath the squid and helps obscure the shadow that the squid might cast on a moonlit night, making it less noticeable to its prey as it hunts.

Unfortunately, not all bacteria are beneficial to the host. While the vast majority of bacteria do not cause human disease, some

do. Bacteria and other organisms that cause disease are known as **pathogens.** Many pathogenic bacteria cause disease by producing toxins that harm their hosts. Such toxins can either be part of the bacterial cell itself or secreted by the bacterium. For example, certain strains of the bacterium *Escherichia coli* secrete a potent toxin that causes bloody diarrhea and even sometimes kidney failure and death in its host (these cases are often associated with the O57:H7 strain of *E. coli* that has been implicated in several foodborne outbreaks). Keeping food refrigerated helps prevent food poisoning by

INFOGRAPHIC 18.6 EXPLORING BACTERIAL DIVERSITY

→ Bacteria live in every imaginable place on Earth and have a diverse array of lifestyles.

Symbiotic

Vibrio fischeri
Lives symbiotically with bioluminous squid

Pathogenic

Treponema pallidum
Causes syphilis in humans

Fermenters

Lactobacillus bulgaricus
Ferments milk, producing yogurt

Motile

Salmonella typhimurium
Moves by flagella through intestines; causes diarrhea and fever

Adhere to surfaces via pili

Escherichia coli
Adheres by pili to intestinal tract; causes urinary tract infections and food poisoning

Avoid defenses via capsule

Klebsiella pneumoniae
Causes pneumonia and wound infections following surgery

Photosynthesizers

Anabaena (cyanobacteria)
Lives in freshwater, photosynthesizes, and fixes nitrogen

Bacteria

Archaea   Eukarya

> **"** *[Lost City] is a good example of what we really don't know and what there is to still discover on the seafloor.* **"**
>
> — GRETCHEN FRÜH-GREEN

slowing the growth of the bacteria and, therefore, production of its toxin.

Not all pathogens produce toxins. Some cause disease by living and reproducing in the body and interfering with its normal processes—an example is the bacterium *Treponema pallidum,* which causes syphilis, a sexually transmitted disease (STD).

Sometimes the line between harmless and harmful bacteria can be blurred. Organisms that can, but don't always, cause disease are known as opportunistic pathogens. For example, most of us have *Staphylococcus aureus* on our skin at many times during our lives. Most of the time, *S. aureus* does not cause any harm, but if it penetrates the skin—through a wound, for example—it can cause a serious infection and even death, as discussed in Chapter 14.

In addition to nutritional and metabolic differences, bacteria display a variety of structural adaptations that suit their various lifestyles. They come in different shapes: spherical (in which case they are known as cocci), rod-shaped (bacilli), and spiral (spirochetes). Many bacteria are equipped with **flagella,** tiny whiplike structures that project from the cell and help it move. For example, the bacterium *Helicobacter pylori,* the most common cause of stomach ulcers, uses its flagella to propel itself through the gastric mucus of the stomach. **Pili** are shorter, hairlike appendages that enable bacteria to adhere to a surface. *Neisseria gonorrhoeae,* the bacterium that causes the STD gonorrhea, uses its pili to remain attached to the lining of the urinary tract. Without pili, the bacteria would be flushed out by the flow of urine.

Other bacteria are surrounded by a **capsule,** a sticky outer layer that helps the cell adhere to surfaces and to avoid the defenses of the host. *Streptococcus mutans,* for example, produces a capsule that allows it to adhere to teeth, where it forms the plaque that can lead to cavities (**INFOGRAPHIC 18.6**).

For all their impressive diversity and abundance, bacteria are far from the totality of prokaryotic life. Moreover, their lifestyles and adaptations may seem tame next to those of the other domain of prokaryotic life: the Archaea.

GOING TO EXTREMES: ARCHAEA

Archaea are similar to bacteria in that archaea are simple cells that lack a nucleus, but genetically they are as different from bacteria as humans are. All those genetic differences add up to a number of unique features that distinguish archaea from bacteria. For example, while bacteria have cell walls made of the molecule peptidoglycan, archaea have cell walls made of other molecules.

Though archaea are found in many run-of-the-mill habitats such as rice paddies, forest soils, ocean waters, and lake sediments, the most well-known species are the so-called extremophiles. Many of these extreme-loving archaea, including those living in Lost City, are hyperthermophiles—organisms that can survive only at extremely high temperatures. Many hyperthermophilic archaea are anaerobic and rely on sulfur instead of oxygen in

▶FLAGELLA (SINGULAR: FLAGELLUM)
Whiplike appendages extending from the surface of some bacteria, used in movement of the cell.

▶PILI (SINGULAR: PILUS)
Short, hairlike appendages extending from the surface of some bacteria, used to adhere to surfaces.

▶CAPSULE
A sticky coating surrounding some bacterial cells that adheres to surfaces.

▶ARCHAEA
One of the two domains of prokaryotic life; the other is Bacteria.

Archaea are sometimes known as "extremophiles." They live in diverse environments, often with very harsh conditions.

Halophiles

Halobacterium
Lives in places with high salt concentrations, such as the San Francisco Bay evaporation ponds

Methanogens

Methanopyrus kandleri
Lives in ocean hydrothermal vents. Produces methane as a by-product of its energy-converting metabolic pathways

Hyperthermophiles

Sulfolobus
Grows at extremely high temperatures (80°–100°C), as in the volcanic Grand Prismatic Spring, Yellowstone National Park

Archaea

Bacteria

Eukarya

metabolism. Sulfur-rich hot springs like those in Yellowstone National Park are home to these archaea.

Other archaea are methanogens, which consume carbon dioxide and hydrogen and produce methane as a by-product in a process called methanogenesis. Because this gaseous meal is completely inorganic, these archaea are considered autotrophs. A methanogen that can survive in an even more extreme environment than Lost City is *Methanopyrus kandleri*, which lives in a type of hydrothermal vent known as a black smoker, where the temperature can be 121°C–thought to be the upper temperature limit of life. Methanogens occupy more mundane environments as well, including the digestive systems of methane-belching cows.

Some archaea are halophiles, or "salt lovers," and prefer a home saturated in salt, which would shrivel most other living things. Their presence is detectable by the colorful pigments they produce–bright reds, yellows, and purples–as seen in salt ponds in San Francisco Bay (INFOGRAPHIC 18.7).

The archaea's affinity for such extreme environments is suggestive of their evolutionarily ancient roots. The early Earth was a lot warmer than it is now, and it's long been a question how living things could withstand the heat that then prevailed. If the archaea at Lost City are any indication, they would have done just fine.

In addition, some researchers suspect that methanogenesis may be among the most ancient forms of metabolism on Earth. Scientists who study the origin of life believe that the first organisms were likely autotrophs that obtained their carbon from carbon dioxide, using hydrogen as an energy source and emitting methane as a by-product–in other words, doing exactly what certain archaea do today.

More recently, scientists studying Lost City have suggested that the archaea found there may survive by consuming, rather than producing, methane. For many years, it was thought that the only organisms capable of "eating" methane were bacteria that required oxygen in order to do so, much as we require oxygen to perform aerobic respiration (see Chapter 6). But scientists now think that archaea can consume methane without oxygen. That's a handy trait to have when you live in an environment lacking oxygen–which is exactly what the ancient Earth was.

LOST CITY: WHERE LIFE BEGAN?

In its early days, more than 4 billion years ago, Earth was mostly warm ocean, and its atmosphere lacked significant oxygen. Photosynthetic organisms did not exist, so there were no living things to harness the energy of sunlight to make organic molecules. Such conditions pose a conundrum for those who study the origin of life. With no photosynthetic organisms to make organic molecules, where did the energy and building blocks necessary to assemble living things come from? Lost City provides an important clue.

The hydrothermal vents at Lost City are driven by a geochemical process called serpentinization, which occurs any time a particular type of rock from Earth's mantle comes in contact with seawater. The rocks are exposed as a result of faulting and uplift of the seafloor. The reaction itself generates a lot of heat, releases hydrogen gas, and–when this hydrogen reacts with carbon from rocks or seawater–produces hydrocarbons such as methane and simple organic molecules. All this happens completely abiotically–that is, without the participation of living things. Such an environment would have been an ideal place for life to begin, says

Coated with a mucusy layer of biofilm, the rocky towers at Lost City "look like they've been sneezed on."

ENERGY FROM THE EARTH FUELS LIFE AT LOST CITY

→ Many prokaryotes thrive in the seemingly inhospitable environment at Lost City. Abiotic reactions at Lost City generate energy-rich and carbon-containing molecules that can sustain the organisms at Lost City. These reactions may be similar to ones that supported life on the early Earth.

Serpentinization
Hydrogen gas is formed when some mantle rock is exposed to seawater. This gas is a source of energy for the organisms living in Lost City.

Hydrogen gas H_2

Hydrogen gas H_2

Carbon dioxide gas CO_2

Seawater

Anaerobic Archaea Make Methane
Archaea on the inside of the Lost City chimneys produce methane from inorganic carbon dioxide with hydrogen gas as an energy source.

Methane CH_4

Hydrogen gas H_2

Carbon in rock and water

Anaerobic Archaea "Eat" Methane
Heterotrophic archaea on the inside of the Lost City chimneys get energy by consuming methane in the absence of oxygen.

Mantle rock

Mantle rock

Methane CH_4

Methane CH_4

Anaerobic Bacteria "Eat" Methane
Heterotrophic bacteria on the outside of the chimneys get energy by consuming methane in the presence of oxygen.

Methane CH_4

Abiotic Synthesis of Carbon Molecules
Methane and other simple organic molecules are formed spontaneously from carbon in the rock and hydrogen gas.

CO_2

Brazelton, of the University of Washington. "You have energy and organic compounds and liquid water all in a warm spot, so that's a great place where you might imagine life could have got started."

Lost City may also help to explain the chicken-and-egg problem posed by the origin of life: organic molecules are needed to build living things, but living things are generally the source of organic molecules. So which came first, the chicken or the egg? "I think a big clue to the chicken-and-egg problem is that you don't need life to make organic compounds," says Brazelton. "We are studying environments right now where the organic compounds are literally pouring out of these chimneys, and they're being made without the help of life" (INFOGRAPHIC 18.8).

Many microbiologists would agree that deep-sea vents like Lost City likely represent some of the oldest habitats for microbial life on Earth. Radiometric dating of the rock layers indicates that Lost City's vents have been pumping strong for at least 100,000 years, and likely for much longer: Lost City sits on a layer of Earth's crust that is at least 1.5 million years old. Back then, rocks from the interior of Earth—from the mantle—were much closer to the surface than they are now, which means their reaction with seawater would have been more common. A journey to Lost City is thus like a journey back in time, to Earth's primordial past.

Lost City may even provide a clue to life beyond Earth. The rocks involved in serpentinization are quite common in the solar system. They are all over the surface of Mars, for example, and researchers suspect that the chemical reaction might be occurring right now beneath the surface of Mars, where recent evidence suggests that methane is being produced. NASA is therefore extremely interested in Lost City as a way to understand potential life on Mars (see Chapter 2).

Ironically, scientists know more about the surface of Mars than they do about the ocean floor of our own planet. Ocean covers 70% of Earth's surface, yet much of it remains unexplored. If Lost City is any indication, many scientific treasures await the patient explorer. "[Lost City] is a good example of what we really don't know and what there is to still discover on the seafloor," says geologist Früh-Green.

If life on Earth did begin at hydrothermal vents like those at Lost City, then it could mean that these extreme-loving prokaryotes are the descendants of the most ancient form of life on Earth. For nearly 2 billion years, these earliest prokaryotic organisms reigned supreme, with no challengers. Not until photosynthetic prokaryotes evolved, some 2.5 billion years ago, did they meet their match. Then, in an instant, geologically speaking, life on our planet underwent a radical and unprecedented change: 2 billion years ago, one of these early prokaryotes engulfed another and the two cells began a symbiotic relationship. That was the birth of the first eukaryote. ■

The tallest tower at Lost City is called Poseidon, after the Greek god of the sea.

CHAPTER 18 Summary

▶ Prokaryotes are unicellular organisms that lack internal organelles and whose DNA is not contained in a nucleus.

▶ Prokaryotes are found in virtually every environment on Earth, even those with seemingly inhospitable conditions, such as hydrothermal vents on the ocean floor.

▶ Genetic analysis has led to the categorization of life into three domains: Bacteria, Archaea, and Eukarya. Each domain of life has a distinct evolutionary history.

▶ Both bacteria and archaea have prokaryotic cells, but they otherwise differ in their structure, biochemistry, and lifestyles.

▶ Bacteria are a diverse group of prokaryotic organisms with many unique adaptations such as flagella and capsules that allow them to live and thrive in many environments.

▶ Some bacteria are disease-causing pathogens, but most are harmless and even beneficial. Cyanobacteria, for example, are responsible for much of the photosynthesis that supports life on Earth.

▶ Often known as "extremophiles," archaea live in some of the most inhospitable conditions on Earth, such as hydrothermal vents. Many archaea flourish in less extreme environments as well.

▶ The harsh conditions of Lost City may resemble the conditions of the early Earth. The prokaryotic inhabitants of Lost City may be metabolically similar to the earliest known life.

▶ The energy that fuels life in Lost City comes from a geological source, rather than from sunlight, making Lost City one of the few communities on Earth that is not powered by photosynthesis.

MORE TO EXPLORE

▶ Lost City Home
http://www.lostcity.washington.edu/

▶ Brazelton, W. J., et al. (2011) Physiological differentiation within a single-species biofilm fueled by serpentinization. *mBio* 2(4): e00127-11.

▶ Brazelton, W. J., et al. (2006) Methane- and sulfur-metabolizing microbial communities dominate the Lost City hydrothermal field ecosystem. *Applied and Environmental Microbiology* 72(9):6257–6270.

▶ Kelley, D. S. (2005) From the mantle to microbes: the Lost City hydrothermal field. *Oceanography* 18(3):32–45. http://www.tos.org/oceanography/archive/18-3_kelley.pdf

▶ Exploring Life's Origins
http://exploring origins.org

CHAPTER 18 Test Your Knowledge

DRIVING QUESTION 1

What are the prokaryotic domains of life?

By answering the questions below and studying Infographics 18.3, 18.4, and 18.5, you should be able to generate an answer for the broader Driving Question above.

KNOW IT

1 Organisms are placed into one or another of the three domains of life on the basis of

  a. cell type.

  b. physical appearance.

  c. evolutionary history as assessed by genetic relatedness.

  d. ability to cause disease.

  e. degree of sophistication, that is, how evolutionarily advanced they are.

2 Describe the major difference(s) between prokaryotic and eukaryotic organisms.

3 The absence of membrane-bound organelles in a cell tells you that the cell must be

  a. from a member of the domain Bacteria.

  b. from a member of the domain Archaea.

  c. from a member of the domain Eukarya.

  d. either a or b

  e. either b or c

USE IT

4 Why were bacteria and archaea originally grouped together?

5 When first discovered, archaea were called "archaeabacteria." Why do you suppose this was? What are the strengths and weaknesses of this earlier term?

DRIVING QUESTION 2

What are the features of bacteria and of archaea?

By answering the questions below and studying Infographics 18.4, 18.6, and 18.7, you should be able to generate an answer for the broader Driving Question above.

KNOW IT

6 The term *prokaryotic* refers to

  a. a type of cell structure.

  b. a domain of life.

  c. a group with a shared evolutionary history.

  d. a type of bacterium.

  e. a type of archaea.

7 If you were looking for a bacterium, where would expect to find one?

  a. on your skin

  b. in soil

  c. in the ocean

  d. associated with plants

  e. any of the above

8 What is the function of flagella?

  a. production of methane

  b. sticking to a surface

  c. motility

  d. luminescence

  e. metabolism

9 If you are unable to culture archaea from an environmental sample, is it safe to conclude that there are no archaea present? Why or why not?

USE IT

10 Can you use cell structure to classify a cell as either bacterial or archaeal? Explain your answer.

11 Many prokaryotic organisms can carry out both photosynthesis and nitrogen fixation. Why are these processes important to humans?

12 If *Neisseria gonorrhoeae* had no pili, would it still be a successful pathogen? Explain your answer.

DRIVING QUESTION 3

What are the challenges faced by organisms living at Lost City, and how do they face them?

By answering the questions below and studying Infographics 18.1, 18.2, and 18.8, you should be able to generate an answer for the broader Driving Question above.

KNOW IT

13 List the features that make Lost City a particularly harsh environment. For each feature, give a brief explanation of why that environment is inhospitable for many organisms.

14 If you were a prokaryotic organism and wanted to be successful at Lost City, what energy source must you be able to use?

    a. sunlight

    b. oxygen

    c. hydrogen gas

    d. electricity

    e. None of the above is available at Lost City.

USE IT

15 What is the significance of methane and other hydrocarbons at Lost City? (Think about both the origin of life and the sustenance of early life.)

16 If methane were not produced abiotically at Lost City, what would be the implications for early life?

17 Would you expect to find photosynthetic organisms at Lost City? Explain your answer.

18 Do you think that the scientists studying Lost City should be concerned about introducing microbial contaminants from their submersibles onto the towers of Lost City? How probable is this, given the conditions at Lost City and on the surface? If such an event could happen, what would be the implications?

INTERPRETING DATA

19 Some of the chimneys at Lost City are actively venting. These chimneys have hot (80°–100°C) interiors that lack oxygen and have a pH range of 9–11. The exterior surfaces of the active chimneys are cooler (~7°C), contain oxygen and have a pH of ~8.

    The inactive chimneys at Lost City are no longer venting hot fluids. Compared to the actively venting chimneys, their interiors are much cooler (7°–20°C), have a pH of 8–10, and lack oxygen. The exteriors of the inactive chimneys are very similar to those of the active chimneys.

    From the properties of the organisms given in the table below, complete the table to indicate where in Lost City each of these organisms is most likely to be found.

ORGANISM	AEROBIC OR ANAEROBIC?	OPTIMUM pH	OPTIMUM TEMPERATURE (°C)	DOMAIN: BACTERIA OR ARCHAEA?	LOCATION?
A	Anaerobic	9	15	Archaea	
B	Aerobic	8	6	Bacteria	
C	Aerobic	8.5	8	Bacteria	
D	Anaerobic	11	90	Archaea	
E	Aerobic	7.5	7	Bacteria	
F	Anaerobic	11	85	Archaea	

MINI CASE

 20 In many ways, the discovery of Lost City is about the discovery of life, as is the voyage of Curiosity to Mars (see Chapter 2).

   a. What features do these two environments, Lost City and Mars, share?

   b. How do the challenges faced by *Curiosity* compare to those faced by *Jason*?

   c. From the way microbes survive at Lost City, what properties might you expect Martian organisms to have?

BRING IT HOME

21 Do you think that, instead of spending time studying microbes from extreme and remote environments, scientists should be studying microbes that are more apparently relevant to humans, such as ones that cause disease? In what ways might understanding the organisms at Lost City be useful to humans?

RAIN FOREST

RICHES

Restoring eukaryotic diversity in Olympic National Park

19

ON A CHILLY JANUARY AFTERNOON IN 2008, A CROWD OF eager onlookers gathered at a snowy campground in Washington State to watch natural history being made. The stars of the show were a trio of rarely seen animals held inside small wooden crates. As photographers craned for a good look and schoolchildren held their breath, the door to the first crate was opened. With a flash of whiskers and brown fur, a weasel-like animal bolted from the box and made a break for the forest. The fisher was finally home.

"There were lots of oooohs and aaaahhs and clapping and cheering," recalls Jeffrey Lewis of the Washington Department of Fish and Wildlife, who helped coordinate the release of the fishers into the Washington forest that day.

It was a long-awaited homecoming. Once plentiful in the state, fishers have been hunted and trapped nearly to extinction; they have not been seen in Washington since the early 20th century. The animal was named a Washington State endangered species in 1998 after a careful investigation failed to find any evidence of a local population. The three animals released that afternoon—two females and one male—were imported from British Columbia, Canada, as part of a coordinated effort to bring fishers back to Washington. They were the first of 90 fishers to be released over 3 years into their new home: Olympic National Park.

A nearly 1-million-acre plot of wilderness occupying the northwest corner of the state, Olympic National Park is attractive to more than just fishers. It

▶▶ DRIVING QUESTIONS

1. What are eukaryotic organisms, and what factors influence their diversity?
2. How are plants defined, and what influences their diversity?
3. How are animals defined, and what influences their diversity?
4. How are fungi defined, and what influences their diversity?
5. What are protists, and what influences their diversity?

The fishers released in 2008 are part of a coordinated effort to bring the species back to Washington State.

▶**EUKARYOTE**
Any organism of the domain Eukarya; eukaryotic cells are characterized by the presence of a membrane-enclosed nucleus and organelles.

is home to a mind-boggling array of different species, in numbers uncommon in most other parks. "It's got amazing biological diversity in a very compressed area," says Patti Happe, who is Wildlife Branch Chief at the park. Among this biodiversity is a seemingly endless variety of **eukaryotes**—the plants, animals, fungi, and unicellular protists making up one of the three main branches of life (**INFOGRAPHIC 19.1**).

Visitors to the park can find some of the world's oldest and tallest Douglas fir and Sitka spruce trees, the country's largest herd of Roo-

sevelt elk, and a number of eukaryotic species found here and nowhere else, including the Olympic marmot, the Olympic pocket gopher, and the Olympic torrent salamander. A designated UNESCO Biosphere Reserve and a World Heritage Site, Olympic National Park is a microcosm of the planet's eukaryotic diversity, and biologists are eager to protect it.

By returning the missing fishers to Olympic, park managers hope to ensure the resiliency of the rain forest. The restoration effort follows the guiding principle of conservation, which holds

INFOGRAPHIC 19.1 TREE OF LIFE: DOMAIN EUKARYA

➜ From genetic evidence, the domain Eukarya is sorted into the Plant, Fungi, and Animal kingdom, as well as multiple groups of protists—single-cell eukaryotes that don't fit neatly into the other kingdoms.

Bacteria
Spirochetes
Gram-positive bacteria
Green filamentous bacteria
Proteobacteria
Cyanobacteria
Planctomyces
Bacteroides
Cytophaga
Thermotoga
Aquifex

Archaea
Methanobacterium
Methanosarcina
Methanococcus
Halophiles
T. celer
Thermoproteus
Pyrodicticum

Eukarya
Animals
Fungi
Plants
Ciliates
Slime molds
Flagellates
Trichomonads
Microsporidia
Diplomonads
— **Protists**

Common Ancestor

INFOGRAPHIC 19.2 THE LANDSCAPE OF OLYMPIC NATIONAL PARK

→ Olympic National Park has an enormous diversity of physical, geological, and climatic conditions, all of which contribute to the huge biological diversity in the park.

1

Old Growth Forest
The Olympic National Park temperate rain forest has tremendous plant diversity, providing habitat and resources for other living organisms.

Olympic National Park

WASHINGTON

5

Hart Lake
A variety of lakes provides a type of aquatic habitat for living organisms.

2

Pacific Coastline
The park includes saltwater beaches.

3

Olympic Mountains
Mount Olympus has a major influence on local climate.

4

Elwha River
Rivers running through the park provide a specific type of aquatic habitat for many species.

that the healthiest ecosystems are ones where all the various parts remain intact. "National parks are here to serve as our national treasures, where we preserve and protect both our cultural and our natural heritage for future generations," says Happe. "Really what we're doing is wise stewardship of our nation's resources."

A GREEN WORLD

Olympic's unique collection of wildlife reflects the geological history and ecology of the region. During the last ice age, approximately 20,000 years ago, the Olympic Peninsula was isolated by glaciers and largely separated from the rest of what is now the United States. Today, it is surrounded by saltwater on three sides and is essentially an ecological island,

distinctive in its geography and topography.

Far from being a single landscape, however, Olympic National Park is more like three parks in one: a glacier-topped mountain region with flowering subalpine meadows, valleys of temperate rain forest threaded with freshwater rivers and lakes, and nearly 60 miles of jagged Pacific coastline. Central to the park's ecology is the towering presence of Mount Olympus, a nearly 10,000-foot peak that traps warm air blowing in from the Pacific, making the western slope one of the wettest spots in the United States; it's doused in more than 12 feet of rain each year. This mosaic of physical, geographic, and climatic conditions provides numerous habitats for the many species of eukaryotes that live here (**INFOGRAPHIC 19.2**).

▶PLANT
A multicellular eukaryote that has cell walls, carries out photosynthesis, and is adapted to living on land.

Each segment of the park is marked by its distinct form of vegetation, or plant life. Within the low-elevation rain forests, for example, stands of giant Sitka spruce and Douglas fir trees form a thick canopy of green, steadily dripping moisture. In dense forest understory, 300 feet below the canopy, plants form a junglelike tangle: nearly every log and tree trunk is coated with a shaggy carpet of mosses, ferns, and lichens (lichens are partnerships between a fungus and a photosynthetic organism), while hanging plants drape on branches like luxurious scarves with long ground-reaching stems.

In all this variety, how can scientists differentiate plants from other eukaryotes, like protists or fungi? At the most basic level, a **plant** is a multicellular eukaryote that possesses cells with cell walls, carries out photosynthesis, and is adapted to living on land. Plants such as those found in Olympic first evolved from water-dwelling algae about 450

INFOGRAPHIC 19.3 EVOLUTION OF PLANT DIVERSITY

→ All plants have evolved from an ancient common ancestor. Different groups of plants have developed different specializations that allow them to be successful on land.

Bryophytes
Example: Cat's tail moss
- Nonvascular, so can't move water throughout their bodies
- Small plants living in damp environments
- Nonflowering; no seeds
- Reproduction relies on water

Ferns
Example: sword fern
- Vascular, so can live in drier environments and move water throughout their bodies
- Nonflowering; no seeds
- Reproduction relies on water and involves spores

Gymnosperms
Example: Douglas fir
- Vascular, so cell walls of vascular cells support plant body to great heights
- Seeds are "exposed" (typically held in cones)

Angiosperms
Example: Big-leaf maple
- Vascular, so cell walls of vascular cells support plant body to great heights
- Flowers are important in reproduction
- Seeds are contained within fruit

Cones

Flowers

Seeds

Vascular

Common Ancestor

million years ago, when life on Earth was confined primarily to the seas. As plants radiated and diversified on land, they evolved a number of adaptations that made them increasingly independent of water.

The earliest plants that made the transition from water to land were small, seedless plants called **bryophytes.** Bryophytes lack roots and tissue for transporting water and nutrients throughout their bodies, and therefore can grow only in damp environments, where they can easily absorb water. One of the wettest places on Earth, the Olympic rain forest is a soggy paradise for bryophytes, such as mosses and liverworts, which appear as squat, spongy mats.

The rain forest is also home to many **vascular plants**—those with specialized tissues for transporting nutrients and water through the plant body. The first true vascular plants were **ferns,** such as the hip-high sword ferns that cover a good portion of the Olympic forest. Like bryophytes, ferns do not produce seeds. Yet unlike those vertically challenged relatives, ferns can stand upright and grow tall, thanks to the vascular tissue that keeps stems rigid and transports water and nutrients from the plant's roots to its leaves. At one time, ferns ruled the plant world, spreading their massive fronds across the entire landscape in the Carboniferous period. But their reign was short lived. Soon, another kind of plant evolved to challenge the ferns' dominance: those with seeds.

Seed plants first emerged about 360 million years ago, during the late Devonian Period. A seed, which envelops a plant's embryo, is an ideal package for withstanding harsh conditions and transporting the embryo to a location where it can grow into a new plant. Seed plants were so successful that they quickly came to dominate forests by the time dinosaurs appeared in the Mesozoic Era.

Today, more than 90% of all living plants are seed plants. **Angiosperms** are flowering plants with seeds contained in a fruit—an apple, say, or an acorn ("angio" is from the Greek for "vessel" or "container"). Olympic National Park is home to many species of angiosperms, including oaks, maples, huckleberry bushes, and willows, as well as hundreds of species of flowers. Plants with exposed seeds, as in a pinecone, are known as **gymnosperms**—spruce, pine, redwood, fir, and other conifers, for example. (*Gymnos* is Greek for "naked," so the name literally means "naked seeds.") **(INFOGRAPHIC 19.3)**

A FOREST FOR FISHERS (AND MORE)

Olympic National Park is an especially good home for the tree-loving fisher (*Martes pennanti*). "Because the park is 95% wilderness area, little of it has been logged, and it contains great expanses of older forest, which provide the large trees, snags, and logs that fishers need," says Lewis. Fishers rest in nooks in the trees, and females use tree cavities as dens in which to birth and nurse their kits. The shy fishers are also attracted to places with dense

▶**BRYOPHYTE**
A nonvascular plant that does not produce seeds.

▶**VASCULAR PLANT**
A plant with tissues that transport water and nutrients through the plant body.

▶**FERN**
The first true vascular plants; ferns do not produce seeds.

▶**ANGIOSPERM**
A seed-bearing flowering plant with seeds typically contained within a fruit.

▶**GYMNOSPERM**
A seed-bearing plant with exposed seeds typically held in cones.

A Pacific fisher (Martes pennanti).

▶**ANIMAL**
A eukaryotic multicellular organism that obtains nutrients by ingesting other organisms.

▶**RADIAL SYMMETRY**
The pattern exhibited by a body plan that is circular, with no defined left and right sides.

▶**BILATERAL SYMMETRY**
The pattern exhibited by a body plan with right and left halves that are mirror images of each other.

▶**VERTEBRATE**
An animal with a bony or cartilaginous backbone.

canopy cover, woody debris, and understory vegetation, all of which provide plentiful hiding places. Many of the tree species found here make for prime fisher habitat, including western Hemlock, Sitka spruce, and Pacific silver fir, which also provide a reliable food source for seed- and insect-eating mammals that fishers stalk as prey, such as squirrels, mice, and shrews.

The fisher is an animal, of course, but what defines an animal? Scientifically, an **animal** is a multicellular eukaryotic heterotroph that obtains nutrients by ingestion—that is, by eating. When we humans think of "animal," we tend to picture mammals, such as the fur-covered fisher. But the term applies to a great variety of creatures, from sponges to worms to insects to humans. To help bring some order to this diversity, biologists sort animals into smaller groups on the basis of shared characteristics and ancestry.

Many features can be used to group and sort animals. Historically, biologists relied mostly on anatomical and embryological evidence, but in recent years it has become more

common to use genetic evidence. From genetic evidence, it is clear that all animals descend from a common ancestor that lived nearly 1 billion years ago and whose descendants diversified into the different forms we see today.

Early in their history, animals branched into three main lineages, and the legacy of that division can be seen in three distinct animal body plans in existence today. The simplest living animals, such as sponges, lack defined tissues or organs and are amorphous—that is, they have no definite shape. These asymmetrical organisms are likely similar to the earliest animals to have populated the oceans. All other animals have defined tissues and fall into one of two broad categories based on the type of body symmetry they possess.

Animals such as jellyfish and coral exhibit **radial symmetry,** meaning that they're shaped like a pizza—circular, with no defined left and right sides. All other animals—everything from worms and insects to fishers and humans—exhibit **bilateral symmetry:** if you cut them down the middle you produce left and right halves that are mirror images of each other.

Bilateral symmetry has become very prevalent in the animal kingdom because it is a useful adaptation for seeking out food, stalking prey, and avoiding predators. For instance, bilaterally symmetrical animals have an eye on each side of the face, enabling them to look straight ahead. In the fisher's case, such bilateral symmetry enables it to climb down trees head first in search of prey.

Fishers, members of the phylum Chordata, are **vertebrates,** meaning they are animals with a backbone. A fisher's backbone is made of bony vertebrae, a feature they share with most other vertebrates. (A few vertebrates, primarily sharks and several other fish, have backbones made of cartilage.) While vertebrates—which include humans—are some of the most easily recognized animals, they represent a sliver of the total animal world. In fact, most animals lack a backbone and are

The banana slug, one of the forest's most voracious inhabitants, is distasteful to predators.

INFOGRAPHIC 19.4 EVOLUTION OF ANIMAL DIVERSITY

All animals have descended from a common ancestor. Many features help classify animals, including body symmetry, type of body support, and the presence or absence of a spinal cord and backbone. In addition, genetic sequencing and the study of embryonic development have informed this phylogenetic tree.

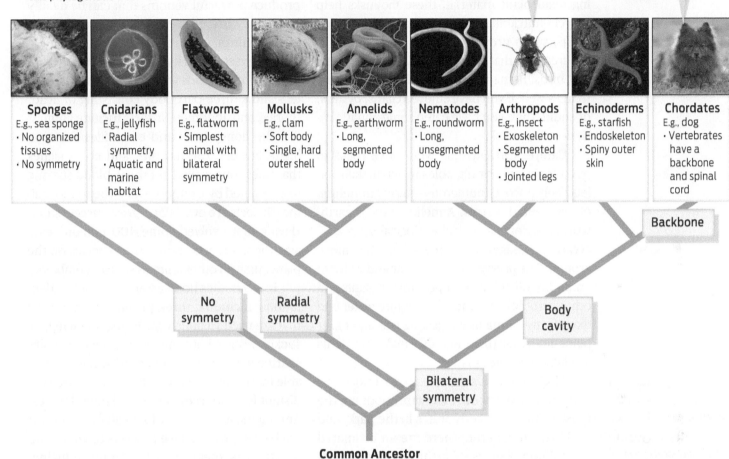

The arthropods are the most abundant animal group.

The chordates are the only group that includes vertebrates.

Sponges
E.g., sea sponge
· No organized tissues
· No symmetry

Cnidarians
E.g., jellyfish
· Radial symmetry
· Aquatic and marine habitat

Flatworms
E.g., flatworm
· Simplest animal with bilateral symmetry

Mollusks
E.g., clam
· Soft body
· Single, hard outer shell

Annelids
E.g., earthworm
· Long, segmented body

Nematodes
E.g., roundworm
· Long, unsegmented body

Arthropods
E.g., insect
· Exoskeleton
· Segmented body
· Jointed legs

Echinoderms
E.g., starfish
· Endoskeleton
· Spiny outer skin

Chordates
E.g., dog
· Vertebrates have a backbone and spinal cord

Backbone

No symmetry

Radial symmetry

Body cavity

Bilateral symmetry

Common Ancestor

---

therefore called **invertebrates.** While invertebrates are often lumped together on the basis of what they lack, the division of the animal world into those with and those without backbones makes about as much sense as dividing the world into sponges and nonsponges—it overlooks a lot of differences, and obscures the fact that most animals—an astounding 95%—are invertebrates (**INFOGRAPHIC 19.4**).

Olympic National Park hosts a squirming, wriggling, buzzing swarm of invertebrates. If you were hiking or camping in the park, you would easily encounter invertebrates from sev-

eral major phyla—some more welcome than others, perhaps.

Sliding quietly amid leaf litter on forest trails are many specimens of the Pacific coast's best known **mollusk,** the brightly colored banana slug (one of the few invertebrates to serve as a university mascot, as it does for the University of California at Santa Cruz). A squishy yellow creature that can grow nearly to the size of its namesake fruit, the banana slug is one of the forest's most voracious inhabitants, eating its way through just about everything in its path, from animal carcasses and droppings to

▶**INVERTEBRATE**
An animal lacking a backbone.

▶**MOLLUSK**
A soft-bodied invertebrate, generally with a hard shell (which may be tiny, internal, or absent in some mollusks).

mushrooms, lichens, and leaves. This mollusk is so successful in the forest in part because it is distasteful to would-be predators, who know by its unmistakable color to avoid eating it.

Slugs—and their shelled cousins, the snails—are often considered garden pests. Yet by digesting dead plant material, these mollusks help recycle nutrients. And with their calcium-rich shells, snails provide this valuable mineral to the creatures that feast on them, such as rodents and birds. Some humans find mollusks a tasty treat as well: if you have enjoyed clams, oysters, or squid, then you have eaten some aquatic varieties of mollusk.

Campers in Olympic who move rocks to pitch their tents or dig holes to act as their toilets likely uncover numerous squirmy **annelids,** or segmented worms. Annelids such as earthworms perform a critical ecological service by creating passageways in the soil as they move around. The passageways allow air and water to enter the soil, which is important for plants and other aerobic organisms that require water and oxygen. By eating and digesting leaf and other plant litter, earthworms also make nutrients available for other plants.

The park is also home to an enormous collection of **arthropods.** Arthropods are the most abundant invertebrates in the park, and on Earth in general. There are an estimated 2-4 million species of arthropods, of which 855,000 have been officially described so far. The number of individual arthropods on the planet is estimated to be more than 10^{18} (that's 1 with 18 zeros after it). They include animals as diverse as water-dwelling crustaceans like crabs and lobsters and terrestrial spiders, millipedes, and flying insects.

Despite their abundance and diversity, all arthropods share some common physical characteristics. They have segmented bodies with jointed appendages such as legs, antennae or pincers, and a hard **exoskeleton,** or external skeleton, made up of proteins and chitin (a type of polysaccharide). An arthropod's exoskeleton serves multiple functions: it

protects the organism from predators, keeps it from drying out, and provides structure and support for movement, just as our internal **endoskeleton** does.

Most arthropods are harmless or even helpful to humans, but some are not. A few produce powerful venoms that can be deadly when they are conveyed to victims through bites or stings.

The vast majority of all arthropods are **insects**—arthropods with three pairs of jointed legs and a three-part body consisting of head, thorax, and abdomen. Insects include animals like the honey bees and butterflies that pollinate flowers, the termites and cockroaches that infest our walls, and those blood-sucking insects hated by campers everywhere: mosquitoes. Insects are evolution's great success story. Having first evolved some 400 million years ago, there are now more insect species on the planet than all other animals species combined.

Insect bodies boast an array of useful adaptations, including three-pronged mouthparts that are used variously for biting, chewing, or sucking. But what really sent insect diversity soaring was the evolution of wings. Wings enable insects to fly away from predators, access distant food sources, and travel to find mates. Among the most successful of all flying insects are beetles, which have two sets of wings and mouthparts specialized for biting, mincing, or chewing. Taxonomists have catalogued approximately 350,000 beetle species so far, and some estimates put the total number of species in the millions.

But even insects with six feet firmly planted on the ground can be remarkably successful—just look at the ants. They can't fly, but they can communicate, split up tasks, solve problems, and shape their local environment—a picnic, for example—for their own needs. Given their complex social behavior and adaptations, it's not surprising to find ants nearly everywhere on the planet; Antarctica, Greenland, Iceland, and Hawaii are some of the few places on Earth believed to harbor no native ant species.

▶**ANNELID**
A segmented worm, such as an earthworm.

▶**ARTHROPOD**
An invertebrate having a segmented body, a hard exoskeleton, and jointed appendages.

▶**EXOSKELETON**
A hard external skeleton covering the body of many animals, such as arthropods.

▶**ENDOSKELETON**
A solid internal skeleton found in many animals, including humans.

▶**INSECT**
A six-legged arthropod with three body segments: head, thorax, and abdomen.

> They are occasionally killed by cougars, coyotes, and eagles, but the fishers' biggest foes, by far, are humans.

THE VALUE OF DIVERSITY

Park officials had long suspected that fishers were missing from the forest, but it wasn't until 1998 that the species was officially designated as endangered in Washington State, after a careful scientific investigation failed to find any. That's when wildlife experts decided it was time to take action. Over the next decade, park managers worked with the Washington Department of Fish and Wildlife to research and devise a restoration plan. The multi-person effort relied upon a careful understanding of the fisher's attributes and its ecological role.

Fishers fall into a class of vertebrates called **mammals,** animals with mammary glands and a body covered with hair. Like many mammals, fishers are predators, hunting mostly other small and midsize mammals such as snowshoe hares, squirrels, mice, and beaver. With their keen sense of sight and smell, sharp teeth, and nonretractable claws, these nocturnal creatures are among the most effective predators on the ground and in trees.

Fishers themselves have few natural predators. They are occasionally killed by cougars, coyotes, and eagles, but the fishers' biggest foes, by far, are humans. For hundreds of years, fur traders trapped and killed the animals for their soft pelts. By 1934, when the practice was outlawed, fishers had largely been hunted to extinction. Fishers were also an unintended target of a massive government-sponsored predator control program in the early 20th century intended to reduce the number of wolves, which posed a danger to livestock. Officials with the U.S. Forest Service set poisoned traps for wolves and in a matter of years slashed the wolf population in the lower 48 states to near zero. Meanwhile, unsuspecting fishers ate the same poison bait and also perished.

Because humans are largely responsible for the decline of the fisher in Washington, many conservationists believe it is our ethical duty to help undo that damage–restoring diversity for diversity's sake, as it were. But there is a good ecological rationale for such action as well. "When you start getting up into the mammalian species, going up the phylogenetic tree," says wildlife manager Happe, "there's fewer species, but one animal has a big effect on the ecosystem." That is especially true of predators, such as the fisher, that act as a natural population control on other species.

For example, in some parts of their range, fishers are one of the few natural predators of porcupines. A healthy fisher presence in a forest helps keep the porcupine population under control and prevents what would quickly become a prickly problem for the trees that the porcupines eat and the loggers who cut them for lumber. Restoring fishers to the ecosystem is therefore necessary to keep the natural web of the environment intact, and to preserve economic goods–lumber. "They fill that niche back up that nothing else can quite totally fill and restore some resilience to the ecosystem," says Happe.

Exactly what effect the reintroduced fishers are having in Olympic, researchers can't say for certain at this point. "I'm sure that it's having repercussions in the ecosystem, it's just that we can't interview the squirrels and the

▶**MAMMAL**
An animal having mammary glands and a body covered with fur.

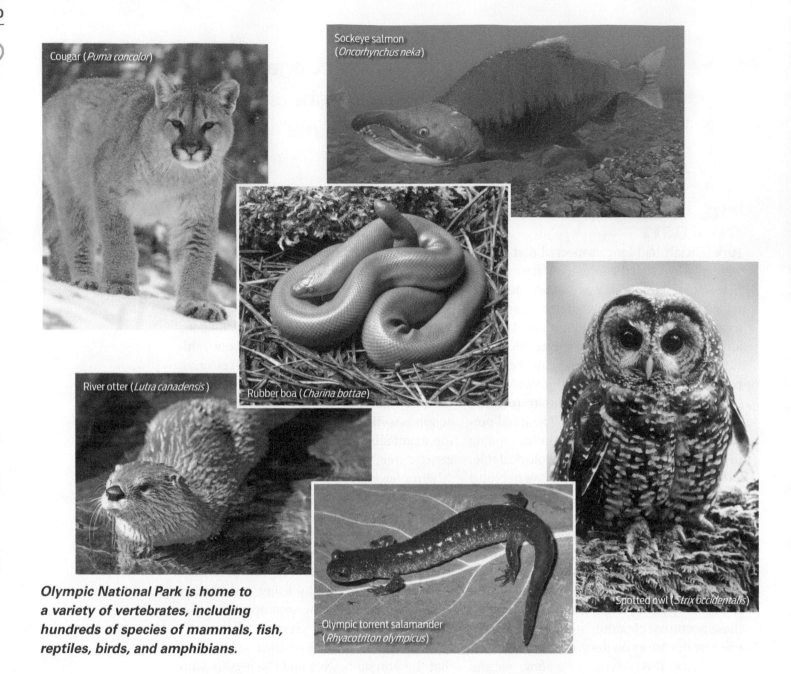

Cougar (*Puma concolor*)

Sockeye salmon (*Oncorhynchus neka*)

Rubber boa (*Charina bottae*)

River otter (*Lutra canadensis*)

Spotted owl (*Strix occidentalis*)

Olympic torrent salamander (*Rhyacotriton olympicus*)

Olympic National Park is home to a variety of vertebrates, including hundreds of species of mammals, fish, reptiles, birds, and amphibians.

rabbits to find out what they think of all this," says Happe.

Restoration efforts such as these are guided by the scientific understanding that the healthiest, most stable ecosystems are ones that have abundant biodiversity. Predators, especially, play an important role in an ecosystem—something that conservation biologists have come to understand in the wake of wolf eradications. Wolves, for example, prey on elk—one of the only species that can do so. By keeping elk herds in check, wolves help keep trees from being overgrazed by the elk (see Chapter 21). When predators go missing, a whole cascade of disruptive effects can result. "When you take those species out, then it's kind of like a tower of building blocks: you have less structure to keep your environment together," says Happe.

With the return of the fisher, the only species that is still missing from Olympic is the gray wolf. There's been talk of restoring wolves to Olympic—as has been done in other national parks, such as Yellowstone—but so far no definite plans have been made.

The restored fishers—90 so far—join a variety of other vertebrates living in Olympic, including other mammals—cougar, black-tailed deer, mountain goat, black bear, river otter, and Douglas squirrel, for example, as well as hundreds of species of fish, amphibians, birds, and reptiles. These various vertebrates are easily recognized by the unique adaptations that allow them to survive and flourish in their particular habitats within the park. Fish can live in the ponds and rivers because of their scales, fins, and gills; amphibians metamorphose from a water-dwelling juvenile form to an air-breathing adult; birds have hollow bones and feathers that enable many of them to fly; and reptiles have a body covered in water-tight scales that equip them for life on dry land. Each of these animals plays an important ecological role in the park, serving either as food for other animals, or as predators themselves.

CYCLES OF LIFE IN THE RAIN FOREST

The rain forest is a place of irrepressible life; it is also one of death. Coyotes kill fishers. Fishers hunt squirrels. Insects eat trees. Death casts a long shadow over life in the park, yet without it there would be no life at all.

When organisms are alive, they store nutrients and chemical building blocks in the fabric of their bodies. When the organisms die and decompose, these nutrients and building blocks are returned to the soil and eventually taken up into new life. Crucial to this cycle of life and death, growth and decomposition, are **fungi,** a third major branch on the eukaryotic tree. It's impossible to put an exact figure on the number of fungi in the forest, but their role can hardly be overestimated.

By breaking down organic matter into smaller particles, fungi help release trapped nutrients. Without fungi, dead trees and animal carcasses would pile up in the forest and smother everything in it. Thanks to the action of fungi, however, the organisms decompose and the elements they contained will nourish many organisms throughout the environment. Many decomposing organisms even provide shelter, such as the tree holes in which fishers and other animals make their dens.

Fungi come in many forms. There are unicellular species, such as molds and yeasts, and multicellular species, such as mushrooms and the shelf fungus you sometimes see growing on a tree trunk. Underlying this physical diversity is a method of obtaining nutrients common to all fungi: they secrete digestive enzymes onto their food source and then absorb the digested products. As one of nature's **decomposers,** fungi can break down just about anything that has organic components, including plant parts that are indigestible to many bacterial decomposers. All fungi have cell walls made of chitin (the same molecule that makes up the exoskeleton of the arthropods), and all modern fungi evolved from a common unicellular ancestor—the same one that gave rise to animals—approximately 1 billion years ago.

Multicellular fungi, such as the mushrooms poking up through leaves in the forest, have a body composed of threadlike structures known as **hyphae.** Each individual hypha is a chain of many cells, capable of absorbing nutrients. Fungal hyphae interweave to form a spreading mass known as a **mycelium.** These structures can vary in size and location: the mold on a slice of bread has a small mycelium that's in plain view. By contrast, the mushrooms you see on the forest floor are merely one aboveground part of what can be a huge, underground fungal mycelium. The largest known organism in the world is a fungus in Oregon with an underground mycelium stretching across nearly 4 square miles.

There are fungi everywhere in Olympic National Park, growing on, in, and under the abundant vegetation. Some species play a critical role in the soil, where they form a symbiotic relationship with the roots of many trees. Their slender hyphae grow into microscopic spaces in the soil where the tree's roots can't fit, greatly enhancing a root's ability to absorb water and nutrients. Trees supply nutrients to

▶**FUNGUS (PLURAL: FUNGI)**
A unicellular or multicellular eukaryotic organism that obtains nutrients by secreting digestive enzymes onto organic matter and absorbing the digested product.

▶**DECOMPOSER**
An organism such as a fungus or bacterium that digests and uses the organic molecules in dead organisms as sources of nutrients and energy.

▶**HYPHA (PLURAL: HYPHAE)**
A long, threadlike structure through which fungi absorb nutrients.

▶**MYCELIUM (PLURAL: MYCELIA)**
A spreading mass of interwoven hyphae that forms the often subterranean body of multicellular fungi.

INFOGRAPHIC 19.5 FUNGI, THE DECOMPOSERS

Fungi are a diverse group of organisms with a variety of reproductive strategies. However, all fungi eat in the same way: they secrete digestive enzymes onto their food, then absorb the digested products.

Yeast
- Single-cell fungi
- Contribute to tasty food and beverages when they ferment sugars
- Many live symbiotically with plants and animals

Spores are reproductive structures

Molds
- Many form mycelia—mats of strands called hyphae
- Hyphae spread throughout the soil, decaying leaf litter and dead organisms
- Many associate with plant roots, supplying and receiving nutrients

Mushrooms
- Hyphae form mycelia in soil and trees
- Fruiting bodies emerge from mycelia as part of the reproductive strategy
- Decay leaf litter and dead organisms
- Source of food for other animals

Mushroom fruiting body

Hyphae making up a mycelium

▶PROTIST
A eukaryote that cannot be classified as a plant, animal, or fungus; usually unicellular.

▶ALGA (PLURAL: ALGAE)
A uni- or multicellular photosynthetic protist.

the fungi, which do not photosynthesize. Many fungi also live in and on animals, as you probably know if you have ever had athlete's foot or a yeast infection (**INFOGRAPHIC 19.5**).

A MICROCOSM IN A DROP OF WATER

The remaining eukaryotes in Olympic National Park are the ones you rarely see—swimming in drops of water or hiding in puddles under the leaf cover. Informally known as **protists,** these varied members of what used to be called kingdom Protista do not fit neatly into one group and are tricky to classify.

Most protists are unicellular, but there are also multicellular varieties, such as some types of **algae.** Multicellular algae photosynthesize like plants, but unlike plants they lack specialized adaptations for living on land, such as roots, stems, and leaves. Other protists are similar to animals in that they are heterotrophic, eating other organisms, but since they are unicellular they are not technically animals. Some protist species have long filamentous bodies resembling fungi, but they are no more related to fungi than animals are. In fact, genetic evidence shows that protists do not form a cohesive evolutionary group; some may be as distinct from one another as plants are from animals. Still a work in progress, our understanding of these diverse organisms is likely to evolve in coming years as we continue to learn more about them (INFOGRAPHIC 19.6).

Despite their diversity, protists do share some common traits. They are all susceptible to drying out, so they are typically found in wet environments: lakes, oceans, ponds, moist soils, and living hosts. Many disease-causing

protists, for example, must spread directly from host to host because otherwise they would dry out. This explains why trichomoniasis, a sexually transmitted infection caused by the protist *Trichomonas vaginalis,* can be spread only through the exchange of bodily fluids during direct sexual contact.

Other protists live in the gastrointestinal systems of animals such as beavers and can be found in pond water where those animals defecate. Unwary campers who drink from a pond may find themselves stricken with an unpleasant diarrheal disease called giardiasis (aka "beaver fever") caused by the protist *Giardia lamblia.*

However small and difficult they are to classify, protists can rightfully claim an important position on the eukaryotic family tree.

INFOGRAPHIC 19.6 THE CHALLENGE OF CLASSIFYING PROTISTS

→ Protists are a diverse group of organisms that are difficult to classify. They share features with animals, plants, and fungi, but are not classified as any one of these. Nor do they have a single unifying characteristic that places them within a single evolutionary group. Most protists are unicellular.

Animal-like, but not animals

Plant-like, but not plants

Fungus-like, but not fungi

River and lake water is teeming with unicellular protists that feed on other organisms (for example, algae), as heterotrophic animals do.

Aquatic environments contain a diversity of unicellular organisms that carry out photosynthesis, similar to plants. Some have a whiplike flagellum that allows them to move, a characteristic of animals.

Dog vomit slime mold grows on the ground in wooded areas. Slime molds are decomposers, like fungi, but may also eat other organisms (for example, bacteria) as food, similar to heterotrophic animals.

Algae float in lakes and rivers, performing photosynthesis, like plants. Many algae form filamentous strands of cells, similar to many fungi.

The Olympic rain forest is a soggy paradise for bryophytes, such as mosses and liverworts, which appear as squat, spongy mats.

> **❝** *National parks are here to serve as our national treasures, where we preserve and protect both our cultural and our natural heritage for future generations.* **❞**
>
> —PATTI HAPPE

They were the first eukaryotes on the planet, evolving long before plants, animals, and fungi were even around.

PROTECTING DIVERSITY

The fisher is but one eukaryotic species whose existence in Washington has been threatened by human actions. The gray wolf and spotted owl are two others. Next to hunting, habitat loss poses the biggest threat to wildlife. Be-tween the 1930s and the early 1990s, the total area of old-growth forest in Washington State was slashed by approximately 70%–down to 3 million acres from more than 9 million–most of it used for lumber. Much of the remaining forest habitat–what is not protected by Olympic National Park–is fragmented by highways, power lines, railroads, and residential development, leaving no place for many species to call home.

reefs—poses a grave threat to countless species. If current rates of habitat destruction continue, we may witness levels of extinction rivaling the greatest extinction events of geological history. Can anything be done to reverse the trend of dwindling biodiversity around the globe?

Though the rapidly expanding human population (see Chapter 24) is gobbling up resources faster than Earth can restore them, there are things we can do to mitigate the destruction. One conservation strategy is to protect those areas that are known to be especially diverse (called "hot spots"), ensuring that they remain so. That means safeguarding habitat and forbidding overhunting. Where possible, it also means taking efforts to restore missing diversity, in an effort to keep ecosystems whole. "We have an obligation to try to keep an ecosystem intact if we can," says Lewis, of the Washington Department of Fish and Wildlife, noting that the best way to protect the environment may be to keep all of its parts in place.

For the fishers, at least, things seem to be looking up. They are dispersing and reproducing in the forest—at least seven females so far have had kits—and park manager Happe says she is "guardedly optimistic" about their chance of survival. Over the next few years, Happe and Lewis will continue to monitor the fishers, which have been equipped with radio collars, to make sure they are adapting to and surviving in their new home. Only then will they be able to label the restoration project a success. Fishers may have returned to the forest, says Happe, but "they're not out of the woods." ■

The challenges facing Washington are not unique. The United Nations Food and Agricultural Organization estimates that the total global area covered by forests shrank by 23 million acres a year during the 1990s—most of it cleared for agriculture. Some experts estimate that only half the acreage of the planet's original rain forest remains.

Forests are only one place biodiversity is in danger. Habitat destruction in ecosystems around the globe—wetlands, ice caps, coral

CHAPTER 19 Summary

▸ Forest ecosystems like Olympic National Park are home to a great variety of eukaryotic organisms.

▸ The domain Eukarya encompasses all eukaryotic organisms—plants, animals, fungi, and the many types of protists.

▸ Plants are multicellular eukaryotes that carry out photosynthesis and are adapted to living on land. All plants have cells with cell walls, but not all have a vascular system, not all produce seeds, and not all produce flowers.

▸ Plants can be subdivided into groups, including the bryophytes, ferns, gymnosperms, and angiosperms, on the basis of their terrestrial adaptations.

▸ Animals are multicellular eukaryotic heterotrophs that obtain nutrients by ingestion.

▸ Most animals are invertebrates (that is, they lack a backbone). The most abundant invertebrates by far are arthropods, and especially insects.

▸ Vertebrates (animals with a backbone) are members of the phylum Chordata. Common vertebrates include mammals such as the fisher, as well as amphibians, reptiles, birds, and fish.

▸ Fungi are decomposers, acquiring their nutrition by breaking down dead organic matter and absorbing the results. There are unicellular and multicellular fungi.

▸ Protists are a diverse group of mostly unicellular eukaryotic organisms that do not cluster on a single branch of the evolutionary tree. They include photosynthetic plantlike algae and animal-like parasites.

▸ Healthy ecosystems rely on a web of relationships among its members. Removing valuable species—predators especially—can cause a cascading chain of events leading to ecosystem disruption.

MORE TO EXPLORE

▸ **National Park Service: Olympic National Park http://www.nps.gov/olym/index.htm**

▸ **Lewis, J. C., et al. (2012) Olympic Fisher Reintroduction Project: Progress Report 2008–2011. http://wdfw.wa.gov/publications/01393/wdfw01393.pdf**

▸ **Washington Department of Fish and Wildlife, 2011 Annual Report: Fisher http://wdfw.wa.gov/conservation/endangered/species/fisher.pdf**

▸ **UNESCO World Heritage Site: Olympic National Park http://whc.unesco.org/en/list/151**

▸ **Beschta, R. L., and Ripple, W. J. (2008) Wolves, trophic cascades, and rivers in the Olympic National Park, USA. *Ecohydrology* 1:118–130.**

CHAPTER 19 Test Your Knowledge

DRIVING QUESTION 1

What are eukaryotic organisms, and what factors influence their diversity?

By answering the questions below and studying Infographics 19.1 and 19.3, you should be able to generate an answer for the broader Driving Question above.

KNOW IT

1 How does the physical landscape diversity of Olympic National Park affect biodiversity in the park?

2 What are the defining features of eukaryotes, members of the domain Eukarya?

3 What do a fisher and a fir tree have in common?

USE IT

4 How do you think the diversity of eukaryotic organisms in each of the following areas would compare to the diversity in Olympic National Park— would there be more or less? Explain the reasons for your answers.

   a. Lake Michigan

   b. the Sonoran Desert in Arizona

   c. the prairies of Kansas

5 If a fungicide were applied throughout Olympic National Park, how might it affect eukaryotes in the park? Explain your answer.

DRIVING QUESTION 2

How are plants defined, and what influences their diversity?

By answering the questions below and studying Infographic 19.3, you should be able to generate an answer for the broader Driving Question above.

KNOW IT

6 Which group of plants was the first to live on land? Why do we find these plants only in particular environments (after all, if they were first, shouldn't they have spread everywhere by now)?

7 A major difference between a fern and a moss is

   a. the presence of seeds.

   b. the presence of flowers.

   c. the presence of cones.

   d. the presence of a vascular system.

   e. the ability to carry out photosynthesis.

USE IT

8 What is an advantage of having seeds? (Think about spreading to new locations and whether or not reproduction relies on water.)

9 What type of seed plant is likely to rely on hungry animals to spread its seeds? Explain your answer.

10 How did the evolution of vascular systems in plants change the landscape?

DRIVING QUESTION 3

How are animals defined, and what influences their diversity?

By answering the questions below and studying Infographic 19.4, you should be able to generate an answer for the broader Driving Question above.

KNOW IT

11 A sand dollar gets its name from its body shape— it resembles a large coin. What type of body symmetry does a sand dollar have?

   a. bilateral

   b. radial

   c. none (sand dollars are amorphous)

   d. hyphae

   e. mycelium

12 What do a backbone and an exoskeleton have in common?

   a. They are found in closely related groups of animals.

   b. They are made of the same substance.

   c. They both help provide support to an animal's body.

   d. They both require an animal to molt in order to be able to grow.

   e. all of the above

13 You and a fisher are both mammals; as such, what are some characteristics you and the fisher have in common?

14 Which of the following statement(s) is/are true about both cockroaches and lobsters?

a. They are invertebrate insects with bilateral symmetry.

b. They are mollusks with an exoskeleton.

c. They are arthropods with segmented bodies and no symmetry.

d. They are arthropods with an exoskeleton.

e. They are mollusks with a segmented body.

USE IT

15 Many characteristics are used to classify animals. Why do we need to use so many different characteristics? Consider the following five animals: woodpecker, human, wasp, ant, and fisher; and the following three characteristics: ability to fly, two-legged, bearing feathers

a. Which of the five animals could be grouped by each characteristic?

b. Would this grouping reflect their real taxonomic relationship?

c. By what feature(s) would you put wasps and ants together in their own group? What about humans and fishers?

16 Judging from their numbers, arthropods are a tremendously successful group. What traits do you think have enabled them to be so successful? Justify your answer with examples.

DRIVING QUESTION 4

How are fungi defined, and what influences their diversity?

By answering the questions below and studying Infographic 19.5, you should be able to generate an answer for the broader Driving Question above.

KNOW IT

17 Consider the "eating habits" of fungi.

a. Can fungi carry out photosynthesis?

b. Can fungi ingest their food?

c. How do fungi obtain their nutrients and energy?

18 Which of the following meals include fungi as food?

a. a bread and blue cheese platter with fruit

b. mushroom risotto

c. a and b

d. a fruit salad

e. yogurt

USE IT

19 A very early classification scheme placed the fungi together with the plants. Why do you think fungi were grouped with plants? What features distinguish them from plants?

DRIVING QUESTION 5

What are protists, and what influences their diversity?

By answering the questions below and studying Infographic 19.6, you should be able to generate an answer for the broader Driving Question above.

KNOW IT

20 What do all members of the informal group known as protists have in common?

a. nothing

b. They are all eukaryotic.

c. They all carry out photosynthesis.

d. They are all human parasites.

e. They are all decomposers.

USE IT

21 Why do scientists no longer consider protists a separate kingdom? How might scientists find new taxonomic "homes" for the protists? Do you think structural features (for example, chloroplasts) or genetic information will be more useful in their classification?

22 Many protists have an organelle called the contractile vacuole that pumps out water that enters the cell by osmosis. Why is this a useful adaptation for a protist? What might happen to a protist if its contractile vacuole stopped working? (Think about where many protists live, and what happens to bacteria whose cell walls are disrupted by antibiotics.)

MINI CASE

23 Reintroducing species to their native habitats is sometimes controversial. One reintroduction effort in particular that has caused quite a stir is the reintroduction of the Mexican gray wolf (*Canis lupus baileyi*) into New Mexico and Arizona. You can read about this project at http://www.fws.gov/southwest/es/mexicanwolf/.

a. Why might it be important to reintroduce species into their native habitats? Answer first in general terms, then specifically for the Mexican gray wolf.

b. What factors could impede the success of such reintroductions? Again, answer in general terms first, then specifically for the Mexican gray wolf.

INTERPRETING DATA

 24 The otter was officially declared extinct in the Netherlands in 1988. A reintroduction program was started in 2002. The reintroduced otters were either captured in Eastern Europe, or bred in captivity in countries including Sweden and Russia. Between 2002 and 2008, a total of 31 otters were released into a specific area in the Netherlands.

DNA from otter "spraint" (fecal material) was used to identify individual otters, and their offspring. The DNA spraint data are summarized in the table below.

   a. Graph these data.

   b. What is the total otter population in the Netherlands by the seventh year of the project?

   c. What appears to be the major contributor to the otter population: introductions of new otters, or breeding of otters in the Netherlands?

The project also monitored genetic diversity in the population and sources of mortality. The DNA analysis also included looking at the total number of alleles for genes in the population. Those data are shown in the table below. Of the 54 otters that died, 80% died in traffic (hit by vehicles), 5 deaths were due to unknown causes, and the remaining deaths were caused by drowning, disease, a bite wound, and traps.

   a. What does the genetic diversity analysis suggest about inbreeding?

   b. From all the data presented, is the otter reintroduction successful?

   c. What recommendations would you make to ensure the long-term survival of otters in the Netherlands?

PROJECT YEAR	NUMBER OF FOUNDER SPRAINTS	NUMBER OF NATIVE-BORN OTTER SPRAINTS
1	15	0
2	7	5
3	8	7
4	6	12
5	7	28
6	7	39
7	6	45

PROJECT YEAR	AVERAGE NUMBER OF ALLELES PER GENE
2	6.5
3	5.5
4	4.5
5	4.2
6	4.3
7	4.7

Source: *Results and perspectives of Otters in the Netherlands.* H. Jansman, et al. Presentation at http://ls.gd/spv5ZY

BRING IT HOME

 25 Many species reintroductions are being carried out across the United States. Do some research to learn about at least one such effort. For the species you research, address the following questions:

   a. What caused it to be lost from its native habitat?

   b. Is its reintroduction important?

   c. Are there are controversies about its reintroduction?

   d. What made you interested in this particular species and its reintroduction? Is it an "attractive" species? Is it being reintroduced near where you live?

SKIN DEEP

Science redefines the meaning of racial categories

WHEN BARACK OBAMA WAS ELECTED FOR HIS FIRST term in 2008, he was hailed as America's first black president. When Tiger Woods won the Masters Golf Tournament in 1997, he was lauded as the first black man to win. When Halle Berry won an Oscar in 2001 for best actress, she was commended as the first black woman to win in that category.

Why was skin color so significant? A 250-year history of slavery and racial discrimination in the United States has left a bitter legacy. Almost 150 years after slavery was legally abolished in the United States, people of color are still underrepresented in positions of power and prestige. Although the reasons for this underrepresentation are complex, the recognition of the achievements of Obama, Woods, and Berry signaled a major change: barriers to social advancement were beginning to come down.

To shoehorn any of these three people into a simple racial category, however, is misleading: Barack Obama was born to a white mother and a black African father; Tiger Woods's background includes African, Chinese, Dutch,

▶▶ **DRIVING QUESTIONS**

1. What contributes to human skin color, and why is there so much variation in skin color among different populations?

2. Where did the earliest humans evolve, and how do we know?

3. What can we learn about human evolution from the fossil record?

INFOGRAPHIC 20.1 HUMANS ARE GENETICALLY SIMILAR

The human genome includes 3 billion nucleotide base pairs. Its entire sequence of A, G, T, and C nucleotide bases would fit into 200 phone books, each 1,000 pages long. When stacked, the phone books would reach to the top of the Washington Monument in Washington, D.C. That's 200,000 pages of genetic code!

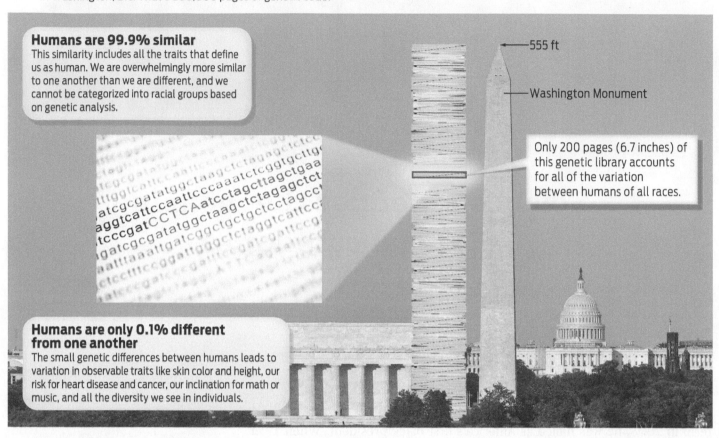

Humans are 99.9% similar
This similarity includes all the traits that define us as human. We are overwhelmingly more similar to one another than we are different, and we cannot be categorized into racial groups based on genetic analysis.

555 ft

Washington Monument

Only 200 pages (6.7 inches) of this genetic library accounts for all of the variation between humans of all races.

Humans are only 0.1% different from one another
The small genetic differences between humans leads to variation in observable traits like skin color and height, our risk for heart disease and cancer, our inclination for math or music, and all the diversity we see in individuals.

and Thai forebears; Halle Berry was born to a white mother and an African-American father. So what does the term "black" mean?

Historically, racial categories were employed by one group to maintain power over another and to justify forms of oppression, including slavery. In the United States, racial categories were reinforced by laws like the "one drop" rule adopted by several states in the 1920s, which held that any American with one drop of African blood was to be considered black. People then continued to use these categories and their connotations to justify racial discrimination and, in some places, racial segregation.

Though social and political attitudes have changed, people continue to invoke racial categories like "black" or "white" for various reasons, including simple physical description. From a biological perspective, however, it is

increasingly clear that racial categories have little meaning. Research on human genetics shows that race is a social, not a biological, category. Groups of people can and do share similar physical characteristics, such as skin color and other features, but these superficial differences obscure how fundamentally similar human beings are.

All humans are members of a single biological species, *Homo sapiens*. Biologically, we are far more similar to one another than we are different. In fact, genetic studies comparing regions of the human genome from person to person show that each person's DNA is 99.9% identical to that of any other unrelated person. It is only in the remaining 0.1% of our DNA that we differ, and it is this tiny fraction that accounts for the diversity of traits we see from person to person (INFOGRAPHIC 20.1).

If people are so fundamentally similar on a biological level, why is there so much variation in human skin tone? And how did these differences come about? The answers lie in the evolution and migration of our earliest ancestors. Humans are a recent species, first walking the Earth a mere 200,000 years ago. Since that time, humans have evolved many traits that helped them survive the environments they encountered as they migrated around the globe. People with African ancestry, for example, carry with high frequency an allele that helps them resist malaria. An allele common among Northern Europeans enables them as adults to digest milk better than other populations do, an indication that at some point in history, dairy products provided an important source of nutrition and those who could digest them were more likely than others to survive. Tibetans carry in high frequency an allele that helps their red blood cells compensate for the low oxygen level in their high-altitude environment.

Skin color is another example of human evolution. For many years, scientists had wondered why humans differ in their skin color. But it wasn't until the work of Nina Jablonski, an anthropologist at Pennsylvania State University, that they found a convincing answer.

THE EVOLUTION OF SKIN COLOR

More than a decade ago, Jablonski and her husband, George Chaplin, a geographic information systems specialist, set out to understand why human populations evolved varying skin tones. They knew that skin tone largely reflects the amount of **melanin** present in the skin. People naturally produce different levels of melanin: more melanin yields darker skin, less melanin lighter skin. Skin also responds to sunlight by producing more melanin and becoming darker temporarily (INFOGRAPHIC 20.2).

Jablonski and Chaplin also knew that, in general, skin tone correlates with geography:

▶MELANIN
Pigment produced by a specific type of skin cell that gives skin its color.

INFOGRAPHIC 20.2 MELANIN INFLUENCES SKIN COLOR

→ Melanocytes are a type of cell located in the epidermis, the outermost layer of skin. Melanocytes make the pigment melanin and deposit it into other cells in the skin. A person's skin color depends largely on the amount and type of melanin that his or her skin melanocytes produce. Sunlight can also temporarily increase the amount of melanin in a person's skin.

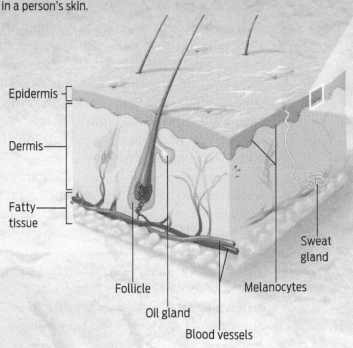

Epidermis

Dermis

Fatty tissue

Follicle

Oil gland

Blood vessels

Melanocytes

Sweat gland

Skin epidermis

Melanin

Melanocyte

▶**FOLATE**
A B vitamin also known as folic acid, folate is an essential nutrient, necessary for basic bodily processes such as DNA replication and cell division.

In certain environments, is there an evolutionary advantage to light or dark skin?

people from regions closer to Earth's poles tend to be lighter-skinned and those from areas closer to the equator tend to have darker skin. Jablonski wanted to understand this, so she searched the scientific literature. Might there be an evolutionary advantage to having light or dark skin in different environments?

Jablonksi (right) and Chaplin examine a map of predicted human skin color based on UV light intensity.

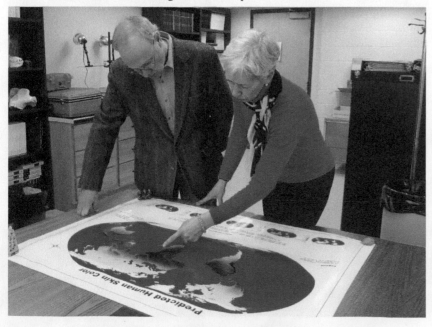

She found her first clue in a 1978 study showing that an hour of intense sunlight can halve the level of an important vitamin known as **folate** in light-skinned people. Folate, also called folic acid, is an essential nutrient, necessary for basic bodily processes like DNA replication and cell division.

Then, at a seminar, Jablonski learned that low folate levels can cause severe birth defects such as spina bifida–a condition in which the spinal column does not close–and anencephaly, the absence at birth of all or most of the brain. She subsequently came across three case studies that linked such birth defects to the mothers' visits to tanning studios, where the women would have been exposed to ultraviolet light. She also learned that folate is necessary for the normal development of sperm.

Taken together, these observations suggested to Jablonski that people with light skin are more vulnerable to folate destruction than are darker-skinned people–presumably because melanin absorbs damaging UV light and dissipates it as heat. Could the need to protect the body's folate stores from UV light have driven the evolution of darker skin shades? The supporting evidence was compelling. But then what was the advantage of having light skin at all, as many populations today do?

Jablonski considered a hypothesis first proposed in the 1960s by biochemist W. Farnsworth Loomis, who suggested that vitamin D might play a role in the evolution of skin color. Unlike folate, which is destroyed by excess sunlight, the production of vitamin D requires ultraviolet light. **Vitamin D** is crucial for good health: it helps the body absorb calcium and deposit it in bones. During pregnancy women need extra vitamin D to nourish the growing embryo. And because vitamin D is so important for healthy bone growth, too little might cause bone distortion,

and a distorted pelvis would make it difficult for a woman to bear children (**INFOGRAPHIC 20.3**).

Building on this previous work, in 2000 Jablonski and Chaplin compared data on skin color in indigenous populations from more than 50 countries to levels of global ultraviolet light as measured by NASA satellites. They found a clear correlation: the weaker the ultraviolet light, the fairer the skin—a compelling suggestion that both dark and light skin are linked to levels of global sunlight. They published their results in the *Journal of Human Evolution*.

The researchers now had a complete hypothesis: light skin evolved in sun-poor parts of

▶VITAMIN D
A fat-soluble vitamin required to maintain a healthy immune system and to build healthy bones and teeth. The human body produces vitamin D when skin is exposed to UV light.

INFOGRAPHIC 20.3 FOLATE AND VITAMIN D ARE NECESSARY FOR REPRODUCTIVE HEALTH

→ Folate, also known as folic acid or vitamin B₉, is abundant in beans, citrus fruit, dark green leafy vegetables, whole grains, poultry, pork, shellfish, and liver. Folate is especially critical during periods of rapid cell division, such as during embryonic and fetal development. UV light destroys the body's folate stores.

UVB rays

Intense UV light destroys stored folate.

Sunlight

Vitamin D is produced in skin exposed to UV light.

UVB rays

→ Human skin produces vitamin D when exposed to UV rays in sunlight. We can also get vitamin D from some foods such as vitamin D–fortified milk, fish, cheese, butter, and fortified cereals. Sufficient vitamin D permits absorption of calcium and phosphate in the small intestine and enhances bone mineralization, among other tasks in the body.

Folate
Folate
Folate
Folate

People with light skin are more vulnerable to folate destruction.

—Skin color

Vitamin D
Vitamin D
Vitamin D
Vitamin D

People with dark skin are more vulnerable to insufficient vitamin D production.

Folate deficiency

Vertebra
Dura mater
Spinal cord
Spinal fluid

Low Sperm Count
Folate is required for sperm to develop normally.

Spina Bifida
The spinal column does not close around the spinal cord before birth.

Anencephaly
The brain and skull are highly under-developed. This condition is always fatal.

Vitamin D deficiency

Risks for pregnant women
• Preeclampsia—elevated blood pressure
• Reduced bone density in pelvis

Risks for children
• Premature birth
• Rickets disease—abnormal bone formation from calcium deficiency
• Multiple sclerosis

INFOGRAPHIC 20.4 HUMAN SKIN COLOR CORRELATES WITH UV LIGHT INTENSITY

Nina Jablonski and George Chaplin used NASA satellite measurements of UVB intensity to predict the amount of skin pigment that would best block harmful UV rays yet still enable the body to produce sufficient vitamin D in populations around the globe. Their predictions closely match actual skin color variations around the world.

Predicted pigmentation of skin based on UVB intensity

SOURCE: Chaplin, G. (2004) Geographic distribution of environmental factors influencing human skin coloration. *Journal of Physical Anthropology* 125:292 -302. Map updated in 2007.

the world because it helped the body produce vitamin D, while dark skin evolved because it helped protect the body's folate stores in people who lived in sunny climates. The body's need to balance levels of these two important nutrients given varied levels of UV light explains why there is so much variation in skin tone around the globe (**INFOGRAPHIC 20.4**).

Since the publication of Jablonski and Chaplin's work, many other scientists have tested their hypothesis, and it is now the most widely accepted explanation for the evolution of human skin color. As Jablonski points out, "It synthesizes the available information on the biology of skin from anatomy, physiology, genetics, and epidemiology, and has not been contradicted by any subsequent data."

OUT OF AFRICA

If light and dark skin tones developed over time, then at some point in history, all humans likely had the same skin tone. This conclusion—a logi-

cal extension of Jablonski's work—is supported by genetic studies suggesting that anatomically modern humans first evolved in Africa.

In 1987, a team of geneticists led by Allan Wilson of the University of California at Berkeley used **mitochondrial DNA (mtDNA)**—genetic material we inherit solely from our mothers—to construct an evolutionary tree of humanity. Mitochondrial DNA is DNA located in the mitochondria in all our cells. Unlike nuclear DNA, which is inherited from both parents in most multicellular organisms (including humans and other animals), and which undergoes recombination during meiosis, mtDNA passes from mothers to offspring essentially unchanged. That's because sperm do not contribute their mitochondria to the newly formed zygote (**INFOGRAPHIC 20.5**).

Like nuclear DNA, mtDNA mutates at a fairly regular rate, although it appears to mutate faster than nuclear DNA. A mother with a mutation in her mtDNA will pass it to all

▶**MITOCHONDRIAL DNA (MTDNA)**
The DNA in mitochondria that is inherited solely from mothers.

her children, and her daughters will pass it to their children in turn. Because these mutations pass down without being combined and rearranged with paternal mitochondrial DNA, mtDNA is a powerful tool by which to track human ancestry back through hundreds of generations.

Wilson and his colleagues Rebecca Cann and Mark Stoneking collected mtDNA from 147 contemporary individuals from Africa, Asia, Australia, Europe, and New Guinea. On the basis of the mtDNA sequence patterns—that is, the number of individual nucleotide differences from one person to the next—researchers were able to determine how closely or distantly related the individuals were: individuals sharing nearly identical sequences are more closely related than individuals with many sequence differences between them. As mitochondrial DNA mutates at a more or less constant rate, the researchers could also tell how long it had been since groups diverged from one another, based on the number of differences. Using this information, the researchers created an evolution-

ary tree, grouping first individuals that shared the most similar mtDNA sequences, and then grouping these groups with respect to one another, until finally the largest group—containing all the others—was reached. A computer program helped them choose the most statistically probable tree.

The researchers found that branches of the tree from all five geographical areas could be traced back to a single female ancestor who lived in eastern Africa some 200,000 to 150,000 years ago. In other words, if every person on the planet were to construct a family tree that listed every female ancestor for thousands of generations back in time, they would all eventually converge at a single common female ancestor. As a newspaper reporter put it, the researchers had found "Mitochondrial Eve."

While catchy, the nickname "Eve" is misleading in a number of respects. This single female common ancestor wasn't the only female living at the time; she was merely one female in a population of many ancient humans. Nor was she the only female to leave descendants.

INFOGRAPHIC 20.5 MITOCHONDRIAL DNA IS INHERITED FROM MOTHERS

When egg and sperm fuse during fertilization, sperm contribute only nuclear DNA to the nucleus of the newly formed zygote. The egg provides all other organelles, including mitochondria. Consequently, only mothers contribute mitochondrial DNA to their children.

Sperm
The mitochondria in a sperm body produce ATP to power its tail to swim. Only the sperm's DNA-containing nucleus fuses with an egg nucleus. Sperm mitochondria do not enter the egg at fertilization.

Egg
Eggs are much larger than sperm. They carry all of the cytoplasm and organelles (including the mitochondria) necessary for a fertilized embryo to begin developing.

Nucleus
Mitochondrion
Head
Midpiece
Sperm cell

Mitochondrial DNA

37 genes

DNA in egg mitochondria is passed to all offspring.

Only the mother's mitochondrial genes are passed on to the offspring.

But Eve's mitochondrial DNA is the only mitochondrial DNA that modern humans still carry today. How can this be? The reason is simple: while other females living at the time also had descendants, the lines of these descendants either died off, perhaps in a bottleneck event (see Chapter 15), or left only sons. When that happened, their mitochondrial DNA lines became extinct. All humans today can trace their mitochondrial DNA back to Mitochondrial Eve (INFOGRAPHIC 20.6).

Interestingly, the tree Wilson generated had two major branches: one that leads to individuals now living in all five of the regions studied—Asia, Australia, Europe, New Guinea,

and Africa—and one that leads only to modern-day Africans. The mtDNA of people on the exclusively African branch had acquired twice as many mutations as the mtDNA of people on the rest of the tree. The most likely interpretation of these data, the scientists reasoned, was that the African mtDNA had had more time to accumulate mutations, and was consequently older, evolutionarily speaking. This would mean that humans likely originated in Africa, where they formed several ancestral populations. After some period of time, one group of Africans left the continent, and their descendants continued to migrate to other continents, eventually becoming the

INFOGRAPHIC 20.6 MODERN HUMAN POPULATIONS SHARE A COMMON FEMALE ANCESTOR

Many women of Eve's generation left descendants, but only Eve's mitochondrial DNA survives among humans today.

200,000 to 150,000 years ago • Other Early females • mtEve • Other Early females • Generational time • Dead End Lineage • Dead End Lineage

Thousands of generations
All modern human populations can trace their mtDNA back to a single mitochondrial ancestor.

INFOGRAPHIC 20.7 OUT OF AFRICA: HUMAN MIGRATION

→ Genetic evidence suggests that the earliest modern humans originated and evolved for thousands of years in Africa before a group of them migrated to the other continents. Because African populations are older, they have more genetic diversity. As migrating descendants only took a fraction of the alleles present in Africa, there is less genetic variation in other regions.

Bering Strait land bridge

40,000 years ago

67,000 years ago

20,000 years ago

"Eve"

200,000–
150,000
years ago

60,000–40,000 years ago

13,000 years ago

More genetic variation

Less genetic variation

Omo fossils
195,000 years old

Herto fossils
160,000 years old

ancestors of modern-day Asians, Australians, and Europeans. Evidence suggests this migration began around 70,000 years ago.

Since Wilson's study, additional evidence continues to back the "out of Africa" hypothesis. Fossil discoveries in Ethiopia in 2003 and 2005 represent the oldest known fossils of modern humans—160,000 and 195,000 years old, respectively—and plug a major gap in the human fossil record. Both sets of remains date precisely from the time Wilson and his colleagues think that Mitochondrial Eve lived in eastern Africa. The fossil discoveries provide evidence that anatomically modern humans were living in that region around the same time that Mitochondrial Eve lived, and provide further evidence that the earliest humans originated in Africa.

This hypothesis is also supported by research that sampled genetic diversity from nuclear DNA. In a 2008 study, for example, Richard Myers, of the Stanford University School of Medicine, and colleagues found less and less genetic variation in people the farther away from Africa they lived—the same pattern of variation that scientists have found in human mtDNA sequences. This finding suggests that as each small group of people broke away to explore a new region, it carried only some of the parent population's alleles. Consequently, genetic diversity decreased in tandem with the distance people traveled away from Africa—a classic example of the founder effect described in Chapter 15 (**INFOGRAPHIC 20.7**).

440

▶HOMINID

Any living or extinct member of the family Hominidae, the great apes—humans, orangutans, gorillas, chimpanzees, and bonobos.

BECOMING HUMAN

A number of lines of evidence peg Mitochondrial Eve as the most recent matrilineal common ancestor of all humans living today. However, she represents merely one branch on the evolutionary tree that includes our species; this tree has several other branches representing other hominid species. A **hominid** is any member of the biological family Hominidae, which includes living and extinct great apes (humans, orangutans, gorillas, chimpanzees, and bonobos).

Humans are grouped with the great apes because fossil evidence shows that modern humans and other present-day great apes evolved from a common ancestor that lived

INFOGRAPHIC 20.8 TRAITS OF MODERN HUMANS REFLECT EVOLUTIONARY HISTORY

➔ *Homo sapiens* is the only surviving lineage in the evolutionary history of humans. In other words, several hominids have existed or coexisted as related but distinct species in the past. The physical traits of modern-day humans, such as skin color and body hair, evolved in response to selective pressures. A species with less hair could better regulate body temperature in hot and sunny environments, for example, but would require darker skin to protect it from high UV light exposure.

13 million years ago. Of the living members of this group, humans and chimpanzees are the most closely related, although it has been about 7 million years since their last shared ancestor lived. During those 7 million years, both humans and chimps have undergone a tremendous amount of evolutionary change, which is why living humans look and behave so differently from chimps—or any other primate species living today.

Scientists haven't yet discovered fossil remains of the last common ancestor of chimps and humans. However, in October 2009 the first analyses of fossil remains of a 4.4-million-year-old hominid, *Ardipithecus ramidus*, nicknamed Ardi, were published. Ardi's remains are among the oldest hominid fossils so far discovered and, as such, give tantalizing clues to early human origins.

Among the defining characteristics of *Homo sapiens* are the ability to walk upright and a big brain. An upright gait meant the hands were free to make and use tools; a big brain enabled *H. sapiens* to develop complex language. Which came first, a big brain or walking upright? Ardi provides the answer: Ardi had a small brain, suggesting that it could not use complex language. But Ardi's bones clearly show that it could walk upright without dragging its knuckles, while still able to maneuver on all fours in trees. The ability to walk upright therefore evolved first and was a major step toward becoming human.

The fossil record after Ardi also reveals some of the major milestones in human evolution. For example, artifacts found at various archeological sites indicate that simple tool use began approximately 2.6 million years ago, most likely when our hominid ancestors began eating meat from large animals. The first tool-users were members of the genus *Australopithecus*. This genus walked upright and appears to have lived on the ground, rather than in trees, as evidenced by the lack of an opposable big toe, which had helped the early hominids grip branches.

Another milestone was the ability to use and control fire, which appeared about 800,000 years ago. Artifacts such as clay shards found at various fossil sites show that *Homo erectus* was likely the first species able to control fire. Using fire enabled *Homo erectus* to cook meat and bone marrow, to stay warm, and probably to fight off predators.

Finally, at some point between 800,000 and 200,000 years ago, hominid brain size began to expand rapidly. Studies of fossilized organisms from ocean sediments show that this was also a time of climate instability. The temperature dramatically shifted from high to low and back several times during this period. Scientists hypothesize that a larger brain would have enabled better communication and problem solving, which would have been very useful to our hominid ancestors as they coped with climate change. This was also around the time that anatomically modern humans like Mitochondrial Eve and our own species, *Homo sapiens*, appeared (INFOGRAPHIC 20.8).

SELECTION FOR SKIN COLOR

That anatomically modern humans originated in Africa suggests that the first humans likely had dark skin. But then how might varying skin tones have later evolved? Nina Jablonski's research has shown that environmental factors likely played a role in the evolution of different skin tones. Environment alone doesn't produce evolution, however. Rather, the environment acts on traits, or phenotypes, increasing or decreasing the frequency of alleles in a population by natural selection. Where did these alleles come from?

Recall that each time a cell replicates, mutations caused by errors in replication can occur. If these mutations occur in germ cells during meiosis (see Chapter 10), they will permanently change the genome of the next generation. This process continually introduces new alleles into the population. Some of these alleles can be negative or harmful, as in the case of hereditary cancer or cystic fibrosis. But new alleles can also be benign or even beneficial. Indeed, sometimes alleles can be so positive and confer such a survival advantage that they become more common in succeeding generations and can eventually become fixed in a population, reaching 100% frequency (**INFOGRAPHIC 20.9**).

Sometimes alleles that are harmful in one environmental context may be beneficial in another. For example, the recessive allele responsible for cystic fibrosis (CF) can cause this serious disease when it occurs in homozygotes, who have two copies of the allele. However, research has suggested that being heterozygous for CF–that is, having only one CF allele–may have reduced the severity of diarrhea caused by cholera or some other infection in times past. Consequently, carrying a CF allele provided an advantage during epidemics. This would help explain why the CF allele became relatively common.

Skin color is another example of a trait that likely conferred an advantage to humans and underwent natural selection at some point in human history. Otherwise, dark or light skin color wouldn't be so common among specific populations. In fact, the dark skin of those *Homo sapiens* who evolved in Africa was probably an early adaptation; it is likely that before dark skin evolved, our earliest ancestors had light skin, just as chimpanzees do today.

Fossil and genetic evidence suggests that about 2 million years ago hominids became "bipedal striders, long-distance walkers, and possibly even runners," according to Jablonski. But to sustain such activities, hominids needed an effective cooling system, a feature they could have developed only by losing excessive body hair and gaining more sweat glands. In contrast, hairy chimpanzees, our closest living animal relatives, can sustain only short bouts of activity without getting overheated. "It's like sweating in a wool blanket," Jablonski explains. "After that blanket gets saturated, you can't lose very much heat."

Eventually, some factor–food scarcity, perhaps–forced ancient hominids out of the forests

INFOGRAPHIC 20.9 NATURAL SELECTION INFLUENCES HUMAN EVOLUTION

 The environment selects for specific genetically determined traits. Different environments select for different traits, and therefore different alleles.

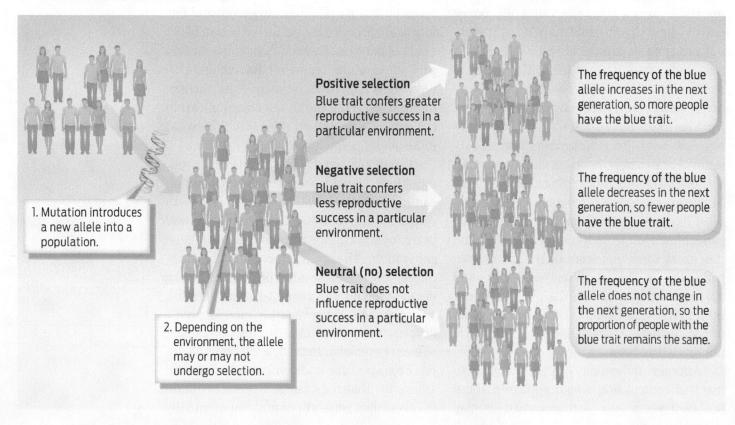

1. Mutation introduces a new allele into a population.

2. Depending on the environment, the allele may or may not undergo selection.

Positive selection
Blue trait confers greater reproductive success in a particular environment.

The frequency of the blue allele increases in the next generation, so more people have the blue trait.

Negative selection
Blue trait confers less reproductive success in a particular environment.

The frequency of the blue allele decreases in the next generation, so fewer people have the blue trait.

Neutral (no) selection
Blue trait does not influence reproductive success in a particular environment.

The frequency of the blue allele does not change in the next generation, so the proportion of people with the blue trait remains the same.

and into the open savannahs to hunt for food. Hominids with less hair and more sweat glands were likely better hunters because they could sustain long bouts of activity without getting overheated. Like modern-day chimpanzees, these hominids likely had fair skin under their hair. Without hair to protect their light skin, they were exposed to the intense African sun. And, scientists hypothesize, exposure to the sun would have reduced their folate levels and thus their fitness in the sun-drenched environment.

Any of these ancient hominids that carried an allele or developed or inherited a mutation that increased their ability to produce more melanin would have been able to spend more time in the sun without the detrimental effects. Darker coloration would have protected their skin, and consequently their folate levels, from the sun, enabling them to hunt and travel in the open fields.

Evidence to support this hypothesis comes from genetics. In 2004, Alan Rogers of University of Utah and his colleagues studied a gene that influences skin color. They discovered that more than a million years ago, an allele that contributes to dark skin became fixed–that is, its frequency approached 100%–in the African population of hominids. "This is critical," Jablonski says. "It shows that darkly pigmented skin became extremely important to us" around the time that hominids became more humanlike.

The allele for darker skin was such an advantage in terms of survival and reproduction that hominids with darker skin left more offspring than their lighter-skinned relatives. Hominids born with light skin weren't able to survive and reproduce in great enough numbers for the trait to persist in the population. The allele for darker skin eventually increased in the population until it essentially reached 100%.

Populations that migrated north, away from the African sun, however, faced a different environment. Folate was not as easily destroyed in this lower-UV-light environment. But the high levels of melanin present in dark skin were a disadvantage; they prevented bodies from producing enough vitamin D. In this low-UV-light environment, fair skin allowed the body to soak up more ultraviolet light and

produce essential vitamin D. In these environments fair-skinned people thus were more fit and left more descendants than did dark-skinned people. Consequently, the frequency of light skin in northern climates increased with each generation.

If the earliest humans migrating from Africa to northern regions all had dark skin, where then did alleles for light skin come from? There are two possibilities. One is that the alleles were introduced as new mutations into the population either shortly before or shortly after humans migrated out of Africa and were subsequently selected by natural selection. The other is that the alleles were still present in the African population, albeit at very low frequencies, and then subsequently became more common under selection in a new environment.

Both explanations may be true. Genetic studies show that the frequency of alleles for light skin increased and swept through populations as they migrated north–most likely more than once. Alleles of at least three different genes have been found to confer light skin in people from northwestern Europe and light skin in people from eastern Asia– suggesting that mutations for light skin arose independently and spread through those two populations separately. These mutations were genetically minor, yet phenotypically significant: single nucleotide changes in two of these genes have been shown to account for a large proportion of the difference in skin color between light- and dark-skinned people–two nucleotides among the 3 billion total in our DNA. Skin color really is skin deep.

AN EVOLVING EXPLANATION

There are a number of other hypotheses for the evolution of skin tone, but of all of them, the folate-vitamin D hypothesis has the most supporting evidence and is consequently "the most reasonable," says Mark Stoneking of the Max Planck Institute for Evolutionary Anthropology in Leipzig, Germany, who collaborated on the Mitochondrial Eve study. In fact, Stoneking says, "Skin tone is one of the best examples of human evolution." It's

INFOGRAPHIC 20.10 THE EVOLUTION OF SKIN COLOR

Human skin color is an example of a trait that has undergone natural selection. Varying levels of UV light have selected for a range of skin tones around the globe. In each case, the amount of melanin represents a compromise between the need to protect folate and the need to make vitamin D.

2. Migration into Low-UV-Light Environment
Individuals with light skin reproduce more successfully
· Lower melanin levels enable sufficient vitamin D levels even with low-UV-light levels.
· Folate is not destroyed in low-UV environments.

1. High-UV-Light Environment
Individuals with dark skin reproduce more successfully
· More melanin protects folate even from high levels of UV light.
· High-UV-light intensity allows even those with more melanin to produce sufficient vitamin D.

3. Migration into High-UV-Light Environment
Individuals with dark skin reproduce more successfully
· More melanin protects folate even from high levels of UV light.
· High-UV-light intensity allows even those with more melanin to produce sufficient vitamin D.

an example in which we can see that genes have definite phenotypic effects, in this case on skin pigmentation, and we can also see that the traits were selected, he explains.

Scientists know that these skin-color genes have been favored by specific environments because they carry genetic signatures of natural selection. To study whether natural selection favored any particular trait, scientists typically look at the amount of allelic diversity that exists for a gene of interest. Fewer alleles than average in a population means that there was some environmental pressure that selected the alleles for that trait (see Chapter 15). Genes that are not being acted on by natural selection show higher amounts of allelic diversity in the population.

Indeed, skin-color genes show this very pattern—they show less allelic diversity than genes for other traits. The particular alleles that predominate in a particular population are ones that originally provided a selective advantage to its members by producing more or less melanin. Dark-skin alleles were favored in high-UV environments, while light-skin alleles were favored in low-UV environments. Skin color is thus a proxy for the geographic origin of our ancestors, but not much else (INFOGRAPHIC 20.10).

Throughout human history, the lines between what we have come to call races have been fluid. Genetic studies show that hardly any population is "pure" in the way that many have thought. As people moved around the globe, they settled and often bore children with people they met along the way, introducing new alleles into the local gene pool. The particular environment people encountered favored some traits over others, and that is why populations that live in similar environments share similar features.

Though people tend to create racial groupings based on obvious physical characteristics, such features can be shared with other groups, says Jablonski. Not all Africans have equally dark skin and not all Europeans are light skinned, for example. And as humans travel more, settle in different areas, and intermarry, Jablonski says, "racial categories will get messier and messier." Perhaps in time the concept of race itself will disappear. ■

CHAPTER 20 Summary

- Humans are 99.9% genetically identical to one another regardless of geographic origin. Biologically distinct human races do not exist.

- All humans are members of a single biological species, *Homo sapiens,* which evolved relatively recently—just 200,000 years ago.

- Physical features shared by people within populations reflect adaptations to specific environments.

- Alleles can be harmful, beneficial, or neutral in their effect on survival and reproduction.

- Skin color most likely evolved in response to environmental UV levels, an example of evolution by natural selection. Alleles for darker skin conferred an advantage in sunnier environments, while alleles for lighter skin conferred an advantage in regions that receive weak sunlight.

- Skin color represents an evolutionary trade-off between the need for vitamin D, which requires adequate sunlight for its production, and the need for folate, which is destroyed by too much sunlight.

- Fossil evidence shows that humans and apes descended from a common ancestor and that walking upright preceded development of a big brain. There were many species that could walk upright before *Homo sapiens* appeared.

- Fossil and DNA evidence shows that anatomically modern humans first emerged in Africa, approximately 200,000 years ago, and subsequently spread to other continents, beginning about 70,000 years ago.

- All modern-day humans can trace a portion of their genetic ancestry back to a single woman, Mitochondrial Eve, who lived 200,000 to 150,000 years ago in Africa.

- Humans evolved from apelike primate ancestors who likely had light skin. Darker skin emerged in tandem with loss of body hair as our hominid ancestors ventured into the hot savannah, while lighter skin emerged as humans migrated farther north.

MORE TO EXPLORE

- Jablonski, N. G., and Chaplin, G. (2010) Human skin pigmentation as an adaptation to UV radiation. *Proceedings of the National Academy of Sciences* 107: 8962–8968.

- Cann R. L., et al. (1987) Mitochondrial DNA and human evolution. *Nature* 325: 31–36.

- Cann, R. L., and Wilson, A. C. (2003) The recent African genesis of humans. *Scientific American* 13(2):54–61.

- Groleau, R. (2002) Tracing ancestry with mtDNA. NOVA Online: http://www.pbs.org /wgbh/nova/neanderthals/mtdna.html.

- White, T. D., et al. (2009) *Ardipithecus ramidus* and the paleobiology of early hominids. *Science* 326(5949):64, 75–86.

- Stringer, C., and McKie, R. (1996) *African Exodus: The Origins of Modern Humanity.* New York: Henry Holt.

CHAPTER 20 Test Your Knowledge

DRIVING QUESTION 1

What contributes to human skin color, and why is there so much variation in skin color between different populations?

By answering the questions below and studying Infographics 20.1, 20.2, 20.3, 20.4, 20.9, and 20.10, you should be able to generate an answer for the broader Driving Question above.

KNOW IT

1 What pigment molecule gives dark skin its color? What cell type produces this pigment?

2 In the course of human evolution, which of the following environmental factors likely influenced whether populations had mostly light-skinned individuals or mostly dark-skinned individuals?

  a. average annual temperature

  b. average annual rainfall

  c. levels of UV light

  d. the vitamin D content of the typical diet

  e. mitochondrial DNA inheritance

3 As hypothesized by Jablonski and Chaplin, darker skin is advantageous in _____ UV environments because darker skin _____.

  a. high-; reduces vitamin D production

  b. high-; protects folate from degradation

  c. high-; increases the rate of folate synthesis

  d. low-; allows more vitamin D to be produced

  e. low-; allows more folate to be produced

USE IT

4 If folate were *not* destroyed by UV radiation, predict the skin color you might find in populations living at the equator; in populations living in Greenland. Explain your answers.

5 Which of the following would help darker-skinned people who live in low-UV environments remain healthy?

  a. folate supplementation

  b. sunscreen

  c. increased production of melanin

  d. vitamin D supplementation

  e. calcium supplements

6 What can you infer about the skin-color genotype and the geographic origins of the ancestors of a light-skinned person and a dark-skinned person?

7 Our closest primate relatives, chimpanzees, have light-colored skin yet live in tropical (high-UV) environments. How would the Jablonski–Chaplin hypothesis explain this observation?

  a. Chimpanzees don't need folate for successful reproduction.

  b. Chimpanzees are not susceptible to skin cancer.

  c. The hair of chimpanzees protects their light skin from UV light.

  d. Chimpanzees require much higher levels of vitamin D than humans do.

  e. In chimpanzees a light-colored pigment offers UV protection.

8 Vitiligo is a disease in which melanocytes are destroyed, with resulting loss of pigmentation. If a dark-skinned person develops vitiligo and therefore lighter-colored skin, would his or her race change? What factors have led people to classify (or misclassify) themselves or others as members of one race or another?

DRIVING QUESTION 2

Where did the earliest humans evolve, and how do we know?

By answering the questions below and studying Infographics 20.5, 20.6, 20.7, and 20.8, you should be able to generate an answer for the broader Driving Question above.

KNOW IT

9 What percentage of DNA sequences do all humans share?

  a. 0%

  b. 25%

  c. 50%

  d. 75%

  e. >99%

10 Why is mtDNA a useful tool in the study of human evolution? (Think about how mitochondrial DNA is inherited.)

11 According to the "out of Africa" hypothesis of human origins and migration, which group of people should show the highest level of genetic diversity?

  a. Africans

  b. Europeans

  c. Asians

  d. South Americans

  e. Australians

USE IT

12 Rank the levels of genetic diversity you would expect to find within the five populations listed in Question 11 from highest to lowest. Justify your ranking.

13 If there were many human females living ~200,000 years ago, why do we find that the mitochondrial DNA in all living humans is all related to a single woman from that time?

What kind of evidence could you look for to test your explanation? (Think about all the human fossils that have been uncovered, and consider that it is possible to extract DNA from fossils.)

MINI CASE

14 A mother with medium skin tone gives birth to a baby with darker skin than she has. Her lighter-skinned husband accuses her of infidelity.

a. How reasonable is this, given the genetics of skin color? (Hint: Refer to Infographic 12.9.)

b. Could an mtDNA analysis be used in a paternity test? Why or why not?

DRIVING QUESTION 3

What we can learn about human evolution from the fossil record?

By answering the questions below and studying Infographics 20.7 and 20.8, you should be able to generate an answer for the broader Driving Question above.

KNOW IT

15 Of the following traits that are associated with being human, which evolved most recently?

a. upright walking   **d.** tool use

b. ability to control fire   **e.** big brain

c. social communication

16 Place the following ancestors in order of most ancient (1) to most recent (5).

_____ *Homo sapiens*

_____ Last common ancestor of chimpanzees and humans

_____ *Australopithecus*

_____ *Ardipithecus ramidus*

_____ *Homo erectus*

USE IT

17 Where would the last common ancestor of gorillas and humans fit into the ordering in your answer to Question 16? Explain.

18 Why would individual Australopithicines who could make and use tools have had a selective advantage (that is, higher fitness) over individuals who could not make or use tools?

19 Ardi was partially arboreal (that is, the species could live in trees). The ability to move around in trees was facilitated by an opposable big toe that would help grip branches. Once ancient hominids moved permanently to a grounded lifestyle, would there have been any selective pressure to maintain an opposable big toe? Explain your answer.

20 Members of the genus *Australopithecus* walked upright, and their fossilized footprints show **no** evidence of an opposable big toe.

a. What foot structure and lifestyle might have been selected if early hominid evolution occurred in a forested environment? In a grasslands environment? Would you predict any differences because of the selective pressures in each environment? Why or why not?

b. What other traits would you expect to be favored in a forested environment? In open grasslands?

INTERPRETING DATA

21 An extensive study of a hominid fossil dating from approximately 2 million years ago was published in 2013. For each of the features described below, consider whether they are closer to an ancestral state or closer to modern humans. On the basis of the features described, where would you place this fossil on the lineage between the chimpanzee–human ancestor and modern humans—what genus is it likely a member of?

- The shoulder structure and very long arms suggest the ability to climb and perhaps hang or swing.
- The spine and other skeletal features suggest an upright stance.
- There is no opposable toe.
- The heel is very narrow and pointed (not flat and wide).
- The skeleton suggests that the gait would have been rolling, the feet rolling inward with each step.
- The skull is very small.
- The chest is not cylindrical but wider at the base and narrow at the shoulders, much like a triangle.

BRING IT HOME

22 The U.S. Census Bureau provides information on classifying race (http://www.census.gov/population/race/about/). What races does the U.S. Census Bureau recognize? What about people of mixed race? What about people who identify themselves as Hispanic? How easy is it for you to identify yourself with respect to race given the racial categories on the U.S. Census?

On the Tracks of Wolves and Moose

Ecologists learn big lessons from a small island

21

JOHN VUCETICH SPENT A FREEZING FEBRUARY DAY IN 2010 trudging through knee-high snow on an island in Lake Superior. He was tracking a young gray wolf he called Romeo. The tracks led to a site in the forest where cracked branches and crimson-stained snow were evidence that a violent struggle had taken place just hours earlier. Later, thawing in his cabin beside a wood-burning stove, Vucetich wrote in his field journal:

> Teeth, hooves, blood, bruises, adrenaline, exhaustion. Romeo killed a moose. Very likely, this is the first moose he'd ever killed. He'd seen his parents, the alpha pair of Chippewa Harbor Pack, do it many times. . . . He'd wounded moose a couple of times this winter, but never killed one.

▸ **DRIVING QUESTIONS**

1. What is ecology, and what do ecologists study?

2. What are the different patterns of population growth?

3. What factors influence population growth and population size?

> **❝** *Isle Royale is not too big and it's not too small and it's not too close and not too far. It's just the right size to have a population of wolves and moose that we can study.* **❞**
>
> — JOHN VUCETICH

▶ECOLOGY
The study of the interactions between organisms and between organisms and their nonliving environment.

For nearly 20 years, Vucetich has been shadowing wolves like Romeo and his kin on Isle Royale, a remote island about 15 miles off the Canadian shore in the northwest corner of Lake Superior. A 200-square-mile slice of roadless wilderness that is accessible only by boat and seaplane, Isle Royale may seem an unlikely place for a scientific laboratory, but that's exactly what it is for Vucetich and his colleagues. Every summer, and for a few weeks every winter, they investigate the island's packs of gray wolves (*Canis lupis*) and the herd of moose (*Alces alces*) that are their prey.

Begun in 1958, the Isle Royale wolf and moose study is the longest-running predator-prey study in the world. For more than 50 years, researchers have studied how these two island inhabitants have interacted and co-existed in this wild place. They are motivated by a simple yet increasingly pressing goal: "to observe and understand the dynamic fluctuations of Isle Royale's wolves and moose, in the hope that such knowledge will inspire a new, flourishing relationship with nature," according to study's annual report.

Islands have long provided biologists with important lessons about nature—think of Darwin and Wallace and their island adventures. But none has ever produced such a mountain of data for researchers interested in **ecology,** the study of the interactions between organisms and between organisms and their nonliving environment. At a moment when wild populations are threatened all over the world, the lessons of Isle Royale couldn't have come at a better time.

IN NATURE'S LABORATORY

A number of features make Isle Royale an ideal place in which to study ecology. Because the island is uninhabited by humans and is protected as a national park, scientists can study moose-wolf interactions in a nearly natural environment, undisturbed by settlement, hunting, or logging.

Isle Royale is also an ideal distance from shore—close enough to the mainland for moose and wolves to have reached it, but far enough away that other animals do not migrate easily to it. Because there are no other predators or prey on the island, the only things eating moose are wolves, and moose are just about the only thing wolves eat. These simplified

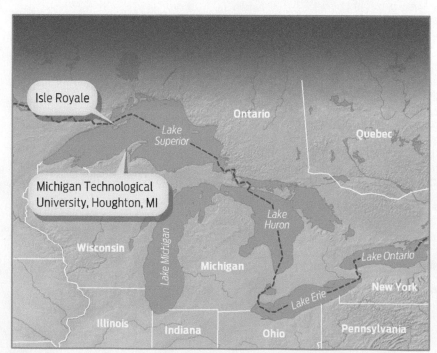

Isle Royale

Ontario

Quebec

Lake Superior

Michigan Technological University, Houghton, MI

Lake Huron

Wisconsin

Lake Michigan

Michigan

Lake Ontario

New York

Illinois

Indiana

Ohio

Lake Erie

Pennsylvania

conditions allow scientists a good look at the two residents' behavior and ecological impact.

Another feature that makes Isle Royale good for research is its size. The island is not so big as to have an unmanageably large population of moose, and not so small as to be unsupportive of a wolf population. "It's a little bit of the Goldilocks thing," says Vucetich, now a professor of ecology at Michigan Technical University. "Isle Royale is not too big and it's not too small and it's not too close and not too far. It's just the right size to have a population of wolves and moose that we can study."

Vucetich began studying wolves as a college student at Michigan Tech in the early 1990s. In 2001, he became coleader of the Isle Royale study, working alongside his former teacher and mentor, Rolf Peterson. It's challenging work at times, but Vucetich says he may have been destined for this career path: "*Vuk*"—the root of his last name—"is the Croatian word for wolf."

Ecologists study organisms at a number of levels. They can look at an individual organism, such as a single moose or wolf, studying how it fares in its surroundings. They may also look at a group of individuals of the same species living in the same place—a herd of moose, or a pack of wolves, for example—watching what happens to this **population** over time. Two or more interacting populations of different species constitute a **community.** Isle Royale, for example, is home to a community of wolves, moose, and the plants the moose feed on.

Finally, ecologists may want to understand the functioning of an entire **ecosystem,** all the living organisms in an area and the nonliving components of the environment with which they interact. When moose eat trees, for example, they reduce the available habitat for other animals, such as birds. However, the heat of summer can reduce the ability of moose to feed, which in turn improves tree growth (**INFOGRAPHIC 21.1**).

Vucetich was initially drawn to ecology as a way to experience the outdoors, his first love. Only later did he realize he was actually quite

INFOGRAPHIC 21.1 ECOLOGY OF ISLE ROYALE

Individual: a single organism of a particular species
· one wolf

Population: a group of individuals of the same species living and interacting in the same region
· a pack of wolves

Community: interacting populations of different species
· wolves prey on moose
· ticks infest moose
· moose feed on trees

Ecosystem: species interacting with other species and the environment
· moose eat the trees, changing the vegetation, which in turn changes the landscape for other animals
· hot summers reduce the ability of moose to feed, affecting their winter survival

good at something ecology has a lot of: math. "As a high school student, I didn't like math at all," he says. But when he saw that math allowed him to spend more time outdoors he became, as he says, "interested and inspired to learn a great deal about math."

Vucetich is a population ecologist, and population ecology is all about numbers. On Isle Royale, the main numbers the researchers are interested in year after year are the numbers of wolves and moose. "In any given season there are more or less of those species and we want to understand why," says Vucetich. Answering the "why" involves a lot of time, patience, and, of course, counting.

▶**POPULATION**
A group of organisms of the same species living and interacting in a particular area.

▶**COMMUNITY**
Interacting populations of different species in a defined habitat.

▶**ECOSYSTEM**
All the living organisms in an area and the nonliving components of the environment with which they interact.

Much of the counting is done from the air. Sitting one in front of the other inside a tiny two-person plane, pilot and observer circle the island scanning for evidence of wolves and moose. Wolves are relatively easy to find and count, especially in the snow: "You follow the wolf tracks until you find the wolves," says Vucetich. The other thing that makes counting wolves easy is that they live in packs: if you find one wolf, you've generally found the others. And since there are usually no more than a couple dozen wolves on the island at any time, it's possible to count every one.

It's a different story with moose. There can be more than a thousand moose on the island—too many to count all at once. Besides, moose are relatively solitary creatures, and their brown coloring makes them harder to spot

A pack of wolves moving in for the kill.

INFOGRAPHIC 21.2 DISTRIBUTION PATTERNS INFLUENCE POPULATION SAMPLING METHODS

Wolves and moose have different lifestyles and distribution patterns on Isle Royale. Determining the size of each population requires a distinct counting strategy.

Individual wolves cluster together in packs, making them easier to spot, track, and count.

2013 Wolf Territories and Kill Locations

West-end Trio
Chippewa Harbor Trio

Moose are largely solitary creatures and are distributed more randomly on the island. Counting them requires a different strategy.

1.8 moose/km²

2013 Moose Distribution

Moose survey plot #87

This is an aerial view of plot #87, outlined in orange. The green dotted line defines the 9 overlapping circles flown as moose are counted. This sampling is done for 91 plots across 20% of the island to gather enough data for an accurate estimate of moose population size and density on the island.

SOURCE: Vucetich, J. A., and Peterson, R. O. (2013) Ecological studies of wolves on Isle Royale, Annual Report 2012–2013.

INFOGRAPHIC 21.3 POPULATION DISTRIBUTION PATTERNS

Different organisms have different distribution patterns. There are three main types, but few organisms in nature fall into strictly one category.

Random

Individuals are equally likely to be anywhere within the area.

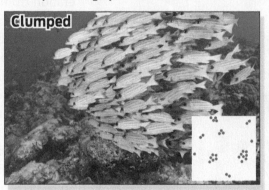

Clumped

High-density clumps are separated by areas of low abundance.

Uniform

Individuals maximize space between them by being uniformly spaced.

against the backdrop of dark evergreen trees. When moose are feeding in the forest–which is much of the time–counting them is, according to Vucetich, "like trying to count fleas on a dog from across the room." It's simply not possible to count them all.

Instead, the team uses a shortcut: they count all the moose in a series of square-kilometer plots representing about 20% of the island, average the number of moose per plot, and then extrapolate to the rest of the island. But even this shortcut requires many careful hours of study in the plane, straining to see the moose through the trees. To help himself concentrate, Vucetich recites a sort of mantra: "Think moose, think moose, look for the moose." **(INFOGRAPHIC 21.2)**

The somewhat random dispersion of individually roaming moose represents one type of **distribution pattern** found in nature. Distribution patterns generally reflect behavioral or ecological adaptation. For moose, being solitary and randomly distributed may help protect them from predation, since single moose are harder to spot in the forest than a large group would be. A random distribution may also allow individuals to maximize their access to resources. Pine trees, for example, have air-blown seeds that are spread far and wide by gusty winds, resulting in a random distribution of trees in the forests on the island.

A truly random distribution is rare in nature; even wind-blown seeds must fall on fertile soil to grow, and this does not always happen. More common is a clustered, or clumped, distribution, which results when resources are unevenly distributed across the landscape, or when social behavior dictates grouping, as it does with the highly social wolf. Clumping has its advantages: for wolves, clumping helps them to gang up on moose; they circle their prey and close in for the kill. Clumping can also be a defense against predation, as it is for a school of fish.

A third distribution pattern found in nature is uniform distribution. In this case, individuals keep apart from one another at regular distances, usually because of some kind of territorial behavior. Birds such as penguins that nest in defined spaces a few feet away from one another are a good example **(INFOGRAPHIC 21.3)**.

POPULATION BOOM AND BUST

Moose have not always roamed Isle Royale. The first antlered settlers likely arrived around 1900, when a few especially hardy individuals swam across the 15-mile-wide channel from Canada. With an abundant food supply and no natural predators on the island, the moose population exploded, growing from a handful of individuals around the turn of the century to more than a thousand by 1920.

▶**DISTRIBUTION PATTERN**
The way that organisms are distributed in geographic space, which depends on resources and interactions with other members of the population.

Moose have a random distribution on Isle Royale.

This rapid increase reflected the population's high **growth rate,** defined as the birth rate minus the death rate. Because it denotes the simple balance between birth and death, the growth rate is also known as the rate of natural increase. When the birth rate of a population is greater than the death rate, the population grows; when the death rate is greater than the birth rate, the population declines; and when the two rates are equal, the result is zero population growth.

In many populations, **immigration,** the movement of individuals into a population, and **emigration,** the movement of individuals out of a population, make substantial contributions to population growth. But because the moose and wolves of Isle Royale are isolated, and individuals neither come to nor go from the island on a regular basis, their population growth rates are due only to births and deaths.

Ecologists describe two general types of population growth. The rapid and unrestricted increase of a population growing at a constant rate is called **exponential growth.** When a population is growing exponentially, it increases by a certain fixed percentage every generation. Thus, instead of a constant number of individuals being added at each generation—say, the population going from 100 to 120 to 140 to 160—the increase is more like credit card interest, with each increase added to the principal (the

population) before the percentage is applied. And so, with an exponential growth rate of 20%, a population of 100 would increase at each generation from 100 to 120 to 144 to 173 to 207. If the population continued to grow exponentially, it would quickly get out of control, not unlike a credit card bill you don't pay on time.

Such unrestricted growth is rarely if ever found unchecked in nature. As populations increase, various environmental factors such as food availability and access to **habitat,** the physical environment where an organism lives and to which it is adapted, limit an organism's ability to reproduce. When population-limiting factors slow the growth rate, the result is **logistic growth,** a pattern of growth that starts rapidly and then slows.

Eventually, after a period of rapid growth, the size of the population may level off and stop growing. At this point, the population has reached the environment's **carrying capacity**—the maximum number of individuals that an environment can support given its space and resources. Carrying capacity places an upper limit on the size of any population; no natural population can grow exponentially forever without eventually reaching a point at which resource scarcity and other factors limit population growth. This is true even of the human population (see Chapter 24) (**INFOGRAPHIC 21.4**).

The size of a population may fluctuate around the environment's carrying capacity, briefly exceeding it and then dropping back. After an initial overshoot of carrying capacity, factors like disease or food shortage will cause the population to shrink. This drop in turn may allow the environment time to recover its food supply, at which point the population may begin to grow again, briefly exceeding carrying capacity, and so on, in a cycle of "boom and bust."

When moose first arrived on Isle Royale, their population grew exponentially. This unchecked proliferation of hungry mouths took a severe toll on the island; by 1929, the moose had munched their way through most of its vegetation. In turn, the reduction of the island's food supply caused the moose population to crash. The moose population had

INFOGRAPHIC 21.4 POPULATION GROWTH AND CARRYING CAPACITY

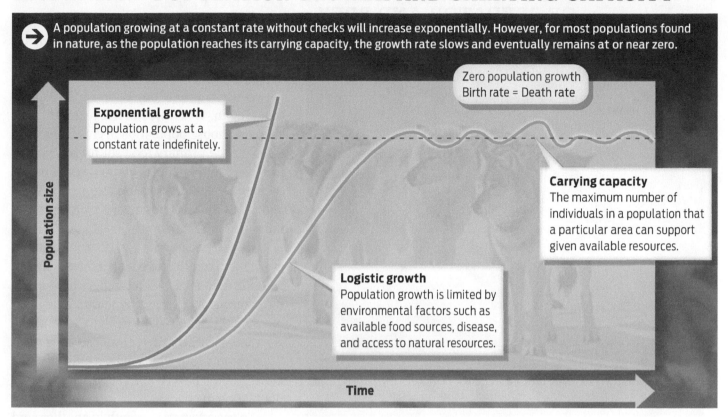

→ A population growing at a constant rate without checks will increase exponentially. However, for most populations found in nature, as the population reaches its carrying capacity, the growth rate slows and eventually remains at or near zero.

Exponential growth
Population grows at a constant rate indefinitely.

Zero population growth
Birth rate = Death rate

Carrying capacity
The maximum number of individuals in a population that a particular area can support given available resources.

Logistic growth
Population growth is limited by environmental factors such as available food sources, disease, and access to natural resources.

Population size

Time

exceeded the island's carrying capacity, and by 1935, it had dwindled to a few hundred starving individuals.

The herd got lucky, though. The next summer, fire consumed 20% of the island, and the scorched areas provided space for new trees to grow. But as soon as the forest recovered, moose numbers again began to explode, ravaging the forests once more.

Then, around 1950, everything changed. One especially cold winter, a pair of gray wolves crossed an ice bridge connecting Canada to Isle Royale, forever altering the ecology of the island. Since then, the fates of the wolves and moose have been inextricably linked.

At the beginning of the Isle Royale study, in 1959, there were about 550 moose and 20 wolves on the island. Moose numbers climbed for about 15 years, reaching a peak of approximately 1,200 animals in 1972, and then declined rapidly, to a low of approximately 700 moose in 1980. As moose numbers fell,

wolf numbers rose—from a low of 17 wolves in 1969 to a high of 50 animals in 1980. These two trends were linked: the wolves were feeding themselves well enough to increase their own

John Vucetich examines a kill site.

INFOGRAPHIC 21.5 POPULATION CYCLES OF PREDATOR AND PREY

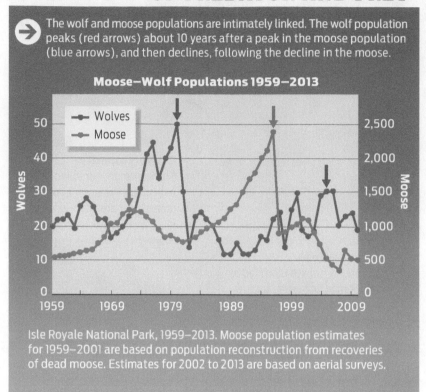

→ The wolf and moose populations are intimately linked. The wolf population peaks (red arrows) about 10 years after a peak in the moose population (blue arrows), and then declines, following the decline in the moose.

Moose–Wolf Populations 1959–2013

Isle Royale National Park, 1959–2013. Moose population estimates for 1959–2001 are based on population reconstruction from recoveries of dead moose. Estimates for 2002 to 2013 are based on aerial surveys.

ECOLOGICAL DETECTIVES

The size of the wolf population affects more than just the size of the moose population. Another pattern that has emerged in the decades of data collected on Isle Royale is a correlation between a large wolf population and vigorous tree growth. When wolves are plentiful, they keep the moose population in check. Because trees are the primary food source for moose, they grow more when fewer moose are eating them. It's therefore possible to follow the rise and fall of the wolf population by monitoring the state of the forest.

One way ecologists can determine forest growth and health is to count and measure the width of tree rings, which reflect how much trees have grown season by season. They also measure the height of the trees. Taller and bigger trees mean that fewer moose have been foraging on them, which in turn indicates that more wolves have been keeping the moose population in check.

Wolves affect tree growth in another, and indirect, way as well. Because the wolves don't always consume the entire carcass of a moose they kill, the remains decay and fertilize the ground where they lie, enriching the soil with nutrients for plant growth. Researchers have found that nitrogen levels are between 25% and 50% higher in these hot spots compared to control sites. This work shows that predators—in this case, wolves—are an important component of a balanced and healthy ecosystem and illustrates what can be lost when predators are eliminated (INFOGRAPHIC 21.6).

Another clue the ecological detectives look at is urine-soaked snow and droppings, or scat. Urine and scat may seem crude objects of scientific study, but they reveal a host of information about the animal that produced them. By analyzing moose scat samples under the microscope, researchers can tell exactly what moose have been eating. During the winter months, for example, moose eat mostly twigs from deciduous (leaf-shedding) plants and needles from balsam fir and cedar trees. Knowing a moose's diet is important because the supply of balsam fir on the island has been

population, but by hunting and killing so many moose they caused the moose death rate to exceed the birth rate. With a negative growth rate, the moose population shrank.

What would happen next? Would the wolf predators simply drive their moose prey to extinction? No one knew. The only thing to do was watch and wait. Eventually, it became clear that the two populations were rising and falling together in a specific pattern, with the size of the wolf population peaking several years after the size of the moose herd and then dropping.

Why does the wolf population fall? Because even for wolves, there's no such thing as a free lunch: they pay a price for predation in the form of a declining food supply. The result is a repeating cycle in the numbers of predator and prey. Rather than growing exponentially and leveling off, the populations cycle through repeated rounds of boom and bust (INFOGRAPHIC 21.5).

steadily dwindling over the years. By monitoring moose scat, researchers will be able to tell if moose are able to switch over to other, more common food items.

Scat also provides important information about an animal's genetics. It is used to obtain DNA profiles, for example, which can be used to confirm population counts and to track which wolves were involved in killing which moose. DNA can also be used to look for diseases or signs of inbreeding. "Through the DNA we can get a good sense of individual wolves—how they live and how they die," says Vucetich. (For more information on DNA profiling, see Chapter 7.)

Yet more clues can come from studying a moose kill site, which is a bit like analyzing a crime scene. Researchers can tell if a moose was killed by wolves because in that case there will often be blood spattered on nearby trees and signs of struggle in the form of broken branches. Wolves also typically scatter bones as they feast, whereas the carcasses of moose that die of starvation may be relatively intact.

At the kill site, researchers gather moose bones. From these bones, the researchers can tell how old a moose was when it died, as well as learn about other aspects of the animal's health, such as whether it had arthritis or osteoporosis. The value of this information goes

INFOGRAPHIC 21.6 PATTERNS OF POPULATION GROWTH

Wolf, moose, and tree populations are all interconnected. Trees provide food for moose, and moose provide food for wolves. Anything that impacts the size of one population will impact the size of the others.

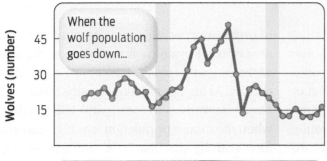

When the wolf population goes down...

The main diet of Isle Royale wolves is moose. Wolf populations grow and diminish in response to the availability of this food resource.

the moose population goes up...

The main diet of moose is trees. A larger moose population means that more tree material is eaten, so tree growth slows.

...resulting in slower tree growth.

Tree growth can be measured by the width of each tree ring. One ring represents the amount of growth in 1 year: the wider the ring, the more the growth in that year.

SOURCE: McLaren, B.E., and Peterson, R.O. (1994) Wolves, moose, and tree rings on Isle Royale. *Science* 266(5190):1555–1558.

INFOGRAPHIC 21.7 MOOSE AND WOLF HEALTH IS MONITORED BY A VARIETY OF DATA

In addition to information about population size, researchers collect other data that are essential for monitoring the physical health of moose and wolf populations.

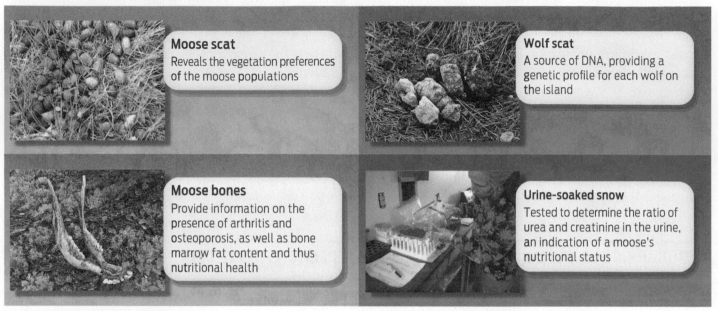

Moose scat
Reveals the vegetation preferences of the moose populations

Wolf scat
A source of DNA, providing a genetic profile for each wolf on the island

Moose bones
Provide information on the presence of arthritis and osteoporosis, as well as bone marrow fat content and thus nutritional health

Urine-soaked snow
Tested to determine the ratio of urea and creatinine in the urine, an indication of a moose's nutritional status

▶POPULATION DENSITY
The number of organisms per unit area.

▶DENSITY-DEPENDENT FACTOR
A factor whose influence on population size and growth depends on the number and crowding of individuals in the population (for example, predation).

▶BIOTIC
Refers to the living components of an environment.

▶DENSITY-INDEPENDENT FACTOR
A factor that can influence population size and growth regardless of the numbers and crowding within a population (for example, weather).

beyond understanding individual animals. It allows researchers to know whether wolves are targeting healthy moose or sickly ones. Killing a healthy moose has a bigger effect on moose population dynamics than killing one that is already near death, because a young, healthy moose might have gone on to reproduce had it lived (**INFOGRAPHIC 21.7**).

TOO CLOSE FOR COMFORT?

Moose are formidable foes of their wolf predators. At 900 pounds, 10 times the weight of a wolf, an adult moose can successfully defend itself against an aggressive pack of wolves with its powerful front legs. For that reason, wolves often attack older and weaker or young moose. They typically target the nose and hindquarters, where they bite and latch onto the flesh like a steel trap. When enough wolves are attached, their collective weight brings down the moose, and the feeding begins.

A number of factors can influence the likelihood that wolves will kill moose. One of the simplest is **population density,** the number

of organisms per unit area. Because the total area of Isle Royale stays the same, as the size of the moose population increases so does its density. At high population density, moose are easier for wolves to locate and kill. Further, when the moose population is at high density, food scarcity can also be a problem, leaving moose hungry and weak and therefore more vulnerable to attack.

Because wolf predation and plant abundance have a greater effect on moose when the moose population is large, these are examples of **density-dependent factors**—factors that exert different degrees of influence depending on the density of the population. As living parts of the environment, they are also examples of **biotic** factors influencing growth.

Some environmental pressures take a toll on a population no matter how large or how small it is. In an exceptionally cold winter with deep snow, for example, moose can die of cold or starvation. The weather can also weaken them so they are easier to hunt and kill. Since cold weather affects moose regardless of population size, it is considered a **density-independent** factor: whether there are 10 moose or 1,000, a harsh winter affects them all.

Wolves are also affected by density-independent factors. "A mild winter is always tough on the wolves," says Peterson, who notes that moose can more easily escape wolves when snow cover is light. Density-independent factors like snowfall can be considered nature's form of bad luck. Most density-independent factors are **abiotic** and include things like temperature, precipitation, and fire (INFOGRAPHIC 21.8).

▶ABIOTIC
Refers to the nonliving components of an environment, such as temperature and precipitation.

INFOGRAPHIC 21.8 ABIOTIC AND BIOTIC INFLUENCES ON POPULATION GROWTH

→ Both nonliving (abiotic) and living (biotic) environmental factors influence the size and growth of populations.

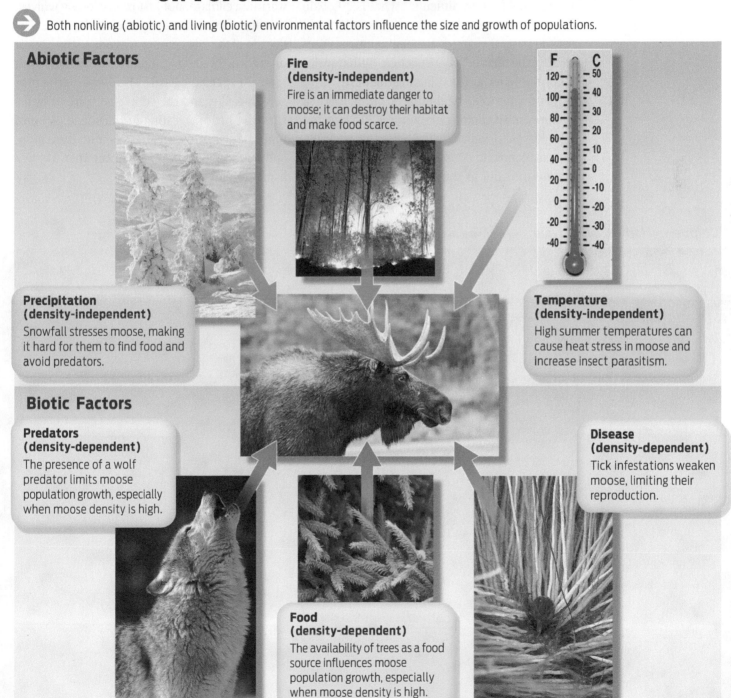

Abiotic Factors

Fire (density-independent)
Fire is an immediate danger to moose; it can destroy their habitat and make food scarce.

Precipitation (density-independent)
Snowfall stresses moose, making it hard for them to find food and avoid predators.

Temperature (density-independent)
High summer temperatures can cause heat stress in moose and increase insect parasitism.

Biotic Factors

Predators (density-dependent)
The presence of a wolf predator limits moose population growth, especially when moose density is high.

Disease (density-dependent)
Tick infestations weaken moose, limiting their reproduction.

Food (density-dependent)
The availability of trees as a food source influences moose population growth, especially when moose density is high.

WATCHING AND WAITING

For the scientists on Isle Royale, population ecology is full of unexpected twists and turns. There is often no sure way to know how various environmental factors will influence the growth of a population. Even on an isolated island with only one large predator and one large prey, population dynamics are never simple. Scientists gather data, look for patterns, and form hypotheses, but predicting what will happen next is much more difficult. "What Isle Royale has shown us—and has shown us convincingly for the past 50 years—is that we're lousy at predicting the future," says Vucetich. "What we're a fair bit better at is explaining the past."

For example, beginning around 1980, a disease called canine parvovirus (CPV) infected Isle Royale's wolves. The disease typically affects domestic dogs and was likely brought unintentionally to the island on the boots of hikers. The disease killed all but 14 of the island's wolves, and over the next 10 years the moose population skyrocketed, suggesting that wolves exert a strong influence on the abundance of their prey. The event was useful from a scientific standpoint—but entirely unexpected. "There's no way that anyone could have predicted that. Not in a million years," says Vucetich.

That wasn't the end of the surprises. In the last 15 years, it's become apparent that a warming climate, not just predation by wolves, is influencing moose population size. The first decade of the 21st century was one of the hottest on record. Sweltering summer temperatures hit moose especially hard. The large herbivores get hot easily, and they don't perspire; they escape the heat by resting in the shade. A lot of time spent resting means less time for eating, and a moose who eats less all summer has less insulation for winter.

Chippewa Harbor, Isle Royale. *(Photo © John and Ann Mahan)*

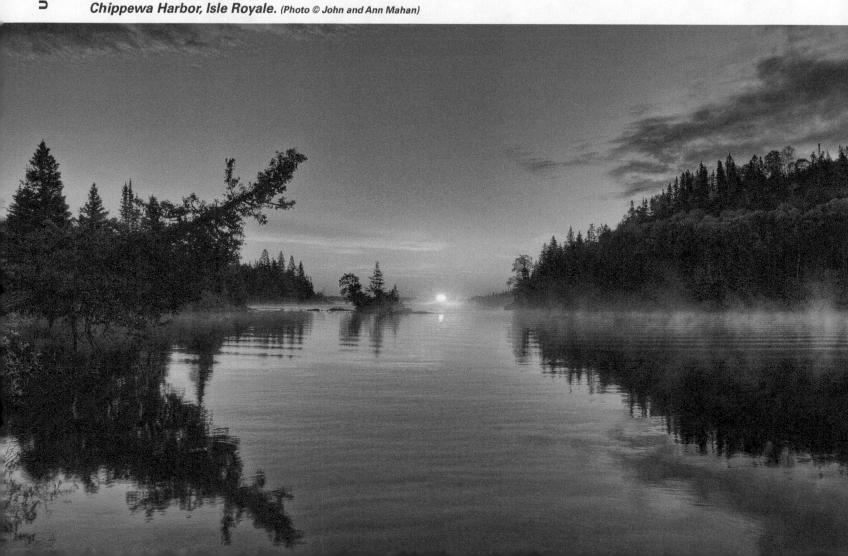

Warmer temperatures have affected moose in a more insidious way as well. About 10 years ago, Vucetich and his colleagues began to notice that a tick parasite was bothering the moose, and that warm weather seems to favor ticks. Ticks suck the moose's blood and cause them to itch. The moose scratch themselves against trees and chew their hair out trying to rid themselves of the itchy freeloaders. Since a single moose may host many thousands of ticks, the combination of tick-related blood loss and heat-induced weight loss can be deadly. In 2004, the average moose had lost more than 70% of its body hair, the result of carrying more than 70,000 ticks.

By 2007, the deadly combination of blood-sucking ticks, hot summers, and relentless predation from wolves had driven the moose population to its lowest point in at least 50 years—385, down from 1,100 in 2002. Predictably, the wolf population followed suit, declining from 30 individuals in 2005 to 21 in 2007. As of 2013, the moose population was back up to about 975 moose, while the wolves declined to just 8 individuals (INFOGRAPHIC 21.9).

Hunted by wolves, preyed on by ticks, dogged by oppressive heat, moose certainly do not have it easy. They can live to be 17 years old, but most moose die before reaching their tenth birthday.

Life is no picnic for wolves, either. While they can live to be 12 years old, most die by age 4. The most common cause of death is starvation. With few available food sources, a wolf may go 10 days without eating. Obtaining a meal on the eleventh day may mean having to wrestle a 900-pound moose on an empty stomach.

The difficulty of finding food is just one obstacle for wolves. They also have a high incidence of bone deformities, which cause back pain and partial paralysis of the hind legs. In the early years of the Isle Royale study, such deformities were rare, but they've become more common in recent years, almost certainly a result of inbreeding. For the last 15 years, every dead wolf on Isle Royale has had such deformities. Vucetich thinks that climate change may be an indirect cause of the inbreeding by lessening the frequency that ice bridges form between

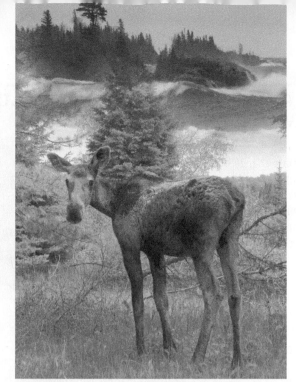

A young moose on Isle Royale (note the patchy fur, a sign of tick infestation).

Canada and Isle Royale, bridges over which new wolves could reach the island and diversify the wolf gene pool.

It's not the first time wolf populations have been in trouble. When colonial settlers first arrived in North America, the gray wolf roamed throughout all of the future 48 contiguous U.S. states. By 1914, hunting and trapping had greatly reduced the population, and survivors were limited to remote wooded regions of Michigan, Wisconsin, and Minnesota. The federal government officially listed the species as endangered in the early 1970s, when it seemed on the verge of extinction. Since then, as a result of conservation efforts, wolves in other areas have started to bounce back: there are an estimated 7,000 to 11,000 gray wolves in Alaska, and approximately 5,000 in the rest of the country. But on Isle Royale, wolves are at their lowest numbers since the study began.

The Isle Royale wolves' latest plight poses an ethical dilemma: should scientists intervene on their behalf—say, by importing wolves from another population to reintroduce genetic diversity—or let nature take its course? It's a question that Vucetich thinks about a lot. The answer, he says, will require balancing a number of competing values—not just the value of

INFOGRAPHIC 21.9 A WARMING CLIMATE INFLUENCES MOOSE AND WOLF POPULATION SIZE

In recent years, climate change has become a significant influence on moose and wolf populations on Isle Royale. Warmer temperatures lead to increased tick infestations of moose, resulting in a weakened and depleted population.

One moose may be home to tens of thousands of ticks at a time.

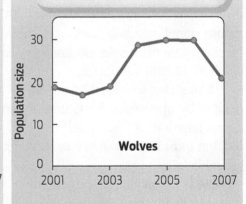

Ticks cause moose to lose their hair, their appetite, and a good deal of blood.

Ticks make moose weak and vulnerable to predation and starvation. So, while ticks have been increasing...

...the moose population has been decreasing.

Moose weakened by ticks are easier for wolves to catch. After an initial population increase in response to an abundance of moose, the wolf population begins to suffer (2007) as the moose population continues to decline.

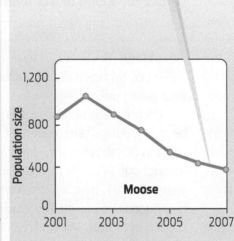

individual populations, but the values of ecosystem health, scientific knowledge, and wilderness. Without wolves, for example, would the moose population once again explode and decimate the island's forest? Would healthier wolves be able to completely overwhelm moose, and drive them to extinction on the island? Does the fact that human-caused climate change has worsened these problems mean that we have a moral obligation to alleviate them? These are some of the difficult questions that wildlife managers will need to consider when debating whether and how to intervene.

The dilemma is a familiar one to conservation biologists. According to Vucetich, these competing values show up in varying degrees in almost any management question that we have in any part of the world. They represent, he says, "this grand question of How should humans relate to nature?" There are no easy or obvious answers. Nevertheless, he believes it is important for people to debate and discuss these issues—not just scientists and experts, but lay people, too, because "every citizen has a stake in this question of how we relate to nature." ■

CHAPTER 21 Summary

- Ecology is the study of the interactions between organisms and between organisms and their nonliving environment.

- Ecologists study these interactions at a number of levels, including population, community, and ecosystem.

- Living organisms may have a clumped, random, or uniform distribution pattern, depending on ecological and behavioral adaptations. Few organisms fall into strictly one category.

- Population growth is an increase in the number of individuals in a population. The growth rate of a population is defined as the birth rate minus the death rate. Growth rate is also known as the rate of natural increase.

- Exponential growth is the unrestricted growth experienced by a population growing at a constant rate. Logistic growth is the slowing of the growth of a population due to environmental factors such as crowding and lack of food.

- Carrying capacity is the maximum population size that an area can support, given its food supply and other life-sustaining resources. Populations cannot grow exponentially forever; eventually, they hit the carrying capacity for the region and stop growing.

- Population growth can be limited by a variety of factors, including biotic (living) and abiotic (nonliving) parts of the environment.

- Density-independent factors, such as a severely cold winter, can affect a population of any size.

- Density-dependent factors, such as the presence of predators, have different impacts on the population, depending on the size and crowding of individuals in the population.

- Populations in a community are interconnected, the fate of one often influencing the fate of the others.

MORE TO EXPLORE

- Wolves and Moose on Isle Royale http://isleroyalewolf.org/

- Vucetich, J. A., and Peterson, R. O. (2013) Ecological studies of wolves on Isle Royale, Annual Report 2012–2013.

- PBS. (2007) *Nature*: In the Valley of the Wolves http://www.pbs.org/wnet/nature/episodes /in-the-valley-of-the-wolves/introduction/212/

- Ripple, W. J., and Beschta, R. L. (2012) Trophic cascades in Yellowstone: the first 15 years after wolf reintroduction. *Biological Conservation* 145:205–213.

- Seabird Monitoring on the Isle of May, Scotland http://www.ceh.ac.uk/sci_ programmes/IsleofMayLong-TermStudy.html

CHAPTER 21 Test Your Knowledge

DRIVING QUESTION 1

What is ecology, and what do ecologists study?

By answering the questions below and studying Infographics 21.1, 21.2, and 21.3, you should be able to generate an answer for the broader Driving Question above.

KNOW IT

1 What is the difference between a community and a population?

2 An ecosystem ecologist might study

a. plant populations.

b. herbivores that eat the plants.

c. predators in the population.

d. the impact of precipitation patterns on the plant populations.

e. all of the above

3 Why do the researchers collect scat as part of their study on Isle Royale?

USE IT

4 How would you explain to a 10-year-old what ecologists do?

5 Your local environmental group wants to determine the population size of squirrels in a nearby nature preserve. What are some methods you could use to estimate the size of the squirrel population? Would the same approaches be as useful in determining the population size of maple trees in the same area? Why or why not?

6 How would you use scat analysis to determine whether an herbivore had a preference for a particular type of vegetation? Be specific about both the type of analysis and what the analysis would reveal for herbivores with or without a preference for a particular type of vegetation.

DRIVING QUESTION 2

What are the different patterns of population growth?

By answering the questions below and studying Infographic 21.4, you should be able to generate an answer for the broader Driving Question above.

KNOW IT

7 Which of the following would cause a population to grow?

a. identical increases in both the birth rate and the death rate of a population

b. a decrease in the birth rate and an increase in the death rate of a population

c. an increase in the birth rate and a decrease in the death rate of a population

d. an increase in the birth rate and a larger increase in the death rate of a population

e. an identical decrease in both the birth rate and the death rate of the population

8 Which of the following statements describes an example of population growth?

a. The average weight of Americans has increased substantially in the past decade.

b. Tropical fish have been found in waters more northerly than their usual habitat.

c. The number of people in a town has increased by 25% in the past 5 years.

d. The number of butterflies in a region has stayed the same from 1950 to 2010.

e. all of the above

9 When a population reaches its carrying capacity, what happens to its growth rate?

USE IT

10 The wolf population on Isle Royale in 2013 was eight. There was no evidence of wolf reproduction in 2013.

a. What is the current birth rate for the wolf population?

b. If the death rate for wolves is 1 per year, what will be the size of the wolf population in 2014, assuming the birth and death rates do not change?

c. If the winter of 2015 is very cold and an ice bridge linking Isle Royale and the mainland forms, what factor could contribute to growth of the wolf population?

INTERPRETING DATA

11 Population Q has 100 members. Population R has 10,000 members. Both are growing exponentially at a 5% annual growth rate.

a. Which population will add more individuals in 1 year? Explain your answer.

b. After 5 years, what will be the size of each population?

c. If the larger population reaches its carrying capacity at the end of the third year, what will its size be after 5 years?

What factors influence population growth and population size?

By answering the questions below and studying Infographics 21.5, 21.6, 21.7. 21.8, and 21.9, you should be able to generate an answer for the broader Driving Question above.

KNOW IT

12 The Mexican gray wolf has been reintroduced into parts of New Mexico and Arizona. There have been several wolf deaths due to shootings and traffic kills. Are these influences on the wolf population biotic or abiotic factors?

13 Drought causes a pond to dry up almost completely. The frog population dropped to 10% of its initial size. What kind of factor—biotic or abiotic—influenced the frog population?

14 Which of the following is a density-dependent factor influencing population growth?

　　a. elevated temperature
　　b. prolonged winters
　　c. a viral disease
　　d. a devastating forest fire
　　e. all of the above

USE IT

15 Why is it important for researchers to determine the cause of death of moose on Isle Royale? Can this information be used to help make predictions about moose and wolf populations? Explain your answer.

16 A group of predatory fish lives in a school in a large lake. If a parasite were introduced to the lake—for example by a vacationing fisherman—would you expect it to have a greater impact on the population if the fish were at high density or at low density? (Assume the parasite is passed from one fish to another through the water but can remain alive in the water only for a very short period of time.) What would happen to this same population if there were a severe drought and very hot summer?

17 Classify each of the following as a biotic or an abiotic factor in an ecosystem. Then predict the impact of each factor on the moose population of Isle Royale. Explain your answers, keeping in mind possible interactions between the various factors and between the moose and wolf populations.

　　a. hot summer temperatures
　　b. ticks that parasitize moose
　　c. declining numbers of balsam fir trees
　　d. a parvovirus in wolves
　　e. deep winter snowfall

18 Assume that a new herbivore is added to Isle Royale that is not a prey for wolves. Predict the effect of this introduction on

　　a. the populations of trees.
　　b. the moose population.
　　c. the wolf population.

19 If the moose population remains stable, what other factors could influence the wolf population on Isle Royale?

MINI CASE

20 The wolves of Isle Royale are suffering from bone deformities, probably as a result of inbreeding in their small population.

　　a. Do you think that humans should intervene to save the wolves? Would your answer be different if the wolves were near human populations or agricultural centers?

　　b. If humans were to intervene, what kinds of strategies might help stabilize or increase the wolf population? Explain your answer.

BRING IT HOME

21 Use your knowledge of ecology to plan a home saltwater aquarium. Think about the factors you need to consider in order to have a healthy community of fish and plants in your aquarium. Consider both biotic and abiotic factors and the population densities of the organisms in your aquarium.

What's Happening To Honey Bees?

A mysterious ailment threatens a vital link in the food chain

DAVE HACKENBERG HAS BEEN KEEPING BEES FOR MORE than 40 years. Every spring, as flowering plants start to bloom, he trucks bees from his home in central Pennsylvania to farms around the country, where they help farmers pollinate local crops–everything from California almonds to Florida melons. In November 2006, as he had done for years, Hackenberg brought his buzzing cargo to his winter base in central Florida. When he dropped them off, his 400 healthy hives were "boiling over" with bees, he says. Three weeks later, when he returned to check on them, the bees had essentially vanished; only 40 healthy hives remained.

Mysteriously, there were no dead bees lying in or near the hives. Nor were there any signs of intruders who might have destroyed the hives in search of honey. The bees were simply gone. It was, as Hackenberg said, a bee ghost town.

"I literally got down on my hands and knees and looked between the stones for dead bees," says Hackenberg, but the beekeeper found none. "I was kind of speechless. And people know I'm not speechless."

▸▸ **DRIVING QUESTIONS**

1. What are keystone species in a community, and why are pollinators considered keystone species?

2. What are food chains and food webs, and how does energy flow through them?

3. What positive and negative interactions occur among members of a community?

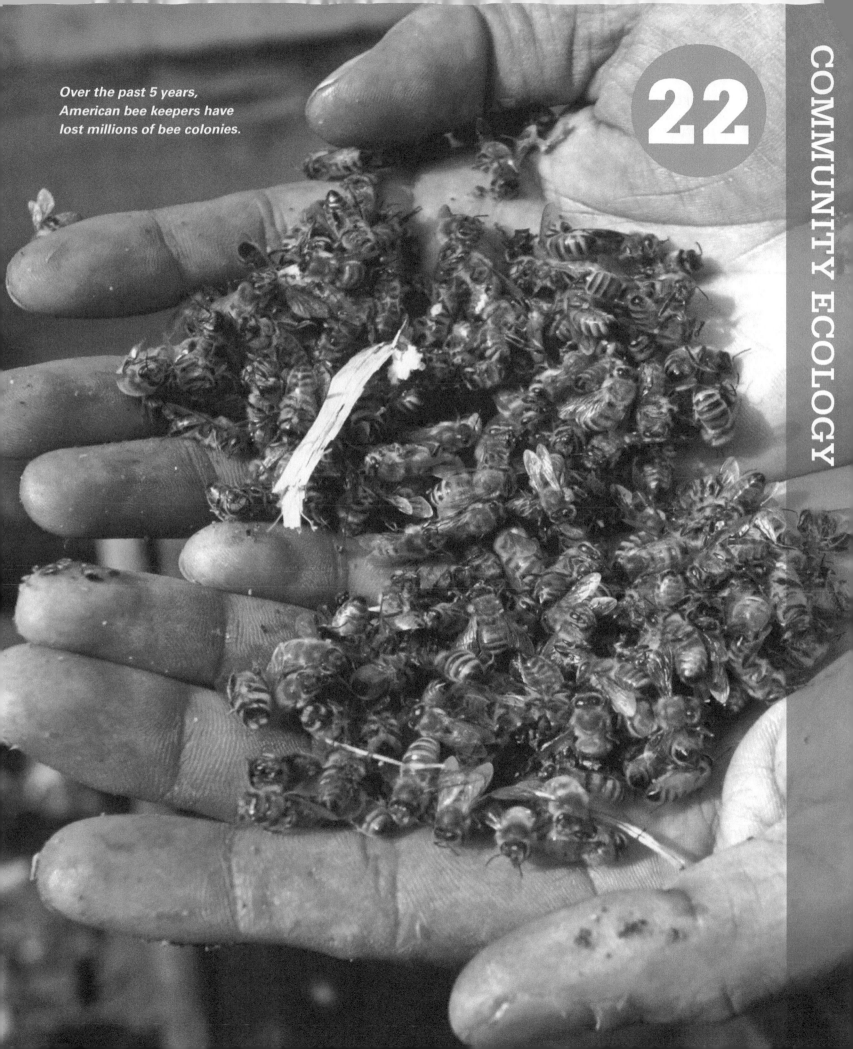

Over the past 5 years,
American bee keepers have
lost millions of bee colonies.

▶POLLEN
Small, thick-walled plant structures that contain cells that develop into sperm.

▶POLLINATION
The transfer of pollen from male to female plant structures so that fertilization can occur.

▶KEYSTONE SPECIES
Species on which other species depend, and whose removal has a dramatic impact on the community.

▶COMMUNITY
A group of interacting populations of different species living together in the same area.

Hackenberg's was the first reported case of what has since become known as colony collapse disorder, or CCD. But Hackenberg was not alone. Surveys conducted in 2007 and 2008 by the U.S. Department of Agriculture and the Apiary Inspectors of America found that beekeepers all across the United States had suffered similar unexplained devastation, losing anywhere from 30% to 90% of their colonies.

Since that first case, some 10 million honey bee colonies across the United States have reportedly been wiped out, with American beekeepers losing an average of 30% to 40% of their colonies every year since 2006. To date, CCD has been documented in 35 states and in Canada, as well as in Europe and Asia.

What's happening to honey bees? No one knows for sure, but it's a predicament that has beekeepers, farmers, and scientists racing to understand and combat the plight of the precious pollinator. At stake are not only billions of dollars' worth of agricultural crops but also the health and diversity of natural ecosystems that rely on the valuable services of bees.

THE DANCE OF POLLINATION

A hundred million years ago, when dinosaurs roamed Earth, the plant world was dominated by cone-bearing conifers such as pine and redwood trees, which spread their **pollen** by the wind; when the pollen reaches a plant, fertilization of an egg can occur. But a new type of **pollination** was evolving at this time, one that would forever change the terrestrial landscape: pollination by insects.

With the arrival of pollinating insects, including bees, on the evolutionary scene, flowering plants blossomed, diversified, and radiated around the globe. Their great success owes everything to the reproductive advantage of relying on insects, rather than wind, to deliver pollen.

Wind pollination is like junk mail: you need to send thousands of letters to hook just one receptive customer—or in this case, to fertilize just one egg. All those wasted pollen grains represent a huge energy loss to the plant. Pollinating insects, on the other hand, are like FedEx: they deliver a pollen package directly to the appropriate recipient, contributing to the reproductive success of the plants.

Of the 250,000 species of flowering plants, or angiosperms, that exist worldwide, more than 75% are dependent on insect pollinators to reproduce. And of the many types of insect pollinators, bees are by far the most important ecologically. In fact, in many natural environments, bees are **keystone species**—those that play a central role in holding the **community** together.

You can think of a keystone species as analogous to the keystone in an archway—it doesn't support as much weight as the other stones, but if it is removed, the doorway collapses (**INFOGRAPHIC 22.1**).

The majority of beekeepers in the United States and Europe cultivate the bee species *Apis mellifera*, the Western honey bee. It is hard to overestimate the importance of this tiny pollinator to modern agriculture. In the United States alone, more than 100 different crops—worth an estimated $15 billion annually—are dependent on honey bee pollination, including apples, oranges, blueberries, melons, pears,

INFOGRAPHIC 22.1 BEES ARE KEYSTONE SPECIES

➔ As keystone species, bees play a fundamental role in supporting the entire community. While bees may not be the most abundant member of the community, their loss has a huge impact on the community and the ecosystem.

Bees are keystone species. Their removal from the ecosystem devastates the community, just as the removal of the keystone from an arch causes it to collapse.

pumpkins, cucumbers, cherries, raspberries, broccoli, avocados, asparagus, clover, alfalfa, and almonds. A 2007 study published in the British journal *Proceedings of the Royal Society B* found that 87 of the leading global food crops, accounting for 35% of global crop production, are dependent upon pollinators, the most important of which are honey bees.

"One in every three bites of food we eat is pollinated directly or indirectly by honey bees," says Dennis vanEngelsdorp, State Apiarist for Pennsylvania's Department of Agriculture. Without honey bees, he says, we wouldn't starve—we would still have wheat, rice, corn, and other crops that are either wind- or self-pollinated—but many of our favorite foods might no longer grace our tables (**INFOGRAPHIC 22.2**).

For bees, flowers are food: they contain the protein-rich pollen and sugary nectar that bees need to nourish themselves and their hives.

Apiarist Dennis vanEngelsdorp examines a honey bee colony.

INFOGRAPHIC 22.2 COMMERCIAL CROPS REQUIRE BEES

→ Many of the crops that we rely on for food, fuel, and fiber rely on bees for their pollination and reproduction.

Bee hives are placed in the field. Forager bees leave the hive in search of nectar and pollen, which they use to make food for themselves and other bees in their hive.

	U.S. crop value in billions (2006)	Percentage pollinated by honey bees	Percentage of crop pollinated by		
			Honey Bees	Other Insects	Other Means
Soybeans	$19.7	5%			
Cotton	5.2	16			
Grapes	3.2	1			
Almonds	2.2	100			
Apples	2.1	90			
Oranges	1.8	27			
Strawberries	1.5	2			
Peanuts	0.6	2			
Peaches	0.5	48			
Blueberries (cultivated)	0.5	90			

Besides insects, other means of pollination include birds, wind, and rainwater.

SOURCES: U.S. Department of Agriculture (crop values); Roger Morse and Nicholas Calderone, Cornell University (insect pollination dependence).

Bees gather nectar from blossoms. During this process, they transfer pollen between flowers.

▶**STAMEN**
The male reproductive structure of a flower, made up of a filament and an anther.

▶**PISTIL**
The female reproductive structure of a flower, made up of a stigma, style, and ovary.

▶**STIGMA**
The sticky "landing pad" for pollen on the pistil.

▶**STYLE**
The tube-like structure that leads from the stigma to the ovary.

With their long tonguelike proboscis, bees are able to reach deep into a flower to draw out the nectar. Being fuzzy and having a slight electrical charge, bees attract pollen as they snuggle up to a flower the way warm socks attract other clothes as they come out of the dryer. The bees can then transfer this pollen to other plants as they continue their hunt for food.

Honey bees are more efficient at pollination than many other types of pollinators, which is why farmers have come to depend on them. The average honey bee will make 12 or more foraging trips a day, visiting several thousand flowers. On each trip, she (the foragers are all female) will confine herself to flowers from a single plant species, thus ensuring delivery of the proper pollen. Because they come and go from a central home base—their hive—honey

bees can be counted on to stay in a fixed area around a crop. And with roughly 40,000 individual bees per hive, this is a versatile workforce that's also easy for beekeepers to transport from crop to crop.

Of course, pollination isn't just about feeding bees and humans; it's also how flowering plants have sex. The insect pollinators that travel among them play an unwitting yet crucial role in the courtship. Masters of seduction, flowers have evolved countless colorful and fragrant adaptations that lure their pollinators to the blossom. In the words of poet Kahlil Gibran, "For to the bee a flower is a fountain of life, and to the flower a bee is a messenger of love."

A flower is the reproductive hub of an angiosperm—it is where its reproductive organs are located, male and female in the same flower

INFOGRAPHIC 22.3 FLOWERING PLANT REPRODUCTION RELIES ON POLLINATORS

Flowering plants attract pollinators with their flowers and nectar. When a pollinator such as a bee visits a flower to collect energy-rich nectar, it also picks up pollen. When it visits another flower, this pollen is transferred to that next flower, resulting in pollination.

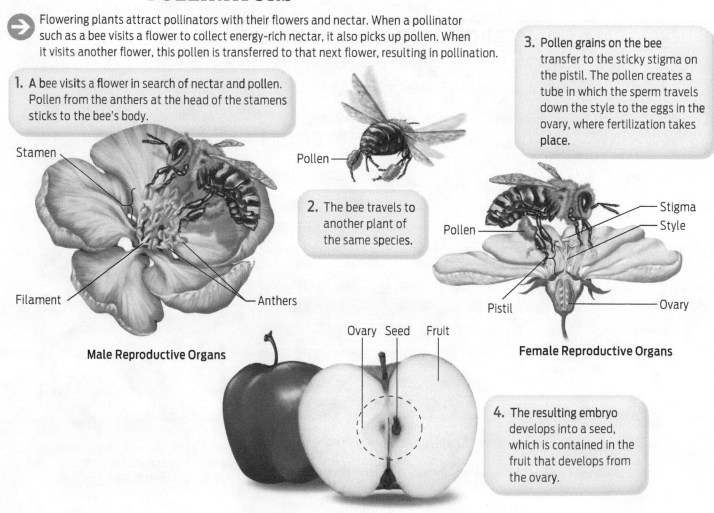

1. A bee visits a flower in search of nectar and pollen. Pollen from the anthers at the head of the stamens sticks to the bee's body.

Stamen

Filament

Anthers

Male Reproductive Organs

Pollen

2. The bee travels to another plant of the same species.

3. Pollen grains on the bee transfer to the sticky stigma on the pistil. The pollen creates a tube in which the sperm travels down the style to the eggs in the ovary, where fertilization takes place.

Pollen

Pistil

Stigma

Style

Ovary

Female Reproductive Organs

Ovary Seed Fruit

4. The resulting embryo develops into a seed, which is contained in the fruit that develops from the ovary.

A wolf hunting a mouse and a deer nibbling on a bush are both examples of predation.

in some species, separate flowers in others. In order for a flowering plant to reproduce, male pollen–containing sperm–must find a way to deliver the sperm to the female eggs of a plant of the same species. The male reproductive organ, called a **stamen,** consists of a stemlike filament topped with a pollen-saturated anther. When a bee lands on or brushes against an anther during her pursuit of nectar and pollen, her furry body picks up pollen grains.

As the bee continues to forage, she carries the pollen to the female reproductive organ of another flower–the **pistil.** The pistil is topped with a sticky "landing pad" called a **stigma.** When a bee lands on the stigma, pollen grains are deposited. The pollen grain creates a tube through which the sperm can travel down the length of the **style**–a tubelike passage into the ovary. At this point, the sperm fertilizes the egg. A fertilized egg will eventually develop into an embryo-containing **seed,** while the surrounding ovary eventually becomes the fruit (**INFOGRAPHIC 22.3**).

A LINK IN THE CHAIN

By helping plants reproduce, bees not only help sustain human food production, they also maintain the integrity and productivity of many natural communities. Without these miniature matchmakers, many flowering plants would become extinct, and many birds and mammals would go hungry. That's because, as keystone species, bees are a critical

link in the **food chain**–the linked sequences of feeding relationships in a community.

Take the tiny blueberry bee (*Osmia ribifloris*), which feeds on the flowers of blueberry plants. This speedy pollinator will visit 50,000 blueberry flowers over the course of a few weeks in spring, helping to produce more than 6,000 blueberries. All those blueberries are an important food source for many wild animals, including bluebirds, robins, fox, mice, rabbits, chipmunks, and deer. In turn, the small animals that eat the blueberries are fed upon by larger carnivorous animals, such as hawks and coyotes. Without the blueberry bee–the keystone in this community–the chain is broken.

Organisms in a food chain can be categorized by who eats whom. At the base of the food chain are **producers**–autotrophs such as plants and algae, which obtain energy directly from the sun and supply it to the rest of the food chain. Organisms higher up the food chain are **consumers**–heterotrophic organisms that eat the producers or eat other organisms lower on the chain to obtain energy.

When one organism feasts on another, that's **predation.** Usually when we think of predators, we think of large, fierce animals such as wolves hunting moose (see Chapter 21). Predation can also take more subtle forms, however. Feeding on plants, for example, is a type of predation known as **herbivory,** an activity that may or may not kill the plant. Predators that feed on plants are known as herbivores–bees

▶**SEED**
The embryo of a plant, together with a starting supply of food, all encased in a protective covering.

▶**FOOD CHAIN**
A linked series of feeding relationships in a community in which organisms further up the chain feed on ones below.

▶**PRODUCERS**
Autotrophs (photosynthetic organisms) that form the base of every food chain.

▶**CONSUMERS**
Heterotrophs that eat other organisms lower on the food chain to obtain energy.

▶**PREDATION**
An interaction between two organisms in which one organism (the predator) feeds on the other (the prey).

▶**HERBIVORY**
Predation on plants, which may or may not kill the plant preyed on.

INFOGRAPHIC 22.4 ENERGY FLOWS UP A FOOD CHAIN

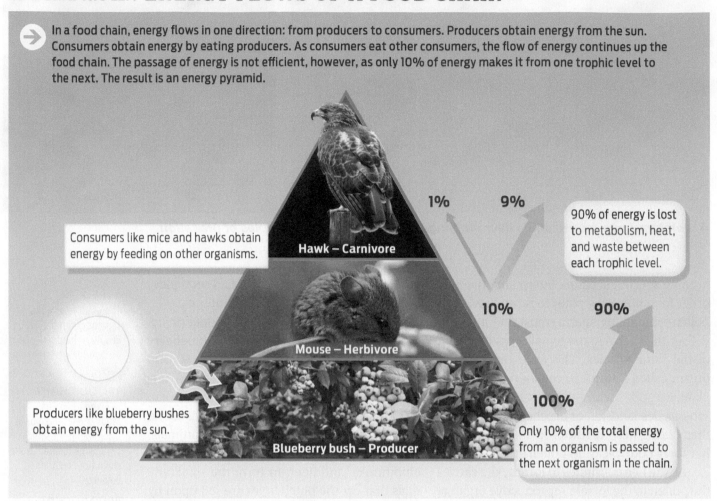

→ In a food chain, energy flows in one direction: from producers to consumers. Producers obtain energy from the sun. Consumers obtain energy by eating producers. As consumers eat other consumers, the flow of energy continues up the food chain. The passage of energy is not efficient, however, as only 10% of energy makes it from one trophic level to the next. The result is an energy pyramid.

Consumers like mice and hawks obtain energy by feeding on other organisms.

Hawk – Carnivore

1%   9%

90% of energy is lost to metabolism, heat, and waste between each trophic level.

Mouse – Herbivore

10%   90%

Producers like blueberry bushes obtain energy from the sun.

100%

Blueberry bush – Producer

Only 10% of the total energy from an organism is passed to the next organism in the chain.

▶TROPHIC LEVELS
Feeding levels, based on positions in a food chain.

feasting on nectar and pollen, for example, or a mouse nibbling on a blueberry bush. Herbivores are an important food source for carnivorous consumers higher up the chain.

At the top of the food chain are those animals known as top consumers–animals such as coyotes, hawks, and wolves (as well as meat-eating humans), who have no natural predators and are not generally eaten by anything else in the community.

While it might seem preferable to be at the top of a food chain rather than at the bottom, there are downsides to being last in line to eat. For one, it's harder to obtain the energy necessary to live. As consumers prey on organisms below them in the chain, energy is transferred up the chain through what are known as **trophic levels** (from the Greek *trophe*, meaning "food"). But not all the energy stored in a lower

level makes it to the level above it; at each step, much of this energy is lost from the chain.

When a mouse feeds on a blueberry bush, for example, most of the energy in the blueberries is either burned (in aerobic respiration) to fuel the mouse's activities and given off as heat, or passed through the mouse as indigestible plant fiber. Only a very small portion (about 10%) of the energy stored in the blueberries goes to putting weight on the mouse (**INFOGRAPHIC 22.4**).

This is the main reason top carnivores like coyotes, hawks, and wolves are scarce on earth, and why there are no predators of these creatures: there's simply not enough energy left in the chain to sustain more of them. (It's also why human vegetarianism is more energetically efficient than meat-eating: the same amount of a crop can feed many more vegetarians than meat-eaters who eat the animals that eat the crop.)

While it's helpful to think of the food chain as a stepwise series, the food-chain concept is an oversimplification. Many organisms are omnivores (that is, they eat both plants and animals) and so occupy more than one position in the chain. The result is a complex, intertwined **food web**—like the one that links bees to the food on your breakfast table. This food web includes not just the fruits and vegetables that bees pollinate directly, but the animals that eat bee-pollinated crops and provide us with meat and dairy products (**INFOGRAPHIC 22.5**).

▶**FOOD WEB**
A complex interconnection of feeding relationships in a community.

INFOGRAPHIC 22.5 A HONEY BEE FOOD WEB

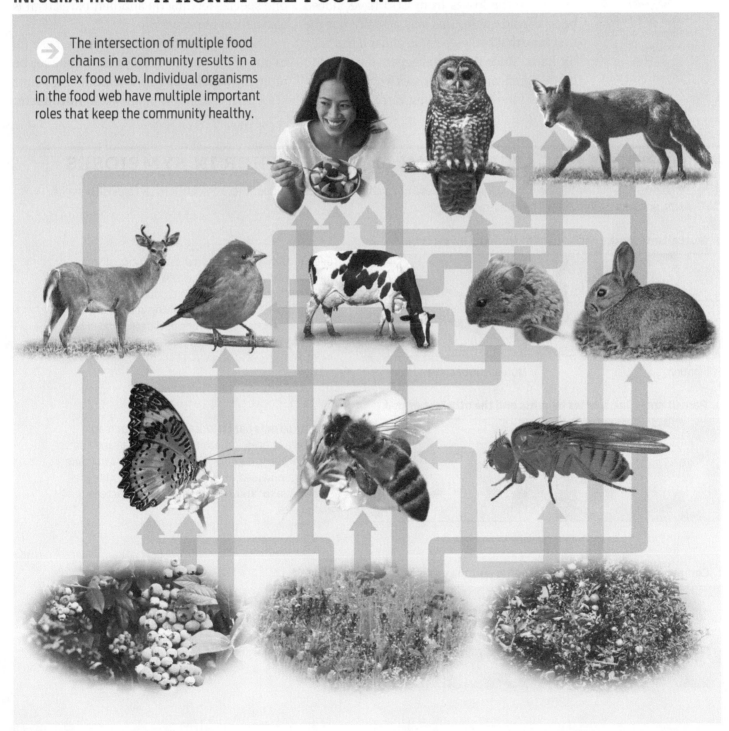

→ The intersection of multiple food chains in a community results in a complex food web. Individual organisms in the food web have multiple important roles that keep the community healthy.

▶**PARASITISM**
A type of symbiotic relationship in which one member benefits at the expense of the other.

▶**SYMBIOSIS**
A relationship in which two different organisms live together, often interdependently.

▶**MUTUALISM**
A type of symbiotic relationship in which both members benefit; a "win–win" relationship.

A SWARM OF PROBLEMS

The United States is home to approximately 1,000 commercial beekeepers, who together cultivate about 2.5 million bee colonies. To a beekeeper—essentially a small business owner—losing 30% of his or her colonies represents an unsustainable financial loss. The sudden bee disappearances of 2006 were a serious worry for Dave Hackenberg and other beekeepers.

While the losses from CCD have indeed been considerable, this was actually not the first time that beekeepers' livelihood has been hit hard. Since 1987, beekeepers have had to battle significant annual losses from an aggressive pest: the blood-sucking varroa mite.

An invasive species, the varroa mite was likely introduced into the United States on the backs of imported bees. The sesame seed-size freeloader is a parasite that feeds on bees' blood, weakening their immune systems and spreading viruses. **Parasitism** is a type of **symbiosis**—a close relationship between two species—in which one species (in this case, the mite) clearly benefits, and one species (the honey bee) clearly loses. Because it involves one species feeding on another, parasitism is also a form of predation.

Not all symbiotic relationships are harmful to one of the partners; they can sometimes be mutually beneficial. Bees and flowering plants are a perfect example of one such **mutualism.**

INFOGRAPHIC 22.6 ORGANISMS MAY LIVE TOGETHER IN SYMBIOSES

 Symbioses are relationships in which different species live together in close association. These associations can provide benefits, harm, or have no effect on the partners involved.

a. Mutualism – Both species benefit from the interaction

Pollination   Mutualistic bacteria

Bees and flowering plants represent a mutualism: bees get nectar and pollen, which they use for food, and pollination allows successful reproduction for the plant. In another mutualism, bacteria living in the bee gut provide protection against bee pathogens, and in turn get a safe place to live and a constant source of food.

b. Parasitism – One species benefits and the other is harmed

 —— Varroa mite ——

The varroa mite has historically been the primary pathogen of bees. It parasitizes both larvae and adult bees, obtaining nutrition and leaving the bee immune system suppressed. Hives infected with the mite are susceptible to fatal infections caused by bacteria and viruses.

c. Commensalism – One species benefits and the other is unharmed

Bees can live in hollows in trees. The bees benefit from the shelter provided by the tree, and the tree is not harmed by the symbiotic association with the bees.

Bees can't survive without the flowers, which provide food, and plants depend on the bees to help them reproduce. Honey bees have other mutualistic symbioses, as well, including with bacteria that live safely inside the bees and benefit their hosts by helping them combat disease.

A third type of symbiotic relationship is **commensalism,** a relationship in which one species benefits while the other is unaffected or unharmed–bees living in a hollowed-out oak tree, for example **(INFOGRAPHIC 22.6)**.

As devastating as the parasitic varroa mite infestation has been, it is unlikely to be the sole or even primary factor responsible for the most recent colony collapses. Research by apiarist vanEngelsdorp, and others has shown that levels of mite infections in collapsing colonies are no higher than they had been in previous years. Moreover, a mite infestation does not explain the most curious aspect of the condition: the sudden disappearance of entire hives.

Honey bees are a colonial species: they live in hives of thousands of individuals, in which worker bees collectively support all the juvenile larvae. In collapsing colonies, the worker bees (all female) abandon the hive. With no workers to help larvae reach maturity, the colony dies.

According to vanEngelsdorp, the worker bees may be practicing what's called "altruistic suicide." "The worker bee knows she's sick," he explains. "She knows, 'Well, I better fly out of here and die away from the hive and maybe preserve my nest mates.'" But that altruistic practice can spiral out of control, he says, and the result is a collapsing colony.

Alternatively, the sick bees may simply have trouble finding their way back to the hive.

Not surprisingly, the sudden disappearances have fueled intense speculation among beekeepers and laypeople alike about what's going on. Hypotheses have included everything from pesticides, viruses, and genetically modified crops to cell phone radiation, global warming, and even alien abduction. It's a baffling who-done-it with many suspects but no smoking gun.

HONEY BEE FORENSICS

Among the first to investigate the die-offs was a team of Pennsylvania State University biologists headed by vanEngelsdorp and Diana Cox-Foster. It was Cox-Foster whom beekeeper Hackenberg called the day his bees went missing.

The team started their investigation by performing autopsies on the few remaining bees in Hackenberg's colonies. When vanEngelsdorp looked through his microscope, he was shocked: "I found a lot of different scar tissue, and [what] looked like foreign organs," he says. There were also signs of multiple infections, including a parasitic fungus called *Nosema ceranae*. The bees' insides were overrun with pathogens.

Though the bees were clearly sick, each colony seemed to suffer from a different spectrum of ailments. "The bees are getting the flu," says vanEngelsdorp. "What we don't understand is the fact that it's not always the same strain of flu." The researchers hypothesized that something had compromised the bees' immune system, making them vulnerable to infections that a healthy colony could normally fend off. Some observers have even likened the condition to "bee AIDS."

The analogy certainly seems fitting. But early attempts to identify a bee-equivalent of HIV (the virus that causes AIDS) were unsuccessful. The initial prime suspect, the varroa mite, was not present at high enough levels to cause a crippled immune system. And all of the other parasites and infections had previously been documented in healthy bee populations, and so were unlikely to have caused the heavy losses.

Hoping to isolate a previously unidentified culprit, in 2007, Cox-Foster and her colleagues enlisted genomics experts from Columbia University to scour genetic material from the hives for evidence of a new invader. After months of intensive work, their efforts seemed to pay off: genetic tests revealed that a virus called Israeli acute paralysis virus (IAPV) was present in 96% of the hives affected with CCD. The researchers thought they had found the smoking gun.

▶**NICHE**
The space, environmental conditions, and resources that a species needs in order to survive and reproduce.

▶**COMPETITION**
An interaction between two or more organisms that rely on a common resource that is not available in sufficient quantities.

Subsequent research, however, showed that not all honey bee colonies that are infected with IAPV have symptoms of CCD, suggesting that the virus alone is not the source of the problem.

More recently, in 2010, researchers from the University of Montana and the U.S. Army's Edgewood Chemical Biological Center presented evidence that another viral culprit–invertebrate iridescent virus (IIV)–was present in essentially all collapsing hives. Whether this virus proves to be the decisive factor in CCD remains to be seen. But since IIV is present in noncollapsing hives as well, it is unlikely to be acting alone.

In fact, there may be no single cause of CCD, but rather a complex combination of causes. "All the evidence so far has really supported the idea that it's likely a combination of factors that are stressing the bees beyond their ability to cope," says Maryann Frazier, a bee researcher at Penn State who is part of Cox-Foster's team.

One factor that almost certainly plays a role in exacerbating the condition is poor nutrition. Just like humans, bees need a well-balanced diet that contains all the essential nutrients to remain healthy. For a number of reasons, honey bees are finding it harder and harder to obtain a nutritious diet.

COMPETING FOR RESOURCES

When European settlers first brought the Western honey bee to the United States in the 1600s, the bees quickly spread from managed colonies into the wild, in some cases displacing native bee species. Honey bees are excellent at

INFOGRAPHIC 22.7 POLLINATORS HAVE DIFFERENT ECOLOGICAL NICHES

→ Some pollinators share similar but not identical ecological niches. They may require the same seasonal temperatures, structures for shelter, yearly rainfall, and food from the nectar and pollen of flowers but prefer flowers of different size, color, and shape.

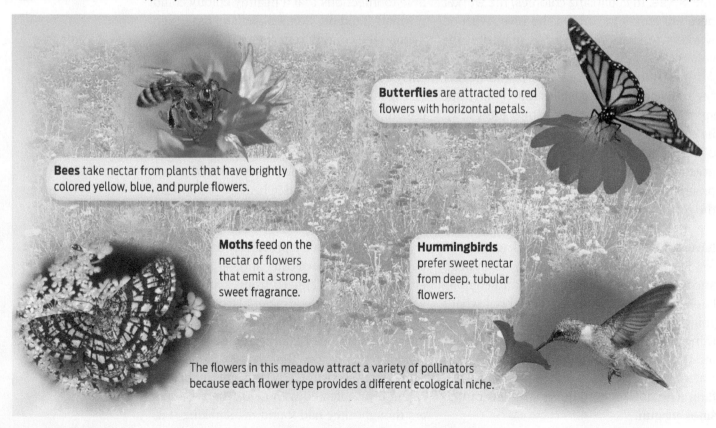

Butterflies are attracted to red flowers with horizontal petals.

Bees take nectar from plants that have brightly colored yellow, blue, and purple flowers.

Moths feed on the nectar of flowers that emit a strong, sweet fragrance.

Hummingbirds prefer sweet nectar from deep, tubular flowers.

The flowers in this meadow attract a variety of pollinators because each flower type provides a different ecological niche.

colonizing new habitats because they are largely generalists when it comes to flower choice.

Though they tend to visit a single species of flower on each foraging trip, honey bees may visit more than 100 species of flowers within a single geographic region over the course of a season. In warm climates, they are active year round, and tend to feed throughout the day and start foraging earlier in the morning than many native bee species. In other words, honey bees have a broad ecological **niche**–the space, environmental conditions, and resources (including other living species) that a species needs in order to survive and reproduce.

Different pollinators generally have different niches, owing to their varying sizes and preferences for different flower types. Bees, for instance, are attracted to brightly colored blossoms–those with yellow, blue, and purple petals, for example, but not red. Butterflies, by contrast, are commonly attracted to red flowers that are large and easy to land on. Moth-pollinated flowers tend to have pale or white petals with no distinctive color pattern but with strong fragrance (**INFOGRAPHIC 22.7**).

When two or more species rely on the same limited resources–that is, when their niches overlap–the result is **competition.** Competition tends to limit the size of competing populations and may even drive one out. In theory, no two species can successfully coexist in identical niches in a community because one would eventually out-compete the other–a concept described by the **competitive exclusion principle.** In reality, however, very few species share *exactly* the same niche, so different species may find a competitive balance by subdividing resources (so-called "resource partitioning").

Some species compete through behavior. African honey bees (*Apis mellifera scutellata*), for example, are a subspecies that was brought from Africa to Brazil in 1956 and which quickly expanded its range to include Central America and the southern United States. When African bees mate with local varieties they produce hybrid species with

a blend of traits: these so-called Africanized honey bees are much more aggressive than Western honey bees. They will chase away other pollinators from food sources, and even swarm and sting animals that get near their hives–behaviors earning them the colloquial name "killer bee" (**INFOGRAPHIC 22.8**).

INFOGRAPHIC 22.8 BEES COMPETE FOR RESOURCES

→ Species with similar niches compete for resources that may be limited because of natural or human influences. Species may out-compete one another or find a balance, depending on their foraging abilities and behaviors.

Blueberry bee
(*Osmia ribifloris*)

Food Partitioning
Native bees like the blueberry bee successfully coexist with non-native bees by specializing on one food (blueberry flowers) that the honey bees can't efficiently utilize.

Honey bee
(*Apis mellifera*)

Generalist Foraging Patterns
Imported honey bees can forage over great distances and feed on a wide variety of flowers. As they feed year round in warmer climates, they are very successful in their competition with other bee species, which may have more limited niches.

Africanized honey bee
(*Apis mellifera scutellata*)

Defensive Behavior
Killer bees compete successfully because of their aggressive defense of food resources. They chase other pollinators away from available food.

▶**COMPETITIVE EXCLUSION PRINCIPLE**
The concept that when two species compete for resources in an identical niche, one is inevitably driven to extinction.

Well-manicured lawns (top) and farms planted with monocrops (center) are "a desert to pollinators." Lawns with wildflowers and diverse plants (bottom) are more welcoming to pollinating creatures.

Some scientists are worried that killer bees from Central America may displace or interbreed with U.S. populations of Western honey bees, with potentially disastrous results for beekeeping and agriculture. This hasn't happened–yet. The much more serious problem for the Western honey bee, it seems, is competition from an even more dangerous species: humans. Humans and their activities have limited the resources used by bees and other pollinators. Agriculture, suburban sprawl, and development, for example, have all decreased bees' natural forage areas, and fragmented their habitat into nonoverlapping zones. Unable to access as many resources in a single foraging trip, bees must compete with each other in the patches that remain.

The ubiquitous well-manicured lawn is particularly problematic. An immense stretch of green grass and no flowers, a lawn is "basically a desert to pollinators," says Frazier. There is literally nothing for them to eat. Likewise, many agricultural areas are planted with monocultures (that is, single crops), which all bloom at the same time, leaving no flowers for the rest of the year. Worse yet, certain genetically engineered pollen-free crops trick bees into thinking they'll find food, only to leave them hungry. And certain non-native plants have floral structures that are inaccessible to indigenous pollinating insects.

As a consequence of these and other human actions, in some geographic regions bees must compete both with one another and with other pollinators for a dwindling supply of food-providing flowers. Many are going hungry and thus are left vulnerable to conditions that can lead to CCD.

HONEY BEE IN THE COAL MINE?

Honey bees aren't the only pollinator in peril. According to a report published in 2007 by the National Research Council, the number and abundance of pollinator species have declined greatly over the last several years. In fact, several bumblebee species are becoming or have become extinct in North America. The concern among researchers and beekeepers is that honey bees may be the "canary in the coal mine," forecasting what's in store for other pollinators. "It's not only the honey bees that are in trouble," says Hackenberg. "All the beneficial insects are in a bad situation."

What's ailing these insects? In addition to a shrinking and fragmented habitat, a disquieting possibility is that they are being poisoned by pesticides. Penn State researcher Frazier and her colleagues have looked at pollen and wax from beehives and found large amounts of many different kinds of pesticides, some of which are approaching toxic levels for the bees. "Pesticides are definitely in the mix and we think they are definitely a player in the stresses that bees are experiencing," says Frazier.

Of particular concern to beekeepers is a class of pesticides known as neonicotinoids, or "neonics" for short. Neonics are an artificial form of nicotine heavily used in commercial agriculture all over the world. (Nicotine, made by tobacco plants, is a natural deterrent to plant-eating insects.) Virtually all corn and most soybean seeds in the U.S. are treated with neonics. Research has shown that neonics can impair honey bees' ability to find and return to their hives, and the U.S. Environmental Protection Agency (EPA) acknowledges that neonics are "highly toxic to honey bees." Yet so far the

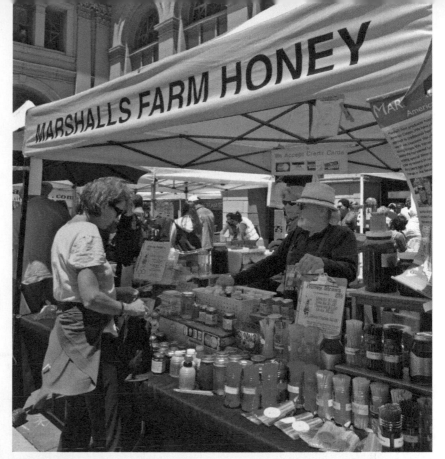

Locally grown honey is, according to vanEngelsdorp, "the most ethical sweetener," taking the least amount of fossil fuel to produce and transport.

EPA has not restricted use of these pesticides in the U.S. A 2012 report produced by the USDA acknowledged that neonics and other pesticides may be a contributing factor to CCD, but stopped short of singling them out as a primary cause. "The bottom line is we have not been able to put together, here in the United States, the evidence that neonicotinoids are causing

> **"** *All the evidence so far has really supported the idea that it's likely a combination of factors that are stressing the bees beyond their ability to cope.* **"**
>
> —MARYANN FRAZIER

the decline in honeybees," says Frazier. "But we are not convinced that they are *not* playing a role."

The situation is a bit different in Europe. In 2013, the European Union imposed a two-year continent-wide ban on neonicotinoids on flowering crops such as corn, rapeseed, and sunflowers that are attractive to bees, after a report by the European Food Safety Authority identified a number of risks posed to bees from the pesticides. The short-term ban will allow researchers the chance to study the effects of neonics more thoroughly—especially at the sublethal doses bees are likely to encounter.

Although researchers have not yet been able to prove that neonics are playing a role in CCD—"the jury is still out," says Frazier—beekeepers like Hackenberg are understandably cautious about what pesticides they expose their colonies to (**INFOGRAPHIC 22.9**).

Because it may involve a complex combination of triggers, there is no easy remedy to CCD. It may require making fundamental changes to our beekeeping and agricultural

INFOGRAPHIC 22.9 WHAT IS CAUSING COLONY COLLAPSE DISORDER?

→ Bees live in a social colony with a single queen and her offspring. The collapse of colonies all over the world is of great concern. The cause of this disorder is likely to be complex and to involve an interplay of several factors.

A healthy bee colony is full of busy adult bees.

A collapsing colony has very few adults, so the developing larvae that depend on them will not survive.

Possible contributors →

Stress and nutrition
Bees need both nectar and pollen for a complete, nutritional diet. When blossoms are scarce, beekeepers feed their colonies sugar mixtures but not pollen supplements. The lack of essential nutrients leaves the colony stressed, which weakens its defenses.

Pesticides
The pesticides used in agriculture can make their way into pollen particles. The amounts measured in pollen have reached toxic levels and may be affecting the health of the bees that eat the pollen.

***Nosema ceranae* parasite**
This intestinal parasite prevents bees from processing food properly and thus can weaken and kill bees.

Other pathogens
Several viruses, including israeli acute paralysis virus, can impair and kill bees.

practices. In particular, we could break up fields of monocultures with varied bee-friendly plants: red clover, foxglove, and bee balm, for example. We could also use pesticides sparingly and avoid spraying at times of day when bees are actively foraging (although this won't help with neonics, which are applied to the seeds themselves before they get planted).

While these individual steps would certainly help matters, apiarist Dennis vanEngelsdorp diagnoses a more systemic problem. In his estimation, we suffer from NDD—"nature deficit disorder." To help bees, he says, we need also to cure ourselves. As treatment, he prescribes reconnecting to nature in a more immediate and local way—"having a meadow or living by a meadow," for example, or becoming a beekeeper oneself.

In addition, says bee expert Frazier, "People need to take more time to understand where their food comes from, what it takes to produce food and have this incredible supply of food available to us."

While the fate of the honey bees remains uncertain, there are signs that a more bee-friendly awareness is beginning to emerge, thanks in part to the concerns raised by CCD. In 2008, Häagen-Dazs, the ice-cream maker, launched a "Help the Honey Bee" campaign, in recognition of the fact that honey bee-dependent products are used in 25 of its 60 flavors. So far, the company has donated more than $700,000 to honey bee research at universities including the University of California-Davis and Penn State University. Burt's Bees, the maker of "Earth-friendly" lip balms

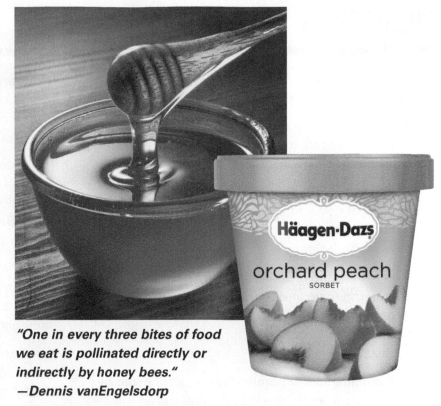

"One in every three bites of food we eat is pollinated directly or indirectly by honey bees."
—Dennis vanEngelsdorp

and other personal products, has created a series of CCD-related public service announcements, viewable on YouTube, including one starring Isabella Rossellini dressed as a honeybee. Even ordinary citizens are catching the bee buzz. From city-dwellers becoming amateur rooftop beekeepers to suburbanites letting more flowers grow in their yards, the ranks of people wanting to make the environment pollinator-friendly has swelled. And that's a cause that just about everyone can get behind—because, as more and more people are coming to realize, a world without honey bees just wouldn't be as sweet. ∎

CHAPTER 22 Summary

- An ecological community is made up of interacting populations of different species.

- Bees are keystone species because they play a fundamental role in supporting the entire community, much like the keystone in an arch.

- Bees are the primary pollinators for many species of flowering plants, which depend on the pollinators to transfer pollen between plants of the same species.

- Flowers are the reproductive hub of a plant, containing male and female reproductive structures. Pollination, the transfer of pollen from male to female structures, results in fertilization.

- The organisms in a community are connected by a food chain. Each player in the chain is an important ecological link in the chain.

- Organisms at the base of the food chain are producers—they obtain energy directly from the sun and supply it to the rest of the food chain; organisms higher up the food chain are consumers—they obtain energy by eating organisms lower on the chain.

- In predation, one organism eats another. Herbivory (eating plants) and parasitism are both types of predation.

- As energy flows up trophic levels in the food chain, some of it is lost to the environment.

- Organisms can have different types of symbiotic relationships. In mutualistic symbioses, both members benefit; in parasitism, one member benefits while the other suffers; and in commensalism, one member benefits while the other is unharmed.

- The space and resources, including other members of the community, that a species uses to survive and reproduce define its ecological niche. Some species have overlapping niches, leading to competition for resources.

- Colony collapse disorder may have no one single cause, but may result from many interacting factors affecting bees, including poor nutrition, pathogens, and pesticides.

- Bees are not the only pollinators in peril. Human development and agriculture have decreased habitat and foraging areas for many natural pollinators, resulting in increased competition among them.

MORE TO EXPLORE

- Bee Informed http://beeinformed.org

- Henry, M., et al. (2012) A common pesticide decreases foraging success and survival in honey bees. *Science* 20:348–350.

- Krupke, C. H., et al. (2012) Multiple routes of pesticide exposure for honey bees living near agricultural fields. *PLOS ONE* 7(1):e29268.

- Dennis vanEngelsdorp (2009) Colony Collapse Disorder: A Descriptive Study. *PLOS ONE* 4(8):e6481.

- Cox-Foster, D. L. (2007) A metagenomic survey of microbes in honey bee colony collapse disorder. *Science* 318(12):283–286.

CHAPTER 22 Test Your Knowledge

DRIVING QUESTION 1

What are keystone species in a community, and why are pollinators considered keystone species?

By answering the questions below and studying Infographics 22.1, 22.2, and 22.3, you should be able to generate an answer for the broader Driving Question above.

KNOW IT

1 How does a community differ from a population?

2 What are keystone species?

3 A rocky shoreline that is covered at high tide but exposed at low tide supports a community of mussels, algae, barnacles, and starfish. An ecologist systematically removes species from different areas of the beach. Removing the mussels doesn't substantially change the community, but removing the starfish dramatically changes the mix of species in the area. Which is the keystone species?

  a. mussels
  b. barnacles
  c. algae
  d. starfish
  e. all of the above

4 Bees transfer pollen from the _____ to the _____.

  a. anther; stigma
  b. stigma; style
  c. filament; ovary
  d. anther; ovary
  e. stigma; anther

USE IT

5 Think about a community of organisms that you are familiar with. From what you know about this community, choose what you think might be a keystone species and defend your choice.

6 If you have pollen allergies, are you more likely to be suffering from the effects of bee-carried pollen or wind-carried pollen? Explain your answer.

DRIVING QUESTION 2

What are food chains and food webs, and how does energy flow through them?

By answering the questions below and studying Infographics 22.4 and 22.5, you should be able to generate an answer for the broader Driving Question above.

KNOW IT

7 In relation to a food chain, what do plants and photosynthetic algae have in common?

  a. nothing.
  b. They are both producers.
  c. They are both first level consumers.
  d. They are both top level consumers.
  e. Their numbers are limited by the energy they take in from heterotrophic food sources.

8 A bear that eats both blueberries and fish from a river can be referred to as

  a. an omnivore.
  b. a heterotroph.
  c. a consumer.
  d. a producer.
  e. all of the above.
  f. a, b, and c
  g. a and c

USE IT

9 Describe a natural food web that includes a terrestrial food chain (including honeybees) and at least one aquatic organism from an aquatic food chain.

10 Explain how a cow can eat so many kilograms of grass but not produce the equivalent amount of energy in the form of meat. What happens to the energy stored in the grass once it is ingested by the cow?

11 Compare the diet of a human who is an herbivore with that of a human who is a top consumer. Consider what each might actually eat; how much energy from a producer is captured in the herbivore human; and how much energy from a producer is captured in the top consumer human.

DRIVING QUESTION 3

What positive and negative interactions occur between members of a community?

By answering the questions below and studying Infographics 22.6, 22.7, 22.8, and 22.9, you should be able to generate an answer for the broader Driving Question above.

KNOW IT

12 What are some important features of a honey bee niche? How is it that other nectar-feeding organisms can coexist with bees as part of a community?

13 Competition is most likely to occur

   a. when one species eats another.

   b. when two species occupy different niches.

   c. when one species helps another.

   d. when two species occupy overlapping niches.

   e. when two species help each other.

14 Which of the following characterizations best describes a symbiotic relationship?

   a. Both organisms benefit.

   b. The organisms live in close association.

   c. Only one organism benefits.

   d. The relationship is mutually harmful.

   e. Neither organism benefits.

15 Would you characterize the relationship between the bacteria that live symbiotically within bees and their bee hosts as a type of competition, parasitism, mutualism, or commensalism? Explain your answer.

USE IT

16 On a rocky intertidal shoreline (the area between the highest and lowest tidelines, so the intertidal zone is alternately exposed and covered by seawater), mussels and barnacles live together attached to rocks where they obtain food by filtering it from ocean water. Since these two species coexist in the same habitat, we predict that they do not have identical niches. What might be separating their niches enough to allow them to occupy the same rocky intertidal zone?

17 If a meadow of wildflowers were converted to a field of corn, would you predict the number and diversity of bees in the community to increase or decrease? Explain your answer.

18 What is the evidence for and against each of the following being responsible for colony collapse disorder (CCD)?

   a. varroa mites

   b. IAPV

   c. neonicotinoids

19 We all have *E. coli* bacteria living in our intestinal tracts. Occasionally these *E. coli* can cause urinary tract infections. From this information, which of the following terms would you say describe(s) the relationship between us and our intestinal *E. coli*? Why did you choose the term(s) you did?

   a. competition

   b. mutualism

   c. parasitism

   d. symbiosis

   e. predator–prey

BRING IT HOME

20 Many people consider bees a stinging nuisance. What could you say to such people to dissuade them from killing all the bees in their backyards?

MINI CASE

21 Farmers often plant large acreage of a single crop in order to maximize yield and simplify harvesting. This is true of almonds in the central valley of California.

   a. From what you have read in this chapter, what are some of the pros and cons associated with monoculture?

   b. Do some online research to develop a specific model for an alternative to monoculture that addresses at least one of the issues you have identified.

INTERPRETING DATA

22 Scientists carried out an experiment to test the hypothesis that a neonicotinoid pesticide called imidacloprid could cause colony collapse disorder (CCD). They had a total of 20 hives (colonies) that were broken into 5 groups (with 4 hives per group). Four groups received imidacloprid at different dosages (400 μg/kg; 200 μg/kg; 40 μg/kg and 20 μg/kg). One group did not receive imidacloprid. Hives were monitored for 23 weeks after the initial dose. The data are summarized in the table below.

a. Graph these data.

b. What patterns do you observe?

c. Do you think that the data support the hypothesis? Why or why not?

DOSE (μg/kg)	NO. OF DEAD HIVES AT 12 WEEKS	NO. OF DEAD HIVES AT 14 WEEKS	NO. OF DEAD HIVES AT 16 WEEKS	NO. OF DEAD HIVES AT 18 WEEKS	NO. OF DEAD HIVES AT 21 WEEKS	NO. OF DEAD HIVES AT 23 WEEKS
400	0	2	2	4	4	4
200	0	0	2	2	3	4
40	0	1	1	2	3	3
20	0	0	0	0	2	4
0	0	0	0	0	1	1

SOURCE: Lu, C., et al. 2012. In situ replication of honey bee colony collapse disorder. *Bulletin of Insectology* 65:99-106.

THE HEAT IS ON

From migrating maples to shrinking sea ice, signs of a warming planet

FOR MORE THAN TWO CENTURIES, BURR MORSE'S family has collected sap from Vermont's maple trees and boiled it to sweetened perfection. If you pour maple syrup over your breakfast pancakes or eat maple-cured ham, you've likely enjoyed the results of their careful craft, or that of other Vermont maple syrup farmers. About one in four trees in the state of Vermont is a sugar maple (*Acer saccharum*), and each year the state produces roughly a million gallons of syrup, making Vermont the number one maple syrup producer in the United States. But what has been a proud family tradition and the economic lifeblood for generations of maple syrup farmers could now very well be in jeopardy.

Sap dripping from a tapped sugar maple tree.

▸▸ DRIVING QUESTIONS

1. What are ecosystems, and how are ecosystems being affected by climate change?
2. What is the greenhouse effect, and what does it have to do with global warming?
3. How do carbon and other chemicals cycle through ecosystems?
4. How are scientists able to compare present-day levels of atmospheric carbon dioxide to past levels, and why would they want to?

INFOGRAPHIC 23.1 U.S. MAPLE SYRUP: A THING OF THE PAST?

 The amount of maple syrup produced in in the United States has been declining, in part because of the shortening of the maple syrup season in the northeastern states. Meanwhile, Canadian production has been increasing because of increased marketing, government subsidies, improved technologies, and, likely, climate change.

Maple Syrup Production in the United States and Canada, 1860–2010

Canadian production has significantly increased in the past 40 years.

U.S. production has significantly decreased over time. Recent U.S. increases are linked to improved technologies for sap extraction and syrup production.

SOURCE: Farrell, M. L., and Chabot, B. F. (2012) Assessing the growth potential and economic impact of the U.S. maple syrup industry. *Journal of Agriculture, Food Systems, and Community Development* 2(2):11 -17.

"In the last 20 years we have had a number of bad seasons and most of those I would attribute to temperature that is a little too warm," says Morse. "For maple sugaring to work right, the nights have to freeze down into the mid-20s, and the days have to thaw up into the 40s. And the nights for those 20 years, it seemed, were not quite getting cold enough."

❝ The total economic impact of maple in Vermont alone is nearly $200 million each year. ❞

—TIM PERKINS

Morse isn't the only one to notice the shift. Maple syrup farmers across New England have noted the changes in temperature and are concerned about their long-term effects.

Warmer winters in New England could have a large economic impact on the region. As ecologist Tim Perkins, director of the Proctor Maple Research Center at the University of Vermont, testified to Congress in 2007, "If the northeast regional climate continues to warm as projected, we expect that the maple industry in the U.S. will become economically untenable during the next 50-100 years." This is not just a drop in the bucket: according to Perkins, "The total economic impact of maple in Vermont alone is nearly $200 million each year."

Before 1900, 80% of the world's maple syrup came from trees in the United States, the rest from Canada. Today, the pattern is reversed, with Canada greatly out-producing the United States. Canada now accounts for about 80% of world maple syrup production. While part of this reversal has to do with marketing, Canadian government subsidies, and improved technologies, scientists believe that climate change is a significant contributing factor, putting New England sugar farmers at a competitive disadvantage (**INFOGRAPHIC 23.1**). **Climate change** is any substantial change in climate that lasts for an extended period of time (decades or more). One contributor to current climate change is **global warming,** a recent and continuing increase in the average global temperature.

New England's maples are not the only ones feeling the heat. Plant and animal species throughout the world–from herbs in Switzerland to starfish in California–are being affected by rising temperatures. Some are shifting their geographic ranges as a result. Many historically subtropical aquatic animals, such as seahorses and turtles, for example, are moving toward the coasts of northern England and Scotland, where ocean temperatures are warmer than they used to be. And fish that were once wholly tropical are turning up in North Atlantic waters. Other organisms that cannot easily relocate, like some plants and mountain-dwelling animals, are being driven to extinction.

Earth's climate changes naturally, and has done so many times in its long history. But scientists now believe that humans are accelerating the pace of change, with potentially dire consequences for life on our planet.

TO EVERYTHING A SEASON

In nature, timing is everything. And for many species temperature is nature's clock, cueing seasonally appropriate tasks like mating or producing flowers. Rising temperatures around the globe are interfering with these natural rhythms. Many plants are flowering earlier now than they once did; animals, such as the yellow-bellied marmot, are emerging from hibernation earlier; and many bird and butterfly species are migrating north and breeding earlier in the spring than they did a few decades ago—all because of slight changes in temperature cues. It's a pattern of change that scientists are seeing around the globe (**INFOGRAPHIC 23.2**).

▸**CLIMATE CHANGE**
Any substantial change in climate that lasts for an extended period of time (decades or more).

▸**GLOBAL WARMING**
An increase in Earth's average temperature.

INFOGRAPHIC 23.2 RISING TEMPERATURES AFFECT PLANT BEHAVIOR

→ Global warming is changing the seasonal behavior of plants and animals. Increased average spring temperatures result in earlier flowering dates for a variety of plant species. For each 1°C rise in mean spring temperature, plants flower an average of 3.2 days earlier in Massachusetts and 4.1 days earlier in south-central Wisconsin.

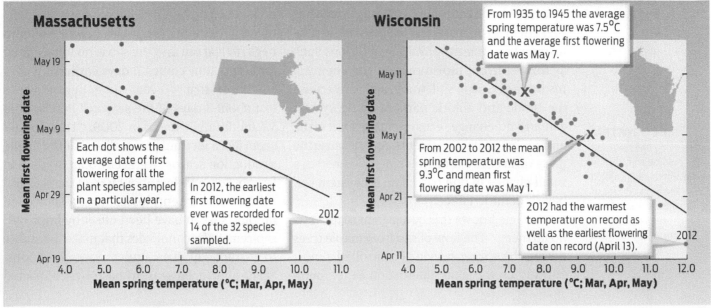

SOURCE: Ellwood, E. R., et al. (2013) Record-breaking early flowering in the eastern United States. *PLOS ONE* 8(1):e53788.

Burr Morse boiling sugar maple sap in his Vermont sugar farm.

Who cares if flowers bloom earlier or marmots come out of hibernation sooner? By themselves, these changes wouldn't necessarily be a big deal. But because living things are exquisitely adapted to their environments, and are interdependent, a change in one part of an ecosystem may upset others. With longer summers, for example, marmots spend more time eating and less time hibernating. As a result, in addition to getting fat, the marmots are putting more pressure on the animals and plants they prey on, and therefore changing the wider ecosystem.

An **ecosystem** is a complex, interwoven network of interacting components. It includes both the community of living organisms present in an area and the features of the non-living environment–physical conditions such as temperature, moisture, and the chemical resources found in soil, water, and air. Because the biotic and abiotic parts of an ecosystem can and do change, ecosystems are not static entities but dynamic systems. And because the parts of an ecosystem are so interconnected, a small change in one part of an ecosystem can have a domino effect.

No one knows this better than maple syrup farmers. "The flow of sap from maple trees during the spring season is controlled almost entirely by the daily fluctuation in tempera-

ture," explains ecologist Perkins. "Small changes in the day-to-day temperature pattern will have large consequences on sap flow."

Historically, trees were tapped in early March when the sap began to flow; the sap was then collected for the next 6 weeks. But about 10 years ago, Perkins started getting calls from syrup producers saying that they were tapping earlier and making syrup earlier. Curious, he and his colleagues decided to investigate. They scoured historical records and surveyed hundreds of maple syrup producers in New England and New York. Their results were startling: over a mere 40 years, between 1963 and 2003, the start of the tapping season had moved up by about 8 days. Even more significant, the end of the season, when maples begin to leaf out and the sap is no longer good for syrup, now comes 11 days earlier.

"Over that 40-year time period we've lost about 3 days of the season," Perkins told Vermont Public Radio in 2009. "That doesn't seem like a lot until you realize that the maple production season averages about 30 days in length. So we've lost about 10% of the season."

To some extent, losses from a shortened tapping season have been offset by improved sap-removal technologies that make it possible to extract sap even under poor conditions. "[The shortened season] hasn't yet impacted yields, because the technology of sap extraction has improved," says Perkins, who points in particular to the use of vacuum tubes to suck out sap. The bigger problem is what will happen if the climate changes so much that New England no longer provides a suitable **habitat** for maple trees.

As Perkins testified in 2007, current climate computer models predict that by the end of the century New England's forests will more closely resemble those of present-day Virginia, North Carolina, and Tennessee, dominated by hickory, oak, and pine rather than maple, beech, and birch. If that happens, not only maple syrup but the brilliant fall foliage New

England is famous for will be a thing of the past **(INFOGRAPHIC 23.3)**.

New England's colorful foliage is part of a distinct **biome** known as temperate deciduous forest. Biomes are large, geographically cohesive regions whose defining vegetation—its plant life–is determined principally by climatic factors like temperature and rainfall. The temperate deciduous forest biome is characterized by having distinct seasons, trees that lose their leaves in the fall, and a total annual precipitation of between 750 and 1,500 millimeters that occurs evenly throughout the year. The sugar maple is right at home in this biome (see **UP CLOSE: BIOMES**).

▸**HABITAT**
The physical environment where an organism lives and to which it is adapted.

▸**BIOME**
A large geographic area defined by its characteristic plant life, which in turn is determined by temperature and levels of moisture.

INFOGRAPHIC 23.3 MAPLE TREE RANGE IS AFFECTED BY RISING TEMPERATURES

 Computer simulation models that assume future increases in greenhouse gases predict major changes in forest makeup in the northeast United States over the next century. Predictions suggest that the currently dominant maple, beech, and birch forests will be replaced with hickory, oak, and pine forests.

White–Red–Jack Pine
Spruce–Fir
Longleaf–Slash Pine
Loblolly–Shortleaf Pine
Oak–Pine
Oak–Hickory
Oak–Gum–Cypress
Elm–Ash–Cottonwood
Maple–Beech–Birch
Aspen–Birch
No data

SOURCE: U.S. Global Change Research Program.

UP CLOSE BIOMES

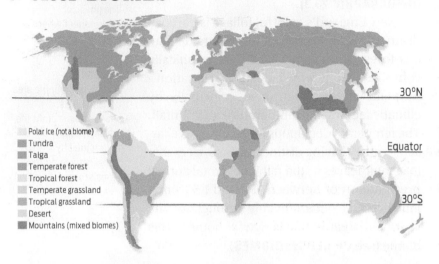

Polar ice (not a biome)
Tundra
Taiga
Temperate forest
Tropical forest
Temperate grassland
Tropical grassland
Desert
Mountains (mixed biomes)

30°N

Equator

30°S

Aquatic: Marine

This biome covers about three-fourths of the earth and includes the oceans, coral reefs, and estuaries.

Aquatic: Freshwater

A biome characterized by having a low salt concentration. Fresh-water biomes include ponds and lakes, rivers and streams, and wetlands.

Tropical Forest

Tropical forests are biomes characterized by warm temperatures and sufficient rainfall to support the growth of trees. Tropical forests may be deciduous or evergreen, depending on the presence or absence of a dry season.

Temperate Forest

Temperate forests are characterized by moderate winters and rainfall. Trees are mostly evergreen or deciduous, dropping their leaves in winter.

Grassland

A biome characterized by perennial grasses and other nonwoody plants. In North America, the prairies are examples of grasslands.

Desert

A biome characterized by extreme dryness. Cold deserts experience cold winters and hot summers, while hot deserts are uniformly warm throughout the year.

Taiga

A biome characterized by evergreen trees, with long and cold winters and only short summers.

Tundra

A biome that occurs in the Arctic and mountain regions. Tundra is characterized by low-growing vegetation and a layer of perma-frost (frozen all year long) very close to the surface of the soil.

Climate change is beginning to redraw the map of biomes around the world. In northern Alaska, where once there was only sparsely vegetated tundra, woody shrubs now grow. When Montana's Glacier National Park was opened in 1910, it held approximately 150 large glaciers; in 2013, there were only 25, and scientists predict that by 2030 there will be no glaciers left. As the vegetation in these landscapes changes, so does the community of organisms that rely on it for food and habitat.

The difference in average global temperatures between today and the last ice age—10,000 years ago —is only about 5°C (9°F).

WARMING PLANET, DIMINISHING BIODIVERSITY

Although temperature swings and shifts in the ranges of organisms are natural phenomena, the amount of warming in recent years is unprecedented, and evidence suggests that the change is not merely part of a natural cycle. From 1880 until 2012, Earth's surface has warmed, on average, by about 0.8°C (1.4°F), according to a 2013 report by NASA's Goddard Institute for Space Studies. That may not sound like a lot. But consider this: the difference in average global temperatures between today and the last ice age—10,000 years ago, when much of North America was buried under ice—is only about 5°C (9° F). Where global temperatures are concerned, even a 1° change is significant.

The rate of warming has increased as well. Ten of the warmest years on record have occurred since 1998. The last decade, from 2000 through 2010, was the hottest decade so far, with 2010 tying 2005 for the title of hottest year on record. For the continental United States, 2012 broke the heat record. Much of this warming is attributable to the **greenhouse effect,** the trapping of heat in Earth's atmosphere. As sunlight shines on our planet, it warms Earth's surface. This heat radiates back to the atmosphere, where it is absorbed by **greenhouse gases** such as carbon dioxide. The heat trapped by greenhouse gases raises the temperature of the atmosphere, and in turn, Earth's surface (**INFOGRAPHIC 23.4**).

The greenhouse effect is a natural process that helps maintain life-supporting temperatures on Earth. Without this greenhouse effect, the average surface temperature of the planet would be a frigid -18°C (0°F). In recent years, however, rising levels of greenhouse gases have increased the strength of the greenhouse effect, a phenomenon known as the enhanced greenhouse effect. As the amount of greenhouse gases in the atmosphere has

INFOGRAPHIC 23.4 **THE GREENHOUSE EFFECT**

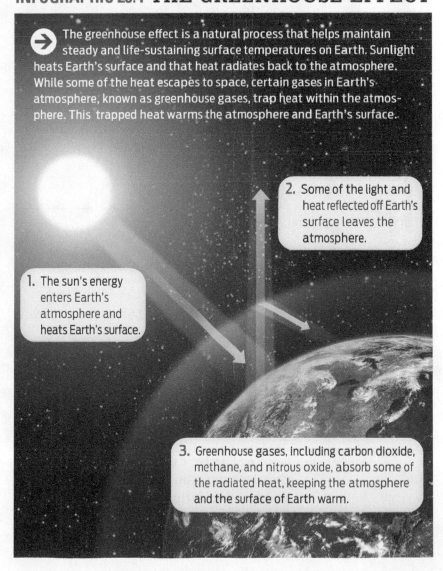

→ The greenhouse effect is a natural process that helps maintain steady and life-sustaining surface temperatures on Earth. Sunlight heats Earth's surface and that heat radiates back to the atmosphere. While some of the heat escapes to space, certain gases in Earth's atmosphere, known as greenhouse gases, trap heat within the atmosphere. This trapped heat warms the atmosphere and Earth's surface.

1. The sun's energy enters Earth's atmosphere and heats Earth's surface.

2. Some of the light and heat reflected off Earth's surface leaves the atmosphere.

3. Greenhouse gases, including carbon dioxide, methane, and nitrous oxide, absorb some of the radiated heat, keeping the atmosphere and the surface of Earth warm.

INFOGRAPHIC 23.5 EARTH'S SURFACE TEMPERATURE IS RISING WITH INCREASES IN GREENHOUSE GASES LIKE CARBON DIOXIDE

As measured directly by thermometers and as documented by historical records and other biological indicators (including tree rings, corals, and ice cores) the temperature on Earth has increased rapidly in the past 140 years. This is paralleled by an increase in greenhouse gases, including carbon dioxide.

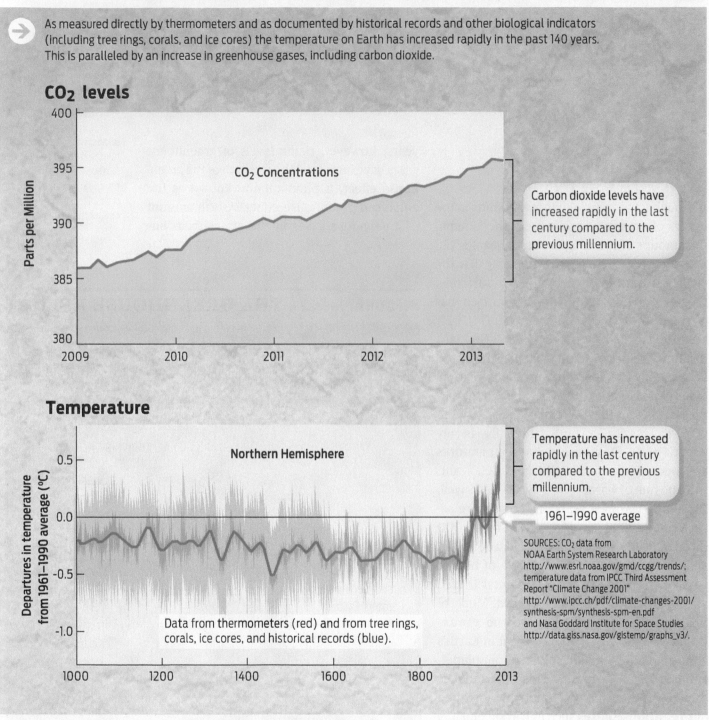

CO₂ levels

CO₂ Concentrations

Parts per Million

Carbon dioxide levels have increased rapidly in the last century compared to the previous millennium.

Temperature

Northern Hemisphere

Departures in temperature from 1961–1990 average (°C)

Temperature has increased rapidly in the last century compared to the previous millennium.

1961–1990 average

Data from thermometers (red) and from tree rings, corals, ice cores, and historical records (blue).

SOURCES: CO₂ data from NOAA Earth System Research Laboratory http://www.esrl.noaa.gov/gmd/ccgg/trends/; temperature data from IPCC Third Assessment Report "Climate Change 2001" http://www.ipcc.ch/pdf/climate-changes-2001/ synthesis-spm/synthesis-spm-en.pdf and Nasa Goddard Institute for Space Studies http://data.giss.nasa.gov/gistemp/graphs_v3/.

increased, so have temperatures. The result is global warming, an overall increase in Earth's average temperature (INFOGRAPHIC 23.5).

For ecologist Hector Galbraith, director of the Climate Change Initiative at the Manomet Center for Conservation Sciences, in Plymouth, Massachusetts, the most worrying thing about climate change is how quickly it is happening and how sensitive species are to the changes. "Most people think of climate change as something that's 30 years out," says Galbraith. But that's simply not true, he notes. "We began seeing responses in ecosystems 20 years ago. The ecosystems knew about it before we did."

Plants, of course, are slower to adapt than animals; they cannot simply get up and move (although they may change their range over time by dispersing seeds into more favorable habitats). But some animals can change their ranges quite quickly. "A bird can simply open its wings, and within 2 hours it's 50 miles farther north," says Galbraith.

What will be the outcome of all these changes? We don't really know. "We're seeing changes to systems that have been relatively stable for thousands of years," says Galbraith.

"The really scary thing about climate change is it's very difficult to predict the ecosystem effects of these changes."

Nevertheless, there are disturbing scenarios. Take the relationship between birds and insects. Many forests are susceptible to insect attacks. Given their insect-rich diet, birds are a natural form of pest control in these forests. If the birds move north, as evidence suggests they are doing, the forests they leave behind will be more vulnerable to threats from insects that thrive in warmer, drier weather. The maple-tree-loving pear thrip and the forest tent caterpillar are just two examples of insects that might be happy to see the birds go. More insects means more dead trees, which in turn means more fuel for forest fires (INFOGRAPHIC 23.6).

INFOGRAPHIC 23.6 RISING TEMPERATURES MEAN WIDESPREAD ECOSYSTEM CHANGE

➔ Climate change is having dramatic impacts on entire ecosystems. The Annual Audubon Christmas Bird Count has shown that 58% of 305 bird species studied have shifted their winter ranges significantly northward, and that this shift correlates with increasing temperature. This has implications for insect control, leaving forests vulnerable to fire when insects are not eaten by birds.

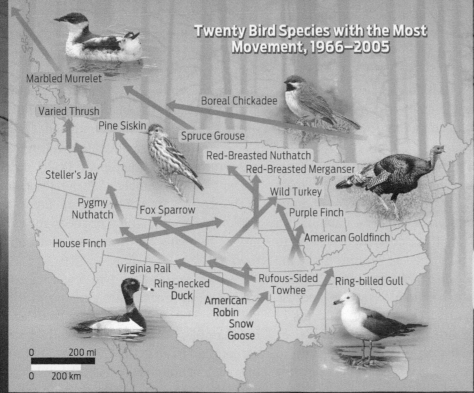

Twenty Bird Species with the Most Movement, 1966–2005

Marbled Murrelet
Varied Thrush
Pine Siskin
Boreal Chickadee
Spruce Grouse
Steller's Jay
Red-Breasted Nuthatch
Red-Breasted Merganser
Wild Turkey
Pygmy Nuthatch
Fox Sparrow
Purple Finch
House Finch
American Goldfinch
Virginia Rail
Ring-necked Duck
American Robin
Snow Goose
Rufous-Sided Towhee
Ring-billed Gull

0   200 mi
0   200 km

The migration of birds north leaves forests vulnerable to insects like the tent caterpillar that voraciously eat, and eventually kill, trees. Dead trees lead to increased forest fires.

SOURCES: Autubon Society; NOAA.

Sap from sugar maples is collected and will later be boiled down to a thicker syrup.

Not all species will be negatively affected by climate change—some may actually benefit. But one species' success in coping with climate change may contribute to another's demise. For example, the adaptable red fox (*Vulpes vulpes*), found throughout the northern hemisphere, is venturing into the range of the endangered Arctic fox (*Vulpes lagopus*), whose habitat—the Arctic tundra—has become warmer. When the two species share a range, the Arctic fox inevitably suffers because the red fox out-competes it for food and also preys on Arctic fox pups.

While some species can adapt to a changing climate by shifting range, future climate change will likely exceed the ability of many species to adapt, as hospitable habitats can no longer be found or accessed. According to a 2011 study published in *Proceedings of the National Academy of Sciences,* 1 in 10 species could be driven to extinction by 2100 because of climate change. The natural residents of mountaintops are especially vulnerable: as temperatures rise, species may move up to higher, colder elevations, but eventually they will have nowhere left to go.

ARCTIC MELTDOWN

Predictably, snow- and ice-covered regions such as the Arctic stand to suffer most immediately from a warming climate, as frozen habitats start to melt. But the situation is worse than one might imagine. As Mark Serreze, director of the National Snow and Ice Data Center at the University of Colorado, Boulder, notes, the Arctic has warmed, on average, twice as much as the rest of the planet. This phenomenon is known among climate scientists as Arctic amplification, and it has to do with the way sea ice affects temperature. As Serreze explains, sea ice both reflects solar radiation and insulates the ocean. As global temperatures rise, ice begins to melt. With less sea ice, more solar radiation is absorbed by the ocean and more of the relatively warm ocean is exposed to air, raising the air temperature even more. It's a positive feedback loop: as additional ice is lost, temperatures rise at an accelerated pace.

According to the extensive Arctic Climate Impact Assessment, the result of 4 years' work by more than 300 scientists around the world, Arctic temperatures are projected to rise by an additional 4°-7°C (7°-13°F) over the next 100 years (INFOGRAPHIC 23.7).

Warming temperatures could spell disaster for species that call the Arctic their home. Polar bears, for example, spend most of the year roaming the Arctic on large swaths of floating sea ice that blanket a good portion of the Arctic Ocean from September through March. The massive mammals travel on sea ice to hunt for seals, which periodically pop up through "whack-a-mole"-like breathing holes in the ice and are nabbed by the bears. The size of this frozen habitat has been shrinking, greatly reducing the bears' ability to obtain food.

Moreover, over the past few decades the ice has been breaking up earlier and earlier in spring. The sea ice in Hudson Bay, Canada, for example, now breaks up nearly 3 weeks earlier than it did in the 1970s. In the absence of sufficient summer sea ice, the polar bears are stuck on land (where there are no seals), or

INFOGRAPHIC 23.7 ARCTIC TEMPERATURES ARE RISING FAST

Current measurements suggest that the Arctic is warming faster than other parts of Earth. The year 2012 was the ninth warmest year on record (since measurements began in 1880). Much of Earth was warmer in 2012 than average temperatures from 1951 to 1980. The Arctic was notably warmer in 2012.

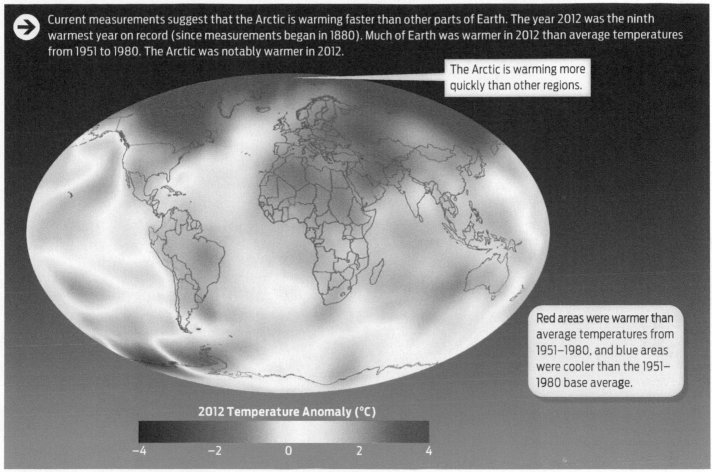

The Arctic is warming more quickly than other regions.

Red areas were warmer than average temperatures from 1951–1980, and blue areas were cooler than the 1951–1980 base average.

2012 Temperature Anomaly (°C)

−4   −2   0   2   4

SOURCE: NASA's Goddard Institute for Space Studies (GISS) http://earthobservatory.nasa.gov/IOTD/view.php?id=80167

are forced to swim long distances to reach sea ice. Some, exhausted by the journey, drown. Those that do survive have fewer opportunities to hunt. Canadian polar bears now weigh on average 55 pounds less than they did 30 years ago, a loss that seriously compromises their reproductive ability.

Scientists have monitored sea ice daily by satellite since 1979. Over the past three decades, the area of Arctic sea ice has shrunk by more than 1 million square miles, an area roughly four times the size of Texas, according to Walt Meier, a research scientist with the National Snow and Ice Data Center. Arctic sea ice hit a record low in September 2012, at the end of the summer melt season, shrinking to a level that climate change models had predicted wouldn't happen until at least 2050.

Scientists now fear that nearly all of the polar bear's summer sea ice could vanish by 2040– possibly sooner (**INFOGRAPHIC 23.8**).

Warming temperatures are also causing glaciers and ice caps on land to melt. Unlike sea ice, which, like an ice cube in a glass of water, doesn't raise the water level as it melts, melting glaciers and ice caps do. How much will seas rise? "By 2100, you're looking at probably about a meter," says Serreze. "Here in Boulder we're at 5,400 feet–we're not worried about that. But if you're living in Miami, this is something that should concern you."

It's important to note that much of the data we have on climate change relates to global, long-term trends. From year to year, there may be slight variations–slightly warmer summers and less sea ice one year, slightly cooler summers

INFOGRAPHIC 23.8 ARCTIC SEA ICE IS MELTING

Rising temperatures have caused the Arctic sea ice to melt and break apart earlier in the season. The reduction in the extent of summer sea ice is threatening the survival of polar bears, which require the sea ice to hunt for seals.

**Median Sea Ice Extent
1979–2000**

2.72 million square miles

**New Record Low
Sept. 16th, 2012**

1.32 million square miles

SOURCE: National Snow and Ice Data Center Sea Ice Index.

and more sea ice the next. And indeed, from a low in 2007, sea ice did indeed bounce back a bit in 2008 and 2009. But the overall trend is still unmistakably downward—toward less sea ice. By 2030 or 2040, says Serreze, there could be no summer ice to speak of. "You could take a ship across the north pole."

FOLLOW THE CARBON

The immediate cause of this planetary warming is a fired-up greenhouse effect. And that, scientists argue, is the result of human activity. As a 2010 statement from the National Academy of Sciences puts it, "There is compelling, comprehensive, and consistent objective evidence that humans are changing the climate in ways that threaten our societies and the ecosystems on which we depend." How did we get to be the culprits in this situation? In short, by pumping more carbon dioxide into the atmosphere.

▶**FOSSIL FUEL**
A carbon-rich energy source, such as coal, petroleum, or natural gas, formed from the compressed, fossilized remains of once-living organisms.

Carbon dioxide is the most notorious player in the greenhouse effect, and scientists believe it is responsible for most of the warming. In fact, atmospheric carbon dioxide concentrations are higher now than they have been in more than 700,000 years.

As discussed in Chapter 2, carbon is a natural ingredient in every living organism, part of the backbone of all organic molecules. Carbon also exists in inorganic forms: as carbon dioxide in the atmosphere, as carbonic acid dissolved in water, as calcium carbonate in limestone rocks. If dead organisms are fossilized before being digested by decomposers, the organic molecules contained within their bodies become trapped below Earth's surface or under the seas. Over time, these compressed organic molecules become **fossil fuels**—coal, oil, and natural gas.

Like other chemical elements, the total amount of carbon on Earth remains essentially constant. In contrast to the way energy flows through an ecosystem in one direction (from the sun to producers to consumers and out to the universe as heat; see Chapter 22), elements such as carbon move in cycles. The movement of carbon through the environment follows a predictable pattern called the **carbon cycle.**

As it cycles through the environment, carbon moves between organic and inorganic forms. For example, animals take in organic carbon when they eat other organisms and release inorganic gaseous CO_2 into the atmosphere as a by-product of cellular respiration. Similarly, when organisms die, decomposers

in the soil use the dead organic material for food and energy, releasing some of the carbon during respiration as CO_2.

Plants, photosynthetic bacteria, and algae take up CO_2 during photosynthesis and convert it into organic sugar molecules, thus reducing atmospheric CO_2 levels. Photosynthesis, respiration, and decomposition form a cycle that keeps carbon dioxide at a relatively stable level in the atmosphere. But human actions, like deforestation and burning fossil fuels, inject carbon dioxide that was not otherwise moving into the cycle (**INFOGRAPHIC 23.9**).

Carbon isn't the only element that cycles through ecosystems. Other elements, such

▶**CARBON CYCLE**
The movement of carbon atoms as they cycle between organic molecules and inorganic CO_2

INFOGRAPHIC 23.9 THE CARBON CYCLE

The carbon cycle involves the movement of carbon atoms as they cycle between organic molecules and inorganic CO_2. Natural processes such as photosynthesis, respiration, and decomposition are responsible for most carbon cycling. Since the 1700s, human activities, including burning fossil fuels and deforestation, have made significant contributions to the carbon cycle, primarily by increasing the amount of carbon in the form of CO_2 (measured in tons of carbon).

| CO_2 produced from natural processes **210** | + | CO_2 produced from human activity **8** | − | CO_2 removed through photosynthesis **214** | = | Net CO_2 released into the atmosphere each year **4** |

Atmospheric CO_2
6.5
1.5
120
122
Photosynthesis
Deforestation/land use
Oceanic photosynthesis and respiration 90 92
Storage in land plants
Respiration
Burning fossil fuels releases carbon.
Carbon enters soil via organic matter. 2000
Coal 2000
Oil 300
Gas
Fossil fuels lock carbon out of the carbon cycle.
Dead marine life becomes sediment. 100,000,000
Fossil carbon
All numbers are billions of tons of carbon.

as nitrogen, phosphorus, and sulfur also follow natural cycles (see **UP CLOSE: CHEMICAL CYCLES**). But it's the carbon cycle that is most relevant to the phenomenon of global warming.

For most of human history, the amount of carbon present in the atmosphere as carbon dioxide has remained fairly constant. But since the 1800s, with the rise of industry and the invention of the internal combustion engine, humans have begun to alter the carbon cycle,

adding increasing amounts of CO_2 to the atmosphere.

Before the industrial revolution, the carbon trapped in fossil fuels was not easily accessible, and therefore it wasn't cycling as part of the carbon cycle. But modern drilling and mining methods have unlocked the deep reserves of this ancient planetary energy. The CO_2 released when humans burn fossil fuels is the largest source of the carbon being added to

UP CLOSE CHEMICAL CYCLES: NITROGEN

Nitrogen is a critical component of the amino acids that make up proteins and the nucleotides of DNA and RNA. Nitrogen atoms cycle between different chemical and biochemical compounds as they move from organisms to the soil, water, and air and back to organisms. A variety of natural processes as well as some human activities contribute to the transformation and movement of nitrogen through the ecosystem.

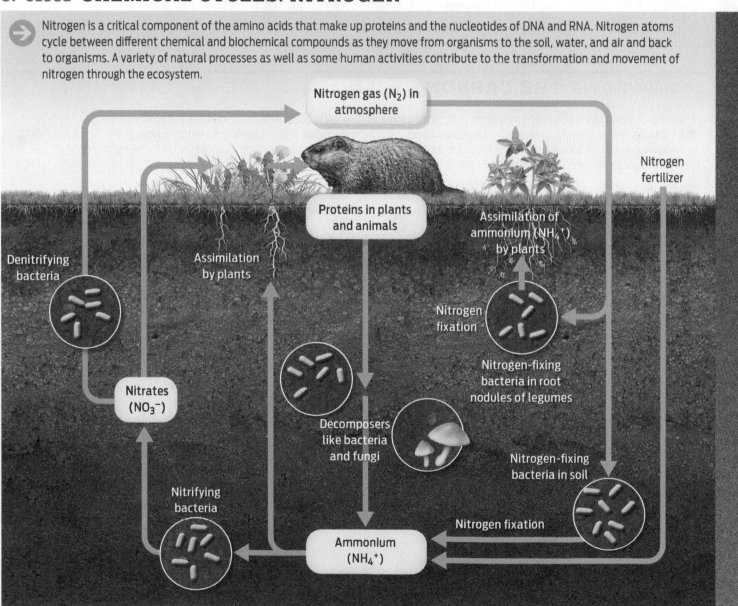

the atmosphere by humans and is a major contributor to the enhanced greenhouse effect.

How do scientists know that carbon dioxide levels are much higher now than in the past? There are two main sources of evidence. Air bubbles trapped in glacial ice from Greenland and Antarctica provide a historical measure of carbon dioxide in the atmosphere. Ice cores drilled at these sites provide data on very long term changes in CO_2 levels (and temperature), going back hundreds of thousands of years. These data show, for example, that levels of CO_2 have cycled in patterns that correlate with major ice ages. Since 1958, scientists have also directly measured CO_2 in the atmosphere—for example, at the Mauna Loa Research Station, which sits atop an inactive volcano in Hawaii. When combined, these two types of data show that atmospheric CO_2 has been rising steadily since the industrial revolution—increasing from

CHEMICAL CYCLES: PHOSPHORUS

Phosphorus is critical for the structure of DNA, RNA, ATP, and phospholipids. it is also critical for bones and teeth. Phosphorus cycles primarily through soil, water, and organisms. It is not a major component of gases in the atmosphere. Phosphorus is generally added to an ecosystem by the weathering of rocks, although human activities can also add phosphorus compounds to soil and water. When taken up by organisms, it is incorporated into organic molecules, then released back to the environment by decomposition.

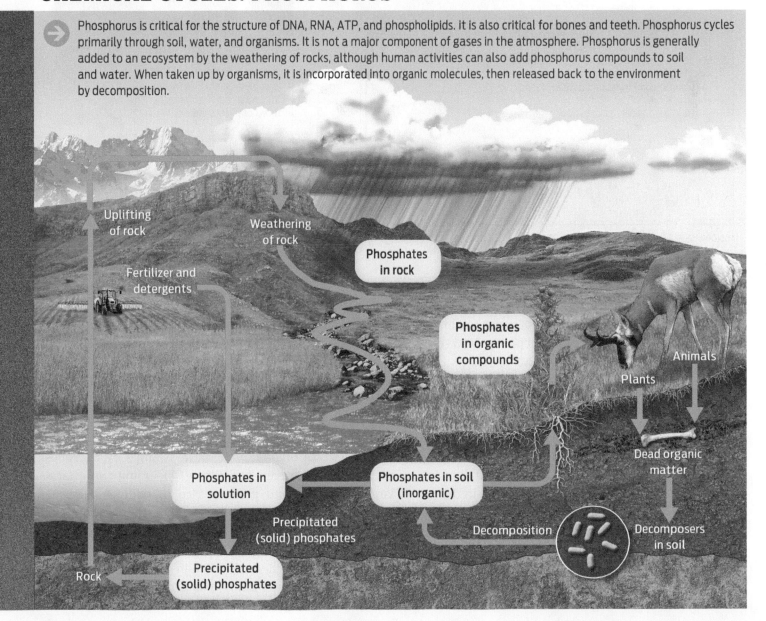

INFOGRAPHIC 23.10 MEASURING ATMOSPHERIC CARBON DIOXIDE LEVELS

Examining data from ancient air bubbles and present-day air measurements, scientists have recorded an approximately 40% increase in atmosphere CO_2 levels since 1800.

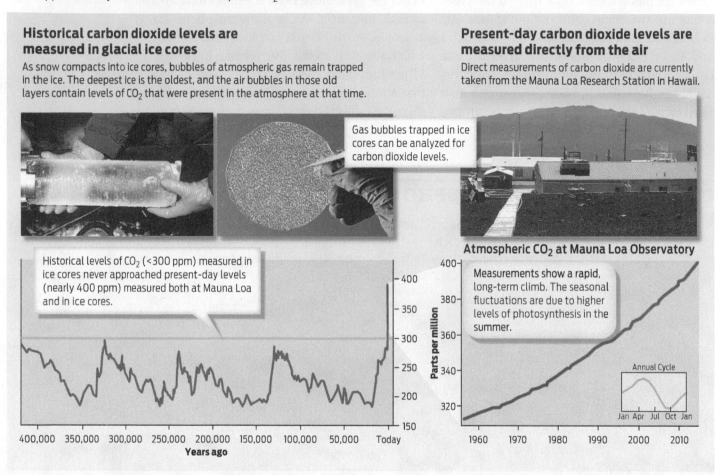

Historical carbon dioxide levels are measured in glacial ice cores

As snow compacts into ice cores, bubbles of atmospheric gas remain trapped in the ice. The deepest ice is the oldest, and the air bubbles in those old layers contain levels of CO_2 that were present in the atmosphere at that time.

Gas bubbles trapped in ice cores can be analyzed for carbon dioxide levels.

Historical levels of CO_2 (<300 ppm) measured in ice cores never approached present-day levels (nearly 400 ppm) measured both at Mauna Loa and in ice cores.

Present-day carbon dioxide levels are measured directly from the air

Direct measurements of carbon dioxide are currently taken from the Mauna Loa Research Station in Hawaii.

Atmospheric CO_2 at Mauna Loa Observatory

Measurements show a rapid, long-term climb. The seasonal fluctuations are due to higher levels of photosynthesis in the summer.

Annual Cycle

SOURCE: Historic levels of CO_2, NOAA National Climactic Data Center]; atmospheric CO_2 at Mauna Loa Observatory, Scripps Institution of Oceanography and NOAA Earth System Research Laboratory.

about 280 parts per million (ppm) in 1800 to nearly 400 ppm in 2013—or more than 40% **(INFOGRAPHIC 23.10).**

Activities that decrease the number of photosynthetic organisms also increase global CO_2 levels. Since photosynthesizers are the only consumers of carbon dioxide in the carbon cycle, removing them not only reduces the amount of carbon dioxide they might have consumed, but also—in the case of large trees and stable populations of algae—eliminates what are in essence long-term storage vessels of carbon. Human activities that reduce the number of photosynthetic organisms on the planet include large-scale slash-and-burn agriculture, development that leads to deforestation, and various forms of pollution. To-

gether, these activities contribute to our **carbon footprint,** a subset of our total ecological footprint, which is discussed in Chapter 24.

Though CO_2 is the major player in global warming, another important greenhouse gas is methane (CH_4). Methane is produced by natural processes, such as the decomposition of organic material in swamps by microbes. However, agriculture, including cattle farming and growing rice in paddies, now accounts for more than half the total methane being pumped into the atmosphere. One of the main sources of methane is the gas produced by archaea that live in the digestive systems of cattle. Emitted as flatulence, it adds an estimated 100 million tons of methane a year to the atmosphere. Although the atmospheric

▶CARBON FOOTPRINT

A measure of the total greenhouse gases produced by human activities.

concentration of methane is far less than that of CO_2, methane is particularly worrisome because it is 30 times more potent than CO_2 as a greenhouse gas.

With this steep rise in greenhouse gases have come steadily rising temperatures around the globe, with most of that warming occurring since the 1970s. Virtually all climate scientists agree that greenhouse gases emitted by human activities—primarily driving gasoline-powered cars and burning coal to generate electricity—have caused most of the global rise in temperature observed since the mid-20th

century. In 2007, the Intergovernmental Panel on Climate Change, composed of hundreds of scientists from around the world, concluded that the global rise in average yearly temperature since the mid-20th century was "very likely" anthropogenic—that is, caused by humans. (In statistics, "very likely" means a certainty of greater than 90%.) (**INFOGRAPHIC 23.11**).

NO TIME FOR FATALISM

The United States is among the world's biggest emitters of greenhouse gases, yet for political reasons it has been reluctant to make

INFOGRAPHIC 23.11 ANTHROPOGENIC PRODUCTION OF GREENHOUSE GASES

→ A variety of human activities are increasing the levels of greenhouse gases in the atmosphere. Power generation and transportation account for the largest proportion of greenhouse gas emission.

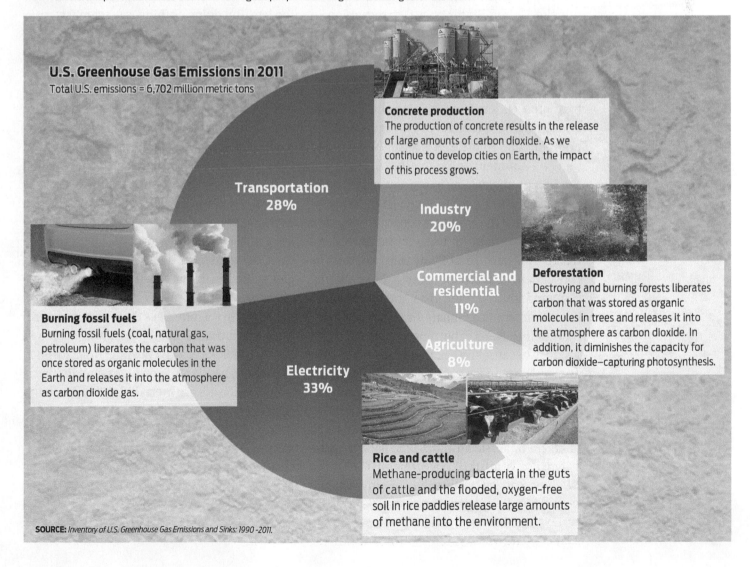

U.S. Greenhouse Gas Emissions in 2011
Total U.S. emissions = 6,702 million metric tons

Transportation 28%

Industry 20%

Commercial and residential 11%

Agriculture 8%

Electricity 33%

Concrete production
The production of concrete results in the release of large amounts of carbon dioxide. As we continue to develop cities on Earth, the impact of this process grows.

Deforestation
Destroying and burning forests liberates carbon that was stored as organic molecules in trees and releases it into the atmosphere as carbon dioxide. In addition, it diminishes the capacity for carbon dioxide–capturing photosynthesis.

Burning fossil fuels
Burning fossil fuels (coal, natural gas, petroleum) liberates the carbon that was once stored as organic molecules in the Earth and releases it into the atmosphere as carbon dioxide gas.

Rice and cattle
Methane-producing bacteria in the guts of cattle and the flooded, oxygen-free soil in rice paddies release large amounts of methane into the environment.

SOURCE: *Inventory of U.S. Greenhouse Gas Emissions and Sinks: 1990 -2011.*

> ❝ *We've got to get beyond the deer in the headlights stage and begin to think as conservation biologists about what we're going to do about this to help to mitigate the impact.* ❞
>
> — HECTOR GALBRAITH

significant reductions. It is one of the few countries that refused to ratify the Kyoto Protocol, a United Nations agreement adopted in 1997 that obligates endorsing countries to reduce carbon dioxide emissions. In 2010, the Obama administration signaled support for the Copenhagen Accord, which would commit the United States to a 17% reduction in greenhouse gases from 2005 levels by 2020, but carrying out those goals depends on Congress passing a climate bill. The Environmental Protection Agency (EPA) could decide to regulate CO_2 levels on its own; in 2007, the Supreme Court ruled that CO_2 qualifies as a pollutant and so the EPA is legally required to regulate it under the Clean Air Act. But so far, the EPA has been slow to act, perhaps fearing a backlash and legal challenges from industry.

Even if all the world's greenhouse gas emissions were turned off today like a faucet, we would still face decades of warming and its consequences because of past emissions—what climate scientists refer to as heat in the pipeline. Basically, the climate has not yet caught up with the effects of burning fossil fuels in past decades because the oceans are slower to heat up than the land (as any beachgoer knows). But eventually, the oceans will catch up, leading to more warming of the atmosphere.

This grim reality could easily lead some to take a fatalistic attitude, but that would be a dangerous mistake, says Hector Galbraith. "We've got to get beyond the deer in the head-lights stage and begin to think as conservation biologists about what we're going to do about this to help to mitigate the impact." These preventative measures are what he and other climate experts call "adaptation."

Adaptation will not be easy. For many species, like Vermont's maples, it may already be too late. But doing nothing, say scientists, risks turning a bad problem into a catastrophic one.

In concrete terms, adaptation means planning for the inevitable: more frequent droughts, heat waves, and severe storms, as well as a rise in sea level. Increasingly, cities and towns all across the country are taking steps to incorporate adaptation into urban and land-use policies—restricting new construction in flood zones, for example, and working to conserve water in drought-prone areas, as well as protecting wetlands.

But these efforts won't do much to stem the tide unless we deal with the underlying cause of rapid climate change—runaway emissions of greenhouse gases. And that means dealing with the practices that collectively produce more than 90% of greenhouse gases in the United States today: burning fossil fuels for electricity, heat, transportation, and industry.

"Fossil fuels are incredibly efficient sources of energy," says Serreze, of the University of Colorado. "We've built our whole infrastructure around that. But what we didn't realize is that it's a trap, and that's what we're coming to grips with now." ■

Summary

- Ecosystems are made up of the living and nonliving components of an environment, including the communities of organisms present and the physical and chemical environment with which they interact.

- Temperature is an important physical feature of any ecosystem and serves as a clock that cues many biological events, such as breeding, blooming, and hibernation.

- Biomes are large, geographically distinct ecosystems, defined by their characteristic plant life, which in turn is determined by temperature and levels of moisture.

- Climate change is a persistent pattern of change in Earth's climate. Global warming is an increase in Earth's average temperature over time.

- Climate change, and especially global warming, is having widespread effects on plant and animal life on the planet—altering seasonal life cycles, shifting ranges, and contributing to species loss by extinction.

- The greenhouse effect is a natural process by which heat radiated from Earth's surface is absorbed by heat-trapping gases in the atmosphere, maintaining a global temperature that can support life. Rising levels of greenhouse gases have led to the enhanced greenhouse effect.

- Elements cycle through ecosystems. The carbon cycle is the movement of carbon atoms through living and nonliving components of the environment by the biotic processes of photosynthesis, cellular respiration, and decomposition, as well as by long-term geological processes.

- Global warming results from an increase in the amount of carbon dioxide and other greenhouse gases in the atmosphere; the warming that has occurred over the past half century is due largely to human activities, such as burning fossil fuels and deforestation.

- Global warming is leading to the melting of sea ice in the Arctic, which is diminishing habitat for the organisms that rely on sea ice and creating a positive feedback loop for increased warming. Melting glaciers and ice caps on land are leading to rising sea levels.

- Methane is a significant greenhouse gas whose levels have increased because of human activities, including raising cattle and farming rice in paddies.

MORE TO EXPLORE

- NASA Goddard Institute for Space Studies http://www.giss.nasa.gov/

- National Snow and Ice Data Center http://nsidc.org/

- NOAA, Earth System Research Laboratory, Trends in Atmospheric Carbon Dioxide (animation) http://www.esrl.noaa.gov/gmd/ccgg/trends/history.html

- Ilya M. D. Maclean, I. M. D., and Wilson, R. J. (2011) Recent ecological responses to climate change support predictions of high extinction risk. *Proceedings of the National Academy of Sciences* 108:12337–12342.

- Oreskes, N., and Conway, E. M. (2010) *Merchants of Doubt: How a Handful of Scientists Obscured the Truth on Issues from Tobacco Smoke to Global Warming.* New York: Bloomsbury Press.

- Kolbert, E. (2006) *Field Notes from a Catastrophe: Man, Nature, and Climate Change.* New York: Bloomsbury Books.

CHAPTER 23 Test Your Knowledge

DRIVING QUESTION 1

What are ecosystems, and how are ecosystems being affected by climate change?

By answering the questions below and studying Infographics 23.1, 23.2, 23.3, and 23.6 and Up Close: Biomes, you should be able to generate an answer for the broader Driving Question above.

KNOW IT

1 Which of the following are parts of an ecosystem?

a. the plant life present in a given area

b. the animals living in a given area

c. the amount of annual rainfall in a given area

d. the soil chemistry in a given area

e. none of the above

f. all of the above

2 List several examples of species discussed in this chapter that have changed their geographic distributions or the timing of events in their life cycle as a result of global climate change.

3 In identifying a biome, for which of the characteristics below would it be most important to have data? (Select all that apply.)

a. monthly rainfall

b. temperatures throughout the year

c. plant life

d. animal life

e. size of the human population in the area

4 Which biome is characterized principally by evergreen trees?

5 Look at Up Close: Biomes. Where in North and South America do you find temperate forest? Tropical forest?

6 If global warming causes Arctic sea ice to melt, what will be the effect on sea levels in a low-lying region like Miami? If large parts of the Antarctic polar ice cap should melt, what would be the effect on sea level?

USE IT

7 Although trees may not be able to walk away from increasingly warm regions, evolutionary adaptations may allow trees to survive in warmer regions. Discuss each of the adaptations listed below and decide if it is likely to be helpful or harmful in a warming environment. (Think about water—water is taken up by the roots of plants, and lost through pores in the leaves; CO_2 levels— CO_2 is taken up by plants through pores in leaves, then used by leaves for photosynthesis; and the movement of other species, for example insects, in response to global warming.)

a. having smaller leaves

b. having a larger number of pores on each leaf

c. having thicker and waxier bark

8 What is a possible risk for humans if insects that carry pathogenic bacteria or viruses expand their range northward?

DRIVING QUESTION 2

What is the greenhouse effect, and what does it have to do with global warming?

By answering the questions below and studying Infographics 23.4, 23.5, 23.7, 23.8, and 23.11, you should be able to generate an answer for the broader Driving Question above.

KNOW IT

9 Which greenhouse gas is emitted every time you breathe out?

a. oxygen

b. carbon dioxide

c. methane

d. nitrogen

e. water vapor

10 Which of the following organisms contributes to reducing atmospheric CO_2 levels?

a. maple trees

b. most algae

c. polar bears

d. pearthrips

e. a and b

f. a, b, and d

11 Could we live in the absence of the greenhouse effect? Explain your answer.

USE IT

12 Explain how each of the following contributes to an elevation of levels of greenhouse gases.

   a. large-scale slash-and-burn agriculture

   b. driving gasoline-fueled cars

   c. producing cattle for beef and dairy products

   d. rice production

> ### DRIVING QUESTION 3

How do carbon and other chemicals cycle through ecosystems?

By answering the questions below and studying Infographic 23.9 and Up Close: Chemical Cycles, you should be able to generate an answer for the broader Driving Question above.

KNOW IT

14 Decomposers _____ CO_2 by the process of _____.

   a. emit; photosynthesis

   b. take up; photosynthesis

   c. emit; cellular respiration

   d. take up; cellular respiration

   e. store; cellular respiration

15 Atmospheric nitrogen is in the form of

   a. N_2.

   b. proteins.

   c. ammonium (NH_4^+).

   d. nitrates (NO_3^-).

   e. any of the above

13 Fill in the blanks in the image below.

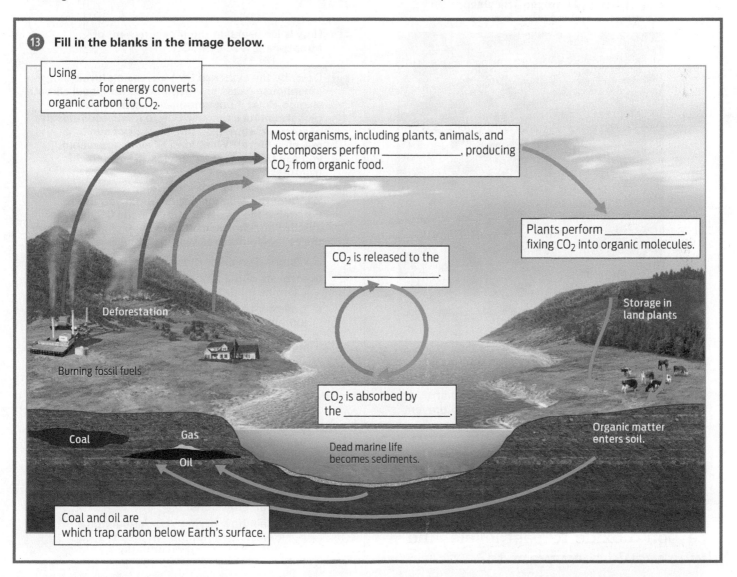

Using _____ _____ for energy converts organic carbon to CO_2.

Most organisms, including plants, animals, and decomposers perform _____, producing CO_2 from organic food.

Plants perform _____, fixing CO_2 into organic molecules.

CO_2 is released to the _____.

Storage in land plants

Deforestation

Burning fossil fuels

CO_2 is absorbed by the _____.

Coal

Gas

Oil

Dead marine life becomes sediments.

Organic matter enters soil.

Coal and oil are _____, which trap carbon below Earth's surface.

MINI CASE

16 For more than 30 years, a system of lakes in northwestern Ontario, Canada, called the Experimental Lakes, has been used to study the impacts of various pollutants in order better to understand how to maintain healthy freshwater ecosystems. A challenge in many aquatic ecosystems is eutrophication—nutrient enrichment that leads to overgrowth of algae. When the algae die, the decomposers use so much oxygen as they feast on the dead algae that other organisms in the ecosystem cannot survive. Eutrophication led to loss of many fish from Lake Ontario in the 1960s and 1970s. In one of the Experimental Lakes, two basins of the lake were separated by a plastic sheet. The basin shown at the bottom of the photo below received nitrogen, phosphorus, and carbon. The basin shown at the top of the photo received only carbon and nitrogen. The green scum covering the water surface is algae.

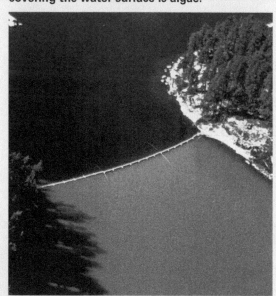

a. What nutrient is most important in the eutrophication process?

b. What are human-derived sources of this nutrient?

c. Do some online research to see what measures were taken in Ontario to prevent eutrophication.

DRIVING QUESTION 4

How are scientists able to compare present-day levels of atmospheric carbon dioxide to past levels, and why would they want to?

By answering the questions below and studying Infographics 23.5, 23.9, 23.10, and 23.11, you should be able to generate an answer for the broader Driving Question above.

KNOW IT

17 Fossil fuels are most immediately derived from

a. organic molecules.

b. CO_2.

c. methane.

d. melting ice caps.

e. photosynthesis.

18 Name at least two human activities that increase CO_2 levels in the atmosphere and two natural processes that contribute CO_2 to the atmosphere.

USE IT

19 How is ice useful in the measurement of atmospheric levels of CO_2?

20 Describe the evidence that increasing levels of greenhouse gases are responsible for global climate change. What if someone suggested to you that global climate change is due to increased intensity of solar radiation (i.e., the amount of sunlight reaching Earth)? What kind of evidence would you ask this person to provide to support this hypothesis?

21 Which of the following data would you use to determine the levels of atmospheric CO_2 in 1750? Justify your choice, including an explanation of why the other alternatives would not be as effective.

a. historical weather records of daily temperatures

b. archives of the Manua Loa Observatory (to examine records from 1750)

c. tree-ring analysis (to look for evidence of extreme fires)

d. ice cores from ice formed in 1750

BRING IT HOME

22 Visit an online carbon footprint or carbon emissions calculator (for example, http://www.epa.gov/climatechange/ghgemissions/ind-calculator.html) and calculate your total carbon emissions.

a. What is your largest source of emissions?

b. What steps can you take to decrease your carbon emissions?

c. Explain how drying your laundry on a clothesline rather than in the dryer can decrease your carbon emissions.

INTERPRETING DATA

23 A 2010 study compared the amount of CO_2 emitted when locally grown broccoli was delivered to Virginia Tech University with the amount emitted when broccoli grown in California was delivered to Virginia Tech. The California broccoli was delivered in shipments containing 768 lbs of broccoli, in a tractor-trailer that traveled 2,786 miles. Tractor-trailer fuel efficiency is 5 miles per gallon, and 20 lbs of CO_2 are released per gallon of fuel burned. The local broccoli was delivered in shipments of 587 lbs of broccoli in a cargo van that traveled 19.1 miles. Cargo van fuel efficiency is 16 miles per gallon, and 20 lbs of CO_2 are released per gallon of fuel burned.

a. Complete the table below to determine the CO_2 emissions associated with delivering 1 lb of local and 1 lb of nonlocal broccoli.

b. Is locally sourced fresh broccoli a year-round option at Virginia Tech?

c. Do some online research to determine approximately what proportion of CO_2 emissions are associated with food delivery vs. food production.

SOURCE OF BROCCOLI	MILES PER SHIPMENT	GALLONS OF FUEL BURNED PER SHIPMENT	CO_2 RELEASED PER SHIPMENT (LBS)	CO_2 RELEASED PER LB OF BROCCOLI DELIVERED
Nonlocal	2786			
Local	19.1			

SOURCE: Schultz, J. and Clark, S. (2010) Foodprint Comparison of Local vs Nonlocal Produce.
http://www.blacksburgfarmersmarket.com/docs/Schultz_Foodprint_Comparison_of_Local_vs_Nonlocal.pdf (accessed 5/4/2013)

Progress OR

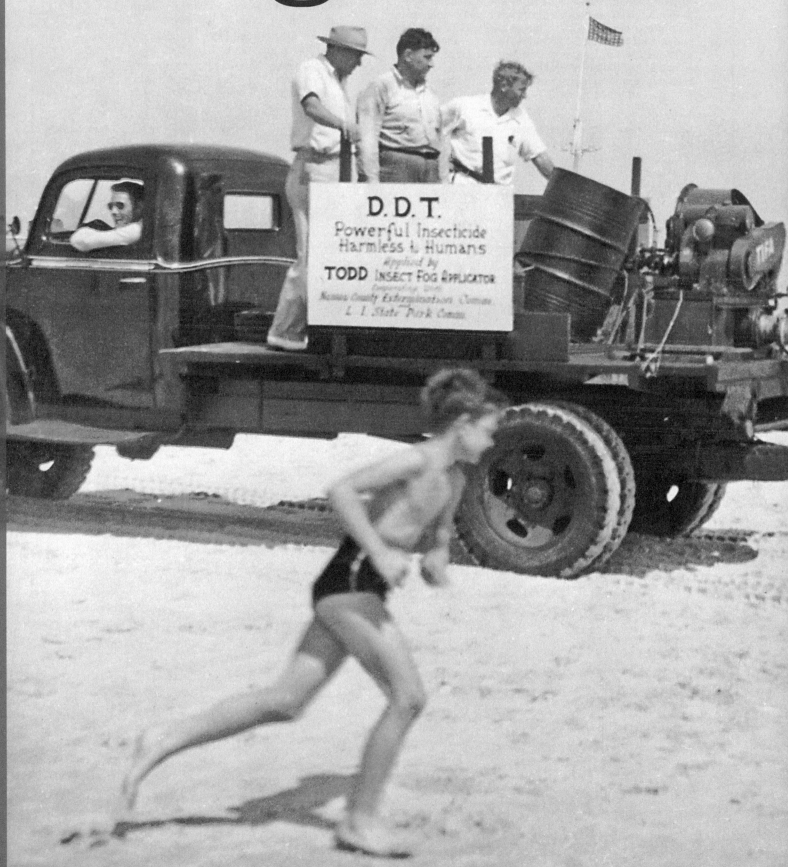

D.D.T.
Powerful Insecticide
Harmless to Humans
Applied by
TODD INSECT FOG APPLICATOR
Cooperating With
Nassau County Extermination Comm.
L. I. State Park Comm.

Poison?

6

Rachel Carson, pesticides, and the birth of the environmental movement

IN 1958, THE BIOLOGIST AND SCIENCE WRITER RACHEL CARSON RECEIVED a disturbing letter from a friend in Massachusetts. The letter described an event that had recently taken place on the friend's property near Cape Cod: an airplane had sprayed a thick cloud of a pesticide called DDT as part of a coordinated campaign to eradicate mosquitoes. In the days following the spraying, the friend noticed many dead songbirds in the area. The birds had clearly suffered: "Their bills were gaping open, and their splayed claws were drawn up to their breasts in agony." Could anything be done to stop these aerial sprayings, the friend wanted to know?

At the time she received the letter, Carson was a well-known science writer who had written some widely popular books about the sea. She cared deeply about nature and was horrified by her friend's report—especially since it wasn't the first time she'd heard about DDT toxicity. Carson had been concerned for a number of years about the effects DDT might be having on beneficial insects, birds, and fish, ever since she had worked as a marine biologist with the U.S. Fish and Wildlife Service, which had conducted studies on the pesticide in 1945. But the letter from her friend was a tipping point.

Carson decided that someone should research and write an article about the dangers of pesticides. She tried at first to persuade other writers to take on the topic. When no one would, she realized she had to tackle it herself. Four years later, the outcome

▸▸ DRIVING QUESTIONS

1. How can a policy designed to help society cause great harm?
2. What is biomagnification, and how does it occur?
3. What properties of DDT permit it to negatively affect organisms at a variety of levels in a food chain?

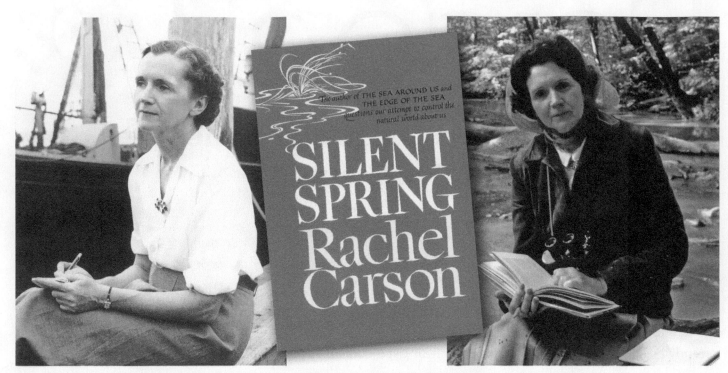

Rachel Carson was a best-selling author and marine biologist before she wrote Silent Spring.

was *Silent Spring*, a book that sparked a national conversation about pesticides and ushered in a new way of thinking about human impacts on the environment.

In the book's famous opening chapter, "A Fable for Tomorrow," Carson asked readers to imagine a town "in the heart of America" where "a strange blight" had crept over the land. The birds that had once greeted the coming of spring with a chorus of song now lay sick, dying, and silent. "It was a spring without voices," she wrote. What had caused this strange blight? "No witchcraft, no enemy action had silenced the rebirth of new life in this stricken world," Carson wrote. "The people had done it themselves."

I WANT MY DDT

DDT, a synthetic chlorinated hydrocarbon, was discovered to be a potent insecticide in 1939 by the Swiss chemist Paul Müller. DDT poisons the nervous system of insects and other animals, and was first widely used during World War II to combat insect-carried diseases, such as typhus and malaria, among U.S. soldiers. (Typhus is carried by lice, which infested many soldiers living in cramped quarters; malaria is carried by mosquitoes, such as those that live in the islands of the South Pacific.) As a public health measure, DDT succeeded marvelously, saving countless lives during the war. For his work Müller was awarded a Nobel prize in 1948.

Paul Muller, a Swiss chemist, showed that DDT was an effective insecticide. As a result of his work, DDT became a widely used pesticide.

The chemical formula for the synthetic chemical DDT

After the war, DDT became available to the public as a commercial pesticide and was widely used by farmers to protect their crops and by public health officials to contain insect-borne disease. The chemical companies that manufactured the potent bug-killer advertised it as an aid to domestic comfort and tranquility. Photos and newsreels from the era document mass spraying of suburban neighborhoods as children played happily in the chemical clouds (INFOGRAPHIC M6.1).

But by the late 1950s, a number of scientists and citizens had become concerned that indiscriminate spraying of pesticides like DDT was doing more harm than good. Efforts by the U.S. Department of Agriculture (USDA) to eliminate gypsy moths and fire ants in a number of states had reportedly killed off more than just these target pests—scores of other insects, fish, and birds had also been wiped out. By 1958, when Carson received the letter from her friend, there were several court cases directed against the USDA in an effort to stop the indiscriminate spraying.

Carson began approaching scientists around the country, including many of her

INFOGRAPHIC M6.1 WIDESPREAD USE OF DDT TO KILL MOSQUITOES AND LICE

→ As an insect neurotoxin, DDT was used to combat a number of diseases associated with insects, including those borne by lice and mosquitoes.

DDT was applied directly to human hair, skin, and clothing to eliminate lice that spread typhus, a microbial disease.

DDT was sprayed freely throughout neighborhoods to eradicate mosquitoes and other insects that carry human pathogens.

INFOGRAPHIC M6.2 UNINTENDED CONSEQUENCES OF USING DDT

→ While DDT was effective against its insect targets, it also made its way into the food chain of many organisms, with unintended consequences.

1. Trees were sprayed with DDT to kill the beetle that spread the fungus that causes Dutch elm disease.

2. Leaves coated in DDT fell to the ground and were eaten by earthworms.

3. Birds, like the American Robin, that ate earthworms died from the accumulated DDT in their systems.

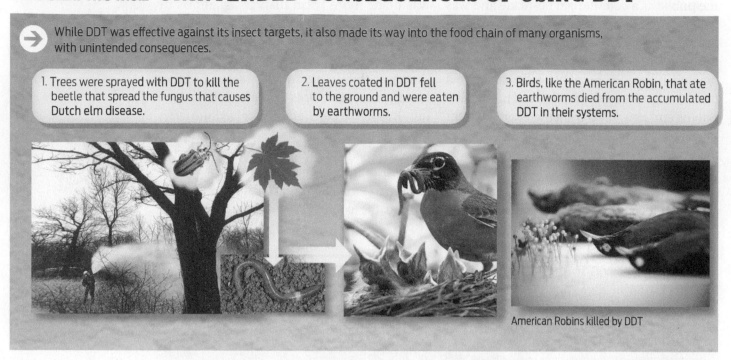

American Robins killed by DDT

former colleagues in the Fish and Wildlife Service, for information about pesticide use and toxicity. "The more I learned about the use of pesticides, the more appalled I became," she later said. Over the next 4 years, she interviewed more scientists, scoured the research literature, and culled through newspaper reports. The result was *Silent Spring*, a comprehensive treatise with 50 pages of footnotes.

Carson's book was filled with example after depressing example of the destruction that synthetic pesticides were wreaking on natural populations of animals, most famously birds. For example, in the mid-1950s, government officials in states across the Midwest attempted to deal with growing problem of Dutch elm disease, which was killing off the popular trees, by spraying the trees with DDT. Dutch elm disease is caused by a fungus, but it is spread from tree to tree by beetles that feed on the trees' leaves. To combat the beetles, scientists sprayed the trees with DDT. In the autumn, leaves coated in DDT fell to the ground, where they were eaten and decomposed by earthworms, which took up the chemical and accumulated it inside their bodies. In the spring, robins mi-

grated to these areas, fed voraciously on the earthworms, and were poisoned by ingesting high amounts of DDT (Infographic M6.2).

Carson presented evidence that aquatic ecosystems are also affected by DDT. Since 1945, coastal waters around the United States had been sprayed with DDT to combat the salt-marsh mosquito. DDT entered the water supply, where it and its breakdown product DDE were taken up into the bodies of small fish and crabs. Larger fish would eat the smaller fish, thus taking in higher quantities of stored DDT and DDE, until, finally, the larger fish were eaten by top predators like eagles, which ingested the highest quantities of all. Because of the interactions of organisms at different trophic levels in a food chain (see Chapter 22), organisms at the highest trophic levels can have high concentrations of DDT in their tissues even if they were not directly exposed to DDT. The process by which environmental toxins accumulate as they move up the food chain is called biomagnification (INFOGRAPHIC M6.3).

The accumulated DDE impaired reproduction in birds of prey: their eggshells became so thin that mothers would crush their eggs when they sat on their nests to incubate them. From numbers in the hundreds, the eagle population in the United States plummeted in the 1950s and appeared on the verge of extinction—an ominous fate for our national emblem.

Besides the threat to wild animals, Carson also pointed to possible dangers to humans. Though DDT was deemed safe for use as an insecticide on humans, there were concerns about long-term effects. "We have to remember that children born today are exposed to these chemicals from birth, perhaps even before birth," Carson said in a 1962 interview on CBS News. "Now what is going to happen to them in adult life as a result of that exposure? We simply don't know." These uncertainties were made more unnerving when DDT was shown to persist in the environment for many years, long after its application. We now know that DDT persists for at least 15 years in soil and much longer in aquatic environments, possibly for hundreds of years. (The EPA currently classifies DDT as a "probable carcinogen"; DDT bioaccumulates in human tissues and in breast milk, and has been associated with neurological disorders in places where it has been used heavily to treat malaria.)

To Carson, the most galling and stupefying aspect of all these stories was the blind faith people put in pesticides—using them even when they had not been tested and when other methods had proved effective. As a means of controlling Dutch elm disease in the Midwest, for example, DDT spraying was a colossal failure; it did not save the trees, and may have actually made the problem worse because it had the unintended effect of killing birds that serve as a natural form of pest control. Yet effective means of controlling the disease—through removal and burning of contaminated wood—had been used successfully for decades in other parts of the country, for example in New York.

How to account for these lapses of scientific judgment? A big part of the problem, Carson argued, was an unquestioned faith in the power of experts to control nature, bend it to

INFOGRAPHIC M6.3 BIOMAGNIFICATION

Chemicals such as DDT are retained in the bodies of organisms that take them up. When those organisms are eaten by others, the concentration of chemicals increases at each level of the food chain.

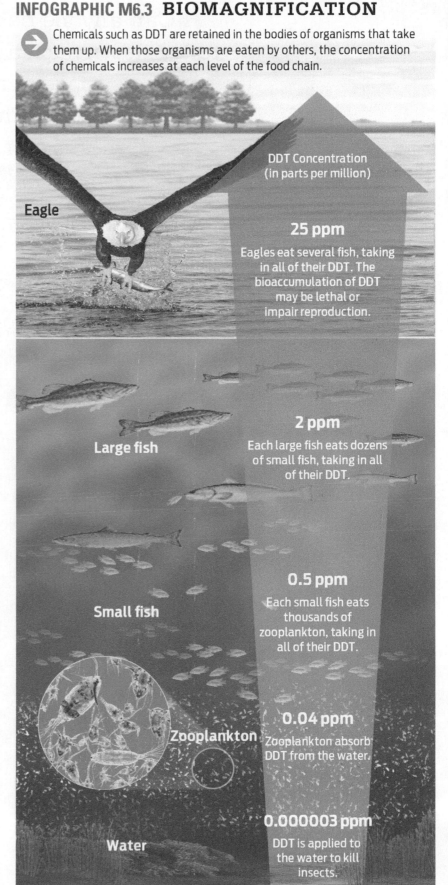

> **"**Can anyone believe it is possible to lay down such a barrage of poisons on the surface of the earth without making it unfit for all life?**"**
>
> —RACHEL CARSON

their whim with technology. In other words, it was as much a mind-set as a specific technology that was the source of the trouble.

"Who has made the decision that sets in motion these chains of poisonings, this ever-widening wave of death that spreads out, like ripples when a pebble is dropped into a still pond?" she asked in her book. Carson minced

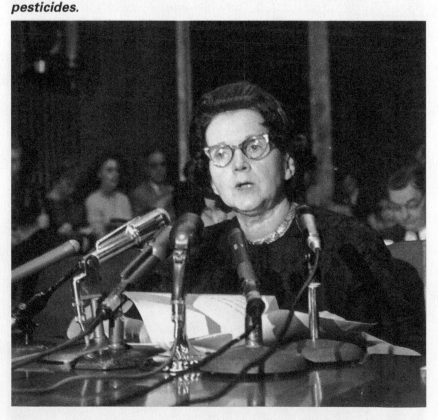

Carson testified before Congress in 1963 on the dangers of pesticides.

no words with her answer: it was shortsighted individuals in positions of power whose capacity for destruction was not matched by an informed awareness of the associated costs.

FROM SILENT SPRING TO NOISY SUMMER

Silent Spring was serialized in *The New Yorker* in the spring of 1962 and published as a book by Houghton Mifflin that fall. It caused an immediate sensation, as people around the country began to question the safety of these omnipresent and unseen chemicals. "'Silent Spring' Is Now Noisy Summer," the *New York Times* headlined its story documenting the furor that erupted in the wake of publication. Though Carson found many supporters and admirers, among them President John F. Kennedy, she also found herself attacked by angry representatives of the chemical industry.

Velsicol Chemical, a maker of DDT and other pesticides, threatened to sue both Houghton Mifflin and *The New Yorker*. A lawyer for the company accused Carson herself of being a communist sympathizer who wanted to shrink the American food supply. Monsanto, another pesticide producer, published and distributed 5,000 copies of a brochure parodying Carson's book titled "The Desolate Year," which argued that without pesticides to help agriculture, food supplies would plummet and millions around the world would suffer

History of Environmental Legislation

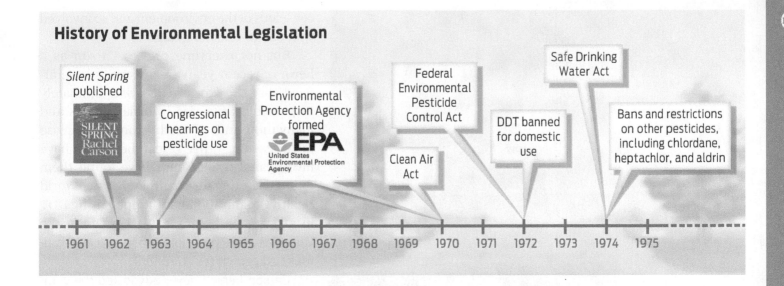

Silent Spring published

Congressional hearings on pesticide use

Environmental Protection Agency formed

EPA
United States Environmental Protection Agency

Clean Air Act

Federal Environmental Pesticide Control Act

DDT banned for domestic use

Safe Drinking Water Act

Bans and restrictions on other pesticides, including chlordane, heptachlor, and aldrin

1961 1962 1963 1964 1965 1966 1967 1968 1969 1970 1971 1972 1973 1974 1975

from hunger and starvation. "If man were to faithfully follow the teachings of Miss Carson," scoffed an executive of American Cyanamid, "we would return to the Dark Ages, and the insects and diseases and vermin would once again inherit the earth."

Some of the criticism Carson faced was clearly because she was a woman. Her credentials as a scientist were routinely attacked (although she had an M.A. in zoology from Johns Hopkins), and she was often referred to as "hysterical." A former U.S. Secretary of Agriculture wondered aloud to the press "why a spinster with no children was so interested in genetics." A physician, writing in a medical journal, stated that reading *Silent Spring* "kept reminding me of trying to win an argument with a woman. It can't be done."

But Carson had done her homework, and her view ultimately carried the day. A commission convened by President Kennedy to investigate the claims made by Carson in her book found them to be sound. Soon thereafter, in 1963, congressional hearings were held at which Carson testified. She argued that a commission should be established to review pesticide issues and make decisions on the basis of broad public interest rather than the profit motives of a few.

Silent Spring is universally credited as the spark that ignited the modern environmental movement. In less than 10 years, what began as one woman's fight grew to enlist an army of activists, including many college students, who together changed the national conversation. The first national Earth Day celebration was held on April 22, 1970. Later that year, President Richard Nixon created the Environmental Protection Agency (EPA) and gave it the authority to set safe levels for chemicals. Shortly thereafter came several pieces of legislation including the Clean Air Act, the Federal Environmental Pesticide Control Act, and the Safe Drinking Water Act. The EPA banned DDT for domestic use in 1972, except for cases of public health (e.g., to prevent malaria). In 1974, it further banned or severely restricted the other pesticides that Carson had written about in her book: chlordane, heptachlor, dieldrin, aldrin, and endrin.

Carson would surely have been gratified to see the enactment of these milestones, but she didn't live long enough to witness them. She died in 1964, at the age of 56, of breast cancer, less than 2 years after *Silent Spring* was published. Though she had been sick for much of the period when she wrote her book, she never mentioned her illness publicly, fearing

The beliefs of the time: An ad from **Life** *magazine in 1948 featured model Kay Heffernon at Jones Beach, NY, demonstrating that DDT was safe.*

that her analysis of the human dangers of pesticides—cancer among them—might be labeled as biased. But her legacy lives on in the many people who credit her with changing their worldview and relationship to nature.

"For me personally," former Vice President Al Gore wrote in an introduction to the 1992 edition of her book, "*Silent Spring* had a profound impact. . . . Indeed, Rachel Carson was one of the reasons that I became so conscious of the environment and so involved with environmental issues."

But not everyone praises Carson as a hero. In recent years, many critics have argued that the ban on DDT was ultimately responsible for a rise in malaria rates and many thousands of deaths from this disease in Africa. But other researchers who have investigated these claims have found them dubious, noting for example that the World Health Organization had stopped using DDT in Africa even before it was banned in the United States, principally because mosquitoes had grown resistant to it.

Carson herself never advocated the complete elimination of chemical pesticides. Rather, she opposed their indiscriminate use without conducting studies of their ecological effects. "It is not my contention that chemical insecticides must never be used. I do contend that we have put poisonous and biologically potent chemicals indiscriminately into the hands of persons largely or wholly ignorant of their potentials for harm," she wrote.

It was the height of human arrogance and, ultimately, of stupidity, she argued, to think that we could inflict great damage on one part of the environment without harming others. The issues she raised are as applicable today as they were then, foreshadowing our current debates over greenhouse gases and endocrine disruptors, as well as a new crop of synthetic pesticides (see Chapter 22).

"Can anyone believe it is possible to lay down such a barrage of poisons on the surface of the earth without making it unfit for all life?" she wrote. ■

519

MILESTONES IN BIOLOGY: PROGRESS OR POISON?

M6

MORE TO EXPLORE

▸ Lear, L. (1997) *Rachel Carson: Witness for Nature.* New York: Henry Holt.

▸ Griswold, E. (2012) How *'Silent Spring'* ignited the environmental movement. *New York Times Magazine.* September 21. http://www.nytimes.com/2012/09/23/magazine/how-silent-spring-ignited-the-environmental-movement.html

▸ Environment & Society: *Rachel Carson's Silent Spring, a Book That Changed the World.* Online Exhibit. http://www.environmentandsociety.org/exhibitions/silent-spring/overview

▸ CBS News. (2007) The Price of Progress. http://www.cbsnews.com/video/watch/?id=2714989n

▸ Oreskes, N., and Conway E. (2010) *Merchants of Doubt: How a Handful of Scientists Obscured the Truth on Issues from Tobacco Smoke to Global Warming.* New York: Bloomsbury.

MILESTONES IN BIOLOGY 6 Test Your Knowledge

1 On the origins of DDT:

a. When was the chemical DDT first widely used, and what was its intended purpose?

b. How effective was it for that purpose?

2 On the mechanism of DDT:

a. How does DDT kill insects?

b. How does DDT harm top predator birds?

3 What is biomagnification?

4 PCBs (polychlorinated biphenyls), a type of chlorinated hydrocarbon, were used for a variety of purposes (including electrical insulation), until their use was banned in 1979. A 2000 survey of top predator fish in the Great Lakes showed that the concentration of PCBs in these fish ranged from 0.8 to 1.6 ppm. The wildlife protection value (the concentration that should not be exceeded in order to protect the safety of wildlife) is 0.16 ppm. How could there be such high levels of PCBs in top predator fish 21 years after PCBs were banned?

5 In 2013, a group of beekeepers launched a lawsuit against the EPA concerning the use of neonicotinoid pesticides and possible unintended impacts on honey bees (see Chapter 22). From what you have read here, what kinds of testing would you want the EPA to require before approving a pesticide applied widely to crops such as corn or soybeans?

The Makings of
A GREEN

One Kansas town reinvents itself sustainable

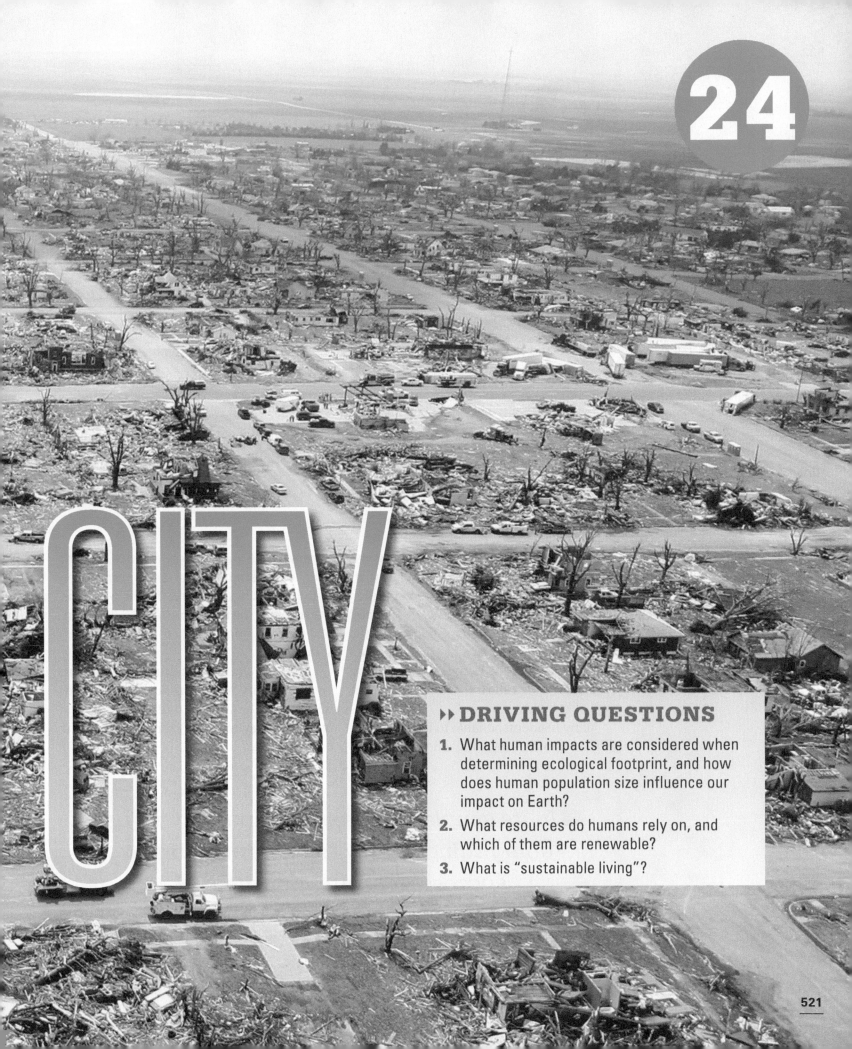

CITY

▶▶ DRIVING QUESTIONS

1. What human impacts are considered when determining ecological footprint, and how does human population size influence our impact on Earth?

2. What resources do humans rely on, and which of them are renewable?

3. What is "sustainable living"?

O N THE NIGHT OF MAY 4, 2007, AT 9:50 P.M., A MAMMOTH EF-5 TORNADO roared through the small town of Greensburg, Kansas. Nearly 2 miles wide—wider than the town itself—with winds topping 200 mph, the twister was one of the largest ever recorded. Once the warning sirens sounded, residents had about 20 minutes to retreat to their storm cellars before the raging monster was upon them.

"The sound was like a jet engine going right over us, about to take off," wrote Megan Gardiner, then a high school senior, in a personal account of the storm published by the *Wichita Eagle*. She and her family huddled in their basement, ears popping, listening helplessly as first windows exploded and then nails were sucked out of walls with a rhythmic high-pitched screech. "Just hearing the house rip into shreds was horrible."

The ordeal lasted 10 long minutes. When the ferocious funnel had finished churning, 11 people had died. People emerged from their storm cellars to find the homes above them gone. Fully 95% of the buildings in town—961 homes and 110 businesses—were leveled. The town of Greensburg had been wiped off the map.

Before the tornado hit, Greensburg was a struggling farming town of about 1,300, its population shrinking every year as people left the town for brighter horizons. (A reporter for the *New York Times* once quipped that the town's biggest export was its young people.) After the disaster, the people of Greensburg were faced with a stark choice: abandon the town or rebuild?

Many residents, unable to face the prospect of another tornado, chose to leave. Others took a wait-and-see attitude. But for town administrators, who had for years been searching for ways to breathe life back into Greensburg, the tornado offered them a rare and unlikely opportunity to start over from scratch.

"We can paint this community any way we want," City Administrator Steve Hewitt, whose own home was destroyed, told the mayor the day after the storm. "It's a blank canvas."

> **"In its simplest sense, sustainability is just doing things today to ensure a vibrant successful future for others."**
>
> — STACEY SWEARINGEN WHITE

The EF-5 tornado that hit Greensburg was wider than the town itself. At right, Doppler radar shows the position and severity of the storm.

GOING GREEN

An ambitious young professional with a wife and son, Hewitt was one of the first to start thinking about rebuilding in a fundamentally different way. He saw in the catastrophe an opportunity to fix problems that had plagued Greensburg before the storm—things like crumbling infrastructure, a lackluster economy and jobs base, and dim prospects for future growth.

In the days immediately following the storm, he met with the mayor and Kansas Governor Kathleen Sebelius to discuss ideas for improving on the old ways. Sebelius listened patiently to what Hewitt was saying about rebuilding smarter, with an eye toward economic growth, and then summarized what she heard in one simple statement: "Steve, you're talking about green."

Hewitt hadn't immediately connected his ideas to a green agenda, but the conceptual beauty was not lost on him. This was *Greensburg*, after all. "You're right," he said. "That is what we're talking about." From there, the idea took off.

A small town in Kansas might seem like a strange place for an environmental movement to emerge. After all, this is a politically conservative area, in a staunchly red state, where climate change is still viewed skeptically by a large proportion of the population. "Green, I thought that was just for people who hug the trees," said one resident at an early town meeting called to discuss the idea of rebuilding green. But Greensburg residents—even those who did not see themselves as environmentalists—eventually warmed to the idea of putting the green in Greensburg. What sold them was the concept of living more sustainably.

Sustainability means different things to different people, and it can be a difficult

▶**SUSTAINABILITY**
The use of Earth's resources in a way that will not permanently destroy or deplete them; living within the limits of Earth's biocapacity.

INFOGRAPHIC 24.1 THE HUMAN ECOLOGICAL FOOTPRINT

→ How much of Earth's resources does your lifestyle require? The ecological footprint is a measure of our demand on nature. Ecologists use 5,400 different measures gathered from government agencies and scientific publications to calculate a footprint, a measure of how much biologically productive land and water area (cropland, forests, grazing lands, fishing area, and built-up land) a human population requires to produce the resources it consumes and to absorb the waste it produces.

How do you live?

Energy
Do you drive a car?
Use a computer?
Cook your food?

Buildings
Do you live in
a house?
Work in an office?
Eat at restaurants?

Timber and Paper
Do you read
magazines?
Print your
assignments?
Use a textbook?

Food and Fiber
Do you eat three
meals a day?
Wear clothes?
Have furniture?

Seafood
Do you eat fish?
Take a fish oil
supplement?
Like shrimp cocktail?

Stored Carbon
Consumption of
fossil fuels derived
from ancient
organisms

Built-Up Land
Once
biologically
productive, now is
space we live on

Forest Land
Trees cut for
consumer products

**Grazing and
Crop Land**
Require a lot of
land, water, and
other resources like
fertilizer

Fishing Grounds
Depletion of wild
stock or resources
used in farming

What natural resources does it require?

thing to define and measure. "In its simplest sense, sustainability is just doing things today to ensure a vibrant successful future for others," says Stacey Swearingen White, a professor of environmental planning at the University of Kansas.

To ecologists, sustainability has to do with the way humans use Earth's resources. We require a variety of **natural resources**

to live: farmland to grow crops and raise cattle, gasoline to power cars, oxygen to fill our lungs, to name just a few. In addition to providing us with natural resources, Earth acts like a sponge, absorbing our wastes. The carbon dioxide we emit, for example, is soaked up and stored by plants; the garbage we produce is decomposed in landfills.

The collective demand we place on these resources is referred to as our **ecological footprint.** This measure is a tally of the land and water area needed to supply a given population with resources and to absorb its wastes. By quantifying the amount of biologically productive area it takes to sustain our lifestyles, the ecological footprint puts a number on our environmental impact (**INFOGRAPHIC 24.1**).

The human ecological footprint is often compared with Earth's **biocapacity**—its ability to sustain human demand given its available natural resources and its ability to absorb waste. If we think of the footprint as our demand on Earth, the biocapacity is the amount of supplies that Earth can produce to meet that demand. The question is: does Earth have a sufficient supply of resources to meet our demands?

Earth's biocapacity is measured in units called **global hectares (gha).** One global hectare represents the biological productivity—both the resource-providing and waste-absorbing capacity—of an average hectare of Earth's surface (including cropland, pasture, forest, and fisheries). A hectare is 10,000 square meters—about the size of a soccer field. In 2008 (the last year for which data are available), Earth's total biocapacity was 12 billion gha, or 1.8 gha per person. That same year, the average ecological footprint was 2.7 gha per person per year. In other words, it takes more than two and a half soccer fields of land and water area to support one average human for 1 year. Since the average human ecological footprint is greater than Earth's total biocapacity, our demand on Earth's resources is currently outstripping its supply (by almost 1 gha per person).

How is this possible when there is only one Earth? You can think of Earth as being like an interest-earning bank account: it has a certain amount of natural capital that is continually renewing itself. Just as it is possible to overdraw a bank account faster than the interest

▸ECOLOGICAL FOOTPRINT
A measure of how much land and water area is required to supply the resources an individual or a population consumes and to absorb the wastes it produces.

▸BIOCAPACITY
The amount of Earth's biologically productive area—cropland, pasture, forest, and fisheries—that is available to provide resources and absorb wastes to support life.

▸GLOBAL HECTARE (GHA)
A unit of measurement representing the biological productivity (both resource-providing and waste-absorbing capacity) of an average hectare of Earth.

The Greensburg grain elevator was one of the few buildings left standing after the tornado.

UNIT 4 • **WHAT MAKES UP OUR ENVIRONMENT? ECOLOGY**

24

INFOGRAPHIC 24.2 THE HUMAN ECOLOGICAL FOOTPRINT IS GREATER THAN EARTH'S BIOCAPACITY

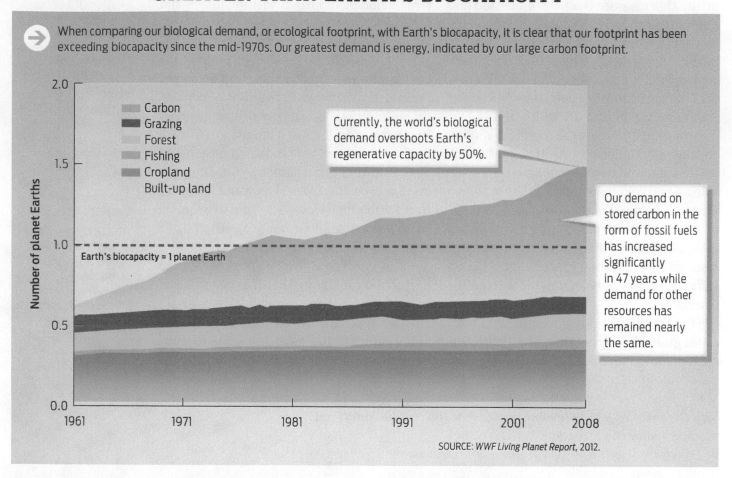

When comparing our biological demand, or ecological footprint, with Earth's biocapacity, it is clear that our footprint has been exceeding biocapacity since the mid-1970s. Our greatest demand is energy, indicated by our large carbon footprint.

Legend:
- Carbon
- Grazing
- Forest
- Fishing
- Cropland
- Built-up land

Y-axis: Number of planet Earths

Earth's biocapacity = 1 planet Earth

Currently, the world's biological demand overshoots Earth's regenerative capacity by 50%.

Our demand on stored carbon in the form of fossil fuels has increased significantly in 47 years while demand for other resources has remained nearly the same.

SOURCE: *WWF Living Planet Report*, 2012.

can regenerate the principal, our demands on Earth's resources are diminishing the natural capital faster than it can be renewed. We're living in the red (**INFOGRAPHIC 24.2**).

These footprint numbers are averages, tallied across the whole globe. But, of course, not everyone on the planet uses resources to the same extent. An average American, for instance, has an ecological footprint of about 7.2 global hectares, while the average Haitian uses just 0.60 global hectares. These are per capita figures, averages for one resident in each of those countries. It's also possible to calculate the ecological footprint of a whole country. China, for example, has a per capita footprint of 2.1 global hectares, much smaller than that

of the United States, but because China's population is so large, its total footprint is larger than that of the United States, which has fewer people. In fact, China has the largest total footprint in the world (**INFOGRAPHIC 24.3**).

According to ecological footprint analysis, if everyone on the planet were to live like the average resident of the United States, it would take about four Earths to support us. By contrast, if everyone in the world lived like the average person in Indonesia, we would need two-thirds of an Earth to sustainably satisfy our demands.

What is it about the U.S. lifestyle that leaves such a heavy footprint compared to other countries? Energy consumption, by far, is the largest culprit. The cars and SUVs we drive, the computers we work and play with, the washers and dryers that clean our clothes, the air conditioners that cool our homes, the food we truck across the country or fly around the world–all these require energy. Globally, the energy component of our ecological footprint increased roughly 700% between 1961 and 2008, accounting for roughly half our total current footprint.

In the United States, as in most parts of the world, most of this energy comes from fossil fuels–oil, coal, and natural gas. As discussed in Chapter 23, burning fossil fuels

INFOGRAPHIC 24.3 COUNTRIES DIFFER IN THEIR ECOLOGICAL FOOTPRINT

 Per capita, the United States has a larger footprint than most other world populations. However, the enormous size of some populations, like China's, means that their total footprint is greater than that of the United States–in fact, China has the largest footprint in the world.

Ecological Footprint, 2008

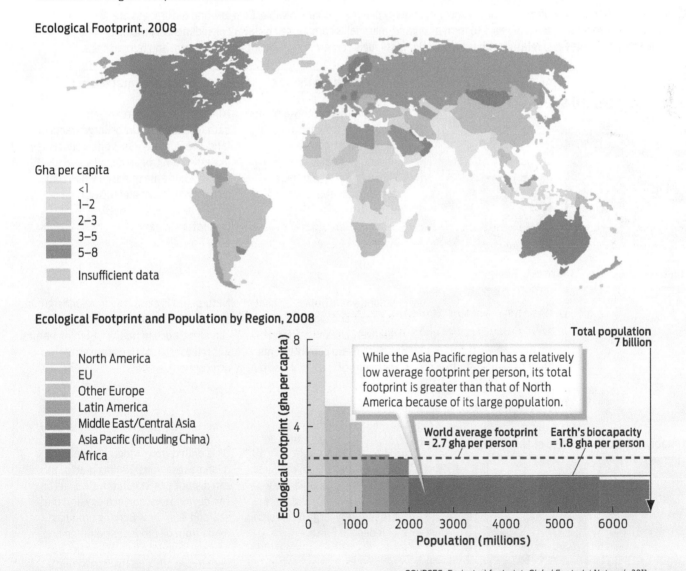

Gha per capita
- <1
- 1–2
- 2–3
- 3–5
- 5–8
- Insufficient data

Ecological Footprint and Population by Region, 2008

- North America
- EU
- Other Europe
- Latin America
- Middle East/Central Asia
- Asia Pacific (including China)
- Africa

Ecological Footprint (gha per capita)

Total population 7 billion

While the Asia Pacific region has a relatively low average footprint per person, its total footprint is greater than that of North America because of its large population.

World average footprint = 2.7 gha per person

Earth's biocapacity = 1.8 gha per person

Population (millions)

SOURCES: Ecological footprint, *Global Footprint Network*, 2011; ecological footprint and population, *WWF Living Planet Report*, 2012.

▶**NONRENEWABLE RESOURCES**
Natural resources that cannot easily be replaced; fossil fuels are an example.

releases carbon dioxide to the atmosphere, contributing to global warming. Therefore, the ecological footprint takes into account the amount of land and water area needed to absorb CO_2. This, combined with increased consumption, is what makes our ecological footprint so large.

Because they take millions of years to form naturally, fossil fuels are considered **nonrenewable resources:** once depleted, they are essentially gone for good. Besides contributing to our carbon footprint, burning these nonrenewable resources also releases pollutants such as sulfur dioxide and nitrogen dioxide, which contribute to acid rain. Coal also contains toxic elements such as arsenic and fluorine, which can cause bone and lung disease when inhaled or consumed. For many reasons, urban planners are eager to wean people off fossil fuels (**INFOGRAPHIC 24.4**).

INFOGRAPHIC 24.4 FOSSIL FUELS ARE NONRENEWABLE

Most of the natural resources we use to supply our energy needs are nonrenewable. Coal, oil, and natural gas are fossil fuels that take millions of years to form as organic material is compressed by layers of sedimentary rock. While plentiful today and relatively cheap to obtain, fossil fuels come with significant environmental and human costs.

Coal: Supplies 20% of U.S. Energy

Why Do We Use It?	Environmental Impact
Coal is burned in power plants to produce steam to turn turbines that generate electricity. Coal is relatively cheap to mine, and there is currently an abundance of it in Earth.	Mining coal from Earth often damages the habitat on large tracts of land. Greenhouse gases and pollutants like arsenic, nitrogen dioxide, and sulfur dioxide are released when coal is burned to make electricity. Coal miners have increased risk of respiratory illness.

Natural Gas: Supplies 25% of U.S. Energy

Natural gas is burned to heat buildings and water. It is relatively cheap to extract, and there are currently large reservoirs of it deep in Earth.	Natural gas is extracted from underground and offshore reservoirs. Drilling platforms can disrupt ocean habitat. Burning natural gas releases greenhouse gases to the atmosphere.

Petroleum Oil: Supplies 36% of U.S. Energy

Oil is used to produce gasoline, petroleum products, and plastics. It is relatively cheap to extract, and there are currently large reservoirs of it deep in Earth.	Oil is drilled from undergound and offshore reservoirs. Drilling platforms can disrupt ocean habitat. Oil spills can devastate ocean ecology and the seafood economy. Burning products made from oil produces pollution and emits greenhouse gases. Plastics do not biodegrade and therefore create a huge amount of landfill waste.

> **"We really believe that the best economic solutions can usually be, and often are, green solutions."**
>
> — LYNN BILLMAN

Hewitt didn't talk about ecological footprints at the first town meeting he called, which was held in a giant striped circus tent 1 week after the tornado and attended by about 700 town residents. But he knew that the idea of sustainability would resonate with this group of midwesterners whose pioneering ancestors had settled the area generations ago. As farmers, many of them had an intrinsic connection to nature and an awareness of the goods it can provide if managed properly.

Environmental concerns are crucial to the concept of sustainability, but they aren't the whole story. "The term sustainability often gets focused exclusively on environmental aspects," says White. "I'm an environmental planner so those are near and dear to my heart, but I think true sustainability also has to bring in the social or the equity concerns as well as economic concerns."

Greensburg residents recognized the importance of these factors, too. As the building plan the town eventually drew up stated, "A truly sustainable community is one that balances the economic, ecological, and social impacts of development."

PLATINUM RATED

Mike Estes, owner of the local John Deere dealership and service shop, was one of the Greensburg residents who initially resisted the notion of rebuilding green. The 60-year-old Kansas native had "only minimal interest in green building" at first. Then he realized how much money he'd save in energy costs over the years and was ultimately persuaded. With technical assistance provided by the National Renewable Energy Laboratory (NREL) of the U.S. Department of Energy, he rebuilt his dealership to include a suite of green technologies, including solar panels and on-site wind turbines. Since rebuilding, he says he's saved $25,000 to $30,000 a year in energy costs.

From the beginning, going green was presented to the residents of Greensburg as an opportunity for economic success rather than a financial burden. "We're very honest and straightforward about getting people the best economic solutions," says Lynn Billman, a research analyst with NREL who worked closely with residents of Greensburg on their building plans. "We really believe that the best economic solutions can usually be, and often are, green solutions."

Yes, rebuilding green would cost more in the short term, but in the long term, the savings in energy costs would more than make up for it. Moreover, becoming a pioneer in the green energy movement promised to attract national attention as well as new business to the area, putting Greensburg back on the map with a claim to fame (besides having the world's largest hand-dug well).

"Before the storm, I couldn't get business to talk to us about relocating to [Greensburg]," Hewitt told CNN. "Once we went green, our phones starting ringing and opportunities put Greensburg in the game."

▶RENEWABLE RESOURCES

Natural resources that are replenished after use as long as the rate of consumption does not exceed the rate of replacement.

One of the first green things the city council did was pass an ordinance, in December 2007, requiring all new public buildings larger than 4,000 square feet to be built to LEED platinum standards. "LEED" stands for "Leadership in Energy and Environmental Design," a certification system run by the U.S. Green Building Council, a nonprofit organization that supports sustainable building prac-

tices. Platinum is the highest rating a building can achieve, above gold and silver.

LEED-certified buildings must be shown to conserve energy and water, reduce waste sent to landfills, lower greenhouse gas emissions, and be healthier and safer for occupants. By committing to platinum standards, the town agreed to design buildings that would reduce energy consumption by

GREENSBURG STYLE: LEED PLATINUM BUILDINGS

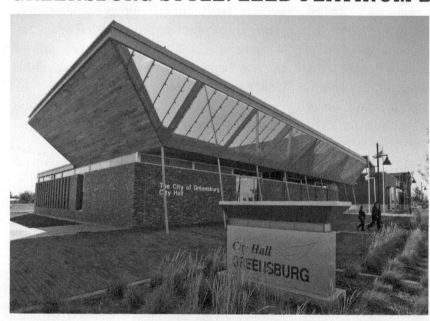

Greensburg City Hall

LEED Platinum Features

▶ Solar panels

▶ Geothermal heating and cooling

▶ Reclaimed brick and wood from buildings destroyed in the tornado

▶ Green roof with vegetation

▶ Rainwater collection and irrigation

▶ 38% more energy efficient

Kiowa, KS, County Schools

LEED Platinum Features

▶ Natural daylight and ventilation

▶ Wind energy generator

▶ Recycling center

▶ Cabinets and lockers made from natural and recycled materials

▶ Spaces for community sharing

▶ Rainwater collection and irrigation

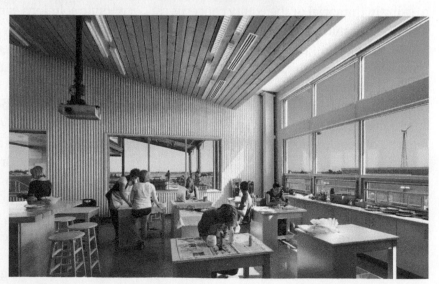

more than 40% over current building code requirements.

"I am so excited about being the first city in the U.S. to adopt this system for a town," City Administrator Hewitt said after the vote.

Greensburg now has one of the highest per capita concentrations of LEED-certified buildings in the country, one per every 130 citizens. Among the more than 15 LEED platinum buildings that have been constructed are a new city hall, public school, hospital, art center, and business incubator. The buildings were designed and constructed by a variety of approaches and technologies to achieve their platinum rating, including extensive use of natural light, geothermal heating, solar panels, and even wind turbines. The goal of all these approaches is to reduce energy use overall and end reliance on fossil fuels.

The city ordinance for platinum buildings applies only to public buildings. On the private side, more than half the residents of Greensburg chose to rebuild their homes using green technologies to reduce energy use. An analysis by NREL revealed that more than 100 of the newly built homes would, according to their design plans, use an average of 40% less energy than before, for an average saving of about $512 per year.

In addition to using less energy overall, the residents of Greensburg also chose to have their electricity be "100% renewable, 100% of the time." They decided to take advantage of a **renewable resource** that Kansas has in abundance: wind.

On the outskirts of town, a stately row of 10 wind turbines, each more than 200 feet tall, turns slowly and reliably in the breeze. Completed in 2010, the wind farm supplies enough electricity to power about 4,000 households—enough for all of Greensburg and dozens of neighboring towns.

Other renewables at work in Greensburg include LED (light-emitting diode) lamps that

GREENSBURG STYLE: NET-ZERO ENERGY

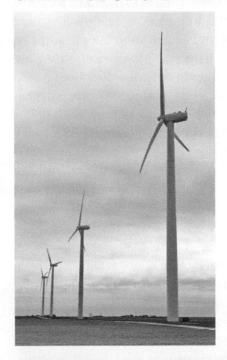

Greensburg Wind Farm

▶ 100% renewable energy

▶ Powers the entire city

▶ Greensburg uses 25%–33% of the power generated from 10 turbines. Remaining energy goes to powering neighboring towns.

100% LED Streetlights

▶ 40% more efficient than traditional fixtures

▶ 70% total energy and maintenance savings

▶ Reduce nighttime light pollution by pointing straight downward

Solar Energy

▶ 100% renewable energy

▶ Many individual homes and buildings have a collection of solar panels.

▶ Additional panels on roofs and walls or in vacant fields

illuminate city streets. Greensburg is the first city in the United States to use LEDs for 100% of its street lighting. LED lamps are 40% more efficient than regular streetlamps, and also help avoid light pollution because they are easier to direct in only one direction: down.

With all these changes, Greensburg has effectively achieved its energy goal. It is a net-zero energy city, generating as much electricity from renewable energy as it uses. "Greensburg has the best infrastructure of any community in the world right now," Hewitt says proudly.

GREEN LIKE US

With a population of about 900 people, what this small town does to conserve resources may seem insignificant. But green becomes meaningful when you consider how lifestyle choices affect resource use when employed by people in a whole state, country, or even the whole planet.

Two thousand years ago, there were roughly 300 million people on Earth, less than the current population of the United States alone. In 2012, there were 7 billion people. Much of that growth has occurred since 1950, thanks in large part to antibiotics and advances in public health that have allowed many of us to live longer. And each hour more than

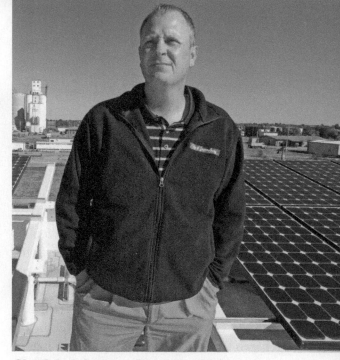

City Administrator Steve Hewitt led Greensburg's efforts to go green.

10,000 new people are added to the planet. By 2050, demographers estimate, we'll hit the 9 billion mark (**INFOGRAPHIC 24.5**).

With every person added to the planet, Earth's pie of resources is divided into smaller and smaller slices. Moreover, as the human population grows, so does our impact on the other species. Ecologists have documented a striking correlation between the rise in human population and loss of **biodiversity** as species

INFOGRAPHIC 24.5 HUMAN POPULATION GROWTH

→ Since the advent of agriculture, the human population has been following an exponential growth pattern, reaching 7 billion people in 2011. Some estimate that the human population will reach 9 billion by 2050.

Domestication of plants and animals

Agriculturally based urban societies

Agricultural Revolution
10, 000—8000 BCE

OLD STONE AGE   NEW STONE AGE   BRONZE AGE

10,000 B.C.   8,000 B.C.   3,000 B.C.   2,000 B.C.

decline in numbers and eventually become extinct. The principal causes of these losses include habitat destruction and overconsumption. Faced with the prospect of 2 billion more people in the coming years, the world is badly in need of new ways of living within Earth's biocapacity (**INFOGRAPHIC 24.6**).

What Greensburg lacks in size, it is hoping to make up for in visibility and daring. It seeks to be a model for other towns and cities across the country, and indeed the world, to emulate.

In a speech to Congress in 2009, President Obama echoed this view, describing Greensburg as "a global example of how clean energy can power an entire community, how it can bring jobs and businesses to a place where piles of bricks and rubble once lay."

Already, the city is attracting international attention. In 2012, a group of architects from tsunami-ravaged areas of Japan came to Greensburg to learn about building sustainable infrastructure. And many of the strategies enacted by Greensburg are being used to design and build ecocities around the world in countries as diverse as Argentina, Australia, Finland, and Vietnam.

A revitalized Greensburg suddenly finds itself serving as the poster child for a sustainability movement that seems at times to be facing an uphill battle.

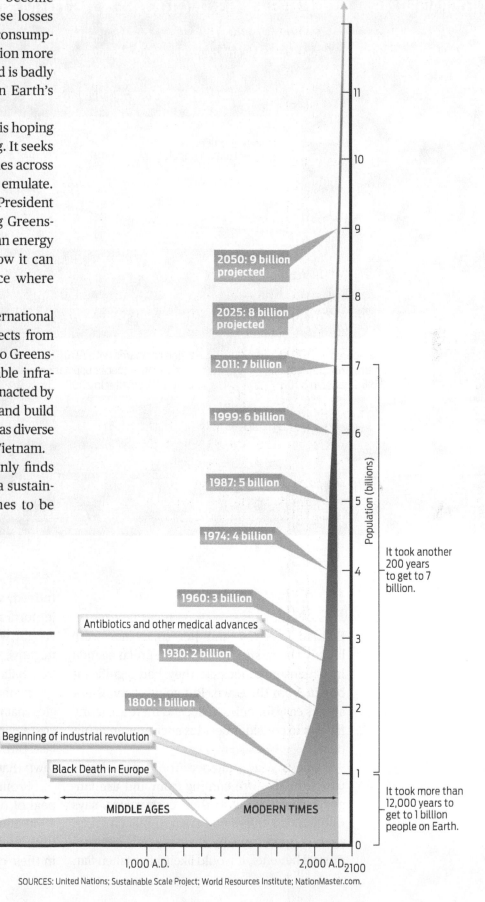

SOURCES: United Nations; Sustainable Scale Project; World Resources Institute; NationMaster.com.

SPECIES LOSS INCREASES AS HUMAN POPULATION GROWS

As the human population increases, so does the number of species lost to extinction. The causes for this include habitat destruction, pollution, spread of disease and overconsumption. Similarly, many wildlife populations are declining.

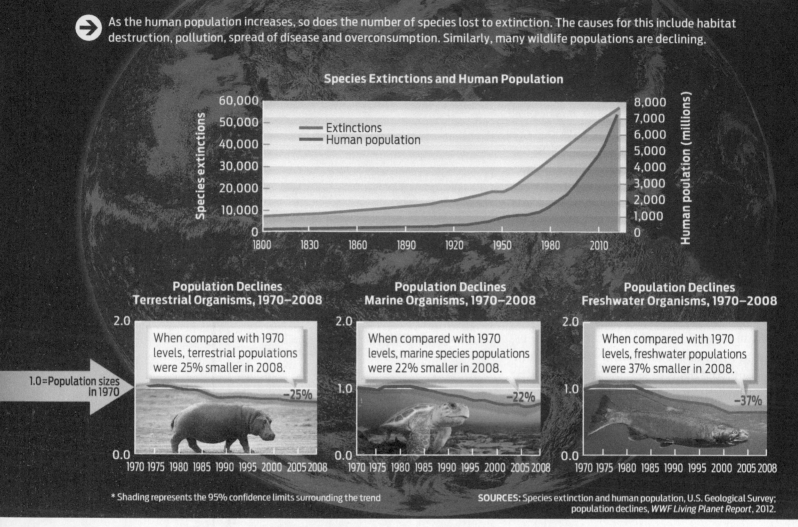

Species Extinctions and Human Population

Species extinctions — *Human population (millions)*
— Extinctions
— Human population

**Population Declines
Terrestrial Organisms, 1970–2008**

When compared with 1970 levels, terrestrial populations were 25% smaller in 2008.

−25%

1.0=Population sizes in 1970

**Population Declines
Marine Organisms, 1970–2008**

When compared with 1970 levels, marine species populations were 22% smaller in 2008.

−22%

**Population Declines
Freshwater Organisms, 1970–2008**

When compared with 1970 levels, freshwater populations were 37% smaller in 2008.

−37%

* Shading represents the 95% confidence limits surrounding the trend

SOURCES: Species extinction and human population, U.S. Geological Survey; population declines, *WWF Living Planet Report*, 2012.

IT AIN'T EASY BEING GREEN

Billman says she and her colleagues were largely successful in realizing a green agenda in Greensburg because they had significant buy-in from the town. But going green is not always easy for cities to do, and there are many factors to consider besides environmental impact.

One is your audience. You have to know where people are coming from and use language that speaks to their core values, says Billman. If leaders had tried to push a green agenda on citizens by stressing only environmental benefits, it would likely have fallen flat.

Indeed, while environmental concerns were important to city leaders, they carried less weight with many ordinary citizens. "It was negative for us to bring up climate change," says Billman. "So we didn't."

Instead, she says, they stressed the two angles that resonated most with this community: cost savings and resiliency–building things to last. Ultimately, the same goal was achieved: a town that saw the value of renewable energy.

From an ecological standpoint, the appeal of renewable sources of energy such as wind and solar is undeniable: they are plentiful, powerful, and environmentally neutral in their carbon emissions. Solar power alone could theoretically provide more than enough clean energy to supply the needs of everyone on the planet many times over—assuming we could adequately and inexpensively harvest it.

The major problem with wind and solar power is that the technologies to harness them are currently much more expensive to build and operate than ones based on oil, coal, and natural gas. What makes fossil fuels such convenient and inexpensive sources of energy is the fact that the difficult work of harvesting the energy of sunlight has already been done by the photosynthetic organisms that were compressed over millions of years into oil, coal, and gas. (In a way, we are already using the energy of sunlight to power our lifestyles, but indirectly.) New natural gas and oil harvesting technologies, like hydraulic fracturing ("fracking"), have upped the ante even more since they deliver products that are cheaper than coal.

Of course, when you consider the environmental and human costs of obtaining and burning fossil fuels—from coal-mine explosions to air and water pollution to oil spills—they aren't actually that cheap. Think of the 2010 disaster in the Gulf of Mexico: an oil rig exploded and sank, spewing 4.9 million barrels of oil into the Gulf and costing 11 lives and more than $40 billion in cleanup. Or consider that hydraulic fracturing can lead to serious water pollution, including contaminated drinking wells that pump out flammable water. These downstream costs of fossil fuels, which are not reflected in their market price, are known among economists as externalities. If externalities were included in the price, as some economists and environmentalists suggest they should be, then the playing field with other forms of energy would be more level.

Moreover, explains Billman, the government subsidizes fossil fuel use in lots of ways that aren't as obvious to the public. According to a 2009 study by the nonpartisan Environmental Law Institute, the U.S. government provided $72 billion in tax breaks and subsidies to the fossil fuel industry over the years 2002-

2008 versus $29 billion to renewable energy companies over the same period. Such hidden subsidies can create disincentives to change.

But as Greensburg shows, there are ways to make renewables more competitive. First and foremost, there's the approach of emphasizing cost savings: when energy savings over time are considered in the math, the economic calculations look pretty good. As Mike Estes, owner of the local John Deere dealership, experienced firsthand, renewables make up for their initial higher price by saving money in the long term.

There are also ways to capitalize on the desire of companies that want to become better stewards of the environment. The Greensburg wind farm, for example, was funded in part with money provided by Vermont-based *Native*Energy, a company that sells "carbon offsets" to businesses that want to counterbalance their fossil fuel use (but may not be able to implement substantial carbon reductions of their own). Current purchasers of *Native*Energy carbon offsets include Ben & Jerry's, Stonyfield Farm, Clif Bar, and Green Mountain Coffee Roasters.

Yet there are factors besides cost that limit how much we can rely on renewables at present. The most significant are technological problems of storing and transmitting energy produced from wind and solar. Threats posed by wind farms and solar panels to native animal habitat also complicate the picture. Given these limitations, says Billman, "wind and solar still are a ways from sweeping over the country." The same goes for algae-based biofuels (see Chapter 5) and other renewables. At least for the next decade, we cannot take fossil fuels out of our energy mix (**INFOGRAPHIC 24.7**).

NOR ANY DROP TO DRINK

As any Kansas farmer will tell you, fossil fuels aren't the only natural resources being overdrawn by a growing human population—water is, too. "We are *so dry* right now," says urban planner White.

INFOGRAPHIC 24.7 NONFOSSIL FUEL RESOURCES REDUCE OUR ECOLOGICAL FOOTPRINT

 While many renewable resources are available to us, economic, technological, and environmental considerations currently limit their use as alternatives to fossil fuels.

	How It Reduces the Ecological Footprint	**Why Don't We Use It More?**
Solar: Supplies 0.09% of U.S. Energy 	Solar energy traps energy from the sun and converts it into electricity and heat with little impact on the environment. As nothing is burned to make the electricity, there are zero polluting emissions from this process.	Solar power is currently much more expensive to produce than nonrenewable options. The mining and processing of the rare earth metals used in the solar panels use hazardous chemicals, and the resulting waste must be properly disposed of.
Wind: Supplies 1.17% of U.S. Energy 	Wind energy is used to turn turbines, producing electricity with little impact on the environment. In the absence of combustion, no pollutants are released to the environment.	Wind power is currently much more expensive to produce than nonrenewable options. Wind generators take up space, either on land or in the water, and must be located in windy areas. Some people don't want a visible wind farm near their homes. The mining and processing of the rare earth metals used in the turbines uses hazardous chemicals. Bird species may be affected as turbines encroach on their air space.
Nuclear: Supplies 8% of U.S. Energy 	Nuclear energy uses concentrated radioactive elements harvested from Earth. As these elements decay, they give off tremendous heat, which is used to produce electricity. As nothing is burned in the process, there are no polluting emissions.	Nuclear reactors are expensive to design and build, and a reactor has a limited life span. Extracting uranium from mines has an environmental impact, and the waste from uranium mines is radioactive. Most important, the waste from nuclear reactors is highly radioactive, making storage complicated. Weapons-grade plutonium can be made from reactor waste, posing a security threat.
Biomass: Supplies 4.32% of U.S. Energy 	Biomass includes biofuels, biomass waste, and wood. When burned, the only CO_2 released to the atmosphere is what the plants and algae took in through photosynthesis, so fossil deposits of carbon are not used.	While being intensively researched, biofuels have not yet become feasible replacements for fossil fuels. In some cases, significant emissions are associated with their production. In other cases, more research and investment is required to optimize the production process. Growing plants for biofuels can also compete with growing crops.
Hydroelectric: Supplies 3.15% of U.S. Energy 	Hydroelectric power relies on the conversion of potential energy (stored in the position of accumulated water behind a dam) to kinetic energy, which can turn a generator. There are no emissions associated with hydroelectric power. Hydro plants have long life spans, and hydro power can potentially power half the projected energy demands of the planet.	To produce hydroelectric power, dams are built to block rivers, creating lakes with immense amounts of potential energy. Building dams destroys habitat, affects local fish populations, and can force human populations to relocate.
Geothermal: Supplies 0.18% of U.S. Energy 	Geothermal energy relies on naturally occurring heat from the magma layer beneath Earth's crust. This is a sizable and sustainable resource that can be tapped to drive generators or directly heat homes and businesses.	Geothermal energy is used extensively in Iceland and in some areas of California. However, it has yet to be fully developed in other areas, primarily because optimal technologies require further development. It can also be very expensive, and in some cases, noxious pollutants are released with the steam from geothermal resources.

For residents of Kansas and other states of the Great Plains, water for drinking and irrigation comes from the large Ogallala Aquifer, a deep underground source of water spanning eight states in the middle of the country. Rain is so scarce in this part of the United States that, were it not for the **aquifer**, it would be a desert. Thanks to the aquifer, it is one of the most fertile agricultural regions in the world.

Studies by the U. S. Geological Survey and others suggest that the Ogallala, also known as the High Plains Aquifer, is being depleted faster than it is being replenished, and some experts predict that it could dry up by 2050. The aquifer is at "historical lows right now," White says.

To help conserve water, many buildings in Greensburg have rooftop rain filtration and storage systems, which collect rainwater and use it to flush toilets and irrigate the grounds. Rain barrels to collect water are also becoming quite common in many parts of Kansas.

The very fact of water shortages may seem counterintuitive. After all, 70% of the globe's surface is covered with water. But of this vast supply, only 2.5% is freshwater, and most of that is locked up in ice caps and glaciers. A meager 1% of the total water on Earth is available for human consumption.

Nevertheless, freshwater is considered a renewable resource because the supply in lakes, rivers, reservoirs, and underground aquifers is continually being replenished by the water cycle. As long as the rate of water withdrawal from these sources is less than the rate of replacement, the supply of freshwater remains relatively constant (**INFOGRAPHIC 24.8**).

▶**AQUIFER**
An underground layer of porous rock from which water can be drawn for use.

INFOGRAPHIC 24.8 WATER IS A RENEWABLE RESOURCE

→ Freshwater is a valuable resource. In addition to its role in keeping us hydrated, it irrigates crops, sustains fisheries, and provides recreational opportunities. Although water is "used," it is not "used up": it is ultimately returned to the global ecosystem as it evaporates to the atmosphere, flows into rivers or streams, or enters underground aquifers.

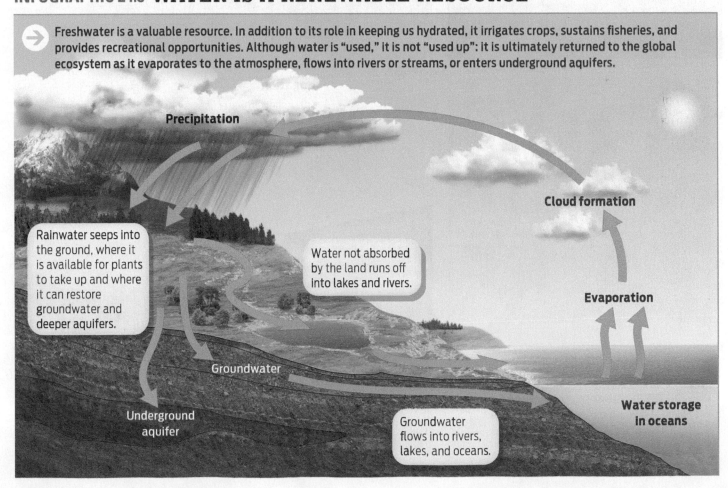

Precipitation

Cloud formation

Rainwater seeps into the ground, where it is available for plants to take up and where it can restore groundwater and deeper aquifers.

Water not absorbed by the land runs off into lakes and rivers.

Evaporation

Groundwater

Underground aquifer

Groundwater flows into rivers, lakes, and oceans.

Water storage in oceans

GREENSBURG STYLE: WATER CONSERVATION

Water Collection

▶ Roofs tilted to channel rainwater

▶ Gutter downspouts collect water in troughs

▶ 10k gallon aboveground cistern stores collected rainwater

▶ 50k gallon belowground cistern stores water runoff

Irrigation

▶ collected water used for irrigation of landscaping

▶ Collected water used to flush water-conserving toilets

▶ Waterless urinals save 0.5 gallons per use

Plants

▶ Native plants are drought tolerant and require less watering

▶ Landscaping minimizes rainwater runoff into city sewers—it filters and channels it for capture

▶ Athletic fields are artificial turf that requires no watering

Although water is not consumed in the same way as coal or oil, our supply of freshwater is being divided among more and more people, so there is less available, on average, for each of us. According to the United Nations, water use increased sixfold during the 20th century, more than twice the rate of population increase. Today, more than half of all the accessible freshwater contained in rivers, lakes, and aquifers is appropriated by humans, most of it for irrigation in agriculture. A striking example of the consequences of increased water use can be seen in Lake Mead, which is fed by the Colorado River. This reservoir, which provides freshwater to much of the southwestern United States, is at historic lows today (**INFOGRAPHIC 24.9**).

Pollution also shrinks the total amount of available clean freshwater on the planet. Agriculture, industry, and cities all play a role. Run-off from streets carries pollutants such as motor oil and sewage; fertilizers, pesticides, and toxic chemicals leach from fields and factories. These substances can eventually reach aquifers, rivers, and oceans, contaminating the water that both humans and wildlife depend on.

Food and lifestyle choices also affect water availability. According to environmental scientist Arjen Y. Hoekstra, it takes 900 liters of water to produce 1 kilogram of corn, but more than 15 times that to produce 1 kilogram of beef.

Some countries experience water scarcity more acutely than others. That's because the

INFOGRAPHIC 24.9 DEPLETION OF FRESHWATER BY A GROWING POPULATION

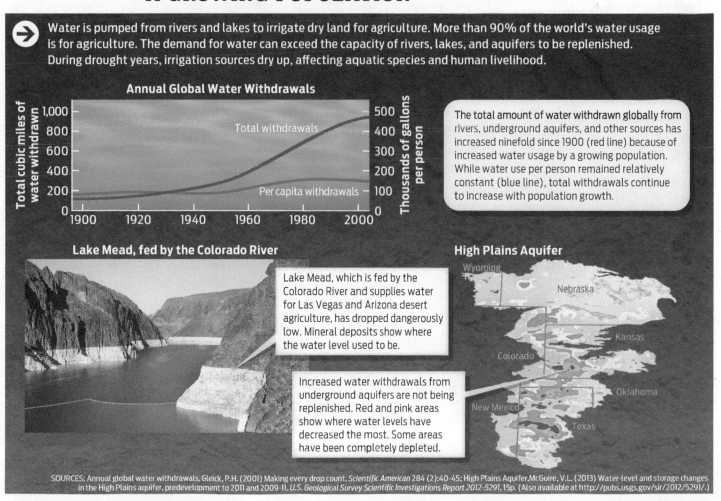

Water is pumped from rivers and lakes to irrigate dry land for agriculture. More than 90% of the world's water usage is for agriculture. The demand for water can exceed the capacity of rivers, lakes, and aquifers to be replenished. During drought years, irrigation sources dry up, affecting aquatic species and human livelihood.

Annual Global Water Withdrawals

Total withdrawals

Per capita withdrawals

The total amount of water withdrawn globally from rivers, underground aquifers, and other sources has increased ninefold since 1900 (red line) because of increased water usage by a growing population. While water use per person remained relatively constant (blue line), total withdrawals continue to increase with population growth.

Lake Mead, fed by the Colorado River

Lake Mead, which is fed by the Colorado River and supplies water for Las Vegas and Arizona desert agriculture, has dropped dangerously low. Mineral deposits show where the water level used to be.

Increased water withdrawals from underground aquifers are not being replenished. Red and pink areas show where water levels have decreased the most. Some areas have been completely depleted.

High Plains Aquifer

Wyoming

Nebraska

Kansas

Colorado

Oklahoma

New Mexico

Texas

SOURCES: Annual global water withdrawals, Gleick, P.H. (2001) Making every drop count. *Scientific American* 284 (2):40-45; High Plains Aquifer, McGuire, V.L. (2013) Water-level and storage changes in the High Plains aquifer, predevelopment to 2011 and 2009-11. *U.S. Geological Survey Scientific Investigations Report 2012-5291*, 15p. (Also available at http://pubs.usgs.gov/sir/2012/5291/.)

INFOGRAPHIC 24.10 WATER AVAILABILITY IS NOT EQUALLY DISTRIBUTED

Freshwater is not evenly distributed across the globe, and its availability does not always follow international borders. In addition, access to even a sufficient water supply may be limited by economic, social, and political circumstances, such as war and ethnic conflict. As the human population continues to grow, and access to clean freshwater continues to decline, these problems are likely to intensify, particularly in areas with existing scarcities of water.

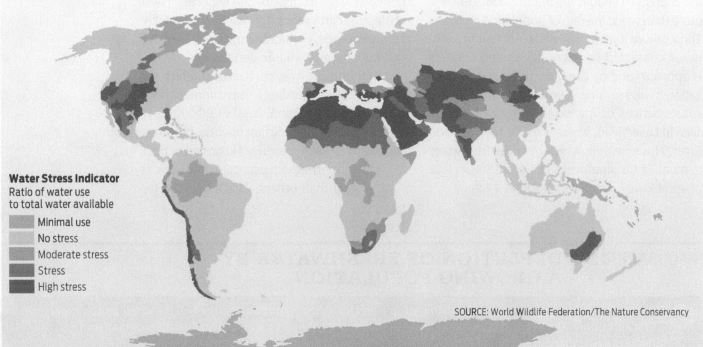

Water Stress Indicator
Ratio of water use
to total water available

- Minimal use
- No stress
- Moderate stress
- Stress
- High stress

SOURCE: World Wildlife Federation/The Nature Conservancy

geographic distribution of freshwater does not match the distribution of the world's population. Canada has just 0.5% of the world's population, but 20% of the global freshwater supply. The United Nations estimates that at least a billion people in the world currently lack access to clean and safe drinking water, and by 2025, two-thirds of the world's population will live in areas of moderate to severe water stress. Climate change, if it changes precipitation patterns, may also affect the global availability of water in unpredictable ways (**INFOGRAPHIC 24.10**).

In their sustainable building plan, Greensburg pledged to "treat each drop of water as a precious resource."

NO PLACE LIKE HOME

When Greensburg decided to go green, its leaders hoped to capitalize on the green message to lure new business to the area. Mayor Bob Dixson was quoted in a brochure produced by NREL: "We'd like to see Greensburg become the ecotourism capital of the world. . . . We want to be a living laboratory."

Unfortunately, the town's green building project was just getting underway when the 2008 financial crisis hit and made businesses wary about making new investments. So far, only one new company has settled in Greensburg: the Canadian-based BTI Wind Energy, a manufacturer of wind turbines for the United States.

> **❝We'd like to see Greensburg become the ecotourism capital of the world. . . . We want to be a living laboratory.❞**
>
> — BOB DIXSON

But town leaders haven't given up hope. They believe more companies will eventually flock to Greensburg. And they are happy and proud of what they have been able to build already: a community that has a future.

City Administrator Hewitt frames the discussion of sustainability in terms of parenting. As citizens and community members, he says, we have the same obligation that we do as parents: to recognize that the choices we make today have consequences, and to make decisions that are going to be good for the long term (INFOGRAPHIC 24.11).

Hewitt says his 4-year-old son, Gunner, was often on his mind while he was working to transform Greensburg into a sustainable city. "By doing it for Gunner," he says, "I'm taking care of a lot of people." ■

INFOGRAPHIC 24.11 WHAT YOU CAN DO TO LIVE MORE SUSTAINABLY

Take Action	Why?	Your Impact!
Drive Less and Invest in Fuel Efficiency	With less than 5% of the world's population, the United States consumes a quarter of the world's oil and emits a quarter of its greenhouse gases, largely from automobiles.	Driving smaller vehicles and those with more fuel efficiency cuts carbon dioxide emissions and reduces dependence on nonrenewable fossil fuels.
Install Compact Fluorescent Lightbulbs (CFLs)	Electricity production is the largest source of greenhouse gas emissions in the United States, and lighting accounts for about 25% of American electricity consumption.	By replacing just four standard bulbs with CFLs, you can prevent the emission of 5,000 pounds of carbon dioxide and reduce your electricity bill by more than $100 over the lives of those bulbs.
Reduce Vampire Energy Waste	Electronics use energy even when they are turned off. This standby "vampire energy" accounts for 5% to 8% of a single family's home electricity use per year.	When you plug your electronics into a power cord that you turn off each night, you will save the equivalent of 1 month's electric bill each year.
Reduce Home Water Use	The average U.S. household uses over 22,000 gallons of water per year for showers and baths. Water is almost always heated, resulting in increased fossil fuel consumption and greenhouse emissions.	If only 1,000 of us install faucet aerators ($2–$5) and efficient showerheads (<$20), we can save nearly 8 million gallons of water and prevent over 450,000 pounds of carbon dioxide emissions each year.
Eat Less Feedlot Beef	Producing 1 pound of feedlot beef in some cases can require 5 pounds of grain and over 2,400 gallons of irrigation water.	Eating lower on the food chain (eating plant-based foods) requires fewer energy and resource inputs.
Recycle	Each year, the U.S. population discards enough glass bottles and jars to fill 12 giant skyscrapers, even though 75% of our trash can be recycled. Recycling materials uses fewer nonrenewable resources, saves energy, results in less air and water pollution, and creates more jobs than making new materials.	Recycling one aluminum can saves enough energy to run a TV for 3 hours. To produce each week's Sunday newspapers, 500,000 trees are cut down. Recycling a single run of the Sunday New York Times would save 75,000 trees. Taking reusable bags on your weekly grocery trip reduces our demand for petroleum for plastic bags.

(24)

CHAPTER 24 Summary

▸ Sustainability refers to the ability of humans to live within Earth's biocapacity—its ability to provide current and future generations with natural resources and to absorb our wastes.

▸ Natural resources include nonrenewable resources such as fossil fuels (oil, coal, and gas) and renewable resources such as sunlight, wind, and water.

▸ Ecologists measure human demand on Earth's resources using ecological footprint analysis, which quantifies the amount of biologically productive land and water needed to support our lifestyles.

▸ The ecological footprint of the current human population is greater than Earth's biocapacity, which means that we are living unsustainably.

▸ The largest component of our footprint is burning fossil fuels, which generates harmful wastes, including greenhouse gases and pollutants.

▸ As the human population grows, so does our ecological footprint. As of 2012, the human population totaled 7 billion people. Some demographers predict the number could hit 9 billion by 2050.

▸ Freshwater is a renewable resource, but the world's supply is not distributed equally, and many people around the world suffer from water scarcity, a problem exacerbated by a rising population, the demands of agriculture, and socioeconomic challenges.

▸ Sustainable practices minimize the consumption of nonrenewable resources by using renewable resources like wind and solar power instead of fossil fuels to generate electricity and heat.

▸ At their current level of development, technologies to harvest renewable energy cannot meet our total energy demands. Fossil fuels cannot yet be taken out of our energy mix.

▸ Individually, we can decrease our ecological footprint by driving less, reducing water and electricity use, consuming less meat, and recycling.

MORE TO EXPLORE

▸ National Renewable Energy Laboratory, Greensburg Deployment Project eere.energy. gov/deployment/greensburg.html

▸ Global Footprint Network, Footprint Calculator www.footprintnetwork.org/en/ index.php/GFN/page/calculators/

▸ World Wildlife Fund (2012) Living Planet Report http://www.footprintnetwork.org/ images/uploads/LPR_2012.pdf

▸ White, S. S. (2010) Out of the rubble and towards a sustainable future: the "greening" of Greensburg, Kansas. *Sustainability* 2:2302–2319.

▸ Planet Green (Discovery Communications) (2010) *Greensburg* (documentary series).

CHAPTER 24 Test Your Knowledge

DRIVING QUESTION 1

What human impacts are considered when determining ecological footprint, and how does human population size influence our impact on Earth?

By answering the questions below and studying Infographics 24.1, 24.2, 24.3, 24.5, and 24.6 you should be able to generate an answer for the broader Driving Question above.

KNOW IT

1. From what you have read in this chapter, explain some of the advances that have permitted the human population to grow exponentially.

2. What is an ecological footprint?

USE IT

3. From your reading of this chapter, how do you think that the ecological footprint of Greensburg compares to that of a small tribal village without electricity in the northern hill regions of Thailand? Describe the points that you considered in your answer.

4. On the outskirts of a small town, a farmer has just sold his 5 acres of cropland to a developer who is planning to build 20 single-family condominium units on that land. Discuss the ways that this transaction will affect the size of the nearby town's population and the ecological footprint of the residents of the nearby town and outskirts.

5. What building considerations could the developer in the Question 4 take into account to minimize the impact of this development on the ecological footprint of the town and outskirts?

DRIVING QUESTION 2

What resources do humans rely on, and which of them are renewable?

By answering the questions below and studying Infographics 24.4, 24.7, 24.8, 24.9, and 24.10 you should be able to generate an answer for the broader Driving Question above.

KNOW IT

6. Which of the following waste products is/are associated with the burning of fossil fuels?
   a. water
   b. carbon dioxide
   c. nitrogen dioxide
   d. all of the above
   e. b and c

7. Mark each of the following natural resources as renewable (R) or nonrenewable (N).

   _____Freshwater   _____Wind
   _____Coal   _____Sunlight
   _____Codfish populations in the North Atlantic

8. If oil is formed from fossilized remains of once-living organisms, and if organisms keep dying, why is oil considered to be a nonrenewable resource?

USE IT

9. The renewability of some resources can depend on human choices and activities. List some such resources, and explain how human activities may lead a renewable resource to become essentially nonrenewable.

10. Think about your local region—for example, do you live in the southwestern desert or on the northeastern ocean shore? Describe the nonrenewable and renewable energy resources that are available in your region, or that your region can harvest. What are some of the challenges that must be overcome in order to tap into the renewable energy resources in your region?

INTERPRETING DATA

11 Wind power has both advantages and disadvantages. Advantages include its sustainability and that it is emission-free. Disadvantages include possible impacts on wildlife.

a. Taking the high end of the estimates given in the figure below, what is the total number of bird deaths attributable to different sources each year in the United States?

BIRD KILLERS

Wind farms kill fewer birds than most other hazards, but pose a particular threat to species including raptors. Estimates of annual fatalities — shown for the United States — are highly uncertain.

Communication towers:
5 million to 6.8 million

Buildings: Strikes to building glass and lighted buildings
100,000 to 1 billion+

Cats (domestic and feral):
365 million to 1 billion

Wind turbines:
100,000 to about 440,000

Automobiles:
60 million to 80 million

Agricultural pesticides:
67 million to 90 million

Power line electrocutions and collisions:
hundreds of thousands to 175 million

b. Taking the high estimates given in the figure, what % of annual bird deaths can be attributed to each of the following?
- Communication towers
- Wind turbines
- Agricultural pesticides
- Cats
- Collisions with buildings

c. The graph below is adapted from two graphs presented in a 2010 report from the National Wind Organization ("Wind Turbine Interactions with Birds, Bats and Their Habitats: A Summary of Research Results and Priority Questions" www.nationalwind.org.) Are the same sites equally responsible for bird and bat deaths? What possible explanations can you think of to account for any differences? What factors should be considered in placing and running wind farms to minimize bird and bat deaths? (You may need to do some research to answer this last question.)

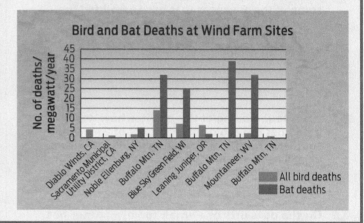

Bird and Bat Deaths at Wind Farm Sites

No. of deaths/megawatt/year

Diablo Winds, CA / Sacramento Municipal Utility District, CA / Noble Ellenburg, NY / Buffalo Mtn, TN / Blue Sky Green Field, WI / Leaning Juniper, OR / Buffalo Mtn, TN / Mountaineer, WV / Buffalo Mtn, TN

All bird deaths
Bat deaths

DRIVING QUESTION 3

What is "sustainable living"?

By answering the questions below and studying Info-graphics 24.6, 24.7, and 24.11 you should be able to generate an answer for the broader Driving Question above.

KNOW IT

12 How would you define "sustainability"? Highlight several features of sustainability in your answer.

13 What aspects of LEED-certified buildings contribute to sustainability?

USE IT

14 If you don't live in Greensburg, what practices can you adopt where you live to reduce your ecological footprint and embrace the philosophy of sustainable living? For each practice that you think of, explain how it would contribute to sustainability and the reduction of your ecological footprint.

15 Many cities have been developed in the hot and dry southwestern states of the United States. What are some of the sustainability implications of living in the desert?

MINI CASE

16 A single mother of two children living in Boston (a city with a decent public transportation system) asks you to help her reduce both her day-to-day living expenses and her ecological footprint. What specific recommendations can you make? What evidence do you have to back up your recommendations? Are any of your recommendations also consistent with a healthier lifestyle? Explain your answer.

BRING IT HOME

17 Explore your school's website to find out about your school's commitment to sustainability. Are there plans for LEED-certified buildings? Are there bike lanes? Bike-parking spots? What about a recycling or composting program? From what you learn, assign a letter grade (A–F) for your school's sustainability plan, and write a short report explaining the reasons for your grade. If you assigned a high grade, what factors contributed to it? If you assigned a low grade, what could be improved at your school?

MAN VS. MOUNTAIN

Physiology explains a 1996 disaster on Everest

AT 1:17 P.M., ON MAY 10, 1996, JON KRAKAUER PLANTED ONE FOOT IN CHINA, the other in Nepal, and stood on the roof of the world. He was at the highest point above sea level that any human has ever reached—short of standing on the moon. Yet he didn't feel like celebrating. It had taken him 6 long weeks to climb to the top of Mount Everest, and now that he was here his toes ached in the subzero cold, his breath came in short, painful bursts, and his head pounded from the altitude. It was a struggle just to stay upright. "I cleared the ice from my oxygen mask, hunched a shoulder against the wind, and stared absently at the vast sweep of earth below," wrote Krakauer in an account of his climb for *Outside* magazine later that year.

Krakauer's journey to Everest was a lifetime in the making. While other kids were idolizing astronaut John Glenn and baseball pitcher Sandy Koufax, Krakauer's childhood heroes were Tom Hornbein and Willi Unsoeld—two men from his hometown in Oregon who, in 1963, became

▸▸ DRIVING QUESTIONS

1. How are the bodies of living organisms organized?
2. How do humans and other organisms regulate body temperature?
3. What physiological systems are regulated by homeostatic feedback loops?

▶**ANATOMY**
The study of the physical structures that make up an organism.

▶**TISSUE**
An organized collection of a single cell type working to carry out a specific function.

▶**ORGAN**
A structure made up of different tissue types working together to carry out a common function.

the first climbers to scale the daunting western ridge of Everest. As a teenager, Krakauer became a skilled climber, vanquishing many of the world's most difficult peaks, and he dreamed of one day climbing Everest himself. By his mid-twenties, though, he had largely abandoned the idea as a boyhood fantasy. But old dreams die hard.

In 1995, Krakauer was working as a journalist when the opportunity came to shadow an Everest climb and report on it for *Outside* magazine. The 42-year-old writer-adventurer jumped at the chance. He would join a team headed by the celebrated climbing guide Rob Hall, whose company, Adventure Consultants, had successfully put 39 amateur climbers on top of Everest. Reaching the summit himself

> At the summit, where windchill temperatures average –53°C (–63°F), freezing to death is a real possibility.

would mean enduring a weeks-long ascent from Base Camp, giving his body time to adjust to the high altitude.

It would also mean risking his life on a daily basis.

The icy tip of Mount Everest sits at 29,035 feet above sea level; cruising altitude of most commercial jetliners is 30,000 feet. A human plucked from sea level and deposited at this altitude would quickly lose consciousness and die. A climber who has spent weeks adjusting to the altitude can function better at the summit, but not very well, and not for very long. Everest is not only the highest place on Earth, it is also one of the coldest. At the summit,

where windchill temperatures average –53°C (–63°F), freezing to death is a real possibility.

Despite these dangers—or perhaps because of them—about 150 fearless men and women try to climb Everest every season. And every season, some of them don't come back. Ten people died on Everest in 2012. In the first half of 2013 alone, nine climbers perished. There are many reasons for these disasters—poor training, unforeseen accidents, raw egotism—but among the most important is basic biology: the human body is not equipped to survive at such extreme altitude, and such extreme temperature, for long.

THE BODY AS MACHINE

Like a car or a computer, a human body is made up of many parts working together in a coordinated fashion. The parts are organized hierarchically, so that smaller components are organized into increasingly larger units, which are themselves organized into more complex systems. The study of all this intricate hardware is called **anatomy.**

The product of millions of years of evolution, human bodies have an anatomical structure that is impressively well adapted to living in certain environments and performing certain functions. Our species evolved in the hot, flat savannahs of Africa where environmental conditions favored big brains, opposable thumbs, and bipedal posture—as well as the ability to keep cool (see Chapter 20). As a result, modern humans excel at grasping a pencil or looking through a microscope; we do less well swimming at the bottom of the ocean or living on mountaintops. Fundamentally, that's because of the way we're put together.

For all living things, the smallest anatomical unit is the cell. Human bodies are made up of trillions of cells, each of which can be classified as one of a few hundred different types. Cells, in turn, are organized into **tissues**—integrated groups of specialized cells, often organized by surrounding proteins, work-

INFOGRAPHIC 25.1 HOW THE HUMAN BODY IS ORGANIZED

Cells
Cells come in many forms, each type performing a specific function.

Tissues
Specialized cells work together to form organized tissues.

Organs
Different tissues coordinate activities in organs.

Organ systems
Multiple organs are organized into systems that perform major tasks for the body.

Muscle cell

Muscle tissue

Connective tissue cell

Connective tissue

Digestive system

Nerve cell

Nervous tissue

Stomach (organ)

Esophagus

Epithelial cell

Epithelial tissue

Large intestine
Small intestine

ing together to execute a particular function. Humans and other animals have four different types of tissue–epithelial, connective, muscle, and nervous–which carry out specific tasks in the **organs** of which they are a part. The stomach, for example, is an organ composed of the four types of tissue organized into a compartment for churning and digesting food. At the highest level of organization, organs interact chemically and physically as part of **organ systems.** For instance, the stomach works with the esophagus, small intestine, and liver to digest and absorb food as part of the digestive system (**INFOGRAPHIC 25.1**).

If the body is like a machine, then physiologists–the scientists who study **physiology**–are interested in how this machine keeps running smoothly. Physiologists want to understand how organ systems cooperate to accomplish basic tasks, such as obtaining energy from food, taking in nutrients to build new molecules during growth and repair, and ridding the body of wastes–in short, how the body functions. To the physiologist, the body is an integrated system for processing inputs and outputs and maintaining **homeostasis**– a relatively stable internal environment even when the external environment changes.

▶**ORGAN SYSTEM**
A set of cooperating organs within the body.

▶**PHYSIOLOGY**
The study of the way a living organism's physical parts function.

▶**HOMEOSTASIS**
The maintenance of a relatively stable internal environment even when the external environment changes.

THERMOREGULATION: THE PHYSIOLOGY OF STAYING WARM

Like many other animals, humans have an optimal operating temperature and are exquisitely sensitive to temperature changes. Although we can tolerate a wide range of external temperatures in our daily lives, we cannot tolerate large fluctuations in our internal temperature. That's because the enzymes that catalyze the chemical reactions in our body function only within a very narrow temperature range. The body works hard to maintain a relatively constant internal temperature compatible with life, fluctuating only about 0.5° C. Through this process of **thermoregulation,** our body temperature is kept at a consistent—and toasty—37°C (98.6°F).

Keeping a consistent body temperature is just one example of how the body tries to maintain homeostasis. "What we're really thinking about with homeostasis are certain set points that your body needs to maintain," explains Robert Kenefick, a research physiologist with the U.S. Army Research Institute of Environmental Medicine in Natick, Massachusetts. The body has a number of such set points, he says, for things like temperature, blood pH, and blood pressure, and it works hard to keep these factors balanced within a very narrow range, even in the face of a changing external environment. The consequences of not maintaining this balance can be deadly: a body temperature increase or decrease of just a few degrees, for example, can be lethal (INFOGRAPHIC 25.2).

As an exercise physiologist, Kenefick has spent his career trying to understand how the human body maintains homeostasis during strenuous activities like hiking and running marathons. He works in the Army Research Institute's Thermal and Mountain Medicine Division, where a main focus of his research is understanding how the body performs in extreme cold.

Staying warm is hard to do when ambient temperatures drop below −50°C (−58°F), as they routinely do on Everest. To seal in heat, mountain climbers wear layers of protective gear designed to trap heat, wick away moisture, and insulate their bodies from the wind and cold. When not hiking, climbers consume copious amounts of hot tea or coffee to warm their insides. But insulated clothes and hot beverages would be of little help without the body's natural way of keeping warm.

Human bodies respond to cold in two main ways: by conserving the heat they already have and by generating more. To conserve heat, the body performs what is called peripheral **vasoconstriction**—a decrease in the diameter of blood vessels just below the surface of the skin. By constricting blood vessels in the skin, peripheral vasoconstriction pushes blood from the skin to the body core, where the internal organs are.

"A lot of people believe this is done to increase the amount of blood that goes to your core to help protect those organs," says Kenefick. There's some truth in that, Kenefick notes, but a more important reason for peripheral vasoconstriction is, he explains, "to decrease the amount of heat loss from your skin to the environment." Like most things in the universe, heat moves along a gradient from higher to lower. "If the temperature is higher in your skin and lower in the air, then you're going to lose heat to the air. By bringing [these temperatures] closer together, you lose much less heat to the environment." The clamping down of blood flow near the skin surface is why hands, feet, and noses are the first to feel cold on a cold day, and it's a sign that your body is trying to retain heat.

The second way the body responds to cold is by trying to generate more heat. It does this by shivering, the involuntary contraction of normally voluntary muscles. "We

▶THERMO-REGULATION
The maintenance of a relatively stable internal body temperature.

▶VASOCON-STRICTION
The reduction in diameter of blood vessels, which helps to retain heat.

INFOGRAPHIC 25.2 THE BODY WORKS TO MAINTAIN HOMEOSTASIS

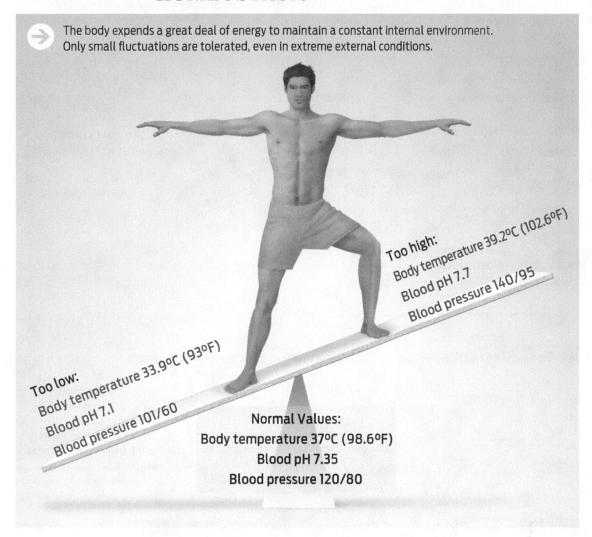

The body expends a great deal of energy to maintain a constant internal environment. Only small fluctuations are tolerated, even in extreme external conditions.

Too high:
Body temperature 39.2ºC (102.6ºF)
Blood pH 7.7
Blood pressure 140/95

Too low:
Body temperature 33.9ºC (93ºF)
Blood pH 7.1
Blood pressure 101/60

Normal Values:
Body temperature 37ºC (98.6ºF)
Blood pH 7.35
Blood pressure 120/80

know that the by-products of cellular respiration—any time cells work, and that includes your muscle cells—are CO_2, heat, and water," says Kenefick. "So when your muscles contract through shivering, they create heat, and that heat helps to warm up the core of your body."

Of course, to maintain a constant temperature, our bodies must not only keep from getting too cold, they must also keep from getting too hot. Two main physiological responses help prevent overheating: peripheral

vasodilation, the expansion of the diameter of blood vessels, which increases blood flow to the skin; and evaporative cooling, or sweating, which cools the body by releasing heat to the air.

In other words, you have a set point for body temperature: if you get too cold, you vasoconstrict and shiver; if you get too hot, you vasodilate and sweat. A precise balance between the two must be maintained to keep tissues healthy. If peripheral vasoconstriction goes on for too long, for example, the result is

▶VASODILATION
The expansion in diameter of blood vessels, which helps to release heat.

INFOGRAPHIC 25.3 THERMOREGULATION IN RESPONSE TO COLD AND HEAT

 When the outside temperature is cold, the body works to maintain a constant internal temperature by conserving the heat it has and generating additional heat as well. When the outside temperature is hot, the body works to bring heat to the surface and release it into the environment.

In cold...

Vasoconstriction Conserves Heat

Skin surface temperature decreases.

Surface blood vessels decrease in diameter, limiting blood flow and thereby limiting heat loss to the environment.

Constricted surface blood vessels

Normal internal blood vessels

Shivering Generates Heat

Heat

Contraction

Skeletal muscles contract involuntarily, generating heat for the body.

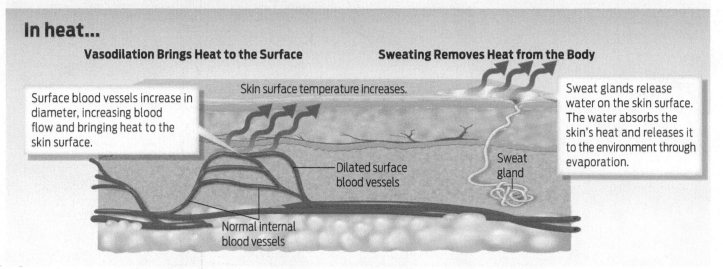

In heat...

Vasodilation Brings Heat to the Surface

Skin surface temperature increases.

Surface blood vessels increase in diameter, increasing blood flow and bringing heat to the skin surface.

Dilated surface blood vessels

Normal internal blood vessels

Sweating Removes Heat from the Body

Sweat gland

Sweat glands release water on the skin surface. The water absorbs the skin's heat and releases it to the environment through evaporation.

frostbite—the death of tissues caused by lack of blood flow (**INFOGRAPHIC 25.3**).

Krakauer and his teammates were no strangers to the cold. After weeks of slowly ascending from camp to camp, they reached the launching pad for the summit, the South Col, at 1 P.M. on May 9. "It is one of the coldest, most inhospitable places I have ever been," Krakauer wrote. A windswept saddle of rock and ice that sits between the peaks of Everest and neighboring Lhotse, the Col rests at 26,000 feet above sea level. Climbers pitch their tents on the relatively flat terrain and try not to think about the fact that they have entered what's known as the death zone.

Conditions were particularly bad on the Col that day. Gale-force winds blew through the camp, limiting visibility. As Krakauer's teammate, Beck Weathers, later recalled, "The weather was so crummy that when we first got in there, I didn't think there was any chance that we were going to climb that night."

But at 7 P.M., conditions improved markedly. It was still cold—26°C (−15°F)—but the wind had abruptly ceased, and by 11 P.M., above their heads, the stars appeared, while a gibbous moon reflected off the mountain snow. It was the perfect night for a climb.

INTO THIN AIR

The 15-member team left camp shortly after 11 P.M. Night climbing is necessary on Everest in order to arrive at the most difficult parts of the climb during daylight hours and still have enough time to get back down to camp before nightfall. Krakauer led the pack that night, along with the team's head Sherpa, Ang Dorjee.

The pair reached the Southeast Ridge, the second to last stop before the summit, at 5:30 A.M., just as the sun was peering over the eastern peaks. By this time, Krakauer's hands and feet felt like unwieldy blocks of ice, nearly useless in performing the delicate work of laying ropes and scaling ice. But it wasn't just the cold he had to deal with. His brain and body were also showing the effects of altitude: "Plodding slowly up the last few steps to the summit," wrote Krakauer in *Into Thin Air,* his 1997 book about the expedition, "I had the sensation of being underwater, of life moving at quarter speed."

At high altitudes, there is lower barometric pressure, which means there are fewer air molecules banging around in the atmosphere—including fewer molecules of oxygen. The percentage of oxygen in the air is the same as at sea level (about 20%), but since there are many fewer air molecules overall, the pressure of oxygen is much less, and therefore fewer oxygen molecules are taken in per breath of air (see Chapter 28). The lower pressure means that fewer oxygen molecules bind to the hemoglobin in blood, which means that blood is less saturated with oxygen, a condition called **hypoxia.** Since all cells require oxygen to function, hypoxia has many bodily consequences. The most serious and immediate occur in the brain. "I've been at 19- and 20,000 feet climbing myself," says physiologist Kenefick, "and I can tell you that doing simple tasks like tying your shoes—even though you've tied your

shoes many times—is much more difficult." For the hikers on Everest, he says, each step would have been a struggle.

To help cope with conditions of low oxygen, climbers spend about 6 weeks **acclimatizing** their bodies to the conditions, spending a few nights at progressively higher elevations. Their bodies respond by increasing the production of red blood cells, the cells that contain hemoglobin and carry oxygen.

▶**HYPOXIA**
A state of low oxygen concentration in the blood.

▶**ACCLIMATIZATION**
The process of physiologically adjusting to an environmental change over a period of time. Acclimatization is generally reversible.

Climbers spend weeks acclimatizing to increasingly higher elevations before attempting to reach the summit from Camp IV on the South Col, pictured below.

The physiological adjustment of acclimatization allows a climber to carry more oxygen than could someone coming straight from sea level. But even well-acclimatized climbers usually need bottled oxygen to reach the summit successfully.

At 1:17 P.M., after more than 12 hours of climbing, Krakauer finally reached the summit. It was smaller than he expected–a patch of ice the size of a picnic table, with Buddhist prayer flags tied to a pole and flapping noisily in the wind. He stood and took in the 360° view. The towering peaks of the surrounding Himalayas, which had once loomed so large, now sat below him, draped in low-lying clouds. Beyond the mountain range, endless miles of continent stretched to the horizon.

Standing on top of the world, Krakauer was surprised by his lack of elation. He had just cleared a huge personal hurdle, yet the victory felt hollow. Partly, he was too exhausted to truly care: he hadn't slept soundly in more than 50 hours, and the only food he had been able to choke down in 3 days was a bowl of ramen soup and some peanut M&Ms (sleep disturbances and digestive difficulties are additional side effects of high elevation). But another thought lurked in his brain: the oxygen tank he had slung on his back to help him breathe was running low, and he still had to get down the mountain.

"With enough determination, any bloody idiot can get up this hill," guide Rob Hall had famously said. "The trick is to get back down alive." Keenly aware of the clock, Krakauer snapped a few perfunctory photos, and within 5 minutes was headed back down the mountain toward Camp IV on the Col.

Fifteen minutes later, after scaling the steep ice fin of the Southeast Ridge, he arrived at the pronounced notch in the mountain known as the South Summit, just below the main peak. As he prepared to rappel over the edge, he caught a glimpse of an alarming sight: a queue of 20 climbers, from three separate expeditions, waiting to come up. They were backed up at the notorious Hillary Step–a 40-foot wall of rock and ice named for Sir Edmund Hillary, who, with Tenzing Norgay, was the first to scale it successfully, in 1954. Getting up the Step requires ropes, so climbers must go up one by one, and on this day there was a traffic jam.

While waiting for his turn to get down the Step, Krakauer peered into the distance and saw something he hadn't noticed before: on the horizon, dark clouds were sweeping in from the south, filling up a corner of what had been a clear blue sky. A storm was brewing.

By this point, it was well past the agreed-upon turnaround time of 1 P.M. set by Hall. The climbers who were still headed up the mountain at this hour were willfully flouting safety rules. Not only that, weather conditions were getting worse. Snow had started to fall, and it had become hard to see where mountain ended and sky began. The lower Krakauer got on the mountain, the worse the weather became.

Krakauer made it back to Camp IV on the Col just before 6 P.M. The bedraggled climber fell into his tent and quickly passed out. He was delirious, shivering uncontrollably, and exhausted. But he was alive.

SENSORS WORKING OVERTIME

Even as he slept, Krakauer's body was working hard to thermoregulate. Like many physiological processes, thermoregulation is not something that requires conscious thought. It is more like the automated response of a home heating system, triggered when the thermostat is tripped.

The body's thermostat is the **hypothalamus,** a grape-size structure that sits at the base of the brain, right above the brain stem. The hypothalamus receives signals from many different **sensors,** specialized cells in the body that detect changes in both the internal and external environment. For cold, the major sensors are thermoreceptors in the skin and in the hypothalamus itself. Information from various sensors is fed to the hypothalamus, where it is integrated

Acting as a thermostat, the hypothalamus has a specific temperature set point below

▶**HYPOTHALAMUS**
A master coordinator region of the brain responsible for a variety of physiological functions.

▶**SENSOR**
A specialized cell that detects specific sensory input like temperature, pressure, or solute concentration.

which a warning message is triggered that body temperature is dropping. When that happens, the hypothalamus tells the body to take corrective action. For example, it can send a signal to blood vessels in the skin, causing them to constrict in peripheral vasoconstriction. It can also send a signal to muscles to start shivering. Both signals are sent from the hypothalamus to their target tissues by way of nerve fibers running from the brain to the rest of the body. The cells, tissues, or organs that respond to such signals are known as **effectors:** they act to cause a change in the internal environment. Once the effectors have raised the body temperature, the sensors detect the changed conditions and the signals are turned off.

This circuit of sensing, processing, and responding is an example of a homeostatic **feedback loop** (INFOGRAPHIC 25.4). In this case, the loop is a negative feedback loop because the output of the loop—an increase in body temperature—feeds back on the sensors to *decrease* the response. As body temperature rises, the responses that would further increase body temperature—shivering and vasoconstriction—are no longer needed and so they are turned off. This loop helps keep the system at the set point.

Not all feedback loops act in a negative fashion. Positive feedback loops occur when the output of a system acts to *increase* the response of the system. An example is the formation of a blood clot when you cut yourself,

▶**EFFECTOR**
A cell or tissue that acts to exert a response on the basis of information relayed from a sensor.

▶**FEEDBACK LOOP**
A pathway that involves input from a sensor, a response via an effector, and detection of the response by the sensor.

INFOGRAPHIC 25.4 HOMEOSTASIS FEEDBACK LOOPS REQUIRE SENSORS AND EFFECTORS

By means of sensors, the body constantly monitors factors like body temperature. The sensors relay temperature information to the hypothalamus. If the temperature is too hot or too cold, the hypothalamus sends signals to effector tissues and organs that work to return the temperature to homeostasis levels.

Sensors in the skin detect a temperature increase.

Hypothalamus detects change and sends signals to effectors.

Vasodilation   Sweating

Effectors carry out a response.

Body too hot

Body cools down.

Homeostasis: Temperature set point at 37ºC (98.6ºF)

Effectors carry out a response.

Body heats up.

Body too cold

Vasoconstriction   Shivering

Hypothalamus detects change and sends signals to effectors.

Sensors in the skin detect a temperature decrease.

UP CLOSE ORGAN SYSTEMS

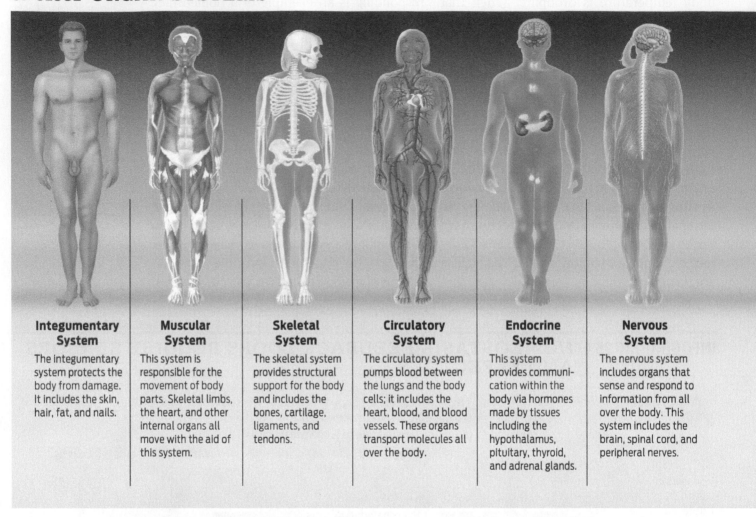

Integumentary System

The integumentary system protects the body from damage. It includes the skin, hair, fat, and nails.

Muscular System

This system is responsible for the movement of body parts. Skeletal limbs, the heart, and other internal organs all move with the aid of this system.

Skeletal System

The skeletal system provides structural support for the body and includes the bones, cartilage, ligaments, and tendons.

Circulatory System

The circulatory system pumps blood between the lungs and the body cells; it includes the heart, blood, and blood vessels. These organs transport molecules all over the body.

Endocrine System

This system provides communication within the body via hormones made by tissues including the hypothalamus, pituitary, thyroid, and adrenal glands.

Nervous System

The nervous system includes organs that sense and respond to information from all over the body. This system includes the brain, spinal cord, and peripheral nerves.

▶**NERVOUS SYSTEM**

The collection of organs that sense and respond to information, including the brain, spinal cord, and nerves.

▶**ENDOCRINE SYSTEM**

The collection of hormone-secreting glands and organs with hormone-secreting cells.

a response critical to preventing blood loss. Blood platelets stick to damaged blood vessels and release molecules that attract even more platelets to the area, which in turn attract even more platelets, and eventually a blood clot forms (see Chapter 8). Positive feedback loops are effective at rapidly amplifying a response, but negative feedback loops tend to be more common in physiology because they help return the body to its set point and ensure that homeostasis is maintained.

The hypothalamus does more than regulate body temperature. In fact, it is the body's main homeostasis control center, regulating many bodily states including hunger, thirst, sexual arousal, and sleep. The hypothalamus is part of what Kenefick calls our "lizard brain"—the evolutionarily ancient parts of the brain, which control our most basic physiological responses through unconscious reflexes.

The hypothalamus is able to play such an important role in homeostasis because it is so well connected to sensors and effectors. The hypothalamus is not only a key part of the **nervous system,** connected to parts of the body through nerves, it is also intimately associated with the **endocrine system,** which produces changes in the body through the action of chemical signals called **hormones,** which travel through the blood. The hypothalamus connects to the endocrine system via a direct circulatory connection to the **pituitary gland,** a pea-size structure that sits right below the hypothalamus. Hormones released by the hypothalamus travel directly to the pituitary gland, signaling it to release more

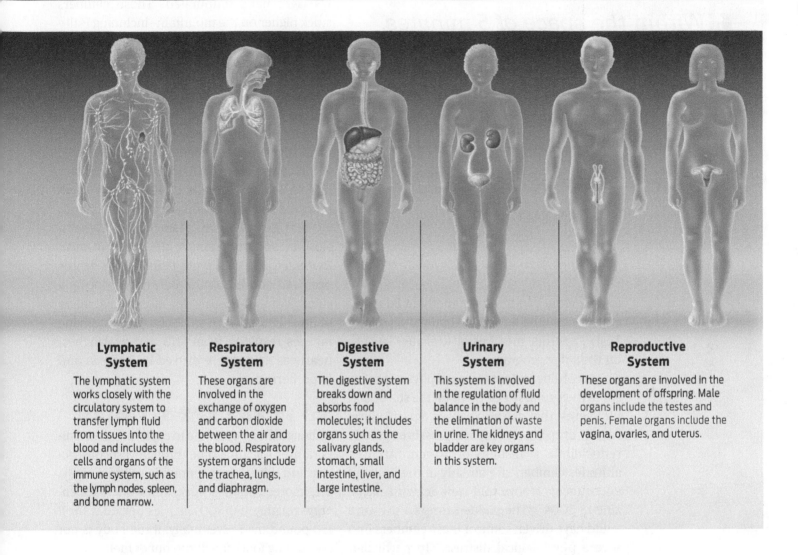

Lymphatic System

The lymphatic system works closely with the circulatory system to transfer lymph fluid from tissues into the blood and includes the cells and organs of the immune system, such as the lymph nodes, spleen, and bone marrow.

Respiratory System

These organs are involved in the exchange of oxygen and carbon dioxide between the air and the blood. Respiratory system organs include the trachea, lungs, and diaphragm.

Digestive System

The digestive system breaks down and absorbs food molecules; it includes organs such as the salivary glands, stomach, small intestine, liver, and large intestine.

Urinary System

This system is involved in the regulation of fluid balance in the body and the elimination of waste in urine. The kidneys and bladder are key organs in this system.

Reproductive System

These organs are involved in the development of offspring. Male organs include the testes and penis. Female organs include the vagina, ovaries, and uterus.

hormones, which in turn travel through the bloodstream and act on many tissues in the body, including other glands. The endocrine system, with its numerous hormone-secreting glands, is just one of 11 organ systems found in the human body that cooperate to perform the tasks necessary for survival—from food intake and waste removal to self defense and reproduction (see **Up Close: Organ Systems** and subsequent chapters in this unit).

During the night, Krakauer was awakened by a teammate who gave him grave news: a number of his teammates, including Rob Hall, had not yet returned to Camp IV. They were still out in the blistering subzero cold somewhere above 26,000 feet. Krakauer's heart sank. He knew the chances of surviving in the cold for that long were slim. By 5 P.M., everyone's oxy-

gen tank would have been empty. It was now midnight. Krakauer feared for the others' lives. But he was also dumbfounded. Hall and the rest of his team had not been far behind him on the mountain. What had gone wrong?

The storm he had spotted on the horizon began as a cyclone in the Bay of Bengal. It came in low from the valley and then rose up the mountain, gaining in ferocity and strength as it climbed. "One minute, we could look down and we could see the camp below. And the next minute, you couldn't see it," recalled Lou Kasischke, a member of Krakauer's team, who was one of 11 people trapped on the Col when the storm hit and who recounted his experience in the PBS documentary *Storm over Everest*. "Within the space of 5 minutes, it changed from really a good day with a little bit of wind

▶**HORMONE**
A chemical signaling molecule that is released by a cell or gland and travels through the bloodstream to exert an effect on target cells.

▶**PITUITARY GLAND**
An endocrine gland in the brain that secretes many important hormones.

" *Within the space of 5 minutes, it changed from really a good day with a little bit of wind to desperate conditions, something I'd never experienced the ferocity of before.* "

—JOHN TASKE

to desperate conditions, something I'd never experienced the ferocity of before," said John Taske, another member of Krakauer's team, on the same program.

According to Kent Moore, a physics professor at the University of Toronto, the storm that hit Everest that day also caused a particularly severe drop in barometric pressure, greatly reducing the availability of oxygen. "At these altitudes climbers are already at the limits of endurance," Moore told *New Scientist* magazine in 2004. "The sudden drop in pressure could have driven some of these climbers into severe physiological distress." In particular, they would have experienced the mental side effects of extremely low oxygen levels, which include confusion and disorientation.

Unable to tell in which direction they were going, and not wanting to take a wrong turn and step off a cliff, the climbers were forced to hunker down in the hurricane-force winds and wait for the storm to abate. Eventually, after 4 long hours, the clouds parted long enough for one of them to see where they were. Six climbers who were able to walk made it back to camp during this lull. An additional three were brought back safely by the efforts of Anatoli Boukreev, a Russian guide who, having descended to Camp IV, went back to search for them.

But others were not so lucky. Two climbers, too weak to make it back to camp, suffered severe frostbite before being rescued. One lost all his fingers and toes; the other had to have

▶**HYPOTHERMIA**
A drop of body temperature below 35°C (95°F), which causes enzyme malfunction and, eventually, death.

▶**GLYCOGEN**
An energy-storing carbohydrate found in liver and muscle.

▶**PANCREAS**
An organ that secretes the hormones insulin and glucagon, as well as digestive enzymes.

▶**INSULIN**
A hormone secreted by the pancreas that causes a decrease in blood sugar.

his right hand amputated. Those climbers stuck higher on the mountain—including Hall—could not be rescued. Trapped without shelter in the subzero temperatures all night, their supplemental oxygen and food long gone, the hikers eventually lost their ability to cope with the cold and succumbed to **hypothermia,** a drop in body temperature below 35°C (95°F). In all, eight climbers died on Everest that day.

This was not the first time that disaster had struck the summit. A 2008 study of all reported Everest deaths from 1921 to 2006, led by researchers at Massachusetts General Hospital and published in the *British Medical Journal,* found that more than 80% of deaths occurred above 26,000 feet, either during a summit attempt or the day after. While many of these deaths were attributable to traumatic injuries resulting from falls and avalanches, nearly as many were caused by hypoxia and hypothermia.

NO FUEL LEFT TO BURN

Although the body is able to cope with cold temperatures for some time through vasoconstriction and shivering, it cannot do so indefinitely. Thermoregulation is work, and work takes energy—roughly 150-300 Calories per hour for a 150-pound man. Eventually, if the body is not consuming food, it will run out of fuel.

The main fuel the body uses in times of intense activity is the sugar glucose, a breakdown product of carbohydrate digestion. When we eat carbohydrates, sugars are released and absorbed into the circulation, and "blood sugar" (the concentration of glucose in the blood) increases (see Chapter 4). Some of this sugar may be used immediately as fuel for aerobic respiration in cells of the body (see Chapter 6). Whatever is not needed right away will be converted into **glycogen,** which is stored in muscles and the liver.

Blood sugar is monitored and controlled by the **pancreas,** a small organ near the stomach that functions in both the endocrine and digestive systems (see **Milestone 7**). In response to high blood sugar, endocrine tissue in the pancreas produces the hormone **insulin,** which acts on liver and muscle cells, signaling them to remove sugar from the blood. The glucose taken up from the blood will either be used immediately for cellular respiration or stored as glycogen. Like many hormones, insulin acts by binding to receptors on target cells, much as a key fits into a lock.

When blood sugar is low, the body first prompts us to eat by sending a signal to the hypothalamus, which cues hunger. If eating isn't an option, the body begins to break down its stored glycogen. The key signal here is the hormone **glucagon,** which is released by the pancreas in response to low blood sugar and triggers liver and muscle cells to break down their stored glycogen into glucose. The liver can release this glucose into the blood, providing fuel for other tissues, particularly the brain. The breakdown of glycogen in

skeletal muscle helps provide energy for the muscle itself.

The amount of glucose in the blood is therefore tightly regulated: when blood- sugar levels are high, excess glucose is stored in cells as glycogen (in response to insulin). The stored glycogen represents a source of glucose to be released into the blood during periods of starvation (in response to glucagon). The opposing effects of insulin (reducing blood glucose) and glucagon (increasing blood glucose) illustrate another example of homeostasis, in this case maintaining a relatively stable blood-sugar level (**INFOGRAPHIC 25.5**).

The trapped climbers hadn't eaten in hours, which meant they were operating on stored energy. Glycogen is the main stored fuel that is tapped during vigorous exercise. But

▶GLUCAGON
A hormone produced by the pancreas that causes an increase in blood sugar.

INFOGRAPHIC 25.5 THE PANCREAS REGULATES BLOOD-GLUCOSE LEVELS

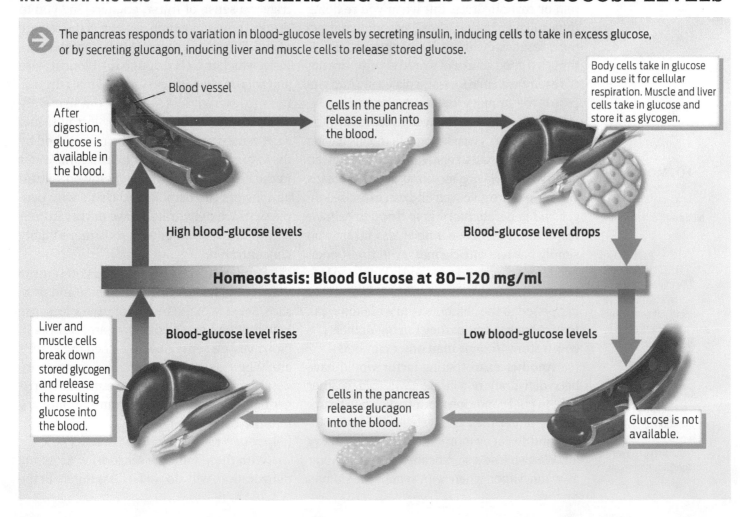

The pancreas responds to variation in blood-glucose levels by secreting insulin, inducing cells to take in excess glucose, or by secreting glucagon, inducing liver and muscle cells to release stored glucose.

Blood vessel

After digestion, glucose is available in the blood.

Cells in the pancreas release insulin into the blood.

Body cells take in glucose and use it for cellular respiration. Muscle and liver cells take in glucose and store it as glycogen.

High blood-glucose levels

Blood-glucose level drops

Homeostasis: Blood Glucose at 80–120 mg/ml

Blood-glucose level rises

Low blood-glucose levels

Liver and muscle cells break down stored glycogen and release the resulting glucose into the blood.

Cells in the pancreas release glucagon into the blood.

Glucose is not available.

the human body can store only so much glycogen. Eventually, after hiking and shivering for many hours, you will exhaust this fuel supply. Without this fuel, your body will struggle to remain active and shiver. And if you can't remain active and shiver, then you can't generate heat and your body temperature will fall. That's when hypothermia can set in.

The average adult has enough stored glycogen to power about 12 to 14 hours of routine activity. When a person is exercising strenuously—say, running or hiking—glycogen stores can be depleted in as little as 2 hours. Marathon runners often refer to this point, which occurs at about mile 20, as "hitting the wall." Then, in order to continue exercising, you must eat something—preferably something with carbohydrates.

"A lot of mountaineering communities think you need fat," says Kenefick. "And that's true—fat has more Calories per gram—9 kcals per gram compared to 4 kcals per gram of protein or carbohydrate. But when you're doing things like shivering, those types of contractions, especially, use a lot of glucose." Fats—though a good source of stored energy—are not as readily available for immediate use. And glucose is the primary fuel for the brain.

Fitness also likely played a role in how the Everest climbers fared. Being fit means having more muscle mass relative to fat for a given body weight. Having more muscle mass means you have more glycogen and can exercise longer and generate more heat through cellular respiration. Someone who is less fit, or who simply has less muscle mass, will tire sooner, need to sit down and rest, and continue to lose heat to the environment. This is likely what happened to the climbers who were too weak to hike back to camp: they ran out of their glycogen stores sooner than other climbers.

Another exacerbating factor would have been dehydration—a little-known cold-weather risk. In cold conditions, our bodies must work harder under the extra weight of heavy clothing, and sweat evaporates quickly in cold, dry air. We also lose a significant amount of water as water vapor when we exhale. The human

body is about 65% water by weight, and when the total amount of water drops by only a few percent, we become dehydrated—which can cause dangerous side effects like delirium, confusion, and convulsions. Kenefick points out that people do not feel as thirsty when it's cold, and thus become even more dehydrated. "We're really tropical animals," says Kenefick. "We came from the Sub-Sahara. We do much better in the heat."

Our body's sense of thirst relies on **osmolarity,** the concentration of dissolved solutes in the blood. Among the solutes dissolved in blood are electrolytes—ions such as sodium and potassium that are critical for nerve signaling and muscle contraction. Osmolarity is monitored by the hypothalamus as part of **osmoregulation.** When you are dehydrated—when you have less fluid in your blood—the concentration of dissolved solutes is higher. If the hypothalamus registers that the concentration of solutes in the blood is high, it will trigger a sense of thirst, encouraging you to drink. At the same time, it triggers the release of antidiuretic hormone (ADH) from the pituitary, which travels through the bloodstream and acts on the **kidneys.** ADH signals the kidneys to excrete less water in the urine. By reducing the amount of water lost in urine, ADH causes more water to be reabsorbed by the kidneys back into the bloodstream. Water in the bloodstream dilutes dissolved solutes and lowers the osmolarity. That's why people who are dehydrated have darker urine—it contains less water and so is more highly concentrated.

Osmoregulation also depends on sensors that detect changes in blood volume and pressure. Sensors in the heart, for example, sense how full the heart's chambers are; sensors in blood vessels sense how stretched the vessels are. When low blood volume and pressure are detected, the hypothalamus responds by triggering the release of ADH from the pituitary into the blood, which acts on the kidneys to help retain water (**INFOGRAPHIC 25.6**).

With these multiple sensors for detecting dehydration, why do we feel less thirsty in the

▶**OSMOLARITY**
The concentration of dissolved solutes in blood and other bodily fluids.

▶**OSMO-REGULATION**
The maintenance of relatively stable volume, pressure, and solute concentration of bodily fluids, especially blood.

▶**KIDNEY**
An organ involved in osmoregulation, filtration of blood to remove wastes, and production of several important hormones.

INFOGRAPHIC 25.6 THE KIDNEYS RESPOND TO CHANGES IN WATER BALANCE

The amount of water in the bloodstream controls the concentration of dissolved molecules in the blood and also determines blood volume and blood pressure. The kidneys control water availability by responding to a variety of signals.

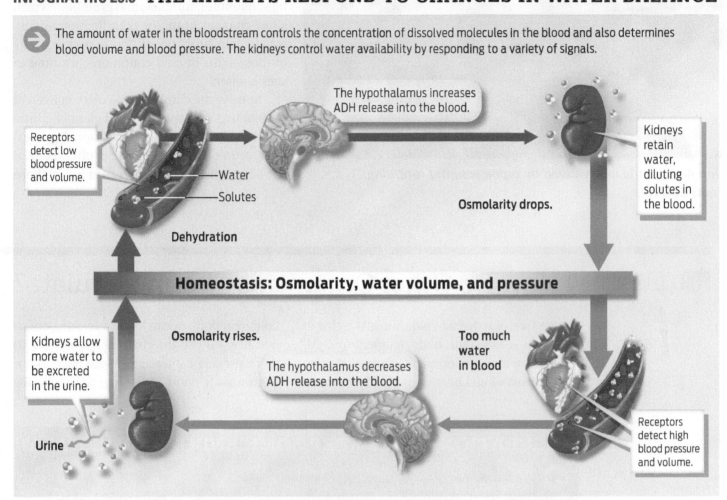

Receptors detect low blood pressure and volume.

Water

Solutes

Dehydration

The hypothalamus increases ADH release into the blood.

Osmolarity drops.

Kidneys retain water, diluting solutes in the blood.

Homeostasis: Osmolarity, water volume, and pressure

Kidneys allow more water to be excreted in the urine.

Osmolarity rises.

The hypothalamus decreases ADH release into the blood.

Too much water in blood

Urine

Receptors detect high blood pressure and volume.

---

cold? The reason, says Kenefick, is that peripheral vasoconstriction pushes blood toward the core. All that blood pushed centrally is sensed by the body as a normal amount of hydration. As a result, the sensation of thirst is reduced, despite the fact that you're dehydrated. This is why it's very important to drink adequate amounts of water in winter, even when you aren't thirsty.

"Because water plays such a large role in cellular function," says Kenefick, "being dehydrated is going to put a greater stress on your body." Dehydration can alter the concentration of electrolytes in the blood, and therefore alter nerve function and muscle contraction. Dehydration also lowers blood pressure and thus makes the heart work harder. Together, these effects can have dangerous consequences, im-

pairing thinking and coordination—two things that matter a great deal when you're navigating the treacherous terrain of the world's tallest mountain during a blizzard.

WARNING SIGNS

Hypothermia isn't only a danger for death-defying mountain climbers: it's a leading cause of death during outdoor recreation like rafting and skiing, and is the number one way to lose your life while outdoors in cold weather. The Centers for Disease Control and Prevention estimate that hypothermia causes more than 1,000 deaths each year in the United States.

Wilderness medicine experts say the best way to prevent hypothermia—in addition to dressing appropriately and carrying plenty of food and water—is to be aware of its signs.

Jon Krakauer is interviewed by reporters after returning from Everest.

In particular, watch for the "umbles": stumbles, mumbles, fumbles, and grumbles, which show changes in motor coordination and altered brain function. If you experience any of these signs in cold conditions, it's time to seek shelter.

None of the climbers who died on Everest in 1996 was an inexperienced climber—three of them, in fact, were professional guides. Why didn't they heed these physiological warning signs? Part of the reason is that there

FOR COMPARISON How Do Other Organisms Thermoregulate?

In the face of extreme cold, humans strive to maintain a constant body temperature. Although we may pile on warm clothes or sit by a fire, most of our heat is coming from metabolic reactions occurring inside our bodies. Because we use internal metabolic heat to thermoregulate, humans are classified as **endotherms** ("endo" meaning "inside"). We

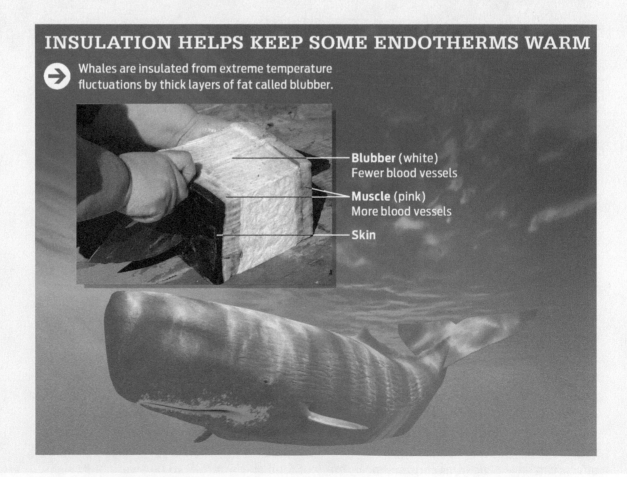

INSULATION HELPS KEEP SOME ENDOTHERMS WARM

→ Whales are insulated from extreme temperature fluctuations by thick layers of fat called blubber.

Blubber (white)
Fewer blood vessels

Muscle (pink)
More blood vessels

Skin

was simply no time. The swift-moving storm made the decision for them. But the climbers had also made questionable choices earlier that affected their fate. Whether from over-confidence or brain-addled thinking, they continued climbing toward the summit even when the hour was late. In the end, the climbers made a fatal wager with biology: in their race to the summit, they pushed themselves beyond the breaking point, overestimating, in Krakauer's words, "the thinness of the margin by which human life is sustained above 25,000 feet."

Not long after the disaster, Krakauer returned to climbing mountains. In a 1997 interview with *Bold Type* magazine, he was asked whether he was fearful of climbing again after the trauma he experienced on Everest. The chastened climber replied: "It wasn't like 'Am I afraid of this?' It was more like 'Is this right? Is it too selfish?' I won't go back to Everest—I'm afraid of that." ■

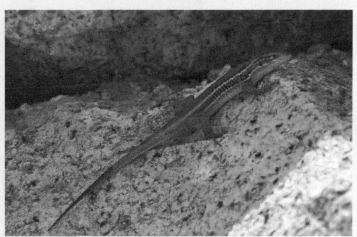

Ectotherms rely on the environment to regulate their body temperatures, seeking sun or shade as necessary.

expend a great deal of energy maintaining a warm body temperature in a cold environment.

Like us, whales are mammals and endotherms. As you can imagine, whales face a great challenge in maintaining a sufficiently warm body temperature in the cold ocean depths. They are protected from this cold by a thick layer of fat tissue called blubber. Not only is blubber an excellent insulator that helps prevent heat loss to the environment, it does not have many surface blood vessels—an adaptation that prevents heat loss from blood to the environment. Many endothermic animals rely on feathers, fur, or fat as insulation.

In contrast to endotherms, other animals must obtain their body heat from the environ-ment and are therefore known as **ectotherms** ("ecto" meaning "outside"). While these animals are commonly referred to as "cold-blooded," their body temperature actually mimics that of their environment. By using behavioral adaptations, many ectotherms also maintain a relatively stable body temperature (although by definition not as stable as that of endotherms'). For example, lizards bask in the sun to warm up and seek out shade or a protected burrow to prevent overheating. Through these behaviors, lizards keep their bodies in a temperature range compatible with their metabolism.

While basking in the sun works well for terrestrial ectotherms, this is not an option for certain fish, such as marlins. The surrounding

▶ENDOTHERM
An animal that can generate body heat internally to maintain its body temperature.

▶ECTOTHERM
An animal that relies on environmental sources of heat, such as sunlight, to maintain its body temperature.

water absorbs most of the heat of the sun, so marlins cannot heat themselves by swimming in sunny waters. Their brains would be dangerously close to freezing if not for specialized "heater tissue" that generates heat to keep their brains warm. In this case, the heater tissue is modified eye muscle that acts to generate heat rather than force or movement. This form of heating is a type of "nonshivering thermogenesis."

Another form of nonshivering thermogenesis, used by bats (and human babies), occurs in a tissue known as brown fat. Brown fat is located in the neck and shoulders areas and is several degrees Celsius warmer than the rest of the body. In brown fat, specialized mitochondria convert energy to heat rather than to ATP, and the many blood vessels in brown fat deliver that heat to other parts of the body.

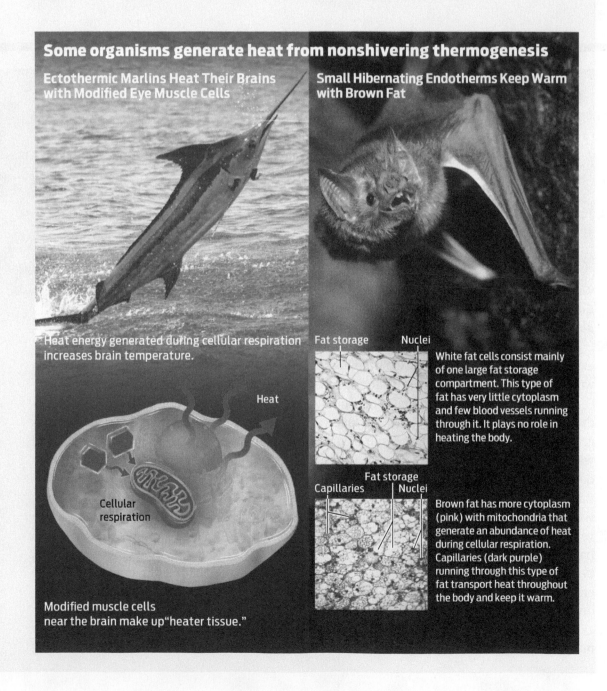

Some organisms generate heat from nonshivering thermogenesis

Ectothermic Marlins Heat Their Brains with Modified Eye Muscle Cells

Heat energy generated during cellular respiration increases brain temperature.

Heat

Cellular respiration

Modified muscle cells near the brain make up "heater tissue."

Small Hibernating Endotherms Keep Warm with Brown Fat

Fat storage   Nuclei

White fat cells consist mainly of one large fat storage compartment. This type of fat has very little cytoplasm and few blood vessels running through it. It plays no role in heating the body.

Fat storage
Capillaries   Nuclei

Brown fat has more cytoplasm (pink) with mitochondria that generate an abundance of heat during cellular respiration. Capillaries (dark purple) running through this type of fat transport heat throughout the body and keep it warm.

CHAPTER 25 Summary

- ▸ Living organisms have an anatomical structure that is adapted to suit their physiological functions.

- ▸ Humans and other multicellular organisms are organized hierarchically: cells assemble to make up tissues; tissues congregate to form organs; organs work together as part of organ systems.

- ▸ Humans have many different organ systems that cooperate to accomplish basic physiological tasks, such as obtaining energy, taking in nutrients to build new molecules during growth and repair, and ridding themselves of wastes.

- ▸ Most organisms cannot tolerate wide fluctuations in their internal environment; their bodies work to maintain homeostasis, a stable internal environment.

- ▸ The process whereby organisms maintain a relatively constant internal temperature is called thermoregulation.

- ▸ The body responds to cold temperatures in two main ways: by conserving the heat it has through vasoconstriction, and by generating more heat through shivering. When overheated, the body releases heat by vasodilation and sweating.

- ▸ Maintaining homeostasis requires both sensors and effectors. Sensors include nervous system receptors that detect changes in a variety of internal states (for example, temperature and blood pressure). Effectors include the glands and muscles that respond to an abnormal state in an effort to correct it.

- ▸ Sensors and effectors work together as part of a circuit or feedback loop. Negative feedback loops are important in homeostasis.

- ▸ Hormones are chemical messengers that travel through the bloodstream, bind to receptors on a target cell, and effect a change in that cell. Insulin and glucagon are hormones that regulate blood-glucose levels.

- ▸ Osmoregulation is the control of water balance in the body. Sensors detect blood pressure, blood volume, and solute concentration. Kidneys are important effectors in maintaining water balance.

- ▸ Maintaining homeostasis is work and requires adequate energy and oxygen to power cellular respiration.

- ▸ Humans (and other mammals) are endotherms: we generate heat internally. Many other organisms, such as reptiles and fish, are ectotherms: they rely on behavior and the environment to maintain a temperature compatible with life.

MORE TO EXPLORE

- ▸ Jon Krakauer (1996) "Into Thin Air," *Outside*. Sept 1.

- ▸ Krakauer, J. (1997) *Into Thin Air: A Personal Account of the Mt. Everest Disaster*. New York: Random House.

- ▸ PBS (2008) Frontline: *Storm over Everest*. http://www.pbs.org/wgbh/pages/frontline /everest/

- ▸ Firth, P. G., et al. (2008) Mortality on Mount Everest, 1921–2006: descriptive study. *British Medical Journal* 337:a2654.

- ▸ Moore, G. W. K., et al. (2006) Weather and death on Mount Everest: an analysis of the *Into Thin Air* storm. *Bulletin of the American Meteorological Society* 87:465–480.

CHAPTER 25 Test Your Knowledge

How are the bodies of living organisms organized?

By answering the questions below and studying Infographic 25.1, you should be able to generate an answer for the broader Driving Question above.

KNOW IT

1 Compare and contrast anatomy and physiology.

2 Organize the following terms on the basis of level of structure, from the simplest (1) to the most complex (4).

_____ Small intestine

_____ Mucus-secreting cell of the small intestine

_____ Digestive system

_____ The layer of muscle that contributes to the function of the small intestine

USE IT

3 An emergency room doctor setting a complex bone fracture is relying primarily on knowledge of

a. anatomy.
b. physiology.
c. thermoregulation.
d. homeostasis.
e. osmoregulation.

4 Is a personal trainer who works with clients to help them lose weight through a combination of diet and exercise focusing primarily on anatomy or physiology? Explain your answer.

5 Why is the heart considered an organ and not a tissue?

How do humans and other organisms regulate body temperature?

By answering the questions below and studying Infographics 25.2, 25.3, and 25.4, you should be able to generate an answer for the broader Driving Question above.

KNOW IT

6 What is homeostasis?

7 How does brown fat contribute to thermoregulation in bats?

a. by providing insulation to retain heat

b. by providing a highly vascularized tissue to release heat to the environment

c. by generating heat through shivering

d. by generating heat via cellular respiration in specialized mitochondria

e. b and d

8 Describe the feedback loop involved in thermoregulation in cold conditions. Use the following terms in your answer: hypothalamus, sensor, muscle, effector, low body temperature, normal body temperature.

USE IT

9 How could damage to the hypothalamus prevent shivering even if the core body temperature drops dramatically?

10 Why is glucagon released as part of the response to a drop in body temperature?

11 Name two or three physiological responses that could help the body dissipate heat during exertion on a hot day. For each mechanism that you propose, explain how it would dissipate heat.

What physiological systems are regulated by homeostatic feedback loops?

By answering the questions below and studying Infographics 25.5 and 25.6, you should be able to generate an answer for the broader Driving Question above.

KNOW IT

12 What internal signals are associated with dehydration?

a. high osmolarity
b. low blood volume
c. low blood pressure
d. all of the above
e. a and b

13 People who are severely dehydrated produce _____ urine that is _____.

a. a high volume of; highly concentrated and dark in color

b. a high volume of; dilute and light in color

c. a low volume of; highly concentrated and dark in color

d. a low volume of; dilute and light in color

e. a normal volume; a normal color (neither very light nor very dark)

14. Insulin is released from the _____ in response to _____.

a. pancreas; elevated blood sugar
b. pancreas; low levels of blood sugar
c. hypothalamus; elevated blood sugar
d. hypothalamus; low levels of blood sugar
e. pituitary; signals from the hypothalamus

USE IT

15. What conditions might cause high levels of insulin in the bloodstream? What events would follow?

16. Tibetan Sherpas, many of whom serve as guides and rescuers on Everest, often do not require bottled oxygen to reach the summit. Why might Tibetans, who have lived at high elevations for many generations, have an easier time than others with hypoxia? (Think about both short-term and long-term changes.)

MINI CASE

17. Hypertension (high blood pressure) has been diagnosed in a 65-year-old woman. She needs medication that will return her blood pressure to normal (and safe) levels. Her doctor tells her that there are two main categories of drugs for hypertension: thiazides, a type of diuretic, and ACE inhibitors, which help relax blood vessels and prevent their constriction. Explain how both of these could reduce blood pressure.

INTERPRETING DATA

18. Jonas and Jennifer have abnormal fasting glucose levels in their blood. They each had a blood test to measure the levels of insulin in their blood (also measured after fasting). Their blood values for both glucose and insulin are shown in the table below. From the data shown, is Jonas or Jennifer more likely to have type 2 diabetes (which is characterized by an inability of cells to respond to insulin)? Which of them is more likely to have type 1 diabetes (which results from a failure of insulin production)? Be sure you can explain all of their blood test results.

	FASTING GLUCOSE LEVEL (mg/dl)	FASTING INSULIN LEVEL (INTERNATIONAL UNITS, μIU/ml)
Jonas	115	1
Jennifer	119	35
Normal Range	70–100	5–25

BRING IT HOME

19. The U.S. National Park Service has to rescue stranded hikers, often at great expense. Do you think that hikers' level of preparation should be a factor in determining whether or not they should bear the cost of their rescue? What factors would you consider to determine whether or not a hiker was adequately prepared? Give a physiological reason for each factor that you propose.

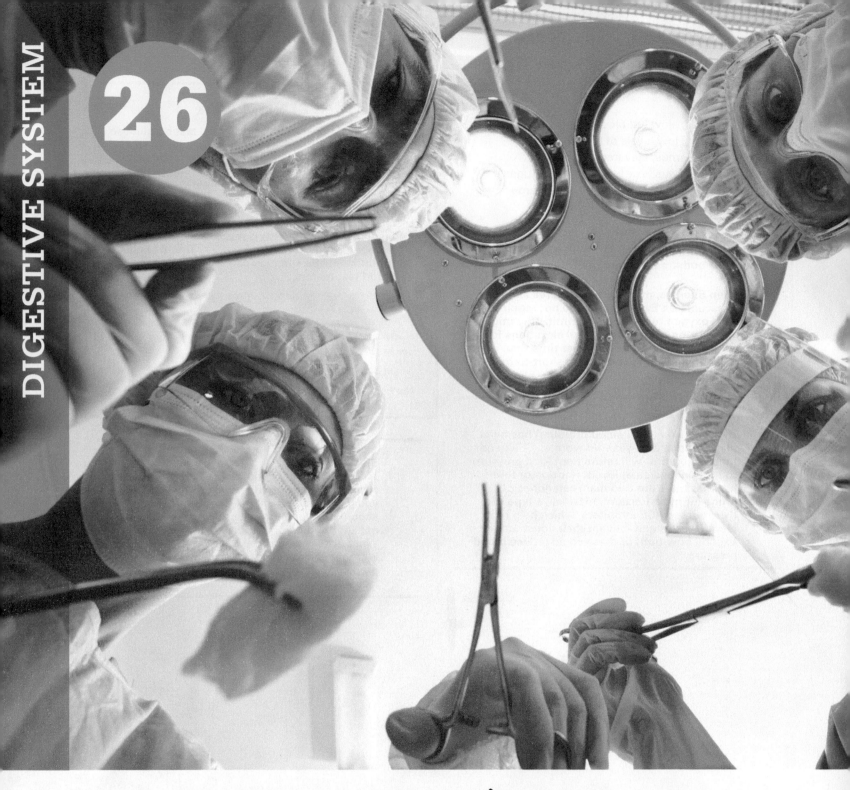

DRASTIC

For the morbidly obese, stomach

AMY JO SMITH HARDLY RECALLS A TIME GROWING UP when her family wasn't dieting. Her parents were both obese, and they were always trying to lose weight.

Smith herself was relatively slender until her senior year in high school, when her weight began to creep up. She grew up on a horse farm in northeast Maryland, and as a teenager spent much of her spare time on the road, taking her horses to shows. She attributes her weight gain to a diet that consisted primarily of fast food. "I was always eating on the run," says Smith, now 38 years old and a computer literacy teacher. But, she says, her growing girth "never stopped me from doing the things I wanted to do." The extra weight did bother her, though, and she tried several diets and diet pills, only to see her weight yo-yo up and down.

In 2004, at a routine checkup, Smith's doctor noticed that she suffered a number of ills that were likely caused by Smith's 264-pound weight. For one thing, she had been suffering from migraines. She also had stress incontinence—a bladder that leaked when stressed by coughing or laughing, for

▸▸ DRIVING QUESTIONS

1. What is the anatomy of the digestive system?
2. How is food broken down and utilized as it moves through the digestive tract?
3. What are the risks and benefits of bariatric surgery?

Measures
-shrinking surgery is a last resort

Studies show that even 10 years after surgery, most patients still weigh 25% to 30% less than they did before the surgery.

example. "I thought that was just normal," she says. And she went months at a time without having a period–a telltale sign of a hormonal imbalance often associated with obesity. Smith's physician suggested that, to lose weight, Smith consider having a surgical procedure that would shrink her stomach to the size of a golf ball–a drastic measure, but one that her doctor believed was warranted given the health complications caused by Smith's weight.

Though we often use the word "fat" casually, obesity–having too much body fat–is actually a medical condition. Obesity is defined as weighing 20% or more than one's "ideal" body weight (based on insurance tables of death rates) or having a body mass index (BMI)

of 30 or higher (see Chapter 6). Morbid obesity–sometimes called clinically severe obesity–is defined as being 100 pounds or more overweight or having a BMI of 40 or higher. Obesity becomes "morbid" when it significantly increases the risk of one or more obesity-related health conditions or serious diseases, such as heart disease and diabetes. At 5′ 2″, and with a BMI of 48, Smith had become morbidly obese.

Even then, Smith had a hard time accepting that she needed such a drastic method to lose weight. "At first I thought he was a quack," she says of her doctor. But she began to think more about the surgery after a friend underwent the stomach-shrinking surgery with dramatic results.

There are several types of bariatric, or weight-loss, surgery. ("Bariatric" comes from the Greek *baros,* meaning "heaviness.") All involve surgically reducing the size of the stomach, either temporarily or permanently. The specific type of surgery recommended depends on the individual patient's medical history and weight-loss goal. Because there are many associated risks, a National Institutes of Health panel of experts has recommended surgery only for people considered morbidly obese–people whose risk of death from diabetes or heart disease because of excess weight is five to seven times greater than for those of average weight.

But bariatric surgery is no miracle cure. "It's sort of barbaric," says Monica Skarulis, director of the Metabolic Clinical Research Unit at the National Institutes of Health. Because the surgery so drastically reduces the size of the stomach and restricts how much a person can eat, it amounts to "forced behavior

Amy Jo Smith in 2004.

control," since patients must live on a strict diet, she says. If they overeat, they suffer nasty side effects such as vomiting and diarrhea. On top of that, some gastric bypass patients suffer mineral and vitamin deficiencies over the long term that cause bone loss and potentially other health impairments. And the surgery itself is risky: as many as 20% of patients suffer complications a year after the surgery that are severe enough to put them back in the hospital.

For some morbidly obese people, however, the risk of dying from obesity-related diseases is higher than the risk of surgical complications. And in terms of weight reduction, the surgery is more effective than lifestyle changes alone. Almost all patients lose 30% to 50% of their excess weight in the first 6 months and 77% of their excess weight after about a year. Studies show that even 10 years after surgery, most patients still weigh 25% to 30% less than they did before the surgery. Consequently, demand for the surgery has soared in the United States since the early 2000s, peaking at about 220,000 surgeries a year.

FOOD FIX

The rationale behind bariatric surgery is simple: by reducing the amount of food the stomach can hold, the surgeries prevent a person from overeating. Reducing the amount of food a person takes in means less food is digested, fewer Calories are absorbed into the body, and the patient eventually loses weight.

The stomach is an easy target for a surgical fix to overeating because it's relatively simple to reduce its size surgically. The two most common types of bariatric surgery are gastric bypass and adjustable gastric banding ("gastric" means "of the stomach"). In gastric bypass, the stomach is surgically made smaller using staples. In adjustable gastric banding, a surgeon wraps an adjustable band around the stomach to make it smaller so that it holds less food.

The stomach is a key part of the **digestive system,** which breaks down food molecules into smaller subunits, absorbs nutrients,

INFOGRAPHIC 26.1 THE DIGESTIVE SYSTEM

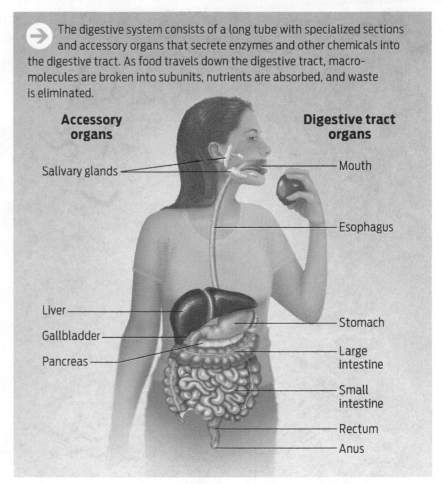

The digestive system consists of a long tube with specialized sections and accessory organs that secrete enzymes and other chemicals into the digestive tract. As food travels down the digestive tract, macromolecules are broken into subunits, nutrients are absorbed, and waste is eliminated.

Accessory organs

Salivary glands

Liver
Gallbladder
Pancreas

Digestive tract organs

Mouth

Esophagus

Stomach
Large intestine
Small intestine
Rectum
Anus

and eliminates waste. The digestive system is composed of a central **digestive tract**—essentially a long tube lined with muscles that extends from the mouth to the anus—as well as accessory organs that assist in digestion. As the muscles of the digestive tract alternately relax and contract, food is pushed along; accessory organs located along the length of the tract secrete enzymes and other chemicals into the tract to help break down food molecules. Through these coordinated actions, the digestive system transforms the food we eat into a form our bodies can use and rids the body of the waste left over once usable nutrients are removed from food we have taken in (**INFOGRAPHIC 26.1**).

Digestion, the process of breaking down food molecules so they can be absorbed by the body, begins as soon as we put food into

▶**DIGESTIVE SYSTEM**

The organ system that breaks down food molecules into smaller subunits, absorbs nutrients, and eliminates waste; it is composed of the digestive tract and accessory organs.

▶**DIGESTIVE TRACT**

The central pathway of the digestive system; it is a long muscular tube that pushes food between the mouth and the anus.

▶**DIGESTION**

The mechanical and chemical breakdown of food into subunits, enabling the absorption of nutrients.

INFOGRAPHIC 26.2 THE UPPER DIGESTIVE TRACT

→ The upper digestive tract includes the mouth, esophagus, and stomach, as well as enzymes and other chemicals secreted by the salivary glands and the stomach.

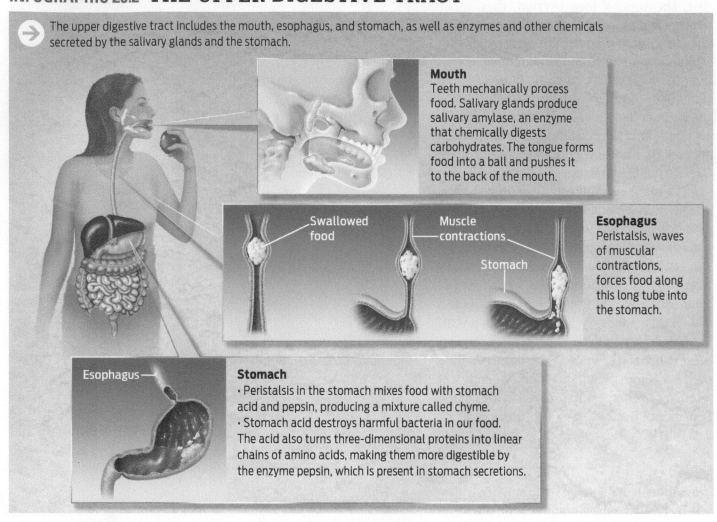

Mouth
Teeth mechanically process food. Salivary glands produce salivary amylase, an enzyme that chemically digests carbohydrates. The tongue forms food into a ball and pushes it to the back of the mouth.

Swallowed food

Muscle contractions

Stomach

Esophagus
Peristalsis, waves of muscular contractions, forces food along this long tube into the stomach.

Esophagus

Stomach
• Peristalsis in the stomach mixes food with stomach acid and pepsin, producing a mixture called chyme.
• Stomach acid destroys harmful bacteria in our food. The acid also turns three-dimensional proteins into linear chains of amino acids, making them more digestible by the enzyme pepsin, which is present in stomach secretions.

▶**INGESTION**
The act of taking food into the mouth.

▶**SALIVARY GLANDS**
Glands that secrete enzymes, including salivary amylase, which digests carbohydrates, into the mouth.

▶**ESOPHAGUS**
The section of the digestive tract between the mouth and the stomach.

▶**PERISTALSIS**
Coordinated muscular contractions that force food down the digestive tract.

our mouths–**ingest** it. The act of chewing mechanically breaks food down into smaller pieces, while **salivary glands** in our mouth secrete enzymes into saliva that chemically dismantle macromolecules into their subunits. Salivary amylase, for example, breaks down carbohydrates into simpler sugars. The tongue compresses the food into a ball and works it to the back of the mouth.

When we swallow, food is propelled along the **esophagus,** the section of the digestive tract between the mouth and the stomach, by rhythmic waves of contracting muscles in a process called **peristalsis.** Food then enters the **stomach,** where stomach acid–which has a pH of close to 1, approximately the same as battery acid–destroys harmful bacteria and

protects us against food-borne diseases. Stomach acid also causes proteins in food to lose their three-dimensional shapes, turning them into linear chains of amino acids. This makes it easier for the enzyme **pepsin,** which is produced in the stomach, to chemically break proteins apart into individual amino acids.

Like the esophagus, the stomach is muscular, expanding and contracting as it accepts food and churns it. Each time it contracts, stomach acid mixes with food, producing a soupy mixture called **chyme** (**INFOGRAPHIC 26.2**).

Stomach acid is strong and can dissolve most foods. But despite so much acid churning inside, the stomach itself remains intact. This is possible because the stomach is lined with a thick layer of protective mucus. Occasionally, this mucus layer is damaged–by a bacterial infection, for example–and the stomach lining becomes more vulnerable to gastric juices; the result is a painful sore called an ulcer.

While the stomach can absorb some substances, such as water, ethanol, and certain drugs, directly into the bloodstream, most of the chyme is pushed farther down the digestive tract, primarily into the small intestine, where it is further processed.

Although the stomach is only a small part of the upper digestive tract, it plays a large part in weight gain. Evolutionarily speaking, the reason we have a stomach in the first place is to enable us to store food. Without a stomach, we would have to eat constantly to fuel our activities. When we eat a large meal, the stomach expands greatly to accommodate and store all that food. It's partly because of this elasticity that we can eat enough to sustain us for hours. But we can also eat more than our bodies need.

STOMACH STAPLE

Because of Smith's weight-related illnesses, her doctor recommended that she have bariatric surgery. The doctor explained that bariatric surgery changes the anatomy of the digestive system to limit the amount of food a person can eat and digest before feeling full. Of the different types of bariatric surgery, gastric bypass is the most commonly performed, and also the most drastic: a surgeon uses staples to seal the stomach except for a small pouch at the top. The pouch is about the size of a golf ball. The surgeon then cuts the **small intestine** and sews part of it directly onto the pouch. Consequently, food is redirected, bypassing most of the stomach and the first section of the small intestine. The rest of digestion continues normally (**INFOGRAPHIC 26.3**).

Bypassing part of the small intestine contributes to weight loss because the small intestine is the primary organ that extracts nutrients

▶**STOMACH**
An expandable muscular organ that stores, mechanically breaks down, and digests proteins in food.

▶**PEPSIN**
A protein-digesting enzyme that is active in the stomach.

▶**CHYME**
The acidic "soup" of partially digested food that leaves the stomach and enters the small intestine.

▶**SMALL INTESTINE**
The organ in which the bulk of chemical digestion and absorption of food occurs.

INFOGRAPHIC 26.3 GASTRIC BYPASS SURGERY

In gastric bypass surgery, the most common type of bariatric surgery, most food is rerouted directly to the small intestine. The surgery reduces the amount of food that a person can ingest and also reduces the amount of nutrients and Calories that a person can absorb.

Before surgery

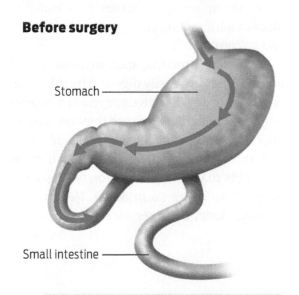

Stomach

Small intestine

Food passes into the stomach, which can stretch from the size of a large sausage (when empty) to hold 3–4 liters of food. Chyme made in the stomach enters the small intestine, where digestion releases food nutrients.

After surgery

New stomach pouch

Staples

Bypassed part of stomach

Small intestine

Unused portion of small intestine

The stomach is reduced to the size of a golf ball, diminishing drastically the amount of food it can hold. Also, stomach contents bypass the upper part of the small intestine, reducing the amount of nutrients (and therefore of Calories) the person can absorb.

▶**DUODENUM**
The first portion of the small intestine; the duodenum receives chyme from the stomach and mixes it with digestive secretions from other organs.

▶**PANCREAS**
An organ that helps digestion by producing enzymes (such as lipase) that act in the small intestine, and by secreting a juice that neutralizes acidic chyme.

▶**LIVER**
An organ that aids digestion by producing bile salts that emulsify fats.

▶**BILE SALTS**
Chemicals produced by the liver and stored by the gallbladder that emulsify fats so that they can be chemically digested by enzymes.

▶**EMULSIFY**
To break up large fat globules into smaller fat droplets that can be more efficiently chemically digested by enzymes.

▶**GALLBLADDER**
An organ that stores bile salts and releases them as needed into the small intestine.

▶**LIPASE**
A fat-digesting enzyme active in the small intestine.

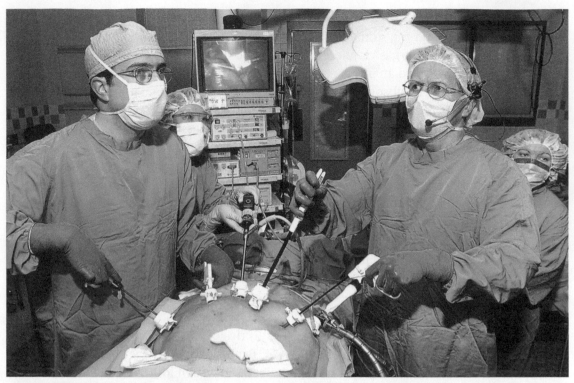

Doctors performing gastric bypass surgery with the help of a tiny camera, called a laparoscope, inserted into the body.

from food. The "small" intestine isn't actually small. It winds through the lower part of the abdomen for about 20 feet. It's called "small" because its diameter is smaller than that of the large intestine, which comes later in the digestive tract. The first part of the small intestine (about 10 inches or so) is the **duodenum,** and this is where chyme from the stomach combines with digestive secretions from several other organs. The **pancreas,** for example, secretes pancreatic juice, a basic fluid (pH greater than 7), into the small intestine to neutralize the acidic chyme (pH less than 7), which would otherwise damage the small intestine. The pancreas also secretes enzymes that help break or-

ganic molecules such as carbohydrates, proteins, and fats into smaller pieces. Enzymes secreted by the small intestine itself further break down macromolecules into building blocks such as amino acids, sugars, fatty acids, and glycerol.

Whereas proteins, carbohydrates, and nucleic acids are easily digested by this powerful mixture of digestive enzymes, fats pose a special challenge. Because they are hydrophobic (see Chapter 2), fats don't mix well with the watery solutions in the small intestine. This makes it difficult for fat-digesting enzymes to break them down. Helping the process along, the **liver** secretes **bile salts,**

The "small" intestine isn't actually small. It winds through the lower part of the abdomen for about 20 feet.

which are chemically suited to dividing large hydrophobic fat globules into smaller droplets—that is, **emulsifying** them. These bile salts pass from the liver into the **gallbladder,** which stores them for future use. When we eat a high-fat meal, bile salts pass from the gallbladder into the small intestine, where they help emulsify the fats. Once the fats are emulsified, **lipase,** a lipid-digesting enzyme secreted by the pancreas, chemically breaks them down to release their constituent fatty acids and glycerol (**INFOGRAPHIC 26.4**).

The liver and gallbladder are crucial organs of the digestive system. A person with no gallbladder or a dysfunctional one—for example, someone who had gallstones and has had the gallbladder removed—cannot digest fats well. Although the liver may still secrete bile salts, they can't be stored. If that person eats a high-fat meal, most of the fat taken in will pass through the body undigested and be eliminated in the feces.

Once digested into their smallest subunits, food molecules are taken up by **epithelial cells** lining the small intestine. This stage of digestion is called **absorption.** The inner lining of the small intestine is folded into finger-like projections called **villi** that are composed of many densely packed epithelial cells. The folds greatly increase the surface area through which the intestine can absorb nutrients. The food molecules then pass into blood vessels of the circulatory system, and the bloodstream transports them throughout the body, where

▶**EPITHELIAL CELLS**
Cells that line organs and body cavities; in the digestive tract they sit in direct contact with food and its breakdown products.

▶**ABSORPTION**
The uptake of digested food molecules by the epithelial cells lining the small intestine.

▶**VILLI (SINGULAR: VILLUS)**
Fingerlike projections of folds in the lining of the small intestine that are responsible for most nutrient and water absorption.

INFOGRAPHIC 26.4 MOST CHEMICAL DIGESTION OCCURS IN THE SMALL INTESTINE

→ The small intestine is the major organ that digests food. Accessory organs secrete enzymes and other substances into the small intestine, which itself also produces digestive enzymes.

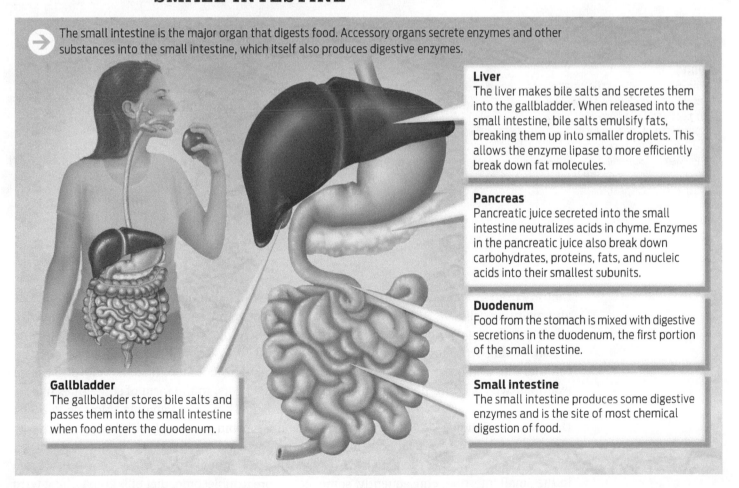

Liver
The liver makes bile salts and secretes them into the gallbladder. When released into the small intestine, bile salts emulsify fats, breaking them up into smaller droplets. This allows the enzyme lipase to more efficiently break down fat molecules.

Pancreas
Pancreatic juice secreted into the small intestine neutralizes acids in chyme. Enzymes in the pancreatic juice also break down carbohydrates, proteins, fats, and nucleic acids into their smallest subunits.

Duodenum
Food from the stomach is mixed with digestive secretions in the duodenum, the first portion of the small intestine.

Gallbladder
The gallbladder stores bile salts and passes them into the small intestine when food enters the duodenum.

Small intestine
The small intestine produces some digestive enzymes and is the site of most chemical digestion of food.

INFOGRAPHIC 26.5 THE SMALL INTESTINE ABSORBS NUTRIENTS

→ The small intestine is the primary organ that absorbs nutrients from food. Nutrients enter the circulatory system via blood vessels connected to the small intestine.

Intestine

Blood vessels

Digested nutrients

Muscle layers

Villi
The inner surface of the small intestine is folded into fingerlike projections called villi, which are composed of many densely packed epithelial cells. Villi increase the surface area of the small intestine, enabling more nutrients to be absorbed.

Blood capillary

Blood vessels

Blood vessels in villi
Nutrients, including digested food molecules and vitamins and minerals released from food, are absorbed by the intestinal lining. On the other side of the lining, nutrients enter blood vessels of the circulatory system, which transports nutrients to the rest of the body.

they are used as a source of nutrients to build and maintain cells (**INFOGRAPHIC 26.5**).

Because most digestion occurs in the small intestine, that structure is a common target for diet pills and other supplements. Alli, for example, which is the first over-the-counter weight-loss pill to be approved by the Food and Drug Administration (FDA), prevents the breakdown of up to 25% of the fat consumed. The drug works by blocking the activity of lipase in the small intestine; consequently, some fat remains undigested and passes out in the feces.

While diet pills can help some people lose weight temporarily, they are largely ineffective in the long term, especially for people who need to lose more than just a few pounds. (They also come with some very unpleasant side effects, in the form of sometimes uncontrollable bowel movements.) Most people who are morbidly obese, Smith included, have already tried various diet pills and lost weight, but not enough to improve their health. Even more troublesome, diet pills aren't a long-term solution. Over time, most people gain all of the lost weight back, and more, after they stop taking the pills. Smith's weight fluctuated over the years, and though she weighed 264 pounds when her doctor recommended surgery, this weight was likely not her heaviest, she says.

The other main type of weight-loss surgery, which Smith also considered, does not involve altering the small intestine. Called adjustable gastric banding, the procedure involves partitioning the stomach into two parts by wrapping an inflatable plastic band around the upper part of the stomach to seal it off and create a small pouch. The principle is similar to that of gastric bypass—making the stomach smaller—except that with an adjustable gastric band food still passes through the entire stomach to the intact small intestine. The pouch holds approximately ½ cup of food, whereas the typical stomach holds about 6 cups. When the upper part of the stomach is filled the brain receives a signal that the entire stomach is full. This "misinformation" helps one feel full for longer periods of time, eat smaller portions, and lose

weight over time. The band can be adjusted or surgically removed if necessary. Banding is becoming more popular because it's simpler and has a lower complication rate than gastric bypass surgery (**INFOGRAPHIC 26.6**). In fact, the FDA recently approved it for more widespread use: now people with a BMI of at least 30 may qualify for the procedure.

Gastric bypass, however, is still the more commonly performed procedure—in part because it's more effective in reducing weight than adjustable gastric banding. That's because gastric bypass fundamentally alters the normal path of digestion through the stomach and small intestine; food bypasses not only most of the stomach but also more than half of the small intestine. Consequently, not as many nutrients and associated Calories are absorbed into the bloodstream. Smith chose gastric bypass surgery over adjustable banding because, she says, she "wanted a permanent fix." She went on: "Bands can always be removed. This was a lifelong decision and commitment for

INFOGRAPHIC 26.6 GASTRIC BANDING SURGERY

→ In this procedure, an adjustable band is used to make the stomach functionally smaller. The upper pouch cannot hold as much food as a normal-size stomach, limiting food intake. Food still passes from the stomach to the upper portion of the small intestine.

Stomach pouch can hold only a small meal.

Food molecules

Only small amounts of food can pass through the gastric band and into the small intestine.

Gastric band

Larger stomach portion

Port

The inflatable pillows that line the gastric band squeeze the stomach to slow the passage of food.

Inflated band

The port is placed just under the skin and is used to inflate or deflate the gastric band to adjust the amount of food that can pass through.

INFOGRAPHIC 26.7 THE LARGE INTESTINE

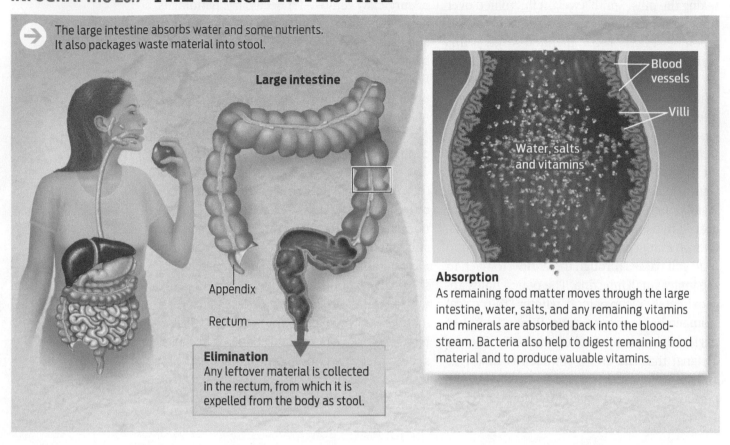

→ The large intestine absorbs water and some nutrients. It also packages waste material into stool.

Large intestine

Appendix

Rectum

Elimination
Any leftover material is collected in the rectum, from which it is expelled from the body as stool.

Blood vessels

Villi

Water, salts and vitamins

Absorption
As remaining food matter moves through the large intestine, water, salts, and any remaining vitamins and minerals are absorbed back into the bloodstream. Bacteria also help to digest remaining food material and to produce valuable vitamins.

▶**LARGE INTESTINE**
The last organ of the digestive tract, in which remaining water is absorbed and solid stool is formed.

▶**COLON**
The first and longest portion of the large intestine; the colon plays an important role in water reabsorption.

▶**STOOL**
Solid waste material eliminated from the digestive tract.

▶**ELIMINATION**
The expulsion of undigested matter in the form of stool.

me so the idea of being able to change your mind was not for me."

Once chyme passes through the small intestine, it moves on to the **large intestine.** The large intestine functions like a trash compactor—it holds and compresses material that the body can't use or digest, such as plant fiber. Within the **colon,** the first section of the large intestine, fiber, small amounts of water, vitamins, and other substances mix with mucus and bacteria that normally live in the large intestine. As this waste travels through the colon, most of the water and some vitamins and minerals are reabsorbed into the body through the colon lining. Bacteria chemically break down some of the fiber to produce nutrients for their own survival and also to nourish cells lining the colon, which is one reason fiber is an important dietary nutrient. As the large intestine expands and contracts, it pushes what ultimately becomes **stool** into the rectum, from which it is expelled

through the anus—a process called **elimination** (INFOGRAPHIC 26.7).

COSTS AND BENEFITS OF SURGERY

After looking into various medical centers and attending informational sessions, Smith decided to have her surgery at Christiana Care Hospital in Wilmington, Delaware, in August 2009. She went through the hospital's presurgical screening program, which entailed a number of medical tests and consultation with a team of doctors that included a cardiologist, a pulmonologist, a psychologist, a dietician, and other specialists to ensure that she was physically and mentally fit for the surgery. Often there are previously undiagnosed medical problems that must be considered or treated before a patient can undergo surgery.

After the surgery Smith lost weight, but it wasn't a smooth ride. The stomach takes time to heal, and so doctors advise patients to ingest only liquids for the first few weeks, puréed foods for the next few weeks, and then gradually progress to solid foods. Because the stomach is made so small, it can carry only about an ounce of food at a time–a handful of crackers or a few broccoli florets. Eating too much at once can cause vomiting or intense stomach pain. But some people find it hard to control their eating habits and wind up suffering.

Moreover, patients must also stick to a special diet after the surgery or suffer other unpleasant consequences. Patients must introduce carbohydrates like breads and pasta into their diet very slowly, says Skarulis, of the National Institutes of Health. If they eat too many simple carbohydrates, the carbs enter the small intestine too quickly. This effect, called gastric dumping, causes nausea and massive diarrhea.

But the surgery does lead to weight loss. In 2007, Swedish researchers published results from a study in which they followed about 2,000 obese patients who had undergone weight-loss surgery–either gastric bypass or surgical banding–over 15 years and compared them to about 2,000 similarly obese people who didn't have surgery but who were counseled in diet and exercise. After 10 years, those who had gastric bypass surgery weighed 25% less than their presurgery weight; those who had stomach-banding surgery were down about 15%. Those who got traditional diet advice lost no more than 2% of their weight. And research suggests that this weight loss, despite the nasty side effects, does more good than harm. There were 129 deaths in the diet-only group, mostly from weight-related heart disease and cancer, and 101 deaths in the surgery group–a large difference statistically. Deaths in the surgery group were also mainly from heart disease and cancer, although there were half the number of heart attack deaths in this group compared with the diet group (INFOGRAPHIC 26.8).

Even more encouraging to scientists is the finding that both major types of weight-loss surgery can reverse type 2 diabetes, a condition in which the body can no longer regulate blood-sugar levels and that over time can damage organs and nerves (see Milestone 7). In 2004, a review in the *Journal of the American Medical Association* of 130 studies of more than 22,000

INFOGRAPHIC 26.8 WEIGHT-LOSS SURGERY SAVES LIVES

A multiyear study of obese people published in 2007 showed that those who underwent weight-loss surgery lost significantly more weight than those given only weight-loss counseling; there were also fewer deaths in the weight-loss group than in the counseling-only group.

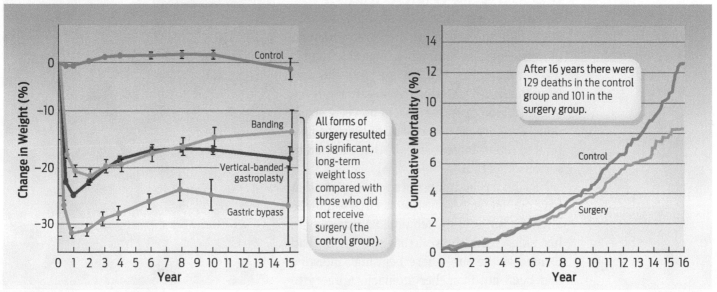

SOURCE: Sjöström L., et al. (2007) Effects of bariatric surgery on mortality in Swedish obese subjects. *New England Journal of Medicine* 357:741-752.

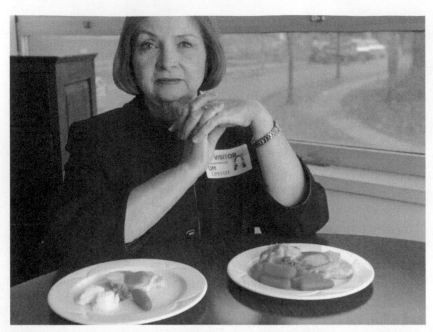

After having gastric bypass, patients must reduce food portion size. Since her surgery, this patient can only comfortably eat about a half cup of food at Thanksgiving (plate on the left).

patients found that 77% of diabetics who have bariatric surgery are cured of diabetes and that 86% are either cured or have their symptoms improve. A 2012 study published in the *New England Journal of Medicine* showed that more than 75% of obese patients who had bariatric surgery achieved remission of their diabetes after 2 years compared to 0% of patients who had conventional medical therapy for diabetes.

NO CURE FOR OBESITY

Bariatric surgery isn't for everyone. As with any major surgery, there are risks. Although bariatric surgery is much safer today than it was 10 years ago, 1 in 200 patients still dies from the surgery. Complications after surgery include blood clots, hernias, or bowel obstructions. Patients can also end up back in the hospital to repair intestinal leaks that can lead to serious infection.

Smith has had a host of complications that have put her back in the hospital. A few months after her surgery, she felt terrible cramping in her side. Tests showed that scar tissue had formed at the site where her small intestine had been cut from her stomach. Surgery to remove the tissue revealed that part of her intestine and stomach had twisted and anchored onto this scar tissue, causing her pain. Even after the surgery, she was still having stomach pains when she ate, so doctors decided to put a feeding tube into her stomach and a catheter into a vein in her arm through which she could take nutrients directly into her bloodstream. Smith spent weeks in and out of the hospital between January and April of 2010.

Smith's health has improved since then, but she is still at risk. Since people who have gastric bypass surgery (as opposed to gastric banding) end up with part of the small intestine bypassed, they absorb fewer of the micronutrients they eat. Patients must take vitamin supplements for the rest of their lives. There may be additional micronutrient deficiencies that scientists haven't yet recognized; only long-term follow-up of these patients will reveal how serious a problem this is. To monitor her micronutrient levels, Smith has a blood test every 3 months.

What's more, the surgery is not a permanent cure for obesity. Most people who have the surgery regain some weight over time. The

Amy Jo Smith in 2011 after her weight-loss surgery.

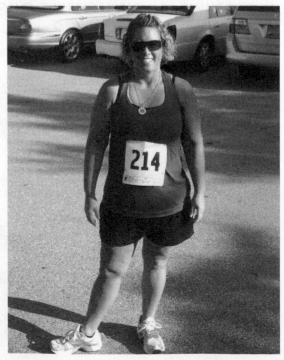

stomach and intestines can sometimes expand to allow greater ingestion of food. And hormonal changes may occur after surgery that alter body metabolism. While scientists are still studying the hormonal mechanisms of weight gain, they do know that if people do not exercise control over their diet and lifestyle, even those who have had surgery can regain significant amounts of weight.

Smith's own experience with bariatric surgery has been mixed. She still struggles with nausea every day; strong smells can cause her to vomit. She also feels pain in her left side, for which she takes medication.

Despite these complications, Smith has had not "one day of regret," she says. On her first "surgiversary"–her surgical anniversary–she wrote a letter to her surgical team in which she said: "I have been blessed with 35 birth-

days but none can compare to my surgiversary. I never imagined in a year that I would lose over 100 lbs, run a 5K the day of my surgiversary . . . sit sideways in a student desk, wear a size 12 pants from a 24-26 . . . be able to sit comfortably in a restaurant booth, and be able to stand on a table or chair without thinking, 'My gosh, am I going to break this?'"

By March 2011, Smith had dropped down to 146 pounds. To maintain her weight loss, she wakes up at 3:30 every morning to get to the gym to exercise, and she adheres to a strict diet that is high in protein and sparse on simple sugars like sweets. "I don't recognize myself anymore," Smith says.

Of her strict regimen, Smith says, "The surgery is merely a tool. If you aren't willing to make a lifestyle change, it's not going to work for you." ■

FOR COMPARISON How Do Other Organisms Digest Food?

Humans and many other animals have what's known as a complete digestive tract–one shaped like a tube, with two openings: a mouth and an anus. Not all organisms have such a tubelike digestive tract. In fact, many organisms have no digestive tract at all and yet are still able to obtain and process food from their environment.

Take fungi, for example. Fungi are eukaryotic organisms. Some, such as yeasts, are unicellular, while the majority are multicellular. Multicellular fungi have bodies that are made up of microscopic filaments called hyphae. Their bodies do not have distinct organs or organ systems, and they do not have digestive tracts. To obtain nutrients, fungi extend hyphae into food–a piece of bread, say, or leaf litter in a forest. Individual hypha cells then release digestive enzymes directly into the food and absorb the digested nutrients directly. Since digestion occurs outside

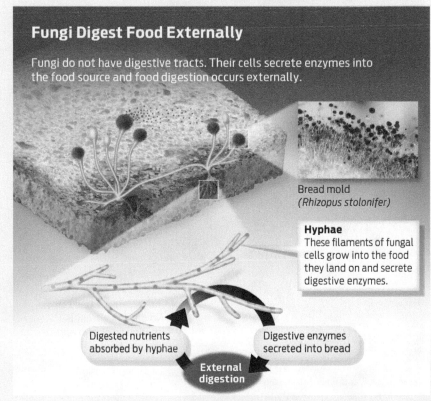

Fungi Digest Food Externally

Fungi do not have digestive tracts. Their cells secrete enzymes into the food source and food digestion occurs externally.

Bread mold
(*Rhizopus stolonifer*)

Hyphae
These filaments of fungal cells grow into the food they land on and secrete digestive enzymes.

Digested nutrients absorbed by hyphae

Digestive enzymes secreted into bread

External digestion

Sea Anemones Have an Incomplete Digestive Tract

While digestion occurs internally in sea anemones, their digestive tracts have only one opening through which food enters and waste exits—the mouth.

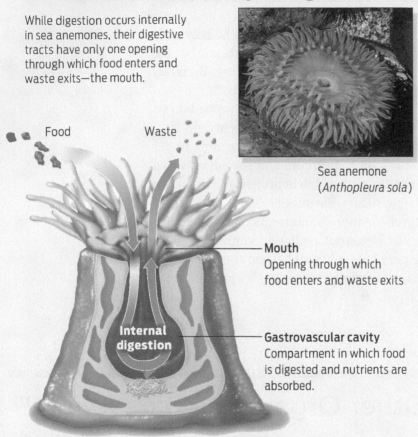

Sea anemone
(*Anthopleura sola*)

Food

Waste

Mouth
Opening through which food enters and waste exits

Internal digestion

Gastrovascular cavity
Compartment in which food is digested and nutrients are absorbed.

Plants Have No Digestive Tract

Because plants rely on photosynthesis to make their own food, they do not have a digestive tract.

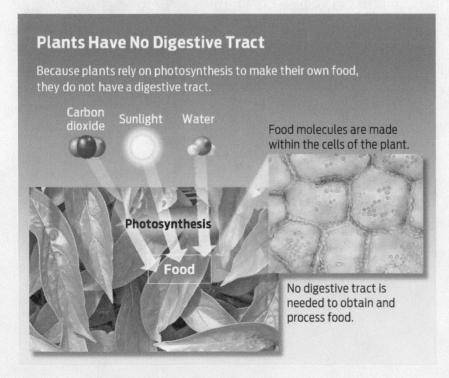

Carbon dioxide

Sunlight

Water

Food molecules are made within the cells of the plant.

Photosynthesis

Food

No digestive tract is needed to obtain and process food.

their bodies, fungi do not need a stomach or a mouth, or even a digestive tract. Rather, each cell can absorb the products of this external digestion directly.

As another example, consider sea anemones, invertebrate animals that live in the oceans, attached to surfaces such as rocks. Sea anemones digest their food internally, but they don't have a digestive tract like humans. Rather, they have a single, multifunctional digestive cavity called a gastrovascular cavity, where digestion and absorption of nutrients take place. Sea anemones capture food and shove it into this cavity using the tentacles that surround their mouths. The gastrovascular cavity has only one opening—the mouth—through which food enters and wastes exit. This is an example of an incomplete digestive tract, in contrast to our own complete digestive tract in which food flows one way from the mouth to the anus.

Photosynthetic organisms such as plants do not need a digestive system because they are autotrophs: they make their own food and therefore don't need to "eat" it (see Chapter 5). Instead, they take up carbon dioxide and use water and the energy of sunlight to convert the atmospheric carbon dioxide to carbohydrates in the plant body.

CHAPTER 26 Summary

▶ The digestive system is composed of a central digestive tract and accessory organs. Its function is to break down food molecules into smaller subunits, absorb nutrients, and eliminate waste.

▶ Digestion begins in the mouth, where teeth chew food (mechanical digestion) and salivary enzymes begin breaking down carbohydrates (chemical digestion).

▶ Food passes from the mouth into the stomach through the esophagus, propelled by waves of muscular contractions called peristalsis.

▶ The stomach is muscular and acidic and contains a protein-digesting enzyme called pepsin. It is elastic and can expand after a large meal to store food for a few hours.

▶ Food processed in the stomach is called chyme. Chyme passes into the small intestine, where enzymes further digest it.

▶ Bariatric surgery reduces the size of the stomach so that it can no longer hold as much food. The surgery may also shorten the small intestine. As a result, fewer Calories are taken in and absorbed, and weight loss occurs.

▶ Enzymes from the pancreas help to digest organic molecules in the small intestine.

▶ Bile salts, produced in the liver and stored in the gallbladder, emulsify fats and help the body digest them.

▶ Epithelial cells that line the small intestine absorb the broken-down products of food; once absorbed, food molecules enter the bloodstream and are transported throughout the body.

▶ The large intestine absorbs water and forms solid stool from indigestible matter in food such as fiber.

▶ Humans and many other animals have a complete digestive tract—one with a mouth and an anus. Not all organisms have a complete digestive tract; many have no digestive tract at all.

MORE TO EXPLORE

▶ Centers for Disease Control and Prevention: Obesity http://www.cdc.gov/obesity/

▶ Mingrone, G., et al. (2012) Bariatric surgery versus conventional medical therapy for type 2 Diabetes. *New England Journal of Medicine* 366(17):1577–1585.

▶ Sjöström L., et al. (2007) Effects of bariatric surgery on mortality in Swedish obese subjects. *New England Journal of Medicine* 357:741–752.

▶ Nestle, M., and Malden, N. (2012) Why Calories *Count: From Science to Politics*. Berkeley: University of California Press.

▶ Brownell, K., and Horgen, K. B. (2004) *Food Fight: The Inside Story of The Food Industry, America's Obesity Crisis, and What We Can Do About It*. New York: McGraw-Hill.

CHAPTER 26 Test Your Knowledge

DRIVING QUESTION 1

What is the anatomy of the digestive system?

By answering the questions below and studying Infographics 26.1, 26.2, 26.3, 26.6, and 26.7, you should be able to generate an answer for the broader Driving Question above.

KNOW IT

1 Place the following structures of the digestive system in order from the entry of food (1) to the exit of waste (5).

_____ Esophagus

_____ Large intestine

_____ Stomach

_____ Mouth

_____ Small intestine

2 Which part of the digestive tract has the most acidic pH?

a. esophagus

b. colon

c. small intestine

d. stomach

e. mouth

3 Why is it helpful to have an expandable stomach?

USE IT

4 What do the gallbladder, liver, and pancreas have in common with respect to the digestive system? How do they differ from the mouth, stomach, and small intestine?

5 Muscle paralysis in the digestive tract would compromise which digestive function?

a. chemical digestion in the stomach

b. chemical digestion in the small intestine

c. absorption in the small intestine

d. chemical digestion in the mouth

e. movement of food from the mouth to the stomach

6 Pepsin is most effective at a pH of about 2. Digestive enzymes in the small intestine are most effective at a pH of or near 7. If the pancreas were unable to secrete bicarbonate (the basic component of pancreatic secretions), what would you predict about the waste eliminated from the large intestine?

DRIVING QUESTION 2

How is food broken down and utilized as it moves through the digestive tract?

By answering the questions below and studying Infographics 26.4 and 26.5, you should be able to generate an answer for the broader Driving Question above.

KNOW IT

7 Where does the majority of chemical digestion take place?

a. small intestine

b. esophagus

c. mouth

d. stomach

e. colon

8 What do pepsin and salivary amylase have in common? How do their activities differ?

9 What organ produces lipase?

USE IT

10 A person who has had the gallbladder surgically removed will have trouble processing

a. fats.

b. carbohydrates.

c. minerals.

d. vitamins.

e. proteins.

11 Compare and contrast the roles of bile salts and lipase.

12 Why would someone with a blocked duct between the pancreas and the small intestine experience pancreatic inflammation (pancreatitis)? Note that in this case inflammation is a response to tissue damage.

13 If you stand on your head, can processed food still pass from your small intestine into your large intestine? Explain your answer.

MINI CASE

14 Alicia has had her gallbladder removed. She must be careful not to eat high-fat meals, or else she is likely to experience greasy diarrhea. Her friend Tammy, who is taking Alli, must also avoid high-fat meals, in order to avoid oily stools, a possible side effect of the drug. Why are both Alicia and Tammy at risk for similar digestive upsets when their situations are different (gallbladder removal, taking Alli)?

DRIVING QUESTION 3

What are the risks and benefits of bariatric surgery?

By answering the questions below and studying Infographics 26.3, 26.6, and 26.8, you should be able to generate an answer for the broader Driving Question above.

KNOW IT

15 Gastric bypass surgery causes the _____ to become _____.

a. stomach; smaller

b. small intestine; larger

c. stomach; less acidic

d. small intestine; less acidic

e. stomach; larger

16 Compare and contrast gastric banding and gastric bypass.

17 Which of the following is the best first option for someone who wishes to lower a BMI of 32?

a. gastric banding

b. gastric bypass

c. dietary modification

d. more exercise

e. all of the above

f. c and d

USE IT

18 Several studies have found that gastric bypass has a greater impact on several conditions (e.g., cardiovascular disease and diabetes) than gastric banding. Why might this be?

BRING IT HOME

19 If a morbidly obese person who is considering gastric bypass surgery asked for your opinion on the procedure, what would you say about its known risks, benefits, and any unknowns? Would you say the same to someone considering the surgery who is simply overweight, not morbidly obese? Explain your answer.

INTERPRETING DATA

20 A 2012 study compared the impacts of medical therapy and bariatric surgery in obese people with uncontrolled type 2 diabetes. Patients were randomly assigned to receive aggressive medical therapy for their diabetes (including medications and diet and lifestyle modifications) or bariatric surgery. Several dependent variables were measured for 1 year. Two of these—average BMI and average number of diabetes medications—are shown in the table below.

a. Draw two graphs, one plotting the average diabetes medications over time for the group receiving medical therapy and for the group receiving surgery, and one plotting the change in BMI from baseline for the two groups (set the baseline values at 0 on the time axis because by definition no change can have taken place yet).

b. From these graphs, how does gastric bypass compare to medical therapy for diabetes management in obese patients with type 2 diabetes?

c. The data shown in the table are for 41 patients who had medical therapy and 50 patients who had gastric bypass. From this additional information and any other limitations you can identify, are these data sufficient to make a recommendation of surgery for diabetes management? Why or why not?

TREATMENT	VARIABLE	BASELINE	TIME AFTER TREATMENT			
			3 MOS.	6 MOS.	9 MOS.	12 MOS.
Medical therapy	Average BMI	36.3	35.4	34.8	34.5	34.4
	Average # of diabetes meds	2.8	3.1	3.1	3.0	3.0
Gastric bypass	Average BMI	37.0	31.8	28.2	26.9	26.8
	Average # of diabetes meds	2.6	1.1	0.6	0.4	0.3

SOURCE: Schauer, P. R., et al. (2012) *New England Journal of Medicine* 366(17):1567–1576.

Beginning with Leonard Thompson, the discovery of insulin was a life-saver for many children.

INSULI
50 Units — 5 C.C.
10 Units per C.
CONNAUGHT LABORAT
UNIVERSITY OF TORO

STUMBLING ON A CURE

Banting, Best, and the Discovery of Insulin

IN DECEMBER 1921, A 14-YEAR-OLD BOY NAMED LEONARD Thompson lay sick and dying in a Canadian hospital bed. Weighing just 65 pounds, he was lapsing in and out of consciousness. Elevated levels of sugar in his blood, caused by the disease diabetes mellitus, were wreaking havoc on his internal organs, and doctors told his parents he would likely not survive more than a month. With nothing to lose, Thompson's parents agreed to let the doctors try something unusual: they injected the boy with a chemical that scientists had recently isolated from the pancreas of a dog. The chemical radically altered Thompson's fate and that of countless others since.

For most of human history, diabetes was a dreaded and deadly disease. People with diabetes have unusually high blood-sugar levels–a sign that the body's cells are not taking up sugar from the blood. Since cells require sugar as fuel to power their activities, this lack of uptake eventually starves the body of nourishment, while the high blood-sugar levels cause a host of problems of their own.

> ▶▶ **DRIVING QUESTIONS**
>
> 1. What is the role of insulin in blood-sugar regulation and diabetes?
> 2. In general terms, what is a hormone?
> 3. What features are shared by type 1 diabetes and type 2 diabetes, and what features are unique to each type?

The word "diabetes" comes from the Greek word meaning "to pass through," and refers to the fact that diabetics tend to urinate excessively. By eliminating excess sugar, urinating restores normal blood-sugar levels. Before there were blood tests, diabetes was diagnosed by testing for the presence of sugar in a patient's urine, which would often attract flies because it was so sweet ("mellitus" means "like honey"). Until 1921, there was no effective treatment for diabetes, and a diagnosis was in essence a death sentence.

The breakthrough came in a laboratory located just footsteps from where Thompson was being treated at the University of Toronto. A team of upstart researchers, working on a shoestring budget, succeeded where many other scientists had failed. With nothing more than a pack of stray dogs and some borrowed laboratory equipment, they discovered a treatment for millions.

HUNTING DOWN A MYSTERY CHEMICAL

In the fall of 1920, a young Canadian surgeon named Frederick Banting was preparing a lecture on the physiology of the pancreas, the carrot-shaped organ that sits next to the stomach. At the time, the pancreas was known to be important for the digestion of food molecules: digestive juices secreted from the pancreas into the small intestine contain enzymes that help break down carbohydrates, proteins, and fats. But scientists were learning that the pancreas has other functions as well. Experiments in 1890 had shown that surgical removal of the pancreas in dogs led quickly and inevitably to all the symptoms of severe diabetes: high blood-sugar levels, sugar in the urine, coma, and death. This finding suggested that the pancreas plays a critical role in the regulation of blood sugar.

INFOGRAPHIC M7.1 THE PANCREAS PRODUCES DIGESTIVE ENZYMES AND INSULIN

→ The pancreas plays distinct roles. Some pancreatic cells secrete digestive enzymes into the small intestine, where those enzymes help break down organic molecules, including carbohydrates, into smaller components. Pancreatic cells in the islets of Langerhans secrete hormones—including insulin—into the circulatory system.

Intestine

Stomach

Insulin and other hormones are secreted into the blood.

Blood

Paul Langerhans

Islets of Langerhans cells produce insulin.

Duct cells secrete digestive enzymes into the intestine.

Islets of Langerhans

Until 1921, there was no effective treatment for diabetes, and a diagnosis was in essence a death sentence.

Scientists were beginning to suspect that the pancreas was essentially two organs in one: one part of the pancreas released digestive enzymes into the intestines through ducts, while another part, made up of discrete islands of cells, released an unidentified substance into the blood that regulated blood-sugar levels. A number of researchers had tried, unsuccessfully, to isolate this mystery chemical from these groups of cells (called islets of Langerhans after Paul Langerhans, who first noticed them under a microscope in 1869). So convinced were researchers that the substance existed that they gave it a name: "insuline"–from the Latin *insula,* meaning "island" (INFOGRAPHIC M7.1).

The notion that certain 'internal secretions' produced by glands in the body and released into the bloodstream could influence physiology was a relatively new idea at the beginning of the 20th century. Before that time, conventional wisdom held that physiological processes, like the regulation of blood sugar, were controlled primarily by the brain through nerves. But evidence was accumulating that blood-borne chemicals called hormones controlled a number of different physiological processes, including blood-sugar regulation. The study of hormones and the glands that produce them would come to be known as endocrinology.

Today we know that the body has numerous hormone-secreting glands, which together coordinate many physiological processes. The adrenal glands, located above the kidneys, for example, secrete the hormone adrenaline, which controls the "fight or flight" response;

the ovaries and testes produce steroid hormones that control reproductive physiology; the thyroid gland, located in the neck, produces thyroid hormones, which regulate metabolism; and the pituitary gland in the brain produces several hormones that control glands and tissues throughout the body. All these hormones work in a similar fashion: they are released by glands into the circulation and carried in the blood to their target cells.

Banting knew about previous attempts to isolate a pancreatic hormone. While preparing his lecture on the physiology of the pancreas, he read an article in a medical journal describing how surgically tying off the main duct leading from the pancreas to the intestine led to the death of the enzyme-secreting cells of the organ, but left the "islands" unharmed. That's when an idea came to him: perhaps other researchers had been unable to isolate insulin because, when they ground up the pancreas to perform the isolation procedure, the digestive enzymes that the pancreas produces destroyed the insulin. Perhaps he could avoid that problem by first performing surgery on an animal to tie off the pancreatic duct and then waiting for the enzyme-secreting part of the organ to die. He jotted down some hasty notes, then went to bed, but was so excited he was unable to sleep.

A few days later, Banting presented his idea to John Macleod, the head of the physiology department at the University of Toronto. Macleod listened patiently to Banting but was skeptical of his ideas, finding them not very well thought out and poorly researched. Macleod was aware that other researchers had tried and failed to isolate insulin and was doubtful about the new approach. But Banting was enthusiastic and persistent; eventually, he convinced Macleod to provide him with some lab space and 10 dogs on which to test his hypothesis. Macleod also agreed to appoint a graduate student, Charles Best, to assist him.

A ROCKY START

Banting and Best began by performing surgery on dogs to tie off the pancreatic duct. Then, in a few weeks, when the enzyme-secreting part of the pancreas had died, they would remove the Langerhans cells and crush them into an extract. This extract would then be injected into other dogs whose pancreases had been surgically removed, to see if the extract would prevent them from getting diabetes. At least, that was the idea. But it was easier said than done.

The initial results were disappointing: 7 of the 10 dogs died of complications from the surgery even before the researchers could perform the rest of the experiment. To obtain more research animals, they resorted to buying dogs off the street for $1, no questions

INFOGRAPHIC M7.2 BANTING AND BEST ISOLATED INSULIN FROM DOG PANCREASES

Banting and Best developed a method of extracting insulin from dog pancreases. The extracted insulin reversed the symptoms of diabetes in diabetic dogs.

Left, Charles Best; right, Frederick Banting

Dog's pancreas is removed.

Dog develops diabetes.

Solution injected into diabetic dog

Dog's diabetes is successfully treated.

Pancreatic duct of healthy dog tied off

Islets of Langerhans isolated from pancreas

Cells crushed into an extract and solids filtered off

Further experiments isolated and purified insulin from the solution.

A page out of Banting's laboratory notebook, documenting glucose levels in dogs.

asked. Banting practiced the surgery on these new dogs. Finally, he managed to keep the dogs alive long enough to extract the "island" cells. Banting and Best were now ready to inject the extract into another dog made diabetic by removing its pancreas. The procedure worked: the extract significantly lowered the dog's blood-sugar levels and also reduced the amount of sugar in the dog's urine. A pancreatic hormone did indeed seem to regulate blood sugar. They reported these results to Macleod, who realized a major breakthrough had been made (INFOGRAPHIC M7.2).

Banting and Best were excited to have discovered the elusive hormone insulin, but it quickly became clear that producing insulin from stray dogs was not a viable way to create a treatment for thousands, and they kept looking for other sources. They discovered that fetal calf pancreas was also an abundant source of insulin. But still there were problems: the insulin prepared from ground-up calf pancreas often produced a severe allergic reaction in the animals who received it. Macleod believed that these reactions were caused because the insulin was impure—mixed with other chemicals of the pancreas. Since neither Banting nor Best was a chemist, Macleod hired a young biochemist, James Collip, to help prepare a more chemically pure version of the drug. Such a cleaned-up version would be needed if the drug were to be used in people.

Eventually, using a method that included alcohol and evaporation, Collip was able to produce a pure form of the drug, and the team was ready to test insulin on humans. They talked to doctors at the hospital across the street at the University of Toronto, who were desperate for a new way to treat patients with diabetes. Leonard Thompson, the 14-year-old diabetic patient who was falling in and out of consciousness, received this purified version on January 23, 1922. Within moments, he was alert and smiling, to the astonishment of his parents. Encouraged by these results, the doctors next went from bed to bed in the diabetes ward, injecting one patient after another. Before they reached the last patient, the first patients to be injected were regaining consciousness. Insulin as a treatment for diabetes was born.

Chemist James Collip (left) helped to make insulin pure enough to inject into patients suffering from diabetes (above).

INSULIN AND DIABETES

Today we know that diabetes is caused by the inability to produce or properly respond to insulin. There are two main types of the condition. Type 1 diabetes, which represents about 5% of all adult cases, is due to pathological destruction of the insulin-producing cells in the islets of Langerhans. People with type 1 diabetes require daily injections of insulin to survive. Type 2 diabetes, which accounts for about 90%-95% of all adult cases, begins when cells in the body don't respond to insulin; researchers don't fully understand why this happens. Having cells that are unresponsive to insulin is as bad as not making insulin at all.

Normally, insulin binds to receptors on target cells and triggers the cells to take up glucose from blood. This glucose is then used by the cells as a source of energy. Without properly functioning insulin (or insulin receptors), cells cannot take up sugar from the blood and so they starve (INFOGRAPHIC M7.3).

The pancreas produces other hormones as well, including glucagon, which raises blood sugar. The two hormones work together to regulate blood sugar through a classic negative feedback loop. (For more on glucose metabolism, including the role of glucagon, see Chapter 25.)

Left untreated, chronic diabetes can cause a suite of health problems: increased risk of

Insulin normally signals cells to take up sugar from the blood, thereby reducing blood-sugar levels. In diabetes, insulin is either not produced, or cells don't respond to it, leading to elevated blood-sugar levels.

1. Glucose is released from carbohydrates by digestive enzymes in the digestive tract. Pancreas produces sufficient amounts of insulin.

2. Glucose and insulin enter the bloodstream.

3. Insulin enables glucose to enter the body tissue cells. Blood-glucose levels return to normal.

	Digestive System	Insulin	Bloodstream	Glucose	Body Tissues
Normal Insulin Response	Glucose is released from carbohydrates by digestive enzymes in the digestive tract. Pancreas produces sufficient amount of insulin.	↑	Glucose and a sufficient amount of insulin enter the bloodstream.	↑	Insulin enables glucose to enter body tissue cells. Blood-glucose levels return to normal.
Type 1 Diabetes	Glucose is released from carbohydrates by digestive enzymes in the digestive tract. Pancreas produces little or no insulin.	↓	Glucose and an insufficient amount of insulin enters the bloodstream.	↑	Very little glucose is able to enter and be used by body tissues. Blood-glucose levels remain high.
Type 2 Diabetes	Glucose is released from carbohydrates by digestive enzymes in the digestive tract. Pancreas produces sufficient amount of insulin.	↑	Glucose and a sufficient amount of insulin enter the bloodstream.	↑	Body tissues are unresponsive to insulin. Very little glucose can enter and be used by body tissues. Blood-glucose levels remain high.

heart disease and stroke, loss of vision, kidney disease, nerve damage—and, eventually, death. An estimated 26 million Americans (roughly 8% of the population) suffer from diabetes, making it the most common endocrine disorder in the United States.

RECEPTION AND CONTROVERSY

Overnight, Banting went from being an unknown surgeon to a world-famous scientist, receiving many accolades. Because of his efforts, Leonard Thompson would live 13 more years, to age 27, when he died of pneumonia.

Ironically, Banting's original hypothesis turned out to be wrong. Contrary to what Banting believed, it was not necessary to remove the digestive enzymes from the pancreas in order to isolate insulin. With Collip's purification method, the researchers were able to extract insulin from whole pancreas, without first performing Banting's surgery. But if Banting hadn't pursued his hypothesis, we might not have the lifesaving drug we know today. Why earlier attempts by other researchers failed to isolate insulin remains a mystery. One possibility is that Collip's use of alcohol and evaporation to extract the chemical somehow made the difference.

If Banting hadn't pursued his hypothesis, we might not have the lifesaving drug we know today.

Frederick Banting.

Some scientists thought Banting's fame and status were undeserved because his initial premise was wrong. But these detractors, it was argued by others, misunderstood the nature of the scientific method, which often involves a good deal of luck.

"Nobody can deny that a discovery of first-rate importance has been made, and, if it proves to have resulted from a stumble into the right road, where it crossed the course laid down by a faulty conception, surely the case is not unique in the history of science," wrote Henry Hallet Dale, a prominent British physiologist, in a letter to the *British Medical Journal* in 1922.

For their work on developing insulin, Banting and Macleod shared the 1923 Nobel Prize in Physiology or Medicine. What about Best and Collip? The Nobel committee deemed their contributions less critical to the discovery, a view that didn't sit well with the participants. When Banting heard that Best had not been recognized, he was furious and refused at first to accept the award. Eventually, he accepted, but shared half his prize money with Best. Macleod, in turn, shared half his prize money with Collip. The earlier researchers who had determined the role of the pancreas in diabetes and deduced the existence of insulin—even gave it a name—got nothing.

Since Banting and his team made their important discoveries, a number of other scientific milestones have occurred. Insulin was determined to be a protein and, in 1952, the British biochemist Frederick Sanger determined its amino acid sequence—the first such sequence to be described. Today, insulin is no longer obtained from pancreas harvested from animal tissues, but is instead synthesized by genetically engineered bacterial cells modified to contain the human insulin gene (see Chapter 8). Insulin remains one of the most important drugs of modern medicine. ■

MORE TO EXPLORE

▸ The Discovery of Insulin. Nobelprize.org
www.nobelprize.org/educational/medicine/insulin/discovery-insulin.html

▸ Rosenfeld, L. (2002) Insulin: discovery and controversy. *Clinical Chemistry* 48 (12):2270–2288.

▸ Henderson, J. (2005) Ernest Starling and 'hormones': an historical commentary. *Journal of Endocrinology* 184:5–10.

▸ Chemical Heritage Foundation: Frederick Banting, Charles Best, James Collip, and John Macleod www.chemheritage.org/discover/online-resources/chemistry-in-history/themes/pharmaceuticals/restoring-and-regulating-the-bodys-biochemistry/banting--best--collip--macleod.aspx

▸ Centers for Disease Control and Prevention: Diabetes Public Health Resource www.cdc.gov/diabetes

MILESTONES IN BIOLOGY 7 Test Your Knowledge

1 What organ produces insulin?

  a. the liver

  b. the stomach

  c. the pancreas

  d. blood

  e. skeletal muscle

2 Someone who cannot produce insulin will likely have blood sugar levels that are

  a. normal.

  b. lower than normal.

  c. higher than normal.

3 In general terms, what is a hormone?

  a. A signaling molecule that is transported by the circulatory system and acts on target cells.

  b. A signaling molecule that regulates blood sugar.

  c. A signaling molecule that is produced by the pancreas.

  d. A signaling molecule that increases the concentration of blood sugar.

  e. A protein that is released by one cell and acts on an adjacent cell.

4 A person has chronically elevated blood-sugar levels. From this information, are you able to say if the person has type 1 diabetes or type 2 diabetes? Why or why not?

5 Insulin is a/an

  a. monosaccharide.

  b. protein.

  c. triglyceride (fat).

  d. phospholipid.

  e. any of the above, depending on the diet

6 List key differences in glucose levels and insulin levels between type 1 and type 2 diabetes. Consider levels during periods of fasting and levels after a carbohydrate-rich meal.

DEATH

IN BOGALUSA

From tragic deaths in a southern town, insight into heart disease

---

▸▸ DRIVING QUESTIONS

1. What structures make up the cardiovascular system, and how does blood flow through the system?

2. What is the structure of the heart and of the different types of blood vessel?

3. What is the composition of blood, and what does blood do?

4. What is cardiovascular disease, and what are some of the risk factors for developing cardiovascular disease?

Gerald Berenson.

SOME DIED IN CAR ACCIDENTS. OTHERS DROWNED, OR WERE STRICKEN with pneumonia. Still others were slain by bullets. The victims were as young as 2 and as old as 39. Together, they transformed our understanding of heart disease.

Some scientific discoveries begin in a laboratory. This one begins in a funeral home, in 1978. Under glaring lights, in the back room of the Cook-Richmond Funeral home in Bogalusa, Louisiana, two pathologists hover over the lifeless body of a young African-American male. There is no morgue in Bogalusa, so an autopsy is being conducted here, with newspapers spread beneath the body. The pathologists slice through skin and muscle with scalpels, looking for the usual suspects—internal bleeding, broken bones—then write up their report for the coroner.

The pathologists aren't quite done, however. Before closing up, they carefully remove the victim's heart, package it in saline, and prepare it for a trip to a medical school in New Orleans. It was an unusual step, not standard for an autopsy. But the pathologists were heeding the instructions of a prominent local physician, who had the blessing of the family.

Gerald Berenson, a cardiologist with the Louisiana State University School of Medicine, had a passionate interest in heart disease. Since 1972, he had spearheaded what was then a novel study: an epidemiological study of heart disease in Bogalusa, a rural town of about 16,000 people located 60 miles north of New Orleans.

The study began with a pretty simple idea: follow a large group of children over a period of time and correlate their physical and lifestyle attributes with their risk for developing heart disease in adulthood. The biggest hurdle was a logistical one—how to enlist the thousands of children necessary to produce a robust data set, and keep them coming back for evaluation year after year.

But Berenson wanted to do more than make statistical correlations. He also wanted to document the progression of heart disease directly. And for that he needed a different type of evidence.

The heart that Berenson obtained in 1978 was one of more than 200 such organs collected from young people in Bogalusa over the next 20 years. In fact, nearly every young person who died, of whatever cause, was autopsied and had the heart removed. From this medical detective work has emerged a detailed understanding of heart disease in young people, and the evidence needed to clinch the case against a cold-blooded killer.

A SILENT EPIDEMIC

Cardiovascular disease (CVD), more colloquially known as heart disease, is the number one killer of men and women in the United States and the developed world. Every year, more people die from cardiovascular disease than from cancer and Alzheimer disease combined. According to the Centers for Disease Control and Prevention (CDC), one in three deaths in the United States is caused by cardiovascular disease—roughly 2,200 deaths per day (**INFOGRAPHIC 27.1**).

CVD claims so many lives each year for two main reasons. The disease is largely "silent"—people don't know they're sick until it's too late. And the system it affects is one of the most important to our survival,

> Every year, more people die from cardiovascular disease than from cancer and Alzheimer disease combined.

so when things go wrong, the result tends to be fatal.

The **cardiovascular system** transports nutrients, gases, hormones, and other critical molecules throughout the body. It consists of the heart, blood vessels, and blood. The cardiovascular system works closely with the respiratory system (see Chapter 28), which moves oxygen from the air into the lungs and removes carbon dioxide waste.

The **heart** is essentially a pump. It is about the size of a fist and very muscular. By repeatedly contracting and relaxing, it pumps blood throughout the body. The heart has four chambers: two paired **atria** above two paired **ventricles**. The atria receive blood into the heart, and the ventricles pump blood out of it.

Blood travels through the body in different types of blood vessel. Blood vessels that carry blood away from the heart are called **arter-**

INFOGRAPHIC 27.1 CARDIOVASCULAR DISEASE IS THE MAJOR KILLER IN THE U.S.

➡ Cardiovascular disease (including heart disease and stroke) continues to be the leading cause of death in both men and women in the United States.

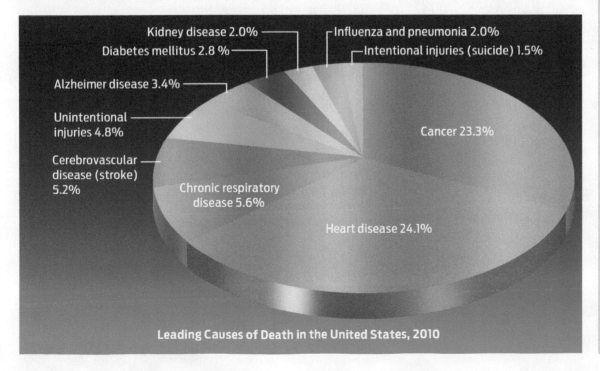

Kidney disease 2.0%
Diabetes mellitus 2.8%
Alzheimer disease 3.4%
Unintentional injuries 4.8%
Cerebrovascular disease (stroke) 5.2%
Chronic respiratory disease 5.6%
Influenza and pneumonia 2.0%
Intentional injuries (suicide) 1.5%
Cancer 23.3%
Heart disease 24.1%

Leading Causes of Death in the United States, 2010

▶CARDIOVASCULAR SYSTEM
The system that transports nutrients, gases, and other critical molecules throughout the body. It consists of the heart, blood vessels, and blood.

▶HEART
The muscular pump that generates force to move blood throughout the body.

▶ATRIA (SINGULAR: ATRIUM)
The chambers of the heart that receive blood. The right atrium receives low-oxygen blood from the body, and the left atrium receives high-oxygen blood from the lungs.

▶VENTRICLES
The chambers of the heart that pump blood away from the heart. The right ventricle pumps blood to the lungs, and the left ventricle pumps blood to the body.

▶ARTERIES
Blood vessels that carry blood away from the heart.

▶VEINS
Blood vessels that carry blood toward the heart.

▶CORONARY ARTERIES
The blood vessels that deliver oxygenated blood to the heart muscle.

▶HEART ATTACK
A restriction of blood flow to the heart, resulting in damage to the heart muscle.

ies, and blood vessels that return blood to the heart are called **veins**.

Blood is composed of cells and liquid. One of its important roles is to carry oxygen to all tissues of the body and ship away carbon dioxide waste **(INFOGRAPHIC 27.2)**.

For decades, doctors have known that cardiovascular disease is essentially a matter of bad plumbing. Fatty deposits develop in the arteries that deliver blood to the body or to the heart muscle itself, cutting off or reducing blood flow to these tissues. This is dangerous because blood carries not only nutrients like sugar needed for energy, but also gases like oxygen needed to carry out aerobic respiration (see Chapter 6). When the arteries that supply blood directly to the heart muscle—the **coronary arteries**—are blocked, the cells of the heart begin to die. This is a **heart attack**.

For the most part, young people do not have heart attacks (and those that do usually suffer from rare genetic conditions). But just because teenagers do not die from heart disease does not mean they can ignore heart health. That's because heart disease can be insidiously gaining a foothold long before we have any obvious symptoms.

This is common knowledge now, but when Gerald Berenson started his study in 1972, it was far from accepted wisdom. Cardiologists—influenced by the prevailing beliefs of the day—focused more on treatment than on prevention. But Berenson had been trained in pediatrics as well as cardiology, and he knew that it was important to understand the beginnings of heart disease as well as its endings. He submitted an application to the National Institutes of Health (NIH) and won

INFOGRAPHIC 27.2 THE CARDIOVASCULAR SYSTEM

The cardiovascular system transports nutrients and gases throughout the body. It consists of the heart, the blood vessels, and the blood.

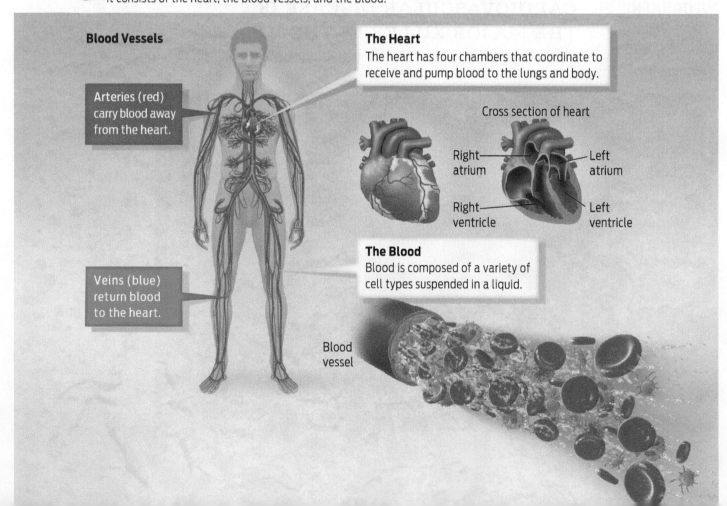

Blood Vessels

Arteries (red) carry blood away from the heart.

Veins (blue) return blood to the heart.

The Heart
The heart has four chambers that coordinate to receive and pump blood to the lungs and body.

Cross section of heart

Right atrium — Left atrium

Right ventricle — Left ventricle

The Blood
Blood is composed of a variety of cell types suspended in a liquid.

Blood vessel

a competitive grant to study heart disease in children. Known as the Bogalusa Heart Study, it would fundamentally change how doctors understand heart disease.

The study would take an epidemiological approach: children and young people ages 5-17 would be enrolled through the school system and monitored twice yearly. Nurses would measure their height, weight, smoking history, blood pressure, and cholesterol levels, among other variables. They would return for evaluation every several years, up to age 45. Later, it was decided that any of the study participants who died would be autopsied.

From small beginnings, the study grew to enroll some 16,000 individuals. It is now one of the longest-running cardiovascular disease studies in the world, the only one with a biracial (black-white) study population, and the first to document that heart disease begins in childhood.

MATTERS OF THE HEART

At 90 years old, Gerald Berenson has been around long enough to see many changes in the practice of medicine. He is currently Director of the Tulane Center of Cardiovascular Health and Principal Investigator of the Bogalusa Heart Study, yet he still retains an air of the old-fashioned family doctor. Over the years, he has even been known to make house calls to many of the participants in the study.

"If there was any kind of medical problems I went and examined them myself and took care of them. I had 4,500 patients to look at initially," Berenson recalled in an interview with Tulane's *Global Health News* in 2012.

It's this personalized approach to medicine, combined with Berenson's brand of southern tenacity, that has ensured success of the study over the years. "I'm often asked 'Why Bogalusa?'" The answer is simple, he says: "It's where I'm from."

Because he grew up there, Berenson knew Bogalusa like the back of his hand. And that firsthand knowledge of this town less than 2

hours north of New Orleans proved crucial in solidifying support for the study. In fact, nearly everyone in the town has participated in the study in some way. Teachers and nurses at local schools serve as study liaisons. The pathologists who conducted the autopsies had Berenson as an instructor in medical school. Even Berenson and the coroner were old friends, and it was because of that relationship that Berenson was able to work out the arrangement that enabled the heart autopsies to be performed.

"Eighty percent of the known deaths in the area we were able to autopsy," says Berenson proudly. "Nobody gets that kind of rate."

The autopsies that Berenson's team performed were geared toward understanding when heart disease first takes hold and begins to threaten heart function. Key to understanding heart function is the fact that blood passes through the heart twice for every trip through the body—it follows a double circuit. One circuit involves the trip of blood between the heart and the lungs, known as the **pulmonary circuit**; the other involves the trip of blood between the heart and the rest of the body, known as the **systemic circuit**.

The two circuits can best be understood by tracing the path of blood as it makes one complete trip through the body. Blood returns to the heart from the body, through veins, into the right atrium. It is then pumped into the right ventricle. From there, the pulmonary artery carries the blood to the lungs, where it picks up oxygen and drops off carbon dioxide. The oxygenated blood then returns to the left atrium of the heart through the pulmonary vein. This completes the pulmonary circuit. Blood is then pumped into the left ventricle, and from there into the **aorta,** the major artery leading to the rest of the body. The aorta carries oxygenated blood to the body's other arteries. The systemic circuit is completed when blood returns to the right atrium.

The heart is equipped with two sets of valves —one valve in each chamber. The coordinated

▶**PULMONARY CIRCUIT**
The circulation of blood between the heart and the lungs.

▶**SYSTEMIC CIRCUIT**
The circulation of blood between the heart and the rest of the body.

▶**AORTA**
The large artery that receives blood from the left ventricle.

▶**ATHEROSCLEROSIS**
A disease process that restricts blood flow through arteries; also known as hardening of the arteries.

opening and closing of these valves ensures a one-way flow of blood through the heart, with no backflow. On average, it takes about 1 minute for blood to make a complete trip around the body (**INFOGRAPHIC 27.3**).

Obviously, for the heart to do its job properly, it has to keep pumping. A group of cells located in the right atrium is the heart's natural pacemaker, setting the heart's tempo—60–90 beats per minute when we are rest. In fact, for as long as you are alive the heart muscle never stops beating, and in that sense it is the only muscle in the body that never gets a rest. The human heart beats more than 2.5 billion times in an average lifetime.

Like any muscle, the heart requires oxygenated blood to power the work of contraction. If a part of the heart loses its blood supply, that portion of the muscle begins to die, and the heart as a whole can begin to beat irregularly, or stop beating altogether. At that point, blood circulation stops. And very quickly, so do you.

Understanding, then, how blood flow becomes obstructed is key to combating heart disease.

UNDER THE KNIFE

William Newman is one of three pathologists on Berenson's team who analyzed the tissues collected from the autopsies. He was also directly involved in harvesting the specimens. Whenever a young person in Bogalusa died, Newman and a colleague would drive the 66 miles up from Louisiana State University School of Medicine in New Orleans to Bogalusa to perform the autopsy for the coroner.

What Newman and his colleagues saw when they opened the hearts of the young people shocked them: fatty streaks lining the coronary arteries that supply the heart muscle with blood, as well as the aorta, the major artery leaving the left ventricle of the heart, ultimately supplying blood to the rest of the body. Virtually all the individuals autopsied had these fatty streaks, and the extent of streaking increased with age.

As Newman explains, a fatty streak is an early form of **atherosclerosis**, a disease that restricts blood flow in blood vessels. He says you can identify these streaks when you cut open a vessel because they "look a bit yellow." They also retain a lipophilic ("fat-loving") dye, which makes them more apparent and easily quantified.

To quantify the extent of atherosclerosis, each pathologist would look at every autopsy

INFOGRAPHIC 27.3 BLOOD FLOW FOLLOWS A DOUBLE CIRCUIT

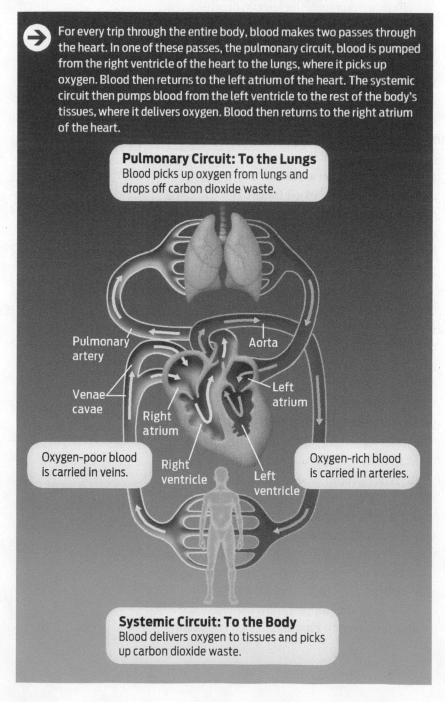

For every trip through the entire body, blood makes two passes through the heart. In one of these passes, the pulmonary circuit, blood is pumped from the right ventricle of the heart to the lungs, where it picks up oxygen. Blood then returns to the left atrium of the heart. The systemic circuit then pumps blood from the left ventricle to the rest of the body's tissues, where it delivers oxygen. Blood then returns to the right atrium of the heart.

Pulmonary Circuit: To the Lungs
Blood picks up oxygen from lungs and drops off carbon dioxide waste.

Pulmonary artery

Aorta

Venae cavae

Left atrium

Right atrium

Oxygen-poor blood is carried in veins.

Right ventricle

Left ventricle

Oxygen-rich blood is carried in arteries.

Systemic Circuit: To the Body
Blood delivers oxygen to tissues and picks up carbon dioxide waste.

specimen and assign the stained fatty streaks a score, and the scores were then averaged together. The study was blind, so the pathologists didn't know the source of the tissues beforehand. Once they had these measurements, statisticians correlated these anatomical measurements with known **risk factors** for heart disease–things like high blood pressure, smoking, and high cholesterol–the young people had when they were alive. The result? "We found that those individuals who had higher levels of known risk factors on the average had more fatty streaks in their coronary arteries and the aortas than individuals who had lower levels," says Newman.

In other words, though many of them were not even old enough to vote, these young people already had telltale signs of heart disease. Some even had evidence of more severe atherosclerosis in the form of fibrous **plaques** that had begun to obstruct

blood flow. Had these young people lived, those with the higher levels of atherosclerosis would have been at risk for a heart attack or other complications of heart disease (INFOGRAPHIC 27.4).

A SERIES OF TUBES

Newman and his colleagues found evidence of atherosclerosis in the aorta and in the heart's coronary arteries. The aorta is the largest artery in the body. Blood leaves the heart through the aorta, which branches out into a series of smaller arteries that deliver oxygen-rich blood to the rest of the body. The coronary arteries that branch off the aorta supply the heart muscle with blood. It is these smaller coronary arteries that are the usual sites of blockages that lead to a heart attack.

If the heart is full of blood, why does it need its own blood supply from arteries leading from

▶**RISK FACTOR**
Anything (e.g., a behavior or exposure) that increases the probability of developing a disease.

▶**PLAQUES**
Raised fatty deposits that accumulate inside arteries, limiting the flow of blood.

INFOGRAPHIC 27.4 ATHEROSCLEROSIS: A COMMON CAUSE OF CARDIOVASCULAR DISEASE

Atherosclerosis begins with fat deposits in the lining of a blood vessel. These deposits cause fatty streaks which, over time, may increase in size and harden, causing an obstructive plaque that restricts blood flow.

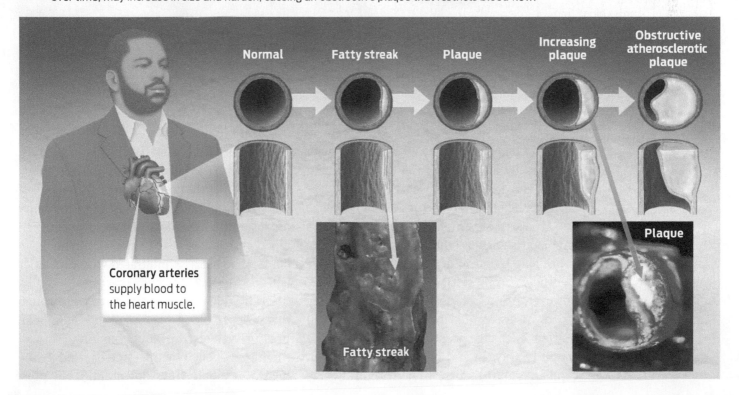

Normal | Fatty streak | Plaque | Increasing plaque | Obstructive atherosclerotic plaque

Coronary arteries supply blood to the heart muscle.

Fatty streak

Plaque

INFOGRAPHIC 27.5 CAPILLARIES ARE SITES OF NUTRIENT AND GAS EXCHANGE

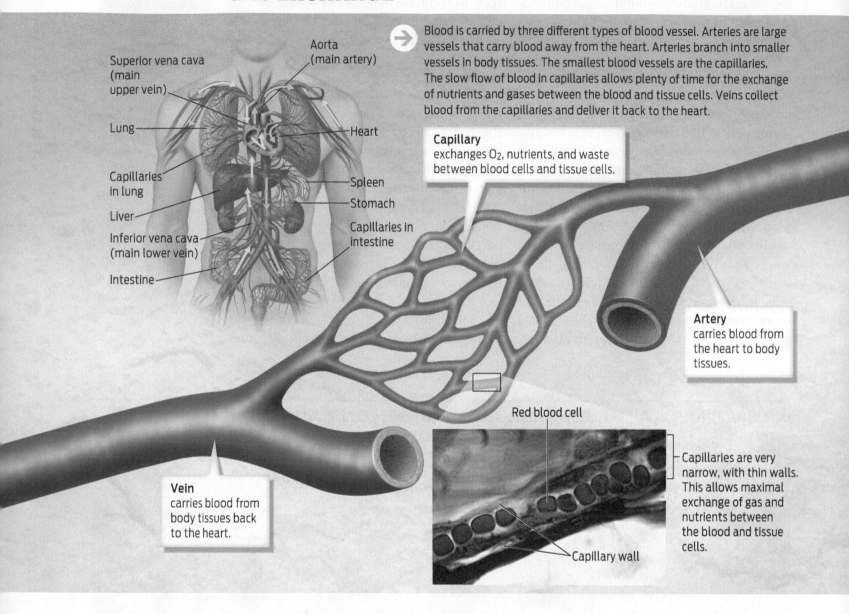

Blood is carried by three different types of blood vessel. Arteries are large vessels that carry blood away from the heart. Arteries branch into smaller vessels in body tissues. The smallest blood vessels are the capillaries. The slow flow of blood in capillaries allows plenty of time for the exchange of nutrients and gases between the blood and tissue cells. Veins collect blood from the capillaries and deliver it back to the heart.

Superior vena cava (main upper vein)

Aorta (main artery)

Lung

Heart

Capillaries in lung

Spleen

Liver

Stomach

Inferior vena cava (main lower vein)

Capillaries in intestine

Intestine

Capillary exchanges O_2, nutrients, and waste between blood cells and tissue cells.

Artery carries blood from the heart to body tissues.

Vein carries blood from body tissues back to the heart.

Red blood cell

Capillaries are very narrow, with thin walls. This allows maximal exchange of gas and nutrients between the blood and tissue cells.

Capillary wall

▶**CAPILLARIES**
The smallest blood vessels. Capillaries are the sites of gas, nutrient, and waste exchange between the blood and tissue cells.

the aorta back into the heart muscle? The answer has to do with how gases like oxygen and carbon dioxide, and nutrients like sugar, are exchanged between blood vessels and tissues.

Arteries and veins are like major highways for blood transport—their main responsibility is bringing blood to and from the body. Situated between these two large types of vessel, however, are capillaries. **Capillaries** are tiny blood vessels, located in tissues, where gas and nutrient exchange occurs. They are where the action is. Oxygen and nutrients diffuse *out* of capillaries into tissues, while carbon dioxide

and wastes diffuse from tissues *into* capillaries. Capillaries are ideally suited to do their job by virtue of their thin walls and their slow rate of blood flow, which allows ample opportunity for gas, nutrient, and waste exchange between blood and tissues. Every tissue in the body, including the heart, is infiltrated with capillaries (INFOGRAPHIC 27.5).

How does blood actually move through this series of blood vessels? Arteries are high-pressure vessels; the blood that flows through them is propelled by the force exerted by the heart. As blood slows down and loses pressure in the capillaries, it enters the veins at low pressure. Therefore, veins are low-pressure vessels; they rely on muscular contractions of the skeletal muscles in which they are found, and on valves, to prevent the backflow of blood and return it to the heart.

The function of this interconnected system of blood vessels is to bring blood in proximity to nearly every cell in the body. What makes blood so vital? Blood is a complex tissue, made up of cells and a liquid known as the plasma, which is composed of water and proteins. For cardio-vascular purposes, the most important cells are the **red blood cells** (or **erythrocytes**), which are specialized for carrying oxygen; they lack a nucleus and have a distinctive concave shape

that allows them to squeeze through capillaries in single file. Oxygen diffuses directly across the membranes of these cells into tissues.

Also present are **white blood cells** (or **leukocytes**), which are involved in many aspects of the immune response (see Chapter 31), and **platelets**, which are cell fragments that play a critical role in blood clotting. Blood also carries dissolved carbon dioxide, the waste product of cellular respiration (see Chapter 6). Finally, blood contains traces of whatever else we put into our bodies–from the food we eat to the drugs we take to the smoke we breathe. Some of the things that circulate in our blood will contribute, over time, to the development of atherosclerosis (**INFOGRAPHIC 27.6**).

▸**RED BLOOD CELLS (ERYTHROCYTES)**
A type of blood cell specialized for carrying oxygen.

▸**WHITE BLOOD CELLS (LEUKOCYTES)**
Leukocytes are involved in the body's defenses (the immune response).

▸**PLATELETS**
Fragments of cells involved in blood clotting.

INFOGRAPHIC 27.6 THE COMPONENTS OF BLOOD

 Blood contains a variety of cells in a liquid called plasma. Red blood cells pick up oxygen in the lungs and deliver it to tissues in the body. White blood cells are critical for immune defense. Platelets are important for blood clotting in areas of injury. The plasma contains many types of dissolved proteins, nutrients, and other important molecules.

Blood vessel

Plasma
is the liquid that carries blood cells and important substances through blood vessels.

Red blood cells
carry oxygen from the lungs to cells in tissues.

White blood cells
mount immune responses against infection.

Platelets
collect at sites of injury and form a plug to help blood clot.

▶BLOOD PRESSURE
The overall pressure in blood vessels, expressed as the systolic pressure over the diastolic pressure.

▶SYSTOLIC PRESSURE
The pressure in arteries at the time the ventricle contracts.

▶PULSE
The detectable force of blood entering arteries, e.g. in the neck or wrist.

▶DIASTOLIC PRESSURE
The pressure in arteries when the ventricle is relaxed.

UNDER PRESSURE

When the Bogalusa Heart Study began, in 1972, Berenson and his colleagues were interested in identifying risk factors that contribute to heart disease, specifically in kids. Other studies had identified risk factors in adulthood, but not in children. Moreover, no other study had ever looked at risk factors among blacks as well as whites (at the time, all major heart disease studies had an exclusively white study population). And that's important, because different populations can have different patterns of risk.

One of the main risk factors the researchers were interested in was elevated **blood pressure**. Blood pressure is expressed as two separate numbers that relate to the action of the heart muscle as it pumps. With every heartbeat, the ventricles contract and blood is forced out of the heart into arteries. This relatively high pressure is the **systolic pressure**. The force of the blood entering the arteries can be felt as the **pulse**–as, for example, in your neck or your wrist. When the ventricles relax, the pressure in the arteries drops; this lower pressure during ventricular relaxation is the **diastolic pressure**. Blood pressure is expressed as systolic pressure over diastolic pressure. The normal measure for blood pressure is systolic pressure less than 120 and diastolic pressure less than 80 (120/80). Chronic elevation of either number increases the risk of cardiovascular disease, including atherosclerosis, heart attack, and stroke.

High blood pressure, or **hypertension**, is dangerous because it can put stress on the

---

INFOGRAPHIC 27.7 HIGH BLOOD PRESSURE CAN RESULT IN ATHEROSCLEROSIS

 When the heart ventricles contract they force blood out of the heart. The force of the blood leaving the heart puts pressure on the walls of the arteries–the systolic pressure. When the ventricles relax there is less pressure on the arteries–the diastolic pressure. Chronically elevated blood pressure can cause small tears in the artery walls. These are sites where fatty deposits can build up.

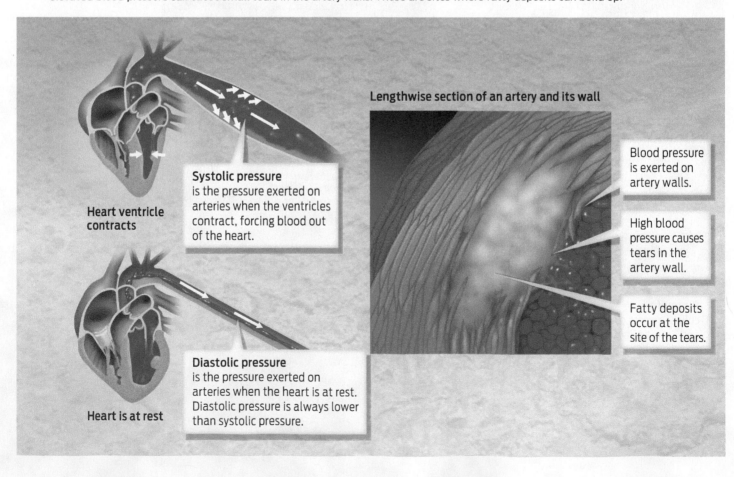

Heart ventricle contracts

Systolic pressure is the pressure exerted on arteries when the ventricles contract, forcing blood out of the heart.

Heart is at rest

Diastolic pressure is the pressure exerted on arteries when the heart is at rest. Diastolic pressure is always lower than systolic pressure.

Lengthwise section of an artery and its wall

Blood pressure is exerted on artery walls.

High blood pressure causes tears in the artery wall.

Fatty deposits occur at the site of the tears.

walls of arteries, damaging them. If the damage is great enough, the blood vessels can burst. Even if they don't burst, the damaged areas can develop scar tissue. The scar tissue provides a site for the buildup of fatty deposits, contributing further to atherosclerosis (INFOGRAPHIC 27.7).

The fatty deposits of atherosclerosis can also promote the formation of blood clots. When plaques rupture, blood clots can form on the surface of the plaque. These clots obstruct blood flow where they form, and can break off and clog other vessels. If blood vessels in the brain are damaged or obstructed, the supply of oxygen to the brain is disrupted, causing a **stroke**.

Hypertension is particularly dangerous because, while it can have deadly consequences, there are typically no symptoms associated with it. This means that people have no way of knowing they are hypertensive unless they have their blood pressure checked and monitored.

What causes hypertension? Researchers are not entirely sure. However, they do know what factors predispose people to it. Some factors are controllable, some aren't. Uncontrollable ones include family history, older age, being African American, and being male; controllable ones include low physical activity, eating a high-salt diet, being overweight or obese, drinking too much alcohol, and stress.

Obviously, when the bodies of the young people arrived at the funeral home in Bogalusa, they no longer had any vital signs, so it was not possible to take their blood pressure. However, because nearly every child in the town had been followed as part of the heart study, researchers were able to correlate the victims' blood pressure when alive with the autopsy results. They found that having had high blood pressure correlated closely with the extent of atherosclerosis found in the arteries of the victims.

High blood pressure wasn't the only risk factor that researchers were able to link with the autopsy results. Other strong links were high body mass index (BMI), high cholesterol

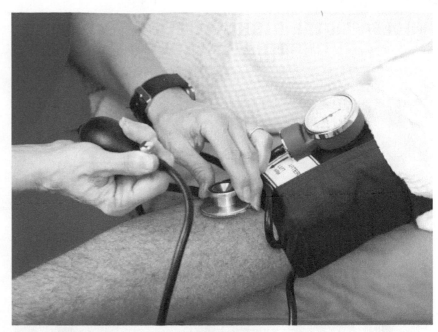

Normal blood pressure is below 120/80.

levels, and smoking cigarettes. When all the known risk factors were considered, a clear pattern emerged: the amount and extent of atherosclerosis increased with age and was strongly correlated with the known risk factors for heart disease. In fact, the young people with the most risk factors when alive were the ones with the most evidence of heart disease at the time of death. For example, those with no risk factors had approximately 20% of their aorta covered with fatty streaks, while those with two risk factors had about 38% of their aorta covered with fatty streaks; those with no risk factors had less than 1% of their coronary arteries covered with fibrous plaques, whereas those with three or four risk factors had more than 7% of their coronary arteries covered with fibrous plaques. The relationship between number of risk factors and extent of heart disease was statistically significant (INFOGRAPHIC 27.8).

"The autopsy studies were really landmark studies," says Peter Katzmarzyk, an epidemiologist at the Pennington Biomedical Research Center in Baton Rouge. Before them, he says, scientists didn't really understand that things like heart attack and stroke have their genesis in childhood. "The Bogalusa Heart Study really put that on the map," he says. And, it made a strong

▶HYPERTENSION
Elevated (high) blood pressure.

▶STROKE
A disruption in blood supply to the brain.

INFOGRAPHIC 27.8 RISK FACTORS AND HEART DISEASE

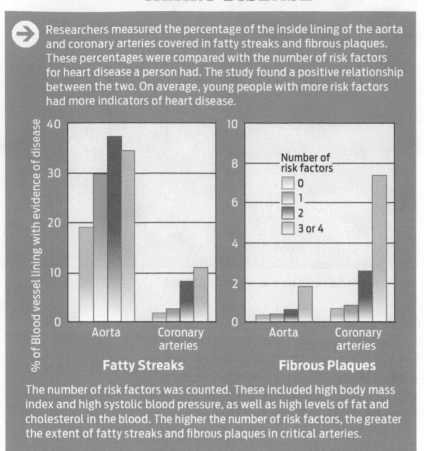

→ Researchers measured the percentage of the inside lining of the aorta and coronary arteries covered in fatty streaks and fibrous plaques. These percentages were compared with the number of risk factors for heart disease a person had. The study found a positive relationship between the two. On average, young people with more risk factors had more indicators of heart disease.

Number of risk factors
0
1
2
3 or 4

Fatty Streaks

Fibrous Plaques

% of Blood vessel lining with evidence of disease

Aorta   Coronary arteries

The number of risk factors was counted. These included high body mass index and high systolic blood pressure, as well as high levels of fat and cholesterol in the blood. The higher the number of risk factors, the greater the extent of fatty streaks and fibrous plaques in critical arteries.

Source: Berenson, G. S., et al., (1998). *New England Journal of Medicine.* 338:1650–1656.

►CHOLESTEROL
A lipid that is transported in the blood by specialized proteins.

►LOW-DENSITY LIPOPROTEIN (LDL)
A form of cholesterol and protein in the blood that contributes to CVD.

►HIGH-DENSITY LIPOPROTEIN (HDL)
A form of cholesterol and protein that is protective.

case that averting severe consequences of heart disease later in life would mean addressing the lifestyle choices we make when we are young.

SOUTHERN DISCOMFORT

If Louisiana is shaped like a boot, then the town of Bogalusa sits in ideal kicking position—right at the easternmost toe. It's a fitting location for a town that has been beaten down over the years, its bruises visible in the form of crumbling playgrounds, dilapidated storefronts, and damage from Hurricane Katrina in 2005. Once known as the Magic City because of its rapid growth around the burgeoning timber industry, Bogalusa has since lost some of its luster, the acrid smell of exhaust from the local paper mill the sole reminder of its former glory days.

Bogalusa is typical of many rural southern towns, in which poverty and poor education

have exacerbated problems of public health. More than 40% of individuals in Bogalusa live below the poverty line (contrasting with 23% for Louisiana as a whole and 15% nationally). And CVD disproportionately affects people of lower socioeconomic status.

Partly, the issue is access to health care. "One of the number one things related to health is socioeconomic status," says epidemiologist Katzmarzyk. "If you're wealthy and you have access to health care, you will be healthier." But there are likely cultural and behavioral factors at work, too.

Take diet, for example. Southern cuisine is perhaps most famous for its fried chicken, which is a staple in Bogalusa. Also popular are fried green tomatoes, shrimp and grits, and collard greens cooked with bacon. In short, there's no shortage of salt and fat in the southern diet.

Eating fatty foods, like meats, butter, and cheeses, can directly contribute to heart disease. That's because these foods are rich in saturated fat (see Chapter 6) and **cholesterol**—both of which can elevate cholesterol levels in the blood, contributing to atherosclerosis.

Cholesterol is not inherently harmful. In fact, it is a critical constituent of cell membranes as well as a precursor to the synthesis of many hormones. However, our bodies can make all the cholesterol they need, and an excess of dietary cholesterol can do more damage than good—especially if it leads to more cholesterol hanging around in our arteries.

Whether cholesterol stays in our arteries or not depends upon the specific form it takes in the blood. As cholesterol is a lipid (see Chapter 2), it is hydrophobic, and so it is not soluble in water-based solutions such as blood. It thus binds with specialized proteins that allow it to be carried in the blood. The resulting protein-cholesterol complexes are known as lipoproteins (from the Greek word *lipos,* meaning "fat") and they come in two main varieties. **Low-density lipoprotein (LDL)** carries cholesterol to body cells, and can accumulate on linings of blood vessels, contributing to atherosclerosis. **High-density lipoprotein (HDL)** helps carry cholesterol out of blood vessels and deliver it to the

Once known as the Magic City, Bogalusa has since lost some of its luster.

liver, where it can be processed into a form that permits its elimination from the body.

In general, HDL is "good" and LDL is "bad" (think "H" for "healthy" and "L" for "lethal"). Too much LDL relative to the amount of HDL in the blood leads to cholesterol deposits in blood vessels. Once cholesterol is deposited, other events lead to fatty streaks. Once these deposits become raised, they are known as plaques. Some plaques harden by accumulating calcium (in the so-called hardening of the arteries) (**INFOGRAPHIC 27.9**).

INFOGRAPHIC 27.9 CHOLESTEROL: GOOD AND BAD

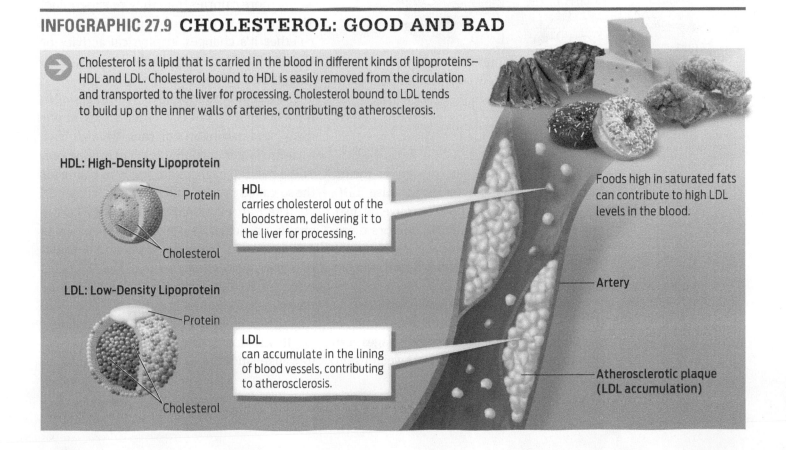

Cholesterol is a lipid that is carried in the blood in different kinds of lipoproteins—HDL and LDL. Cholesterol bound to HDL is easily removed from the circulation and transported to the liver for processing. Cholesterol bound to LDL tends to build up on the inner walls of arteries, contributing to atherosclerosis.

HDL: High-Density Lipoprotein

Protein

Cholesterol

HDL carries cholesterol out of the bloodstream, delivering it to the liver for processing.

LDL: Low-Density Lipoprotein

Protein

Cholesterol

LDL can accumulate in the lining of blood vessels, contributing to atherosclerosis.

Foods high in saturated fats can contribute to high LDL levels in the blood.

Artery

Atherosclerotic plaque (LDL accumulation)

INFOGRAPHIC 27.10 OBESITY IN THE UNITED STATES IS ON THE RISE

→ The percentage of individuals in the U.S. who are overweight (BMI≥25) and obese (BMI≥30) has been steadily increasing.

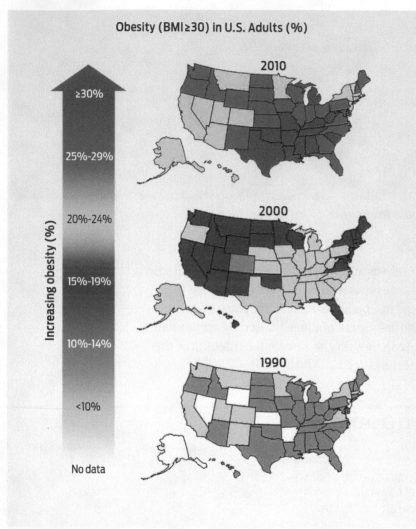

Obesity (BMI≥30) in U.S. Adults (%)

Increasing obesity (%)

≥30%

25%-29%

20%-24%

15%-19%

10%-14%

<10%

No data

2010

2000

1990

Source: CDC

▶**DIABETES**
A disease characterized by chronically elevated levels of blood sugar.

That's because obesity–having a BMI of 30 or above–is an independent risk factor for cardiovascular disease. Being obese makes the heart work harder, and can lead to high blood pressure. Obesity is also the number one risk factor for type 2 **diabetes**, which, in turn, is a leading cause of CVD. In fact, 65% of people with diabetes die of CVD, according to the American Heart Association.

One of the most alarming trends revealed by recent epidemiological research is that obesity rates in the United States have skyrocketed since the 1980s. Nearly 36% of adults and 17% of children in the United States are obese (BMI ≥ 30), according to the CDC (**INFOGRAPHIC 27.10**). In rural parts of the country, such as Bogalusa, the problem is even worse. Since 1973, when the Bogalusa study began, the proportion of children and young people ages 5-17 who are obese has increased more than fivefold, from 6% in 1973 to 31% today.

What's behind this dramatic rise in obesity? Epidemiologists aren't entirely certain.

"This is a big argument among scientists," says epidemiologist Katzmarzyk. "We do know it's energy balance–so definitely the kids are eating more calories than they're expending. The problem is, there's this argument around whether it's changes in physical activity or changes in diet." Quite likely, he says, it's a combination of the two.

As a culture, we spend more time sitting in front of our TVs, our computer games, at our desks, and driving in our cars. We also have access to larger portion sizes than ever before. These cultural changes are partly to blame for the so-called obesity epidemic (see Chapter 6).

There are also regional differences. Louisiana, in particular, is one of a series of states (those located between Texas and Florida) that epidemiologists refer to as the "diabetes belt," "stroke belt," or "obesity belt." For some reason–no one knows exactly why–cardiovascular diseases seem to cluster in this region.

In addition to diet and other lifestyle factors, like smoking, it may partly have to do

Eating saturated fats (found in meats, butter, and cheeses) tends to raise LDL, whereas unsaturated fats (from olive oil, fish, and nuts) tend to raise HDL; doctors therefore recommend that people limit their overall fat intake, and substitute unsaturated fats for saturated ones.

But it's not just eating lots of saturated fat that is the problem. Consuming too much of anything–including sugary beverages–can be harmful if it leads to dramatic weight gain.

with climate. "It's so hot for large parts of the year, people tend not to exercise outside as they would in the north," says Katzmarzyk.

Quite likely there is no one single cause, but rather a confluence of factors that have collectively created "a perfect storm" for cardiovascular disease in this part of the country (INFOGRAPHIC 27.11).

AN OUNCE OF PREVENTION

Given the observed link between risk factors for heart disease and the extent of atherosclerosis in young people, it stands to reason that reducing one's risk factors will reduce one's risk for heart disease. Eating better, quitting smoking, and starting or continuing to exercise will almost certainly reduce

INFOGRAPHIC 27.11 SOUTHERN STATES HAVE THE HIGHEST AVERAGE INCIDENCE OF CARDIOVASCULAR DISEASE

→ States in the Southeast and the Southwest have the highest levels of heart disease in the United States. Societal factors such as diet, activity level, and smoking can influence the risk for developing cardiovascular disease.

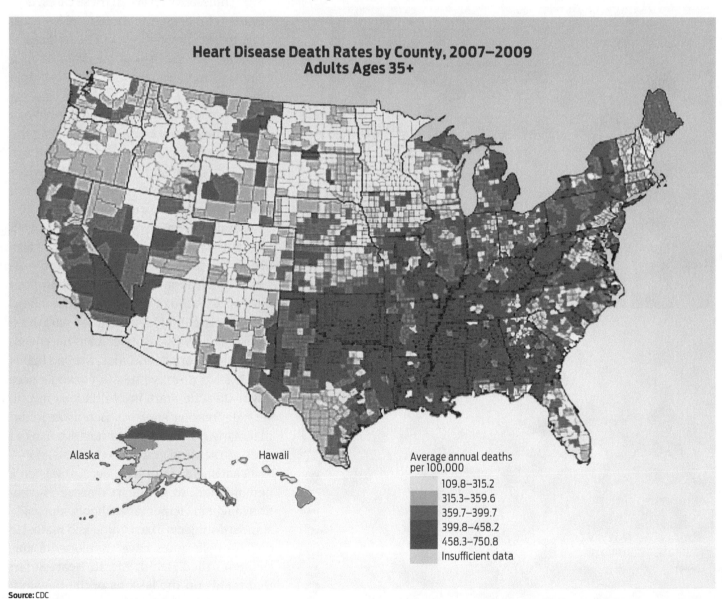

**Heart Disease Death Rates by County, 2007–2009
Adults Ages 35+**

Alaska Hawaii

Average annual deaths per 100,000

	109.8–315.2
	315.3–359.6
	359.7–399.7
	399.8–458.2
	458.3–750.8
	Insufficient data

Source: CDC

27

To reduce your risk of heart disease, follow the ABCs of heart health: Avoid tobacco, Be more active, and Choose good nutrition.

Avoid Tobacco

Be More Active

Choose Healthy Foods

your chances of developing heart disease. In fact, one 2012 study found that eating more fruits and increasing one's activity level can positively affect cholesterol levels and reduce BMI. So that is the good news—there are things you can do to reduce your risk of heart disease: avoid tobacco, get more physical activity, and choose healthy foods.

Unfortunately, for some people, especially those with a family history of heart disease or a genetic predisposition to it, lifestyle changes may not be enough. Some people, for example, naturally have higher "bad" cholesterol levels. In these cases, doctors may also prescribe cholesterol-lowering drugs to keep down the risk of heart disease.

Heart disease is a good example of a complex, modern disease with multiple causes. Dealing effectively with heart disease means not only understanding the biology of atherosclerosis, but also addressing the socioeconomic conditions that create unhealthy environments. In that sense, when it comes to heart disease, your zip code can matter as much as your genetic code.

Gerald Berenson knows this very well. In addition to heart health, Berenson is also deeply concerned with education and worries as much about the high school dropout rate—40% in Bogalusa—as he does about heart disease. Education is a key part of establishing healthy lifestyles, says Berenson, and it starts young. Likewise, the bad habits that lead to negative health consequences down the line start in childhood. In one very depressing statistic, Berenson found that many kids in Bogalusa start smoking as early as the third grade.

That's why, as of late, Berenson is on a new mission: to get heart disease prevention taught in elementary schools, alongside standard subjects like reading and math. He and his colleagues have developed a curriculum called Health Ahead/Heart Smart that builds on the lessons of the Bogalusa Heart Study and attempts to apply them in practical ways to help *prevent* heart disease.

Heart health starts young.

Berenson believes strongly in this approach, which has been used successfully in local schools, but so far he's had trouble garnering national attention for it. "It's a wonderful program, and I don't know how to get it out," he says.

But he hasn't given up just yet. Berenson is rightly proud of his brainchild, and he says he'll never stop trying to convince others of its continued relevance. "There's no other study in the world like Bogalusa," he says. ∎

FOR COMPARISON Open vs. Closed Circulatory Systems (and Other Variations)

All organisms must exchange gases, nutrients, and wastes with their environment. For unicellular organisms, such as bacteria or yeast, this exchange occurs directly across the cell membrane, by the process of **diffusion** (see Chapter 3). For example, oxygen can diffuse in and carbon dioxide can diffuse out, allowing the individual cells to meet their gas exchange needs. Most multicellular organisms are too large to rely solely on this direct transfer with their environment, and so require a cardiovascular system to deliver nutrients and gases to body cells. All cardiovascular systems have three main common elements: a pump, a series of tubes, and a circulating fluid. But there are also important differences among different organisms.

Take the sea squirt, for example. The sea squirt is an invertebrate animal also known as an ascidian. It has a single-chambered heart that is surprisingly similar to the human heart in the way in which it contracts from one end to the other. But, in an interesting twist, the direction of pumping switches regularly in the sea squirt heart. The biggest difference between the two systems, however, is that ascidian blood is not entirely enclosed within the heart and blood vessels. Instead,

▶**DIFFUSION**
The movement of dissolved substances from an area of high concentration to an area of low concentration.

the sea squirt has an **open circulatory system,** in which the heart pumps blood into vessels, which then empty into its central body cavity, where exchange of gases and nutrients occurs. The blood then makes its way back to the heart by vessels. Most mollusks also have an open circulatory system, in which hemolymph (the circulating fluid in mollusks) is pumped through vessels to various parts of the body, where it then leaves the open vessels to mingle with the body cells, eventually returning to the heart.

Unlike the open circulatory system of sea squirts and mollusks, in which the circulatory fluid is not always contained in a vessel, humans and most other animals have a **closed circulatory system,** in which the blood is always contained in the heart or blood vessels, never directly mixing with other body fluids. While many animals share this overall structure, there are many variations on the theme. For example, segmented worms, such as annelids (earthworms) have multiple hearts. Other animals have a single heart, but not every animal heart has four chambers like the human one. Most fish, for example, have a two-chambered heart, while amphibians have a three-chambered heart.

The two-chambered fish heart contains a single atrium and a single ventricle. This means that the blood is pumped in a single circuit around the body. Blood returns from the body and enters the atrium, then is pumped to the ventricle. The ventricle pumps the blood to the gills (the fish equivalent of lungs), where it picks up oxygen. Instead of returning to the heart for a second circuit as it does in a human, the blood then continues directly to the body, delivering oxygen that it picked up in the gills. It is remarkable that the blood can make it all the way through the body, given that a lot of the force of the ventricular contraction is lost as the blood passes through the gills. However, the system clearly works for the fish. Finally, deoxygenated blood then returns to the heart to repeat its circuit through the gills and body.

Amphibians have a three-chambered heart with two atria and a single ventricle. You might think that this would be disastrous: oxygenated blood returning from the lungs (and skin) would mix with deoxygenated blood returning from the body, and that single ventricle

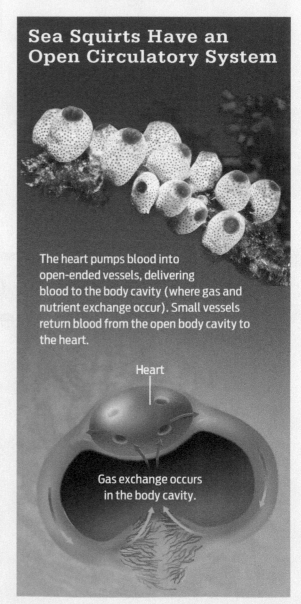

Sea Squirts Have an Open Circulatory System

The heart pumps blood into open-ended vessels, delivering blood to the body cavity (where gas and nutrient exchange occur). Small vessels return blood from the open body cavity to the heart.

Heart

Gas exchange occurs in the body cavity.

would be pumping out to the body blood that was not fully oxygenated. So how do amphibians survive with this counterintuitive model of circulation? Well, there is some mixing of oxygenated and deoxygenated blood, but not as much as you might think. Anatomical features of the vessel leaving the single ventricle help direct deoxygenated blood from the right atrium preferentially to the pulmonary circuit, and similarly preferentially direct oxygenated blood from the left atrium to the systemic (body) circuit.

When an amphibian is submerged underwater it doesn't really need to send as much blood to the lungs (as there won't be as much oxygen to pick up there), so it can redirect more blood to the body, where oxygen can diffuse from the water through the skin directly into the blood. This wouldn't be possible with two completely separate circuits (left and right). ■

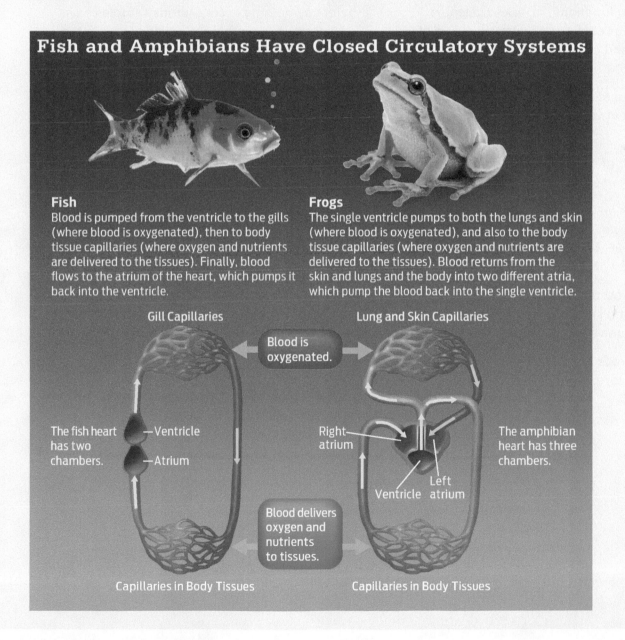

Fish and Amphibians Have Closed Circulatory Systems

Fish
Blood is pumped from the ventricle to the gills (where blood is oxygenated), then to body tissue capillaries (where oxygen and nutrients are delivered to the tissues). Finally, blood flows to the atrium of the heart, which pumps it back into the ventricle.

Frogs
The single ventricle pumps to both the lungs and skin (where blood is oxygenated), and also to the body tissue capillaries (where oxygen and nutrients are delivered to the tissues). Blood returns from the skin and lungs and the body into two different atria, which pump the blood back into the single ventricle.

Gill Capillaries

Lung and Skin Capillaries

Blood is oxygenated.

The fish heart has two chambers.
—Ventricle
—Atrium

Right atrium

The amphibian heart has three chambers.

Ventricle
Left atrium

Blood delivers oxygen and nutrients to tissues.

Capillaries in Body Tissues

Capillaries in Body Tissues

CHAPTER 27 Summary

▸ The cardiovascular system transports nutrients, oxygen, hormones, and other substances throughout the body. It consists of the heart, blood vessels, and blood.

▸ The heart acts as a pump, moving blood in two circuits between the heart and lungs (pulmonary circuit) and between the heart and the rest of the body (systemic circuit).

▸ Different types of blood vessel have different structures and functions. Arteries transport blood away from the heart, veins transport blood to the heart, and capillaries are the sites of nutrient and gas exchange in tissues.

▸ Blood is composed of cells and fluid. Red blood cells (erythrocytes) are the oxygen-carrying cells.

▸ Atherosclerosis, or hardening of the arteries, is a common form of cardiovascular disease, which can lead to heart attack.

▸ A heart attack results when the blood supply to the heart muscle is cut off and the muscle begins to die; atherosclerosis in coronary arteries is a common cause of a heart attack.

▸ High blood pressure (hypertension), obesity, high cholesterol, and smoking are known risk factors for atherosclerosis.

▸ Blood pressure is expressed as systolic pressure/diastolic pressure. Normal blood pressure is ~120/80.

▸ Hypertension is a dangerous condition if left untreated. There are both controllable and uncontrollable causes of hypertension.

▸ Cholesterol is a normal constituent of cells, but too much LDL cholesterol in the blood can contribute to cardiovascular disease through the accumulation of plaque in arteries. Eating saturated fats elevates LDL levels.

▸ Cardiovascular disease begins early in life, long before symptoms occur. Autopsy studies have documented that atherosclerosis occurs in the arteries of young people and correlates with risk factors during life.

MORE TO EXPLORE

▸ American Heart Association www.heart.org

▸ Centers for Disease Control and Prevention: Heart Disease www.cdc.gov/heartdisease

▸ Berenson, G. S., et al., for the Bogalusa Heart Study. (1998) Association between multiple cardiovascular risk factors and atherosclerosis in children and young adults. *New England Journal of Medicine* 338:1650–1656.

▸ Broyles, S., et al. (2010) The pediatric obesity epidemic continues unabated in Bogalusa, Louisiana. *Pediatrics* 125:900–905.

▸ May, A. L., et al. (2012) Prevalence of cardiovascular disease risk factors among US adolescents, 1999–2008. *Pediatrics* 129:1035–1041.

▸ HBO. (2012) *Weight of the Nation*.

CHAPTER 27 Test Your Knowledge

What structures make up the cardiovascular system, and how does blood flow through the system?

By answering the questions below and studying Infographics 27.2 and 27.3, you should be able to generate an answer for the broader Driving Question above.

KNOW IT

1 List the three components of the cardiovascular system.

2 Blood enters the systemic circuit from

 a. the lungs.

 b. the left side of the heart.

 c. the right side of the heart.

 d. the body.

 e. both b and c

3 Which of the following is the correct order of structures that blood flows through as it passes through the cardiovascular system?

 a. right side of heart→body→left side of heart→lungs

 b. left side of heart→body→right side of heart→lungs

 c. left side of heart→right side of heart→body→lungs

 d. right side of heart→left side of heart→lungs→body

USE IT

4 In which circuit (pulmonary or systemic) does blood pick up nutrients, such as sugar?

5 Are the coronary arteries part of the pulmonary or the systemic circuit? Explain your answer.

6 During heart surgery, patients are often placed on a heart–lung machine (known as a cardiopulmonary bypass). This machine circulates and oxygenates blood during the operation, taking over for the patient's heart and lungs. At what points would the patient's blood have to be diverted into the machine and back into the body in order to properly bypass the flow of blood through the heart and lungs?

What is the structure of the heart and of the different types of blood vessel?

By answering the questions below and studying Infographics 27.2 and 27.3, you should be able to generate an answer for the broader Driving Question above.

KNOW IT

7 Blood returning to the heart from the lungs enters the _____ .

 a. right atrium

 b. right ventricle

 c. left atrium

 d. left ventricle

 e. coronary arteries

8 Which chamber of the heart pumps oxygen-rich blood to the body?

 a. the right atrium

 b. the right ventricle

 c. the left atrium

 d. the left ventricle

 e. both the right atrium and the left atrium (to the right and left sides of the body, respectively)

9 How does the heart obtain oxygen?

 a. from the blood in its chambers

 b. from the coronary arteries

 c. from the systemic arteries

 d. from the coronary veins

 e. from the systemic veins

10 What is the defining feature of an artery?

 a. It carries blood away from the heart.

 b. It has very thin walls.

 c. It has valves.

 d. It carries oxygenated blood.

 e. both a and d

USE IT

11 Atrial fibrillation is a condition in which the atria contract very rapidly and irregularly. The ventricles also contract faster than normally, but their contraction is not coordinated with the atrial contraction. Explain how this can disrupt blood supply to the body.

12 Which of the following is most likely to be associated with left-side heart failure?

 a. fluid backing up in the feet and legs, causing swelling

 b. increased systolic pressure in the aorta

 c. reduced oxygenation of the blood leaving the left ventricle

d. fluid backing up in the lungs, causing shortness of breath

e. increased oxygenation of the blood returning to the right atrium

13 Some babies are born with a congenital heart defect in which the septum, or wall, between the left and right ventricles has an opening in it, and the aorta therefore can accept blood from both the left and right ventricles. These babies are known as "blue babies." Why?

DRIVING QUESTION 3

What is the composition of blood, and what does blood do?

By answering the questions below and studying Infographics 27.5 and 27.6, you should be able to generate an answer for the broader Driving Question above.

KNOW IT

14 Describe the function of each of the following cellular components of blood: white blood cells; red blood cells; platelets.

15 Which of the following is *not* true of capillaries?

a. Blood moves through them slowly.

b. They have a very small diameter.

c. They have very thin walls.

d. They occur between arteries and veins.

e. They have valves.

16 What is happening in a capillary in your big toe?

a. Oxygen is diffusing from the blood into your toe tissue.

b. Carbon dioxide is diffusing from your toe tissue into the blood.

c. Nutrients are diffusing from the blood into your toe tissue.

d. all of the above

e. a and c

USE IT

17 TV crime dramas often analyze arterial spray patterns of blood at crime scenes. Why do you never hear about venous spray patterns?

18 Why are soldiers standing at attention for long periods of time advised to contract their calf muscles to avoid becoming light-headed?

DRIVING QUESTION 4

What is cardiovascular disease, and what are some of the risk factors for developing cardiovascular disease?

By answering the questions below and studying Infographics 27.4, 27.7, 27.8, and 27.9, you should be able to generate an answer for the broader Driving Question above.

KNOW IT

19 Match the disease (left column) with the letter corresponding to its underlying cause (right column).

___ heart attack   A. elevated systolic and diastolic pressure

___ hypertension   B. fatty streaks in arteries

___ stroke   C. blockage of a coronary artery

___ atherosclerosis   D. blockage of blood vessels in the brain

20 List at least four risk factors for cardiovascular disease. For each, state whether or not it is modifiable.

21 Why does atherosclerosis increase the risk of stroke?

22 Which type of cholesterol is dangerous with respect to cardiovascular disease?

a. HDL

b. LDL

c. both

d. neither—cholesterol is a healthy lipid

e. HDL is a risk for hypertension and LDL is a risk for atherosclerosis.

USE IT

23 The data generated from the Bogalusa study show a correlation between atherosclerosis and risk factors for cardiovascular disease. Does this mean that everyone with a risk factor (e.g., high BMI) will develop atherosclerosis? Explain your answer.

MINI CASE

24 Steven is 14 years old. He is an ace goalie for his soccer team, which practices twice during the week and plays a game every weekend. He is also an ace online gamer, holding the highest player skill level in World of Warcraft. After many long afternoons of playing, Steven is also a great graphic designer and his mother pays him to design computer-generated fliers for her business. She prepares healthy meals with lots of whole grains, fruits, and dairy (milk and cheese), and she gets regular checkups to monitor her elevated blood pressure. Steven spends some of his money on chocolate bars and po-

tato chips. He is a big guy, and in the heaviest 5% of his age group. From what you've read in this chapter, what cardiovascular risk factors do you identify in Steven? What could you say to Steven and his mom about Steven's potential risk and reducing that risk?

BRING IT HOME

26 Evaluate your own health and lifestyle with respect to risk for cardiovascular disease. Consider what kinds of changes you might want to talk to your physician about.

INTERPRETING DATA

25 Refer to Infographic 27.10 and record your answers to the following questions in the table below.

a. Look at the 2010 data for Texas, Arizona, California, South Dakota and your home state. For each state, record the average percentage of adults with BMI ≥30.

b. Now determine the total population of each state (you can find census data at quickfacts.census.gov/qfd).

c. Now determine the total number of people in each of those states who experienced these cardiovascular events.

State	Adults with BMI ≥30 (%)	Total State Population	Number of Adults with BMI ≥30
Texas			
Arizona			
California			
South Dakota			
Your home state			

PEAK PERF

An inside look at altitude training among elite athletes

Michael Phelps taking a breath.

ORMANCE

621

▸▸ DRIVING QUESTIONS

1. What structures make up the respiratory system?

2. How do the respiratory and cardiovascular systems cooperate to deliver oxygen to body cells and remove carbon dioxide from tissues?

3. What factors influence the oxygen-carrying capacity of blood and breathing rate?

4. How can scientific knowledge of the respiratory system be used to design training regimens for elite athletes?

WITH A WINGSPAN OF 6 FEET 7 INCHES, FEET THE SIZE OF FLIPPERS, and just 4% body fat, swimmer Michael Phelps is built for speed. He holds seven current world record titles, is an 18-time Olympic gold medalist, and is universally recognized as the greatest swimmer of all time. With stats like these, it's hard to imagine how he could possibly improve his performance—short of growing gills. Yet the 27-year-old athlete proved he still has a few tricks up his swimsuit when he announced in June 2012 that he sleeps in a "contraption" designed to give him an athletic boost.

"When I walk into my bedroom, I still have to open another door to get to my bed," explained the swimmer to Anderson Cooper on *60 Minutes*. "It's kind of a boy in bubble sort of thing."

Phelps wouldn't show the device on TV, but he later tweeted a picture of the glass-walled structure to his followers. Known as a hypoxic chamber, the device creates an artificial low-oxygen environment, equivalent to that at 8,000 feet above sea level. The swimmer claims that sleeping in the chamber every night while he's training helps his body heal faster. In fact, it was a central part of his training regimen as he prepared for the summer 2012 Olympics in London, at which he broke the world record for most medals won by an Olympian—22 in all.

Phelps isn't the only elite athlete to resort to a technical fix for improved performance. Manufacturers of hypoxic chambers have seen an increased interest in their product from athletes such as tennis pro Novak Djokovic and triathlete Jonathan Brownlee. Even entire Olympic sports teams are getting in on the action.

But what does sleeping in a low-oxygen chamber have to do with athletic performance? Does it work? And is it ethical?

According to Randall Wilber, a sports physiologist with the U.S. Olympic Training Center in Colorado Springs, Colorado, the use of hypoxic chambers is a variation on the theme of natural altitude training. Athletes spend several weeks living at high altitude, typically defined as 6,000 feet or more above sea level, and then return to sea level to train or compete. The widely used practice is based on the idea that the body will acclimatize to higher elevation by becoming better able to take up and transport oxygen, ultimately improving performance.

"If you look at the number of athletes who win medals in winter sport, summer sport, across many endurance-based sports, I would say in my estimation 90-95% of those athletes

A hypoxic chamber like the one in Michael Phelps's bedroom.

at minimum are using altitude training in some way or another," says Wilber.

But with more and more athletes using technical means to enhance performance, many in the sports field are taking a closer look at the practice, and some are asking questions about fairness.

LIVE HIGH, TRAIN LOW

Altitude training isn't new. According to Wilber, altitude training first became popular after the 1968 Olympic Games, held in Mexico City, elevation 7,350 feet. A number of strange things happened at those Games that drew the attention of scientists and athletes alike. World records were set in a number of short-distance track and field events–like the 100- and 200-m dash–but athletes performing in endurance events, like the 1,500-m run, fared worse than usual.

"From 400 meters on down, including the long jump–think of Bob Beamon–all those people were just smashing world records left and right," says Wilber. "But from 800 meters on up, it went the other way: people were running slower than the world records." These results make perfect sense, he says. At altitude, the air is thinner, its molecules less densely packed, and so there is less air resistance, which explains why records were set in sprinting events. But at high altitude, there is also less oxygen, and so endurance events requir-

Winners at the 1968 Olympic Games in Mexico City.

ing prolonged exertion, like distance running, predictably suffered.

The question that emerged was, what would happen if you did the opposite–lived at high elevation and then competed at lower altitude? Would the body respond differently?

There was tantalizing evidence that it would. In Mexico City, athletes coming from higher-altitude countries like Kenya and Ethiopia in general performed better. They did not experience a dip in performance as

> **"If you look at the number of athletes who win medals in winter sport, summer sport, across many endurance-based sports, I would say in my estimation 90–95% of those athletes at minimum are using altitude training in some way or another."**
>
> — RANDALL WILBER

▶RESPIRATORY SYSTEM
The organ system that allows us to take in oxygen and unload carbon dioxide.

▶LUNG
The major respiratory organ in humans, the site of gas exchange between air and the blood.

▶PHARYNX
The throat.

▶LARYNX
The opening to the lower respiratory tract; also known as the voicebox.

▶TRACHEA
A large airway leading into the lower respiratory tract.

▶BRONCHI
Two airways that branch from the trachea; one bronchus leads into each lung.

> In an adult human, that respiratory surface is enormous: if all the alveoli were splayed out flat, they would occupy an area about the size of a tennis court.

other athletes did, and thus had a competitive advantage. The race was on to understand these results and incorporate the lessons into effective training regimens.

As a sports physiologist with the U.S. Olympic Committee, Wilber serves as a consultant to Olympic athletes, helping them to design appropriate training regimes using the latest science. He has worked with an impressive roster of elite athletes, including cyclist Kristin Armstrong and speed skaters Apolo Ohno and Christine Witty, as well as swimmer Michael Phelps. One of Wilber's main areas of research

is how altitude affects the **respiratory system** of athletes.

The respiratory system is what allows us to take in oxygen from the air and release carbon dioxide. When we breathe, air first enters the nose and mouth and then moves through a series of branched tubes on its way to the **lungs.** It first passes through the throat or **pharynx;** then through the **larynx,** which houses our vocal cords; then through the **trachea,** or windpipe. At that point, air is diverted into two major **bronchi,** one traveling to each lung. Much like an inverted tree, the tubes continue to

INFOGRAPHIC 28.1 THE HUMAN RESPIRATORY SYSTEM

 The respiratory system allows us to take in oxygen from the air and eliminate carbon dioxide waste. Air travels through a series of tubes to small sacs in the lungs called alveoli. The alveoli have a close association with capillaries, permitting the uptake of oxygen into blood and the release of carbon dioxide from blood.

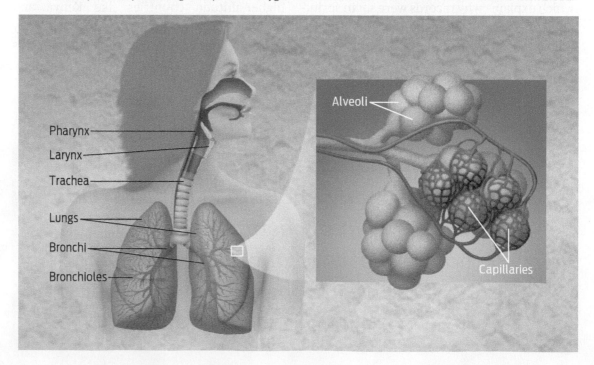

INFOGRAPHIC 28.2 GAS EXCHANGE AND TRANSPORT

→ Oxygen-rich air enters alveoli and oxygen enters the bloodstream. Carbon dioxide from the blood enters the alveoli and is breathed out. The oxygen-rich blood eventually reaches tissues, where oxygen is delivered, and carbon dioxide produced by aerobic respiration enters the blood to be carried back to the lungs.

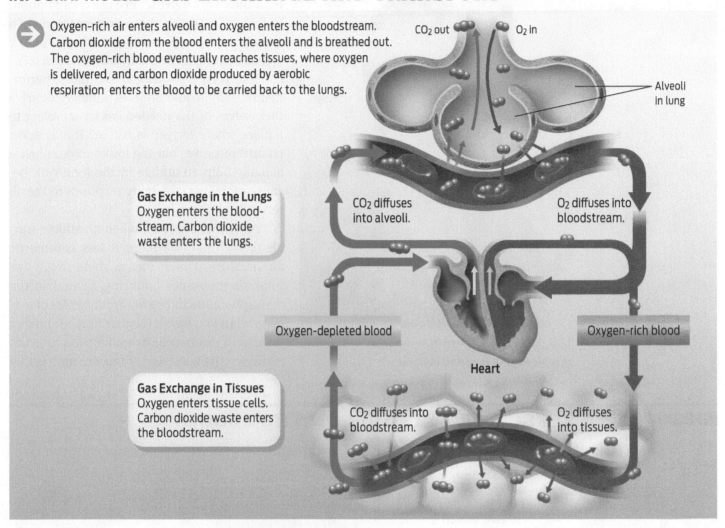

CO₂ out   O₂ in

Alveoli in lung

Gas Exchange in the Lungs
Oxygen enters the bloodstream. Carbon dioxide waste enters the lungs.

CO₂ diffuses into alveoli.

O₂ diffuses into bloodstream.

Oxygen-depleted blood

Oxygen-rich blood

Heart

Gas Exchange in Tissues
Oxygen enters tissue cells. Carbon dioxide waste enters the bloodstream.

CO₂ diffuses into bloodstream.

O₂ diffuses into tissues.

---

branch and split into smaller and smaller **bronchioles** until finally air hits the many tiny sacs called **alveoli,** located deep within the lungs. Alveoli are closely associated with capillaries (see Chapter 27), permitting oxygen and carbon dioxide to be exchanged between the air and the blood (**INFOGRAPHIC 28.1**).

Each lung's **respiratory surface** is made up of millions of gas-exchanging alveoli. In an adult human, that respiratory surface is enormous: if all the alveoli were splayed out flat, they would occupy an area about the size of a tennis court (although the one-cell-thick layer would be so thin you wouldn't be able to see it).

This large respiratory surface ensures that our bodies are able to take in enough oxygen

to power the work of cells. In fact, oxygen is required by every cell in the body in order to perform aerobic respiration to make ATP (see Chapter 6). Yet not every cell in the body is in contact with the respiratory surface (at least not in large, multicellular organisms like humans). How then does oxygen get from the lungs to the rest of the body?

The alveoli are the sites of **gas exchange** between the air and blood: they provide the means by which oxygen enters the blood. Once oxygen has diffused from the air into the blood, and carbon dioxide has diffused from the blood into the airspaces in the lungs, blood vessels carry the oxygenated blood to the heart, where it will be pumped around the body to its various tissues (**INFOGRAPHIC 28.2**).

▶**BRONCHIOLES**
Smaller airways that branch from the bronchi.

▶**ALVEOLI**
Air sacs in the lung across which gases diffuse between air and blood.

▶**RESPIRATORY SURFACE**
A surface across which oxygen enters and carbon dioxide leaves.

▶**GAS EXCHANGE**
The process of taking up and releasing oxygen and carbon dioxide.

U.S. Olympic Committee sports physiologist Randall Wilber.

The respiratory and cardiovascular systems thus work together in close coordination. In fact, it's probably more useful to think of a "cardiorespiratory system" than two separate systems acting individually (see Chapter 27).

If oxygen is so critical to athletic performance, why would athletes willingly deprive themselves of the needed gas by traveling to a place where oxygen is scarce? It may seem counterintuitive, but the lower oxygen levels actually help an athlete in the long run, because of the way the body responds to the altered conditions.

As Wilber explains, when an athlete travels to high altitude there is less barometric air pressure, which means there are fewer total air molecules bouncing around in the atmosphere—including fewer molecules of oxygen. "Thus you have a tougher time running or walking or climbing or swimming or whatever," he says. "The good side of the situation is that

INFOGRAPHIC 28.3 EFFECT OF ALTITUDE ON THE BLOOD

The amount of available oxygen decreases with increasing altitude. Under these conditions, RBC production is stepped up to increase the oxygen-carrying capacity of the blood.

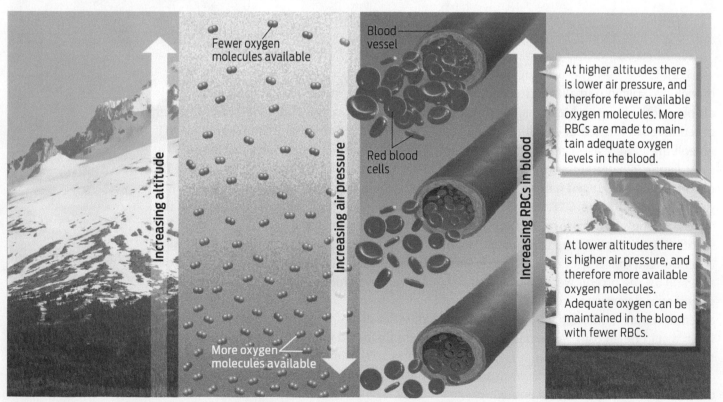

Fewer oxygen molecules available

Blood vessel

Red blood cells

Increasing altitude

Increasing air pressure

Increasing RBCs in blood

At higher altitudes there is lower air pressure, and therefore fewer available oxygen molecules. More RBCs are made to maintain adequate oxygen levels in the blood.

At lower altitudes there is higher air pressure, and therefore more available oxygen molecules. Adequate oxygen can be maintained in the blood with fewer RBCs.

More oxygen molecules available

Michael Phelps training at the U.S. Olympic Training Center in Colorado Springs, Colorado.

when you come to [high] altitude your body senses, *Hey, there's not enough oxygen here, I need to do something about it to compensate!*"

As scientists have learned, the body indeed compensates by increasing the number of oxygen-carrying **red blood cells (RBCs)** in the blood. With more RBCs, the body is able to take in roughly the same amount of needed oxygen, despite the altered conditions, thus maintaining homeostasis (**INFOGRAPHIC 28.3**). When the athlete returns to sea level, after a period of acclimatization to altitude, he will have a competitive advantage.

The most popular form of altitude training in use today is known as "live high, train low": athletes live for a period of time at higher elevations and then return to lower altitudes to train or compete. This approach is better than living and training full time at high elevation, say some experts, since at high elevations one can't exercise or train as rigorously as at sea level–thus the benefits of acclimatization are canceled out.

BOY IN A BUBBLE

Like many athletes on the U.S. Olympic team, Phelps typically travels to Colorado Springs anywhere from three to five times a year, for 4 to 5 weeks at a time. Phelps says he had his personal hypoxic chamber installed in his bedroom at his home in Baltimore, Maryland, in 2011, after noticing that he recovered faster while living at high elevation.

"We've been able to realize after going to Colorado Springs so many times that it is something that helps me recover," Phelps told the Associated Press in 2012. "That's something that is so important to me now being older. I don't recover as fast as I used to."

Hypoxic chambers simulate altitude by artificially reducing the amount of oxygen in the air. By reducing oxygen concentration from about 21% (in untreated air) to about 15%, the chambers reduce oxygen's **partial pressure,** simulating an altitude of approximately 8,000 feet. Partial pressure

▶**RED BLOOD CELLS (RBCs)**
Blood cells specialized for transporting oxygen throughout the body.

▶**PARTIAL PRESSURE**
The proportion of total air pressure contributed by a given gas.

▶ALTITUDE
SICKNESS

An illness that can occur as a result of an abrupt move to an altitude with a reduced partial pressure of oxygen.

and barometric pressure are measured in units called millimeters of mercury (mmHg) **(INFOGRAPHIC 28.4).**

The lower oxygen concentration Phelps experiences while he sleeps stimulates his body to produce RBCs, while the regular oxygen concentration during the day helps him train at his normal, vigorous levels. It's a simulated version of the "live high, train low" scenario.

Phelps may be unusual among athletes in the degree to which he has used a hypoxic chamber, but he is certainly not alone in using altitude training. In fact, according to Wilber, nearly all elite athletes do **(INFOGRAPHIC 28.5).**

Are there any downsides to sleeping in hypoxic conditions? At least over the short term, there don't appear to be. In fact, many humans, including some Tibetan and Andean peoples, live easily at much higher elevations (above 10,000 feet) with even less oxygen. However, there is evidence that these populations have adapted over thousands of years to the hypoxic conditions by becoming better able to absorb and transport oxygen. Someone coming from sea level, who lacked these adaptations, would have a hard time thriving at these elevations, and may in fact succumb to **altitude sickness.** Wilber notes that some athletes find that the cham-

INFOGRAPHIC 28.4 THE RELATIONSHIP BETWEEN ALTITUDE AND OXYGEN PRESSURE

 As altitude increases, overall air pressure (barometric pressure) decreases. This means the pressure contributed by oxygen (its partial pressure) also decreases.

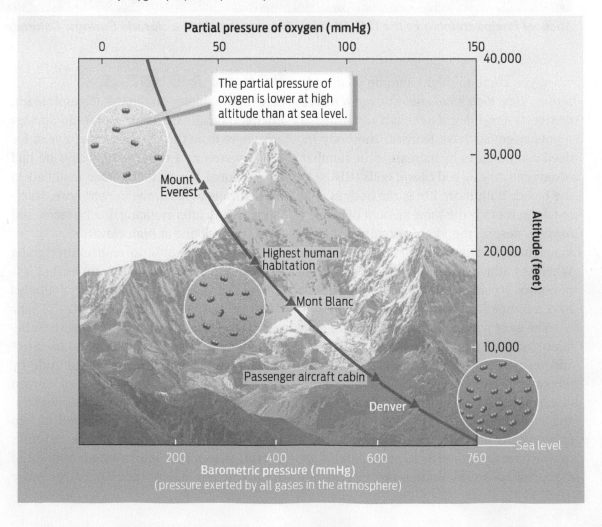

Partial pressure of oxygen (mmHg)

0   50   100   150

40,000

The partial pressure of oxygen is lower at high altitude than at sea level.

30,000

Mount Everest

Highest human habitation

20,000

Mont Blanc

Altitude (feet)

10,000

Passenger aircraft cabin

Denver

Sea level

200   400   600   760

Barometric pressure (mmHg)
(pressure exerted by all gases in the atmosphere)

INFOGRAPHIC 28.5 ALTITUDE TRAINING HYPOTHESIS

 Both hypoxic chambers and natural altitude training are based on the same principle. By living or sleeping at lower partial pressures of oxygen, athletes will elevate their RBC count. This enhances the oxygen-carrying capacity of the blood when competing at sea level.

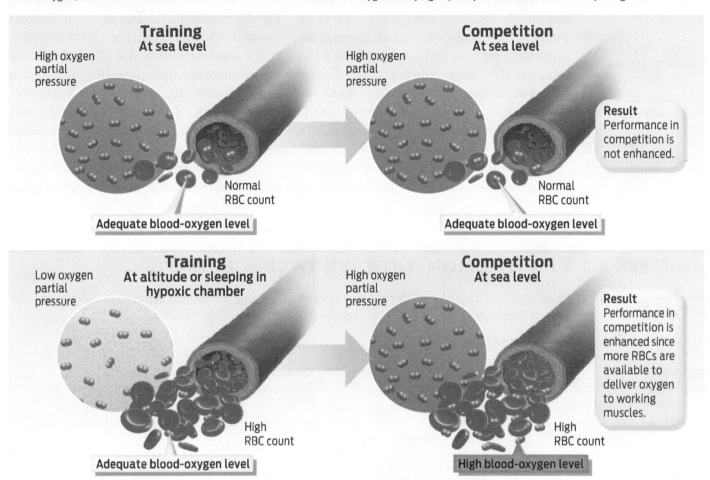

Training
At sea level

High oxygen partial pressure

Normal RBC count

Adequate blood-oxygen level

Competition
At sea level

High oxygen partial pressure

Normal RBC count

Adequate blood-oxygen level

Result
Performance in competition is not enhanced.

Training
At altitude or sleeping in hypoxic chamber

Low oxygen partial pressure

High RBC count

Adequate blood-oxygen level

Competition
At sea level

High oxygen partial pressure

High RBC count

High blood-oxygen level

Result
Performance in competition is enhanced since more RBCs are available to deliver oxygen to working muscles.

bers give them headaches or that they have trouble sleeping.

Phelps's biggest complaint is more mundane: "The worst thing is trying to watch TV in it. I've got to watch it through Plexiglas—it's blurry," Phelps told the Associated Press.

These are complaints that nonendurance athletes don't have to worry about. Typically, nonendurance athletes don't use altitude training. That's because sports such as weightlifting and sprinting require only short bursts of energy, and thus do not rely as much on aerobic respiration, and so having extra RBCs doesn't really help these athletes (see the discussion of fermentation in Chapter 6). "If you wanted to, you could run a 100-meter dash holding your breath," notes Wilber.

The Ladakhi people live at more than 3,500 meters in elevation with no adverse health effects.

EVERY BREATH YOU TAKE

Obviously, to take in oxygen, we need to breathe. But what exactly is breathing and how does it occur? Breathing is the way we **ventilate** our lungs. With every breath, we bring in a new volume of air for gas exchange at the respiratory surface, and then expel the waste products. If our lungs didn't ventilate, the process of gas exchange would be much less efficient because there would be no way to move fresh air to the respiratory surface and get stale air out of the way.

On a mechanical level, breathing is the effect of a series of muscular movements of the **diaphragm** working in coordination with other muscles connected to the rib cage. As these muscles contract, the diaphragm moves downward and the rib cage moves outward, thereby increasing the size of the chest cavity overall. This sudden increase in the volume of the chest cavity has the effect of decreasing air pressure there, creating suction that draws air into the lungs. This is inhalation. As the muscles relax, the chest cavity contracts, and air is expelled. This is exhalation **(INFOGRAPHIC 28.6)**.

Athletes in general have greater lung capacity than the average person, which is partly

VENTILATION
The process of moving air in and out of the lungs.

DIAPHRAGM
A sheet of muscle that contributes to breathing by contracting and relaxing.

INFOGRAPHIC 28.6 VENTILATION: HOW WE TAKE A BREATH

The contraction and relaxation of the diaphragm and rib cage muscles ventilate the lungs. Contraction decreases pressure in the lungs and air flows in. Relaxation of the muscles forces air out of the lungs.

Inhalation
The diaphragm and other muscles contract. This increases the volume of the chest cavity, thereby reducing the pressure and creating vacuumlike suction. Air flows into the lungs.

Exhalation
The diaphragm and other muscles relax. This decreases the volume of the chest cavity, forcing air out.

The diaphragm is like the plunger in a syringe. When it is pulled down, the volume inside the syringe increases, the pressure decreases and fluid rushes in. Pressing the plunger in decreases the volume, increases the pressure, and forces the liquid out.

Air inhaled

Rib cage expands as rib muscles contract.

Lungs expand.

Diaphragm down

Liquid in

Liquid out

Plunger down

Plunger up

Volume of cavity increases, air pressure decreases.

Volume of cavity decreases, air pressure increases.

Air exhaled

Rib cage gets smaller as rib muscles relax.

Lungs shrink.

Diaphragm up

why they can exercise longer, and harder, without getting as winded as someone who is not as fit. And while you can't increase the size of your lungs, you can train the muscles of the chest to work more effectively, as many musicians and singers, as well as athletes, do. Exercising in a pool, for example, makes your chest muscles work harder, which strengthens them, ultimately enhancing lung capacity. Some populations of humans who live at high elevations (like the Bods, whose home, Ladakh, is at 11,500 feet in the Himalaya mountains), have wider chests with stronger chest muscles, permitting them to draw in more air in order to get sufficient oxygen.

Although getting oxygen is an essential part of breathing, the rate of our breathing is not actually determined by the demand for oxygen in our body. It's actually carbon dioxide that is more important. When we exercise, we are using more oxygen and producing more carbon dioxide. The increased concentration of carbon dioxide in the blood causes a decrease in blood pH; when CO_2 dissolves in the water in the blood—the plasma—it produces carbonic acid, which has a low pH (see Chapter 2). The brain senses this drop in pH and responds by sending a signal to the lungs to increase breathing rate. The increased breathing rate increases the rate of gas exchange, bringing more oxygen into the body while unloading excess carbon dioxide, and thus raising blood pH back to normal.

If for some reason a person is unable to unload carbon dioxide from the lungs, the concentration of CO_2 in the blood can rise to dangerous levels, creating a potentially fatal condition known as **acidosis.**

DOES IT WORK?

The logic of altitude training seems airtight: increase the concentration of RBCs in the blood and you will have greater oxygen-carrying capacity, more aerobic respiration, and better performance. But actually proving that altitude training improves performance is harder than it may seem. Even Wilber, who literally wrote the book on the subject, acknowledges that there

A hypoxic workout chamber at a popular gym chain.

is still a fair amount of controversy surrounding the science of altitude training, with some studies concluding that it works, and others concluding it doesn't.

Partly, the issue is the difficulty of designing properly controlled studies. Any gains in performance are likely to be small, so how can you be sure the gains were due to high altitude and not some other aspect of training? Likewise, if athletes don't show consistent improvement, is it because altitude training doesn't work, or because not everyone responds to altitude in the same way?

For his part, Wilber is persuaded by the available evidence that, when pursued

▶ACIDOSIS
A dangerous condition in which blood is too acidic.

INFOGRAPHIC 28.7 ALTITUDE TRAINING IMPROVES PERFORMANCE

→ In a study of 22 elite distance runners, maximal oxygen uptake and 3,000-m race performance were measured before and after a 4-week "live high, train low" altitude-training regimen (living at 2,500 m and training at 1,250 m). Performance and oxygen uptake were improved after the altitude training. Similar performance results were seen for collegiate athletes.

Source: Stray-Gunderson, *J Appl Physiol* 91:1113-1120,2001.

methodically, altitude training does indeed improve performance (**INFOGRAPHIC 28.7**).

Additional (but anecdotal) evidence that it works, he says, comes from the loyalty of athletes and coaches themselves. "When you see a Michael Phelps and his coach and other athletes coming back to altitude year after year after year after year, you know that it's working for them."

At the same time, Wilber cautions that altitude training requires consistent and long-term practice. For that reason, ordinary gym-goers are unlikely to see any significant results from periodic workouts in the numerous "hypoxic rooms" that have sprung up in commercial gyms all over the world.

PICKING UP, DROPPING OFF

If an athlete experiences hypoxia regularly, for 12-20 hours a day, for at least 3 weeks, according to the research he or she will start to see benefits upon returning to sea level. The biggest benefit will be the ability to carry more oxygen in the blood. But unlike carbon dioxide, which travels easily in the blood plasma as carbonic acid, oxygen does not readily dissolve in the watery plasma. How, then, does oxygen make its way through the blood? Oxygen is carried inside RBCs, bound to a molecule called **hemoglobin.** In fact, RBCs are essentially bags of hemoglobin. They lack a nucleus, and are streamlined for doing one thing very well: carrying oxygen.

Hemoglobin is a highly dexterous molecule, composed of multiple interworking parts. The oxygen-binding part is the iron-containing **heme group.** Each hemoglobin molecule has four heme groups, so each hemoglobin molecule can carry up to four molecules of O_2. (Note that because the heme group of hemoglobin contains iron, a dietary deficiency of iron can limit the production of hemoglobin, and therefore reduce the overall oxygen-carrying capacity of blood, leading to a condition known as iron-deficient anemia.)

Key to understanding how hemoglobin functions in oxygen transport is the fact that it binds oxygen reversibly. It can "pick up" (bind) oxygen in the lungs and "drop off" (release) oxygen in the tissues.

What determines whether or not hemoglobin is picking up or dropping off oxygen, and therefore that oxygen is being delivered to the right place at the right time? There are a few important factors. The main one is the partial pressure of oxygen (P_{O_2}). Fresh air entering the lungs has a relatively high partial pressure of oxygen, so the hemoglobin will become saturated with oxygen. In tissues, the partial pressure of oxygen tends to be lower, because cells in tissues are consuming oxygen as they carry out aerobic respiration, so hemoglobin tends to give up some of its oxygen to tissues.

Another important factor is temperature. Hemoglobin is more likely to release oxygen as temperature increases—as, for example, it does in an actively contracting muscle that is burning energy.

Lastly, pH plays a role. At lower pH, hemoglobin releases oxygen. As we've seen, one of the main things that can lower pH is the concentration of CO_2. Since CO_2 is produced during aerobic respiration, the pH of muscle goes down, becoming more acidic during exercise. In turn, the lower pH causes the hemoglobin to give up more of its oxygen, thus ensuring a continuous supply of the critical gas for the active muscle (**INFOGRAPHIC 28.8**).

So hemoglobin has an extremely important role to play in respiration. If there were

> ▸**HEMOGLOBIN**
> A protein found in red blood cells specialized for transporting oxygen.

> ▸**HEME GROUP**
> Iron-containing structures on hemoglobin, the sites of oxygen binding.

INFOGRAPHIC 28.8 HEMOGLOBIN CARRIES OXYGEN IN RED BLOOD CELLS

Hemoglobin reversibly binds O_2. When the partial pressure of O_2 (P_{O_2}) is high, hemoglobin will bind O_2. At low P_{O_2}, hemoglobin will release O_2. The binding is influenced by temperature and pH.

Lungs

High P_{O_2}

Loaded hemoglobin

O_2 diffuses into red blood cells and binds to hemoglobin.

Heme group

Unloaded hemoglobin

Tissues

Low P_{O_2}

Hemoglobin releases O_2, which diffuses into tissue cells.

pH
At any P_{O_2} in tissues, hemoglobin gives up more O_2 as pH decreases.

pH 7.6
pH 7.2

Hemoglobin binds more O_2 at higher pH.

Hemoglobin binds less O_2 at lower pH.

Hemoglobin saturation (%)

Partial pressure of O_2 (mmHg)

Temperature
At any P_{O_2} in tissues, hemoglobin gives up more O_2 as temperature increases.

32°C
42°C

Hemoglobin binds more O_2 at lower temperature.

Hemoglobin binds less O_2 at higher temperature.

Hemoglobin saturation (%)

Partial pressure of O_2 (mmHg)

some way to increase the amount of hemoglobin in blood, then an athlete would be at a competitive advantage in terms of conducting aerobic respiration.

HIGH-TECH DOPING?

When an athlete (or anyone) spends time at altitude, his or her kidneys respond by secreting a hormone called **erythropoietin (EPO),** which stimulates RBC production in the bone marrow. By producing more RBCs—including, within a single cell, hundreds of thousands of hemoglobin molecules—the body increases its ability to absorb and transport oxygen. This is a natural process, which goes on any time the body needs to make more RBCs—for example, after blood loss from a wound.

Because there is such a clear relationship between the concentration of RBCs in the blood and the ability to perform aerobic exercise, many athletes in elite sports in the past resorted to blood doping—artificially increasing the concentration of RBCs in their blood by undergoing blood transfusions or by injecting themselves with synthetic EPO (**INFOGRAPHIC 28.9**).

The International Olympic Committee officially banned blood doping by transfusion

▶**ERYTHROPOIETIN (EPO)**
A hormone that stimulates red blood cell production.

INFOGRAPHIC 28.9 EPO AND BLOOD DOPING

Erythropoietin (EPO) stimulates the production of red blood cells in the bone marrow. After EPO treatment, more immature red blood cells are detectable, and maximum oxygen uptake is enhanced.

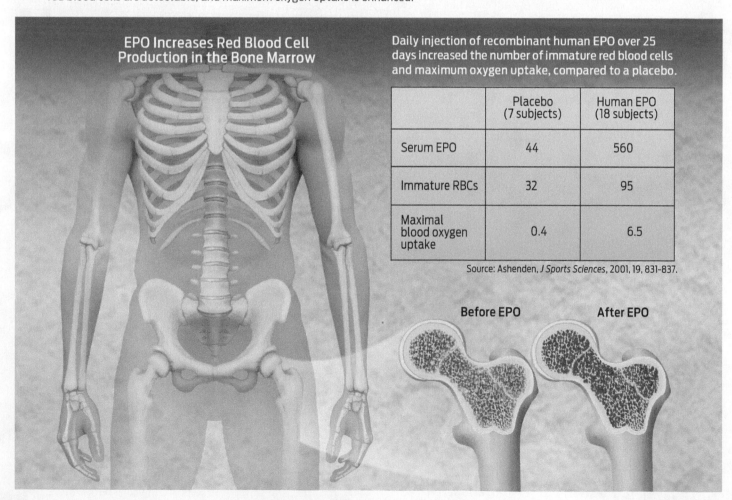

EPO Increases Red Blood Cell Production in the Bone Marrow

Daily injection of recombinant human EPO over 25 days increased the number of immature red blood cells and maximum oxygen uptake, compared to a placebo.

	Placebo (7 subjects)	Human EPO (18 subjects)
Serum EPO	44	560
Immature RBCs	32	95
Maximal blood oxygen uptake	0.4	6.5

Source: Ashenden, *J Sports Sciences*, 2001, 19, 831-837.

Before EPO   After EPO

in 1986 and by EPO in 1990. However, the difficulty in testing made enforcement difficult. (Since EPO is a natural hormone, testing for its presence in the blood is not enough to indicate doping.)

Blood doping became endemic in cycling and caused a major scandal in 1998, when it was discovered that many cyclists were using EPO. That scandal led to the formation of the World Anti-Doping Agency (WADA) in 1999, which has worked to develop tighter and more effective testing programs. One way the agency tests for doping is to measure the percentage of RBCs in the blood; a value over 50% is generally considered evidence of doping, and is grounds for disqualification.

Of course, for testing to be effective, it has to be done systematically and rigorously. Cyclist and seven-time Tour de France winner Lance Armstrong, who admitted to doping and had his titles stripped in 2012, managed to avoid being tested numerous times by simply not answering the door when testers arrived, according to the testimony of his fellow cyclists.

There was speculation during the 2008 Olympics in Beijing that Phelps was doping—how else to account for his superhuman performance, eight gold medals in eight events—but repeated tests proved that he was clean.

Unlike doping with EPO, the use of hypoxic chambers is not currently banned by WADA. Even though the underlying goal of blood doping and the use of hypoxic chambers is the same—increasing the production of RBCs—leaders in the sporting world have so far deemed the use of such chambers a variation on natural training regimens. So for now, it is not illegal to use them. However, it does raise the question of whether is it ethically proper or goes against the spirit of sport.

Controversy surrounding the use of hypoxic chambers reached fever pitch in 2006, when WADA announced that it planned to include these devices on its list of banned sub-

Lance Armstrong admitted to doping during his winning Tours de France.

stances and methods. However, backlash from athletes, coaches, and sports scientists in the end pressured the agency to reconsider its decision. "It doesn't mean we approve it," said Dick Pound, head of WADA, at the time.

Part of WADA's concern was safety. Blood doping with EPO can have dangerous side effects, including abnormal blot clots and stroke. As EPO doping literally thickens the blood by increasing the concentration of red blood cells, the heart has to work harder to pump it, and the thickened blood is more likely to cause a blockage in a vessel. A few athlete deaths have been attributed to the practice. But spokespeople on the other side countered that using hypoxic tents is really no different from using common "live high, train low" methods, and it would be impossible to try to regulate that practice.

Is it fair that some athletes—especially those with lucrative sponsorships, such as Phelps—can afford to buy a hypoxic tent for $7,000–$10,000, while other athletes—especially those from poorer countries—cannot?

Michael Phelps celebrates after winning one of his 18 Olympic gold medals.

Then there were questions of ethics. Is it fair that some athletes—especially those with lucrative sponsorships, such as Phelps—can afford to buy a hypoxic tent for $7,000-$10,000, while other athletes—especially those from poorer countries—cannot?

As Wilber notes, "You're not going to find altitude houses or chambers in South Sudan, or a majority of the countries in the world, but yet in the U.S. or Finland or China, yeah, you're going to find them there. And you're going to find elite athletes using them."

Ethical or not, it is easy to see the appeal of such practices among elite athletes. When fractions of a second can mean the difference between winning gold and coming in fourth, every little ounce of help can matter a great deal. This is a dilemma that Phelps knows all too well, having lost to South African swimmer Chad le Clos by less than a quarter of a second in the 200-m butterfly at the 2012 Summer Games in London. Even with the potential added advantage of a few more RBCs, he was still not able to capture the gold in that race, having to settle instead for silver.

But the swimmer isn't complaining. After all, he's won more Olympic medals than anyone in history, and is recognized as one of the greatest athletes of all time. He says that 2012 will be his last Olympic Games. Maybe now he can retire his bubble as well. ■

How Do Creatures Without Lungs "Breathe"?

In humans and other mammals, the lungs are the primary sites of gas exchange–where oxygen is picked up and carbon dioxide is dropped off. But not all animals have lungs, or any kind of centrally located respiratory surface. Insects, for example, have a vast network of air-filled tubes that bring in air from the body surface to individual cells. This system of air tubes is known as a **tracheal system.** The individual tubes are called tracheae. Each trachea is an inner extension (that is, an invagination) of the body surface. The opening to the air (at the body surface) is called a spiracle. The spiracles are adjustable, to balance the intake of air and the loss of water, and to keep out dust. The tracheae branch into smaller and smaller tubes, resulting in a highly branched network that brings air very close to every cell in the body. Oxygen and carbon dioxide diffuse directly from the air tubes to the body cells, which means there is no need for a circulatory system to shuttle gases to and from cells.

In a process similar to lung ventilation, some insects generate air flow with their tracheal system by compressing and expanding their body wall. This action changes the volume and pressure of the air tubes and pushes air through the tracheal system.

Lungs are specialized structures for extracting gases from air. Not surprisingly, most fish do not have lungs. Instead they have **gills**–delicate tissues that allow the fish to extract oxygen from water. As water flows over gills, oxygen diffuses directly into the blood and is carried through the body. Fish move water over their gills in a one-way direction. Water enters through the mouth, is forced over the gills, and then leaves the gill compartment by passing the gill cover (the operculum). In most fish, the opening and closing of the mouth and coordination of the opening of the

▶**TRACHEAL SYSTEM**
A series of air-filled tubes in insects that deliver air to cells.

▶**GILLS**
The gas exchange surface in fish.

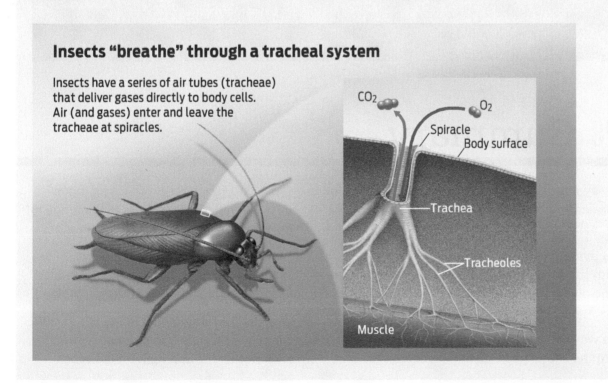

Insects "breathe" through a tracheal system

Insects have a series of air tubes (tracheae) that deliver gases directly to body cells. Air (and gases) enter and leave the tracheae at spiracles.

CO_2

O_2

Spiracle

Body surface

Trachea

Tracheoles

Muscle

operculum is sufficient to force water over the gills. Some large and active fish, such as tuna, instead use "ram ventilation" to keep water moving over their gills. In this case, they swim with their mouths wide open, and the force of their movement through the water keeps water moving over their gills. The more water they can move over their gills, the more oxygen they can extract from the water, and the more oxygen can be carried to cells for use in cellular respiration. Since water is the fish's source of oxygen, a fish out of water will die.

Fish "breathe" with their gills

Most fish use gills for gas exchange between water and blood. As water flows over the gills, oxygen diffuses into the blood and carbon dioxide diffuses out of the blood.

Blood vessel

Oxygen-poor blood

Carbon dioxide is eliminated in water.

Oxygen-rich blood

Direction of water flow

Gill filaments

Oxygen-rich water flows over gills.

Blood flow

CHAPTER 28 Summary

- ▶ The respiratory system is a means of taking in oxygen and eliminating carbon dioxide waste. It consists of the respiratory tract and the lungs.

- ▶ The respiratory system works in close conjunction with the cardiovascular system. Capillaries in the alveoli of the lungs are the sites of gas exchange between air and blood.

- ▶ Breathing is the way we ventilate our lungs and bring fresh air to the site of gas exchange. Ventila-

tion requires a coordinated set of muscle movements that create negative pressure in the lungs to draw air in, and positive pressure to expel air out.

- ▶ Breathing rate is controlled by the brain and relies on sensors that detect pH, which is directly related to the concentration of carbon dioxide in the blood.

- ▶ Oxygen is carried by hemoglobin in red blood cells. Hemoglobin binds reversibly to oxygen, picking it up from the lungs and dropping it off in tissues.

▸ Extended exposure to low oxygen pressure, such as occurs at high altitude, stimulates the production of red blood cells in the bone marrow. The growth signal comes from the hormone erythropoietin (EPO), which is secreted by the kidneys.

MORE TO EXPLORE

▸ World Anti-Doping Agency http://www.wada-ama.org/

▸ Wilber, R. (2004) *Altitude Training and Athletic Performance.* Champaign, IL: Human Kinetics Publishers.

▸ Stray-Gundersen, J., et al. (2001) "Living high-training low" altitude training improves sea level performance in male and female elite runners. *Journal of Applied Physiology* 91:1113–1120.

▸ Ashenden, M. J., et al. (2001). A comparison of the physiological response to simulated altitude exposure and r-HuEpo administration. *Journal of Sports Sciences* 19: 831-837

▸ Beall, C. M., et al. (2010) Natural selection on EPAS1 (HIF2a) associated with low hemoglobin concentration in Tibetan highlanders. *Proceedings of the National Academy of Sciences* 107:11459–11464.

▸ Shermer, M. 2008. "The Doping Dilemma." *Scientific American.* March 31.

▸ Matson, J. 2012. "Rope a Dope: Drug Testing in Sports Enters a More Aggressive Era." *Scientific American.* July 30.

CHAPTER 28 Test Your Knowledge

What structures make up the respiratory system?

By answering the questions below and studying Infographics 28.1 and 28.6, you should be able to generate an answer for the broader Driving Question above.

KNOW IT

1 Add the names of the structures indicated by the labels A–F in the diagram below.

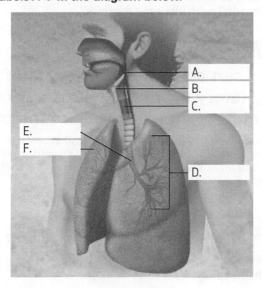

A.
B.
C.
E.
F.
D.

2 Which part of the respiratory system is the site of exchange of gases between blood and air?

a. alveoli   c. trachea   e. bronchi
b. bronchioles   d. pharynx

3 Inhalation is accompanied by

a. muscular relaxation and a decrease in lung volume.

b. muscular relaxation and an increase in lung volume.

c. muscular contraction and a decrease in lung volume.

d. muscular contraction and an increase in lung volume.

e. muscular contraction and no change in lung volume.

USE IT

4 Emphysema is a form of chronic obstructive pulmonary disease (COPD). In this disease, the alveoli have been damaged, most often by cigarette smoking. The alveolar damage results in overinflation of the alveoli that can't exchange their "spent air," and causes an affected person to exhale through pursed lips.

a. From this description, explain the following symptoms of emphysema: breathlessness; inability to exercise.

b. Explain why supplemental oxygen is often a treatment for emphysema.

5 Asthma is a disease that causes swelling and constriction of the airways. Compare the predicted symptoms of asthma with the predicted symptoms of emphysema. Can you think of any treatments for asthma that might be different from treatments for emphysema?

DRIVING QUESTION 2

How do the respiratory and cardiovascular systems cooperate to deliver oxygen to body cells and remove carbon dioxide from tissues?

By answering the questions below and studying Infographics 28.2 and 28.3, you should be able to generate an answer for the broader Driving Question above.

KNOW IT

6 How is O_2 transported throughout the body?

a. dissolved in the plasma of blood
b. bound to hemoglobin in plasma
c. bound to hemoglobin in white blood cells
d. bound to hemoglobin in red blood cells
e. dissolved in the cytoplasm of red blood cells

7 What can cause a drop in blood pH?

a. a decrease in O_2
b. an increase in O_2
c. a decrease in CO_2
d. an increase in CO_2
e. b or d

USE IT

8 If blood pH drops, what happens to the breathing rate? Explain your answer.

9 Oxygen diffuses from the air in alveoli to the blood in lung capillaries. Diffusion occurs rapidly over short distances, but decreases dramatically with increases in distance. Pneumonia is an accumulation of fluid in the alveolar air spaces. Why does pneumonia cause shortness of breath and give a bluish tint to the skin and nails?

10 Breathing in and out of a paper bag will _____ pH and therefore _____ ventilation.

a. not change; not change
b. increase; increase
c. increase; decrease
d. decrease; decrease
e. decrease; increase

DRIVING QUESTION 3

What factors influence the oxygen-carrying capacity of blood and breathing rate?

By answering the questions below and studying Infographics 28.3, 28.4, 28.8, and 28.9, you should be able to generate an answer for the broader Driving Question above.

KNOW IT

11 Relative to a tissue at rest, actively exercising tissues have

a. higher temperature, higher P_{O_2}, and higher pH.
b. higher temperature, lower P_{O_2} and lower pH.
c. higher temperature, higher P_{O_2} and lower pH.
d. lower temperature, higher P_{O_2} and higher pH.
e. lower temperature, lower P_{O_2} and lower pH.

12 What is the particular feature of altitude that increases the oxygen-carrying capacity of the blood?

a. the actual height (elevation)
b. the reduced atmospheric (barometric) pressure
c. the reduced partial pressure of oxygen
d. the increased atmospheric (barometric) pressure
e. the decreased relative humidity

USE IT

13 Hemoglobin releases O_2 at low pH. Give two reasons why a tissue would have a low pH.

14 Carbon monoxide (CO) binds to hemoglobin more tightly than oxygen does. In fact, CO can displace oxygen from hemoglobin. Predict the symptoms of CO poisoning, and provide an explanation.

15 Why do people "suck wind" (that is, breathe very heavily) with vigorous exercise?

DRIVING QUESTION 4

How can scientific knowledge of the respiratory system be used to design training regimens for elite athletes?

By answering the questions below and studying Infographics 28.5, 28.7, and 28.9, you should be able to generate an answer for the broader Driving Question above.

KNOW IT

16 Which of the following mimics high altitude?

a. sleeping in a high-O_2 chamber
b. sleeping in a low-O_2 chamber

c. transfusing RBC into the circulation

d. a and c

e. b and c

17 **What does EPO do?**

a. stimulates RBCs to release stored O_2

b. stimulates RBC production

c. increases the number of heme groups per molecule of hemoglobin

d. increases ventilation rate.

e. b and c

18 **Design an experiment to determine whether hypoxic chambers confer an advantage relative to altitude training. Consider how you will set up your experiment, including appropriate controls, and the variables that you will consider and measure.**

USE IT

19 **WADA must be able to test for a variety of banned substances. How could WADA test for each of the following? Rank them in order of easiest to detect (1) to hardest to detect (4).**

_____ EPO doping

_____ Transfusion of whole blood

_____ Transfusion of RBC

_____ Use of a hypoxic chamber

INTERPRETING DATA

20 Look at Infographic 28.7.

a. Start with the 3000-m performance data for the elite athletes. How many seconds did the men improve by? How many seconds did the women improve by? Now determine that improvement as a % of the pre-altitude race time. What % improvement did altitude training confer? In your opinion, is this method of training worth it for that % improvement?

b. Now look at the graph with both the collegiate and the elite athletes. Even if the graph were not labeled, how could you know which group was which? (Hint: Look at their pre-altitude race times.) Did every athlete experience an improvement in race time? Was the change in race time identical for every athlete? Given the variability in results in this small sample size of conditioned athletes, what do you think you can extrapolate about altitude training in a larger population of active people?

BRING IT HOME

21 Your friend wants to join a new gym. The gym has many amenities, including personal trainers (for an additional fee) and a hypoxic chamber for "performance training." There is a steep sign-up fee and high monthly dues at this gym. Another gym is offering a no-fee sign-up and lower monthly dues. This gym has the same cardiovascular equipment and access to personal trainers (at an additional fee comparable to that at the new gym). Your friend is fit. She runs local 10K races, and often places in the top 10 in her age group in your small town. Given the costs of the two gyms and your friend's fitness level and goals, would you advise her to join the new gym or the old one? Explain your reasoning.

Smoke
ON THE BRAIN

Neuroscience explains why nicotine and other drugs are hard to kick

JACK WARD THOUGHT HE COULD RESIST PICKING UP A cigarette, but the temptation was too strong. He had successfully quit smoking in 2006. But when he walked into a smoke-filled poker room 3 years later at a friend's house in Brooklyn, New York, the scene before him seemed to run in slow motion. He watched intently as smokers sucked smoothly on their cigarettes, deeply drawing in each puff and then exhaling with sighs of contentment. He resisted that night. But poker night became a weekly event, and finally he gave in: he began smoking again.

Ward knew that smoking is risky. He had had countless arguments with his wife, who had insisted he stop smoking to protect his own health and hers

---

▸▸ DRIVING QUESTIONS

1. How is the nervous system organized?

2. How do cells in the nervous system transmit signals?

3. Why are some drugs (and some behaviors) addictive?

Despite public health campaigns, smoking remains a common practice.

as well, which is partly why he quit in the first place. But during poker night, none of that seemed to matter.

Ward had started smoking casually in high school. By the time he went to graduate school, he was smoking a pack and a half a day. "I was surrounded by people who smoked," Ward recalls. "It never seemed unusual to smoke so much."

Cigarette smoking is highly addictive: smokers can develop a physical and psychological need to smoke. And while anyone might be able to smoke one cigarette or even several and not become addicted, most people find it extremely difficult to stop if they have smoked for an extended length of time.

For many years, addiction was seen as a personal or moral failing—the result of a weak will or a set of bad choices. Increasingly, however, researchers are looking to biology to explain addiction's awesome power. Brain scientists now know that cigarettes, alcohol, cocaine, and other drugs of abuse stimulate the brain's reward system, a complex circuit of brain cells that evolved to make us feel good after eating or having sex—activities we must engage in if we are to survive and pass along our genes. Without a feeling of pleasure from these activities that ensure our survival, we might never seek out food or sex, and our spe-

cies would die out. Drugs of abuse stimulate the same pleasure pathways, which is why they make people feel so good, and also why they can be so hard to resist.

Nonchemical addictions can do the same. Certain behaviors—gambling, shopping, sex—can start out as habits but slide into addictions. Even food can be addictive. Studies of obese people, for example, have shown that the brains of compulsive eaters are hyperactive in areas that respond to food. For these people, the mere thought of eating floods the brain with pleasure. In other words, almost anything deeply enjoyable can become an addiction.

But the pleasure comes at a massive cost. After prolonged use, drugs of abuse change the structure and function of the brain in ways that can wreak havoc on users' lives. Severely addicted people may stop eating, stop working, in fact stop all activities because nothing matters except the drug—and they will do anything to get it. Without it, they have intense cravings, obsessive thoughts, and are deeply depressed.

Once established, addiction is difficult to treat. The stories of celebrities who have been in and out of rehab and the deaths of superstars such as Michael Jackson and Heath Ledger from drug overdoses are testament to just how difficult. But scientists today are gaining a better understanding of how physiological changes in the brain cause addiction, and that knowledge is leading to better treatments, which will help addicts reclaim their brains and their lives.

ADDICTION AND THE BRAIN

Despite a massive public health campaign aimed at curbing the dangerous habit, cigarette smoking remains a common practice in this country. Today, in 2013, about 22% of Americans smoke tobacco regularly, according to the National Institute on Drug Abuse (NIDA), down from about 45% in 1955. These numbers testify to tobacco's addictive potential.

The component in tobacco that makes cigarettes addictive is the chemical nicotine. When a person smokes, nicotine floods the lungs, where it is absorbed into the bloodstream. Once in the bloodstream, nicotine travels quickly throughout the body, reaching the **brain** within 8 seconds after tobacco smoke is inhaled. Chewing tobacco also contains nicotine, which the body absorbs through the mucous membranes that line the mouth.

Nicotine is addictive because it stimulates feelings of pleasure in the brain. The brain is the master coordinator of the body, controlling virtually all of the body's activities, including sensation, movement, intellect, and just about all involuntary actions. The brain is also the seat of memory. Pleasurable experiences recorded in our memories serve as motivation; we tend to seek out the same pleasurable experiences over and over. Ultimately, the decision to smoke a cigarette, or to carry out any other behavior, is made in the brain (INFOGRAPHIC 29.1).

The brain doesn't act in isolation: it communicates with the rest of the body, sending out and receiving signals to coordinate our many activities. When we decide to reach for a cigarette, for example, our brain sends a signal to our arm. These signals travel along **nerves**—long fibers made up of specialized cells and supportive tissue that transmit signals through the nervous system. A major collection of nerves is the **spinal cord,** which extends from the base of the brain down to the lower back and is contained

▶**BRAIN**
An organ of the central nervous system that integrates and coordinates virtually all functions of the body.

▶**NERVE**
A bundle of specialized cells that transmit information.

▶**SPINAL CORD**
A bundle of nerve fibers, contained within the bony spinal column, that transmits information between the brain and the rest of the body.

INFOGRAPHIC 29.1 NICOTINE STIMULATES PLEASURE IN THE BRAIN

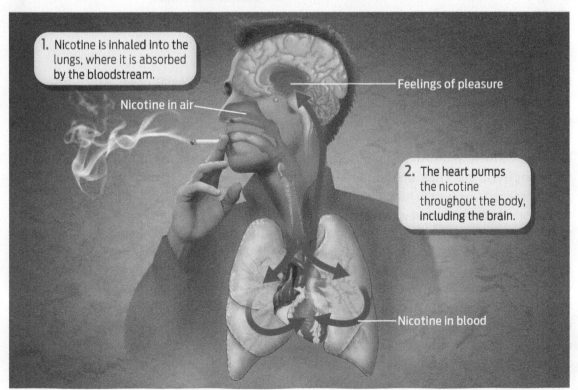

1. Nicotine is inhaled into the lungs, where it is absorbed by the bloodstream.

Nicotine in air

Feelings of pleasure

2. The heart pumps the nicotine throughout the body, including the brain.

Nicotine in blood

INFOGRAPHIC 29.2 THE CENTRAL AND PERIPHERAL NERVOUS SYSTEMS WORK TOGETHER

The nervous system has two main branches: the central nervous system (yellow) and the peripheral nervous system (orange). The central nervous system consists of the brain and spinal cord. The rest of the nervous system is the peripheral nervous system, which connects the CNS to the rest of the body.

Brain

Spinal cord

Peripheral nerves

Central nervous system
Peripheral nervous system

2. The pain signal is transmitted to the spinal cord. The spinal cord can initiate a reflex, and send a signal to the peripheral nerves of the other arm. The spinal cord can also send signals to the brain.

1. A mosquito bites the skin on your arm. Sensory receptors of the **peripheral nervous system** detect pain at the site.

3. Your brain can detect pain, and can also send signals to the peripheral nerves of the other arm.

4. Muscle tissues (the **effector cells**) in the arm and hand respond to slap the mosquito.

▶**CENTRAL NERVOUS SYSTEM (CNS)**

The brain and the spinal cord.

▶**PERIPHERAL NERVOUS SYSTEM · (PNS)**

All the nervous tissue outside the central nervous system which transmits signals from the CNS to the rest of the body.

in and protected by our bony spinal column. Together, the brain and the spinal cord make up the **central nervous system (CNS).** Nerves that travel from spinal cord to distant body sites, like fingers, toes, and heart muscle, compose the **peripheral nervous system (PNS).** The peripheral nervous system includes all nervous tissue outside the brain and spinal cord leading to and from our limbs and organs.

The CNS and the PNS work together to sense and respond to stimuli in the environment. Sensory receptors of the PNS in our ears, our eyes, and in our skin, for example, enable us to sense changes in our environment—a sudden loud noise, for example, or the sting of a mosquito. These stimuli are transmitted to

the brain and spinal cord, which assess the incoming signals and trigger an appropriate response, such as a reflex. The reflex is carried out by effector tissues such as muscles that allow us to respond—by jumping or swatting our arm, for example. Peripheral nervous tissue in other organs, including the lungs, heart, and digestive organs, keeps our bodies operating without conscious thought. The nervous system allows us to perceive and understand the world around us and to translate thought into action (**INFOGRAPHIC 29.2**).

EXCITABLE CELLS

Addiction is a complex psychological and physiological process that affects many parts of the nervous system, but it begins at the level of the cell. Fundamentally, the brain–like any organ–is a collection of cells. The cells of the brain are organized into networks that receive, process, and send information. Highly specialized cells called **neurons** are the individual units of this elaborate network.

Neurons have a unique structure that enables them to send and receive signals: a neuron consists of a large **cell body** with branched extensions called **dendrites** that receive signals and a single large **axon** that carries signals away from the neuronal cell body. The end of the axon, called the **axon terminal,** transmits the signals to the next cell in the network, which could be another neuron, a muscle cell, or an endocrine cell (**INFOGRAPHIC 29.3**).

There are different types of neuron. **Sensory neurons,** such as sensory receptors in the eyes or skin, receive information from the external world and transmit it to the CNS. **Motor neurons** transmit information from the CNS to muscle cells, signaling them to contract or relax.

The signals sent along a neuron are electric currents. Dendrites and axons are like electrical cords that help collect and transmit information in the nervous system. In fact, a nerve is just a bundle of axons of neurons. The axon of a single motor neuron can be over a meter long.

Neurons can conduct electrical signals because of the special properties of their cell membrane. Like other cells in the body, neurons have a slight difference in electrical charge across the membrane: the inside of the neuron has a negative charge relative to the charge outside the cell. But unlike most other cells, neurons can dramatically change the charge difference across the membrane, resulting in an electrical signal. Ion channels in the axon membrane open and close to allow specific ions to cross. When axons of neurons allow positive ions to cross the cell membrane, the charge across the membrane changes. This sudden change of charge initiates an **action potential,** a coordinated pattern of ion flow across the membrane that is conducted down an axon. Like the fuse leading to

▶**NEURONS**
Specialized cells of the nervous system that generate electrical signals in the form of action potentials.

▶**CELL BODY**
The part of a neuron that contains most of the cell's organelles, including the nucleus.

▶**DENDRITES**
Branched extensions from the cell body of a neuron, which receive incoming information.

▶**AXON**
The long extension of a neuron that conducts action potentials away from the cell body toward the axon terminal.

▶**AXON TERMINAL**
The tip of an axon, which communicates with the next cell in the pathway.

INFOGRAPHIC 29.3 NEURONS ARE HIGHLY SPECIALIZED

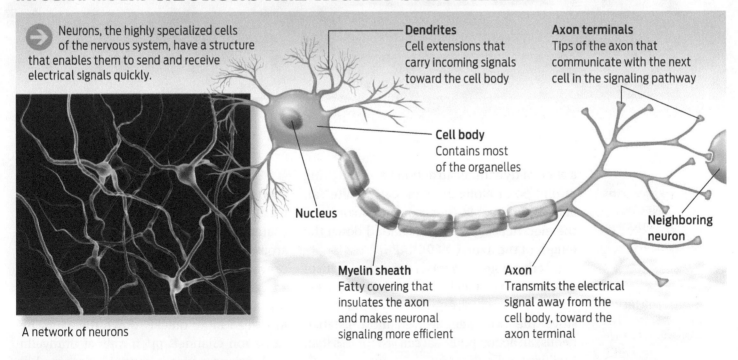

→ Neurons, the highly specialized cells of the nervous system, have a structure that enables them to send and receive electrical signals quickly.

Dendrites
Cell extensions that carry incoming signals toward the cell body

Axon terminals
Tips of the axon that communicate with the next cell in the signaling pathway

Cell body
Contains most of the organelles

Nucleus

Neighboring neuron

Myelin sheath
Fatty covering that insulates the axon and makes neuronal signaling more efficient

Axon
Transmits the electrical signal away from the cell body, toward the axon terminal

A network of neurons

INFOGRAPHIC 29.4 ELECTRICAL SIGNALS ARE TRANSMITTED ALONG AXONS

Neurons conduct electrical signals called action potentials, which are generated as ions flow across the cell membrane. When a neuron "fires", sodium (Na^+) and potassium (K^+) ions enter and leave the axon in a characteristic pattern, creating an action potential. In an axon the firing of an action potential triggers the firing of another farther down the axon. In this way, signals pass along the axon.

1. In response to a stimulus, sodium (Na^+) channels open, generating an action potential here. Na^+ rushes in. The cell becomes more positive inside.

More-positive charge outside the axon membrane

Less-positive charge inside

Na^+ Na^+ Na^+ Na^+ Na^+ Na^+ Na^+ Na^+ Na^+

2. The sodium channels close as potassium (K^+) channels open to release positive charge, returning the inside of the cell to its resting charge.

3. The positive sodium ions that entered the axon (in step 1) trigger sodium channels to open farther down the axon, generating an action potential there.

K^+ K^+ K^+ K^+

K^+ K^+ K^+ K^+

Na^+ Na^+ Na^+ Na^+ Na^+ Na^+ Na^+ Na^+

4. The action potential continues to travel in this way down the length of the axon.

K^+ K^+ Na^+ Na^+ Na^+
K^+ Na^+ Na^+ Na^+
K^+ K^+ Na^+
K^+

▶SENSORY NEURONS
Cells that convey information from both inside and outside the body to the CNS.

▶MOTOR NEURONS
Neurons that control the contraction of skeletal muscle.

a stick of dynamite, an action potential firing at one point along an axon can "ignite" an action potential farther along the axon, and the electrical signal is transmitted down the length of the axon (INFOGRAPHIC 29.4).

Action potentials travel down the length of a neuron very quickly. Just think how quickly your arm moves in response to touching a hot stove—the movement seems almost instantaneous; an action potential can last for less than 1 millisecond.

Electrical signals can travel so quickly down neurons because, much like an electrical cord, neurons are insulated: they are coated with a sheath of **myelin** that wraps around the axon at repeating intervals and prevents the electrical charge from leaking across the membrane. Myelin speeds nerve transmission by allowing electrical charges to spread more quickly down the axon. Because ion channels open only at unmyelinated regions, action potentials appear to hop from one unmyelinated region of the axon to another, and it takes fewer action potentials to conduct the signal down the length of the axon. Myelin is critical to nerve transmission: without it, action potentials weaken, losing strength as they pass along the axon, and can peter out before reaching the axon terminal. Diseases in which myelin degenerates over time, such as multiple sclerosis, can cause progressive paralysis, primarily because motor neurons lacking myelin can't efficiently conduct electrical signals and ultimately can't transmit those signals to muscles. The myelin sheath is produced by a type of supportive cell called a **glial cell.**

THE ANATOMY OF ADDICTION

Neurons communicate not only to move muscles that control limbs but also to solidify thoughts and lay down memories. This constant chatter within the nervous system is the way in which the brain grows and adapts to new environments and learns new tasks. The great capacity that humans have for conscious thought is a function of the sheer number of neurons found in the human brain: approximately 100 billion.

But the brain isn't just an undifferentiated mass of neurons; like any organ, it has specialized regions that perform distinct functions. For example, one part of the brain processes visual information, another specializes in hearing.

Scientists typically divide the brain into four major regions: cerebellum, brain stem, diencephalon, and cerebrum. The **cerebellum,** located in the rearmost portion of the brain, controls movement, coordination, and balance. The **brain stem,** at the base of the brain, coordinates involuntary functions like reflexes, heart rate, and breathing. The **diencephalon,** located above the brain stem, regulates homeostatic functions like body temperature, hunger, thirst, and the sex drive. The **cerebrum** is the largest part of the brain, sitting right on top. Its outer layer, the **cerebral cortex,** is the seat of our more advanced brain functions, including percep-

tion and thinking, and gives us our distinct personalities and most human characteristics. Its inner portion transmits signals, in the form of action potentials, from the cortex to various brain regions and to other parts of the body. The cerebrum is made up of about 10 billion neurons and is divided into left and right hemispheres, with each hemisphere divided into four lobes. Each lobe processes a variety of functions, including smell, hearing, speech, and vision.

Within these large regions are subregions that are further specialized for carrying out certain functions. The diencephalon, for example, includes the thalamus and hypothalamus, subregions important for regulating homeostasis (see Chapter 25). The cerebrum includes the **hippocampus** and **amygdala**—two important components of what is sometimes called the limbic system, the primary seat of our emotions and memories. ("Limbic" comes from the Latin *limbus,* which means "border"; the structures in this system lie along the border of the cerebrum and diencephalon.) Along with two nearby regions, the nucleus accumbens and the ventral tegumental area (VTA), the limbic system, which is stimulated during pleasurable activities, is a major region involved in addiction. It is this system that researchers sometimes refer to as the pleasure center or reward system of the brain.

In 1954, psychologists James Olds and Peter Milner of McGill University provided the first evidence for the existence of such a "pleasure center" in the brain. They implanted electrodes into the brains of rats—specifically in the VTA. The rats were then trained to press a lever that would administer a jolt of electricity through the electrode. The rats apparently found the experience intensely pleasurable because they would continue to press the lever without rest for up to 24 hours, as many as 5,000 times per hour!

Scientists now know that neurons in the VTA are one of the main sources of dopamine in the brain. When a smoker lights up, nicotine that reaches the brain stimulates the release of

►**ACTION POTENTIAL**
An electrical signal within a neuron caused by ions moving across the cell membrane.

►**MYELIN**
A fatty substance that insulates the axons of neurons and facilitates rapid conduction of action potentials.

►**GLIAL CELLS**
Supporting cells of the nervous system.

►**CEREBELLUM**
The part of the brain that processes sensory information and is involved in movement, coordination, and balance.

►**BRAIN STEM**
The part of the brain that is closest to the spinal cord and which controls vital functions such as heart rate, breathing, and blood pressure.

►**DIENCEPHALON**
A brain region located between the brain stem and the cerebrum that includes the thalamus and hypothalamus and regulates homeostatic functions like body temperature, hunger, thirst, and the sex drive.

►**CEREBRUM**
The region of the brain that controls intelligence, learning, perception, and emotion.

►**CEREBRAL CORTEX**
The outer layer of the cerebrum, the cerebral cortex is involved in many advanced brain functions.

►**HIPPOCAMPUS**
A subregion of the brain involved in learning and memory.

►**AMYGDALA**
A subregion of the brain that processes emotions, especially fear and anxiety, and is the seat of emotional memories.

INFOGRAPHIC 29.5 FUNCTIONAL REGIONS OF THE BRAIN

Cerebrum
Controls higher-level functions like thinking and determines personality and behavior

Brain stem
Coordinates involuntary functions like heart rate and respiration

Cerebellum
Controls movement, coordination, and balance

Diencephalon
Regulates homeostatic functions like temperature, hunger, and the sex drive

Spinal cord

Dopamine response to nicotine
Neurons in the ventral tegmental area release dopamine to other parts of the brain (shown here in red), resulting in feelings of pleasure.

Prefrontal cortex

Nucleus accumbens

Limbic system

Ventral tegmental area

Hippocampus and amygdala

▶**DOPAMINE**
A neurotransmitter that is involved in conveying a sense of pleasure in the brain.

▶**NEURO-TRANSMITTER**
A chemical signaling molecule released by a neuron to transmit a signal to a neighboring cell.

memories of the drug. The tinkling of a whiskey glass, meeting an old drinking buddy, being in a bar or a room filled with smokers puffing away can be reminders powerful enough to lure former addicts back to their old ways.

CHEMICAL MESSENGERS

For Ward, who turned 40 in 2013, it wasn't just the smoking at the weekly poker game that rekindled his desire to smoke, it was a combination of events. For one thing, he says he was "extremely stressed" about work. He had started working as a freelance writer, and he was having trouble juggling his deadlines. In the past, he would often rely on smoking to calm himself. "I'm a nervous worker," he says. "Some people shake their legs, or pace, or exercise to work off stress. For me it was smoking." And then a difficult family situation arose. Up to that point, he had managed to avoid smoking, even in the presence of smokers, because, he says, he had learned to think of himself as a nonsmoker: holding to that self-concept was a way to persuade himself not to pick up a cigarette. But in the end, the combination of stressful events and easy access to cigarettes pushed him to smoke again.

As he took a long-awaited drag, the dopamine released by neurons in the VTA excited neurons in the pleasure center of his brain. Dopamine is one type of chemical messenger, or **neurotransmitter,** that cells of the nervous system use to communicate with one another, and with other cells such as muscles. Neurotransmitters are chemical signals released by neurons at the end of their axons when an action potential reaches the tip of the axon. These chemical

dopamine from neurons in the VTA. This dopamine floods other areas of the brain's reward system, giving the smoker a pleasurable sensation (INFO-GRAPHIC 29.5).

That drugs can affect the brain's pleasure centers explains why they can be so hard to kick. It also explains why even addicts who successfully stop taking drugs for decades can relapse simply by being in an environment that conjures up

Regions deep within the cerebral cortex, illustrated here, may be involved in many human experiences, from emotional response to drug use.

signals transmit information to another cell. The site of transmission of a signal from a neuron to another cell is called the **synapse,** which consists of the axon terminal, a small gap between the two cells, and receptors on the receiving cell.

When the electrical signal of an action potential reaches a neuron's axon terminal, it stimulates the neuron to release neurotransmitters from its axon terminal into the space between the neuron and the cell with which it is communicating. Neurotransmitters diffuse across this space, called the **synaptic cleft,** to carry messages from one cell to another. Each neurotransmitter fits into a receptor on the surface of the cell receiving the signal as a key fits into a lock. The act of a neurotransmitter binding to its receptor initiates a response in the receiving cell. If the receiving cell is another neuron, the binding of neurotransmitter may spark another action potential, thereby perpetuating the signal. Signaling between the two cells is terminated when the neurotransmitter is removed from the cleft. Neurotransmitter may be degraded by enzymes in the synaptic cleft, or taken back up into the signaling neuron via reuptake receptors (**INFOGRAPHIC 29.6**).

▶**SYNAPSE**
The site of transmission of a signal between a neuron and another cell; the synapse includes the axon terminal of the signaling neuron, the space between the cells, and receptors on the receiving cell.

▶**SYNAPTIC CLEFT**
The physical space between a neuron and the cell with which it is communicating.

INFOGRAPHIC 29.6 NEURONS USE CHEMICAL SIGNALS TO COMMUNICATE

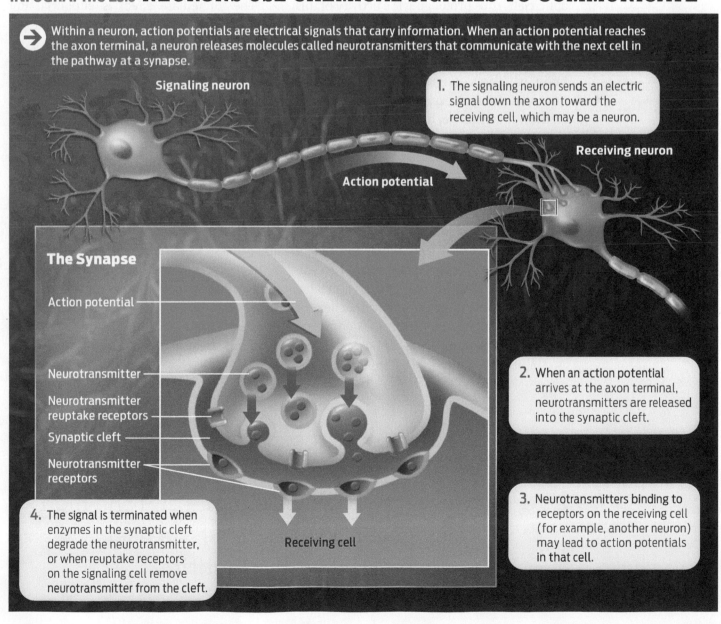

Within a neuron, action potentials are electrical signals that carry information. When an action potential reaches the axon terminal, a neuron releases molecules called neurotransmitters that communicate with the next cell in the pathway at a synapse.

Signaling neuron

1. The signaling neuron sends an electric signal down the axon toward the receiving cell, which may be a neuron.

Receiving neuron

Action potential

The Synapse

Action potential

Neurotransmitter

Neurotransmitter reuptake receptors

Synaptic cleft

Neurotransmitter receptors

Receiving cell

2. When an action potential arrives at the axon terminal, neurotransmitters are released into the synaptic cleft.

3. Neurotransmitters binding to receptors on the receiving cell (for example, another neuron) may lead to action potentials in that cell.

4. The signal is terminated when enzymes in the synaptic cleft degrade the neurotransmitter, or when reuptake receptors on the signaling cell remove neurotransmitter from the cleft.

Cells in the nervous system use several different neurotransmitters to communicate different messages. For example, the neurotransmitter serotonin, which is active in the CNS and the gastrointestinal tract, regulates anxiety, appetite, and sleep. Another important neurotransmitter is acetylcholine, which is involved in learning, memory, and muscle contraction.

Many drugs—including both drugs of abuse and therapeutic medicines—interact with neurotransmitters. The class of antidepressant drugs known as selective serotonin reuptake inhibitors (SSRIs), which includes Prozac and Zoloft, helps to treat depression by preventing the reuptake of serotonin from the synaptic cleft, thus prolonging the activity of this mood-elevating neurotransmitter. Nicotine binds to receptors on neurons and triggers the release of multiple neurotransmitters in the brain, which accounts for the range of sensations that smokers experience: a reduction in anxiety, for instance, as well as an increased ability to concentrate. And nicotine dramatically increases the release of dopamine from certain neurons.

TOO MUCH OF A GOOD THING

In normal circumstances, dopamine's main job is to convey information related to elation and pain. The joy we get from a meal, a promotion, a winning poker hand, or sex—or anything that gives us pleasure—is conveyed in part by dopamine. Drugs of abuse stimulate dopamine production, which is why they can become so addictive.

But these drugs cause such a high that over time they can alter the dopamine signaling system so that the pleasures of everyday life become meaningless compared to the pleasure of the drug. Normally the brain produces dopamine at a relatively constant rate, and dopamine occupies only a portion of dopamine receptors at any given time. But when a person smokes, snorts cocaine, or takes heroin, for example, dopamine levels in the synaptic cleft increase dramatically. With so much dopamine available, practically all of the brain's dopamine receptors become activated simultaneously.

The immediate effect is euphoria. But there is a downside. Because so much dopamine is produced, the brain becomes overwhelmed and tries to dampen the drug's effect by switching off some of its dopamine receptors (yet another example of homeostasis; see Chapter 25). When the drug wears off, fewer receptors are functioning—bringing down mood. Shutting down receptors that would otherwise be available to receive information is called down-regulation.

The low can be so low that normal pleasures such as eating or socializing become dull and listless affairs. The user's mood may now be even lower than it was before taking the drug. As dopamine receptors shut down, ever-larger quantities of the drug are required to produce a high—and the high may never be as high as the user experienced the first time.

> As dopamine receptors shut down, ever-larger quantities of the drug are required to produce a high—and the high may never be as high as the user experienced the first time.

INFOGRAPHIC 29.7 ADDICTIVE DRUGS ALTER DOPAMINE SIGNALING

 Addictive drugs increase dopamine release, causing initial feelings of pleasure. Continued use of the drug alters dopamine signaling, requiring more drug to achieve the same high. Attempts to quit can be difficult because of symptoms of withdrawal, the result of diminished dopamine signaling.

a. Normal Dopamine Signaling

Dopamine — Receptor

Cells release moderate amounts of dopamine into the synapse and not all dopamine receptors are occupied.

b. After Drug Use

More dopamine

Certain drugs cause massive dopamine release into the synapse. Many more dopamine receptors become occupied. The result is an intense feeling of pleasure.

c. After Repeated Drug Use

More dopamine — Fewer receptors

Dopamine overstimulation causes the receiving cell to down-regulate (that is, shut down) some dopamine receptors. Because the user now has fewer receptors available, more of the drug is required to feel high.

d. After Drug Withdrawal

Less dopamine — Fewer receptors

Removing the drug reduces the amount of dopamine released. In combination with fewer dopamine receptors, the user may feel sick and depressed.

As they come down from a high, addicts will likely feel even more unhappy and depressed as the dopamine response system is dampened. Eventually, many addicts need to take drugs simply to feel normal. Without drugs, they suffer the physical symptoms of withdrawal, which may include depression, anxiety, and intense cravings—for a dopamine fix. The specific symptoms and their intensity vary depending on the drug (INFOGRAPHIC 29.7).

A down-regulated dopamine system isn't the only structural brain change that scientists have observed in drug addicts. Researchers have also shown that drugs such as cocaine can change the shape of neurons in specific parts of the brain and consequently may impair their ability to transmit signals. Brain-imaging studies have also shown that addicts consistently have lower than normal levels of blood flow in the frontal regions of

TABLE 29.1 POTENTIALLY ADDICTIVE DRUGS AND THEIR EFFECTS

DRUG	MODE OF ACTION	EFFECT
Cocaine	Causes a large release of dopamine into the synapse and inhibits its removal by reuptake receptors; causes an amplified signal between neurons.	Highly addictive; produces exaggerated feelings of well-being and confidence and loss of interest in life activities. High doses lead to paranoia, anxiety, and increased blood pressure.
Heroin	Mimics endorphins that are normally sent into a synapse to relieve pain; binds opiate receptors in the limbic system, brain stem, and spinal cord, artificially affecting mood, respiration, and pain response.	Rush of euphoria followed by a foggy feeling. Powerful withdrawal symptoms make this drug extremely addictive. Overdose slows breathing to dangerous levels. Associated with transmission of HIV through contaminated needles.
Caffeine	Mimics adenosine and binds to adenosine receptors in the forebrain and blocks the natural sleep response. Stimulated cells also signal the production of adrenaline and dopamine, further supporting an alert and pleasurable response.	Inhibits sleepiness and increases alertness. Withdrawal symptoms include headaches, jittery feelings, and increased anxiety; can interfere with deep-sleep cycles, leading to exhaustion and depression. The adrenaline produced constricts blood vessels, affecting heart rate.
Nicotine	Increases levels of dopamine in the synapse of neurons primarily in the ventral tegumental area and prefrontal cortex.	Enhanced short-term feelings of pleasure, relaxation, increased concentration; very addictive. Other chemicals in inhaled smoke are also toxic.
Ecstasy	Causes excessive release of serotonin in parts of the brain that control the senses, sleeping, and emotions; destroys nerve cells that produce serotonin.	Intense euphoria, followed by depressive "crash." Side effects include paranoia, anxiety, confusion, and difficulty concentrating. Long-term use causes difficulty differentiating reality and fantasy.
Marijuana	Mimics the chemical anandamide; artificially stimulates the anandamide receptors in areas of the brain that affect memory, emotion, and sensory perception.	Relaxation, mild euphoria, and appetite stimulation. Can cause paranoia. Weakens short-term memory and can block the production of long-term memory; weakens problem-solving ability and coordination.
Inhalants (glue, hair spray, paint thinner, etc.)	Vapors destroy the myelin sheath that insulates the axons of neurons and thus slow down or stop nerve signaling; damages cells in the brain, lungs, heart, liver, kidneys, and bones.	Causes headaches, nausea, and disorientation and can diminish the ability to learn, remember, and solve problems; may make the heart more sensitive to brain signaling.
Methamphetamine	Causes release of dopamine and norepinephrine into the synapse. Dopamine signaling creates feelings of pleasure; norepinephrine signaling increases blood pressure and heart rate.	Creates feelings of pleasure and euphoria, paranoia, and hallucinations; may alter dopamine-producing neurons connected with Parkinson disease.
Ritalin	Causes increased release of dopamine into the synapse of nerve cells in the prefrontal cortex; strengthens dominant signals in the brain and weakens background signals that are less organized.	At lower doses increases the ability to focus and concentrate; at higher doses can inhibit formation of new nerve pathways, interfering with cognition and brain development. Prescribed for attention deficit disorder, but has become a common street drug.

> Tobacco, for example, is responsible in some way for one out of every five deaths in the United States.

the cerebrum, a region involved in decision making, during withdrawal from cocaine, and higher than normal levels while they are on the drug. In a variety of tests, drug addicts seem to make poorer decisions, with detrimental consequences.

Even more troubling, adolescents who take drugs may be preventing their brains from developing normally. A 2004 study by researchers at University of California, Los Angeles, and the National Institute of Mental Health (NIMH) that imaged the brains of 13 children and young people age 4 to 21 over 10 years showed that that some parts of the brain, such as the prefrontal cortex, are not fully developed until the mid-twenties or so. Taking drugs at an early age may hinder normal development of this region.

And that's not the end of the story. There are several other neurotransmitters, in addition to dopamine, involved in addiction, says Joe Frascella, director of the division of Clinical Neuroscience and Behavioral Research at the National Institute on Drug Abuse, and scientists are just starting to study how drugs of abuse affect them. While scientists have shown that almost every drug of abuse affects the dopamine system in varying degrees, Frascella says, "It's certainly more complex than just dopamine." Scientists have only just scratched the surface when it comes to learning exactly how long-term drug use affects the brain (TABLE 29.1).

BORN ADDICTS?

Not everyone who takes drugs becomes addicted. Exactly why some people seem to be more at risk than others isn't clear. But researchers have a few hypotheses based on existing evidence. For some of us, it may be a matter of biology: a predisposition to addiction caused by a shortage of dopamine or other types of receptors. Some people may just have been born with fewer receptors, or their brains may have lost receptors over time because of difficult life experiences. Consequently, drugs provide these people a high that they can't get from any other stimulus. And the drug feels too good to stop.

And just as some of us may be biologically predisposed to addiction, others may be predisposed to avoid it: some people's brains may simply be better at overriding the pleasure-seeking impulse.

A 2011 study by researchers at the Scripps Research Institute Florida, for example, identified a brain pathway involved in nicotine addiction. The researchers found that the number of a nicotine receptor called alpha-5 influences how susceptible mice are to nicotine addiction. When given the opportunity to self-administer nicotine, normal mice will stop after reaching a certain dose. Mice with no alpha-5 receptors, however, won't stop until they've taken a much higher dose. Humans also have alpha-5 receptors in varying numbers, and scientists hypothesize that people with fewer receptors are less sensitive to nicotine and may become more easily addicted.

Regardless of the reason people become addicted, addiction is a serious public health problem. Tobacco, for example, is responsible in some way for one out of every five deaths in the United States, according to NIDA. Smokers have a higher incidence of both heart and lung disease. And smoking causes cancer. Tobacco use is the leading preventable cause of death and disease in the United States, killing some 440,000 people a year.

INFOGRAPHIC 29.8 DRUGS DIMINISH MEMORY AND MOTOR SKILLS

Methamphetamine has been shown to decrease the number of dopamine receptors in the brain, impairing motor skills and memory.

Meth users with more dopamine receptors

Dopamine receptor availability (B_m/K_d)

Meth users with fewer dopamine receptors

Motor Skills Test

Time (sec)

Individuals with fewer dopamine receptors take more time walking straight lines.

Memory Test

Number of words

Individuals with fewer dopamine receptors have more difficulty memorizing and recalling lists of unrelated words after being distracted.

SOURCE: Volkow, N., et al. (2001) Association of dopamine transporter reduction with psychomotor impairment in methamphetamine abusers. *American Journal of Psychiatry* 158:377-382.

Drug use also affects health in ways that diminish the quality of life. For example, deficits in the dopamine system caused by drug use weaken memory and motor skills. Nora Volkow, head of NIDA, has shown that methamphetamine (crystal meth) users not only have fewer dopamine receptors than nonusers, they also have poorer short-term memory and score much lower on tests of motor skills, such as quickly walking a straight line (NFOGRAPHIC 29.8).

KICKING THE HABIT

Research on the neurobiology of drug addiction is informing new and better ways to treat addiction. Volkow and other researchers have shown that after a period of abstinence from drugs, the dopamine system can repair itself—but it generally takes longer than a year. And while studies have shown that functional skills such as short-term memory and motor

control do come back, it's not clear whether they come back completely.

Given the nature of many of the drug-induced brain changes, experts now see addiction as a chronic disease. Like heart disease or diabetes, diseases for which patients require long-term treatment, patients with addictions require long-term treatment plans. And the occasional relapse is only a predictable setback, not a failure of the treatment, says Volkow.

Experts also now know that the best treatments should target addictive behaviors in several ways. They should, for example, decrease the reward value of the drug, according to Volkow, perhaps by counseling addicts to seek out other pleasurable experiences and to repeat them to reinforce their value in the brain. Avoiding the drug and focusing on other pleasurable experiences will, over time, weaken conditioned memories of the drug and drug-related stimuli.

Research on the neurobiology of drug addiction is informing new and better ways to treat addiction.

A better understanding of the effects on the brain of substance abuse is also informing efforts to develop medications that can help a person kick the habit. Two popular antismoking drugs, Chantix and Zyban, for example, work by competing with nicotine for binding sites on neurons. By partially activating dopamine release, the drugs help reduce nicotine cravings while a person is trying to quit smoking. They also make smoking less pleasurable because nicotine from cigarettes cannot bind to nicotine receptors while the drugs are present. As a result, smoking becomes much less enjoyable, and therefore easier to stop. Research has shown these medications to be two to three times more effective at aiding smoking cessation compared to placebo. However, there have also been reports of serious side effects of these drugs, such as suicidal thoughts, so use of the medications should be considered carefully.

Scientists are also studying medications that work in different ways. For example, Nabi Biopharmaceuticals, based in Rockville, Maryland, is testing a nicotine vaccine in late-stage clinical trials. According to Nabi, this inhalable vaccine stimulates the immune system to produce antibodies that bind to nicotine, preventing nicotine from reaching receptors in the brain. Smoking then no longer feels pleasurable and smokers should more easily be able to quit. The vaccine has side effects, however, and preliminary results suggest that it is no more effective than Chantix at smoking cessation. Despite these setbacks, researchers haven't yet given up hope for this approach.

In the end, a combination of behavioral strategies and medications that target specific neurotransmitters or brain circuits will likely work the best to help addicts kick the habit, says Frascella, of NIDA.

OLD HABITS DIE HARD

Before Ward quit smoking in 2006, he was up to a pack and a half a day. Smoking had increasingly become a point of tension between him and his nonsmoking wife. She desperately wanted him to kick the habit, both for the sake of his health and because of the risks to herself of heart disease and cancer from secondhand smoke. A medical examination showed that Ward already had risk factors for heart disease, including high blood pressure. His doctor told him to stop smoking. "She didn't ask me to stop," Ward says. "She simply said, 'You are going to stop smoking in two weeks.'" She prescribed a nicotine patch—a skin patch that delivers nicotine to the bloodstream and enables smokers to gradually kick smoking by easing nicotine withdrawal symptoms—and other medication. Within a few months, Ward had stopped smoking.

But after that first poker night, Jack Ward went back to smoking. This time, instead of his old pack-and-a-half-a-day habit, he managed to cut down to only a few cigarettes a week, all of them smoked on his weekly poker night.

Two years after Ward started smoking again, he moved. The weekly poker nights were gone, and gone, too, were the social cues that had tempted him to pick up the habit again. There are more antismoking social influences in his life now than ever. Most of his friends do not smoke, he exercises more, and perhaps most important of all, he wants to be a good role model for his 4-year-old daughter: "I don't want her to see me smoking." ■

FOR COMPARISON Size Matters—Evolution of the Brain

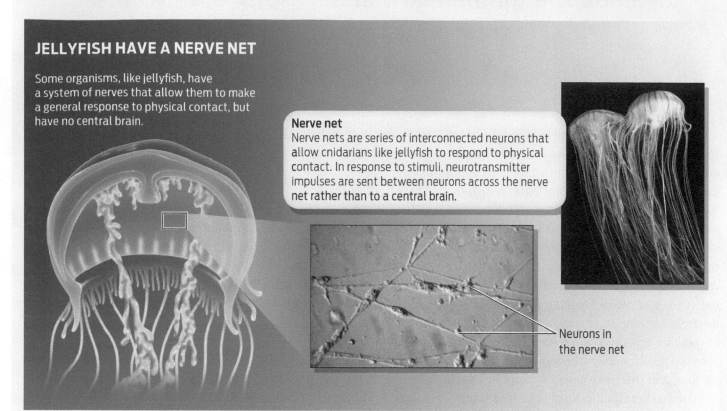

JELLYFISH HAVE A NERVE NET

Some organisms, like jellyfish, have a system of nerves that allow them to make a general response to physical contact, but have no central brain.

Nerve net
Nerve nets are series of interconnected neurons that allow cnidarians like jellyfish to respond to physical contact. In response to stimuli, neurotransmitter impulses are sent between neurons across the nerve net rather than to a central brain.

Neurons in the nerve net

Humans have a sophisticated central nervous system that consists of a brain and spinal cord. Our distinctively large brain (relative to our body size) is partly what endows us with the traits we think of as most human: language, self-consciousness, and the use of tools to shape our environment. But even the simplest organisms have ways of sensing and responding to their environments, and don't need a brain to do so.

Invertebrate animals like jellyfish and sea anemones, for example, have neurons that sense and send information. But these neurons are not organized into a brain. By definition, a brain is a grouping of neurons located in the head of an organism that acts as a control center. Jellyfish and sea anemones exhibit radial symmetry–they are round with no clear front or back–and so they have no head. They therefore do not have a brain. Instead, they have what are called nerve nets, which are collections of sensory neurons, motor neurons that control muscles, and neurons that connect the sensory and motor neurons. Nerve nets enable these animals to respond to stimuli with behaviors such as swimming and eating–but the response is more like a reflex than a thought process.

The first animals to have primitive brains were flatworms, invertebrates that evolved

550 million years ago and include several human parasites such as tapeworms and flukes. Not coincidentally, these organisms are also some of the first animals to exhibit bilateral symmetry—they have clear right and left halves—and also a defined head end. The primitive brain of the flatworm consists of a collection of neurons located at the head end of the organism. The forward location of the flatworm brain allows the organism to sense its environment as it moves forward, and also permits the coordinated movement of its right and left halves. Flatworms also have two large nerves, called nerve cords, that extend along the body and sprout smaller nerves that reach the body tissues—a simple version of a peripheral nervous system. The brain and nerve cords work together to coordinate movement and other responses to the environment.

As animals evolved over time, their brains became larger and more complex, allowing them to perform more complex functions. In vertebrate animals such as fish, frogs, birds,

FLATWORMS HAVE A PRIMITIVE BRAIN

Flatworms were the first group of organisms to have both a central and a peripheral nervous system.

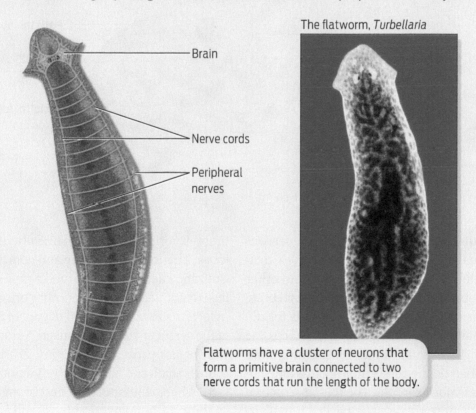

Brain

Nerve cords

Peripheral nerves

The flatworm, *Turbellaria*

Flatworms have a cluster of neurons that form a primitive brain connected to two nerve cords that run the length of the body.

VERTEBRATES HAVE LARGE AND COMPLEX BRAINS

All vertebrates have both a peripheral and a central nervous system. They have brains that vary in size and complexity.

Cerebrum size and complexity

Human

Bird

Frog

Fish

Mammal brains have complex organization

Left hemisphere
This hemisphere controls the right side of the body; it is responsible for linear, rational, and verbal processes.

Right hemisphere
This hemisphere controls the left side of the body; it is responsible for artistic, spatial, and musical processes.

Frontal lobe
Reasoning, behavior, motor control, memory

Temporal lobe
Sound, language, emotion

Parietal lobe
Senses, cognition, speech

Occipital lobe
Vision processing

and humans, brains and peripheral nervous systems are larger and more complex than the primitive brains of flatworms and other invertebrates. But the brains of vertebrate organisms vary, too, which accounts for differences in computational and processing ability. The human cerebrum contains more folds than do the cerebrums of birds and amphibians, for example. The folded structure permits more tissue to be packed into a given amount of space. In addition, the human brain is organized into hemispheres and lobes. This folded structure and complex organization account for the intellectual advantage that humans have over other organisms. In fact, the relative size and degree of folding of the cerebrum of any organism is associated with higher processing abilities. A fish cerebrum is small and has a relatively smooth surface, whereas mammalian cerebrums, such as those of humans, dolphins, and whales, are larger and contain many more folds.

CHAPTER 29 Summary

- The nervous system senses and responds to signals from the environment and coordinates bodily actions, including moving, breathing, and thinking.

- The central nervous system (CNS) consists of the brain and the spinal cord, a thick bundle of nerves extending from the base of the brain to the lower back; the peripheral nervous system (PNS) consists of all the nerves extending from the spinal cord to the limbs and internal organs.

- The PNS senses and responds to information both inside and outside our bodies. It includes the sensory receptors of our sensory organs, such as the eyes, ears, and skin, as well as the effectors, such as the muscles, that respond to signals sent from the CNS.

- Neurons are specialized cells that consist of a cell body, branched dendrites, and a long axon. Neurons conduct electrical signals known as action potentials.

- The coordinated movement of positively charged ions across the axon membrane initiates an action potential. When an action potential reaches the end of a neuron, it causes the neuron to release neurotransmitters.

- Neurotransmitters are chemical signaling molecules released from neuron axon terminals into the synaptic cleft of a synapse and bind to the receptors of other cells (for example, neurons, muscle cells, and endocrine gland cells). Important neurotransmitters include dopamine, serotonin, and acetylcholine.

- Dopamine is a neurotransmitter that produces feelings of pleasure. It is one of the primary neurotransmitters involved in addiction.

- Neurotransmitters bind to receptors on their target cells. The number of receptors can be down-regulated in response to persistently high levels of a neurotransmitter.

- Different parts of the brain coordinate different functions. The four main regions of the brain are the cerebellum, brain stem, diencephalon, and cerebrum. The brain's limbic system, composed of subregions in the diencephalon and cerebrum, includes the brain's "pleasure center" and has been implicated in addiction.

- All organisms have ways of sensing and responding to their environment, but not all have a central nervous system with a brain and spinal cord.

MORE TO EXPLORE

- National Institute of Drug Abuse http://www.drugabuse.gov/

- Volkow, N. D., et al. (2011) Addiction: beyond dopamine reward circuitry. *Proceedings of the National Academy of Sciences* 108(37):15037–15042.

- Kringelbach, M. L., and Berridge, K. C. (2010) The functional neuroanatomy of pleasure and happiness. *Discovery Medicine* 9(49):579–587.

- Proctor, R. N. (2011) *Golden Holocaust: Origins of the Cigarette Catastrophe and the Case for Abolition.* Berkeley: University of California Press.

- Olds J. (1956) Pleasure centers in the brain. *Scientific American* 195:105–116.

CHAPTER 29 Test Your Knowledge

DRIVING QUESTION 1

How is the nervous system organized?

By answering the questions below and studying Infographics 29.2 and 29.5, you should be able to generate an answer for the broader Driving Question above.

KNOW IT

1 For each of the following, indicate whether it is a part of the CNS or of the PNS.

Light-detecting receptor in the eye _____

Amygdala _____

Pain receptor in the skin _____

Spinal cord _____

Thalamus _____

2 Which part of the brain coordinates movement? Which part of the brain maintains body temperature?

3 Rank the complexity of the cerebrums of the following vertebrates from 1 (most complex) to 4 (least complex).

Bird _____

Chimpanzee _____

Salamander (an amphibian) _____

Fish _____

USE IT

4 Is information flow in the spinal cord one way or two way? Explain your answer.

5 How does multiple sclerosis cause muscle weakness? Does multiple sclerosis directly affect muscles? Which part of the nervous system is affected?

6 A brain injury (caused by a blow to the head, for example) that results in the loss of the ability to speak most likely affected the

a. cerebellum.   d. brain stem.

b. cerebrum.   e. hypothalamus.

c. diencephalon.

DRIVING QUESTION 2

How do cells in the nervous system transmit signals?

By answering the questions below and studying Infographics 29.3, 29.4, and 29.6, you should be able to generate an answer for the broader Driving Question above.

KNOW IT

7 Neurons receive information through their

a. axons.

b. axon terminals.

c. cell bodies.

d. dendrites.

e. nuclei.

8 Action potentials are a type of _____ signaling that relies on _____.

a. electrical; neurotransmitters

b. electrical; charged ions

c. electrical; electrons

d. chemical; neurotransmitters

e. chemical; charged ions

9 Neurons release neurotransmitters from their

a. cell bodies.

b. dendrites.

c. axon terminals.

d. all of the above

e. b and c

10 What happens when a neurotransmitter is released into a synaptic cleft?

11 Compare and contrast electrical and chemical signaling by neurons.

USE IT

12 Gatorade and other sports drinks contain replacement electrolytes (ions necessary to enable muscles to continue to contract, especially the ions lost during sweating). Gatorade contains sodium and potassium ions. Other than in the muscle, where else might these ions be crucial during sustained exercise?

13 Botox is a chemical treatment injected into skin to prevent wrinkling. It is primarily a bacterial toxin that prevents certain neurons from releasing the neurotransmitter acetylcholine. Acetylcholine is normally released by motor neurons to signal muscles to contract. Does Botox paralyze muscles in a relaxed state or a contracted state?

14 Is more or less of the neurotransmitter acetylcholine released by the axon terminals of neurons in patients with multiple sclerosis compared to those in people who do not have multiple sclerosis? Explain your answer.

DRIVING QUESTION 3

Why are some drugs (and some behaviors) addictive?

By answering the questions below and studying Infographics 29.1, 29.5, 29.7, and 29.8, you should be able to generate an answer for the broader Driving Question above.

KNOW IT

15 **Addictive substances confer a sense of pleasure because they**

   a. decrease the amount of dopamine in synaptic clefts.

   b. increase the amount of dopamine in synaptic clefts.

   c. increase the number of dopamine receptors on the axon terminals of cells that release dopamine.

   d. increase the number of dopamine receptors on dendrites of cells that release dopamine.

   e. c and d

16 **Why do drug users need to take ever-increasing amounts of drugs to get the same high?**

USE IT

17 **Cocaine prevents dopamine from being removed from the synapse. Why does this cause feelings of pleasure?**

18 **Would you expect a person born with a relatively low number of dopamine receptors to be happier or sadder than the average? Explain your answer.**

MINI CASE

19 **Parkinson disease is caused primarily by a gradual loss of dopamine-producing neurons in the brain. Why is depression often among the debilitating symptoms of Parkinson disease? There are a variety of medications available to treat people with Parkinson disease. Do some internet research to match each medication listed below with its probable mechanism of action.**

Mirapex	Can be used by neurons to make dopamine
Eldepryl	Binds to and activates the dopamine receptor
Levodopa	Inhibits an enzyme that breaks down dopamine

How can drugs with different mechanisms of action all help treat Parkinson disease? (Hint: What do all the underlying mechanisms have in common in terms of their effect?)

INTERPRETING DATA

20 **Study Infographic 29.8.**

   a. What are the independent and dependent variables in each experiment?

   b. The performances of 14 participants are plotted in each graph. Calculate the average time (in seconds) these participants took to walk the straight line, and the average number of words recalled by these participants.

   In the right-hand graph in Infographic 29.8, you will note that there are a few outliers (two are circled in red in the copy of the graph shown below).

   c. For the outlier in the lower left, is this participant recalling more or fewer words than would be predicted based on his or her availability of dopamine receptors? Explain your answer.

   d. For the outlier on the top right, is this participant recalling more or fewer words than would be predicted based on his or her availability of dopamine receptors? Explain your answer.

Memory Test

BRING IT HOME

21 **Replicate the motor skills experiments shown in Infographic 29.8 with some students. On a flat surface, set up a start line and, 10 yards away, a finish line. Instruct each subject to walk in a straight line from the start to the finish and back as fast as they can without running. Start timing when they cross the start line, and stop timing when they cross the start line again at the end of the trial. Have each participant do the trial three times, and record the average of the three trials as the participant's final time. Now calculate the average time in your set of participants. How does this compare to the 14 methamphetamine users whose performances are plotted in Infographic 29.8?**

SOURCE: Method from Robertson, K. R., et al. (2006) Timed gait test: normative data for the assessment of the AIDS dementia complex. *Journal of Clinical and Experimental Neuropsychology* 28:1053–1064.

Nadya Suleman with her 14 children, all of them conceived with assisted reproductive technology.

Too Many Multiples?

The birth of octuplets raises questions about the fertility business

> Scientists estimate that the incidence of natural triplets is 1 in 6,000 to 8,000 births— of quadruplets, 1 in 500,000 births.

THE LIVE BIRTH OF OCTUPLETS IS AN EXTREMELY RARE EVENT, HAVING occurred only once in recorded U.S. history before the year 2000. So the arrival of a second set in a California hospital in January 2009 was greeted with fanfare. Headlines read "Octuplets Stun Doctors" and "Eight Babies!"

But days after news of the miracle multiple birth spread worldwide, the public reception turned sour when it came to light that the 33-year-old mother, Nadya Suleman, already had six children all under the age of 7 who were, like the octuplets, conceived using a form of assisted reproductive technology called in vitro fertilization.

Even more disturbing to some, Suleman was an unemployed single mother on welfare. The public outcry was fierce. How could she support her children? Was she psychologically disturbed? And why had her doctor agreed to give her fertility treatments when she already had six children?

The case cast a spotlight on the business of fertility treatment: in the United States fertility clinics are largely unregulated. Although the American Society of Reproductive Medicine (ASRM) issues guidelines on how doctors should administer fertility services, in most states there are no laws regulating what doctors can and cannot do in this regard. Though critics called Suleman's doctor irresponsible, he had not violated any laws.

Infertility treatment wouldn't be nearly as controversial if it didn't increase the odds that a woman will conceive more than one child during the treatment. "Multiples," as these babies are called,

---

▸▸ DRIVING QUESTIONS

1. What is the anatomy of the male and female reproductive tracts?

2. What hormones are involved in reproduction, and how do they work?

3. What are the different types of assisted reproduction, and how do they work?

▶**OVARIES**
Paired female reproductive organs; the ovaries contain eggs and produce estrogen and progesterone.

▶**ESTROGEN**
A female sex hormone produced by the ovaries.

▶**PROGESTERONE**
A female sex hormone produced by endocrine cells in the ovaries (particularly in the cells of the corpus luteum) that prepares and maintains the uterus for pregnancy.

▶**OVIDUCT**
The tube connecting an ovary and the uterus in females. Eggs are ovulated into and fertilized within the oviducts.

▶**UTERUS**
The muscular organ in females in which a fetus develops.

▶**ENDOMETRIUM**
The lining of the uterus.

▶**TESTES (SINGULAR: TESTIS)**
Paired male reproductive organs, which contain sperm and produce androgens (primarily testosterone).

▶**SCROTUM**
The sac in which the testes are held.

▶**TESTOSTERONE**
The primary male sex hormone, which stimulates the development of masculine features and plays a key role in sperm development.

are often born prematurely. Consequently, they are underweight and have underdeveloped organs, and so are at risk for birth defects such as cerebral palsy and infant respiratory distress syndrome. Carrying multiples also increases the risk that the mother will develop dangerously high blood pressure, diabetes, vitamin deficiencies, or one or another of several other medical conditions during her pregnancy that may affect her health or that of her unborn children. Even women who have twins have a higher risk of medical complications during pregnancy, and their babies are at a higher risk of premature birth than singletons. Suleman's eight children are the only surviving set of octuplets ever.

Multiple births without fertility treatment are extremely rare. Scientists estimate that the incidence of natural triplets is 1 in 6,000 to 8,000 births–of quadruplets, 1 in 500,000 births. The incidence of higher-order multiple births is even rarer. But because the use of assisted reproductive technologies has skyrocketed, so, too, have the number of multiple births. The number of triplet births rose 400% between 1980 and 1998, according to statistics compiled by the Centers for Disease Control and Prevention. The medical complications associated with such pregnancies place a large burden on the health care system.

Although there have been efforts to improve assisted reproductive technology and therefore reduce the likelihood of multiple births, extreme cases such as Suleman's have drawn lawmakers to the scene. Many states are considering legislation that would place restrictions on fertility doctors. Working from another perspective, some groups are fighting for legislation that would require health insurance providers to pay for fertility treatments. Because many patients pay out of pocket, patients with limited financial resources can, because of their circumstances, pressure doctors to be aggressive with treatment, despite the health risks associated with carrying multiples. The goal of this legislation is to make assisted reproduction safer for everyone, not just for those who can afford it.

FERTILITY AND INFERTILITY

Suleman, who grew up as an only child, said she had always wanted a large family to help

Multiples are often born prematurely, and underweight, as were these septuplets born in Alexandria, Egypt, in 2008.

INFOGRAPHIC 30.1 FEMALE REPRODUCTIVE SYSTEM

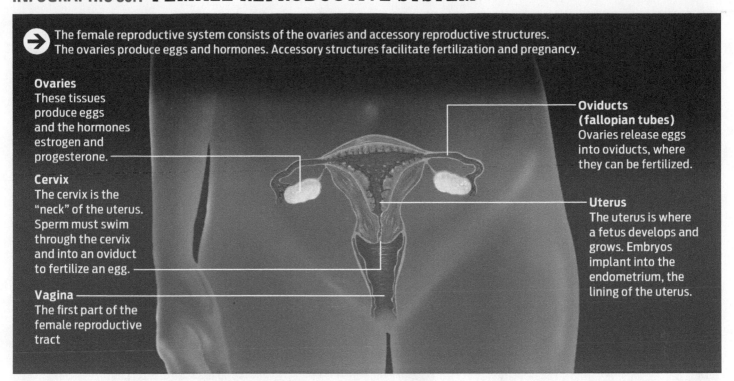

→ The female reproductive system consists of the ovaries and accessory reproductive structures. The ovaries produce eggs and hormones. Accessory structures facilitate fertilization and pregnancy.

Ovaries
These tissues produce eggs and the hormones estrogen and progesterone.

Cervix
The cervix is the "neck" of the uterus. Sperm must swim through the cervix and into an oviduct to fertilize an egg.

Vagina
The first part of the female reproductive tract

Oviducts (fallopian tubes)
Ovaries release eggs into oviducts, where they can be fertilized.

Uterus
The uterus is where a fetus develops and grows. Embryos implant into the endometrium, the lining of the uterus.

---

make up for her lonely childhood. She told Ann Curry on NBC's *Today* show in 2009, after the octuplets were born, that she had tried to become pregnant for several years but had been unable to. That's when she turned to a fertility doctor for help.

Suleman claimed she suffered from a medical condition that prevented her from conceiving a child naturally. If she did, she wasn't alone. In the United States, an estimated one out of eight couples–about 7.3 million women and their partners–experience infertility, which is defined as the inability to conceive within a year or to bring a pregnancy to term. Many things can cause infertility: advanced age, infections, hormonal imbalances, chromosomal abnormalities, and physical blockage of reproductive passages. Men can suffer from a low sperm count or have abnormal sperm. In many cases, the reason for a couple's infertility remains unknown, and most couples are unaware that they have a fertility problem until they begin trying to conceive a child and can't.

According to the ASRM, modern medicine can offer treatment to 90% of infertile couples. But fertility isn't an exact science, and treatment isn't always effective. There are many organs and hormones involved in human reproduction, and communication among them is a highly orchestrated process; even under the best circumstances, successful pregnancies require a bit of lucky timing.

The female reproductive system consists of the **ovaries** and accessory structures that facilitate fertilization and pregnancy. The ovaries contain eggs and produce the sex hormones **estrogen** and **progesterone,** the major female sex hormones. Each month, one ovary typically releases one egg into an adjacent organ called the **oviduct** (also known as the fallopian tube). The egg travels via an oviduct to the **uterus,** an elastic muscular compartment where a fetus may develop should an egg be fertilized by sperm. The tissue that lines the uterus is called the **endometrium,** which becomes enriched with blood vessels to support a potential pregnancy **(INFOGRAPHIC 30.1).**

▶**ANDROGEN**
A class of sex hormones, including testosterone, that is present in higher levels in men than in women and causes male-associated traits like deep voice, growth of facial hair, and defined musculature.

▶**SEMINIFEROUS TUBULES**
Coiled structures that constitute the bulk of the testes and in which sperm develop.

▶**EPIDIDYMIS**
Tubes in which sperm mature and are stored before ejaculation.

▶**VAS DEFERENS**
Paired tubes that carry sperm from the testes to the urethra.

INFOGRAPHIC 30.2 MALE REPRODUCTIVE SYSTEM

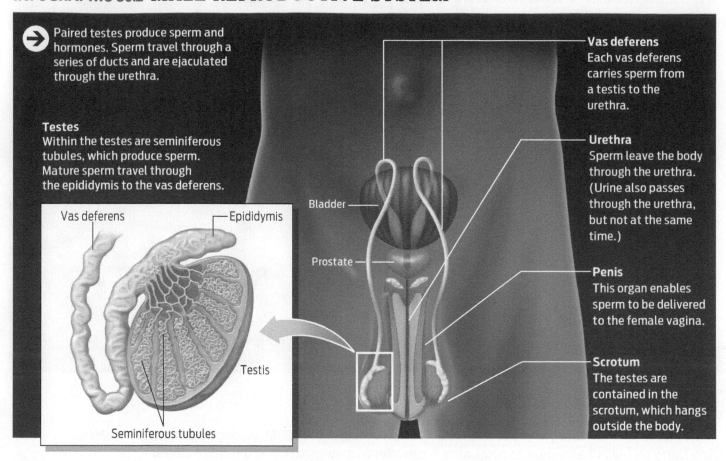

→ Paired testes produce sperm and hormones. Sperm travel through a series of ducts and are ejaculated through the urethra.

Testes
Within the testes are seminiferous tubules, which produce sperm. Mature sperm travel through the epididymis to the vas deferens.

Vas deferens

Epididymis

Testis

Seminiferous tubules

Bladder

Prostate

Vas deferens
Each vas deferens carries sperm from a testis to the urethra.

Urethra
Sperm leave the body through the urethra. (Urine also passes through the urethra, but not at the same time.)

Penis
This organ enables sperm to be delivered to the female vagina.

Scrotum
The testes are contained in the scrotum, which hangs outside the body.

▶**URETHRA**
A tube that connects the bladder to the genitals and carries urine out of the body. In males, the urethra travels through the penis and also carries sperm.

▶**SEMEN**
The mixture of fluid and sperm that is ejaculated from the penis.

▶**VAGINA**
The first part of the female reproductive tract, extending up to the cervix; also known as the birth canal.

In males, paired glands called **testes** (or testicles) produce sperm. The testes are contained in a sac of skin called the **scrotum** that hangs outside the body (an arrangement that keeps the testes slightly cooler than body temperature, ensuring proper sperm development). The testes produce **testosterone,** the primary male sex hormone. Testosterone is one of the **androgens**—a class of hormones that are present in higher amounts in males than in females. Each testis contains tightly coiled **seminiferous tubules** within which sperm develop. Remarkably, each testis contains approximately 250 meters of seminiferous tubules. Sperm travel through the seminiferous tubules and enter the **epididymis,** where they mature and are stored until ejaculated. From the epididymis,

sperm travel through paired tubes called the **vas deferens** and exit the body through the **urethra.** Along the way, the prostate and other glands add fluid to sperm that helps the sperm survive in the female reproductive tract. The fluid contains the sugar fructose as an energy source, alkaline bases that help buffer the acidic pH in the vagina and protect the sperm DNA from being damaged, and mucus that helps keep the sperm mobile in the female reproductive tract. Ejaculated sperm and the accompanying fluid are called **semen (INFOGRAPHIC 30.2).**

During sex, when a man ejaculates into a woman's **vagina,** sperm swim up the reproductive tract, through the opening of the uterus, called the **cervix** (so called from the Latin word for "neck"), and into the oviducts.

The oviducts are where **fertilization**–the fusion of egg and sperm–occurs. Only a single sperm will be successful in fertilizing an egg (**INFOGRAPHIC 30.3**).

That's the way normal conception happens–but several things can go wrong along the way. Physical damage to the reproductive organs can prevent fertilization. In some cases, a woman's oviducts can be blocked or damaged, preventing eggs from entering the uterus or sperm from getting into the oviduct, where fertilization normally takes place. The most common cause of blocked tubes is pelvic inflammatory disease, which can be caused by sexually transmitted diseases such as chlamydia, a bacterial infection. An untreated infection can cause scar tissue to build up in the oviducts and block them. In interviews, Suleman has said that she suffered from fibroids–benign tumors that grow in muscle–that left her oviducts scarred. In men, any obstructions in the vas deferens or epididymis will block sperm transport. Varicose veins in the testicles and sexually transmitted bacterial infections such as chlamydia or gonorrhea can also cause blocked tubes.

Fertility specialists can test for physical blockages and, in some cases, surgically correct them. When surgery isn't an option, **in vitro fertilization (IVF)** is often recommended. In IVF, hormones are administered to a woman to promote egg development, and then eggs are extracted from her ovaries through a needle inserted through the vagina. Sperm are extracted from ejaculate or, in cases of physical blockage, from the epididymis. The sperm and eggs are combined outside the body in a petri dish to allow the sperm to fertilize the eggs. The fertilized eggs begin to divide and the resulting **embryos** are then

▶**CERVIX**
The opening or "neck" of the uterus, where sperm enter and babies exit.

▶**FERTILIZATION**
The fusion of an egg and a sperm.

▶**IN VITRO FERTILIZATION (IVF)**
A form of assisted reproduction in which eggs and sperm are brought together outside the body and the resulting embryos are inserted into a woman's uterus.

▶**EMBRYO**
An early stage of development, an embryo forms when a zygote undergoes cell division.

INFOGRAPHIC 30.3 FERTILIZATION OCCURS IN THE FEMALE OVIDUCT

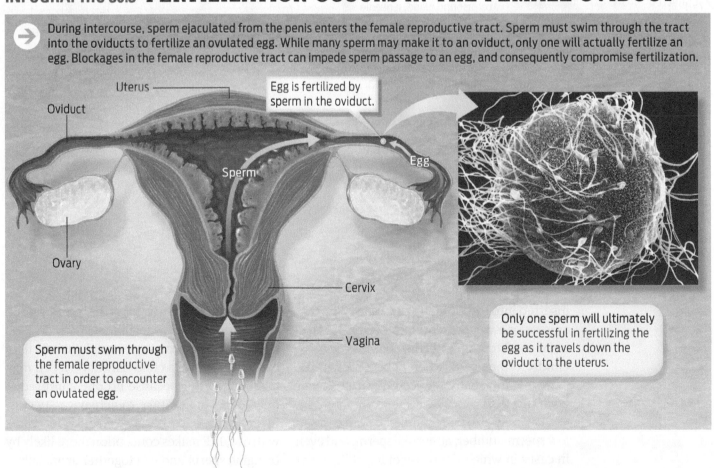

→ During intercourse, sperm ejaculated from the penis enters the female reproductive tract. Sperm must swim through the tract into the oviducts to fertilize an ovulated egg. While many sperm may make it to an oviduct, only one will actually fertilize an egg. Blockages in the female reproductive tract can impede sperm passage to an egg, and consequently compromise fertilization.

Uterus

Oviduct

Egg is fertilized by sperm in the oviduct.

Sperm

Egg

Ovary

Cervix

Vagina

Sperm must swim through the female reproductive tract in order to encounter an ovulated egg.

Only one sperm will ultimately be successful in fertilizing the egg as it travels down the oviduct to the uterus.

INFOGRAPHIC 30.4 IN VITRO FERTILIZATION (IVF)

In vitro fertilization involves extracting eggs and sperm and mixing them together outside the body to allow fertilization. The resulting embryos are then inserted back into a woman's uterus in hope that at least one of them will implant into the uterus and grow into a fetus.

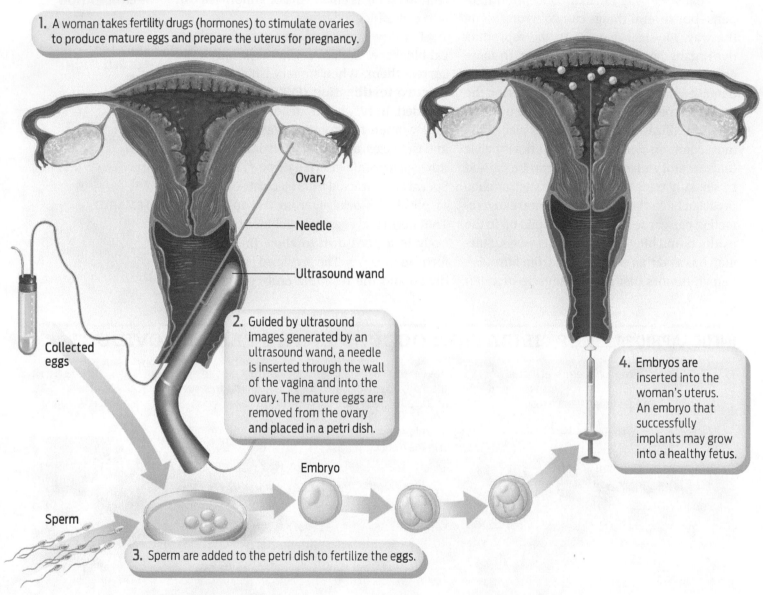

1. A woman takes fertility drugs (hormones) to stimulate ovaries to produce mature eggs and prepare the uterus for pregnancy.

Ovary

Needle

Ultrasound wand

Collected eggs

2. Guided by ultrasound images generated by an ultrasound wand, a needle is inserted through the wall of the vagina and into the ovary. The mature eggs are removed from the ovary and placed in a petri dish.

4. Embryos are inserted into the woman's uterus. An embryo that successfully implants may grow into a healthy fetus.

Embryo

Sperm

3. Sperm are added to the petri dish to fertilize the eggs.

inserted into the woman's uterus in hope that at least one will develop into a fetus. IVF has been used to help infertile couples conceive children since 1978, when Louise Brown, the first "test-tube baby" conceived with IVF, was born (**INFOGRAPHIC 30.4**).

IVF may also be recommended in cases of low sperm number, abnormal sperm, and even in cases in which the cause of infertility can't be determined. In such cases the term "infertility" can actually be a misnomer. Many couples with defective sperm or unexplained infertility can still conceive a child naturally—it just may take longer. But because no one can predict how long it might take to achieve a successful pregnancy and because fertility decreases with age, IVF makes conception more likely by bringing sperm and egg together artificially.

HORMONES AND PREGNANCY

Suleman's doctor told the press that he agreed to give Suleman fertility treatments because she had been trying to become pregnant for several years without success. She had her first round of IVF in 2000, using sperm donated by a friend. Her doctor explained to her that the procedure would begin with a round of hormones to stimulate her ovaries so that her eggs could be harvested. This hormone treatment, he explained, would mimic what happens naturally in a woman's body to trigger her reproductive cycle.

Hormones are responsible for regulating the production of gametes, both sperm and egg. In females, estrogen and progesterone are the key reproductive hormones that support egg maturation. In males, testosterone is the primary hormone that stimulates sperm to develop.

As we saw in Chapter 25, hormones are produced by endocrine glands, which secrete hormones into the circulation. These hormones then travel through the bloodstream to reach their target cells. Hormones interact with their target cells by binding to specific receptors on the target cells, influencing a range of physiological processes.

In females, estrogen and progesterone drive the menstrual cycle, a reproductive cycle that repeats roughly once every 28 days after the onset of puberty. During each cycle, estrogen and progesterone levels rise and fall, triggering the ovaries to release an egg and prepare a woman's uterus for pregnancy should an egg be fertilized.

The brain's **hypothalamus** controls levels of estrogen and progesterone in the body and thus is the ultimate regulator of fertility. The hypothalamus works closely with another endocrine gland called the **anterior pituitary,** which sits just below the hypothalamus in the brain. The hypothalamus secretes hormones that act on the anterior pituitary, causing it to produce two hormones of its own. These hormones, called follicle-stimulating hormone and luteinizing hormone, travel through the bloodstream and directly stimulate the ovaries.

Follicle-stimulating hormone (FSH) acts on structures in the ovaries called **follicles,** each of which contains an immature egg. FSH signals follicles in the ovary to enlarge and to produce estrogen. Estrogen has several effects, one of which is to cause the endometrium to start to thicken. Estrogen also stimulates eggs within the ovaries to mature.

In most women, estrogen levels rise between 10 and 14 days after menstrual bleeding begins (considered the start of the menstrual cycle). This rise in estrogen triggers the brain to release a large amount of **luteinizing hormone (LH).** In turn, this LH surge triggers **ovulation**–the release of an egg from a follicle into the oviduct. After the egg has been ovulated, the remaining follicle becomes a structure called the **corpus luteum,** which secretes progesterone. One of the most important roles of progesterone is to promote the continued thickening of endometrium. The thickened endometrium contains blood vessels and nutrients and is prepared to receive an embryo if the egg is fertilized.

Although both ovaries can release eggs during the same cycle, they typically take turns and only one egg is released per cycle. In about 1% of cycles, however, there are multiple ovulations, in which case fraternal twins, triplets, or higher multiples can develop. (Identical twins occur when a single egg is released and fertilized and then splits into two embryos early in embryonic development.)

Ovulation presents a crucial time window during which a woman can become pregnant. Sperm must swim through the cervix and uterus and into the oviduct containing the released egg in order to fertilize it. Because sperm can survive in the female reproductive tract anywhere from 3 to 7 days, a woman can become pregnant even if she has sex before she ovulates. Sperm can in effect wait in the oviduct for an egg to be ovulated. Once the egg leaves the oviduct, however, the odds that it will be fertilized are extremely small.

▶HYPOTHALAMUS
A master control region in the brain that regulates a variety of physiological functions, including hunger, thirst, temperature regulation, and reproduction.

▶ANTERIOR PITUITARY
The gland in the brain that secretes luteinizing hormone (LH) and follicle-stimulating hormone (FSH).

▶FOLLICLE-STIMULATING HORMONE (FSH)
A hormone secreted by the anterior pituitary. In females, FSH triggers eggs to mature at the start of each monthly cycle.

▶FOLLICLE
The part of the ovary where eggs mature.

▶LUTEINIZING HORMONE (LH)
A hormone secreted by the anterior pituitary. In females, a surge of LH triggers ovulation.

▶OVULATION
The release of an egg from an ovary into the oviduct.

▶CORPUS LUTEUM
The structure in the ovary that remains after ovulation and secretes progesterone.

INFOGRAPHIC 30.5 HORMONES REGULATE THE MENSTRUAL CYCLE

A complex interplay of hormones from the hypothalamus, anterior pituitary gland, and ovaries drives the monthly female reproductive cycle.

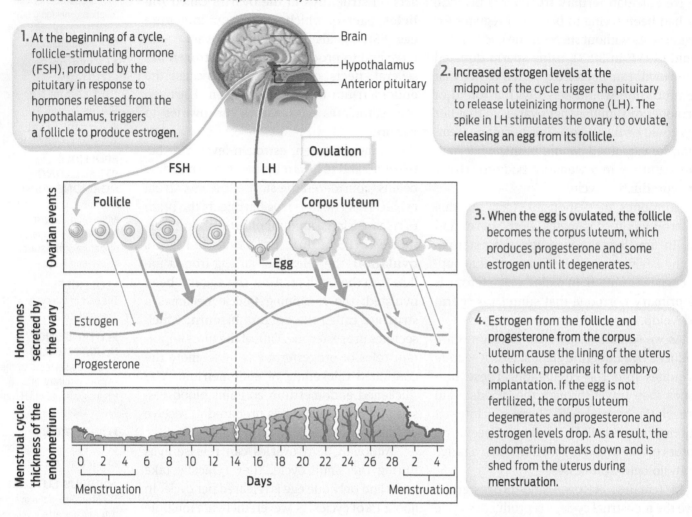

1. At the beginning of a cycle, follicle-stimulating hormone (FSH), produced by the pituitary in response to hormones released from the hypothalamus, triggers a follicle to produce estrogen.

Brain
Hypothalamus
Anterior pituitary

2. Increased estrogen levels at the midpoint of the cycle trigger the pituitary to release luteinizing hormone (LH). The spike in LH stimulates the ovary to ovulate, releasing an egg from its follicle.

FSH   LH

Ovulation

Ovarian events   Follicle   Corpus luteum

Egg

3. When the egg is ovulated, the follicle becomes the corpus luteum, which produces progesterone and some estrogen until it degenerates.

Hormones secreted by the ovary
Estrogen
Progesterone

4. Estrogen from the follicle and progesterone from the corpus luteum cause the lining of the uterus to thicken, preparing it for embryo implantation. If the egg is not fertilized, the corpus luteum degenerates and progesterone and estrogen levels drop. As a result, the endometrium breaks down and is shed from the uterus during menstruation.

Menstrual cycle: thickness of the endometrium
0 2 4 6 8 10 12 14 16 18 20 22 24 26 28 2 4
Days
Menstruation   Menstruation

▶MENSTRUATION
The shedding of the uterine lining (the endometrium) that occurs when an embryo does not implant.

▶ZYGOTE
A fertilized egg.

▶HUMAN CHORIONIC GONADOTROPIN (HCG)
A hormone produced by an early embryo that helps maintain the corpus luteum until the placenta develops.

If an egg is not fertilized within 24 hours of ovulation, it is no longer viable. The corpus luteum degenerates at about day 26 of the cycle, progesterone levels drop, and the uterine lining sloughs off, leading to **menstruation** (INFOGRAPHIC 30.5).

If the egg is fertilized it becomes a **zygote**. As the zygote travels to the uterus it begins dividing and developing into an embryo. In ideal circumstances, the embryo implants in the endometrium of the uterus about 1 week after the egg is fertilized. Once implanted in the uterus, the embryo secretes a hormone called **human chorionic gonadotropin (hCG),** which signals the corpus luteum to continue producing progesterone, which sup-

ports the thickening endometrium. This is the hormone that, in effect, tells the reproductive system that pregnancy has begun (hCG is the hormone that most pregnancy tests use as a marker to detect a pregnancy).

Once the embryo implants, embryonic and maternal endometrial tissues interact to form the **placenta,** a disc-shaped structure that provides nourishment and support to the developing **fetus**. In addition to delivering oxygen, nutrients, and other key molecules like antibodies from the mother that help protect the embryo against infections, the placenta eventually takes over from the corpus luteum the task of producing estrogen and progesterone (INFOGRAPHIC 30.6).

Even if a woman never becomes pregnant, her hormonal cycles will continue until the ovaries stop responding to follicle-stimulating hormone and luteinizing hormone at the time of menopause, which in American women occurs at about age 51; at this time women stop ovulating and stop having monthly reproductive cycles, and are therefore no longer able to conceive.

Because hormones play such a crucial role in pregnancy and reproduction, many types of **contraception** are designed to interfere with the normal female hormone cycle (**TABLE 30.1**). Most birth control pills contain both estrogen and progesterone at levels that prevent the anterior pituitary from releasing follicle-

stimulating and luteinizing hormones. This prevents ovulation and consequently a woman taking this pill does not release eggs. Progesterone in birth control pills prevents successful pregnancy in other ways, too: it thickens the cervical mucus, blocking sperm from entering the uterus and oviducts, and also reduces endometrial thickening–a process that is necessary to support an embryo.

Men do not have a monthly hormone cycle, but beginning at puberty, sperm go through recognizable stages of development. The seminiferous tubules house precursor sperm cells that go through meiosis (see Chapter 11) and specialization to produce mature sperm cells. It takes approximately 6 weeks for sperm

▶PLACENTA
A structure made of fetal and maternal tissues that helps sustain and support the embryo and fetus.

▶FETUS
After the eighth week after fertilization, the embryo is referred to as a fetus.

▶CONTRACEPTION
The prevention of pregnancy through physical, surgical, or hormonal methods.

INFOGRAPHIC 30.6 HORMONES SUPPORT PREGNANCY

Estrogen and progesterone support the implanted embryo as it develops. Early in pregnancy, the embryo secretes human chorionic gonadatropin (hCG), which signals the corpus luteum to continue to produce these hormones. The placenta, once it has formed, takes over estrogen and progesterone production.

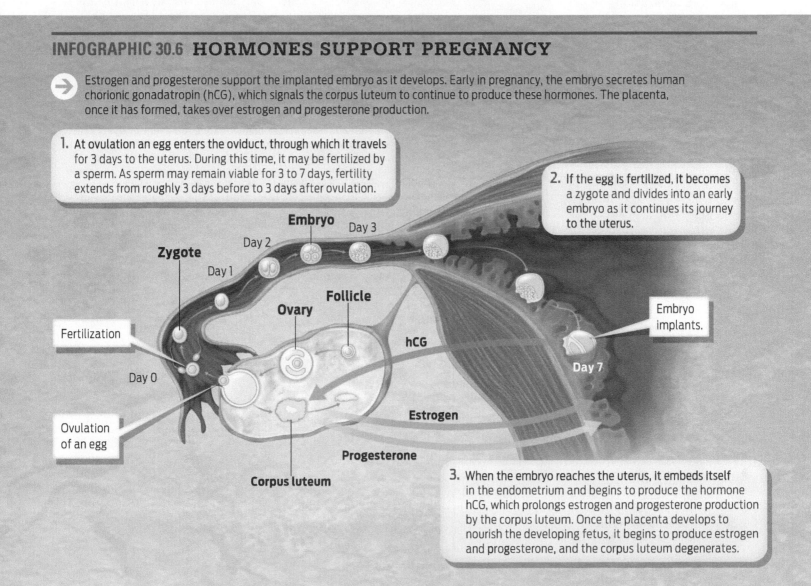

1. At ovulation an egg enters the oviduct, through which it travels for 3 days to the uterus. During this time, it may be fertilized by a sperm. As sperm may remain viable for 3 to 7 days, fertility extends from roughly 3 days before to 3 days after ovulation.

2. If the egg is fertilized, it becomes a zygote and divides into an early embryo as it continues its journey to the uterus.

Embryo Day 3

Day 2

Zygote

Day 1

Follicle

Ovary

hCG

Embryo implants.

Fertilization

Day 0

Day 7

Ovulation of an egg

Estrogen

Progesterone

Corpus luteum

3. When the embryo reaches the uterus, it embeds itself in the endometrium and begins to produce the hormone hCG, which prolongs estrogen and progesterone production by the corpus luteum. Once the placenta develops to nourish the developing fetus, it begins to produce estrogen and progesterone, and the corpus luteum degenerates.

TABLE 30.1 CONTRACEPTION

There are many ways to prevent conception. Currently available contraceptives include behavioral methods, physical and chemical barriers, hormones, and surgery.

METHOD	DESCRIPTION
Abstinence: 0% failure rate	No sexual intercourse
Intrauterine device: 0.2%–0.8% failure rate	An intrauterine device (IUD) is a small T-shaped device that typically contains copper or plastic. It is inserted into the base of the uterus through the cervix. The IUD is a long-term contraceptive option, lasting from 3 to 10 years. It prevents pregnancy by thickening cervical mucus and consequently impeding sperm from entering the uterus. It also weakens the endometrium, making it less able to support an embryo.
Sterilization surgery (tubal ligation, vasectomy): 0.5%–0.15% Failure rate	Surgical options include a vasectomy for men and tubal ligation for women. In a vasectomy, the vas deferens is cut, so sperm can no longer be ejaculated. In tubal ligation, the oviducts are cut and tied off, so sperm can no longer reach an ovulated egg. Both surgeries are essentially permanent, although they can be reversed with limited success. Those who do not want to have children, or any more children, typically choose surgical sterilization.
Hormones (implant, shot, ring, patch, pill): 0.3%–8% failure rate	Female hormonal contraceptives contain a combination of synthetic estrogen and progesterone or progesterone only. Women can take these hormones in the form of combination pills, a skin patch, a cervical ring, a minipill, a regular injection, or an implant. All hormonal methods prevent pregnancy in three major ways: they thicken the cervical mucus, making it less likely that sperm will be able to swim through it and get into the oviducts; they prevent ovulation; and they thin the endometrium, so it cannot support the implantation of an embryo.
Barriers (condom, diaphragm, spermicide): 2%–21% failure rate	The male and female condom, the diaphragm, and the cervical cap. All of these prevent sperm from entering the uterus and are typically used with spermicidal foams or jellies that contain sperm-inactivating chemicals.
Rhythm method and withdrawal: 3%–27% failure rate	The rhythm method (avoiding intercourse around the time a woman ovulates); withdrawal (the male withdraws his penis before ejaculation)
No contraception 85% failure rate	Sexual intercourse without any method of contraception

SOURCE: Trussel, J. (2007) In Hatcher, R. A., et al., *Contraceptive Technology,* 19th revised ed. New York: Ardent Media.

INFOGRAPHIC 30.7 SPERM DEVELOP IN THE TESTES

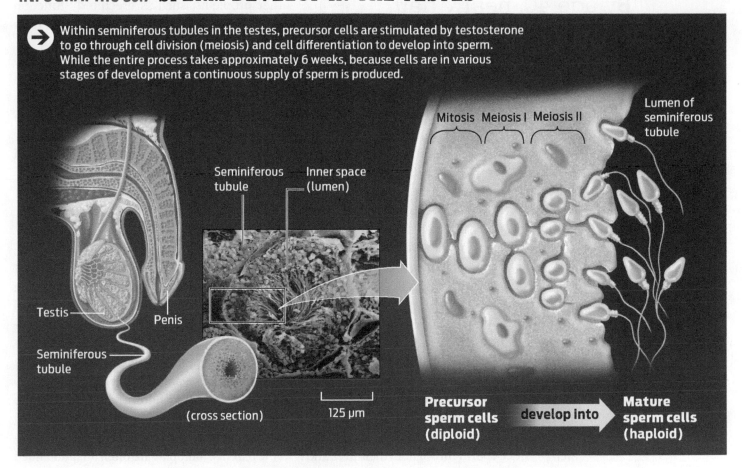

→ Within seminiferous tubules in the testes, precursor cells are stimulated by testosterone to go through cell division (meiosis) and cell differentiation to develop into sperm. While the entire process takes approximately 6 weeks, because cells are in various stages of development a continuous supply of sperm is produced.

Testis

Penis

Seminiferous tubule

(cross section)

Seminiferous tubule

Inner space (lumen)

125 µm

Mitosis   Meiosis I   Meiosis II

Lumen of seminiferous tubule

Precursor sperm cells (diploid)   develop into   Mature sperm cells (haploid)

to mature. Maturation is stimulated by testosterone, and although men produce testosterone continuously throughout their adult lives, they produce slightly less of the hormone as they age and consequently sperm production declines over time (INFOGRAPHIC 30.7).

THE MANY CAUSES OF INFERTILITY

Given how important hormones are to egg maturation and ovulation, it's not surprising that hormonal imbalances are a common cause of female infertility. Low levels of luteinizing hormone and follicle-stimulating hormone, for example, may prevent ovulation or make it erratic. Even slight irregularities in the hormone system can prevent the ovaries from releasing eggs. Specific causes of such hormonal imbalances include injury, tumors, excessive exercise, and starvation. Some medications are as-

sociated with ovulation disorders, and some studies have shown that stress can negatively affect fertility, as can poor nutrition.

Some women suffer from polycystic ovary syndrome, a condition defined by the presence of multiple ovarian cysts that impair ovulation, and associated with the production of excessive amounts of androgens (the "male" sex hormones, typically present in lower amounts in females). This syndrome is one of the most common hormonal disorders, affecting an estimated 10% of women of reproductive age. In addition to impairing ovulation, the condition is associated with irregular menstrual cycles, diabetes, and obesity.

One of the most common causes of infertility in women, especially those living in Western industrialized societies, is advanced maternal age (typically defined as over 35). Women often have a harder and harder time becoming

> By age 40, nearly half of all women will experience difficulty in conceiving.

pregnant as they age, and after age 35 they may choose to consult a fertility specialist to assist in becoming pregnant. Age-related declines in fertility occur in part because both the quantity and quality of a woman's eggs decline with age. At birth, a woman has about 1 million follicles in her ovaries. By puberty, she has about 300,000 follicles. Of these, only about 300 will ever ovulate an egg. Fertility in women peaks around age 25 and declines thereafter. As a result, according to the Centers for Disease Control, by age 40, nearly half of all women will experience difficulty in conceiving.

Men, too, experience declines in fertility as they age, as testosterone levels fall, but usually not as quickly or as dramatically as women. Men produce sperm throughout their lives and therefore can still father children in their 50s, 60s, and 70s. However, many men suffer from other problems, such as erectile dysfunction and testicular varicose veins, that can interfere with sperm reaching their target.

About a third of all cases of infertility are caused by reproductive impairments in women, and about a third are caused by impairments in men. In the remaining cases, the impairments affect both men and women or the cause of infertility cannot be determined (INFOGRAPHIC 30.8).

THE PROBLEM OF MULTIPLES

While Suleman's case made numerous headlines and generated a strong backlash against what many saw as the selfish and irresponsible "Octomom," the lessons of the case extend well beyond this one woman's unusual story. Increasingly, people with all kinds of fertility difficulties—from polycystic ovary syndrome to low sperm count to advanced maternal age—are turning to assisted reproduction for help conceiving a child. Lesbian and gay couples, too, are seeking help from fertility specialists in their quest to become parents. Multiples are often an unintended consequence of the technologies used.

INFOGRAPHIC 30.8 CAUSES OF INFERTILITY

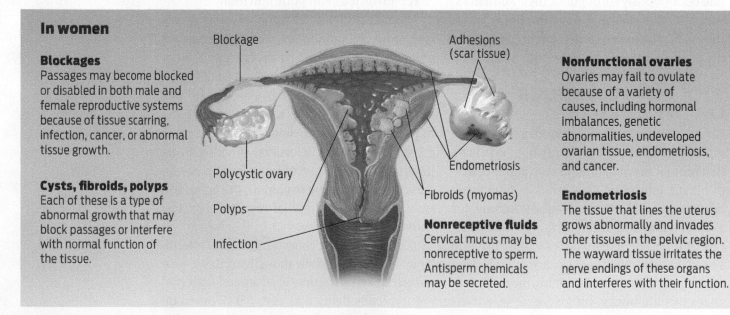

In women

Blockages
Passages may become blocked or disabled in both male and female reproductive systems because of tissue scarring, infection, cancer, or abnormal tissue growth.

Cysts, fibroids, polyps
Each of these is a type of abnormal growth that may block passages or interfere with normal function of the tissue.

Blockage

Polycystic ovary

Polyps

Infection

Adhesions (scar tissue)

Endometriosis

Fibroids (myomas)

Nonreceptive fluids
Cervical mucus may be nonreceptive to sperm. Antisperm chemicals may be secreted.

Nonfunctional ovaries
Ovaries may fail to ovulate because of a variety of causes, including hormonal imbalances, genetic abnormalities, undeveloped ovarian tissue, endometriosis, and cancer.

Endometriosis
The tissue that lines the uterus grows abnormally and invades other tissues in the pelvic region. The wayward tissue irritates the nerve endings of these organs and interferes with their function.

Assisted reproduction takes a number of forms. Besides IVF, another popular method is **intrauterine insemination (IUI),** in which sperm are injected directly into the uterus. This approach, also known as artificial insemination, is used most often when infertility can be traced to low sperm count, or to sperm that are "slow swimmers." It is also commonly used when a woman is trying to conceive with donor sperm or if the cause of infertility cannot be determined. Ultrasound technology is often used to monitor ovulation, so the insemination can be performed when chances are highest that one or more eggs have been released into the oviduct.

Commonly, both IVF and IUI start with a course of fertility drugs in the weeks before the procedure. These are usually FSH and LH, the hormones that stimulate ovarian follicles to develop and eggs to ovulate. The treatment works well for promoting follicle development and for controlling the precise time of ovulation, but there is a catch: fertility drugs almost always cause multiple eggs in multiple follicles to develop. This effect is desirable during IVF, in which the number of eggs that are fertilized outside the body and implanted into the uterus can be controlled. But this advantage turns into a liability with intrauterine insemination. Since sperm are inserted into the uterus during insemination, it's difficult to control the number of eggs that are fertilized. Multiple births result when more than one egg is fertilized (each by a different sperm), leading to more than one embryo implanting in the uterus and developing into a fetus.

This is what happened to Jon and Kate Gosselin, the famous couple who starred in the TLC television reality series *Jon & Kate Plus 8* from 2007 to 2011. The couple first had twins and then later sextuplets, all of whom were conceived through intrauterine insemination.

The book *Multiple Blessings,* by the Gosselins and Beth Carson, tells the story. Kate's first treatment consisted of "painful injections" of fertility drugs to stimulate her ovaries. She ended up having twins. A year later, Kate longed for another baby. This time, Kate saw a different specialist. Kate and her husband told their doctor that they did not want multiples but would not selectively "reduce," or abort a fetus. She received another round of fertility drugs.

> **▶INTRAUTERINE INSEMINATION (IUI)**
> A form of assisted reproduction in which sperm are injected directly into a woman's uterus.

In men

Blockages
Passages may become blocked or disabled in both male and female reproductive systems because of tissue scarring, infection, cancer, or abnormal tissue growth.

Erectile dysfunction
Genetic abnormalities, neurological problems, hormonal imbalances, and physical blockages inhibit the ability of blood to flood the penis tissue to support an erection.

Blockages

Enlarged prostate

Varicose veins

Normal sperm cell

Prostatitis
Sperm pass through and receive fluid from the prostate on their way into the urethra. An enlarged prostate can block the passage of semen.

Testicular varicose veins
Valves in the veins that keep blood flowing in one direction deteriorate, causing blood to back up and pool. These enlarged veins can interfere with sperm production and transport.

Sperm abnormalities
Men may have low numbers of healthy sperm and/or physically abnormal sperm.

INFOGRAPHIC 30.9 ASSISTED REPRODUCTIVE TECHNOLOGIES CAN RESULT IN MULTIPLE BIRTHS

A hazard of assisted reproduction is a high probability of multiple births. Babies born as multiples are more likely to be born underweight and premature, putting them at risk for a variety of serious health conditions.

In Vitro Fertilization (IVF)

A woman is given fertility drugs to stimulate egg development and maturation. Eggs are then retrieved and fertilized in a dish. If multiple embryos are injected, more than one of them may implant in the uterus.

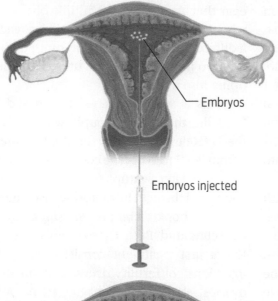

Embryos

Embryos injected

Intrauterine Insemination (IUI)

When a woman has been given fertility drugs to stimulate ovulation, several eggs may mature and ovulate simultaneously. Sperm injected into the uterus may fertilize multiple eggs and create multiple embryos.

Mature eggs

Sperm

Sperm injected

Two developing fetuses

Multiple births: 43% of births by IVF are twins; 3.6% are triplets or more.

When Kate had an ultrasound examination during the treatment, the Gosselins saw three, possibly four, developing eggs. Their doctor told them that they were unlikely to end up with three or four babies, but to be sure to avoid multiples he suggested that they forgo intrauterine insemination then and try again later. They decided to go ahead with the insemination anyway. When the sonogram showed seven developing embryos (one would later disappear), Jon dropped to his knees, Kate recalled in the book (**INFOGRAPHIC 30.9**).

COSTLY CARE

In the United States, only 15 states currently mandate that health insurance policies must offer coverage for infertility treatment, and those that do often do not cover the more expensive treatments, like IVF. As a consequence, many patients must pay out of pocket. Because the costs of treatment are so high, many doctors find themselves under pressure to be aggressive with treatment with patients who have limited resources. Therefore, financial considerations can dictate a course of treatment, even when the

> IVF has a higher success rate—
> about 40%, according to ASRM—but if
> the woman must undergo multiple rounds
> before becoming pregnant, the $12,400
> price tag can be prohibitive.

treatment is unlikely to be successful or to be too successful, as in Suleman's and Gosselin's cases.

As of 2013, a single insemination treatment (including fertility drugs and monitoring) can cost as much as $1,200—steep, but barely one-tenth the cost of an average in vitro fertilization cycle. Many couples may choose insemination over in vitro fertilization simply because it is less expensive, even though the procedure has a low success rate. On average, only 10% to 20% of all insemination treatments result in a live birth. But younger women and women who take fertility drugs and, as a result, have multiple ovulations, tend to have higher success rates. The multiple ovulations increase the odds of a successful pregnancy by insemination.

IVF has a higher success rate–about 40%, according to ASRM–but if the woman must undergo multiple rounds before becoming pregnant, the $12,400 price tag can be prohibitive. This financial burden can cause couples to hedge their bets by choosing to implant more embryos at one time than is typically recommended. The more embryos that are transferred into a woman's uterus during each round of IVF, the higher the odds of a pregnancy.

Kate Gosselin and her eight children.

UP CLOSE PRENATAL DEVELOPMENT

Pregnancy is divided into three trimesters. A normal pregnancy lasts 38 weeks from the time of fertilization, which is 40 weeks from the last menstrual period.

1st Trimester: development of tissue layers and vital organs

Embryonic stage

Month 1   Month 2   Month 3

The first trimester includes the embryonic stage of development and the early fetal stage. The embryo and fetus grow rapidly and critical organs develop.

2nd Trimester: growth and gender determination

Month 4   Month 5   Month 6

During the second trimester, the fetus continues to grow and develop features, including external genitalia.

3rd Trimester: weight gain and organ system development

Month 7   Month 8   Month 9

During the third trimester the fetus becomes fully developed and continues to gain weight. A fetus is technically "full term" at 35 weeks after fertilization (or 37 weeks after the last menstrual period), but most pregnancies last 38 weeks after fertilization (or 40 weeks after the last menstrual period).

In the wake of the revelation that Suleman had received multiple rounds of IVF, many people wanted to know how an unemployed 33-year-old with six children afforded to pay for the costly procedures. Suleman has said in interviews that she saved money for several years while working at a state mental hospital and later supported her family with money from student loans. The Associated Press and various newspapers also reported after the octuplets were born that Suleman received at least $165,000 in disability benefits from the state of California between 2002 and 2008 for an on-the-job injury she sustained in 1999. In addition, Suleman has said she received unemployment insurance and food stamps, as well financial support from the state for her other children, some of whom have disabilities. But even with this governmental support, given her large family and lack of an income, her financial resources would have been limited.

Medical details revealed during a court case brought by the California Medical Board in 2009 against Suleman's doctor, Michael Kamrava, show that Kamrava created 14 embryos and implanted a dozen of them—six times the recommended number for a patient Suleman's age. Eight embryos survived, and the babies were delivered 9 weeks prematurely by cesarean section (C-section). The Associated Press reported that during the hearing Kamrava said he regretted implanting the 12 embryos and "would never do it again."

Kamrava further stated that Suleman was adamant about using all 12 embryos, even though he suggested implanting only 4. "She just wouldn't accept doing anything else with those embryos. She did not want them frozen, she did not want them transferred to another patient in the future," he said, according to the AP story. Kamrava said that he consented only after Suleman agreed to have a fetal abortion if necessary to reduce risk. After the implantation, however, he only heard from Suleman after the birth of her octuplets, despite his numerous attempts to contact her.

From a medical perspective, the birth of multiples from any assisted reproduction procedure carries risks. According to the March of Dimes, 50% of twins and 90% of triplets are born prematurely, as are virtually all quads and quintuplets. A premature baby is defined as one born before 37 completed weeks of pregnancy. When babies are born prematurely, their organs may not be fully developed, which can lead to medical problems. In particular, their lungs are often immature and so the babies must be hooked up to mechanical breathing ventilators, which sometimes scar the lungs so that these children will for the rest of their lives be prone to asthma, pneumonia, chronic lung disease, and other respiratory disorders. And because their brains aren't fully developed, premature babies are susceptible to brain hemorrhages and to developmental difficulties, including learning disabilities (see **UP CLOSE: PRENATAL DEVELOPMENT**).

These medical conditions impose a huge burden on the health care system. Experts who have considered Suleman's case have stated that the medical bill for the delivery and postnatal care of her eight babies likely totaled $1.5 to $3 million. Kaiser Permanente, where Suleman delivered her babies, has asked Medi-Cal, the tax payer-funded state health insurance system, to foot the bill.

To reduce the birth of multiples and prevent the associated health problems, in 2006 the ASRM recommended transferring no more than two embryos in women under 35 years of age and no more than five in women over 40, because as a woman ages the success rate tends to drop. These guidelines appear to have been helped: in 2009, 3.6% of all infants born via assisted reproduction were triplets, as compared to 6.4% in 2003. But assisted reproduction still results in a high number of twins—in 2009, 43% percent of

Suleman (right) with some of her 14 children.

> *"If nothing else, these high profile cases have served as a wake-up call."*
>
> —BARBARA COLLURA

births from IVF alone were twins, compared to 3% of all births.

Despite the drop in multiple births, some lawmakers want more control. The risk that doctors may act against the ASRM guidelines, with negative consequences for women and their babies, remains.

After a lengthy investigation, the California Medical Board revoked Kamrava's medical license, effective July 1, 2011. The Board claimed he was negligent not only in Suleman's case, but also in the cases of two other women who suffered serious medical complications because of aggressive fertility treatments.

Lawmakers are concerned that there is nothing to prevent cases like these from happening again. Several states are considering legislation that would regulate doctors. A Missouri bill, for example, would require doctors not to exceed the ASRM's embryo-transfer guidelines, while a Georgia bill would put a strict limit on the number of embryos that can be transferred. But experts fear that regulation would hinder good care. Fertility doctors need flexibility to tailor treatment to a couple's individual condition of fertility, they argue. Such legislation "seeks to substitute the judgment of politicians for that of physicians and their patients," ASRM president R. Dale McClure said in a statement evaluating the Georgia bill.

Instead of putting controls on doctors, some groups favor regulation that would re-

quire health insurers to cover infertility diagnostics and treatment, for example. The 2010 Affordable Care Act, passed by Congress and signed into law by President Obama, does not include infertility treatments in the "Essential Health Benefits" that health insurers will be required to cover beginning in 2014. However, it does make changes to the tax code that would allow individuals to write off a larger portion of the cost of these procedures on their income taxes.

As long as fertility treatment continues to be a financial burden on couples, fertility doctors will face pressure from patients to give them the most for their money, says Barbara Collura, executive director of RESOLVE, an infertility advocacy group. Broad health insurance coverage would eliminate the cost factor. The result might be that couples would be able to forgo treatments such as insemination that have high risks with low success rates and skip directly to IVF when appropriate.

"If nothing else, these high profile cases have served as a wake-up call," says Collura. The community of health care workers is examining its own procedures and methods because it would rather self-regulate than have regulation imposed on it from the outside, she adds. "After Suleman, the community is really looking into how it [the birth of octuplets] happened and how it can prevent it from happening again." ■

FOR COMPARISON Why Do Organisms Have Sex?

Reproduction is a fundamental fact of life: all species must reproduce if they are to survive. But not all organisms reproduce in the same way, and not all have sex.

Sexual reproduction–reproduction that involves the combination of eggs and sperm from female and male individuals occurs in most eukaryotic organisms: animals, plants, fungi, and protists. In humans and other mammals, fertilization of an egg by sperm occurs internally, in a female oviduct. This approach to sex allows the fertilized egg to be protected as it grows but limits the number of eggs that can be fertilized at any one time. Internal fertilization also requires that the male and female physically copulate in order to achieve fertilization. In humans and other mammals, the penis serves as a sperm delivery system to place sperm in a location where they can easily swim to the egg.

Many nonmammalian species also rely on internal fertilization, but use a very different sexual anatomy. Internal fertilization in birds and reptiles, for example, occurs in an organ called the cloaca (plural: "cloacae"). The cloaca is the shared opening for solid waste, urine, and the reproductive system. When birds have sex, they briefly touch their cloacae together in what's called a cloacal kiss. This speedy kiss-and-run is sufficient to transfer sperm from the male to the female.

Interestingly, human female embryos have a cloaca, but the single cloacal opening divides into separate openings during embryonic development. In very rare cases, baby

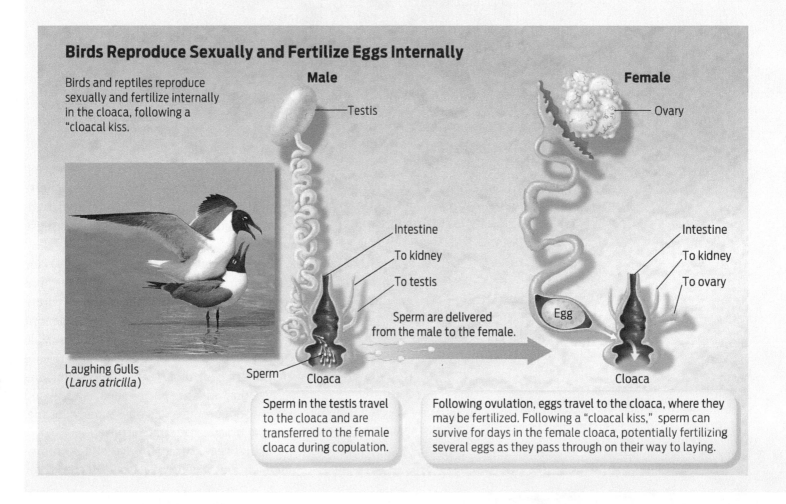

Birds Reproduce Sexually and Fertilize Eggs Internally

Birds and reptiles reproduce sexually and fertilize internally in the cloaca, following a "cloacal kiss.

Male

Testis

Intestine

To kidney

To testis

Sperm are delivered from the male to the female.

Sperm

Cloaca

Laughing Gulls (*Larus atricilla*)

Female

Ovary

Intestine

To kidney

To ovary

Egg

Cloaca

Sperm in the testis travel to the cloaca and are transferred to the female cloaca during copulation.

Following ovulation, eggs travel to the cloaca, where they may be fertilized. Following a "cloacal kiss," sperm can survive for days in the female cloaca, potentially fertilizing several eggs as they pass through on their way to laying.

Fish Reproduce Sexually and Most Fertilize Eggs Externally

Organisms like fish reproduce sexually, with two parents each contributing a gamete. Some fish fertilize internally, but most lay eggs that are fertilized externally.

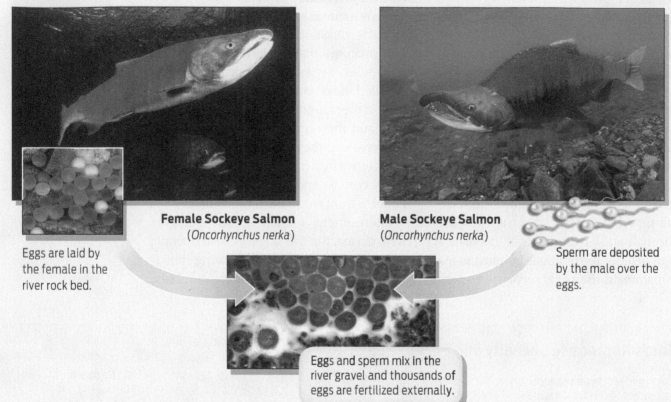

Female Sockeye Salmon
(*Oncorhynchus nerka*)

Male Sockeye Salmon
(*Oncorhynchus nerka*)

Eggs are laid by the female in the river rock bed.

Sperm are deposited by the male over the eggs.

Eggs and sperm mix in the river gravel and thousands of eggs are fertilized externally.

Bacteria Reproduce Asexually

Bacteria reproduce asexually, producing offspring from only a single parent by binary fission. Offspring are genetically identical to each other, as well as to the single parent cell.

Parent cell

Binary fission

Two genetically identical offspring

girls are born with an intact cloaca–having only one opening, rather than the usual three. Surgical correction to form distinct anal, vaginal, and urethral openings is possible in these cases.

Not all sex takes place inside the body. Many fish and amphibians rely on external fertilization–their gametes fuse in the outside environment. Female salmon, for example, deposit their eggs in gravel nests in streambeds. Male salmon then swim over the eggs and release sperm, fertilizing thousands of eggs simultaneously. Other aquatic species such as coral and hydra also reproduce this way. Some plant species also practice external fertilization.

Many very successful creatures reproduce without having sex at all. Bacteria, for example, reproduce asexually–without additional genetic input from another individual. In asexual reproduction, a single parent cell simply divides to produce identical offspring. Some organisms, such as certain types of fungus, can reproduce both sexually and asexually. Baker's yeast (*Saccharomyces cerevisiae*) is a fungus that is commonly used to make bread rise. *S. cerevisiae* can produce gametes called spores, and these gametes can fuse to generate zygotes that develop into unicellular yeast. But both the spores and the yeast can also make exact copies of themselves by mitotic cell division, resulting in identical populations of cells.

If some organisms can reproduce on their own, without the involvement of a partner, why bother having sex at all? Biologists don't fully understand why sex evolved, but they do have some good ideas. One hypothesis has to do with the evolutionary advantage of combining genetic information from two individuals. The resulting genetic diversity may enable offspring to better survive and adapt to changing

Yeast Reproduce Both Sexually and Asexually

Yeast can reproduce asexually by going through mitosis. They can also reproduce sexually by producing spores.

Asexual

A parent goes through mitosis, producing two genetically identical offspring. The parent cell forms a bud during this process, which will form one of the new daughter cells.

Parent cell

Offspring

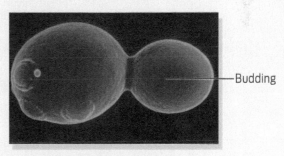

Budding

Saccharomyces cerevisiae

Sexual

First, a diploid parent cell forms four haploid spores during meiosis:

Parent cell

Haploid spores

Haploid spores

Then, two haploid spores from the same or different parent cells combine, producing a diploid offspring that reproduces asexually:

Haploid spores

Diploid offspring

Offspring budding

environments–better than they would if each individual were identical to the parent.

But there are downsides to sex, too. Sexual reproduction generally takes longer than asexual reproduction. Certain bacteria, for example, can reproduce in as little as 20 minutes, which is a great evolutionary advantage in the race of natural selection. And sexual reproduction requires more energy and investment, especially on the part of the female. Internal fertilization, in particular, places a large demand on the mother. In the case of birds and reptiles, she has to produce an energy-rich egg in which the embryo will develop, and then in some cases protect it while it hatches. In most mammals, the maternal investment is even greater: she must supply all nutrients to a growing fetus and also provide a protective environment within her body during the long time it takes for a fetus to grow into a baby. However, this maternal investment, while leaving the mother vulnerable during the pregnancy, often results in the successful birth of a baby from each egg that is fertilized, compared to the exposed embryos that result from external fertilization, which are often eaten by predators before they ever mature.

Over time, different organisms have evolved different reproductive strategies that ensure that their offspring survive in the particular environment each organism occupies.

CHAPTER 30 Summary

▶ The female reproductive system consists of paired ovaries, which produce eggs and secrete the hormones estrogen and progesterone, and accessory structures that enable fertilization and support pregnancy.

▶ The male reproductive system consists of paired testes, which produce sperm, the hormone testosterone, and accessory structures that enable fertilization.

▶ Fertilization of an egg by a sperm occurs in a female oviduct. Only one sperm can fertilize an egg at one time.

▶ The female reproductive cycle is coordinated by a complex balance of hormones that is controlled by the brain.

▶ Hormones from the hypothalamus trigger the anterior pituitary to release follicle-stimulating hormone (FSH) and luteinizing hormone (LH), which in turn stimulate eggs to develop and the ovary to secrete estrogen and progesterone.

▶ Ovulation, the monthly release of an egg from an ovarian follicle, is caused by a spike in production of LH.

▶ Estrogen and progesterone stimulate eggs to develop and the endometrium to thicken and prepare for a possible pregnancy.

▶ Upon fertilization, the zygote divides and travels to the uterus, where it implants in the nutrient-rich endometrium.

▶ The implanted embryo secretes human chorionic gonadotropin hormone (hCG), which acts on the corpus luteum to maintain progesterone secretion.

▶ If an egg is not fertilized, progesterone levels fall and the endometrium sloughs off during menstruation.

▶ Sperm develop in the seminiferous tubules of the testes. The process is stimulated by testosterone.

▶ There are many approaches to contraception, including barriers between sperm and eggs and manipulation of reproductive hormones.

▶ Infertility has many causes, including blocked passageways caused by infection and scar tissue, genetic abnormalities, hormonal deficiencies, and advanced age.

▶ Assisted reproduction involves artificially bringing sperm and egg together, either inside the body (in IUI) or outside the body (in IVF).

▶ Fertility drugs increase the number of eggs that mature and are ovulated by a female at one time. Multiple pregnancies result when sperm fertilize more than one available egg.

▶ Humans and other mammals, as well as birds and reptiles, reproduce sexually using internal fertilization. Other sexually reproducing animals, such as fish, reproduce using external fertilization. Some organisms are asexual and have no sex at all.

MORE TO EXPLORE

▶ NBC News, *Dateline* (2009) http://www.nbcnews.com/id/29130147/ns /dateline_nbc-newsmakers/t/everything-i-do-revolves-around-my-children

▶ American Society for Reproductive Medicine http://reproductivefacts.org/

▶ RESOLVE (National Infertility Association) http://resolve.org/

▶ Sunderam, S., et al. (2012) Assisted reproductive technology surveillance—United States, 2009. Surveillance Summaries. *Morbidity and Mortality Weekly Report* 61(SS7):1–23 http://www.cdc.gov/mmwr/ preview/mmwrhtml/ss6107a1.htm

▶ Mundy, L. (2008) *Everything Conceivable: How the Science of Assisted Reproduction Is Changing Our World.* New York: Anchor Books.

CHAPTER 30 Test Your Knowledge

What is the anatomy of the male and female reproductive tracts?

By answering the questions below and studying Infographics 30.1, 30.2, 30.3, and 30.7, you should be able to generate an answer for the broader Driving Question above.

KNOW IT

1 Sperm develop in

a. the epididymis.

b. the vas deferens.

c. the seminiferous tubules.

d. the urethra.

e. the penis.

2 Why can untreated pelvic inflammatory disease lead to infertility?

a. because it prevents ovulation

b. because it scars and blocks the oviducts

c. because it scars and blocks the cervix

d. because it interferes with estrogen production by the ovaries

e. because it interferes with FSH and LH production by the anterior pituitary gland

3 Describe the relationship between the uterus and the cervix, and between the uterus and the endometrium.

USE IT

4 List the structures that sperm must pass through to reach and fertilize an egg. Begin with the seminiferous tubules.

5 A woman can become pregnant if she has intercourse 5 days before ovulating or on the day she ovulates. What does this suggest about sperm and egg?

a. Sperm can survive for up to 6 days in the female reproductive tract.

b. Fertilization can occur in the ovary.

c. Human eggs cannot survive for very long after ovulation.

d. all of the above

e. a and c

6 A friend tells you that her boyfriend has been diagnosed with gonorrhea, a sexually transmitted infection. She isn't worried for herself because she doesn't have any symptoms of infection. What can you tell her about the invisible risks of an untreated sexually transmitted bacterial infection?

What hormones are involved in reproduction, and how do they work?

By answering the questions below and studying Infographics 30.5 and 30.6, you should be able to generate an answer for the broader Driving Question above.

KNOW IT

7 What is the source—testes, ovaries, anterior pituitary gland, or embryo—of each of the following hormones?

Luteinizing hormone (LH) _____

Follicle-stimulating hormone (FSH) _____

Testosterone _____

Estrogen _____

Progesterone _____

hCG _____

8 The hormone hCG is an indicator of pregnancy; it also

a. signals the corpus luteum to keep producing progesterone.

b. triggers ovulation.

c. acts on the anterior pituitary gland, causing it to release a surge of LH.

d. acts on the endometrium, causing it to thicken.

e. attracts sperm.

USE IT

9 Which of the following would most directly cause reduced levels of estrogen production?

a. an anterior pituitary tumor that increases secretion of LH

b. an increase in hypothalamus hormones that target the anterior pituitary

c. an anterior pituitary tumor that increases secretion of FSH

d. a decrease in hypothalamus hormones that target the anterior pituitary

e. anterior pituitary damage that prevents synthesis and release of FSH

10 In an episode of *Law & Order: Special Victims Unit*, a blood sample from a crime scene was found to have extremely low levels of FSH and LH. From this information, detectives determined that the blood came from a prepubescent girl, not a woman of reproductive age. Explain how they reached this conclusion.

11 As discussed in this chapter, oral contraceptives (such as the combination birth control pill, which contains both estrogen and progesterone) are designed to block ovulation in women. As males do not ovulate, a male hormonal contraceptive would have to target sperm development. Which hormone would have to be blocked to prevent sperm development in males? What would be a likely undesired consequence of this type of male contraception?

INTERPRETING DATA

12 Emergency contraception works by delaying or preventing ovulation. To be effective, it must be taken before LH levels start to rise.

a. Will emergency contraception be effective in preventing a pregnancy after unprotected intercourse on the day that a woman ovulates?

b. Different forms of emergency contraception have different efficacy rates, expressed as the percent reduction in the number of pregnancies that would have otherwise occurred. For example, if 80 women out of 1,000 become pregnant after unprotected intercourse, and an emergency contraceptive is 75% effective, then one would expect the use of the emergency contraceptive to reduce the number of pregnancies by 75% of 80 pregnancies, to 20 pregnancies. Predict the number of pregnancies with each of the following emergency contraceptives (based on 80 of 1,000 pregnancies as a baseline): the pill marketed as Plan B (progestin—a synthetic progesterone—only), 89% effective; combined progestin and estrogen, 74% effective.

DRIVING QUESTION 3

What are the different types of assisted reproduction, and how do they work?

By answering the questions below and studying Infographics 30.4, 30.8, and 30.9, you should be able to generate an answer for the broader Driving Question above.

KNOW IT

13 Which of the following could interfere with ovulation?

a. blocked oviducts

b. chronically low levels of LH

c. excessive production of cervical mucus that blocks the cervix

d. presence of sperm in the oviduct

e. low levels of hCG

14 Compare and contrast in vitro fertilization (IVF) and intrauterine insemination (IUI).

USE IT

15 Assume that there is an array of diagnostic methods available to you, including blood tests to determine hormone levels and ultrasound to visualize internal structures. What results might confirm each of the following infertility-associated conditions? Be specific.

a. a blocked epididymis

b. polycystic ovary syndrome

c. menopause

d. oviduct scarring

16 Why does IUI have a higher risk of multiple births than IVF?

MINI CASE

17 A young couple has been trying to have a baby for over a year, but so far they have not had any luck. Analysis of the man's semen reveals a normal sperm count and no evidence of high rates of abnormal sperm. A physical exam and ultrasound revealed blockages in both of the woman's oviducts (fallopian tubes). From this information, describe two forms of fertility treatments (including assisted reproduction) that the couple could consider, and two that would not be helpful. For each, describe what is involved in the treatment, and why it would or would not be a helpful strategy for this couple.

BRING IT HOME

18 From the perspective of a fertility specialist, how would you respond to a congressional representative about a proposed increased regulation of fertility clinics? In order to make a convincing argument, include both pros and cons, medical and scientific considerations, and a description of the patient population that this specialist serves.

THE FORGOTTEN PLAGUE

After nearly a century, scientists learn what made the 1918 influenza virus so deadly

31

IN THE FALL Of 1918, WORLD WAR I WAS COMING TO AN END—but another, insidious, threat was surfacing. In communities around the world, a deadly illness was gaining momentum, spreading like wildfire from person to person.

At first the disease seemed much like a bad case of influenza, or "flu." High fevers and body aches were common. But some patients also developed more severe symptoms, such as coughing so intense that they spat up blood. Many died so quickly after falling ill that doctors began to suspect they were witnessing something far worse than ordinary flu.

Stories were told of victims who were completely healthy in the evening but were dead by morning. An army physician stationed at Camp Devens in Massachusetts wrote to a friend that patients with what seemed like ordinary flu would rapidly "develop the most vicious type of pneumonia that has ever been seen" and later would "struggle for air until they suffocated." Patients "died struggling to clear their airways of a blood-tinged froth that sometimes gushed from their nose and mouth," recalled Isaac Starr, a physician who was a third-year medical student in Philadelphia during the war and who published his recollections of the illness in the *Annals of Internal Medicine* in 1976.

Between 1918 and 1920, the disease swept around the globe. Had the world not been at war, the pandemic might never have happened. Troops carried the illness with them across Europe and back to the United States, infecting relatives, friends, and strangers along the way.

▶▶ DRIVING QUESTIONS

1. What is the structure of a virus, and how do viruses cause disease?
2. What is innate immunity?
3. What is adaptive immunity, and how does vaccination rely on adaptive immunity?
4. What are specific features of influenza virus that allow it to cause worldwide outbreaks?

Police in Seattle, Washington, keeping the peace during the 1918 flu pandemic.

The result was one of the deadliest pandemics in human history. According to historian John Barry, the 1918 flu pandemic killed more people in 1 year than the Black Death killed in the 14th century; it killed more people in 25 weeks than AIDS has killed in 25 years. By the time it was over, between 50 and 100 million people around the world had perished—more than the total number killed in World War I.

Why this particular strain of flu inflicted so much more devastation than any pandemic that came before or after, scientists of the time couldn't say. For decades after the pandemic had subsided, it seemed it would never be possible to explain the flu's severity, since the virus itself was lost to history. But recent discoveries have made it possible to peer into the viral past, and scientists have begun to uncover what made the 1918 virus so deadly.

SILENT INVADER

Each year in the United States about 36,000 people die from flu-related complications, and more than 200,000 people are hospitalized from flu-related illnesses. Most of these deaths strike young children and the elderly, whose bodies are less able to fight off the infection. The 1918 flu, by contrast, struck down people in the prime of life: people age 20 to 40 were the hardest hit. The death toll in this age group

> The 1918 flu pandemic killed more people in 1 year than the Black Death killed in the 14th century; it killed more people in 25 weeks than AIDS has killed in 25 years.

was so severe that it reduced the average U.S. life expectancy for that year by 12 years.

At the time, doctors did not know what caused influenza. Some suspected that it was caused by a bacterium–as were many other known diseases, such as cholera and tuberculosis. But research in 1933 showed that influenza is caused not by a bacterium but by a **virus.**

The influenza virus, like all viruses, is a particle made up of nucleic acid surrounded by a protein shell. The nucleic acid serves as the virus's genetic material, while the protein shell (called a capsid) allows the virus to bind to and enter cells. Viruses are considered to be nonliving because even though all viruses have genes, viruses are not made of cells. Because they don't have cells, viruses can reproduce only by infecting host cells. They then use the host cell's machinery to replicate and make more viruses. Eventually they destroy the host cell (**INFOGRAPHIC 31.1**).

Unlike cells, which rely on DNA as their hereditary material, viruses can store their genetic information in the form of either DNA or RNA (see Chapter 2). RNA viruses include the influenza virus and the viruses that cause colds, measles, and mumps. RNA viruses also

▸**VIRUS**
An acellular infectious particle consisting of nucleic acid surrounded by a protein shell.

INFOGRAPHIC 31.1 VIRUSES INFECT AND REPLICATE IN HOST CELLS

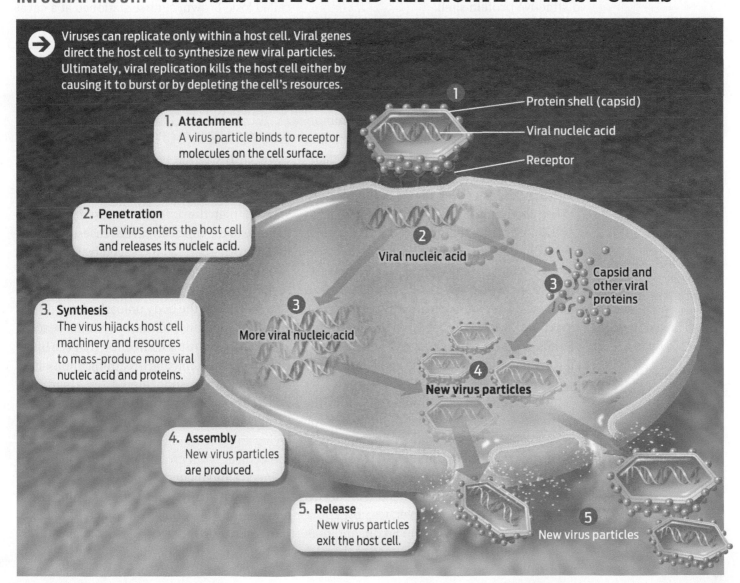

→ Viruses can replicate only within a host cell. Viral genes direct the host cell to synthesize new viral particles. Ultimately, viral replication kills the host cell either by causing it to burst or by depleting the cell's resources.

1. **Attachment**
A virus particle binds to receptor molecules on the cell surface.

2. **Penetration**
The virus enters the host cell and releases its nucleic acid.

3. **Synthesis**
The virus hijacks host cell machinery and resources to mass-produce more viral nucleic acid and proteins.

4. **Assembly**
New virus particles are produced.

5. **Release**
New virus particles exit the host cell.

Protein shell (capsid)

Viral nucleic acid

Receptor

Viral nucleic acid

Capsid and other viral proteins

More viral nucleic acid

New virus particles

New virus particles

> **"** *We didn't take temperatures;*
> *we didn't even have time to take blood pressure.*
> *We would give them a little hot whisky toddy;*
> *that's about all we had time to do.* **"**
>
> — JOSIE BROWN, a nurse stationed at a naval base in Illinois in 1918

▶**IMMUNE SYSTEM**
A system of cells and tissues that acts to defend the body against foreign cells and infectious agents.

▶**PATHOGENS**
Infectious agents, including certain viruses, bacteria, fungi, and parasites. Many pathogens trigger an immune response.

▶**IMMUNITY**
The resistance to a given pathogen conferred by the activity of the immune system.

▶**INNATE IMMUNITY**
Nonspecific defenses, such as physical and chemical barriers and specialized white blood cells that are present from birth and require little or no time to become active.

cause more serious diseases, like AIDS and polio. Diseases caused by DNA viruses include hepatitis B, chicken pox, and herpes. Each type of virus has a characteristic shape and a distinctive protein shell.

Some viruses cause more damage than others. The severity of any illness typically depends on how quickly the **immune system** responds and how well the infected tissue can repair itself. Most people recover completely from the rhinovirus that causes the common cold because the respiratory tract contains rapidly dividing cells that quickly replace damaged ones. Poliovirus, however, attacks nerve cells that almost never divide, which is why polio infection may cause paralysis.

Like the virus that causes common colds, the flu virus spreads from person to person through coughs and sneezes. Viruses are extremely small and are easily aerosolized. One sneeze can send millions of virus particles out into the air, where they are free to infect other victims who are near enough to inhale them. Only one or two particles are enough to cause an infection.

Scientists aren't sure where the 1918 flu outbreak started. Newspaper coverage of the carnage first occurred in Spain, which was a neutral country in World War I and one where news was not censored. For that reason, the pandemic became known at the time as "Spanish flu." However, we now know that the virus did not start in Spain, and that some of the earliest reported cases occurred in the United States, among soldiers housed in

close quarters. Such U.S. military bases were also where Spanish flu first demonstrated its deadly power.

"We didn't have the time to treat them," recalled Josie Brown, a nurse stationed at a naval base in Illinois in 1918. "We didn't take temperatures; we didn't even have time to take blood pressure. We would give them a little hot whisky toddy; that's about all we had time to do."

A MICROSCOPIC BATTLEGROUND

Some early clues as to what made the 1918 flu infection so deadly came from examining victims' bodies. Autopsies performed at the time showed that the infection caused extensive damage to the lungs. In severe cases, the lungs looked torn apart, as if ravaged by shrapnel.

Today we know that victims' lungs were severely damaged not by the virus itself but by all the defenses the body unleashed to fight the virus. The victims' immune systems reacted so strongly against the infection that the lungs filled with fluid, making breathing impossible. Victims literally suffocated from their own fluids. Though disastrous, this response was actually a sign that the immune system was trying to fight the infection. The lungs were the central battleground of the war between influenza and its human host.

Like an army, the body's immune system defends the body from different kinds of **pathogens**—infectious agents such as viruses, bacteria, and parasites. The immune system basically reacts against anything it encounters

INFOGRAPHIC 31.2 THE IMMUNE SYSTEM DEFENDS AGAINST A VARIETY OF PATHOGENS

Viruses

Influenza virus
Flu

Human immunodeficiency virus
HIV

Varicella zoster virus
Chicken pox

Bacteria

Escherichia coli
Food poisoning

Treponema pallidum
Syphilis

Streptococcus pyogenes
Strep throat, scarlet fever

Parasites

Trychophyton mentagrophytes (fungus)
Athlete's foot

Plasmodium falciparum (worm)
Malaria

Giardia lamblia (protist)
Gastroenteritis

as foreign, or "nonself," and tries to destroy it **(INFOGRAPHIC 31.2).**

The human immune system has two primary lines of defense that coordinate to protect us from pathogens, providing us with **immunity.** The first includes physical and chemical barriers such as the skin and mucous membranes, which block invaders from entering the body. It also includes defensive proteins and a variety of white blood cells that serve as sentries and engulf invaders. Because we are born with these defense mechanisms, they are referred to as **innate immunity.** Innate defenses are effective but nonspecific

INFOGRAPHIC 31.3 INNATE AND ADAPTIVE IMMUNITY

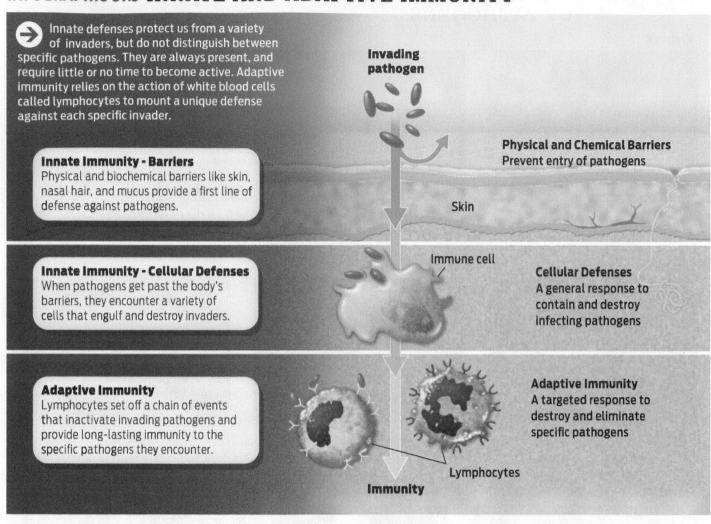

→ Innate defenses protect us from a variety of invaders, but do not distinguish between specific pathogens. They are always present, and require little or no time to become active. Adaptive immunity relies on the action of white blood cells called lymphocytes to mount a unique defense against each specific invader.

Innate Immunity - Barriers
Physical and biochemical barriers like skin, nasal hair, and mucus provide a first line of defense against pathogens.

Innate Immunity - Cellular Defenses
When pathogens get past the body's barriers, they encounter a variety of cells that engulf and destroy invaders.

Adaptive Immunity
Lymphocytes set off a chain of events that inactivate invading pathogens and provide long-lasting immunity to the specific pathogens they encounter.

Invading pathogen

Physical and Chemical Barriers
Prevent entry of pathogens

Skin

Immune cell

Cellular Defenses
A general response to contain and destroy infecting pathogens

Adaptive Immunity
A targeted response to destroy and eliminate specific pathogens

Lymphocytes

Immunity

▶ADAPTIVE IMMUNITY
A protective response, mediated by lymphocytes, that confers long-lasting immunity against specific pathogens.

▶LYMPHOCYTE
A specialized white blood cell of the immune system. Lymphocytes are important in adaptive immunity.

defenses, in that they attack a wide variety of invaders but do not discriminate among them. These defenses are always present and require little or no time to become active.

The second line of defense is called **adaptive immunity** and includes the coordinated actions of specialized white blood cells called **lymphocytes.** Adaptive immunity takes time to develop, but it is targeted and discriminating: it "learns" to respond to specific pathogens and substances. This phenomenon of learning is important because there are many types of pathogens and they evolve quickly, so our immune system must be capable of adapting to new invaders if it is to adequately protect us (**INFOGRAPHIC 31.3**).

INNATE DEFENSES

The innate immune system starts defending at sites where the body is exposed to the outside world. Enzymes in saliva destroy some pathogens; nasal hair and the mucus that lines the throat trap others. Under the layer of mucus is a layer of cells with hairlike projections called cilia that sweep away pathogens to prevent them from taking hold in the throat. When pathogens do successfully breach these physical barriers, the body tries to flush them out with more fluid: runny noses, watery eyes, coughs, and sneezes are your body's way of expelling the invaders.

When pathogens overcome these physical and chemical defenses, they trigger additional innate responses. One of the most important things the innate immune system does is promote **inflammation** in response to tissue damage or infection. Tissues that have been damaged by injury or by infection release chemicals such as **histamine** that cause blood vessels to expand and leak fluid into surrounding tissues. These chemical signals also attract various white blood cells to the region. The redness, heat, swelling, and pain associated with inflammation are the result of increased blood flow to the area (causing redness and heat), release of fluids and white blood cells from the leaky blood vessels (causing swelling), and release of chemical signals from damaged tissue (causing pain). The fluid at inflamed sites also contains clotting proteins that stop bleeding and prevent pathogens from spreading to neighboring tissues.

Several types of white blood cell contribute to our innate defenses. Most important are **phagocytes,** which include **macrophages,** found in tissues, and **neutrophils,** which patrol the bloodstream. When phagocytes come into contact with pathogens, they bind to and engulf the invaders in a process called phagocytosis. Some phagocytes enhance the inflammatory response, and others help initiate an adaptive response against the specific invader. Still other white blood cells, called **natural**

Sample blocks containing lung and brain tissue from victims of the 1918 "Spanish flu" sit atop a list of child victims.

killer cells, attack virus-infected cells—and even cancer cells—and kill them.

The body has other nonspecific defenses. These include proteins that either attack pathogens directly or prevent them from reproducing. Virus-infected cells can send out "SOS" signals in the form of **interferon.** Interferon proteins impede viral replication in neighboring cells and also signal other immune cells to become active. Other defensive

▶**INFLAMMATION**
An innate defense that is activated by infection or tissue damage; characterized by redness, heat, swelling, and pain.

▶**HISTAMINE**
A molecule released by damaged tissue and during allergic reactions.

▶**PHAGOCYTE**
A type of white blood cell that engulfs and ingests damaged cells and pathogens.

▶**MACROPHAGE**
A phagocytic cell that resides in tissues and plays an important role in the inflammatory response.

▶**NEUTROPHIL**
A phagocytic cell in the circulation that plays an important role in the inflammatory response.

▶**NATURAL KILLER CELL**
A type of white blood cell that acts during the innate immune response to find and destroy virally infected cells and tumor cells.

▶**INTERFERON**
Antiviral proteins produced by virally infected cells that help protect adjacent cells from becoming infected.

A CDC researcher works with samples of recreated 1918 flu virus.

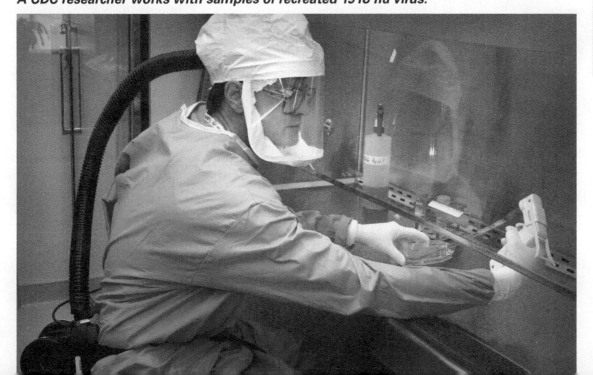

▶**COMPLEMENT PROTEINS**
Proteins in blood that help destroy pathogens by coating or puncturing them.

proteins, called **complement proteins,** coat the surface of pathogens, making them more easily destroyed by phagocytes, or directly destroy some bacteria by punching holes in the cell membrane (**INFOGRAPHIC 31.4**).

Though the innate immune system provides some protection against influenza, occasionally some virus particles manage to evade the body's physical and chemical barriers and take up residence in the upper respiratory tract: the mouth, the nose, and the throat. Within minutes, these virus particles attach to epithelial cells that line the respiratory tract and slip inside the body. The virus then hijacks the host cells' cellular machinery to replicate its own genetic material. Generally, about 10 hours after the virus invades a cell, the cell begins to release newly synthesized viral particles, sending out between 1,000 and 10,000 viruses capable of infecting other cells. Eventually the infected cells die, weakening the respiratory tract.

INFOGRAPHIC 31.4 SOME IMPORTANT FEATURES OF INNATE IMMUNITY

Physical barriers

Skin (on the outside of the body) and mucous membranes (lining the inside of the body) have layers of tightly packed cells that prevent pathogens from entering the body. The mucus that coats mucous membranes traps foreign substances.

Inflammation

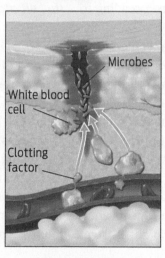

Microbes

White blood cell

Clotting factor

1. Pathogens get past physical barriers.
2. Microbes and damaged cells release molecules that increase blood flow and attract white blood cells to infected areas.
3. Blood vessels leak, causing surrounding tissue to swell with fluid that contains clotting factors and white blood cells.
4. White blood cells ingest pathogens and also trigger an adaptive response. Clotting reactions contain the infection.

Phagocytes

Macrophage

Yeast cell

Macrophage engulfs yeast cells.

Phagocytes (including macrophages and neutrophils) can recognize, bind to, and ingest pathogens. Phagocytes also trigger inflammation and adaptive immune responses.

Antimicrobial chemicals

Acid in the stomach kills many of the microorganisms that we ingest.

Tears and **saliva** contain an enzyme that breaks down the cell walls of bacteria, causing them to burst.

Complement proteins

Complement proteins in blood puncture holes in bacterial cells or coat the cell's surface, making them more easily destroyed by phagocytes.

> What killed many people was not
> the virus itself but the overly aggressive
> inflammatory response mounted against it.

Normally our bodies are able to mount a sufficient immune response to contain the infection, so a typical course of flu lasts only about a week and is not generally fatal. Why the 1918 flu was so deadly, researchers at the time could not say. More perplexing still was why the virus tended to kill people in the prime of their lives—those with the healthiest immune systems. It would be more than 75 years before researchers could even begin to answer those questions.

VIRAL TIME CAPSULE

In 1995, Jeffery Taubenberger, a virologist at the U.S. Armed Forces Institute of Pathology, became interested in the mystery of the 1918 flu, and wondered if advances in molecular biology might help him solve it. He realized that he could use the polymerase chain reaction (PCR; see Chapter 7) to reconstruct the viral genome—if only he had access to some original tissue samples from victims. He knew that the army had routinely saved tissue samples from soldiers killed during wartime and decided to check the institute's archives. Right away, he was able to locate tissues from several U.S. servicemen who had died of flu in 1918. Over the next several months, Taubenberger and his colleagues used PCR to recover viral RNA from these samples and then were able to sequence a few of the virus's genes. But the results were incomplete, and the team eventually ran out of tissue samples from which to obtain virus. Taubenberger published these preliminary results in 1997 in the journal *Science*.

Later that year, Johan Hultin, a Swedish pathologist and physician, came across Taubenberger's paper and realized he had a solution to the problem of finding more tissue samples. Back in the 1950s, while doing research for his dissertation, Hultin had discovered a mass grave of Spanish flu victims outside a remote village in Alaska. The village had been hit especially hard from the pandemic—85% of villagers had died from flu. Because the victims were buried in permafrost, their tissues might still be intact, he reasoned. Hultin wrote to Taubenberger and offered to take him to the site.

With permission from village elders, the scientists dug into the permafrost and exhumed dozens of bodies. Most were not much more than bone, but the body of one obese woman was still relatively intact, preserved by layers of insulating fat. From her tissues, the scientists were able to obtain enough viral RNA to resurrect the complete 1918 flu virus sequence—a landmark breakthrough. From this sequence, published in 2005, researchers were able to determine the identity of the virus and analyze its genetic relationship to other known strains of flu virus. But that's not all. Knowing the sequence of all eight of the virus's genes allowed researchers to begin asking which genes were responsible for the virus's extreme virulence.

A few years later, in 2008, Yoshihiro Kawaoka, at the University of Wisconsin, Madison, and his colleagues used this information to determine what made the 1918 virus so deadly: four genes with alleles specific to the 1918 virus allowed flu virus to penetrate deep into the lungs and cause infection there. While less lethal flu viruses

INFOGRAPHIC 31.5 LUNG INFLAMMATION CAN KILL

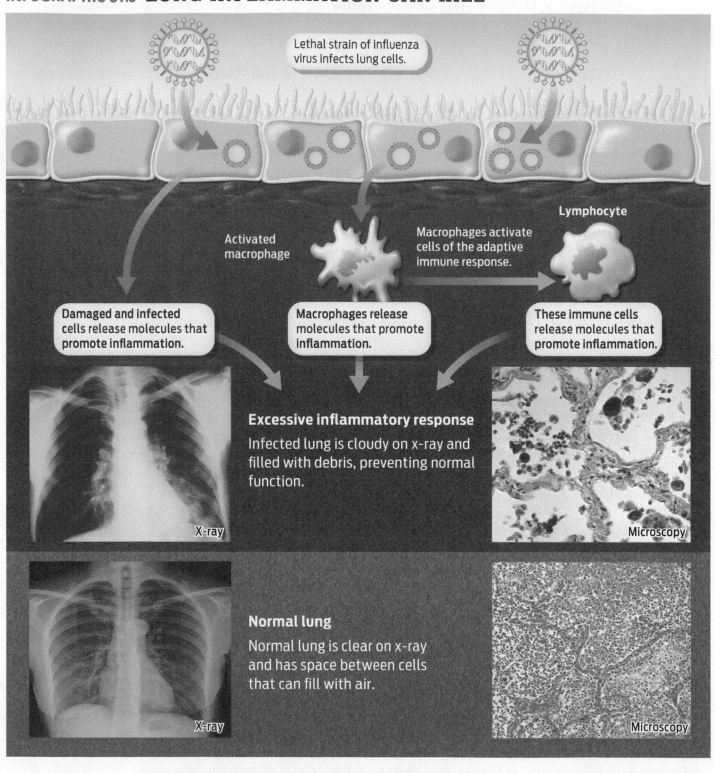

Lethal strain of influenza virus infects lung cells.

Activated macrophage

Macrophages activate cells of the adaptive immune response.

Lymphocyte

Damaged and infected cells release molecules that promote inflammation.

Macrophages release molecules that promote inflammation.

These immune cells release molecules that promote inflammation.

Excessive inflammatory response

Infected lung is cloudy on x-ray and filled with debris, preventing normal function.

X-ray

Microscopy

Normal lung

Normal lung is clear on x-ray and has space between cells that can fill with air.

X-ray

Microscopy

replicate primarily in the upper respiratory tract, the versions of genes in the 1918 virus allowed it to cruise past the mouth, nose, and throat and to invade the lungs. Replication of the virus in the lungs triggered a massive inflammatory response.

The inflammatory response is generally a good thing: it is a very effective way to contain and destroy invaders. But the balance between defense and destruction is delicate. In unusual cases, the inflammatory response can go into overdrive and destroy the very organ it is trying to save. This is what researchers now think happened with the 1918 flu. In the face of such an aggressive flu virus, the inflammatory response of victims was so massive that it damaged the very tissue it was supposed to protect. As fluids leaked out of blood vessels, they impaired oxygen uptake in the lungs; and as phagocytes produced chemicals to kill pathogens, those same chemicals destroyed lung tissue. What killed many people was not the virus itself but the overly aggressive inflammatory response mounted against it. This aggressive immune response also explains why victims with the strongest immune systems–people age 20 to 40–were the ones who suffered most (**INFOGRAPHIC 31.5**).

Other patients likely died from secondary infections occurring after flu infection. Because influenza weakened the defenses of the respiratory tract, other pathogens, such as bacteria, had unimpeded entry into the lungs. And because antibiotics were not yet discovered (see Chapter 3), doctors had no way of treating these infections. Bacterial pneumonia was responsible for at least half of all deaths during the 1918 pandemic, according to Taubenberger.

IMMUNOLOGICAL MEMORY

Not everyone who became infected with the 1918 flu died, despite its virulence. In fact, while there were approximately 50 to 100 million deaths, experts estimate that some 525 million people were infected–about 30% of the entire world population. How so many people were able to fight the infection while others died remains mostly a mystery. But scientists do have some clues.

Of those who became infected and then recovered, some may have had partial immunity from an earlier infection. Such long-lasting immunity is conferred by the adaptive immune system.

Whereas the innate immune system is always ready to fight, the adaptive immune system must be primed over time. From birth into adulthood, our bodies are continuously assaulted by a barrage of pathogens. With repeated exposure, our bodies develop a memory of every pathogen we encounter that gets past our innate defenses. Should we confront the same pathogen again, immunological memory helps our bodies fight off infection before it can take hold.

Two types of lymphocyte play crucial roles in adaptive immunity: B cells and T cells. Like all blood cells, B and T cells develop from precursor cells made in the bone marrow. Some immature lymphocytes in bone marrow become **B cells** (think "B" for "bone"). Other lymphocytes migrate from the bone marrow to the **thymus,** a gland in the chest, where they become **T cells** (think "T" for "thymus"). Both B cells and T cells eventually travel to the **lymph nodes** and other organs of the **lymphatic system,** where they hunt for pathogens (**INFOGRAPHIC 31.6**).

B and T cells cooperate to produce two main types of adaptive immunity. One type, called **humoral immunity,** targets free-floating pathogens in body fluids ("humoral" refers to the "humors," an old name for the liquids in the body). Humoral immunity produces **antibodies**–proteins that circulate in body fluids and help to fight infections. Antibodies are produced by B cells, but specialized T cells called **helper T cells** are needed to jumpstart

A U.S. Public Health Service ad that appeared in local newspapers in 1918.

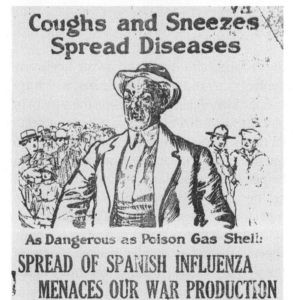

▶B CELLS
White blood cells that mature in the bone marrow and produce antibodies during the adaptive immune response.

▶THYMUS
The organ in which T cells mature.

▶T CELLS
White blood cells that mature in the thymus and can destroy infected cells or stimulate B cells to produce antibodies, depending on the type of T cell.

▶LYMPH NODES
Small organs in the lymphatic system where B and T cells may encounter pathogens.

▶LYMPHATIC SYSTEM
The system of vessels and organs that works with the immune system, allowing B and T cells to respond to pathogens.

▶HUMORAL IMMUNITY
The type of adaptive immunity that fights free-floating pathogens in the blood and lymph fluid.

▶ANTIBODY
A protein produced by B cells that binds to antigens and either neutralizes them or flags other cells to destroy pathogens.

▶HELPER T CELL
A type of T cell that helps activate B cells to produce antibodies.

➔ The immune system is made up of specialized cells called lymphocytes that develop, mature, and act in a variety of tissues, including the bone marrow, thymus, spleen, and lymph nodes, which are part of the lymphatic system.

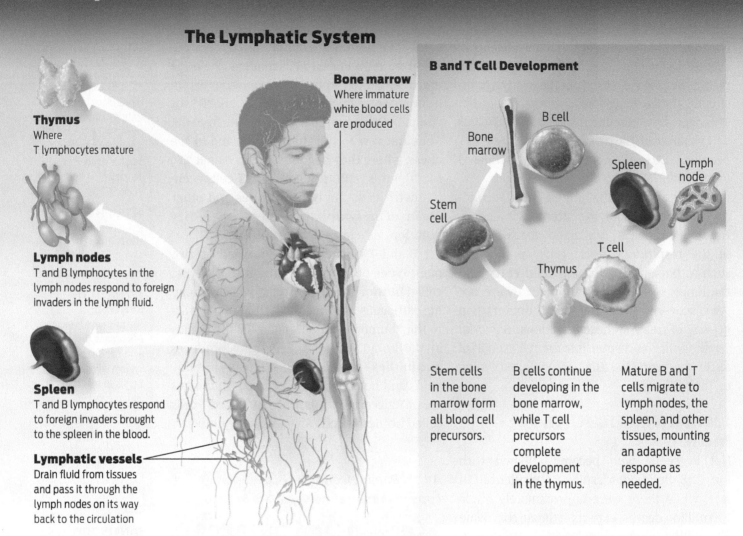

The Lymphatic System

Bone marrow
Where immature white blood cells are produced

Thymus
Where T lymphocytes mature

Lymph nodes
T and B lymphocytes in the lymph nodes respond to foreign invaders in the lymph fluid.

Spleen
T and B lymphocytes respond to foreign invaders brought to the spleen in the blood.

Lymphatic vessels
Drain fluid from tissues and pass it through the lymph nodes on its way back to the circulation

B and T Cell Development

Bone marrow

B cell

Stem cell

Spleen

Lymph node

Thymus

T cell

Stem cells in the bone marrow form all blood cell precursors.

B cells continue developing in the bone marrow, while T cell precursors complete development in the thymus.

Mature B and T cells migrate to lymph nodes, the spleen, and other tissues, mounting an adaptive response as needed.

▶**ANTIGEN**
A specific molecule (or part of a molecule) to which specific antibodies can bind, and against which an adaptive response is mounted.

▶**PLASMA CELL**
An activated B cell that divides rapidly and secretes an abundance of antibodies.

the process. Both B cells and helper T cells recognize bacterial and viral **antigens,** which are specific identifying structures on the surface of pathogens. Antigens are like unique molecular fingerprints, alerting the immune system to the presence of invaders. When a B cell and a helper T cell recognize the same antigen, the B cell becomes activated. That B cell will divide repeatedly to create an army of **plasma cells**—cells that secrete many copies of an antibody specific to that particular antigen. Some of the plasma cells will become **memory cells,** which remain the bloodstream and "remember" the infection; they can be called on in the future to secrete antibodies in response to an infection.

Antibodies fight infections in several ways. First, by binding to antigens, they may physically block the infectious organism from infecting other cells, rendering it harmless. Second, they can flag a pathogen to be digested by phagocytes. Finally, they can mark the pathogen so that complement proteins will bind to it and destroy it. In other words, antibodies that bind to antigens are like giant red flags, alerting the body's immune system to take action.

Each B cell produces only one type of antibody. And antibodies produced against one antigen are usually ineffective against any other antigen—they are highly specific. But because the body can produce more than a billion unique B cells, there are that many different kinds of potential antibodies—a very large war chest indeed (**INFOGRAPHIC 31.7**).

▶**MEMORY CELL**
A long-lived B or T cell that is produced during an immune response and that can "remember" the pathogen.

INFOGRAPHIC 31.7 HUMORAL IMMUNITY: B CELLS AND ANTIBODY PRODUCTION

➡ B cells and helper T cells recognize antigens on pathogens that are free-floating between cells. When a B cell and a helper T cell recognize the same antigen, the T cell can activate the B cell to divide and become plasma cells. Plasma cells secrete antibodies that bind specifically to antigens on pathogens to inactivate and destroy them. Specific memory B cells are also produced and spring into action if the same pathogen is encountered again in the future.

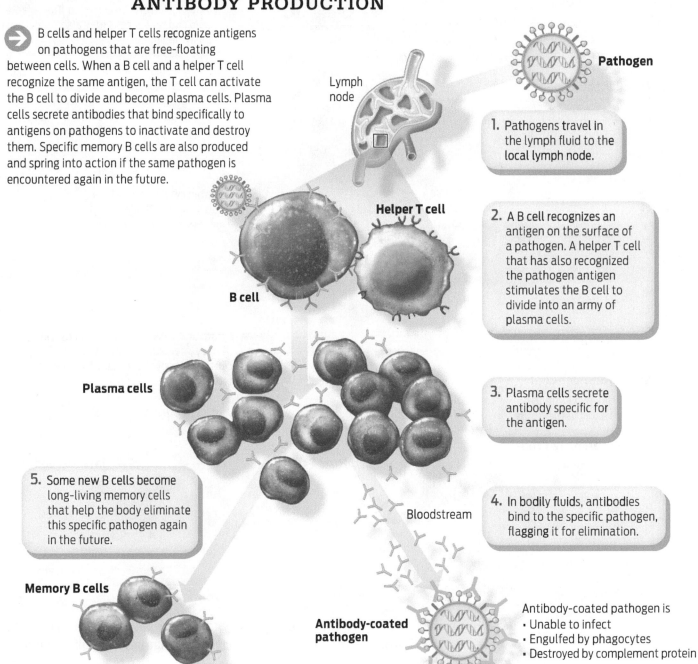

Pathogen

Lymph node

Helper T cell

B cell

Plasma cells

Memory B cells

Bloodstream

Antibody-coated pathogen

1. Pathogens travel in the lymph fluid to the local lymph node.

2. A B cell recognizes an antigen on the surface of a pathogen. A helper T cell that has also recognized the pathogen antigen stimulates the B cell to divide into an army of plasma cells.

3. Plasma cells secrete antibody specific for the antigen.

4. In bodily fluids, antibodies bind to the specific pathogen, flagging it for elimination.

5. Some new B cells become long-living memory cells that help the body eliminate this specific pathogen again in the future.

Antibody-coated pathogen is
· Unable to infect
· Engulfed by phagocytes
· Destroyed by complement protein

INFOGRAPHIC 31.8 CELL-MEDIATED IMMUNITY: CYTOTOXIC T CELLS KILL INFECTED CELLS

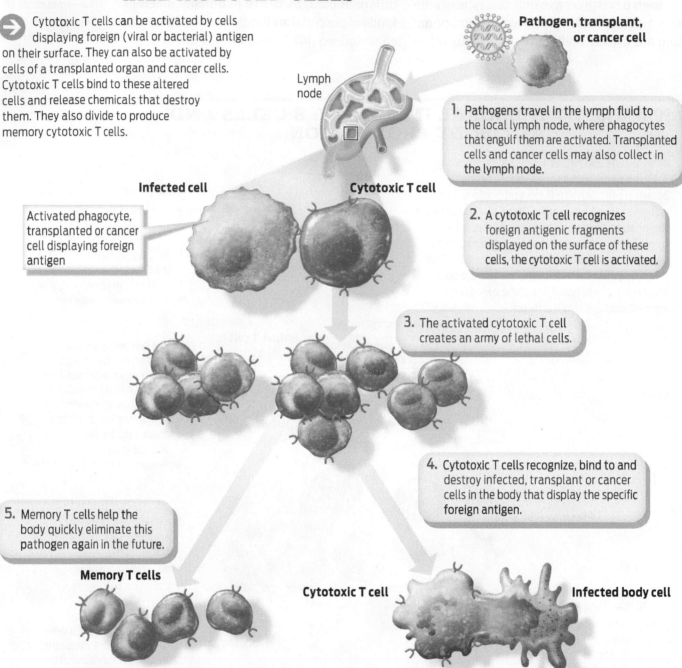

Cytotoxic T cells can be activated by cells displaying foreign (viral or bacterial) antigen on their surface. They can also be activated by cells of a transplanted organ and cancer cells. Cytotoxic T cells bind to these altered cells and release chemicals that destroy them. They also divide to produce memory cytotoxic T cells.

Pathogen, transplant, or cancer cell

Lymph node

1. Pathogens travel in the lymph fluid to the local lymph node, where phagocytes that engulf them are activated. Transplanted cells and cancer cells may also collect in the lymph node.

Infected cell

Cytotoxic T cell

Activated phagocyte, transplanted or cancer cell displaying foreign antigen

2. A cytotoxic T cell recognizes foreign antigenic fragments displayed on the surface of these cells, the cytotoxic T cell is activated.

3. The activated cytotoxic T cell creates an army of lethal cells.

4. Cytotoxic T cells recognize, bind to and destroy infected, transplant or cancer cells in the body that display the specific foreign antigen.

5. Memory T cells help the body quickly eliminate this pathogen again in the future.

Memory T cells

Cytotoxic T cell

Infected body cell

The second type of adaptive immunity, called **cell-mediated immunity,** targets infected or altered body cells. This response requires a type of T cell called a **cytotoxic T cell.** Cytotoxic T cells are activated by for- eign (viral or bacterial) antigens on the surface of phagocytes that have engulfed pathogens. Helper T cells can also help activate cytotoxic T cells. The activated cytotoxic T cells then pa- trol the body, on the lookout for infected cells displaying these distinguishing antigens. In addition to infected body cells, cytotoxic T cells also target foreign cells (from a transplanted organ or tissue, for example) and even cancer cells that the body recognizes as altered. The activated cytotoxic T cells bind to antigens on these target cells and release cytotoxic chemicals that cause the rogue cells to self-destruct (INFOGRAPHIC 31.8).

Note that while both humoral and cell-mediated immune responses destroy invading pathogens, there is a key difference between them: humoral immunity acts by releasing antibodies that bind to antigens on free-floating pathogens in lymph and blood; in a cell-mediated immune response, cytotoxic T cells bind to and destroy infected or altered cells in body tissues.

With its billions of potential lymphocytes, each adapted to recognize a particular antigen, the adaptive immune system enables our bodies to fight off countless numbers of pathogens. Occasionally, however, for reasons that are not entirely clear, the system runs amok and immune cells become active even against antigens that aren't harmful to us, or worse, they begin attacking healthy cells in the body.

When the immune system attacks harmless antigens from outside the body, like dust or certain types of food, an **allergy** is the result. Allergies are quite common, often accompanied by a runny nose, watery eyes, and sneezing—all triggered by histamine, the same chemical involved in inflammation. When the immune system attacks the body's own healthy cells, a more serious **autoimmune disease** can result. Multiple sclerosis, lupus, and rheumatoid arthritis are all autoimmune diseases, caused by a body at war with itself.

BUILDING A LINE OF DEFENSE

The advantage of adaptive immunity is that it learns to respond to new invaders and targets them specifically. The downside is that it takes a while for it to kick in. First-time exposure to a pathogen will almost certainly cause illness, because the adaptive response takes 7 to 10 days to develop. Over time an exposed individual will recover as T and B cells are activated and antibody levels increase. This initial slow response is the **primary response.** As B and T cells are churned out, some of them become memory cells. These memory cells remain in the bloodstream and "remember" the infection. The next time the same pathogen is encountered, memory B and T cells become active, dividing rapidly and leading to the production of very high levels of antibodies. They fight the specific pathogen so quickly that the illness usually doesn't occur a second time. This rapid reaction is called the **secondary response** (INFOGRAPHIC 31.9).

The secondary response is what confers immunity to a particular infection. It's also how **vaccines** work. Vaccines are essentially dead or weakened versions of a pathogen administered to a patient to elicit an immune response. The goal is to create a primary response in the body that's strong enough to create memory cells, yet weak enough so as not to cause disease symptoms. Thus, if the pathogen is subsequently encountered naturally, the secondary response is already primed. Vaccination is like being infected with a pathogen but not having the disease.

Similarly, people exposed to a pathogen that is similar to a pathogen with which they were previously infected may be partly protected from the disease caused by the new pathogen. Memory B and T cells may still respond, although only partially—in which case the illness may occur, but mildly.

Evidence suggests that such partial immunity may have helped some of those infected survive the 1918 flu pandemic. Statistical data from the time show that people over 65 accounted for the fewest influenza cases, suggesting that they might over the years have acquired immunity or partial immunity from earlier infection. Partial immunity might have helped these people fight off the virus before it dug deep into the lungs.

▶CELL-MEDIATED IMMUNITY
The type of adaptive immunity that rids the body of infected, cancerous, or foreign cells.

▶CYTOTOXIC CELL
A type of T cell that destroys infected, cancerous, or foreign cells.

▶ALLERGY
A misdirected immune response against environmental substances such as dust, pollen, and foods that causes discomfort in the form of physical symptoms.

▶AUTOIMMUNE DISEASE
A misdirected immune response in which the immune system attacks healthy cells.

▶PRIMARY RESPONSE
The adaptive response mounted the first time a particular antigen is encountered by the immune system.

▶SECONDARY RESPONSE
The rapid and strong response mounted when a particular antigen is encountered by the immune system subsequent to the first encounter.

▶VACCINE
A preparation of killed or weakened microorganisms or viruses that is administered to people or animals to generate an immune response.

INFOGRAPHIC 31.9 MEMORY CELLS MOUNT AN AGGRESSIVE SECONDARY RESPONSE

The adaptive immune system's primary humoral response is slow and produces low levels of antibodies. Upon subsequent exposures, memory cells produced during the primary response respond quickly and trigger B cells to produce high levels of antibodies. This secondary response is so rapid and effective that illness does not occur. The secondary response of cell-mediated immunity and cytotoxic T cells is also quick and effective.

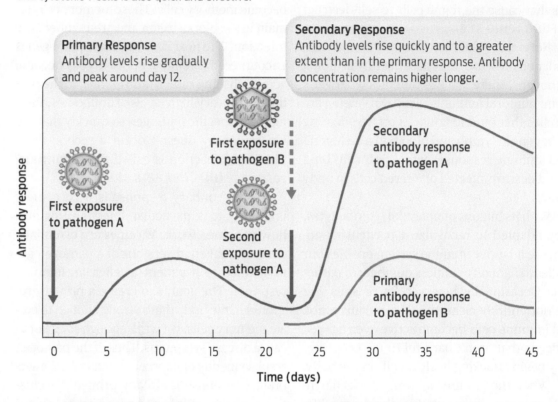

Primary Response
Antibody levels rise gradually and peak around day 12.

Secondary Response
Antibody levels rise quickly and to a greater extent than in the primary response. Antibody concentration remains higher longer.

First exposure to pathogen B

First exposure to pathogen A

Second exposure to pathogen A

Secondary antibody response to pathogen A

Primary antibody response to pathogen B

Antibody response

Time (days)

AN EVOLVING ENEMY

When Kawaoka and his colleagues discovered that the 1918 flu virus carried specific alleles of four genes that enabled the virus to replicate in the lower respiratory tract, thus making the disease so deadly, a piece of the 75-year-old mystery was solved. But where did these alleles come from?

New influenza viruses are constantly being produced by two mechanisms: mutation and gene swapping. Because influenza viruses replicate their genetic material so rapidly and don't "proofread" the replicated copies, mistakes often occur, leading to mutations. **Antigenic drift** is the gradual accumulation of mutations that causes small changes in the antigens on the virus surface. Antigenic drift explains why there can be different types, or strains, of influenza circulating at the same time.

Two important antigens on the influenza virus are hemagglutinin (which binds to receptors on host cells and enables the virus to enter host cells) and neuraminidase (which helps new viruses exit host cells). The host immune system mounts an adaptive response specifically to these two antigens. When there is a change in hemagglutinin or neuraminidase or both, the memory cells no longer recognize them. These new antigens prompt a new and slow primary immune response.

An influenza virus strain can also swap genes with other strains of influenza, including strains that infect animals such as birds and pigs. While every influenza strain contains the same set of genes, the particular alleles of these genes that are present differ from strain to strain. The exchange of alleles between two strains that have infected the same cell does not simply create

▶**ANTIGENIC DRIFT**
Changes in viral antigens caused by genetic mutation during normal viral replication.

a small change in viral gene sequence: it introduces an entirely new allele, and therefore an entirely new antigenic protein. This process, called **antigenic shift,** is responsible for pandemic outbreaks of flu. Because people have no existing immunity to protect against infection by emerging strains of flu, these strains can spread rapidly throughout the human population. The severity of the 1918 flu is believed to have resulted in part from antigenic shift. Antigenic shift is also thought to have played a role in the emergence of avian flu in the 1990s and swine flu (H1N1) in 2009, both of which originated in animals.

Together, antigenic drift and antigenic shift create an increasing variety of strains over time until one of the variants is able to infect human cells so efficiently that it sweeps through the population and causes a pandemic. These strains are named by the type of H (hemagglutinin) and N (neuraminidase) proteins they carry, with some antigenic combinations more deadly than others (**INFOGRAPHIC 31.10**).

Tracking the strains as they move through the population helps public health officials predict how severe a coming flu season will be. But since viruses can mutate every time

▶**ANTIGENIC SHIFT**
Changes in antigens that occur when viruses exchange genetic material with other strains.

INFOGRAPHIC 31.10 ANTIGENIC DRIFT AND SHIFT CREATE NEW INFLUENZA STRAINS

➡ Mutation and gene exchange are two mechanisms by which influenza viruses can change. Mutations that accumulate gradually can cause variations in surface antigen. This process is called antigenic drift. Different strains can also swap genes and cause surface antigens to change more dramatically. This is called antigenic shift. Drift is responsible for annual seasonal variation in influenza; shift is responsible for dramatic pandemics.

Antigenic Drift
Gradual change
Caused by point mutations that occur when the virus replicates

Antigenic Shift
Rapid change
Caused by gene exchange between two different viruses that simultaneously infect the same cell

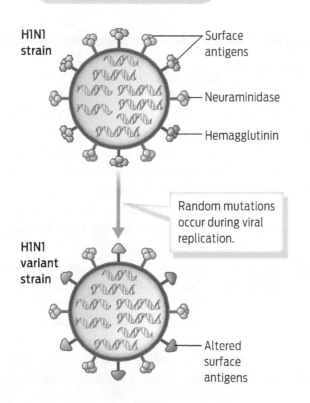

H1N1 strain

Surface antigens

Neuraminidase

Hemagglutinin

Random mutations occur during viral replication.

H1N1 variant strain

Altered surface antigens

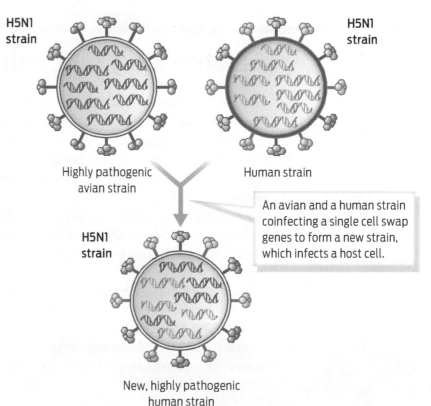

H5N1 strain

H5N1 strain

Highly pathogenic avian strain

Human strain

H5N1 strain

An avian and a human strain coinfecting a single cell swap genes to form a new strain, which infects a host cell.

New, highly pathogenic human strain

they reproduce, they continue to evolve during epidemics. This unpredictability is what makes influenza so frightening: an apparently mild outbreak can suddenly become deadly.

PREVENTING PANDEMICS

Because influenza viruses can gene swap so easily, public health officials closely monitor the types of influenza viruses that infect animals. Viruses that infect animals may accumulate mutations or pick up genes that enable them to infect human cells. Severe acute respiratory syndrome (SARS), for example, is a respiratory illness caused by a virus. In 2003, more than 8,000 people worldwide became sick with SARS, according to the World Health Organization (WHO), and 774 died. Experts think the virus jumped from animals to people, although they have not yet determined the identity of the animal host.

They do, however, suspect that the source was a species of bird. Birds are a primary source of potentially pandemic viruses. Wild birds are routinely infected with viruses that live in their gastrointestinal tracts but cause little or no disease in the bird. Sometimes, however, bird flu viruses can jump from wild birds to poultry or other farm animals and become lethal. And since many human populations maintain close contact with poultry, pigs, and other domestic animals, the risk of an animal flu infecting humans is high.

In the 1990s, a lethal strain of the avian flu virus (H5N1) crossed over into humans from domesticated birds, but it was not easily transferred from person to person. However, scientists continue to monitor this bird strain, not only because it can be deadly to the commercial poultry business, but also because the 1918 flu carried a variation of the H5N1 bird flu genes. In fact, most human influenza pandemics have been caused by flu viruses that carry variations of bird flu genes (TABLE 31.1).

To help people ward off flu infection, public health officials offer flu shots every year at the beginning of flu season, which typically runs from October to February in the northern hemisphere. A flu shot is a vaccine: a weakened version or part of a flu virus is injected into the body in the hope of generating a primary immune response and memory cells against that

FOR COMPARISON Defending Oneself Is a Fact of Life

The adaptive immune system that protects humans and other vertebrates from infection is a relatively recent adaptation. Evidence suggests that adaptive immunity first arose in fishes, some 500 million years ago. The innate immune system, by contrast, is much older and is shared by creatures as diverse as sea sponges, fruit flies, and humans. In fact, all organisms have evolved ways to defend themselves from threats, both large and small.

Some organisms are able to repel invaders and predators by physical and chemical means. The invertebrate animals called sea sponges, for example, combine physical defenses with poisonous secretions to ward off danger. An internal support system made of collagen fibers and stiffened with hard spikes called crystalline spicules acts as a physical defense against predators like fish, who may want to take a bite. In order to deter such predators from getting close enough to nibble, sponges also secrete chemicals that are toxic to many potential predators.

Many bacteria are able to protect themselves from a class of viruses known as bacteriophage–"phage" for short. Phage infection causes a bacterial cell to burst, or lyse–a life-ending event for that bacterium. Bacteria defend themselves against phage by producing enzymes that act like scissors, cutting up the DNA that phage inject strain of virus. However, since influenza viruses mutate so frequently, a yearly flu shot may not protect us from getting the flu the following year. In fact, a flu shot may not confer protection for the duration of a single season. Scientists create a new vaccine each year by tracking which strains of influenza are circulating worldwide and studying the antigens on their surfaces. But their predictions can be wrong. If public health officials decide to vaccinate against a strain of influenza with one variant of hemagglutinin antigen but a strain with a different variant of the antigen strikes, those who are vaccinated may become ill anyway.

This is one reason why public health officials were so concerned about the 2009 H1N1 influenza virus, or swine flu. H1N1 was a new virus strain that was first detected in people in the United States in April 2009 and became a worldwide pandemic in 2010, the first pandemic in 40 years. It was called swine flu because many of its genes were similar to those in influenza viruses that normally occur in North American pigs. Further study has shown, however, that this new virus is actually a mix of genes from flu viruses that normally circulate in European and Asian pigs, birds, and humans.

There are other reasons for concern. Whereas 90% of deaths from seasonal flu occur in people over 65 years old, H1N1 causes most severe disease in people under 25. The 1918 virus also preferentially affected young people, suggesting that swine flu, like Spanish flu, has the potential to become a more virulent infection that can kill within hours.

TABLE 31.1 KNOWN FLU PANDEMICS

NAME	DATE	VIRUS ANTIGEN SUBTYPE	NUMBER OF DEATHS
Asiatic (Russian) flu	1889–1890	H2N2	1 million
1918 (Spanish) flu	1918–1919	H1N1	20–100 million
Asian flu	1957–1958	H2N2	1–1.5 million
Hong Kong flu	1968–1969	H3N2	0.75–1 million
Swine flu	2009–2010	H1N1	150,000–575,000

Sea sponges employ physical and chemical defenses

Crystalline spicules are a painful defense against fish that try to take a bite.

Toxins released by sponges keep predators from getting too close.

Bacterial cells that try to infect a sponge are destroyed by molecules that punch holes in their cell walls.

By the end of the 2009-2010 flu season, a total of 2,125 deaths in the United States were confirmed as influenza deaths. A total of 344 children died, four times higher than in the previous four years.

In 2009, Kawaoka reported in the journal *Nature* that the H1N1 virus could replicate in the lungs of mice, ferrets, and monkeys much better than a seasonal flu virus could. That finding "suggests the virus can cause serious respiratory illness in many people," Kawaoka stated in the paper. He also reported that some people born before 1918 have antibodies to the H1N1 virus, suggesting it shares some antigens in common with the 1918 virus.

Today, scientists have better surveillance, better communication, and can make more-informed decisions about whether and how to quarantine groups to prevent a dangerous virus from spreading—even when extensive global travel might enable a new pandemic to spread rapidly.

into the bacterial cell. By destroying the phage DNA, bacteria prevent phage from replicating. A bacterium's own DNA is protected from these powerful molecular scissors by a mechanism that chemically modifies its DNA, ensuring that it won't be cut up by these enzymes.

Many invertebrate animals have an innate immune system armed with specialized cells

Bacteria use enzymes to defend against infection

Bacterium does not produce the enzyme

Bacterium does produce the enzyme

Bacteriophage

Infection

Infection

Enzyme

Bacterial cell

Phage

Bacterial DNA

Bacterium

Phage-infected bacterium

1. Phage attach to the bacterial host cell and inject nucleic acid.

1. Phage nucleic acid is cut by the enzyme. Bacterial DNA is not affected.

2. Phage make more nucleic acid and phage proteins.

2. The bacterial cell is not infected.

3. Newly assembled phage burst from the cell as it lyses.

The 1918 flu pandemic taught scientists a lot about how viruses can become lethal in a matter of days and demonstrated that developing vaccines can be like taking aim at a moving target. But by finding influenza genes common to strains that mutate less frequently than others, researchers may be able to develop a universal flu vaccine that would be effective over a number of years.

There are now at least two universal flu vaccines that have been shown to provide protection against lethal strains of influenza, including H5N1 and H1N1, in laboratory animals. Some companies are scaling up their vaccine manufacturing efforts, while others are boosting research and development of new universal vaccines. Some researchers are also working on different antiviral drugs to help reduce the impact and spread of the influenza virus. The hope is that such measures will suppress the next big flu outbreak, whenever and wherever it happens. ■

that attack or neutralize invaders. Sea stars are a good example. It has long been known that sea stars very rarely develop bacterial infection. It turns out that their ability to resist bacterial pathogens comes from a class of cells called amoebocytes, which act much like the macrophages of the vertebrate innate immune system. Amoebocytes circulate in the body cavity fluid of sea stars. When bacteria are injected into the body cavity of sea stars, the amoebocytes actively engulf and destroy the bacteria within 10 minutes. This rapid response helps sea stars remain infection free.

Fruit flies also have several innate defense mechanisms. Phagocytic cells called plasmatocytes act like macrophages, ingesting and destroying foreign cells by phagocytocis. Fruit flies also have cells called lamellocytes, which can coat and essentially wall off foreign objects, such as the eggs of parasitic wasps, which are too big to be phagocytosed. This walling off, or encapsulation, helps destroy the foreign object. Humans carry out a similar process in response to lung infections, such as those caused by tuberculosis. In humans, immune cells surround and wall off the pathogen in a structure called a granuloma.

That many organisms from many different phyla share similar kinds of immune responses is evidence that these defensive strategies evolved early in the history of life and were a crucial part of the organisms' survival and reproduction.

Sea stars have innate cellular defenses

Bacterium   Amoebocyte

Phagocytosis
Amoebocytes in the body cavity fluid ingest and destroy invading bacteria.

Fruit flies have multiple physical, chemical, and cellular defenses

External Barriers
Fruit flies have a hard exoskeleton and specialized cells that line the airways and digestive system.

Encapsulation
Lamellocytes (green) surround and wall off infecting parasites (black) inside living *Drosophila* larvae.

Gr
Pl

Phagocytosis
Phagocytic cells known as plasmatocytes (labeled **p**) ingest and destroy invaders.

CHAPTER 31 Summary

- The immune system defends the body against infection by pathogens.

- Pathogens are infectious agents that cause an immune response; they include certain bacteria, viruses, fungi, and parasites.

- Viruses are acellular; they consist of a genome of nucleic acid (DNA or RNA) contained within a protein shell.

- The immune system has two main components: innate immunity and adaptive immunity. Innate immunity is the first defense against invaders; adaptive immunity comes into play once innate defenses are breached.

- The innate immune system includes defenses with which we are born, and which are always active: they include barriers such as skin and mucous membranes; antimicrobial chemicals in tears, saliva, and other secretions; and phagocytic cells that engulf and digest pathogens.

- The inflammatory response, part of the innate immune system, is triggered by tissue infection or injury. During an inflammatory response, blood vessels swell and leak, marshaling phagocytic cells and protective molecules to the area to contain the infection.

- Adaptive immunity is conferred by specialized lymphocytes called B and T cells.

- B cells produce antibodies that specifically recognize antigens on pathogens and mark the pathogens for destruction.

- T cells destroy infected, foreign, and cancer cells and stimulate B cells to produce antibodies.

- The immune response can go awry, causing allergies, which are responses to intrinsically harmless substances (for example, pollen), and autoimmune conditions that result when the immune system mounts a response against the body's own cells and tissues.

- At the first exposure to a particular pathogen, a primary response is generated that takes time to become fully effective; the primary response also produces memory cells. Memory cells remain in the body and become active at the time of a subsequent exposure to the same pathogen, producing a more rapid and vigorous secondary response that fights the pathogen and usually prevents the associated illness.

- Vaccination elicits a primary response. Memory cells produced during the primary response protect against illness following subsequent exposure to the actual pathogen.

- The adaptive response is highly specific for particular pathogens. If the pathogen changes, by antigenic shift or drift, the body must mount a separate response for each strain of the pathogen.

- All organisms have ways of defending themselves against infection. The innate immune system is evolutionarily ancient.

MORE TO EXPLORE

- National Institute of Allergy and Infectious Diseases: Immune System www.niaid.nih.gov/topics/immunesystem

- PBS, *American Experience*, Influenza 1918. www.pbs.org/wgbh/americanexperience/films/influenza/

- Watanabe, T. (2009) Viral RNA polymerase complex promotes optimal growth of 1918 virus in the lower respiratory tract of ferrets. *Proceedings of the National Academy of Sciences* 106(2):587–591. www.pnas.org/content/106/2/588

- Taubenberger, J., et al. (2007) Discovery and characterization of the 1918 pandemic influenza virus in historical context. *Antiviral Therapy* 12(4 Pt B):581–91. www.ncbi.nlm.nih.gov/pmc/articles/PMC2391305/

- Barry, J. M. (2005) *The Great Influenza: The Story of the Deadliest Pandemic in History*. New York: Penguin Books.

- Crosby, A. (2003) *America's Forgotten Pandemic: The Influenza of 1918*. Cambridge, UK: Cambridge University Press.

CHAPTER 31 Test Your Knowledge

DRIVING QUESTION 1

What is the structure of a virus, and how do viruses cause disease?

By answering the questions below and studying Infographics 31.1 and 31.5, you should be able to generate an answer for the broader Driving Question above.

KNOW IT

1 Which of the following is found in all viruses?

  a. DNA

  b. RNA

  c. a membranous envelope

  d. a protein shell

  e. a cell membrane

2 Explain how viruses replicate within humans.

3 Why does poliovirus cause long-lasting damage, whereas those infected by influenza virus typically make a full recovery?

USE IT

4 Both viruses and bacteria can be human pathogens. Describe some key differences between them. (Hint: You may wish to refer to Chapter 14.)

5 Why do poliovirus, influenza virus, and HIV infections cause different symptoms?

DRIVING QUESTION 2

What is innate immunity?

By answering the questions below and studying Infographics 31.3 and 31.4, you should be able to generate an answer for the broader Driving Question above.

KNOW IT

6 Name three components of the innate immune system. For each, provide a brief description of how it offers protection.

7 What do macrophages and neutrophils have in common?

USE IT

8 From what you know about innate immunity, would you predict different or identical innate responses to infections from *E. coli* (a bacterium) and *S. aureus* (another bacterium)? Explain your answer.

9 Neutropenia is a deficiency of neutrophils. Would you expect someone with neutropenia to be able to mount an effective inflammatory response? Explain your answer.

10 Why might someone taking anti-inflammatory drugs be more susceptible than others to bacterial infections?

DRIVING QUESTION 3

What is adaptive immunity, and how does vaccination rely on adaptive immunity?

By answering the questions below and studying Infographics 31.3, 31.6, 31.7, 31.8, and 31.9, you should be able to generate an answer for the broader Driving Question above.

KNOW IT

11 Compare and contrast the features of innate and adaptive immunity.

12 B cells, plasma cells, and antibodies are all related. Describe this relationship, using words, a diagram, or both.

USE IT

13 Anti–hepatitis C antibodies present in a patient's blood indicate

  a. that the patient is mounting an innate response.

  b. that the patient has been exposed to HIV.

  c. that the patient has been exposed to hepatitis C within the last 24 hours.

  d. that the patient has been exposed to hepatitis C at least 2 weeks ago.

  e. that the patient has hepatitis.

14 Vaccination against a particular pathogen stimulates what type of response?

  a. innate   **d.** autoimmune

  b. primary   **e.** b and c

  c. secondary

15 Will someone who has been exposed to seasonal influenza in the past

  a. have memory B cells?

  b. still be at risk for seasonal influenza next year? Why?

  c. still be at risk for H1N1 (swine flu)? Why?

16 *Staphylococcus aureus* can cause a bacterial skin infection that can become very serious.

  a. Why does the body exhibit innate and adaptive responses to *Staphylococcus aureus* but not to its own skin cells?

  b. Will the innate response to *Staphylococcus aureus* be equally effective against *Streptococcus pyogenes*, another bacterium that can cause skin infections? Explain your answer.

  c. Will the adaptive response to *Staphylococcus aureus* be equally effective against *Streptococcus pyogenes*? Explain your answer.

17 HIV is a virus that infects and eventually destroys helper T cells. Why do people with AIDS (that is, with advanced HIV infections) often die from infections by other pathogens?

MINI CASE

18 In 2008, there was an outbreak of measles in San Diego. The first patient was a 7-year-old unvaccinated boy who had returned home from a family trip to Switzerland. He began to develop a cough, sore throat, and fever, but continued to attend school. He was taken to his pediatrician when he developed a rash, and then was sent to the emergency room because of a very high fever. Blood tests revealed antimeasles antibodies. Eleven other children ended up developing measles; none had been vaccinated. The other cases were the 2 siblings of the first patient, 4 children in his school and 5 children who were in the pediatrician's office at the same time as the patient. Of the 11 additional cases, 3 children were less than 1 year old. (Data are from Outbreak of measles—San Diego, California, January—February 2008. (2008) *Morbidity and Mortality Weekly Report* 57(8): 203–206.)

  a. Why was the presence of antimeasles antibodies in the first case an important finding?

  b. What can you infer about how easily measles spreads?

  c. What does this case suggest about the importance of measles vaccinations?

  d. Were all the unvaccinated children necessarily behind on their vaccination schedule? (Hint: Look up the recommended measles vaccination schedule on the CDC website www.cdc.gov.)

BRING IT HOME

19 Almost 10% of the children in the school attended by the original patient described in the Mini Case were unvaccinated because their parents had filed Personal Belief Exemptions stating that they did not want to vaccinate their children. What is your local school district or state policy on vaccinations for enrolled students? This information is typically available at the school district website or the state's Department of Health website. From the policies in place (and whether they permit any exemptions), do you think it is possible that a measles outbreak could occur in a local school?

DRIVING QUESTION 4

What are specific features of influenza virus that allow it to cause worldwide outbreaks?

By answering the questions below and studying Infographics 31.5 and 31.10, you should be able to generate an answer for the broader Driving Question above.

KNOW IT

20 What is the difference between antigenic shift and antigenic drift?

21 A strain of influenza can infect and replicate in birds without causing them disease. The same strain can be transmitted from birds to humans, causing severe illness in humans. What else would this strain need to be able to do in order to become pandemic?

USE IT

22 What processes are responsible for the emergence of pandemic influenza strains such as H1N1 swine flu? Explain how these strains can spread so successfully through the human population.

23 In 2013, the CDC advised people traveling to China to avoid visiting poultry farms or bird markets. Why did the CDC make this recommendation?

24 Why are those with influenza infections susceptible to bacterial pneumonia?

INTERPRETING DATA

25 The table below gives the approximate cumulative rate of influenza hospitalizations in the United States in newborns to 4-year-olds and in people 65 years old and older for the 2009–2010 flu season and the 2012–2013 flu season. As the flu season runs from the fall of one year through the spring of the next, the data are reported by week (36th week of the first year to 12th week of the next).

a. Graph the cumulative incidence rates by week for each age group in each flu season. In plotting the data on the graph, note that the rates in the two seasons are reported differently—per 10,000 and per 100,000—and be sure your graph accounts for this.

b. Examine each season and compare and contrast the age-specific hospitalization rates in the two seasons.

c. Which of the two flu seasons is more typical? Why?

Age Group and Season	Cumulative Incidence Rate							
	Week 36	Week 40	Week 44	Week 48	Week 52	Week 4	Week 8	Week 12
Birth to 4 years 2009–2010 (per 10,000)	0	2	8	8.5	9	9.5	10	10.5
65 years and older 2009–2010 (per 10,000)	0	0	1.5	1.5	1.5	1.5	1.5	1.5
Birth to 4 years 2012–2013 (per 100,000)	0	0	0	5	20	40	45	50
65 years and older 2012–2013 (per 10,000)	0	0	0	5	50	125	150	160

SOURCE: Update: influenza activity—United States, 2009–10 season. (2010) *Morbidity and Mortality Weekly Report 59*(29):901–908; Influenza activity—United States, 2012–13 season and composition of the 2013–14 influenza vaccine. (2013) *Morbidity and Mortality Weekly Report 62*(63):473–479.

Q&A Plants

Exploding seeds, carnivorous flowers, and other colorful adaptations of the plant world

PLANT STRUCTURE

 How do plants stand up?

ANSWER: From tiny 3-inch-tall crocus flowers to massive redwood trees standing 300 feet, plants come in a variety of shapes and sizes. Yet almost all of them share the same basic design: a belowground **root system** for absorbing water and nutrients and an aboveground **shoot system** made up of stems and leaves. This body arrangement works well for plants, allowing them to simultaneously reach for the sky and anchor themselves safely in the soil—all without the help of bones, tendons, or muscles. In the absence of such a musculoskeletal system, what keeps a plant from flopping over?

Bugs are drawn to what look like dewdrops on this carnivorous plant, the Australian sundew. The "dewdrops" are actually sticky mucus bubbles that trap insects the plant will consume.

▶**ROOT SYSTEM**
The belowground parts of a plant, which anchor it and absorb water and nutrients.

▶**SHOOT SYSTEM**
The aboveground parts of a plant, including the stem and photosynthetic leaves.

▸▸ **DRIVING QUESTIONS**

1. How is a plant body structured, and how do plants obtain water and grow?

2. How do plants obtain nutrients?

3. How do plants reproduce, respond to stimuli, and protect themselves?

▶**CELL WALL**
A rigid layer surrounding the cell membrane of some cells, providing shape and structure. In plant cells, the cell wall is made of cellulose, a complex carbohydrate.

▶**CHLOROPLAST**
The organelle in plant cells in which photosynthesis takes place.

Like animals and other eukaryotes, plants are made of cells packed with organelles, including a nucleus, endoplasmic reticulum, and mitochondria. Plant cells contain a few plant-specific parts as well, including a supportive **cell wall,** made of a complex carbohydrate called cellulose; **chloroplasts,** the sites of photosynthesis; and a **central vacuole,** essentially a large water balloon occupying the center of the cell. When filled with water, vacuoles create **turgor pressure** against the cell wall, keeping a plant body rigid and upright. When the vacuoles are less than full, turgor pressure is reduced, and the plant wilts.

Cell walls contribute to a plant's stiffness in another way. Plant cells are packed tightly together, much like bricks in a wall; carbohydrate molecules associated with the cell walls act like glue, helping adjacent cells stick together. With many cells held together in this way, plant tissues are exceptionally strong, which is why they make such durable ropes and fabrics.

Some plant tissues, such as those that make up the stem, are made of cells with two cell walls: an outer cell wall containing cellulose, and a second, inner cell wall containing cellulose and **lignin.** Lignin is a hard, durable material, which lends added support and strength to plant tissues; it is what makes

INFOGRAPHIC 32.1 ALL PLANTS SHARE THE SAME BASIC DESIGN

Trees and other plants are anchored in the soil by a root system and extend their shoot system, including stems and leaves, skyward. A variety of features, including rigid cell walls and a central vacuole, allow the stems to remain upright.

Plant Cell

Chloroplast
Photosynthetic organelle in leaf cells

Nucleus
Compartment for DNA

Cell Wall
Rigid cellulose-based structure surrounding the cell membrane

Central Vacuole
Fills with water, creating turgor pressure against the rigid cell wall and thus keeps a plant from wilting

Leaves
The site of photosynthesis

Stem
Rigid, upright support for the plant

Roots
Anchor the plant and absorb water and minerals from the soil

Taproot
Extends deep into the ground to anchor plants and reach underground sources of water

Root Hairs
Extensions of root cells that increase the surface area for enhanced absorption of water and minerals from the soil

> The tallest plants on Earth are the California redwoods (*Sequoia sempervirens*), which can reach heights of more than 350 feet.

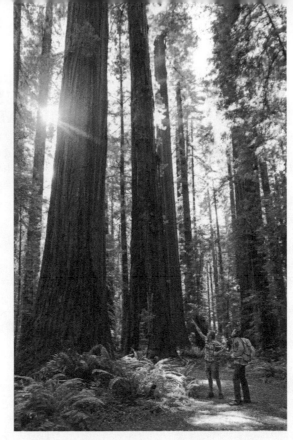

them "woody." A plant with a thick, woody stem (or trunk, in the case of a tree) can grow tall and still not topple over. The tallest plants on Earth are the California redwoods (*Sequoia sempervirens*), which can reach heights of more than 350 feet.

Plants would not be able to grow so tall were it not for the extensive system of roots that anchors them in the ground. Some plants have root systems that are much deeper than the plants are tall, and some plants' roots extend out horizontally to about three times the branch spread. A vast number of tiny **root hairs** increase the surface area of the root, enhancing its ability to absorb water and nutrients from the soil. When root hairs are included, a plant's total root length can be hundreds of miles long. In addition, some plants produce very long **taproots**–deep-reaching roots that help the plant reach the underground water table. Taproots also store water and carbohydrates for the plant, making them appealing and nutritious to animals. Carrots and turnips are actually taproots (**INFOGRAPHIC 32.1**).

 Why don't plants bleed?

ANSWER: Like animals, plants have a **vascular system** for transporting fluids. Instead of blood, however, a plant's vascular system transports nutrients like water and sugar throughout the plant body. Plants need water to live and grow (you may have learned this the hard way if you've ever killed a plant with neglect). Among other things, water is crucial for photosynthesis in leaves. But you don't water a plant's leaves; you water the roots by pouring water in the soil. How does water get from the roots to the rest of the plant?

A plant's water-carrying tissue is known as **xylem** (pronounced ZYE-lum). Xylem tissue is made of cells arranged into long, stiff tubes; the cells have holes in each end and are stacked one on top of the next. Water moves up from the roots through these tubes to the aboveground stems, and eventually into the leaves, where it is used during photosynthesis to make sugar. Plant cells use the newly synthesized sugar as food, so the sugar must be transported out of the leaves and back down through the plant. Another series of tubes, known as **phloem** (pronounced FLO-um; think "f" for "food"), transports sugar through the plant. Phloem supports two-way transport: sugar moves down to the roots, where some of it is stored; later, sugar moves up to the shoot system, where it provides nutrition and energy for growing fruit, buds, and leaves.

▶CENTRAL VACUOLE
A fluid-filled compartment in plant cells that contributes to cell rigidity by exerting turgor pressure against the cell wall.

▶TURGOR PRESSURE
The pressure exerted by the water-filled central vacuole against the plant cell wall, giving a stem its rigidity.

▶LIGNIN
A stiff strengthening agent found in secondary cell walls of plants.

▶ROOT HAIRS
Tiny extensions of root cells that increase the surface area of roots and enhance their ability to absorb water and nutrients.

▶TAPROOT
A long, straight root produced by some plants that stores water and carbohydrates.

▶VASCULAR SYSTEM
Tube-shaped vessels and tissues that transport nutrients throughout an organism's body.

▶XYLEM
Plant vascular tissue that transports water from the roots to the shoots

▶PHLOEM
Plant vascular tissue that transports sugars throughout the plant.

INFOGRAPHIC 32.2 PLANTS TRANSPORT WATER AND SUGAR THROUGH THEIR VASCULAR SYSTEM

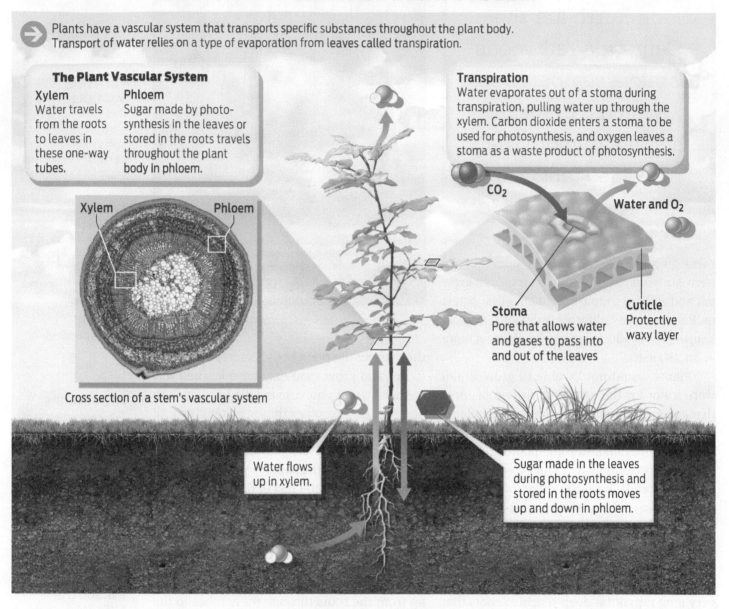

Plants have a vascular system that transports specific substances throughout the plant body. Transport of water relies on a type of evaporation from leaves called transpiration.

The Plant Vascular System

Xylem
Water travels from the roots to leaves in these one-way tubes.

Phloem
Sugar made by photosynthesis in the leaves or stored in the roots travels throughout the plant body in phloem.

Xylem   Phloem

Cross section of a stem's vascular system

Transpiration
Water evaporates out of a stoma during transpiration, pulling water up through the xylem. Carbon dioxide enters a stoma to be used for photosynthesis, and oxygen leaves a stoma as a waste product of photosynthesis.

CO_2

Water and O_2

Stoma
Pore that allows water and gases to pass into and out of the leaves

Cuticle
Protective waxy layer

Water flows up in xylem.

Sugar made in the leaves during photosynthesis and stored in the roots moves up and down in phloem.

▶**TRANSPIRATION**
The loss of water from plants by evaporation, which powers the transport of water and nutrients through a plant's vascular system.

Moving water and other nutrients up to the top of a 300-foot tree against the force of gravity is no easy task—hundreds of pounds of water must be lifted up through what is essentially a long water pipe. And unlike animals, which have a heart to pump blood along, plants have no such mechanical help. Instead, plants rely on evaporation of water from the leaves to draw water up the plant, a process called **transpiration.** Because water is a polar molecule that can form hydrogen bonds with other water mole-cules, it has great cohesive strength (see Chapter 2). As water evaporates from leaves into the air, water in the xylem is pulled up to replace it. The cohesive strength of water is enough to coun-teract the force of gravity and pull water up to astonishing heights.

For a plant, transpiration is life sustaining: it is the force that carries water through the plant. But losing too much water can be dan-gerous. On a hot, dry, or windy day, a large tree can lose hundreds of gallons of water vapor from its leaves. To control the amount of water lost by transpiration, a plant's leaves are coated with a waxy **cuticle** that functions somewhat like a rubber suit, sealing in moisture. At regular intervals, the cuticle is punctuated by pores, called **stomata,** which open and close. When the stomata are open, water vapor leaves freely and other gases enter and exit—specifically, carbon dioxide and oxygen. When the stomata are closed, water and gases are sealed in. Many plants keep their stomata open during the day, letting in carbon dioxide for photosynthesis. At night, they close the stomata, conserving water (**INFOGRAPHIC 32.2**).

Eventually, if a plant loses too much water, it will die. During an extreme drought, if you put a stethoscope to a tree and listen closely, you may hear clicks. This is the sound of the water column in the xylem breaking because there is no water in the soil to replace the water being lost by transpiration.

The presence of specialized tissues for transporting water is what distinguishes plants from their water-dwelling ancestors, the algae. The evolution of this vascular tissue is what allowed plants to colonize nearly every part of the land, from valley to mountaintop. A few primitive land plants, such as mosses and liverworts, also lack true roots and shoots containing vascular tissue. Without specialized tubes for transporting water, these nonvascular plants are limited to environments that are saturated with water, where they grow close to the ground in squat, spongy mats.

PLANT GROWTH

 What are tree rings?

ANSWER: Plants experience two kinds of growth: primary and secondary. Primary growth is growth in length—a tree getting taller. Trees and other plants also grow wider as well—this is secondary growth. Tree trunks increase in diameter because they add a new layer of xylem tissue—otherwise known as **wood**—each year. In temperate regions—as in most of the United

> ## On a hot, dry, or windy day, a large tree can lose hundreds of gallons of water vapor from its leaves.

States—xylem growth is dormant in the winter. In the spring, xylem growth starts again. The first xylem cells to grow in the spring are usually larger in diameter and thinner-walled than xylem cells produced later in the summer. The boundary between the smaller xylem cells from one summer and the larger xylem cells from the next spring appears as a ring in the cross section of a tree trunk. You can determine the age of a tree by counting the number of rings. Tree rings vary in width from year to year because differences in temperature and rainfall affect the amount of xylem tissue that is produced in any one season.

Each dark ring represents the end of one year of growth. This tree was 37 years old.

▶**CUTICLE**
The waxy coating on leaves and stems that prevents water loss.

▶**STOMATA (SINGULAR: STOMA)**
Pores on leaves that permit the exchange of oxygen and carbon dioxide with the air and allow water loss.

▶**WOOD**
Secondary xylem tissue found in the stem of a plant.

PLANT NUTRITION

Q What do plants eat?

ANSWER: This is a trick question, because plants don't "eat" in the same sense that animals do. Plants are autotrophs, making carbohydrates from inorganic CO_2 by photosynthesis. As discussed in Chapter 5, photosynthesis can be represented by the following equation:

Sunlight + Water + Carbon dioxide → Oxygen + Glucose

Give a plant some sunshine, carbon dioxide, and water–plus a few soil nutrients–and you will have a happy plant, capable of feeding and nourishing itself. From the air, plant leaves absorb carbon dioxide and sunlight, which they convert into sugar through photosynthesis. Sugar is a source of energy and building blocks for the plant. From the soil, plant roots absorb both water for photosynthesis and other nutrients–nitrogen, phosphorus, and sulfur–that are required to make plant proteins and other molecules.

Although nitrogen gas is plentiful in air pockets in the soil, it exists there in a form

INFOGRAPHIC 32.3 PLANTS OBTAIN NUTRIENTS IN A VARIETY OF WAYS

→ Plants obtain carbon from atmospheric CO_2 during photosynthesis. They obtain minerals and nutrients such as nitrogen from the soil, or sometimes through other means.

Air
CO_2 gas in the atmosphere provides the carbon and sunlight provides the energy to build organic molecules during photosynthesis.

Sunlight

Carbon dioxide
Gas from the air enters stomata in the leaves.

Glucose
Organic food is made during photosynthesis for plant energy and building blocks.

Photosynthesis:
Sunlight + Water + CO_2 → Oxygen + Glucose

Soil
Water and inorganic nutrients and minerals for plant growth enter plant roots from the soil.

Nitrogen
Sulfur
Phosphorus
Water

Symbiosis
Some plants get their nitrogen by "partnering" with nitrogen-fixing bacteria that live in association with their roots.

Rhizobium
A nitrogen-fixing bacterium

Clover plant roots with nodules containing nitrogen-fixing bacteria

Predation
Carnivorous plants in nutrient-poor soil obtain nutrients such as nitrogen by digesting insects.

Pitcher plants attract insects with bright colors and sweet nectar.

Insect trapped by an Australian sundew plant

that is essentially unusable by plants. Luckily, bacteria within the soil are able to convert nitrogen into a form that can be used by plants, a process called **nitrogen fixation.** Some of these bacteria, like members of the genus *Rhizobium,* can even live symbiotically with a plant's roots, forming lumpy structures called root nodules on the roots of legumes. By making plant growth possible, nitrogen-fixing bacteria play a critical role in supporting nearly all life on the planet (for more on the nitrogen cycle, see Chapter 23).

Fertile soil naturally contains nitrogen-fixing bacteria and therefore provides adequate supplies of usable nitrogen that plants can absorb and use to grow. And soil can be supplemented with fertilizer, giving plants a boost of artificial nitrogen and other nutrients. But in certain natural environments, such as bogs or rock outcroppings, it's hard for plants to obtain the nitrogen they need. The acidity of a bog, for example, prevents organic matter from breaking down, so nutrients are recycled more slowly. In these environments, plants have evolved novel ways to obtain scarce nitrogen—some of which would put animal carnivores to shame.

Trumpet pitchers (*Sarracenia*), for example, lure insects with brightly colored flowers and nectar "bribes." But the rim of the plant's trumpet-shaped flower is slippery. Unsuspecting trespassers climb onto the rim, lose their grip, and tumble into a deep cavity filled with digestive juices. Prevented from escape by downward-pointing spikes, the tiny prisoners drown and are slowly dissolved. The resulting insect soup—a rich source of nitrogen—is absorbed by the plant.

The Venus flytrap (*Dionaea muscipula*) takes an even more dramatic approach. The plant's "flower" is actually a spring-loaded trap that snaps shut around unsuspecting prey. Tiny hairs inside the flower act as sensors; when the sensors are tripped by a moving insect, the trap slams shut and the feasting begins.

Even when plants obtain nitrogen in this "carnivorous" way, they must still perform photosynthesis to make sugar. The plant body is composed of complex carbohydrates, such as cellulose, which the plant makes by stringing sugar molecules together. And the starting material for photosynthesis is carbon dioxide gas. Thus the air, rather than the soil, is where a plant obtains the material to put on weight (INFOGRAPHIC 32.3).

 ## Can plants photosynthesize at night?

ANSWER: By definition, there is a crucial part of photosynthesis that can occur only during daylight hours—the light-absorbing "photo" part. So technically, the answer is, no—plants can't photosynthesize at night. However, some plants are still hard at work at night, especially those living in hot, dry climates.

Because daylight hours tend to be the time when it's hottest and driest, plants lose a lot of water during the day. To conserve water, plants can close their stomata. But closing stomata also prevents carbon dioxide—the raw material of photosynthesis—from entering the leaf. In many plants, wheat and rice, for example, the result of this trade-off is reduced output of plant food—sugar.

Some plants have adapted to sun-scorched surroundings. Corn and sugar cane, for example, are able to thrive in hot, sunny climates by keeping their stomata closed as much as possible. They use a molecule called PEP carboxylase to capture CO_2 even when it is present only at low concentrations in air pockets in the leaf when the stomata are closed. Once the CO_2 has been captured, it can be "fed" into the synthesis part of photosynthesis to generate the sugar that the plants rely on.

Still other plants have adapted to hot, dry conditions by splitting up two parts of photosynthesis that usually occur together: taking in CO_2 and making sugar. Pineapples as well as cacti and other succulent plants—juicy aloe and jade plants, for instance—conserve water by keeping their stomata closed during the day when it's hot. But at night, when it's cooler, they open their stomata, allowing carbon dioxide in. The CO_2 that is

INFOGRAPHIC 32.4 BEATING THE HEAT: DOING PHOTOSYNTHESIS WITHOUT DRYING OUT

➔ All plants must perform the "photo" reactions of photosynthesis during the day. But many plants experience reduced levels of photosynthesis in hot and dry conditions. Other plants are adapted to live in hot and dry climates, and have different strategies to take in CO_2 while minimizing water loss.

Stomata open all day: too much water loss

CO_2 in H_2O out

When stomata are open, carbon dioxide can enter the leaves for photosynthesis. However, this increases transpiration and therefore water loss.

Stomata closed all day: too little sugar produced

CO_2 can't get in

Very little water loss

In dry climates, some plants keep stomata closed during the hot hours of the day. This conserves water, but inhibits the uptake of carbon dioxide for photosynthesis. These plants make less sugar food.

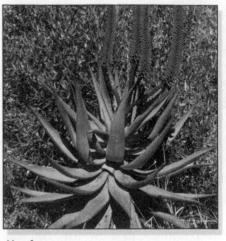

Aloe ferox

To combat this problem, some plants in dry climates:

1. Conserve water by having thick, waxy succulent leaves.
2. Extract CO_2 from airspaces within the plant even when stomata are mostly closed to conserve H_2O.
3. Open stomata at night to capture CO_2, while reducing water loss by transpiration.

---

captured at night isn't used right away but instead is stored. During the day, these plants can leave their stomata closed and use the stored CO_2 to complete photosynthesis and make sugar. By separating in time the steps of photosynthesis, these well-adapted plants can thrive in conditions that would wither their less physiologically adapted cousins (**INFOGRAPHIC 32.4**).

Like other eukaryotic organisms, plants use sugar to perform aerobic respiration to make ATP (see Chapter 6). All plants respire both during the day and at night. Because the total amount of carbon dioxide given off by plants during cellular respiration is less than the total amount taken in for photosynthesis, plants are carbon dioxide sinks that ultimately lower the amount of carbon dioxide in the atmosphere.

▶**CHLOROPHYLL**
The dominant pigment in photosynthesis, which makes plants appear green.

Q Why do leaves change color in the fall?

ANSWER: For most of the year, leaves are photosynthesis factories, making sugar from sunlight, water, and carbon dioxide. A key component of the photosynthetic machinery is a pigment called **chlorophyll,** which absorbs red and blue wavelengths of light and reflects green wavelengths. Chlorophyll is the reason plants appear green. To keep photosynthesis running, plants make abundant chlorophyll in spring and summer. But they also produce a smaller amount of other pigments—yellow-reflecting xanthophyll and orange-reflecting carotene. In the leaves, these pigments capture additional wavelengths of light and therefore expand the range of light that is useful for photosynthesis. You can't see these other pigments in leaves during spring and summer because leaves are chock full of green chlorophyll, camouflaging the other hues (although you can see them elsewhere in some plants: in the flesh of a pineapple, for example, or the taproot that is a carrot).

After a summer of intense sugar stockpiling, trees, bushes, and other deciduous plants that seasonally drop their leaves start to settle in for the winter and begin to shut down their photosynthesis machinery. During the winter months in temperate regions, there isn't enough water to drive photosynthesis; water in the ground is frozen and can't be absorbed. As temperatures cool, plants turn off the production of their light-absorbing pigments. Of all the pigments, chlorophyll is the most chemically unstable and therefore the shortest lived: levels fall quickly once production stops. By contrast, xanthophyll and carotene linger. As green fades, the other colors peek through—mostly yellow and orange, which have been there all along, but hidden.

The really intense colors—the fiery reds and deep purples—that some trees and bushes turn in autumn are the result of a fourth pigment, called anthocyanin. This distinctive chemical, which is the same one that gives apples and acai berries their color, is produced in leaves in response to high sugar concentration. Why does sugar concentration rise in the fall? To conserve water in winter, a tree develops a corklike membrane between the leaf stem and the branch, sealing off the leaf and preventing transpiration. This corky membrane also prevents phloem from transporting sugar out of the leaf. Unable to move out of the leaf, sugar begins to collect. The more sunny days there are in fall, the more photosynthesis occurs, the more sugar is made and trapped in the leaves. More sugar means more anthocyanin is produced, yielding the bright colors we associate with fall.

Some trees naturally produce lots of anthocyanin—maple and oak, for example—which explains why these trees produce such brilliant colors. Others, such as aspen and

INFOGRAPHIC 32.5 PLANTS PRODUCE MULTIPLE LIGHT-CAPTURING PIGMENTS

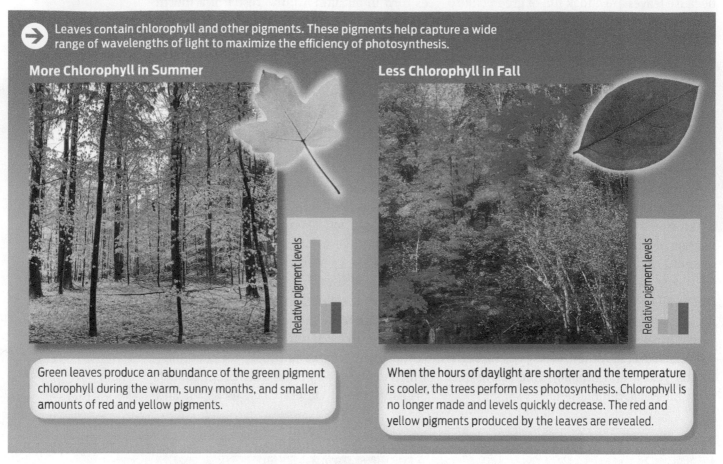

→ Leaves contain chlorophyll and other pigments. These pigments help capture a wide range of wavelengths of light to maximize the efficiency of photosynthesis.

More Chlorophyll in Summer

Relative pigment levels

Green leaves produce an abundance of the green pigment chlorophyll during the warm, sunny months, and smaller amounts of red and yellow pigments.

Less Chlorophyll in Fall

Relative pigment levels

When the hours of daylight are shorter and the temperature is cooler, the trees perform less photosynthesis. Chlorophyll is no longer made and levels quickly decrease. The red and yellow pigments produced by the leaves are revealed.

poplar, produce less anthocyanin and turn yellow or orange in fall (**INFOGRAPHIC 32.5**).

Eventually the corky membrane that seals off the leaf from the tree causes the leaf to dry out. With a slight gust of wind, the leaf flutters to the ground. The tree is now prepared for winter.

PLANT REPRODUCTION

 Do plants have sex?

ANSWER: They may not go on blind dates or join online dating sites, but plants are among the most "flirtatious" and sexually active organisms on earth. The many bright colors, provocative shapes, and alluring fragrances of flowers are the plant equivalents of lipstick, muscle shirts, and perfume–evolutionary novelties designed to attract pollinating partners.

Like other sexually reproducing creatures, plants have distinct male and female reproductive organs that produce haploid gametes: sperm and egg. In flowering plants, male and female sexual organs are housed within a plant's flowers. The male sexual organ, called the **stamen,** consists of a stalklike filament that supports a structure, the anther, that contains **pollen.** Pollen grains contain cells that will develop into sperm. The female structure, called a **pistil,** consists of a tubelike structure (called the style) topped with a sticky stigma, a "landing pad" for pollen. At the base of the pistil is a plump ovary stuffed with egg-containing ovules. Transfer of pollen from a male stamen to a female pistil results in **pollination,** which may then lead to fertilization. To fertilize eggs, sperm from the pollen must travel down the pistil to the ovary, where the eggs are located.

▶**STAMEN**
The male sexual organ of a flower.

▶**POLLEN**
Small, thick-walled structures that contain cells that will develop into sperm.

▶**PISTIL**
The female reproductive organ of a flower.

▶**POLLINATION**
The transfer of pollen from a male stamen to a female pistil.

When a haploid sperm fuses with a haploid egg, the result is a diploid plant zygote, which divides to form the embryo.

In some flowering plant species, male and female organs are found on separate male and female plants. Ginkgo trees and holly bushes are examples of such unisexual plants. More commonly, however, flowering plants—everything from dandelions to lilies—are hermaphrodites: their flowers contain both male and female reproductive parts. A few such hermaphroditic plants can self-pollinate—transferring pollen from stamen to pistil within the same flower or plant. But most hermaphroditic plants have evolved ways to prevent self-pollination and to encourage out-crossing instead. They may have pistils and stamens of different

heights, for example, or their pollen may be chemically rejected by eggs of the same plant.

Plants have evolved many elaborate ways to spread their pollen from plant to plant. For many flowering plants, pollinators such as bees and hummingbirds transfer pollen grains from the anther of one flower to the stigma of another. To these animal pollinators, a flower's scent is a potent aphrodisiac, irresistibly attracting them to the blossom. Other plants rely on wind to transfer pollen (**INFOGRAPHIC 32.6**).

Not all plants are so sexually adventurous. Some are asexual, or have asexual phases, and can produce new individuals through underground root runners or bulbs that develop into whole new plants. Asexual reproduction, in the form of cuttings and grafting, is

INFOGRAPHIC 32.6 PLANTS REPRODUCE SEXUALLY

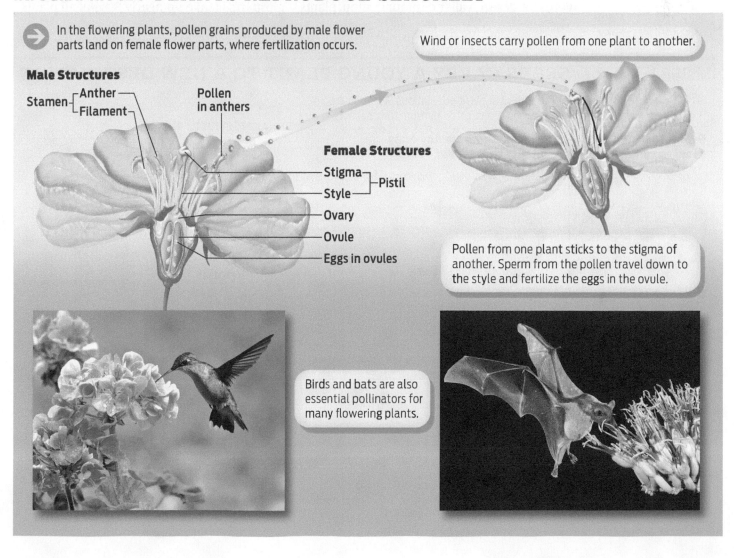

In the flowering plants, pollen grains produced by male flower parts land on female flower parts, where fertilization occurs.

Wind or insects carry pollen from one plant to another.

Male Structures

Stamen — Anther — Filament

Pollen in anthers

Female Structures

Stigma — Pistil
Style —

Ovary

Ovule

Eggs in ovules

Pollen from one plant sticks to the stigma of another. Sperm from the pollen travel down to the style and fertilize the eggs in the ovule.

Birds and bats are also essential pollinators for many flowering plants.

▶**SEED**
An embryonic plant contained in a protective structure.

▶**GYMNOSPERMS**
Cone-bearing seed plants.

▶**ANGIOSPERMS**
Flowering plants.

especially important in agriculture. A cutting taken from one plant can be planted in soil to form a new plant; sugar cane and pineapples are often reproduced this way. A cutting taken from a plant can also be surgically grafted to the root system of a different plant—a procedure commonly used to perpetuate vineyard grapes.

Q What spins like a helicopter, shoots like a rocket, and contains its own parachute?

ANSWER: The answer to this biological riddle: a seed. Unlike animals, land plants cannot move to seek out more hospitable living conditions for their offspring when resources are scarce or the neighborhood gets too crowded. Forest or desert, valley or mountaintop, plants

are stuck where nature put them. Their solution to this enforced sedentary existence is to disperse their offspring far and wide through seeds. A **seed** is a small embryonic plant contained within a sac of stored nutrients—essentially, a plant starter kit. It develops from a fertilized egg and is a perfect package for delivering an immature plant to its new home and protecting it from harsh conditions. It's also a tremendously successful evolutionary adaptation, which is why two types of seed-bearing plants dominate the plant world: **gymnosperms,** which include seed-cone-producing conifers, and **angiosperms,** the flowering plants. *Gymnos* is Greek for "naked," so the name literally means "naked seeds"; in a pinecone, for example, seeds sit nestled under the cone's scales. In angiosperms, seeds are located inside a fruit or nut: "angio" is derived from the Greek for "vessel" or "container." Angiosperms make up nearly 90% of all plants—

INFOGRAPHIC 32.7 SEEDS CARRY A YOUNG PLANT TO A NEW DESTINATION

→ Seeds are important for the spread of plant populations. As seeds are dispersed, offspring reach new locations, often far from their parents.

Gymnosperms produce seeds in cones

Angiosperms are flowering plants that produce seeds in fruit

Basic seed structure

Embryo

Germination

Stored nutrients

Seedling

Seeds disperse in a variety of ways

Burrs hitch a ride on animal fur.

Small dandelion seeds fly for miles in the wind.

Coconuts float, relocating seeds from one shore to another.

Animals eat fruits, dispersing the seeds in their feces.

The squirting cucumber shoots its seeds out with water.

including most of our agricultural crops (see Chapter 22).

There are many shapes and sizes of seeds, and they are dispersed in myriad ways. Some are small and hitch a ride on fur or clothing by means of tiny hooks or burrs. Some, like coconuts, are large and buoyant and can float across oceans to reach distant beaches. The delicate parachutes of dandelions, the spinning helicopters of maples and pine trees, and other lightweight seeds sail on the wind. Cottonseeds are little more than hairy specks of dirt, but on a steady breeze they can windsurf for miles. Other seeds are packaged inside fruit. Tempted by the fruit's bright color and sugary content, animals eat the fruit and then deposit the seeds in feces some distance away from the original plant.

Some seeds are dispersed through a ballistic mechanism. The seedpod of the squirting cucumber (*Ecballium elaterium*), for example, fills with slimy juice as it ripens. Eventually, the mounting pressure of the increased volume of juice causes the cucumber to shoot off the plant like a rocket, trailing a plume of seeds and slime in its wake. At the slightest touch, the bulging seedpods of the aptly named touch-me-not plant explode, spraying seeds like bullets. When a seed lands in favorable conditions, with enough water, it will **germinate** and grow into a young plant, or seedling (INFOGRAPHIC 32.7).

PLANT HORMONES

 ## Can plants see?

ANSWER: Observe an old building and you'll likely see ivy scaling up its walls. Keep houseplants next to a window and you'll find them bending toward the sunlight. How does a plant know where it's going? While plants don't have eyes and therefore cannot see, they are quite adept at sensing and responding to their environment, which they do through various kinds of tropism (from the Greek *tropos*, meaning "turn").

The growth of a plant shoot toward light is called **phototropism.** This is how leaves get

An example of gravitropism: leaf-bearing stems bend upward away from gravity.

the sunlight they need for photosynthesis, and it's why ivy climbs up walls and houseplants lean into the window. Shoots grow toward the light because of **auxin,** a plant hormone that promotes cell elongation as one of its effects. When light hits one side of a plant shoot, auxin moves to the shaded side, creating a gradient of the hormone in the stem. The side receiving the most direct sunlight contains the least auxin, while the shaded side contains the most. Auxin promotes elongation of cells on the shady side of the stem. This causes the shaded side to elongate faster than the sunny side, pushing the stem toward the sun. The whole stem doesn't sense light and produce auxin, though–just the tip of the stem. Cover the tip of a young plant and it won't turn toward the light, but will instead grow straight up because, in the absence of sunlight, auxin is not differentially localized to one side of the stem.

Other mechanisms help a plant sense where it is and orient itself in space. **Gravitropism** is the growth of plants in response to gravity–roots grow downward, with the force of gravity, and shoots grow upward, against it. Auxin is again the main player in this mechanism. When a plant is placed on its side, more auxin is sent to the down side of the stem, in the direction of gravity. This causes the cells on the down side to elon-

▶**GERMINATION**
The process in which a seed comes out of dormancy and a new plant begins to grow.

▶**PHOTOTROPISM**
The growth of the stem of a plant toward light.

▶**AUXIN**
A plant hormone that causes elongation of cells as one of its effects.

▶**GRAVITROPISM**
The growth of plants in response to gravity. Roots grow downward, with gravity; shoots grow upward, against gravity.

THIGMOTROPISM
The response of plants to touch and wind.

gate more. Stems begin to curve away from gravity. Root cells, however, respond the opposite way to auxin: more auxin on the gravity side of roots inhibits root cell elongation on the down side, so roots bend toward gravity. Gravitropism allows a planted seed to send its shoots toward the light and its roots toward the soil.

A plant's sense of touch is called **thigmotropism;** it's how vines sense their way around poles or trellises and carnivorous plants sense their prey. Touch-sensitive growth is also controlled by auxin. Vine cells touching a pole, for example, get less auxin and consequently elongate less, while cells on the other side elongate more. The result is another kind of lopsided growth in the shoot, which eventually causes the vine to coil around whatever

it's touching. A plant's sense of touch can be exquisitely sensitive—more sensitive than a human's. A human can detect the presence of a thread weighing 0.002 mg laid across the arm. By contrast, the feeding tentacle of the insectivorous sundew plant can sense a thread of less than half that weight. The legs of a single gnat are enough to trigger the tentacle into swift action (**INFOGRAPHIC 32.8**).

Q Does one bad apple really spoil the bunch?

ANSWER: This old adage is true, and its truth can be shown empirically: put a ripe apple in a bowl of unripe ones and the unripe

INFOGRAPHIC 32.8 PLANTS SENSE AND RESPOND TO THEIR ENVIRONMENT

Plants can respond to a variety of stimuli, including light, touch, and gravity. Many of these responses are mediated by a plant hormone called auxin. Unequal distribution of auxin leads to a corresponding unequal pattern of cell elongation in stems and roots, and thus growth in a particular direction.

Phototropism: Plants respond to light.

Gravitropism: Plants respond to gravity.

More auxin on the gravity side of stems results in cell elongation. The stems always bend against gravity. For roots, the reverse is true. More auxin on the gravity side causes root cells to stop elongating, resulting in roots bending toward the gravitational force.

Stem cells
(More auxin = more elongation = longer cells on bottom)

Root cells
(More auxin = less elongation = longer cells on top)

Auxin equal on both sides of stem. Cells elongate equally on each side.

Light

Light

More auxin sent down the shaded side of the stem. Cells elongate more on the shaded side, bending the stem toward the light.

Thigmotropism: Plants respond to touch.

Vine cells touching this stem get less auxin and therefore elongate less, and cells on the other side of the vine elongate more. This causes the vine to bend toward the stem, allowing it to coil as it grows.

neighbors will quickly ripen. That's because ripe fruit—bananas and apples, especially—emit **ethylene,** a gaseous plant hormone, one effect of which is to promote ripening. In a confined space, the ethylene gas collects and causes nearby fruit to ripen through the loss of chlorophyll and the breakdown of cell walls. The result is the conversion of a hard, green fruit to a soft, ripe one. Commercial fruit growers take advantage of the action of ethylene when they ship fruit to distributors. Often, fruit is picked while still green and then exposed to natural or synthetic ethylene just before arrival in grocery stores to hasten ripening. This method works especially well with tomatoes, avocados, bananas, and cantaloupe. In some cases, fruit growers remove ethylene from storage containers in order to prevent ripening so that fruits can be stored for a long time—apples that are picked in fall and sold in summer, for example (**INFOGRAPHIC 32.9**).

As a plant hormone, ethylene does more than ripen fruit. It is responsible for a number of different aging effects in plants—including leaf dropping in autumn.

Q Can plants take hormones to improve their performance?

ANSWER: In a manner of speaking, yes. **Gibberellins** are plant hormones that promote growth. Scientists have identified more than 100 different types of gibberellin. One effect of gibberellins is to provide the chemical cue for seeds to germinate and grow. When conditions are right for a seed to germinate—when rising temperatures begin to melt frost, for example—these growth-promoting hormones give the green light for a seedling to grow.

Applying gibberellins to a young plant can increase the length of its stem, which also indirectly increases the size of its fruits—a fact that makes them very useful in agriculture. Gibberellins are commonly used to increase the size of seedless grapes, for example. On an untreated seedless grape plant, the stem

INFOGRAPHIC 32.9 ETHYLENE PROMOTES FRUIT RIPENING

→ Ripe fruit produces ethylene, which can speed the ripening process in neighboring fruit. Ethylene is used commercially to ripen fruit that is picked and transported while still green.

Ethylene gas

Commercial fruit growers often harvest fruit before it is ripe to ensure it doesn't spoil as it travels long distances to market. Ethylene gas is pumped into ripening chambers to ripen fruit before it is placed on grocery store shelves.

remains relatively short, so the bunches of grapes growing on the stem are clustered densely together, resulting in small grapes. When sprayed with gibberellins, the stems grow longer, giving the grapes more room to grow. Because seeds are the normal source of gibberellins, seedless grapes have no source of gibberellins to help them grow naturally, which is why farmers need to spray them in the first place.

In many species of plant, seed germination is controlled by the balance between the growth-promoting properties of gibberellins and the growth-inhibiting effects of another hormone, **abscisic acid (ABA).** ABA is the plant hormone that keeps seeds dormant—for example, over the cold winter months, when conditions for growth are not ideal. With appropriate environmental cues—the right combination of temperature, moisture, and light—ABA is degraded and gibberellins are

▶**ETHYLENE**
A gaseous plant hormone that promotes fruit ripening as one of its effects.

▶**GIBBERELLINS**
Plant hormones that cause stem elongation and cell division.

▶**ABSCISIC ACID (ABA)**
A plant hormone that helps seeds remain dormant.

INFOGRAPHIC 32.10 HORMONES TRIGGER PLANT GROWTH AND DEVELOPMENT

→ Certain plant hormones regulate growth and development of plants. When humans apply these hormones to crops, plant growth can be dramatically enhanced.

Gibberellin hormones enhance plant growth

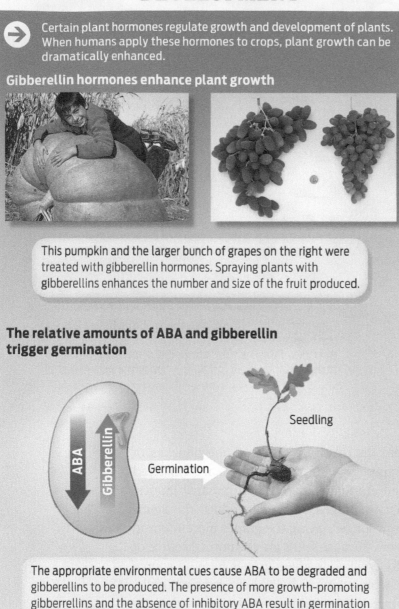

This pumpkin and the larger bunch of grapes on the right were treated with gibberellin hormones. Spraying plants with gibberellins enhances the number and size of the fruit produced.

The relative amounts of ABA and gibberellin trigger germination

Seedling

ABA

Gibberellin

Germination

The appropriate environmental cues cause ABA to be degraded and gibberellins to be produced. The presence of more growth-promoting gibberrellins and the absence of inhibitory ABA result in germination and seedling growth.

synthesized, leading to the end to dormancy and the start of germination. You can think of ABA as the brakes and gibberellins as the gas pedal: as the brakes on growth are lifted, the gas pedal is pressed, ushering in a new season's blooms (**INFOGRAPHIC 32.10**).

Q Why are some plants poisonous?

ANSWER: From juicy peaches and succulent strawberries to zesty basil and peppery arugula, many plants are incontestably delicious. Humans aren't the only ones who think so. Many herbivores such as insects, birds, rodents, and other small mammals find plant parts tasty and irresistible. This is both helpful and hurtful to a plant. On the one hand, plants rely on animals to eat their fruits and disperse their seeds. On the other hand, plants must ensure that only noncrucial parts of the plants are eaten by other organisms. Eating a plant's fruit is one thing; eating all of its leaves is quite another.

Plants have evolved many defenses to protect their important parts from an herbivore's chomping. Some defenses are mechanical: the stems of a raspberry plant are covered in prickly spines to prevent unwanted chewing; holly leaves are waxy and difficult for insect jaws to grasp. Other defenses are chemical: leaves of the tobacco plant produce nicotine, which is toxic to insects; the bark of the South American evergreen cinchona tree produces quinine, an extremely bitter substance that many animals find distasteful (except certain humans, who use it in their gin and tonics). Such antiherbivory chemicals are a highly effective way to deter pests from eating a plant's leaves.

While a plant's fruits are often tasty and meant to be digested, seeds generally are not. The seed contains a new plant, and therefore it must be protected. Many seeds are encased within an indigestible shell that prevents them from being destroyed by an animal's stomach juices. The unlucky animal that succeeds in breaking open the shell and eating the seed is in for an unwelcome surprise. Seeds are sources of some of the most potent poisons on Earth, including ricin, cyanide, and strychnine. Ricin, found in castor beans, can be lethal to animals in quantities as little as two beans. Cyanide, which is found in small doses in the seeds of peaches, apricots, and apples, kills by interrupting cellular respiration in mitochondria; unable to make ample amounts of the short-term energy-storage molecule ATP, nerve and muscle tissues quickly shut down.

Plants also use chemicals to combat other plants. Blue gum eucalyptus trees (*Eucalyptus globulus*), for example, secrete a sticky gumlike substance that acts as a deterrent to the germination and growth of noneucalyptus plants in the nearby vicinity.

Perhaps the most interesting method of deterrence that plants use is a kind of unwitting alliance. Some plants, such as wild tobacco, emit potent vapors when they are eaten by insect pests. The vapors, in turn, attract natural predators of the insects, which are thus enlisted in the plant's defense. This "enemy of my enemy" approach is an example of mutualism (see Chapter 22) (**INFOGRAPHIC 32.11**).

Individual plants can even communicate with other individuals of their species and unite against a common enemy. At the first sign of grazing by a hungry giraffe or antelope, for example, African acacia trees release a bad-tasting poison into their leaves. At the same time, they release a gaseous chemical—the versatile hormone ethylene—which drifts out of the stomata of their leaves. Other acacias in a 50-yard radius detect the gas and are prompted to start releasing poison in their leaves, too.

Although antiherbivory chemicals complicate an herbivore's life, they are often quite useful to humans. Some of our most important medicines are extracts of plant chemical defenses, including aspirin, morphine, digitalis, and the anticancer drug Taxol, which was originally obtained from the bark of the Pacific yew tree (see Chapter 8). ∎

INFOGRAPHIC 32.11 PLANTS DEFEND THEMSELVES

Physical Defenses

Raspberry thorns can impale hungry insects.

The jaws of insects that try to feed on holly leaves may not be strong enough to penetrate their waxy coat.

Chemical Defenses

The leaves of the tobacco plant contain nicotine which is toxic to insects.

The bark of the cinchona tree contains the bitter chemical quinine, which deters insects from feasting on it.

Peach pits have a tough exterior that protects the embryo inside. If an animal is successful in cracking the seed open, it is met with lethal toxins like cyanide.

When a hornworm feeds on a wild tobacco plant, the plant releases chemicals that both repel the worm and attract the worm's predators.

CHAPTER 32 Summary

- Nearly all plants share the same basic structure: an aboveground shoot system that includes the stem and photosynthetic leaves, and an underground root system that anchors the plant in the soil and absorbs water and nutrients.

- All plant cells are surrounded by a cell wall made of the complex carbohydrate cellulose. A central water-filled vacuole creates turgor pressure against the cell wall and helps a plant stand up.

- Some plant tissues have cells with an additional inner cell wall made of cellulose and lignin. Lignin is extremely durable and is what makes plant tissues woody.

- Plants have a vascular system. Xylem transports water and dissolved nutrients from the roots to the shoots. Phloem transports dissolved sugars throughout the plant body.

- Water transport is powered by evaporation through stomata in leaves, a process known as transpiration. Plants have mechanisms to control the amount of water lost by transpiration.

- Plants undergo primary (vertical) and secondary (horizontal) growth. Secondary growth is mostly growth of xylem tissue. Tree rings reflect the growth of xylem tissue from season to season.

- Plants are autotrophs, able to make their food from sunlight and carbon dioxide absorbed from the air by photosynthesis. They also require a few additional nutrients from the soil, including water, nitrogen, phosphorus, and sulfur.

- Bacteria in the soil convert nitrogen into a usable form that plants can absorb. In nutrient-poor soils, some plants obtain nitrogen by trapping and digesting insects.

- Some plants have adapted to hot, dry climates, keeping their stomata closed during the day and collecting carbon dioxide for photosynthesis primarily at night.

- A variety of pigments assist with photosynthesis. Chlorophyll is the most predominant and is responsible for the green color of plant leaves. Other pigments contribute to photosynthesis and are visible in the fall, after chlorophyll has broken down.

- Many plants reproduce sexually and have male and female sexual organs that produce haploid gametes.

- Some plants rely on pollinators such as insects to deliver male pollen to female reproductive structures. Other plants depend on the wind to deliver pollen to the female structures.

- Seed-bearing plants (gymnosperms and angiosperms) produce embryonic plants encased within a protective seed that can disperse a great distance from the parent plant.

- Plants respond to their environment by various tropisms. Phototropism is growth toward light; gravitropism is growth in response to gravity; thigmotropism is growth in response to touch or wind.

- Plants produce hormones that contribute to growth and development: ethylene contributes to fruit ripening; gibberellins and ABA regulate germination and stem growth; auxin controls cell elongation.

- Plants have physical defenses (e.g., thorns) and chemical defenses (e.g., toxins) that fend off herbivores. Humans make use of chemical defenses in developing pharmaceuticals.

MORE TO EXPLORE

- Chamovitz, D. (2013) *What a Plant Knows: A Field Guide the Senses.* New York: Scientific American/Farrar, Straus, Giroux.

- Coursera, What a Plant Knows (and other things you didn't know about plants) https://www.coursera.org/course/plantknows

- U.S. Forest Service, Ethnobotany http://www.fs.fed.us/wildflowers/ethnobotany/

- PBS, *Nature*, What Plants Talk About (2013) http://www.pbs.org/wnet/nature/episodes/what-plants-talk-about/video-full-episode/8243/

- PBS, *Nature*, The Seedy Side of Plants (1999) http://www.pbs.org/wnet/nature/episodes/the-seedy-side-of-plants/introduction/1268/

CHAPTER 32 Test Your Knowledge

DRIVING QUESTION 1

How is a plant body structured, and how do plants obtain water and grow?

By answering the questions below and studying Infographics 32.1 and 32.2, you should be able to generate an answer for the broader Driving Question above.

KNOW IT

1. Draw and label a diagram of two adjacent plant cells. Include key intracellular structures.

2. Which of the following statements represents a true distinction between xylem and phloem?

   a. Xylem provides support only; phloem provides transport.

   b. Xylem provides water and nutrient transport; phloem provides sugar transport.

   c. Xylem transports materials from shoots to roots; phloem transports materials in either direction.

   d. Xylem transports sugars in either direction; phloem transports water from roots to shoots

   e. all of the above

3. What is the function of the cuticle?

   a. It enables neighboring cells to stick together.

   b. It provides rigidity to the cell wall.

   c. It is toxic to many herbivorous insects.

   d. It prevents water loss.

   e. It is sticky and helps pollen stick to a plant during pollination.

USE IT

4. Paper is made from wood that is broken down to pulp. Why are lignin-digesting enzymes included in the pulping process? Would these enzymes have to be included in the pulping process if paper were made from green leaves? Explain your answer.

DRIVING QUESTION 2

How do plants obtain nutrients?

By answering the questions below and studying Infographics 32.3, 32.4, and 32.5, you should be able to generate an answer for the broader Driving Question above.

KNOW IT

5. Plants are autotrophs and can make sugar from CO_2. How do they obtain CO_2?

   a. through stomata

   b. by absorption through the root system

   c. by digesting insects

   d. by breaking down carbon-rich carbohydrates stored in roots

   e. a and b

6. When stomata are open, what is happening?

   a. O_2 is entering the plant for photosynthesis.

   b. CO_2 is entering the plant for photosynthesis.

   c. H_2O is entering the plant for photosynthesis.

   d. H_2O is leaving the plant.

   e. a, b, and c

   f. b and d

7. What is found in root nodules?

   a. PEP carboxylase

   b. CO_2

   c. bacteria

   d. phosphorus

   e. stored glucose

USE IT

8. Describe the "conflict" that plants face with respect to opening and closing their stomata.

9. Are trumpet pitchers strict autotrophs? Explain your answer.

10. If you applied to the soil around your plants a chemical that kills bacteria (but not plants), why might your plants die?

11. Why may the bright coloration of a trumpet pitcher have a different function than do the bright colors of yellow or orange squash blossoms? (Hint: Squash are pollinated by bees.)

12. If a plant could not make chlorophyll, would you expect it to survive? Why or why not?

INTERPRETING DATA

13 Scientists carried out an experiment to examine the effect of CO_2 concentrations on plant growth in a semi-arid (that is, a dry) grassland environment in Colorado. They set up several plots in the field, consisting of chambers that allowed the concentration of CO_2 to be controlled. One set of plots (A) was kept at ambient CO_2 concentration, and one set (B) was kept at elevated CO_2 concentration (two times ambient). In mid-July the total plant mass in each plot set was recorded. The data for three consecutive years are shown in the table. (In order to establish a baseline, the CO_2 levels in 1996 were not manipulated.)

Plot set	Average Plant Mass (g/m^2), 1996	Average Plant Mass (g/m^2), 1997	Average Plant Mass (g/m^2), 1998
A (ambient CO_2)	110	108	145
B (elevated CO_2)	112	145	205

a. Graph these data.

b. Are there any differences between different plot sets in any given year? If so, describe the differences observed.

c. Are there any differences in the same plot set between years? If so, describe the differences and propose an explanation.

d. What are the implications of this study for grassland productivity (at least in Colorado) with rising CO_2 levels?

SOURCE: Morgan, J. A., et al. (2001) Elevated CO_2 enhances water relations and productivity and affects gas exchange in C_3 and C_4 grasses of the Colorado shortgrass steppe. *Global Change Biology* 7:451–466.

DRIVING QUESTION 3

How do plants reproduce, respond to stimuli, and protect themselves?

By answering the questions below and studying Infographics 32.6, 32.7, 32.8, 32.8, 32.10, and 32.11, you should be able to generate an answer for the broader Driving Question above.

KNOW IT

14 What characterizes a dormant seed?

a. the presence of sperm

b. the presence of eggs

c. the presence of an embryonic plant

d. the absence of ABA

e. the presence of several pollen grains

15 Describe how fertilization follows pollination in an angiosperm. What has to happen, and what plant structures are involved?

16 Which of the following pigments are present in green leaves late in the summer?

a. chlorophyll   d. all of the above

b. carotene   e. a and b

c. xanthophyll

17 Which of the following hormones helps fruit to ripen?

a. auxin

b. ethylene

c. gibberellins

d. estrogen

e. anthocyanin

18 If you wanted a plant to grow really tall, which hormone should you apply?

a. auxin

b. ethylene

c. gibberellins

d. anthocyanin

e. ABA

USE IT

19 Why do seedless grapes need hormone treatment to develop big clusters of big grapes, while seeded varieties can develop large fruits without exogenous (that is, externally applied) hormones?

20 Nopales are cactus pads (the large, thick "leaves" of the prickly pear cactus) and make a delicious salad. What antiherbivory mechanism fails when we succeed in making ensalada de nopales—prickly pear salad?

MINI CASE

21 The Natural Products Branch of the National Cancer Institute looks for defensive compounds produced by plants, microbes, and marine organisms that may have anticancer activity. Once compounds have been isolated, they can be chemically modified to enhance their activity. Several drugs have come out of this program, including eribulin mesylate, a chemically modified compound originally purified from a sea sponge. This drug has been approved for women with metastatic breast cancer whose disease thus far has not been responsive to treatment. In a clinical trial, patients taking eribulin mesylate had a significantly longer survival time (13.1 months) than patients taking chemotherapy regimens prescribed by their oncologists (10.6 months). (Data are from http://www.cancer.gov/ncicancerbulletin/041911/page5.)

The first step in developing drugs such as these is to prepare extracts by grinding up the natural product. Design a procedure to test such extracts for anticancer activity. Consider what tests you will use to determine if an extract has an anticancer activity, and what variables you will measure and manipulate in order to find promising candidates to advance along the drug discovery pipeline.

BRING IT HOME

22 Many everyday products contain compounds derived from plants. A few of these are:

Caffeinated sodas or energy drinks

Camphor and menthol-containing vapor rubs and steams (for congestion and muscle aches)

Nicotine

Tea tree oil for minor skin irritations and infections

Quinine-containing tonic water

You, members of your family, or your friends may have used some or all of these. For each item listed, do some internet research to identify the plant (find both the scientific and common name) and something else about it that interests you (e.g., where is the plant found, whether it has historically been used for medicinal purposes, are there other sources of the compound).

Answers

See below for answers to the "Know It" questions. Sign on to LaunchPad to see the answers to the "Use It" questions, Mini Cases, and Interpreting Data problems.

Chapter 1

Driving Question 1
1. c
2. e
3. b
4. c
5. e

Driving Question 2
9. a
10. e

Driving Question 3
15. e
16. b

Chapter 2

Driving Question 1
1. e
2. a
3. six

Driving Question 2
6. c
7. Homeostasis is the ability to maintain a relatively stable internal environment, even when the external environment changes. It is important for a cell or organism to maintain a relatively constant internal environment because many of the molecules and processes of life cannot function outside a narrow range of, for example, pH and temperature. Homeostasis helps maintain the conditions necessary for life.
8. A polymer is a molecule made up of smaller and typically repeating subunits. Examples include proteins, which are made up of amino acids; complex carbohydrates, which are made up of simple sugars or monosaccharides; and nucleic acids, which are made up of nucleotides.
9. a

Driving Question 3
13. b
14. e

Driving Question 4
17. Olive oil does not mix with water, so it is hydrophobic; it is made up of hydrophobic triglycerides. Salt will dissolve in water, so it is hydrophilic. When salt dissolves, the charged Na^+ and Cl^- ions can interact with the partial charges on the water molecules.

18. a
19. a: The solvent in coffee and tea is water. b: The main solute is the dissolved sugar molecules, but there are also additional compounds in coffee and tea that dissolve in the water as the beverage is prepared. c: Because the sugar dissolves in water, sugar is hydrophilic.
20. c
21. a
22. Both ionic bonds and covalent bonds are strong bonds when dry. Covalent bonds involve the sharing of electrons (pairs of electrons). Ionic bonds involve an electrical attraction between two oppositely charged ions.

Chapter 3

Driving Question 1
1. The cell theory states that all living organisms are made of cells, and that all cells arise from existing cells.
2. d
3. prokaryotic and eukaryotic
4. d
5. d
6. e

Driving Question 2
9. e
10. c
11. a: They both involve the movement of a solute across a membrane from an area of higher solute concentration to an area of lower solute concentration, without the input of any additional energy. b: They both require a transport protein in order to transport a solute across a membrane.
12. b

Driving Question 3
17. b
18. Phospholipids of the cell membrane would not be a good target for an antibiotic because human as well as bacterial cells have phospholipids in their cell membranes, and thus the proposed antibiotic would harm both bacterial and human cells.

Driving Question 4
24. a: Mitochondria are rod-shaped organelles. They are surrounded by a double membrane (they have both an inner and outer membrane). They are important in the reactions that extract energy from food and convert it to an immediately usable form. b: The nucleus is a large organelle that stores the genetic instructions (DNA). The nucleus has a double membrane (the nuclear envelope) that has pores that permit the movement of molecules into and out of the nucleus. c: The endoplasmic reticulum is an

extensive network of membrane tubes. The endoplasmic reticulum is connected to the nucleus and has a variety of critical functions in the cell, including protein synthesis, lipid synthesis, and drug detoxification. d: Chloroplasts are organelles found in plant cells. They are enclosed by a double membrane and appear green because of the pigments involved in photosynthesis. Photosynthesis is the process by which energy from sunlight and carbon dioxide from the air are used to synthesize sugars.

25. a
26. d

Milestones in Biology 1: Scientific Rebel

1. a: Mitochondria are cellular "power plants." They carry out the reactions of cellular respiration to generate usable energy. b: Chloroplasts are the photosynthetic organelles in plant cells. They convert atmospheric carbon dioxide to organic carbohydrates (sugars).

2. Chloroplasts have their own DNA; they have their own ribosomes; they are about the same size as bacteria and replicate in the same way as bacteria; and mitochondria share these traits with chloroplasts.

3. a: cyanobacteria; b: *Rickettsia*

4. a

5. a: You could not live without your mitochondrial symbionts. Humans require ongoing cellular respiration to sustain life. b: If plants did not have mitochondria, they would die (much as humans would without mitochondria). That would reduce the food supply for humans (directly or indirectly because of the animals that many humans eat that in turn in eat plants). Furthermore, if plants lacked chloroplasts, they would not be able to carry out photosynthesis. They would die (which would be very bad for humans). Additionally, when plants carry out photosynthesis, oxygen is released, and humans rely on this oxygen.

Chapter 4

Driving Question 1

1. c
2. f
3. b
4. a

Driving Question 2

7. b
8. e

Driving Question 3

12. e
13. Both help enzymes speed up their reaction rates. Cofactors are typically inorganic metals, while coenzymes are organic molecules, such as vitamins.
14. b
15. c

Driving Question 4

18. a: Vitamin C is water soluble, so any excess is removed from the body in the urine. Vitamin E is fat soluble, so excess is stored in fatty tissues in the body, possibly accumulating to toxic levels. b: High levels of vitamin C supplementation will not lead to storage of vitamin C in the body. Any excess over what is needed is excreted from the body, dissolved in the urine.
19. e

Chapter 5

Driving Question 1

1. They are all photosynthetic–that is, they carry out photosynthesis, using the energy from sunlight and carbon dioxide from air to synthesize sugars.
2. No. Animals cannot carry out photosynthesis. However, they rely on photosynthesis, as they eat plant material that is the product of photosynthesis, and animal material that is sustained by eating plants.
3. c
4. The main photosynthetic pigment is chlorophyll, which reflects green wavelengths.
5. Photosynthetic algae rely on photosynthesis for the production of sugars that can be used for short-term energy needs, or for longer-term energy storage. Animals cannot make their own sugars (or other energy-rich organic molecules) from scratch, so must obtain them from their diet.

Driving Question 2

8. b
9. c
10. c

Driving Question 3

12. a
13. c
14. oxygen (O); carbon dioxide (I); photons (I); glucose (O); water (I)
15. a

Driving Question 4

19. g
20. e

Chapter 6

Driving Question 1

1. Her BMI would be ~27, which would place her in the overweight category, according to the CDC.
2. e

Driving Question 2

8. d
9. d
10. e

Driving Question 3

12. c
13. b
14. a
15. b
16. d

Driving Question 4

19. d
20. c
21. b
22. a

Chapter 7

Driving Question 1

1. e
2. c
3. b
4. d
5. c

Driving Question 2

8. c
9. a

Driving Question 3

12. a
13. d
14. d
15. lane B

Driving Question 4

19. a: Suspect B's profile matches the evidence at all markers tested. Suspect B is more than likely the source of the DNA evidence at the crime scene. b: Suspect B is likely to be related to the victim, as suspect B shares at least one band with the victim at every allele tested. Suspect B could be the parent or child of the victim.

Milestones in Biology 2: The Model Makers

1. c
2. d
3. Franklin carried out the key X-ray diffraction studies, as did Wilkins, who recognized that Franklin's data might be helpful to Watson and Crick. Watson and Crick synthesized existing information to build models that were consistent with the observed data. Franklin's experimental observations were critical in assembling the correct model, the one that accurately represented the structure of DNA.
4. The DNA stretched out in the presence of water, suggesting that the water must be interacting with and coating the molecule. As the charged phosphate groups are hydrophilic, and as the water was interacting with DNA on the surface of the DNA, the phosphate groups had to be on the outside of the molecule—a key insight.

5. The DNA double helix is arranged as antiparallel strands. Each strand has its phosphate groups oriented along the external backbone of the molecule, and the nucleotide bases oriented toward the interior of the helix. The bases from one strand pair with bases on the other strand according to specific rules: A pairs with T and G pairs with C. The helix has a constant diameter along its length.

Chapter 8

Driving Question 1

1. b
2. e

Driving Question 2

5. This statement is not accurate. Chromosomes contain genes, and genes encode proteins. A correct statement would be: "A chromosome contains many genes. Each gene encodes one or more proteins."
6. c
7. RNA polymerase (N); ribosome (C); tRNA (C); mRNA (C) (mRNA is made by transcription in the nucleus, but it acts in translation in the cytoplasm.)
8. c
9. d

Driving Question 3

13. e
14. a: The construct is present in every cell, as it was introduced into the embryo. b: The construct is expressed only in mammary cells, because its regulatory region is specific for expression in mammary cells.
15. The beta-casein gene (because of its regulatory sequence) is expressed only in mammary cells. Because this regulatory sequence directs expression only in mammary cells, it will direct expression of the protein to be released in milk.

Driving Question 4

18. In the case of goat cheese, the goat's milk does not include a foreign protein encoded by a transgene. The production of cheese from goat's milk does not require genetic modification.
19. Therapeutic proteins that have to be isolated from blood products or animal organs have several disadvantages. There is a risk of contamination with blood-borne pathogens or zoonotic viruses (those found in other animals that can infect humans)—this is a safety concern. The process often takes a huge volume of blood or organs, requiring many donors or the sacrifice of many animals, and these resources may not always be available in sufficient quantity, affecting the ability of the company to produce sufficient quantity of the drug. Proteins from other animals may not be exactly identical to the human version, so may not be as effective as the human version. Often, the purification process is very elaborate and lengthy, making the purification and the final product very expensive. Thus genetic engineering that establishes

transgenic animals that produce large quantities of the human protein in their milk is attractive, because this method circumvents many of these concerns. Of course, some may object to genetically engineering another organism for human purposes, and there are regulations concerning the process that companies must comply with.

Milestones in Biology 3: Sequence Sprint

1. d
2. Because the Venter–Celera approach required piecing together (in the right order) massive numbers of genome fragments. An enormous amount of data had to be processed, requiring the development of new programs and algorithms.
3. Shotgun sequencing is the breaking up the genome into random small fragments, then sequencing these shorter fragments and putting the sequences back together. The NIH team used a hierarchical approach: they first fragmented individual chromosomes into large pieces, mapped the pieces onto their physical location on the chromosome, then carried out shotgun sequencing on these larger pieces. The Celera team "shotgunned" the entire genome, without first mapping any sections of DNA onto a physical location. They then assembled the entire genome sequence from the small shotgun fragments.
4. b
5. Answers will vary.

Chapter 9

Driving Question 1

1. e
2. c
3. cell division (mitosis and cytokinesis)
4. anaphase of mitosis
5. S phase

Driving Question 2

9. b
10. b

Driving Question 3

14. No. Surgery is not a viable option. Now that the cancer has spread, there is no single tumor that can be surgically removed. There are too many tumors in too many tissues to be able to remove all of them surgically.
15. b
16. Chemotherapy interferes with all dividing cells, not just cancerous ones. Cells lining the digestive tract normally divide rapidly, and when they are affected by chemotherapy, nausea and diarrhea can occur. Similarly, cells at the base of hair follicles normally divide rapidly, and when chemotherapy drugs interfere with their division, the result is hair loss.

Driving Question 4

20. a: the Pacific Yew (*Taxus brevifolia*); b: the northern spotted owl
21. e

Chapter 10

Driving Question 1

1. b
2. c
3. No. Some mutations may have no effect on the encoded protein (silent mutations). Other mutations may be beneficial.

Driving Question 2

5. In their nonmutant states, both tumor suppressor genes and (proto-)oncogenes are important for proper cell cycle progression. In their mutant states, both contribute to the development and progression of cancer. Un-mutated tumor suppressor genes act to prevent the cell cycle from progressing inappropriately (e.g., when there is rampant DNA damage). When mutated, tumor suppressor genes can no longer pause the cell cycle when necessary, and cells with DNA damage may continue to divide. Normal proto-oncogenes act to promote cell division in response to appropriate signals to divide. When proto-oncogenes are activated by mutation (to oncogenes), they continuously "push" cells to divide, even in the absence of growth-promoting signals.
6. *BRCA1* is a tumor suppressor gene that encodes a protein involved in DNA repair of mutations.
7. e

Driving Question 3

10. b
11. e

Chapter 11

Driving Question 1

1. a
2. a: There are 46 chromosomes in each liver cell. b: There are two alleles of each gene in each liver cell.
3. a: A human gamete has 23 chromosomes. b: There is one allele of each gene present in a gamete.

Driving Question 2

7. c
8. f
9. See Infographics 11.4 and 11.5

Driving Question 3

13. b
14. a: no CF; b: no CF; c: CF
15. A person with CF has two copies of the CF-associated (recessive) alleles of the CF gene in his or her lung cells. A carrier is heterozygous, having one CF-associated (recessive) allele and one non-CF allele. A homozygous dominant person has no CF-associated alleles (both alleles are un-mutated).
16. c

Driving Question 4

22. CF cannot be diagnosed prenatally by chromosome analysis because the mutations associated with CF are too small to be detected by chromosome analysis.

Milestone 4: Mendel's Garden

1. a: It could be homozygous dominant (*TT*) or heterozygous (*Tt*). b: You would cross it with a homozygous recessive (true-breeding short) plant (*tt*). If the tall plant is true-breeding (homozygous), then 100% of the offspring will be *Tt* (tall). If the tall plant is heterozygous, 50% of the offspring will be *Tt* (tall) and 50% will be *tt* (short).

2. a: If both parents are true breeding for both traits, then 100% of the offspring will be tall with purple flowers. b: In this case, all the offspring will be homozygous dominant (*TTPP*) (as both their parents were homozygous dominant). If all the plants are homozygous dominant, then they will be true breeding.

3. This is explained by Mendel's law of segregation. The two alleles will segregate from one another at meiosis I.

4. a: Mendel's law of independent assortment would be violated, as the alleles of each gene do not appear to be assorting independently of each other. b: These results would suggest that the *T* and the *P* alleles are closely linked on one homologue, and that the *t* and the *p* alleles are closely linked on the other homologue.

Chapter 12

Driving Question 1

1. c

2. Males have only one X chromosome, while females have 2. This means that if a male inherits an X-linked recessive allele (on his mother's X chromosome), that is the only allele of that gene he has–there is no dominant allele on another X chromosome to mask the recessive. In this case, the male will develop the disease. If a female inherits an X-linked recessive allele on one of her two X chromosomes, she will not develop the disease if her other X chromosome has the dominant allele to mask the recessive allele.

3. No. Sons cannot inherit X-linked conditions from their fathers because sons inherit their father's Y chromosome, not the X. The son will inherit his X chromosome from his mother.

Driving Question 2

8. Many genes contribute to height. As there are multiple genes contributing to the phenotype, height has a polygenic component.

9. d

10. In polygenic inheritance, multiple genes influence the phenotype. Multifactorial traits are those on which environment has an influence.

11. In incomplete dominance, heterozygotes have a phenotype that is intermediate between the phenotypes of the homozygous dominants and the homozygous recessives. In codominance, heterozygotes display traits of both alleles present.

12. If you are type A-positive, then you can donate to other A-positive people, as well as to AB-positive people. If you are A-positive, you can receive type O-negative, O-positive, A-positive, and A-negative blood.

Driving Question 3

18. a: 23; b: 23; c: 46

19. Karyotype analysis can be used to detect trisomy 21, as an extra chromosome is easily visible at this level. However, cystic fibrosis is caused by mutations that change the nucleotide sequence of the gene–these cannot be detected by simply looking at the chromosomes.

Chapter 13

Driving Question 1

1. f

2. Tissues are made up of different cell types that work together. Neurons and glial cells are different cell types that work together to allow electrical impulses to travel faster through the nervous system.

3. d

Driving Question 2

8. d

9. Muscle cells and skin cells have different functions. Keratin is not necessary for muscle cell function, so muscle cells do not express the keratin gene (and so do not have the keratin protein).

10. c

Driving Question 3

13. Embryonic stem cells can differentiate into almost any cell type and are found in early embryos. Somatic stem cells are more limited with respect to the cell types they can differentiate into and are found in tissues in the body.

14. c

Chapter 14

Driving Question 1

1. d

2. a

3. In colonization, the bacteria are growing on or in the body without causing disease. Infections are associated with disease.

4. c

Driving Question 2

7. a: A sensitive strain of *S. aureus* will eventually burst (lyse) and die because of its weakened cell wall. b: A resistant strain of *S. aureus* will not be affected by penicillin and will continue to grow.

8. d

Driving Question 3

11. c

12. d

13. Fitness, in an evolutionary context, describes the ability of an individual to survive and reproduce in a given environment. Individuals that are more fit leave more offspring and more of their alleles in the next generation relative to individuals that are less fit.

14. a

15. c

Milestones in Biology 5: Adventures in Evolution

1. That although the giant sloth was an extinct animal, it resembled modern-day sloths in Argentina. This observation led him to consider that the modern-day animals might be descendants of the ancient giant sloths.

2. Malthus wrote about factors (such as hunger) that would limit the growth of population. Darwin realized that individual organisms must therefore compete for access to resources, and that any small variation that gave an individual an advantage would lead to its success over others. If the variations were successful, then individuals with them would survive and reproduce to a greater extent than individuals without these successful variants. Over time, these variations, or traits, would become more common in the population.

3. Wallace focused on Malthus's writings about disease, and how disease limited the growth of populations. As he himself was suffering from malaria, Wallace realized that disease would eliminate the weakest members of a population, leaving the strongest (the most fit) individuals to survive and reproduce. He reasoned that this would lead to changes leading to adaptations, and even to new species.

4. Both Darwin and Wallace had "aha!" moments inspired by observations while on their voyages. Darwin had read Lyell's work, and when he observed a bed of seashells in a cliff well above sea level and the effects of an earthquake he truly appreciated how much the geology of Earth changed. His observation of the extinct sloth and smaller but similar modern sloths led him to consider how resemblances suggested ancestral relationships. Wallace's observations of distinct yet similar species on either side of a physical separation (e.g., a river or canyon) helped him think about his "closely allied" species and led him to the idea each species was somehow related to a pre-existing species. Without actually seeing these fossils, organisms, and events, it is less likely that either would have been able to develop his understanding of evolution.

Chapter 15

Driving Question 1

1. 3,200 mice have 6,400 alleles, of which 2,200 are *g*. 3,200/6,400 gives an allele frequency of 0.34.

2. Population B appears to have the highest level of genetic diversity, based on allele frequencies and genotype frequencies in the population.

Driving Question 2

7. d
8. c
9. a: The allele frequencies for every allele except for *H* is now 0. The allele frequency for *H* is 1.0 (100% of the alleles are *H*). b: As a random event affecting allele frequencies, this is an example of genetic drift–specifically, it has created a bottleneck.

Driving Question 3

14. e
15. Evolution has occurred. The allele frequencies of the *TUB* gene have changed over many generations (in this case in response to a change in the environmental conditions).
16. Inbreeding can result in matings between relatives that produce offspring with two (detrimental) recessive alleles. Over time, inbreeding reduces the frequency of heterozygotes, and produces homozygotes that have two deleterious alleles.

Driving Question 4

19. d
20. When populations are geographically isolated, they do not exchange alleles. This means that if a mutation arises in one population and not the other, the mutation will be present only in one of the two populations. As the number of different mutations accumulates in each isolated population, and the resulting phenotypes are acted on by natural selection, the two populations could diverge enough so that they cannot successfully interbreed if they come into contact with each other.

Chapter 16

Driving Question 1

1. c
2. The newest fossils will be in the layers closest to the surface.
3. A fossil with four limbs and digits would be the fossil of a more recent organism than a "standard" fish. Thus, the "standard" fish fossil would be in deeper layers, and the four-limbed fossil would be in layers above that fish fossil.
4. This fossil seems to be similar to modern-day bony fishes. It is possible to observe fins with rays, suggesting that this organism was aquatic and able to swim. There does not appear to be a distinct neck, consistent with a fish, and there could be a gill cover present, again consistent with an aquatic organism. This may have been a predatory fish, as there appear to be teeth present.

Driving Question 2

8. c
9. long and sturdy ribs (to help support the body) and pectoral fins that have wrists and can bear weight

Driving Question 3

13. The skeletal anatomy of an eagle wing and that of a human arm are very similar. All major bones are present in each, and in the same locations relative to other bones. In the human, the most distal bones, like the tips of digits, are longer and arranged in a way that permits fine manipulation of objects with hands and fingers. Birds do not need to carry out this fine-scale manipulation, as their wings are specialized for flying.
14. middle ear bones in humans; gills in adult bony fish
15. You could make pairwise comparisons between the sequences, counting the number of nucleotide differences

between them. The more differences there are, the less similar the sequences are. More-similar sequences suggest closer relationships, and less-similar sequences suggest more distant relationships. Ideally, you would compare a large number of genes before coming to a conclusion.

Chapter 17

Driving Question 1

1. They are all radioactive isotopes that decay into other elements at constant rates.
2. You would use uranium-238, which has the longest half-life of the three (4.5 billion years). Isotopes with shorter half-lives may no longer be present in very ancient samples, having completely decayed.
3. (1) the first prokaryotes (-3 billion years ago); (2) an increase of oxygen in the atmosphere (-2.5 billion years ago); (3) the first multicellular eukaryotes (-1.2 billion years ago); (4) the Cambrian explosion (-545 million years ago); (5) the first animals (-540 million years ago); (6) the Permian extinction (-248 million years ago); (7) the extinction of dinosaurs (-65 million years ago)

Driving Question 2

8. No. While they may be closely related, they may also represent convergent evolution, in which unrelated groups of organisms share common characteristics because of independent natural selection in similar environments.
9. See Infographic 17.5. Several of the continental landmasses were much closer together (e.g., North America and Eurasia). Over time, through plate tectonics, the landmasses have moved to their present locations. As the landmasses moved, organisms moved with them and were subjected to changing environments, which influenced the evolution of organisms.

Driving Question 3

13. f
14. (1) domain; (2) kingdom; (3) phylum; (4) genus; (5) species
15. d

Chapter 18

Driving Question 1

1. c
2. The key difference is that prokaryotic cells do not contain membrane-enclosed organelles, particularly a nucleus. Eukaryotic cells are defined by the presence of membrane-enclosed organelles, in particular the nucleus.
3. d

Driving Question 2

6. a
7. e
8. c
9. No. You cannot conclude that there are no archaea present in that environment or sample. Many archaea are very difficult to grow (that is, culture) in the lab. This is likely because we don't understand enough about their growth requirements to be able to provide the appropriate conditions to culture them successfully in the lab. They may be present in the environmental sample, and we may just be unable to culture them.

Driving Question 3

13. High temperatures: most organisms are not adapted to high temperatures, and if they encounter temperatures higher than their optimum temperatures, their proteins will denature (that is, lose their shape and therefore their function) and their membranes will destabilize. High pressure: most organisms do not have adaptations that allow them to survive at high pressure and would be crushed by the pressure at Lost City. High pH: the high pH at Lost City is outside the pH optimum of most organisms, and would lead to their death.
14. c

Chapter 19

Driving Question 1

1. There are a variety of habitats in the park, including freshwater lakes and streams, marine environments at the ocean coastline, glacier-topped mountains, and temperate rain forest. A wide variety of organisms with different adaptations for these different habitats can live in the park.
2. Eukaryotes are defined by having cells with membrane-bound organelles, including a nucleus.
3. They are both eukaryotic and live in Olympic National Park.

Driving Question 2

6. The bryophytes were the first plants to live on land. As they do not have a vascular system to transport water throughout their bodies, they live in damp environments. They do not have adaptations that would enable them to live in other, drier environments.
7. d

Driving Question 3

11. b
12. c
13. As mammals, both fishers and humans have backbones, mammary glands, and body hair.
14. d

Driving Question 4

17. a: Fungi are not photosynthetic; they are heterotrophs. b: They do not ingest their food. c: They obtain nutrients and energy by secreting digestive enzymes onto their food. The food is digested into smaller subunits, which are then absorbed by the fungi.
18. c

Driving Question 5

20. b

Chapter 20

Driving Question 1

1. melanin, which is produced by melanocytes
2. c
3. b

Driving Question 2

9. e
10. Mitochondrial DNA is inherited only from the mother and does not undergo recombination. Thus, it is transmitted intact from mothers to all their children and so can be used to trace a lineage. As mitochondrial DNA mutates at a constant rate, it is possible to use it as a molecular clock to estimate the time since different lineages diverged.
11. a

Driving Question 3

15. e
16. (1) last common ancestor of chimpanzees and humans; (2) *Ardipithecus ramidus;* (3) *Australopithecus;* (4) *Homo erectus;* (5) *Homo sapiens*

Chapter 21

Driving Question 1

1. A population is a group of interbreeding organisms of the same species living in a particular geographic area. A community includes all the organisms (and populations) in a geographic area. Populations represent a single species, communities represent multiple species.
2. d
3. Scat reveals information about the organism that produces it. Information that can be obtained from scat analysis includes dietary preferences of the organism, as well as its DNA. The DNA can be used to identify individual organisms, as well as to examine genetic relationships with other organisms and genetic evidence of inbreeding.

Driving Question 2

7. c
8. c
9. At carrying capacity the population growth rate is 0.

Driving Question 3

12. Shootings and traffic kills are directly related to human activities, and are thus biotic factors.
13. The drying up of the pond due to drought is an abiotic factor.
14. c

Chapter 22

Driving Question 1

1. A community is a collection of interacting populations of different species in a given geographic area. A population is a group in interbreeding organisms of the same species in a given geographic area.
2. Keystone species are species that have a strong influence on the community, without necessarily having a high abundance.

3. d
4. a

Driving Question 2

7. b
8. f

Driving Question 3

12. The honey bee niche includes flowers that have a particular shape or color that attracts bees and facilitate the transfer of pollen. The niche will have a variety of flowering plants that will flower at different times during the seasons in which the bees are active. Ideally, the flowers will not have been treated with insecticides or pesticides that could have a negative impact on the honey bees. Because of different flowering times and different flower shapes and other attributes, different organisms can specialize on different sources of nectar.
13. d
14. b
15. This relationship is a mutualism: the bacteria get nutrients and a safe place to live and the bees get protection from infectious diseases.

Chapter 23

Driving Question 1

1. f
2. Maple trees are producing less sap because of fewer days of optimal temperature. Tropical fish are moving into northern oceans. Flowering plants are flowering earlier in the spring. Marmots are emerging earlier from hibernation. Polar bears are facing starvation because of melting sea ice.
3. Monthly rainfall and temperatures are very important, as are the defining plants in the area.
4. Coniferous forest biome
5. Temperate deciduous forest is in eastern North America; tropical forest is in Central America and northern South America.
6. Melting sea ice will not change sea levels. However, melting ice caps will cause a rise in sea levels, which can put low-lying cities (such as Miami) at risk for flooding.

Driving Question 2

9. b
10. e
11. No. The greenhouse effect helps trap heat in the environment. In the absence of the greenhouse effect, the temperature of the planet would be too cold to support life as we know it.

Driving Question 3

13. Using *fossil fuels* for energy converts organic carbon to CO_2. Most organisms, including plants, animals, and decomposers perform *cellular respiration*, producing CO_2 from organic food.

CO_2 is released to the *atmosphere*.

CO_2 is absorbed by the *oceans*.

Plants perform *photosynthesis*, fixing CO_2 into organic molecules.

Coal and oil are *fossil fuels*, which trap carbon below the surface.

14. c

15. a

Driving Question 4

17. a

18. Burning fossil fuels (e.g., by driving cars) and eating red meat will increase atmospheric CO_2 levels. Decomposition increases CO_2 levels in the atmosphere, as do forest fires.

Milestones in Biology 6: Progress or Poison?

1. a: DDT was intended to combat insect-borne diseases, specifically typhus and malaria, among U.S. soldiers during World War II. b: It was very effective, saving the lives of countless soldiers.

2. a: DDT is toxic to the nervous systems of insects. b: It accumulated in organisms up the food chain, and the high concentrations in top predator birds caused reproductive failures–the thinning of eggshells to the point that the eggs were easily crushed before the chicks hatched.

3. The process by which chemicals (particularly toxic chemicals) increase in concentration with each trophic level. Organisms at the lowest trophic levels have the lowest concentrations and organisms at the highest trophic levels have the highest concentrations.

4. Like DDT, PCBs are very stable in the environment, and degrade only very slowly. Even though they have not been used for decades, they are still present in the environment, and can still magnify up trophic levels.

5. Some of the questions to consider and test include: Do the pesticides leave a residue on the plants or in the environment? Are humans, other animals, or nontarget insects affected by exposure to these pesticides even at low levels? How long does it take for the pesticides to degrade? Do the pesticides accumulate in tissues of nontarget insects, wildlife, or humans? If the pesticides accumulate, what are their impacts on the health and behavior of nontarget insects, wildlife, and humans, in both the short term and the long term? If there are any toxic effects, are these reversible with treatment?

Chapter 24

Driving Question 1

1. Medicine has contributed substantially, particularly in the development of antibiotics and vaccines to treat and prevent infectious disease. Agricultural advances have increased food production to support a larger population. Public health measures to provide clean drinking water to a greater proportion of the population have decreased the incidence of many diseases.

2. An ecological footprint is the amount of land and water required to produce the resources we need and to help absorb our wastes. Sustainable living practices reduce the ecological footprint.

Driving Question 2

6. e

7. freshwater (R); coal (N); codfish populations (R); wind (R); sunlight (R)

8. Oil takes a very long time (millions of years) to form from organismal remains, so it is not renewable on a useful time scale.

Driving Question 3

12. "Sustainability" refers to living in a way that does not lead to enhanced depletion of resources. This includes minimizing use of nonrenewable resources, as well as choosing to use products that generate less waste.

13. LEED certified buildings are designed to conserve energy, conserve water, send less waste to landfills, produce only limited amounts of greenhouse gases and generally be safe and healthy for the occupants (e.g., by using less toxic paints and furnishings).

Chapter 25

Driving Question 1

1. Anatomy is the study of the structure of living organisms. Physiology is the study of how organisms function, particularly with respect to maintaining homeostasis.

2. (1) mucus-secreting cell of the small intestine; (2) the layer of muscle that contributes to the function of the small intestine; (3) small intestine; (4) digestive system

Driving Question 2

6. The maintenance of a relatively stable internal environment, even when the external environment changes.

7. d

8. In cold conditions, *sensors* will detect the *low body temperature* and send a signal to the *hypothalamus*. The hypothalamus will send signals to *effectors* (including *muscles* and blood vessels), which will exert their effects (e.g., shivering and vasoconstriction), bringing about a *normal body temperature*.

Driving Question 3

12. d

13. c

14. a

Chapter 26

Driving Question 1

1. (1) mouth; (2) esophagus; (3) stomach; (4) small intestine; (5) large intestine

2. d

3. By having an expandable stomach you can eat more at a time than you can immediately process. The stomach will expand to store the extra food until there is room in the small intestine. If the stomach were not expandable, we would have to continuously eat small amounts of food throughout the day.

Driving Question 2

7. a
8. They are both digestive enzymes. Pepsin digests proteins in the stomach, an environment with an acidic pH. Salivary amylase digests carbohydrates in the mouth, an environment with a neutral pH.
9. Lipase is produced by the pancreas.

Driving Question 3

15. a
16. Gastric banding reduces the size of the stomach by an adjustable band, producing a small pouch. This pouch cannot hold much food, restricting the amount that a person can eat. The food still passes from the stomach pouch to the small intestine in a normal fashion. In gastric bypass, a small pouch at the top of the stomach is surgically created, and surgery reroutes the passage of food directly from the stomach to the portion of the small intestine that is past the duodenum.
17. f

Milestones in Biology 7: Stumbling on a Cure

1. c
2. c
3. a
4. It is not possible to distinguish between type 1 diabetes and type 2 diabetes on the basis of blood-sugar levels: both types lead to elevated levels of blood sugar. In order to distinguish between them, one would need to monitor insulin levels. In type 1 diabetes insulin levels would be essentially at zero. In type 2 diabetes they would be low between meals and high after meals, providing a critical clue that blood glucose is remaining high after meals, even in the presence of high insulin level.
5. b
6. In both types there are elevated levels of blood glucose, both immediately after a carbohydrate-rich meal and also for extended periods after the meal. In type 1 diabetes, no insulin is produced, so insulin will remain low or at zero at all times. In type 2 diabetes, the body can make and release insulin normally, so (at least in early stages of the disease) insulin will be low between meals, then rise after a carbohydrate-rich meal. In type 2 diabetes, the blood-glucose will not come down in response to insulin.

Chapter 27

Driving Question 1

1. the blood, blood vessels, and the heart
2. b
3. b

Driving Question 2

7. c
8. d
9. b
10. a

Driving Question 3

14. Different types of white blood cell have different specific functions, but overall, white blood cells are important

for the immune response. Red blood cells carry oxygen and deliver it to cells. Platelets are important in blood clotting.

15. e
16. d

Driving Question 4

19. heart attack–C; hypertension–A; stroke–D; atherosclerosis–B
20. There are many possibilities. Some include family history/genetics (not modifiable); diet (modifiable); exercise (modifiable); race (not modifiable); smoking (modifiable); obesity (modifiable to a certain extent). Type 2 diabetes and hypertension are risk factors that have both modifiable and unmodifiable aspects; in both cases, proper management after diagnosis is important and modifiable.
21. The plaques of atherosclerosis can rupture, promoting blood clot formation. If these clots travel to and block a blood vessel in the brain, a stroke can occur.
22. b

Chapter 28

Driving Question 1

1. (A) pharynx; (B) larynx; (C) lung; (D) bronchioles
2. a
3. d

Driving Question 2

6. d
7. d

Driving Question 3

11. b
12. c

Driving Question 4

16. b
17. b
18. There are many possibilities. The independent variable will be the treatment: altitude training, hypoxic chambers, or no oxygen manipulation. The dependent variable could be a variety of factors relative to athletic performance–speed, endurance, strength–and there are a variety of ways to measure these. You would also have to consider whether any advantage was to be measured at sea level or at altitude. You could also consider the relative timing of the treatment and the duration of the treatment before assessing athletic advantage.

Chapter 29

Driving Question 1

1. light-detecting receptor in the eye (PNS); amydgala (CNS); pain receptor in the skin (PNS); spinal cord (CNS); thalamus (CNS)
2. movement–cerebellum; body temperature–hypothalamus (diencephalon)
3. (1) chimpanzee; (2) bird; (3) salamander; (4) fish

Driving Question 2

7. d

8. b

9. c

10. The neurotransmitter diffuses from the signaling cell (the presynaptic cell) to the receiving cell (the postsynaptic cell). It then binds receptors on the postsynaptic cell.

11. Electrical signaling takes place along the length of axons and involves the movement of sodium and potassium ions across the membrane of axons. Chemical signaling involves the release of chemical signaling molecules (neurotransmitters) into synapses, where they communicate information between neurons and their effector cells (other neurons, glands, muscles).

Driving Question 3

15. b

16. Because of the initial surge of dopamine, receptors respond by down-regulating–that is, reducing their numbers. With fewer receptors, the drug cannot elicit excessive feelings of pleasure unless more drug is ingested to overcome the effect of down-regulation.

Chapter 30

Driving Question 1

1. c

2. b

3. a: The uterus is the organ in which embryos and fetuses develop. The cervix is the opening into the uterus from the vagina. b: The endometrium is the lining of the uterus. The embryo implants into the endometrium.

Driving Question 2

7. luteinizing hormone (LH)–anterior pituitary; follicle-stimulating hormone (FSH)–anterior pituitary; testosterone–testes; estrogen–ovary; progesterone–ovary; hCG–embryo

8. a

Driving Question 3

13. b

14. In both IUI and IVF, drugs are administered to stimulate egg production in the woman. In IVF, the eggs are surgically removed, then fertilized in vitro (that is, in a dish) and embryos are placed in the woman's uterus, with the hope that at least one will successfully implant. In IUI, the timing of ovulation is carefully monitored so that sperm can be inserted in the uterus just before ovulation (this procedure generally takes place in a doctor's office or a fertility clinic). Multiple eggs may be ovulated, and the hope is that at least one of them will be fertilized by one of the sperm inserted in the uterus.

Chapter 31

Driving Question 1

1. d

2. Viruses must enter their specific host cell in the human, thus infecting it. Once they have entered the host cell, they use host cell resources and direct the host cell to replicate the virus.

3. Poliovirus infects and damages nerve cells. As nerve cells do not replicate, they cannot repair themselves. Influenza virus infects cells lining the respiratory tract, which are replaced at high frequency.

Driving Question 2

6. Skin provides a barrier to pathogen entry. Enzymes in tears and saliva digest components of pathogens. Phagocytes ingest and destroy pathogens.

7. They are both specialized white blood cells (macrophages are found in tissues, neutrophils circulate in the blood), and they are both phagocytes.

Driving Question 3

11. Innate immunity is present since birth, always active and nonspecific. It does not have "memory." Adaptive immunity is specific for a particular pathogen and must be turned on when the specific pathogen is encountered–it is not always "on"). Adaptive immunity exhibits "memory," strengthening with repeated exposures to the same pathogen.

12. B cells are lymphocytes that are activated during an adaptive response. Upon activation, they specialize into antibody-producing plasma cells. The antibodies produced then specifically bind to and inactivate their targets.

Driving Question 4

20. Antigenic shift is the result of the exchange of segments of viruses, so a virus can end up with a genomic segment (and corresponding genes) from a completely different virus. These viruses tend to be very different than either starting virus. Antigenic drift is the result of small changes–mutations–in the viral genome.

21. To become pandemic this virus would need to be able to spread easily from human to human, thus spreading widely throughout the human population.

Chapter 32

Driving Question 1

1. See Infographic 32.1.

2. b

3. d

Driving Question 2

5. a

6. f

7. c

Driving Question 3

14. c

15. The pollen grain lands on the stigma (pollination). The pollen grain produces a tube that grows down the style. The sperm from the pollen grain travel in this tube to the ovule, where an egg is fertilized.

16. d

17. b

18. c

Glossary

abiotic Refers to the nonliving components of an environment, such as temperature and precipitation.

abscisic acid (ABA) A plant hormone that helps seeds remain dormant.

absorption The uptake of digested food molecules by the epithelial cells lining the small intestine.

acclimatization The process of physiologically adjusting to an environmental change over a period of time. Acclimatization is generally reversible.

acid A substance that increases the hydrogen ion concentration of solutions.

acidosis A dangerous condition in which blood is too acidic.

action potential An electrical signal within neurons caused by ions moving across the cell membrane.

activation energy The energy required for a chemical reaction to proceed. Enzymes accelerate reactions by reducing their activation energy.

active site The part of an enzyme that binds to a substrate.

active transport The energy-requiring process by which solutes are pumped from an area of lower concentration to an area of higher concentration with the help of transport proteins.

adaptation The process by which populations become better suited to their environment as a result of natural selection.

adaptive immunity A protective response, mediated by lymphocytes, that confers long-lasting immunity against specific pathogens.

adaptive radiation The spreading and diversification of organisms that occur when they colonize a new habitat.

adenosine triphosphate (ATP) The molecule that cells use to power energy-requiring functions; the cell's energy "currency."

adhesion Water molecules sticking to other surfaces through hydrogen bonding.

adult (somatic) stem cells Stem cells located in tissues that help maintain and regenerate those tissues.

aerobic respiration A series of reactions that occurs in the presence of oxygen and converts energy stored in food into ATP.

alga (plural: algae) A uni- or multicellular photosynthetic protist.

allele frequency The relative proportion of an allele in a population.

alleles Alternative versions of the same gene that have different nucleotide sequences.

allergy A misdirected immune response against environmental substances such as dust, pollen, and foods that causes discomfort in the form of physical symptoms.

altitude sickness An illness that can occur as a result of an abrupt move to an altitude with a reduced partial pressure of oxygen.

alveoli Air sacs in the lung across which gases diffuse between air and blood.

amino acid The building block, or monomer, of a protein.

amniocentesis A procedure that removes fluid surrounding the fetus to obtain and analyze fetal cells to diagnose genetic disorders.

amygdala A subregion of the brain that processes emotions, especially fear and anxiety, and is the seat of emotional memories.

anabolic reaction Any chemical reaction that combines simple molecules to build more-complex molecules.

anatomy The study of the physical structures that make up an organism.

androgen A class of sex hormones, including testosterone, that is present in higher levels in men and causes male-associated traits like deep voice, growth of facial hair, and defined musculature.

anecdotal evidence An informal observation that has not been systematically tested.

aneuploidy An abnormal number of one or more chromosomes (either extra or missing copies).

angiosperm A seed-bearing flowering plant with seeds typically contained within a fruit.

animal A eukaryotic multicellular organism that obtains nutrients by ingesting other organisms.

annelid A segmented worm, such as an earthworm.

anterior pituitary The gland in the brain that secretes luteinizing hormone (LH) and follicle-stimulating hormone (FSH).

antibiotics Chemicals that either kill bacteria or slow their growth by interfering with the function of essential bacterial cell structures.

antibody A protein produced by B cells that binds to antigens and either neutralizes them or flags other cells to destroy pathogens.

anticodon The part of a tRNA molecule that binds to a complementary mRNA codon.

antigen A specific molecule (or part of a molecule) to which specific antibodies can bind, and against which an adaptive response is mounted.

antigenic drift Changes in viral antigens caused by genetic mutation during normal viral replication.

antigenic shift Changes in antigens that occur when viruses exchange genetic material with other strains.

aorta The large artery that receives blood from the left ventricle.

apoptosis Programmed cell death; often referred to as cellular suicide.

aquifer Underground layers of porous rock from which water can be drawn for use.

Archaea One of the two domains of prokaryotic life; the other is Bacteria.

arteries Blood vessels that carry blood away from the heart.

arthropod An invertebrate having a segmented body, a hard exoskeleton, and jointed appendages.

atherosclerosis A disease process that restricts blood flow through arteries; also known as hardening of the arteries.

atom The smallest unit of an element that cannot be chemically broken down into smaller units.

atria (singular: atrium) The chambers of the heart that receive blood. The right atrium receives low-oxygen blood from the body, and the left atrium receives high oxygen blood from the lungs.

autoimmune disease A misdirected immune response in which the immune system mistakenly attacks healthy cells.

autosomes Paired chromosomes present in both males and females; all chromosomes except the X and Y chromosomes.

autotrophs Organisms such as plants, algae, and certain bacteria that capture the energy of sunlight by photosynthesis.

auxin A plant hormone that causes elongation of cells as one of its effects.

axon The long extension of a neuron that conducts action potentials away from the cell body toward the axon terminal.

axon terminal The tip of an axon, which communicates with the next cell in the pathway.

B cells White blood cells that mature in the bone marrow and produce antibodies during the adaptive immune response.

Bacteria One of the two domains of prokaryotic life; the other is Archaea.

base A substance that reduces the hydrogen ion concentration of solutions.

benign tumor A noncancerous tumor that will not spread throughout the body.

bilateral symmetry The pattern exhibited by a body plan with right and left halves that are mirror images of each other.

bile salts Chemicals produced by the liver and stored by the gallbladder that emulsify fats so that they can be chemically digested by enzymes.

binary fission A type of asexual reproduction in which one parental cell divides into two.

biocapacity The amount of Earth's biologically productive area–cropland, pasture, forest, and fisheries–that is available to provide resources and absorb wastes to support life.

biodiversity The number of different species and their relative abundances in a specific region or on the planet as a whole.

biofuels Renewable fuels made from living organisms (e.g., plants and algae).

biogeography The study of how organisms are distributed in geographical space.

biological species concept The definition of a species as a population whose members can interbreed to produce fertile offspring.

biome A large geographic area defined by its characteristic plant life, which in turn is determined by temperature and levels of moisture.

biotic Refers to the living components of an environment.

blastocyst The stage of embryonic development in which the embryo is a hollow ball of cells. Researchers can derive embryonic stem cell lines during the blastocyst stage.

blood pressure The overall pressure in blood vessels, expressed as the systolic pressure over the diastolic pressure.

blood vessels The components of the cardiovascular system that transport blood throughout the body.

body mass index (BMI) An estimate of body fat based on height and weight.

bottleneck effect A type of genetic drift that occurs when a population is suddenly reduced to a small number of individuals, and alleles are lost from the population as a result.

brain An organ of the central nervous system that integrates and coordinates virtually all functions of the body.

brain stem The part of the brain that is closest to the spinal cord and which controls vital functions such as heart rate, breathing, and blood pressure.

bronchi Two airways that branch from the trachea; one bronchus leads into each lung.

bronchioles Smaller airways that branch from the bronchi.

bryophyte A nonvascular plant that does not produce seeds.

Calorie 1,000 calories or 1 kilocalorie (kcal); the capital "C" in Calorie indicates "kilocalorie." The Calorie is the common unit of energy used in food nutrition labels.

calorie The amount of energy required to raise the temperature of 1 g of water by 1°C.

cancer A disease of unregulated cell division: cells divide inappropriately and accumulate, in some instances forming a tumor.

capillaries The smallest blood vessels. Capillaries are the sites of gas, nutrient, and waste exchange between the blood and tissue cells.

capsule A sticky coating surrounding some bacterial cells that adheres to surfaces.

carbohydrate An organic molecule made up of one or more sugars. A one-sugar carbohydrate is called a monosaccharide; a carbohydrate with multiple linked sugars is called a polysaccharide.

carbon cycle The movement of carbon atoms as they cycle between organic molecules and inorganic CO_2.

carbon fixation The conversion of inorganic carbon (e.g., CO_2) into organic forms (e.g., sugars).

carbon footprint A measure of the total greenhouse gases produced by human activities.

carcinogen Any chemical agent that causes cancer by damaging DNA. Carcinogens are a type of mutagen.

cardiovascular disease (CVD) A disease of the heart or blood vessels or both.

cardiovascular system The system that transports nutrients, gases, and other critical molecules throughout the body. It consists of the heart, blood vessels, and blood.

carrier An individual who is heterozygous for a particular gene of interest, and therefore can pass on the recessive allele without showing any of its effects.

carrying capacity The maximum population size that a given environment or habitat can support given its food supply and other natural resources.

catabolic reaction Any chemical reaction that breaks down complex molecules into simpler molecules.

catalysis The process of speeding up the rate of a chemical reaction (e.g., by enzymes).

cell The basic structural unit of living organisms.

cell body The part of a neuron that contains most of the cell's organelles, including the nucleus.

cell cycle The ordered sequence of stages that a cell progresses through in order to divide during its life; stages include preparatory phases (G_1, S, G_2) and division phases (mitosis and cytokinesis).

cell cycle checkpoint A cellular mechanism that ensures that each stage of the cell cycle is completed accurately.

cell division The process by which a cell reproduces itself; cell division is important for normal growth, development, and repair of an organism.

cell membrane A phospholipid bilayer with embedded proteins that forms the boundary of all cells.

cell theory The concept that all living organisms are made of cells and that cells are formed by the reproduction of existing cells.

cell wall A rigid structure enclosing the cell membrane of some cells that helps the cell maintain its shape.

cell-mediated immunity The type of adaptive immunity that rids the body of altered (that is, infected, cancerous, or foreign) cells.

cellular differentiation The process by which a cell specializes to carry out a specific role.

central nervous system (CNS) The brain and the spinal cord.

central vacuole A fluid-filled compartment in plant cells that contributes to cell rigidity by exerting turgor pressure against the cell wall.

centromere The specialized region of a chromosome where the sister chromatids are joined; critical for proper alignment and separation of sister chromatids during mitosis.

cerebellum The part of the brain that processes sensory information and is involved in movement, coordination, and balance.

cerebral cortex The outer layer of the cerebrum, the cerebral cortex is involved in many advanced brain functions.

cerebrum The region of the brain that controls intelligence, learning, perception, and emotion.

cervix The opening or "neck" of the uterus, where sperm enter and babies exit.

chemical energy Potential energy stored in the bonds of biological molecules.

chemotherapy The treatment of disease, specifically cancer, by the use of chemicals.

chlorophyll The pigment present in the green parts of plants that absorbs photons of light energy during the "photo" reactions of photosynthesis.

chloroplast The organelle in plant and algae cells where photosynthesis occurs.

cholesterol A lipid that is transported in the blood by specialized proteins.

chromosome A single, large DNA molecule wrapped around proteins. Chromosomes are located in the nuclei of most eukaryotic cells.

chyme The acidic "soup" of partially digested food that leaves the stomach and enters the small intestine.

citric acid cycle A set of reactions that takes place in mitochondria and helps extract energy (in the form of high-energy electrons) from food; the second stage of aerobic respiration.

closed circulatory system A circulatory system in which the blood remains in the blood vessels at all times.

coding sequence The part of a gene that specifies the amino acid sequence of a protein. Coding sequences determine the identity, shape, and function of proteins.

codominance A form of inheritance in which both alleles contribute equally to the phenotype.

codon A sequence of three mRNA nucleotides that specifies a particular amino acid.

coenzyme A small organic molecule, such as a vitamin, required to activate an enzyme.

cofactor An inorganic substance, such as a metal ion, required to activate an enzyme.

cohesion Water molecules sticking to water molecules through hydrogen bonding.

colon The first and longest portion of the large intestine; the colon plays an important role in water reabsorption.

commensalism A type of symbiotic relationship in which one member benefits and the other is unharmed.

community A group of interacting populations of different species living together in the same area.

competition An interaction between two or more organisms that rely on a common resource that is not available in sufficient quantities.

competitive exclusion principle The concept that when two species compete for resources in an identical niche, one is inevitably driven to extinction.

complement proteins Proteins in blood that help destroy pathogens by coating or puncturing them.

complementary Fitting together; two strands of DNA are said to be complementary in that A always pairs with T, and G always pairs with C.

conservation of energy The principle that energy cannot be created or destroyed, but can be transformed from one form to another.

consumers Heterotrophs that eat other organisms lower on the food chain to obtain energy.

continuous variation Variation in a population showing an unbroken range of phenotypes rather than discrete categories.

contraception The prevention of pregnancy through physical, surgical, or hormonal methods.

control group The group in an experiment that experiences no experimental intervention or manipulation.

convergent evolution The process by which organisms that are not closely related evolve similar adaptations as a result of independent episodes of natural selection.

coronary arteries The blood vessels that deliver oxygenated blood to the heart muscle.

corpus luteum The structure in the ovary that remains after ovulation and secretes progesterone.

correlation A consistent relationship between two variables.

covalent bond A strong chemical bond resulting from the sharing of a pair of electrons between two atoms.

cuticle The waxy coating on leaves and stems that prevents water loss.

cytokinesis The physical division of a cell into two daughter cells.

cytoplasm The gelatinous, aqueous interior of all cells.

cytoskeleton A network of protein fibers in eukaryotic cells that provides structure and facilitates cell movement.

cytotoxic T cell A type of T cell that destroys infected, cancerous, or foreign altered cells, including virally infected cells.

decomposer An organism such as a fungus or bacterium that digests and uses the organic molecules in dead organisms as sources of nutrients and energy.

dendrites Branched extensions from the cell body of a neuron, which receive incoming information.

deoxyribonucleic acid (DNA) The molecule of heredity, common to all life forms, that is passed from parents to offspring.

dependent variable The measured result of an experiment, analyzed in both the experimental and control groups.

descent with modification Darwin's term for evolution, combining the ideas that all living things are related and that organisms have changed over time.

diabetes A disease characterized by chronically elevated levels of blood sugar.

diaphragm A sheet of muscle that contributes to breathing by contracting and relaxing.

diastolic pressure The pressure in arteries when the ventricle is relaxed.

diencephalon A brain region located between the brain stem and the cerebrum that includes the thalamus and hypothalamus and regulates homeostatic functions like body temperature, hunger, thirst, and the sex drive.

differential gene expression The process by which genes are "turned on," or expressed, in different cell types.

diffusion The movement of dissolved substances from an area of high concentration to an area of low.

digestion The mechanical and chemical breakdown of food into subunits so that nutrients can be absorbed.

digestive system The organ system that breaks down food molecules into smaller subunits, absorbing nutrients, and eliminating waste; composed of the digestive tract and accessory organs.

digestive tract The central pathway of the digestive system; a long muscular tube that pushes food between the mouth and the anus.

diploid Having two copies of every chromosome.

directional selection A type of natural selection in which organisms with phenotypes at one end of a spectrum are favored by the environment.

distribution pattern The way that organisms are distributed in geographic space, which depends on resources and interactions with other members of the population.

diversifying selection A type of natural selection in which organisms with phenotypes at both extremes of the phenotypic range are favored by the environment.

DNA polymerase An enzyme that "reads" the sequence of a DNA strand and helps to add complementary nucleotides to form a new strand during DNA replication.

DNA profile A visual representation of a person's unique DNA sequence.

DNA replication The natural process by which cells make an identical copy of a DNA molecule.

domain The highest category in the modern system of classification; there are three domains–Bacteria, Archaea, and Eukarya.

dominant allele An allele that can mask the presence of a recessive allele.

dopamine A neurotransmitter that is involved in conveying a sense of pleasure in the brain.

double helix The spiral structure formed by two strands of DNA nucleotides bound together.

duodenum The first portion of the small intestine; the duodenum receives chyme from the stomach and mixes it with digestive secretions from other organs.

ecological footprint A measure of how much land and water area is required to supply the resources a person or population consumes and to absorb the wastes it produces.

ecology The study of the interactions between organisms and between organisms and their nonliving environment.

ecosystem The living and nonliving components of an environment, including the communities of organisms present and the physical and chemical environment with which they interact.

ectotherm An animal that relies on environmental sources of heat, such as sunlight, to maintain its body temperature.

effector A cell or tissue that acts to exert a response on the basis of information relayed from a sensor.

electron A negatively charged subatomic particle with negligible mass.

electron transport chain A process that takes place in mitochondria and produces the bulk of ATP during aerobic respiration; the third stage of aerobic respiration.

element A chemically pure substance that cannot be chemically broken down; each element is made up of and defined by a single type of atom.

elimination The expulsion of undigested matter in the form of stool.

embryo An early stage of development reached when a zygote undergoes cell division to form a multicellular structure.

embryonic stem cells Stem cells that make up an early embryo and which can differentiate into nearly every cell type in the body.

emigration The movement of individuals out of a population.

emulsify To break up large fat globules into small fat droplets that can be more efficiently chemically digested by enzymes.

endocrine system The collection of hormone-secreting glands and organs with hormone-secreting cells.

endometrium The lining of the uterus.

endoplasmic reticulum (ER) A network of membranes in eukaryotic cells where proteins and lipids are synthesized.

endoskeleton A solid internal skeleton found in many animals, including humans.

endosymbiosis The scientific theory that free-living prokaryotic cells engulfed other free living prokaryotic cells billions of years ago, forming eukaryotic organelles such as mitochondria and chloroplasts.

endotherm An animal that can generate body heat internally to maintain its body temperature.

energy The ability to do work, including building complex molecules.

enzyme A protein that speeds up the rate of a chemical reaction.

epidemiology The study of patterns of disease in populations, including risk factors.

epididymis Tubes in which sperm mature and are stored before ejaculation.

epithelial cells Cells that line organs and body cavities; in the digestive tract they sit in direct contact with food and its breakdown products.

erythropoietin (EPO) A hormone that stimulates red blood cell production.

esophagus The section of the digestive tract between the mouth and the stomach.

essential amino acids Amino acids the human body cannot synthesize and must obtain from food.

essential nutrients Nutrients that cannot be made by the body, so must be obtained from the diet.

estrogen A female sex hormone produced by the ovaries.

ethylene A gaseous plant hormone that promotes fruit ripening as one of its effects.

eukaryote Any organism of the domain Eukarya; eukaryotic cells are characterized by the presence of a membrane-enclosed nucleus and organelles.

eukaryotic cells Cells that contain membrane-bound organelles, including a central nucleus.

evolution Change in allele frequencies in a population over time.

exoskeleton A hard external skeleton covering the body of many animals, such as arthropods.

experiment A carefully designed test, the results of which will either support or rule out a hypothesis.

experimental group The group in an experiment that experiences the experimental intervention or manipulation.

exponential growth The unrestricted growth of a population increasing at a constant growth rate.

extinction The elimination of all individuals in a species; extinction may occur over time or in a sudden mass die-off.

facilitated diffusion The process by which large or hydrophilic solutes move across a membrane from an area of higher concentration to an area of lower concentration with the help of transport proteins; facilitated diffusion does not require an input of energy.

falsifiable Describes a hypothesis that can be ruled out by data that show that the hypothesis does not explain the observation.

feedback loop A pathway that involves input from a sensor, a response via an effector, and detection of the response by the sensor.

fermentation A series of chemical reactions that takes place in the absence of oxygen and converts some of the energy stored in food into ATP. Fermentation produces far less ATP than does aerobic respiration.

fern The first true vascular plants; ferns do not produce seeds.

fertilization The fusion of an egg and a sperm.

fetus After the eighth week after fertilization, the embryo is referred to as a fetus.

fitness The relative ability of an organism to survive and reproduce in a particular environment.

flagella (singular: flagellum) Whiplike appendages extending from the surface of some bacteria, used in movement of the cell.

folate A B vitamin also known as folic acid, folate is an essential nutrient, necessary for basic bodily processes such as DNA replication and cell division.

follicle The part of the ovary where eggs mature.

follicle-stimulating hormone (FSH) A hormone secreted by the anterior pituitary. In females, FSH triggers eggs to mature at the start of each monthly cycle.

food chain A linked series of feeding relationships in a community in which organisms further up the chain feed on ones below.

food web A complex interconnection of feeding relationships in a community.

fossil fuel A carbon-rich energy source, such as coal, petroleum, or natural gas, formed from the compressed, fossilized remains of once-living organisms.

fossil record An assemblage of fossils arranged in order of age, providing evidence of changes in species over time.

fossils The preserved remains or impressions of once-living organisms.

founder effect A type of genetic drift in which a small number of individuals leaves one population and establishes a new population; by chance, the newly established population may have lower genetic diversity than the original population.

fungus (plural: fungi) A unicellular or multicellular eukaryotic organism that obtains nutrients by secreting digestive enzymes onto organic matter and absorbing the digested product.

gallbladder An organ that stores bile salts and releases them as needed into the small intestine.

gametes Specialized reproductive cells that carry one copy of each chromosome (that is, they are haploid). Sperm are male gametes; eggs are female gametes.

gas exchange The process of taking up and releasing oxygen and carbon dioxide.

gel electrophoresis A laboratory technique that separates fragments of DNA by size.

gene A sequence of DNA that contains the information to make at least one protein.

gene expression The process of using DNA instructions to make proteins.

gene flow The movement of alleles from one population to another, which may increase the genetic diversity of a population.

gene pool The total collection of alleles in a population.

gene therapy A treatment that aims to cure human disease by replacing defective genes with functional ones.

genetic code The set of rules relating particular mRNA codons to particular amino acids.

genetic drift Random changes in the allele frequencies of a population between generations; genetic drift tends to have more dramatic effects in smaller populations than in larger ones.

genetic engineering The process of assembling new genes with novel combinations of regulatory and coding sequences.

genetically modified organism (GMO) An organism that has been genetically altered by humans.

genome One complete set of genetic instructions encoded in the DNA of an organism.

genotype The particular genetic makeup of an individual.

germination The process in which a seed comes out of dormancy and a new plant begins to grow.

germ-line mutation A mutation occurring in gametes; passed on to offspring.

gibberellins Plant hormones that cause stem elongation and cell division.

gills The gas exchange surface in fish.

glial cells Supporting cells of the nervous system.

global hectare A unit of measurement representing the biological productivity (both resource-providing and waste-absorbing capacity) of an average hectare of Earth.

global warming An increase in Earth's average temperature.

glucagon A hormone produced by the pancreas that causes an increase in blood sugar.

glycogen A complex animal carbohydrate, made up of linked chains of glucose molecules, that stores energy for short term use.

glycolysis A series of reactions that breaks down sugar into smaller units; glycolysis takes place in the cytoplasm and is the first stage of both aerobic respiration and fermentation.

Golgi apparatus An organelle made up of stacked membrane enclosed discs that packages proteins and prepares them for transport.

gonads Sex organs: ovaries in females, testes in males.

Gram-negative Refers to bacteria with a cell wall that includes a thin layer of peptidoglycan surrounded by an outer lipid membrane that does not retain the Gram stain.

Gram-positive Refers to bacteria with a cell wall that includes a thick layer of peptidoglycan that retains the Gram stain.

gravitropism The growth of plants in response to gravity. Roots grow downward, with gravity; shoots grow upward, against gravity.

greenhouse effect The normal process by which heat is radiated from Earth's surface and trapped by gases in the atmosphere, helping to maintain Earth at a temperature that can support life.

greenhouse gas Any of the gases in Earth's atmosphere that absorb heat radiated from Earth's surface and contribute to the greenhouse effect; for example, carbon dioxide and methane.

growth rate The difference between the birth rate and the death rate of a given population; also known as the rate of natural increase.

gymnosperm A seed-bearing plant with exposed seeds typically held in cones.

habitat The physical environment where an organism lives and to which it is adapted.

half-life The time it takes for one-half of a sample of a radioactive isotope to decay.

haploid Having only one copy of every chromosome.

Hardy-Weinberg equation A mathematical formula that calculates the frequency of genotypes and phenotypes one would expect to find in a nonevolving population.

Hardy-Weinberg equilibrium The principle that, in a nonevolving population, both allele and genotype frequencies remain constant from one generation to the next.

heart The muscular pump that generates force to move blood throughout the body.

heart attack A restriction of blood flow to the heart, resulting in damage to the heart muscle.

heat The kinetic energy generated by random movements of molecules or atoms.

helper T cell A type of T cell that helps activate B cells to produce antibodies during humoral responses.

heme group Iron-containing structures on hemoglobin, the sites of oxygen binding.

hemoglobin A protein found in red blood cells specialized for transporting oxygen.

herbivory Predation on plants, which may or may not kill the plant preyed on.

heterotrophs Organisms, such as humans and other animals, that obtain energy by eating organic molecules that were produced by other organisms.

heterozygous Having two different alleles.

high-density lipoprotein (HDL) A form of cholesterol and protein that is protective.

hippocampus A subregion of the brain involved in learning and memory.

histamine A molecule released by damaged tissue and during allergic reactions.

homeostasis The maintenance of a relatively stable internal environment, even when the external environment changes.

hominid Any living or extinct member of the family Hominidae, the great apes—humans, orangutans, gorillas, chimpanzees, and bonobos.

homologous chromosomes A pair of chromosomes that both contain the same genes. In a diploid cell, one chromosome in the pair is inherited from the mother, the other from the father.

homology Anatomical, genetic, or developmental similarity among organisms due to common ancestry.

homozygous Having two identical alleles.

hormone A chemical signaling molecule that is released by a cell or gland and travels through the bloodstream to exert an effect on target cells.

human chorionic gonadotropin (hCG) A hormone produced by an early embryo that helps maintain the corpus luteum until the placenta develops.

humoral immunity The type of adaptive immunity that fights free-floating pathogens infections and other foreign substances in the circulation and lymph fluid.

hydrogen bond A weak electrical attraction between a partially positive hydrogen atom and an atom with a partial negative charge.

hydrophilic "Water-loving"; hydrophilic molecules dissolve in water.

hydrophobic "Water-fearing"; hydrophobic molecules will not dissolve in water.

hypertension Elevated (high) blood pressure.

hypertonic Describes a solution surrounding a cell that has a higher concentration of solutes than the cell.

hypha (plural: hyphae) A long, threadlike structure through which fungi absorb nutrients.

hypothalamus A master coordinator region of the brain responsible for a variety of physiological functions.

hypothermia A drop of body temperature below 35°C (95°F), which causes enzyme malfunction and, eventually, death.

hypothesis A tentative explanation for a scientific observation or question.

hypotonic Describes a solution surrounding a cell that has a lower concentration of solutes than the cell.

hypoxia A state of low oxygen concentration in the blood.

immigration The movement of individuals into a population.

immune system A system of cells and tissues that acts to defend the body against foreign cells and infectious agents.

immunity The resistance to a given pathogen conferred by the activity of the immune system.

in vitro fertilization (IVF) A form of assisted reproduction in which eggs and sperm are brought together outside the body and the resulting embryos are inserted into a woman's uterus.

inbreeding Mating between closely related individuals. Inbreeding does not change the allele frequency within a population, but it does increase the proportion of homozygous individuals to heterozygotes.

inbreeding depression The negative reproductive consequences for a population associated with having a high frequency of homozygous individuals possessing harmful recessive alleles.

incomplete dominance A form of inheritance in which heterozygotes have a phenotype that is intermediate between homozygous dominant and homozygous recessive.

independent assortment The principle that alleles of different genes are distributed independently of one another during meiosis.

independent variable The variable, or factor, being deliberately changed in the experimental group.

induced pluripotent stem cell A pluripotent stem cell that was generated by manipulation of a differentiated somatic cell.

inflammation An innate defense that is activated by infection or local tissue damage; characterized by redness, swelling, and pain.

ingestion The act of taking food into the mouth.

innate immunity Nonspecific defenses, such as physical and chemical barriers and phagocytic cells that are present from birth and are always active.

inorganic Describes a molecule that lacks a carbon-based backbone and C-H bonds.

insect A six-legged arthropod with three body segments: head, thorax, and abdomen.

insulin A hormone secreted by the pancreas that causes a decrease in blood sugar.

interferon Antiviral proteins produced by virally infected cells to help protect adjacent cells from becoming infected.

interphase The stage of the cell cycle in which cells spend most of their time, preparing for cell division. There are three distinct sub-phases: G_1, S, and G_2.

intrauterine insemination (IUI) A form of assisted reproduction in which sperm are injected directly into a woman's uterus.

invertebrate An animal lacking a backbone.

ion An electrically charged atom, the charge resulting from the loss or gain of electrons.

ionic bond A strong electrical attraction between oppositely charged ions formed by the transfer of one or more electrons from one atom to another.

isotonic Describes a solution surrounding a cell that has the same solute concentration as the cell.

karyotype The chromosomal makeup of cells. Karyotype analysis can be used to detect trisomy 21 prenatally.

keystone species Species on which other species depend, and whose removal has a dramatic impact on the community.

kidney An organ involved in osmoregulation, filtration of blood to remove wastes, and production of several important hormones.

kinetic energy The energy of motion or movement.

kinetochore Proteins located at the centromere that provide an attachment point for microtubules of the mitotic spindle.

large intestine The last organ of the digestive tract, in which remaining water is absorbed and solid stool is formed.

larynx The opening to the lower respiratory tract; also known as the voicebox.

light energy The energy of the electromagnetic spectrum of radiation that is visible to the human eye.

lignin A stiff strengthening agent found in secondary cell walls of plants.

lipase A fat-digesting enzyme active in the small intestine.

lipids Organic molecules that generally repel water.

liver An organ that aids digestion by producing bile salts that emulsify fats.

logistic growth A pattern of growth that starts off fast and then levels off as the population reaches the carrying capacity of the environment.

low-density lipoprotein (LDL) A form of cholesterol and protein in the blood that contributes to cardiovascular disease.

lung The major respiratory organ in humans, the site of gas exchange between air and the blood.

luteinizing hormone (LH) A hormone secreted by the anterior pituitary. In females, a surge of LH triggers ovulation.

lymph nodes Small organs in the lymphatic system where B and T cells may encounter pathogens.

lymphatic system The organ system of vessels and organs where B and T lymphocytes develop and that works with the immune system, allowing B and T cells to respond to pathogens.

lymphocyte A specialized white blood cell of the immune system.

lysosome An organelle in eukaryotic cells filled with enzymes that can degrade worn-out cellular structures.

macromolecules Large organic molecules that make up living organisms; they include carbohydrates, proteins, and nucleic acids.

macronutrients Nutrients, including carbohydrates, proteins, and fats, that organisms must ingest in large amounts to maintain health.

macrophage A phagocytic cell that resides in tissues and plays an important role in the inflammatory response.

malignant tumor A cancerous tumor that spreads throughout the body.

malnutrition The medical condition resulting from the lack of any essential nutrient in the diet. Malnutrition is often, but not always, associated with starvation.

mammals Members of the class Mammalia; all members of this class have mammary glands and a body covered with hair.

mass extinction An extinction of between 50% and 90% of all species that occurs relatively rapidly.

matter Anything that takes up space and has mass.

meiosis A specialized type of nuclear division that generates genetically unique haploid gametes.

melanin Pigment produced by a specific type of skin cell that gives skin its color.

memory cell A long-lived B or T cell that is produced during an immune primary response and that can be activated rapidly in a secondary response.

menstruation The shedding of the uterine lining (the endometrium) that occurs when an embryo does not implant.

messenger RNA (mRNA) The RNA copy of an original DNA sequence made during transcription.

metabolism All the chemical reactions taking place in the cells of a living organism that allow it to obtain and use energy.

metastasis The spread of cancer cells from one location in the body to another.

micronutrients Nutrients, including vitamins and minerals, that organisms must ingest in small amounts to maintain health.

microtubules Hollow protein fibers that are key components of the cytoskeleton and make up the fibers of the mitotic spindle.

mineral An inorganic chemical element required by organisms for normal growth, reproduction, and tissue maintenance; examples are calcium, iron, potassium, and zinc.

mitochondria (singular: mitochondrion) Membrane-bound organelles responsible for important energy conversion reactions in eukaryotes.

mitochondrial DNA (mtDNA) The DNA in mitochondria that is inherited solely from mothers.

mitosis The segregation and separation of duplicated chromosomes during cell division.

mitotic spindle The structure that separates sister chromatids during mitosis.

molecule Atoms linked by covalent bonds.

mollusk A soft-bodied invertebrate, generally with a hard shell (which may be tiny, internal, or absent in some mollusks).

monomer One chemical subunit of a polymer.

monosaccharide The building block, or monomer, of a carbohydrate.

motor neurons Neurons that control the contraction of skeletal muscle.

multifactorial inheritance An interaction between genes and the environment that contributes to a phenotype or trait.

multipotent Describes a cell with the ability to differentiate into a limited number of cell types in the body.

mutagen Any chemical or physical agent that can damage DNA by changing its nucleotide sequence.

mutation A change in the nucleotide sequence of DNA.

mutualism A type of symbiotic relationship in which both members benefit; a "win-win" relationship.

mycelium (plural: mycelia) A spreading mass of interwoven hyphae that forms the often subterranean body of multicellular fungi.

myelin A fatty substance that insulates the axons of neurons and facilitates rapid conduction of action potentials.

NAD$^+$ An electron carrier. NAD$^+$ can accept electrons, becoming NADH in the process.

natural killer cell A type of white blood cell that acts during the innate immune response to find and destroy virally infected cells and tumor cells.

natural resources Raw materials that are obtained from the earth and are considered valuable even in their relatively unmodified, natural form.

natural selection Differential survival and reproduction of individuals in response to environmental pressure that leads to change in allele frequencies in a population over time.

nerve A bundle of specialized cells that transmit information.

nervous system The collection of organs that sense and respond to information, including the brain, spinal cord, and nerves.

neurons Specialized cells of the nervous system that generate electrical signals in the form of action potentials.

neurotransmitter A chemical signaling molecule released by neurons that communicates with neighboring cells.

neutron An electrically uncharged subatomic particle in the nucleus of an atom.

neutrophil A phagocytic cell in the circulation that plays an important role in the inflammatory response.

niche The space, environmental conditions, and resources that a species needs in order to survive and reproduce.

nitrogen fixation The conversion of atmospheric nitrogen into a form that plants can use for growth.

nonadaptive evolution Any change in allele frequency that does not by itself lead a population to become more adapted to its environment; the causes of nonadaptive evolution are mutation, genetic drift, and gene flow.

nondisjunction The failure of chromosomes to separate accurately during cell division; nondisjunction in meiosis leads to aneuploid gametes.

nonrenewable resources Natural resources like fossil fuels that cannot easily be replaced.

nuclear envelope The double membrane surrounding the nucleus of a eukaryotic cell.

nucleic acids Organic molecules made up of linked nucleotide subunits; DNA and RNA are examples of nucleic acids.

nucleotide The building block, or monomer, of a nucleic acid.

nucleus (atomic) The dense core of an atom.

nucleus (eukaryotic) The organelle in eukaryotic cells that contains the genetic material.

nutrients Components in food that the body needs to grow, develop, and repair itself.

obese Having 20% more body fat than is recommended for one's height, as measured by a body mass index equal to or greater than 30.

oncogene A mutated and overactive form of a proto-oncogene. Oncogenes drive cells to divide continually.

open circulatory system A circulatory system in which the blood is not entirely contained within blood vessels.

organ A structure made up of different tissue types working together to carry out a common function.

organ system A set of cooperating organs within the body.

organelles The membrane-bound compartments of eukaryotic cells that carry out specific functions.

organic Describes a molecule with a carbon-based backbone and at least one C-H bond.

osmolarity The concentration of dissolved solutes in blood and other bodily fluids.

osmoregulation The maintenance of relatively stable volume, pressure, and solute concentration of bodily fluids, especially blood.

osmosis The diffusion of water across a semipermeable membrane from an area of lower solute concentration to an area of higher solute concentration.

ovaries Paired female reproductive organs; the ovaries contain eggs and produce estrogen and progesterone.

overweight Having a BMI between 25 and 29.9.

oviduct The tube connecting an ovary and the uterus in females. Eggs are ovulated into and fertilized within the oviducts.

ovulation The release of an egg from an ovary into the oviduct.

paleontologist A scientist who studies ancient life by examining the fossil record.

pancreas An organ that secretes the hormones insulin and glucagon, as well as digestive enzymes.

parasitism A type of symbiotic relationship in which one member benefits at the expense of the other.

partial pressure The proportion of total air pressure contributed by a given gas.

pathogen Infectious agents including certain viruses, bacteria, fungi, and parasites. Many pathogens trigger an immune response.

pedigree A visual representation of the occurrence of phenotypes across generations.

peer review A process in which independent scientific experts read scientific studies before they are published to ensure that the authors have appropriately designed and interpreted the study.

pepsin A protein-digesting enzyme that is active in the stomach.

peptidoglycan The macromolecule found in all bacterial cell walls that confers rigidity.

peripheral nervous system (PNS) All the nervous tissue outside the central nervous system. The PNS collects sensory information and transmits instructions from the CNS.

peristalsis Coordinated muscular contractions that force food down the digestive tract.

pH A measure of the concentration of H^+ in a solution.

phagocyte A type of white blood cell that engulfs and ingests damaged cells and pathogens.

pharynx The throat.

phenotype The visible or measurable features of an individual.

phloem Plant vascular tissue that transports sugars throughout the plant.

phospholipid A type of lipid that forms the cell membrane.

photons Packets of light energy, each with a specific wavelength and quantity of energy.

photosynthesis The process by which plants and other autotrophs use the energy of sunlight to make energy-rich molecules using carbon dioxide and water.

phototropism The growth of the stem of a plant towards light.

phylogenetic tree A branching diagram of relationships showing common ancestry.

phylogeny The evolutionary history of a group of organisms.

physiology The study of the way a living organism's physical parts function.

pili (singular: pilus) Short, hairlike appendages extending from the surface of some bacteria, used to adhere to surfaces.

pistil The female reproductive structure of a flower, made up of a stigma, style, and ovary.

pituitary gland An endocrine gland in the brain that secretes many important hormones

placebo A fake treatment given to control groups to mimic the experience of the experimental groups.

placenta A structure made of fetal and maternal tissues that helps sustain and support the embryo and fetus.

plant A multicellular eukaryote that has cell walls, carries out photosynthesis, and is adapted to living on land.

plaques Raised fatty deposits that accumulate inside arteries, limiting the flow of blood.

plasma cell An activated B cell that divides rapidly and secretes an abundance of antibodies.

plate tectonics The movement of Earth's upper mantle and crust, which influences the geographical distribution of landmasses and organisms.

platelets Fragments of cells involved in blood clotting.

pluripotent Describes a cell with the ability to differentiate into nearly any cell type in the body.

polar molecule A molecule in which electrons are not shared equally between atoms, causing a partial negative charge at one end and a partial positive charge at the other; for example, water.

pollen Small, thick-walled plant structures that contain cells that develop into sperm.

pollination The transfer of pollen from male to female plant structures so that fertilization can occur.

polygenic trait A trait whose phenotype is determined by the interaction among alleles of more than one gene.

polymer A molecule made up of individual subunits, called monomers, linked together in a chain.

polymerase chain reaction (PCR) A laboratory technique used to replicate, and thus amplify, a specific DNA segment.

population A group of organisms of the same species living and interacting in a particular area.

population density The number of organisms per unit area.

population genetics The study omf the genetic makeup of populations and how the genetic composition of a population changes.

potential energy Stored energy.

predation An interaction between two organisms in which one organism (the predator) feeds on the other (the prey).

primary response The adaptive response mounted the first time a particular antigen is encountered by the immune system.

prion An infectious agent made only of protein.

producers Autotrophs (photosynthetic organisms) that form the base of every food chain.

progesterone A female sex hormone produced by endocrine cells in the ovaries (particularly in the cells of the corpus luteum) that prepares and maintains the uterus for pregnancy.

prokaryote A usually unicellular organism whose cell lacks internal membrane-bound organelles and whose DNA is not contained within a nucleus.

prokaryotic cells Cells that lack internal membrane-bound organelles.

protein An organic molecule made up of linked amino acid subunits.

protist A eukaryote that cannot be classified as a plant, animal, or fungus; usually unicellular.

proton A positively charged subatomic particle in the nucleus of an atom.

proto-oncogene A gene that codes for a protein that helps cells divide normally.

pulmonary circuit The circulation of blood between the heart and the lungs.

pulse The detectable force of blood entering arteries, e.g., in the neck or wrist.

punctuated equilibrium Periodic bursts of species change as a result of sudden environmental change.

Punnett square A diagram used to determine probabilities of offspring having particular genotypes, given the genotypes of the parents.

radial symmetry The pattern exhibited by a body plan that is circular, with no defined left and right sides.

radiation therapy The use of ionizing (high-energy) radiation to treat cancer.

radiometric dating The use of radioactive isotopes as a measure for determining the age of a rock or fossil.

randomized clinical trial A controlled medical experiment in which subjects are randomly chosen to receive either an experimental treatment or a standard treatment (or a placebo).

recessive allele An allele that reveals itself in the phenotype only if a masking dominant allele is not present.

recombinant gene A genetically engineered gene.

recombination An event in meiosis during which maternal and paternal chromosomes pair and physically exchange DNA segments.

red blood cells (erythrocytes) A type of blood cell specialized for carrying oxygen.

regulatory sequence The part of a gene that determines the timing, amount, and location of protein production.

relative dating Determining the age of a fossil from its position relative to layers of rock or fossils of known age.

renewable resources Natural resources that are replenished after use as long as the rate of consumption does not exceed the rate of replacement.

reproductive isolation Mechanisms that prevent mating (and therefore gene flow) between members of different species.

respiratory surface A surface across which oxygen enters and carbon dioxide leaves.

respiratory system The organ system that allows us to take in oxygen and unload carbon dioxide.

ribosome A complex of RNA and proteins that carries out protein synthesis in all cells.

risk factor Anything (e.g., a behavior or exposure) that increases the probability of developing a disease.

RNA polymerase The enzyme that carries out transcription. RNA polymerase copies a strand of DNA into a complementary strand of mRNA.

root hairs Tiny extensions of root cells that increase the surface area of roots to enhance their ability to absorb water and nutrients.

root system The belowground parts of a plant, which anchor it and absorb water and nutrients.

salivary glands Glands that secrete enzymes, including salivary amylase, which digests carbohydrates, into the mouth.

sample size The number of experimental subjects or the number of times an experiment is repeated. In human studies, sample size is the number of participants.

saturated fat An animal fat, such as butter; saturated fats are solid at room temperature.

science The process of using observations and experiments to draw conclusions based on evidence.

scientific theory An explanation of the natural world that is supported by a large body of evidence and has never been disproved.

scrotum The sac in which the testes are held.

secondary response The rapid and strong response mounted when a particular antigen is encountered by the immune system subsequent to the first encounter.

seed The embryo of a plant, together with a starting supply of food, all encased in a protective covering.

semen The mixture of fluid and sperm that is ejaculated from the penis.

semiconservative DNA replication is said to be semiconservative because each newly made DNA molecule has one original and one new strand of DNA.

seminiferous tubules Coiled structures that constitute the bulk of the testes and in which sperm develop.

sensor A specialized cell that detects specific sensory input like temperature, pressure, or solute concentration.

sensory neurons Cells that convey information from both inside and outside the body to the CNS.

sex chromosomes Paired chromosomes that differ between males and females, XX in females, XY in males.

shoot system The aboveground parts of a plant, including the stem and photosynthetic leaves.

short tandem repeats (STRs) Sections of a chromosome in which DNA sequences are repeated.

simple diffusion The movement of small, hydrophobic molecules across a membrane from an area of higher concentration to an area of lower concentration; simple diffusion does not require an input of energy.

sister chromatids The two identical DNA molecules that make up a duplicated chromosome following DNA replication.

small intestine The organ in which the bulk of chemical digestion and absorption of food occurs.

solute A dissolved substance.

solution The mixture of solute and solvent.

solvent A substance in which other substances can dissolve; for example, water.

somatic mutation A mutation that occurs in a body (nongamete) cell; not passed on to offspring.

speciation The genetic divergence of populations, leading over time to reproductive isolation and the formation of new species.

spinal cord A bundle of nerve fibers, contained within the bony spinal column, that transmits information between the brain and the rest of the body.

stabilizing selection A type of natural selection in which organisms near the middle of the phenotypic range of variation are favored by the environment.

stamen The male reproductive structure of a flower, made up of a filament and an anther.

starch A complex plant carbohydrate made of linked chains of glucose molecules; a source of stored energy.

statistical significance A measure of confidence that the results obtained are "real" and not due to chance.

stem cells Immature cells that can divide and differentiate into specialized cell types.

stigma The sticky "landing pad" for pollen on the pistil.

stomach An expandable muscular organ that stores, mechanically breaks down, and digests proteins in food.

stomata (singular: stoma) Pores on leaves that permit the exchange of oxygen and carbon dioxide with the air and also allow water loss.

stool Solid waste material eliminated from the digestive tract.

stroke A disruption in blood supply to the brain.

style The tubelike structure that leads from the stigma to the ovary.

substrate A molecule to which an enzyme binds and on which it acts.

sustainability The use of Earth's resources in a way that will not permanently destroy or deplete them; living within the limits of Earth's biocapacity.

symbiosis A relationship in which two different organisms live together, often interdependently.

synapse The site of communication between a neuron and another cell; the synapse includes the axon terminal of the communicating neuron, the space between the cells, and receptors on the receiving cell.

synaptic cleft The physical space between a neuron and the cell with which it is communicating.

systemic circuit The circulation of blood between the heart and the rest of the body.

systolic pressure The pressure in arteries at the time the ventricle contracts.

T cells White blood cells that mature in the thymus and can destroy infected cells or stimulate B cells to produce antibodies, depending on the type of T cell.

taproot A long, straight root produced by some plants to store water and carbohydrates.

taxonomy The process of identifying, naming, and classifying organisms on the basis of shared traits.

testable Describes a hypothesis that can be supported or rejected by carefully designed experiments or observational studies.

testes (singular: testis) Paired male reproductive organs, which contain sperm and produce androgens (primarily testosterone).

testosterone The primary male sex hormone, which stimulates the development of masculine features and plays a key role in sperm development.

tetrapod A vertebrate animal with four true limbs, that is, jointed, bony appendages with digits. Mammals, amphibians, birds, and reptiles are tetrapods.

thermoregulation The maintenance of a relatively stable internal body temperature.

thigmotropism The response of plants to touch and wind.

thymus The organ in which T cells mature.

tissue An organized group of different cell types that work together to carry out a particular function.

totipotent Describes a cell with the ability to differentiate into any cell type in the body.

trachea A large airway leading into the lower respiratory tract.

tracheal system A series of air-filled tubes in insects that deliver air to cells.

trans fat A type of vegetable fat that has been hydrogenated, that is, hydrogen atoms have been added, making it solid at room temperature.

transcription The first stage of gene expression, during which cells produce molecules of messenger RNA (mRNA) from the instructions encoded within genes in DNA.

transfer RNA (tRNA) A type of RNA that transports amino acids to the ribosome during translation.

transgenic Refers to an organism that carries one or more genes from a different species.

translation The second stage of gene expression, during which mRNA sequences are used to assemble the corresponding amino acids to make a protein.

transpiration The loss of water from plants by evaporation, which powers the transport of water and nutrients through a plant's vascular system.

transport proteins Proteins involved in the movement of molecules across the cell membrane.

triglycerides A type of lipid found in fat cells that stores excess energy for long-term use.

trisomy 21 Carrying an extra copy of chromosome 21; also known as Down syndrome.

trophic levels Feeding levels, based on positions in a food chain.

tumor A mass of cells resulting from uncontrolled cell division.

tumor suppressor gene A gene that codes for proteins that monitor and check cell cycle progression. When these genes mutate, tumor suppressor proteins lose normal function.

turgor pressure The pressure exerted by the water-filled central vacuole against the plant cell wall, giving a stem its rigidity.

unsaturated fat A plant fat, such as olive oil; unsaturated fats are liquid at room temperature.

urethra A tube that connects the bladder to the genitals and carries urine out of the body. In males, the urethra travels through the penis and is also used to carry sperm.

uterus The muscular organ in females in which a fetus develops.

vaccine A preparation of killed or weakened microorganisms or viruses that is given to people or animals to generate a memory immune response.

vagina The first part of the female reproductive tract, extending up to the cervix; also known as the birth canal.

vas deferens Paired tubes that carry sperm from the testes to the urethra.

vascular plant A plant with tissues that transport water and nutrients through the plant body.

vascular system Tube-shaped vessels and tissues that transport nutrients throughout an organism's body.

vasoconstriction The reduction in diameter of blood vessels, which helps to retain heat.

vasodilation The expansion in diameter of blood vessels, which helps to release heat.

veins Blood vessels that carry blood toward the heart.

ventilation The process of moving air in and out of the lungs.

ventricles The chambers of the heart that pump blood away from the heart. The right ventricle pumps blood to the lungs, and the left ventricle pumps blood to the body.

vertebrate An animal with a bony or cartilaginous backbone.

vestigal structure A structure inherited from an ancestor that no longer serves a clear function in the organism that possesses it.

villi (singular: villus) Fingerlike projections of folds in the lining of the small intestine that are responsible for most nutrient and water absorption.

virus An infectious agent made up of a protein shell that encloses genetic information.

vitamin An organic molecule required in small amounts for normal growth, reproduction, and tissue maintenance.

vitamin D A fat-soluble vitamin required to maintain a healthy immune system and to build healthy bones and teeth. The human body produces vitamin D when skin is exposed to UV light.

white blood cells (leukocytes) Cells involved in the body's defenses (the immune response).

wood Secondary xylem tissue found in the stem of a plant.

X-linked trait A phenotype determined by an allele on an X chromosome.

xylem Plant vascular tissue that transports water from the roots to the shoots.

Y-chromosome analysis Comparing sequences on the Y chromosome to examine paternity and paternal ancestry.

zygote A fertilized egg.

Photo Credits

Cover: Masa Ushioda/Robert Harding

FRONT MATTER p. vi: Andrea Gawrylewski. **p. x:** *(bottom left)* Kaspri/Dreamstime.com.

CHAPTER 1 pp. 0-1: Steve Bronstein/Getty Images. **p. 2:** Imaginechina/Corbis. **p. 3:** *Infographic 1.1 (left to right)* PLoS Med 2 (2). Image Credit: Krista Steinke, 28 May 2009, Vol. 338, Issue 7706. Courtesy British Medical Journal, Aleksej Vasic/iStockphoto. **p. 4:** *Infographic 1.2 (left)* 28 May 2009, Vol. 338, Issue 7706. Courtesy British Medical Journal; *(right)* PLoS Med 2 (2). Image Credit: Krista Steinke. **p. 5:** *(left)* SCIENCE Vol. 324, no. 5935, 26 June 2009. Reprinted with permission from AAAS; *(middle)* Cell Metabolism, March 4, 2009. Cover Illustration by Chris Lange. Copyright Elsevier, 2009; *(right)* Reprinted by permission from Macmillan Publishers Ltd: Nature Chemical Biology Feb 2013 Volume 9, Number 2; Cover art by Erin Dewalt, based on imagery from Joseph Schine. Copyright 2013. **p. 8:** *Infographic 1.5 (left)* Jessica Peterson/Photolibrary; *(right top)* Nancy Nehring/iStockphoto, *(right middle)* Science Source, *(right bottom)* G. Lasley/VIREO, FLPA/David Hosking/age fotostock. **p. 11:** AP Photo/Peter Morgan. **p. 13:** Brendan Hoffman for National Breast Cancer Coalition. **p. 15:** *Infographic 1.8* Tony West/Alamy.

CHAPTER 2 pp. 20-21: NASA/JPL-Caltech. **p. 22:** NASA/JPL-Caltech. **p. 23:** UPI/Brian van der Brug/pool/LANDOV. **p. 25:** *Infographic 2.1 (top to bottom)* Blend Images/SuperStock, Juniors Bildarchiv/age fotostock, globestock/iStockphoto, Kazuo Ogawa/ age fotostock, Andres Rodriguez/age fotostock. **p. 26:** *Infographic 2.2 (background)* NASA/JPL-Caltech/ESA/DLR/FU Berlin/MSSS; *(inset)* NASA/JPL-Caltech. **p. 28:** NASA. **p. 32:** NASA. **p. 35:** *Infographic 2.7 (top to bottom)* AgeFotostock/Superstock, iStockphoto/Thinkstock, igorad1/Shutterstock. **p. 38:** *Infographic 2.10 (top left)* Eye of Science/Science Source; *(bottom left)* Theodore Clutter/Science Source; *(inset)* SIPA USA/SIPA/Newscom; *(bottom right)* VERENA TUNNICLIFFE/AFP/Newscom; *(inset)* Beatty, T, et al. An obligately photosynthetic bacterial anaerobe from a deep-sea hydrothermal vent. PNAS (2005) vol. 102 no. 26 9306-9310. © 2005 National Academy of Sciences, U.S.A.

CHAPTER 3 pp. 44-45: Bettmann/Corbis. **p. 46:** *Infographic 3.1 (left to right)* David M. Phillips/Science Source, Science Museum/Science & Society Picture Library; *(last two)* Biophoto Associates/Science Source. **p. 47:** *Infographic 3.2 (top left to right)* Scenics & Science/Alamy, Roland Birke/Getty Images, Kage-Mikrofotografie/age fotostock; *(bottom left to right)* Ed Reschke/Getty Images, *(inset)* David Toase/Photolibrary, Ed Reschke/Getty Images, *(inset)* Martin Shields/Science Source, Michael Abbey/Science Source, *(inset)* A. & F. Michler/Getty Images. **p. 50:** SSPL via Getty Images. **p. 51:** *(top)* Richard J. Green/Science Source; *(bottom)* Advertising Archive/Courtesy Everett Collection. **p. 52:** Fleming, Alexander. 1929. On the Antibacterial Action of Cultures of a Penicillium, with Special Reference to Their Use in the Isolation of *B. influenzae*. British Journal of Experimental Pathology, Vol. 10, pp. 226-236. Fig. 2. By permission of John Wiley & Sons Ltd. **p. 57:** *(top)* SPL/Science Source; *(bottom)* National Library of Medicine. **p. 60:** Ocean/Corbis.

MILESTONE 1 pp. 66-67: SPL/Science Source. **p. 68:** *(left)* Sagan, L., "On the origin of mitosing cells," Journal of Theoretical Biology, (1967), volume 3, 255-74. Reproduced with permission of Elsevier; *(right)* Nancy R. Schiff/Getty Images. **p. 69:** *Infographic M1.1 (top left)* Medical-on-Line/Alamy; *(top middle)* Dr. Gary Gaugler/ Science Source: *(top right)* ISM/Phototake; *(bottom left)* Dr. Jeremy Burgess/Science Source; *(bottom right)* Keith R. Porter/ Science Source. **p. 70:** *Infographic M1.2 (top left to right)* Dr. D. P. Wilson/Science Source, Stock.xchng, Zefiryn/Fotolia, Ximinez/Fotolia, Pasieka/Science Source; *(bottom left)* Medical-on-Line/Alamy; *(bottom right)* Eye of Science/Science Source. **p. 71:** Dr. Edwin P. Ewing, Jr/CDC/Public Health Image Library. **p. 72:** *(clockwise from top)* Peter Zijlstra/ Shutterstock, CNRI/Science Source, DWaschnig/Shutterstock, Janice Carr/CDC/Public Health Image Library. **p. 73:** Paul Hosefros/The New York Times/Redux.

CHAPTER 4 pp. 74-75: CHAMUSSY/Sipa Press. **p. 76:** GIANLUIGI GUERCIA/AFP/Getty Images. **p. 78:** *Infographic 4.2 (top left to right)* Denis Pepin/Featurpics, margouillat photo/Shutterstock, AgeFotostock/Superstock; *(middle left to right)* Jeffrey Coolidge/Getty Images, Juanmonino/iStockphoto, margouillat photo/Shutterstock; *(bottom left to right)* Juanmonino/iStockphoto, AgeFotostock/Superstock, Tetra Images/Getty Images. **p. 80:** Colin Carson/age fotostock. **p. 81:** Photo by Jeffrey Davis, Courtesy Project Peanut Butter. **p. 86:** *Infographic 4.6 (left)* Michael Klein/Getty Images; *(right)* Fancy Photography/Veerp. **p. 87:** *(left)* © 2011 Jon Warren/World Vision; *(right)* © 2011 World Vision. **p. 89:** Project Peanut Butter. **p. 90:** *Infographic 4.8* Shutterstock

CHAPTER 5 pp. 94-95: Pascal Goetgheluck/Photo Researchers, Inc. **p. 95:** REX USA/Phil Hart/Solent News/Rex; *(inset)* FLPA/Alamy. **p. 97:** MCMULLAN CO/SIPA/Newscom. **p. 98:** *Infographic 5.2 (top to bottom)* Simon Owler/iStockphoto, Årni Torfason/iStockphoto, ImageState, Elena Elisseeva/Dreamstime.com, Bernd Lang/iStockphoto, Corbis. **p. 99:** *Infographic 5.3 (top left to right)* Kimberly Deprey/iStockphoto, mihtiander/FeaturePics, Biosphoto/Claudius Thiriet; *(middle)* Eric Draper/The New York Times/Redux; *(bottom left to right)* Gerd Guenther/Science Source, Photo168/Dreamstime.com, Yobro10/Dreamstime, AP Photo/Seth Perlman, File, Gudella/FeaturePics. **p. 100:** *Infographic 5.4 (left)* Corbis/Superstock; *(middle and right)* technotr/iStockphoto. **p. 101:** *Infographic 5.5 (left to right)* Pixelgnome/Dreamstime.com, Vasily Smirnov/Dreamstime.com, Roger Harris/Science Source, Shevelartur/Dreamstime.com, Aleksandr Lazarev/iStockphoto. **p. 102:** *Infographic 5.6 (top left to right)* Value Stock Images/Fotosearch, jeff waibel/Dreamstime.com, NNehring/iStockphoto; *(bottom left)* Value Stock Images/Fotosearch; *(top right)* Mark Hamblin/age fotostock; *(bottom right)* Michael Abbey/Science Source. **p. 105:** Michael Macor/San Francisco Chronicle/Corbis. **p. 106:** *Infographic 5.8 (left)* Brad Calkins/Dreamstime.com; *(right)* Maxrale/iStockphoto. **p. 108:** Courtesy Jim Sears/A2BE Carbon Capture LLC. **p. 109:** PNWL/Alamy.

CHAPTER 6 p. 115: Masterfile (Royalty-Free Div.). **p. 116:** Courtesy Paul Rozin; *Infographic 6.1 (left)* Mauricio Anton/Science Source; *(middle)* Douglas Johns/the food passionates/Corbis; *(right)* OECD

(2012), OECD Obesity Update http://www.oecd.org/health/49716427. pdf. **p. 118:** *Infographic 6.3* Jose Luis Stephens/Radius Images/Getty Images. **p. 119:** Sergio Pitamitz/DanitaDelimont.com.
p. 120: *Infographic 6.4* Ilena Eisseva/Featurepics. **p. 124:** *Infographic 6.6 (top)* apcuk/iStockphoto; *(middle)* EricGerrard/iStockphoto; *(bottom)* bluestocking/iStockphoto; *(top right)* DNY59/iStockphoto. **p. 128:** *(left)* Konstik/Dreamstime; *(right)* Erik Isakson/Getty Images. **p. 130:** *(row 1)* Julia Nichols/iStockphoto, Lauri Patterson/iStockphoto; *(row 2)* Rowan Butler/iStockphoto, John E. Kelly/Getty Images; *(row 3)* Jill Chen/iStockphoto, janeff/iStockphoto; *(row 5)* Flashon Studio/ iStockphoto; *(row 6)* Givaga/iStockphoto, Lee Rogers/iStockphoto. **p. 131:** AP Photo/Harry Cabluck.

CHAPTER 7 p. 136: JUPITERIMAGES/Brand X/Alamy.
p. 138: Tyler Hicks/The New York Times/Redux. **p. 139:** Uli Holz.
p. 142: Scanpix Sweden/Sipa USA. **p. 145:** Tek Image/Science Source.
p. 147: *Infographic 7.6* Dr. Gopal Murti/Science Source. **p. 148:** Kevin Rivoli/The New York Times/Redux. **p. 149:** STEPHEN D. CANNERELLI/ The Post-Standard /Landov.

MILESTONE 2 pp. 154-155: James D. Watson Collection, Cold Spring Harbor Laboratory Archives. **p. 157:** *Infographic M2.2 (top)* Science Source; *(bottom, both)* King's College London. **p. 158:** *Infographic M2.3* Courtesy of Columbia University Medical Center. **p. 159:** *Infographic M2.4* SSPL via Getty Images.

CHAPTER 8 p. 163: Courtesy rEVO Biologics. **p. 164:** *(top)* Nigel Cattlin/Alamy; *(bottom)* Courtesy rEVO Biologics. **p. 170:** AP Photo/ Lisa Poole. **p. 171:** *Infographic 8.6 (left to right)* MedicalRF.com/age fotostock, Lauritzasoare/Dreamstime.com, jennyt/shutterstock, Courtesy of rEVO Biologics. **p. 177:** *Infographic 8.10* Courtesy of rEVO Biologics. **p. 181:** Dan Reynolds/CartoonStock.

MILESTONE 3 pp. 182-183: Alex Wong/Getty Images. **p. 188:** *(top)* JOYCE NALTCHAYAN/AFP/Newscom. *(bottom left)* Reprinted by permission from Macmillan Publishers Ltd: Nature Feb 15, 2001, Volume 409, Number 6822, copyright 2001; *(bottom right)* Cover from Science Vol. 291, no. 5507, 16 February 2001. Reprinted with permission from AAAS. Illustration by Ann Cutting: ann@cutting. com. **p. 189:** Sinclair Stammers/Science Source. **p. 190:** *Infographic M2.4* Genome Management Information System, Oak Ridge National Laboratory, U.S. Department of Energy.

CHAPTER 9 p. 192: *(top left)* AP Photo/Wally Santana; *(top right)* Philippe Psaila/Science Source; *(bottom left)* Alan Sirulnikoff/Science Source; *(bottom right)* Philippe Psaila/Science Source. **p. 194:** Inga Spence/Science Source. **p. 195:** *Infographic 9.2* Targeting Connexin43 Expression Accelerates the Rate of Wound Repair. (2003). Cindy Qiu, Petula Coutinho, Stefanie Frank, Susanne Franke, Lee-yong Law, Paul Martin, Colin R. Green and David L. Becker. Current Biology, 13(19), 1697-1703. **p. 197:** Alan Sirulnikoff/Science Source. **p. 199:** Courtesy Dr. Susan Band Horwitz, Albert Einstein College of Medicine of Yeshiva University. **p. 200:** *(top left)* Dr. Alexey Khodjako /Science Source; *(bottom left)* Pr. G Giménez-Martin/Science Source; *(top middle)* Dr. Alexey Khodjakov/Science Source; *(top right)* Dr. Alexey Khodjakov/ Science Source; *(bottom right)* Pr. G Giménez-Martin/Science Source.
p. 201: *(top left and top right)* Dr. Alexey Khodjakov/Science Source; *(bottom left and bottom right)* Pr. G Giménez-Martin/Science Source.
p. 202: *Infographic 9.3* from Cancer Cell, Vol. 8, Issue 1, Weaver,

B., Cleveland, D., Decoding the links between mitosis, cancer and chemotherapy: The mitotic checkpoint, adaptation, and cell death, 7-12, Copyright 2005, with permission from Elsevier. **p. 205:** Michael Sewell, Visual Pursuit /Peter Arnold/Getty Images. **p. 206:** Department of Chemistry, Florida State University. **p. 207:** *(left)* Jim West/age fotostock; *(right)* Mark Kostich/Getty Images; *(bottom)* Steve Gschmeissner/Science Source.

CHAPTER 10 p. 213: Bill Greenblatt/UPI/Newscom. **p. 215:** *Infographic 10.1 (left)* Stockbyte; *(right)* Ingram Publishing/Photolibrary. **p. 217:** Du Cane Medical Imaging Ltd./Science Source. **p. 220:** *Infographic 10.5 (top left)* Lepas/Dreamstime; *(top middle)* Larry Jordan/FeaturePics; *(top right)* Photodisc; *(bottom left)* Royalty-Free/Corbis; *(bottom middle)* Royalty-Free/Corbis; *(bottom right)* Mary Lane/FeaturePics.
p. 221: Glow Wellness/SuperStock. **p. 224:** Courtesy Lorene Ahern.

CHAPTER 11 p. 228: Carrie Acosta. **p. 232:** *Infographic 11.2* ISM/ Phototake. **p. 236:** *Infographic 11.6* Simon Fraser/Science Source.
p. 237: *(top)* Jeffrey Sauger; *(bottom)* Courtesy Emily Schaller.
p. 242: 2 Snaps Up. **p. 243:** AP Photo/Carlos Osorio.

MILESTONE 4 p. 248: Biophoto Associates/Getty Images.
p. 250: *Infographic M4.1 (left)* Science Source; *(right)* Wellcome Library, London. **p. 250:** Authenticated News/Getty Images.
p. 251: Guy Croft/Alamy. **p. 253:** Trevor Sims/ Garden World Images/ age fotostock.

CHAPTER 12 p. 257: Sebastian Kaulitzki/Alamy. **p. 258:** REUTERS/ Landov; *Infographic 12.1* ISM/Phototake. **p. 259:** Biophoto Associates/ Science Source. **p. 263:** Courtesy of Mason County Museum, Maysville, Kentucky. **p. 266:** AP Photo/Leslie Close. **p. 267:** *(top left to right)* Ludo Kuipers, OzOutback.com.au, ImageSource/age fotostock, Yuri Arcurs/Fotolia, Tatiana Morozova/FeaturePics; *(middle left to right)* PhotosIndia/Alamy, Jacob Langvad/Getty Images, Denis Pepin/ FeaturePics; *(bottom left to right)* leaf/FeaturePics, Dmitriy Shironosov/ Dreamstime.com, Thomas Cockrem/Alamy. **p. 268:** *Infographic 12.6 (top)* leaf/FeaturePics; *(left)* ImageSource/age fotostock; *(right)* Ryan McVay/Getty Images. **p. 269:** Aizuddin Saad/ZUMA Press/Newscom.
p. 271: Courtesy Peter Morenus. **p. 273:** Courtesy J. Dee Higley; *Infographic 12.10* National Institute of Mental Health. **p. 275:** Markus Moellenberg/Corbis. **p. 277:** *Infographic 12.13* ISM/Phototake.

CHAPTER 13 p. 283: Wake Forest Institute for Regenerative Medicine.
p. 286: *Infographic 13.2 (top)* Ed Reschke/Getty Images; *(middle)* Robert Knauft/Biology Pics/Science Source; *(bottom)* Carolina Biological Supply Company/Phototake. **p. 287:** Wake Forest Baptist Medical Center. *Infographic 13.3* AP Photo/Brian Walker. **p. 288:** Wake Forest Institute for Regenerative Medicine. **p. 291:** *Infographic 13.5 (top)* Wake Forest Institute for Regenerative Medicine; *(bottom left)* Rebecca Hale/ National Geograhic Creative; *(bottom right)* Andre Blais/Shutterstock.
p. 293: AFP/GETTY IMAGES/Newscom. **p. 294:** *Infographic 13.7* Deco Images II/Alamy, AP Photo/PA/Files. **p. 295:** *Infographic 13.8* LookatSciences/Phototake. **p. 296:** Hyun-Wook Kang, Wake Forest University Institute of Regenerative Medicine. **p. 297:** *(top)* Hyun-Wook Kang, Wake Forest University Institute of Regenerative Medicine; *(bottom)* Rex Features via AP Images.

CHAPTER 14 p. 302: David M. Phillips/Science Source.
p. 305: *Infographic 14.1 (top left)* Science Source; *(bottom left)* BSIP/

Biochemistry and Molecular Biology, The Pennsylvania State University, Eye of Science/Science Source, Martin Oeggerli/Science Source, Greg Wanger Ph.D. & Gordon Southam Ph.D. **p. 399:** *(top)* Photo © Woods Hole Oceanographic Institution; *(bottom)* Courtesy of IFE, URI-IAO, Lost City science party, and NOAA. **p. 400:** *Infographic 18.6 (clockwise left to right)* E. Nelson and L. Sycuro, provided courtesy of the Vibrio fischeri Genome Project, Dr. Kari Lounatmaa/Science Source, Kwangshin Kim/Science Source, Michael Abbey/Science Source, John Walsh/Science Photo Library, CNRI/Science Source, Science Photo Library/Photolibrary. **p. 402:** *Infographic 18.7 (top inset)* K. O. Stetter and R. Rachel, Univ. Regensburg, Germany; *(left)* NASA, NSF, Woods Hole Oceanographic Institution/ Chris German, WHOI; *(top)* Kevin Kemmerer/Getty Images; *(top inset)* Eye of Science/ Science Source; *(bottom)* Peter Walker/Corbis; *(bottom inset)* Kenneth M. Stedman, Ph.D., NASA Astrobiology Institute-Center for Life in Extreme Environments, Portland State University. **p. 403:** Image courtesy of Kelley, D. S., University of Washington and IFE, URI-IAO, UW, Lost City science party, and NOAA. **p. 405:** Image courtesy of the University of Washington and the Lost City Science team, IFE, URI-IAO, and NOAA.

CHAPTER 19 p. 410: Michael Wheatley/age fotostock. **p. 412:** *(both)* National Park Service, Olympic National Park. Photo by Janis Burger. **p. 413:** *Infographic 19.2* (1) Patrick Robbins/Dreamstime.com, (2) National Park Service, (3) Marcopolo/FeaturePics, (4) Courtesy Peter Wigmore, (5) Courtesy Bob Wightman. **p. 414:** *Infographic 19.3 (left)* Fotogal/FeaturePics; *(left inset)* Mariya Bibikova/iStockphoto; *(middle left)* Mark Turner/Getty Images; *(middle left inset)* George Bailey/Dreamstime.com; *(middle right)* Michael P. Gadomski/Science Source; *(middle right inset)* ray roper/iStockphoto; *(right)* Mark Turner/Getty Images; *(right inset)* Courtesy Greg Rabourn. **p. 415:** Courtesy Northwest Trek. **p. 416:** David Gomez/iStockphoto; **p. 417:** *Infographic19.4b (left to right)* Courtesy of Brooke et al., NOAA-OE, HBOI; Anky10/Dreamstime.com; Ed Reschke/Getty Images; Photolibrary/Alamy; Manipulateur/Fotolia; London Scientific Films/Getty Images; irin-k/Shutterstock; U.S. National Park Service; Karen Arnold/Dreamstime.com. **p. 420:** *(clockwise from top left)* Outdoorsman/Dreamstime.com, Mark Conlin/Alamy, Stone Nature Photography/Alamy, Gary Nafis; *(left)* Jan Gottwald/iStockphot; *(center)* Chris Mattison/Alamy. **p. 422:** *Infographic 19.5 (clockwise from top left)* London Scientific Films/Getty Images, Mike Norton/Dreamstime.com, Alexander Makarov/iStockphoto, Steve Gschmeissner/Science Source, Eye of Science/Science Source; *(center)* Ed Reschke/Getty Images. **p. 423:** *Infographic 19.6 (left to right)* Gary Retherford/Science Source, Roland Birke/Getty Images, Nick Kurzenko/Alamy, Science PR/Getty Images. **p. 424:** Cornelia Doerr/age fotostock/Robert Harding.

CHAPTER 20 p. 430: Courtesy of American Anthropological Association. **p. 432:** *Infographic 20.1* Joe Ravi/Shutterstock; *(inset)* Thinkstock. **p. 433:** *Infographic 20.2* Carolina Biological Supply Company/Phototake. **p. 434:** *(top)* Mark Wilson/Getty Images; *(bottom)* Nina Jablonski. **p. 435:** *Infographic 20.3 (top)* Juergen Berger/ Science Source; *(bottom left)* Elena Rostunova/Alamy; *(bottom right)* Biophoto Associates/Science Source. **p. 436:** *Infographic 20.4* Chaplin G., Geographic Distribution of Environmental Factors Influencing Human Skin Coloration, American Journal of Physical Anthropology 125:292,Äì302, 2004; map updated in 2007. Designer: Emmanuelle Bournay, UNEP/GRID-Arendal. http://maps.grida.no/go/graphic/ skin-colour-map-indigenous-people. **p. 437:** *Infographic 20.5 (left)* SPL/Custom Medical Stock Photo—All rights reserved. *(right)* CNRI/

Science Source. **p. 439:** *Infographic 20.7 (left)* ©2001 David L. Brill/ Brill Atlanta; *(right)* The Natural History Museum, London/The Image Works. **p. 444:** *Infographic 20.10* Chaplin G., Geographic Distribution of Environmental Factors Influencing Human Skin Coloration, American Journal of Physical Anthropology 125:292,Äì302, 2004; map updated in 2007. Designer: Emmanuelle Bournay, UNEP/GRID-Arendal.http://maps.grida.no/go/graphic/skin-colour-map-indigenous-people.

CHAPTER 21 p. 448: John Vucetich. **p. 451:** *Infographic 21.1 (top to bottom)* U.S. Fish & Wildlife Service, AP Photo/Michigan Technological University, John Vucetich, Accent Alaska.com/Alamy, sherwoodimagery/iStockphoto. **p. 452:** John Vucetich; *Infographic 21.2 (top left)* Flirt/SuperStock; *(top right)* JA Vucetich & RO Peterson, www.isleroyalewolf.org; *(middle left)*Tom Hansch/Dreamstime.com; *(middle right)* JA Vucetich & RO Peterson, www.isleroyalewolf.org; *(bottom, both)* John Vucetich; *Infographic 21.3 (left)* Ed Reschke/Getty Images; *(middle)* Stubblefield Photography/Shutterstock; *(right)* Marcel Krol/ Dreamstime.com. **p. 454:** Stachecki, James J./Animals Animals–All rights reserved. **p. 455:** John Vucetich. **p. 457:** *Infographic 21.6 (top)* John Luke/Getty Images; *(middle)* Steve Kazlowski/DanitaDelimont. com; *(bottom)* Zastolskiy Victor/Shutterstock. **p. 458:** *Infographic 21.7 (top left)* imagebroker/Alamy; *(top right)* Cliff Keeler/Alamy; *(bottom left)* Andrew McLachlan/All Canada Photos/Getty Images; *(bottom right)* John Vucetich. **p. 459:** *Infographic 21.8 (clockwise from top left)* Photobac/Dreamstime.com, Peter J. Wilson/Shutterstock, Terry Morris/iStockphoto.com, John Vucetich, Oksana Churakova/ Dreamstime.com, jim kruger/iStockphoto; *(center)* James Mattil/age fotostock. **p. 460:** John and Ann Mahan. **p. 461:** JA Vucetich & RO Peterson, www.isleroyalewolf.org. **p. 462:** *Infographic 21.9 (left)* John Vucetich; *(right)* Courtesy Sandy Updyke.

CHAPTER 22 p. 466: Julia Kumari Drapkin/Tampa Bay Times/ ZUMAPRESS.com. **p. 467:** Danish Ismail/Reuters/Landov. **p. 469:** © 2010 Ellen Harasimowicz. **p. 469:** *Infographic 22.2 (top)* Dennis MacDonald/age fotostock; *(bottom)* Ragnar/FeaturePics. **p. 470:** *Infographic 22.3 (bottom)* Olga Demchishina/iStockphoto. **p. 471:** *(left)* Gail Shumwa/Getty Images; *(right)* John Serrao/Science Source. **p. 472:** *Infographic 22.4 (top)* David Kay/Dreamstime.com; *(middle)* Rui Miguel da Costa Neves Saraiva/iStockphoto; *(bottom)* Kyu Oh/iStockphoto. **p. 473:** *Infographic 22.5b (row 1 left to right)* Blend Images/Superstock, Michael Sewell/Getty Images, Benny Rytter/iStockphoto; *(row 2 left to right)* James Phelps Jr/Dreamstime. com, Abdolhamid Ebrahim/iStockphoto, Lunamarina/Dreamstime. com, Rui Miguel da Costa Neves Saraiva/iStockphoto, Krys Bailey/ Alamy; *(row 3 left to right)* Juniors Bildarchiv GmbH/Alamy, Lukrecja/ FeaturePics, Hermann Eisenbeiss/Science Source; *(row 4 left to right)* Kyu Oh/iStockphoto, brytta/iStockphoto, Roy T. Free/age fotostock. **p. 474:** *Infographic 22.6 (top left)* Jon Yuschock/Fotolia; *(top right)* Courtesy Donald Stahly; *(middle right)* Crown Copyright courtesy of Central Science Laboratory/Science Source; *(bottom)* Debbie Steinhausser/Shutterstock. **p. 476:** *Infographic 22.7 (clockwise from top left)* Willi Schmitz/iStockphoto, Jim McKinley/Getty Images, Steve Byland/iStockphoto, Nigel Downer/age fotostock; *(background)* Joanne Green/iStockphoto. **p. 477:** *Infographic 22.8 (top)* Alliance Images/Alamy; *(middle)* ElementalImaging/iStockphoto; *(bottom)* Nic Bothma/epa/Corbis. **p. 478:** *(top)* Curt Pickens/iStockphoto; *(middle)* mrolands/Featurepics.com; *(bottom)* Maigi/Dreamstime.com. **p. 479:** Ian Shaw/Alamy. **p. 480:** *Infographic 22.9 (top and bottom left)* Custom Life Science Images/photographersdirect.com; *(right top to bottom)*

AP Photo/Sun Journal, Russ Dillingham, Dušan Kostić/iStockphoto, Courtesy Mariano Higes and Raquel Martin-Hernandez, Courtesy Beeologics and Professor Ilan Sela (Maori et al, IAPV, a bee-affecting virus associated with Colony Collapse Disorder can be silenced by dsRNA ingestion, Insect Molecular Biology (2009) 18(1), 55-60.) **p. 481:** *(left)* Valentyn75/Dreamstime.com; *(right)* © HDIP, Inc.

CHAPTER 23 p. 486: America/Alamy **p. 487:** AP Photo/Toby Talbot. **p. 490:** Cheryl Senter/The New York Times/Redux. **p. 492:** *(row 1)* Krzysztof Odziomek/iStockphoto, Jacka/Dreamstime.com, STILLFX/Shutterstock, Andoni Canel/Getty Images; *(row 2)* ricardoreitmeyer/Shutterstock, Richard Walters/iStockphoto, malerapaso/iStockphoto, Art33art/Dreamstime.com. **p. 495:** *Infographic 23.6 (clockwise from top left)* USFWS Pacific Region/Flickr, Nick Saunders/age fotostock, Rita Robinson/age fotostock, Rod Planck/Science Source, Wayne Lynch/Getty Images, Jim Zipp/Science Source; *(background)* Evgeny Dubinchuk/Dreamstime.com; *(right)* MichaelGrant/Alamy. **p. 496:** REUTERS/Mathieu Belanger. **p. 498:** *Infographic 23.8 (left and right)* NASA/Goddard Space Flight Center Scientific Visualization Studio. The Blue Marble data is courtesy of Reto Stockli; *(top right)* WorldFoto/Alamy. **p. 502:** *Infographic 23.10 (left)* AFP/Getty Images; *(middle)* British Antarctic Survey/Science Source; *(right)* U.S. Dept. of Commerce, NOAA, Earth System Research Laboratory. **p. 503:** *Infographic 23.11 (clockwise from top left)* Justin Kase ztwoz/Alamy, Stringer/epa/Corbis, Matt Meadows/Getty Images, olyniteowl/iStockphoto, Anna Lubovedskaya/iStockphoto, ssuaphoto/age fotostock. **p. 508:** D. W. Schindler.

MILESTONE 6 p. 510: Bettmann/Corbis. **p. 512:** *(left)* Courtesy of the Linda Lear Center for Special Collections & Archives, Connecticut College; *(middle)* Cover from *Silent Spring* by Rachel Carson. Copyright © 1962 by Rachel L. Carson and renewed 1990 by Roger Christie. Reprinted by permission of Houghton Mifflin Harcourt Publishing Company. All rights reserved. *(right)* Alfred Eisenstaedt/Time Life Pictures/Getty Images; *(bottom left)* Basement Stock/Alamy; *(bottom right)* Bettmann/CORBIS. **p. 513:** *Infographic M6.1b (clockwise from top left)* George Konig/Keystone Features/Getty Images, Eye of Science/Science Source, Photo Researchers/Getty Images, George Silk/Time Life Pictures/Getty Images. **p. 514:** *Infographic M6.2 (left to right)* Lee Balterman/Time Life Pictures/Getty Images, Graphic Science/Alamy, Valerie Giles/Getty Images, Kenneth C. Zirkel/Getty Images, Greg L. Kohuth, Michigan State University. **p. 515:** *Infographic M6.3 (top)* Ken Canning/Getty Images; *(bottom)* M. I. Walker/Science Source. **p. 516:** AP Photo. **p. 517:** *(left)* Cover from *Silent Spring* by Rachel Carson. Copyright © 1962 by Rachel L. Carson and renewed 1990 by Roger Christie. Reprinted by permission of Houghton Mifflin Harcourt Publishing Company. All rights reserved. *(right)* U.S. Environmental Protection Agency. **p. 518:** George Silk/Time Life Pictures/Getty Images.

CHAPTER 24 p. 520: Jaime Oppenheimer/Wichita Eagle/MCT via Getty Images. **p. 523:** *(left)* Eric Nguyen/Corbis; *(right)* Courtesy WeatherTap.com. **p. 524:** *Infographic 24.1c (top left to right)* Capricornis/Dreamstime.com, tank_bmb/FeaturePics, Photong/Dreamstime.com, Dimaberkut/Dreamstime.com, hfng/FeaturePics; *(bottom left to right)* Gary Whitton/Dreamstime.com, surpasspro/FeaturePics, Borut Trdina/iStockphotos, stu99/FeaturePics, Tommy Schultz/Dreamstime.com. **p. 525:** *(left)* Larry W. Smith/epa/Corbis; *(inset)* Mike Theiss/Ultimate Chase/Corbis. **p. 528:** *Infographic 24.4 (top left)* Frank Roeder/Dreamstime.com; *(top right)* Airwolf01/Dreamstime.

com; *(middle left)* Dennis Macdonald/Photolibrary/Getty Images; *(middle right)* Lukasz Koszyk/iStockphoto; *(bottom left)* Hanhanpeggy/Dreamstime.com; *(bottom center)* Shaun Lowe/iStockphoto; *(bottom right)* Dmitro Tolokonov/iStockphoto. **p. 530:** *(top)* Wichita Eagle/MCT/LANDOV; *(bottom)* Farshid Assassi/Courtesy BNIM. **p. 531:** *(top)* Native Energy, Inc./NREL; *(middle)* Jorgen Gulliksen/ZUMA Press/Corbis; *(bottom)* Stefan Falke/laif/Redux. **p. 532:** Stefan Falke/laif/Redux. **p. 534:** *Infographic 24.6 (left)* Chris Fourie/Dreamstime.com; *(middle)* Michael Patrick O'Neill/Alamy; *(right)* Mark Conlin/Alamy; *(background)* NASA/Corbis. **p. 536:** *Infographic 24.7 (top to bottom)* Aerial Archives/Alamy, Ron and Patty Thomas Photography/iStockphoto, Petr Nad/iStockphoto, LIU XIN/Xinhua /Landov, Pavle Marjanovic/Dreamstime.com, Rob Broek/iStockphoto. **p. 538:** *(top left)* Farshid Assassi/Courtesy BNIM; *(top right, middle, and bottom left)* conflictedracer; *(bottom right)* Joah Bussert/NREL. **p. 539:** *Infographic 24.9* Andrde Babiak/age fotostock. **p. 541:** *Infographic 24.11 (top to bottom)* Dimitri Vervitsiotis/Digital Vision/Getty Images, AP Photo/Damian Dovarganes, rockphoto/FeaturePics, Don Nichols/iStockphoto, stuartbur/iStockphoto, okea/FeaturePics, Yuliyan Velchev/Dreamstime.com, Dmitro Tolokonov/iStockphoto. **p. 544:** Reprinted by permission from Macmillan Publishers Ltd: Nature News, "The trouble with turbines: An ill wind," 2012, Jun 20.

CHAPTER 25 p. 546: Jake Norton /Aurora/Getty Images. **p. 553:** *(top)* DOD Photo/Alamy; *(bottom)* Christian Kober AWL Images/Getty Images. **p. 562:** *(top)* DEVENDRA M. SINGH/AFP/Getty Images; *(bottom)* Dennis Scott/Corbis; *(inset)* Gerrit Vyn. **p. 563:** *(left)* jimbo/FeaturePics; *(right)* Clément Philippe/age fotostock. **p. 564:** *(left)* AfriPics.com/Alamy; *(top right)* Michael Lynch/Alamy; *(middle right and bottom right)* Biophoto Associates/Science Source.

CHAPTER 26 p. 568: Digital Property Ltd/Image Source/Corbis. **p. 570:** Courtesy Amy Jo Smith. **p. 574:** AP Photo/The Daily Reflector, Rhett Butler. **p. 576:** *Infographic 26.5* Biophoto Associates/Science Source. **p. 577:** *Infographic 26.6* Peter Dazeley/Getty Images. **p. 580:** *(top)* The Post-Standard; *(bottom)* Courtesy Amy Jo Smith. **p. 581:** Gregory G. Dimijian/Science Source. **p. 582:** *(top)* Tom Campbell/agefotostock; *(bottom left)* Lessadar/FeaturePics; *(bottom right)* Nancy Nehring/iStockphoto.

MILESTONE 7 p. 586: *(left and middle)* © Copyright Eli Lilly and Company. All Rights Reserved. Photo courtesy of Eli Lilly and Company Archives. *(right)* Sanofi Pasteur Limited (Connaught Campus) Archives, Toronto. **p. 588:** *Infographic M7.1 (top)* Wellcome Library, London; *(bottom)* Biophoto Associates/Science Source. **p. 590:** *Infographic M7.2* Science Source. **p. 591:** The Thomas Fisher Rare Book Library, University of Toronto. **p. 592:** *(left to right)* SSPL/The Image Works; The Thomas Fisher Rare Book Library, University of Toronto; The Thomas Fisher Rare Book Library, University of Toronto; The Thomas Fisher Rare Book Library, University of Toronto. **p. 594:** The Granger Collection, New York.

CHAPTER 27 p. 596: Kevin Curtis/Science Source. **p. 598:** Greg Miles. **p. 603:** *Infographic 27.4 (left)* Biophoto Associates/Science Source; *(right)* VEM/Science Source. **p. 604:** *Infographic 27.5* Ed Reschke/Peter Arnold/Getty Images. **p. 607:** Lisa F. Young/iStockphoto. **p. 609:** The Daily News, Bogalusa, LA. *Infographic 27.9 (clockwise from top)* Mykola Velychko/iStockphoto, Donald Erickson/iStockphoto, spaxiax /iStockphoto, Robyn Mackenzie/iStockphoto. **p. 612:** *(top)* Sascha Burkard/iStockphoto; *(middle)* Christopher

Index

Note: Page numbers followed by f indicate figures; those followed by t indicate tables.